The Developmental Biology of Plants and Animals

The Developmental Biology of Plants and Animals

EDITED BY

C. F. GRAHAM

Lecturer in Zoology and Fellow of
St. Catherine's College, University of Oxford

AND

P. F. WAREING DSc, FRS

Professor of Botany,
University College of Wales, Aberystwyth

W. B. SAUNDERS COMPANY

PHILADELPHIA TORONTO

Western Hemisphere
distribution rights assigned to

W.B. SAUNDERS COMPANY

West Washington Square, Philadelphia, Pa. 19105
833 Oxford Street, Toronto M8Z 5T9, Canada

Library of Congress Catalog Card Number: 76-3086

SBN 0-7216-4205-5

© 1976 Blackwell Scientific Publications

Printed in Great Britain

Contents

The Authors

K. Beckingham-Smith, *National Institute for Medical Research, Mill Hill, London, U.K.*

J.T. Bonner, *Biology Department, University of Princeton, New Jersey U.S.A.*

I. Craig, *Genetics Department, Oxford University, U.K.*

M.J. Evans, *Anatomy Department, University College, London, U.K.*

P.J. Ford, *Molecular Biology Department, Edinburgh University, U.K.*

C.F. Graham, *Zoology Department, Oxford University, U.K.*

B. Gunning, *Research School of Biological Sciences, The Australian National University, Canberra, Australia*

J.B. Gurdon, FRS, *MRC Laboratory of Molecular Biology, Cambridge, U.K.*

M.A. Hall, *Department of Botany and Microbiology, University College of Wales, Aberystwyth, U.K.*

T. Horder, *Human Anatomy Department, Oxford University, U.K.*

B.E. Juniper, *Botany Department, Oxford University, U.K.*

R.B. Knox, *Botany Department, University of Melbourne, Australia*

A. Miller, *Zoology Department, Oxford University, U.K.*

J.D. Pitts, *Biochemistry Department, University of Glasgow, U.K.*

L. Saxen, *Pathology Department III, University of Helsinki, Finland*

H. Smith, *School of Agriculture, University of Nottingham, Sutton Bonnington, Loughborough, U.K.*

H.E. Street, *School of Biological Sciences, University of Leicester, U.K.*

J.R. Tata, F.R.S. *National Institute for Medical Research, Mill Hill, London, U.K.*

P.F. Wareing, FRS, *Botany and Microbiology Department, University College of Wales, Aberystwyth, U.K.*

J. Wartiovaara, *Pathology Department III, University of Helsinki, Finland*

Preface

This book has been written to meet the need for a unified account of plant and animal development for advanced undergraduate and graduate courses. There have been relatively few attempts to produce such a general treatment of developmental biology in the past and, so far as we know, none at the advanced level. The traditional subject divisions into botany, zoology, biochemistry and medicine have resulted in largely independent approaches to developmental problems in plants, animals and micro-organisms and there has been remarkably little communication between workers in these various fields. Indeed, the language and approaches used by plant and animal developmental biologists are in many respects so different that communication between them is often difficult, as is very apparent on the rare occasions when they meet at interdisciplinary conferences. The differences in approach are partly due to the very different nature of the biological material they are dealing with and the consequent different opportunities for experimental approaches which plants and animals offer. Thus, the motility of animal cells offers experimental opportunities which are denied to the plant biologist, while the remarkable capacity for regeneration of many plants offers possibilities envied by the animal biologist. Nevertheless, despite these differences in approach, there is more common ground in the nature of the problems posed and in the concepts developed than is commonly realized, and as we hope will become apparent from the present book. Our aim, therefore, was to combine a variety of biological disciplines in the study of development and thereby to achieve a unification and synthesis of ideas and concepts. We believe that such a unified approach to plant and animal development is fruitful and that the study of one illuminates the other. For instance, the study of pollen−stigma interactions may give a good model of cell recognition in animal development, while knowledge of gene expression in amphibians is helpful in understanding plant differentiation. We therefore consider it blinkered to study one without the other.

In selecting the material to be included in the book we were governed by two main considerations: firstly, that the background knowledge of the student readers would be very diverse; secondly, that it would be impossible and unnecessary to give complete coverage of the whole field to an advanced level, within a reasonable space. Accordingly, we have had to assume that the reader has a general elementary knowledge of

biology, including development, and a reasonable grounding in biochemistry and molecular biology. However, Part 1 is intended to provide a general introduction, particularly for readers who have a less detailed knowledge of biochemistry and molecular biology or who may not be well acquainted with plant development. The remaining chapters require rather more concentration as the book progresses, but it was important that all the book should be readily comprehensible. To this end, the first draft of each chapter was kindly read through by several students and where authors were found to be less than lucid they rewrote and clarified.

We considered it more important, in an advanced text book, that the topics covered should be treated *in depth* than that a comprehensive coverage should be given at a more general level. The criteria we adopted in selecting topics have therefore been, firstly, that they should illustrate or have a bearing on general themes of plant and animal development; and secondly, that there should be a well-established body of information on each of the topics selected. Hence, for example, the selection of photomorphogenesis to illustrate environmental control of plant development.

One of the results of this policy has been that a number of topics which appear in standard texts of plant or animal development have been given very cursory treatment or omitted altogether. On the other hand, some topics, such as cell junctions and short range interactions in plant cells, which have been included are not normally dealt with in standard textbooks of developmental biology, but are included here because they have a bearing on a general developmental problem.

It was important that the chapters be written by authorities who were aware of recent advances in their respective fields. To integrate their diverse contributions we have introduced and summarized the Parts, extensively cross-referenced, and provided a large index. These aids should be used. For instance there is no section entitled 'Developmental genetics' but the index shows that a full treatment of the genes which control development can be found in Chapters 1.3, 2.1, 3.6 and 5.3; again no chapter is called 'Biochemistry of development' but a quick look at the Conclusions at the end of each Part of the book would draw the reader's attention to Parts 4, 5 and 6 where this subject is widely discussed.

We would like to thank the authors who have devoted so much time and effort to produce excellent and original discussions of developmental problems and who have been tolerant of our editing.

Many of the illustrations have been borrowed from other books, review articles, and research papers and we acknowledge the help of all those who have kindly given permission for their inclusion here.

Finally we have been assisted by all those who have read and commented on particular chapters. We would like to thank: E.D. Adamson, A. Colman, M. Dziadek, G.J.C. George, J.B. Gurdon FRS, H. Gutzeit, J. Heslop-Harrison FRS, A. McLaren FRS, and V.E. Papaioannou. Mr Robert Campbell of Blackwell Scientific Publications has sustained our efforts with his enthusiastic support.

Part 1
The Origin of
Cell Heterogeneity
in Early Development

Chapter 1.1
Introduction—
Problems of Development

The term *development* is used in various senses in biology, but in this book we shall apply it in the broadest sense to the complex changes which an individual organism undergoes in its life cycle from the fertilization to death. Development normally involves both quantitative changes, i.e. *growth*, and qualitative changes leading to increased specialization in the various cells and tissues and organs of the body, which we refer to as *differentiation*. The processes leading to and determining the form and structure of the organism are included in the general term *morphogenesis*.

Differentiation can be recognized at various levels of organization, namely at the cell, tissue and organ levels. The use of the term in this way is quite logical since differentiation between tissues and organs simply reflects differentiation which occurred at the cellular level at an earlier stage of development. Development is normally a very orderly process in which one stage follows another in the proper sequence, and at each stage cells which hitherto had showed a common lineage diverge into alternative pathways of differentiation. If such divergence occurs early in development it may lead to subsequent differentiation of whole organs or other major parts of the whole organism, while later the divergence may lead to tissue or cell differentiation.

A central problem of development concerns the nature of the control mechanisms whereby cells of common lineage are caused to diverge into alternative pathways; that is, the problem of the origin of cell heterogeneity, which is the subject of Part I. Once a group of cells has become committed to a particular pathway of differentiation, it normally becomes very difficult to divert them into another pathway; thus, the developmental potentiality of the cells has become restricted and they are said to have become *determined* along a particular pathway. It might be thought that this restriction of developmental potential reflects a loss of genetic material, i.e. of parts of the genome, during the course of development, but it is now clear that the nucleus remains 'pluripotential' and retains the capacity to regenerate a whole organism under certain conditions (Part 2).

The progressive cell specialization which arises by divergence at successive stages of development would not, of itself, lead to an integrated, whole organism, but to a mass of cells which were different in structure and function in different regions, but which did not relate their activities to each other. To achieve integration there is clearly a need for cells and tissues to be able to communicate and interact with each other in controlled ways. Thus, morphogenesis depends upon cell communication and interaction. Moreover, before there can be communication there must be the capacity for recognition between cells. This capacity for recognition is particularly important in animals, in which the cells are potentially motile and if disaggregated can ultimately segregate out again, so that cells of the same type become reassociated together. In plants, most of the cells are non-motile and the need for cell recognition properties is not so evident, but nevertheless we shall see that the capacity for recognition is well developed in certain types of plant cell (Part 3).

Communication between cells can take various forms. Some of these involve short-range interactions, including the induction of differentiation by one type of cell in contiguous cells (Part 3). Such short-range interactions must play a vital role in ensuring the orderly sequence of differentiation in both space and time.

Long-range communication is primarily effected through hormones produced in one part of the organism and acting in a distant part (Part 4). We shall see that hormones probably do not act directly as inducers of differentiation but rather as evocators of the processes leading to differentiation in cells which are already predetermined along certain pathways.

Although for many purposes it is convenient to regard the cell as the unit of differentiation, it is important to recognize that at the subcellular level organelles, such as mitochondria and chloroplasts, show their own cycles of development, which are described in detail in Part 5. Moreover, it is self-evident that structural and functional differences between cells must have a molecular basis and that it must be our ultimate aim to describe and understand differentiation in molecular terms.

There are several different aspects of differentiation at the molecular level to be considered. On the one hand, development can be regarded as a problem of programmed sequential gene expression (Part 5). In so far as gene action is expressed through the synthesis of specific structural and enzyme proteins, this is clearly one field of molecular biology which is vital to a full understanding of development. We know little about the control of gene expression in eucaryotes, but clearly the mechanism whereby the right genes are activated in the right cells at the right time is central to the problem of development. On the other hand, not all aspects of cell organization at the molecular level are enzyme-mediated and therefore directly controlled by the genome. For example, some cellular structures, such as ribosomes and membranes, exhibit 'self-assembly' and do not require enzymes as catalysts. The extent to which self-assembly plays a role in cell development is also discussed in Part 5.

We shall see that many of these problems of development are common to both plants and animals, particularly at the molecular and sub-cellular levels. At higher levels of organization the problems are not always the same in plants and animals, since the plant cell is enclosed in a rigid wall and is non-motile, whereas this is not the case for animals cells. This basic difference has had profound effects upon the patterns of development of plants and animals. Thus, whereas most animals show a distinct embryonic phase which is terminated by the attainment of the structure of the adult, in plants development is open-ended or 'indeterminate', in that the growing points (apical meristems) of the shoot and root often remain permanently embryonic and capable of further growth and differentiation throughout the life of the plant. On the other hand, the development of organs such as the leaves and fruits is 'closed' and much more analogous to the development of organs in animals. Thus when allowance has been made for these differences in organization of the plant and animal bodies, many of the problems of development are closely analogous in the two groups of organism.

Chapter 1.2
Nucleus and Cytoplasm

1.2.1 GENE EXPRESSION AND DEVELOPMENT

J.B.S. Haldane first pointed out as long ago as 1932 [9] that many genes have a specific time of action during development. At that time nothing was known about the manner in which gene action is controlled or as to how the correct sequence of gene expression during development is achieved. Today we still have little understanding of the latter problem, but as a result of advances in molecular biology we know a great deal as to how the information encoded in the DNA is expressed in the structure of the proteins of the cell. Thus, the concept has arisen that the cell specialization which occurs during development and differentiation involves the activation of specific groups of genes which, in turn, control the synthesis of the enzymes and other proteins characteristic of such specialized cells. This is the so-called 'variable gene activity theory' of cell differentiation.

Our current thinking on gene activation and repression is based upon and conditioned by information gained from studies on micro-organisms. These studies led to the well-known theory of Jacob and Monod, according to which the activity of genes which code for specific enzyme proteins, the so-called *structural genes,* is controlled by other genes referred to as *regulator genes.* A regulator gene forms a cytoplasmic protein called the *repressor.* It is postulated that the initiation of mRNA synthesis on the structural gene is controlled by a neighbouring section of DNA called the *operator,* and that the repressor formed by a given gene binds with specific operators, thereby blocking the initiation of mRNA synthesis on the structural gene. A repressor is held to have the property of binding also with certain smaller molecules (metabolites) called the *effectors.* Regulation of mRNA synthesis may be effected by induction or by repression. In induction, the inducing metabolite combines with the repressor to block its interaction with the operator, so that mRNA synthesis on the structural genes can be initiated. In enzyme repression, combination of the repressing metabolite with the repressor *enhances* interaction with the operator (Fig. 1.2.1).

It is not clear how far these concepts derived from studies on bacteria can be applied to the control of gene activity in eukaryotic organisms, about which little firm information is available (Part 5). However, the idea that development involves programmed, selective gene activation and repression is an attractive one and provides a useful framework for thinking

about selective gene expression. Thus, we are led to the idea that differentiation involves the selective transcription of specific genes, with the release of the corresponding mRNAs into the cytoplasm, where the information they contain is translated into protein synthesis. Selective gene expression may therefore be achieved by control of gene activation at the transcriptional level. However, we shall see that there is evidence that in certain

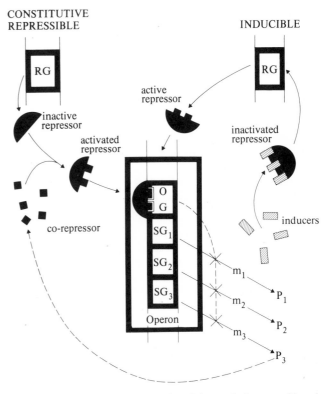

Fig. 1.2.1 A representation of the action of the constitutive repressible and inducible systems of gene regulation. The active and activated repressors produced by the regulator genes (RG) are represented as fitting into the operator gene (OG) of the operon. They then block the transcription of the structural genes (SG) of the operon in producing the mRNAs (m_1, m_2, m_3) which would specify the synthesis of the protein enzymes (P_1, P_2, P_3). One of the enzymes is concerned in activating a repressor in the repressible system. In the inducible system, the inducers are represented as inactivating the repressors by preventing them from fitting the operator genes. From Clowes & Juniper [6].

instances the mRNAs may not be translated until some considerable time after their release into the cytoplasm. Thus, the *time* of gene expression may be controlled at the translational level.

Although it is clear that the role of the nucleus in differentiation is paramount in that the genes are located there and that the coded instructions for protein synthesis they contain is released into the cytoplasm in the form of mRNA, nevertheless the cytoplasm also plays a primary role in determining the time and place of expression of specific genes, i.e. it controls the selective gene expression underlying differentiation.

Some of the best evidence that the cytoplasm may regulate gene expression is provided by studies on the localization within the cytoplasm of certain animal eggs of factors which determine the subsequent pattern of differentiation of the cells derived from the various regions of the egg during cleavage, and which are considered in more detail later in this chapter. The phenomenon of 'unequal division' in plant cells (p. 29) also shows that the activity of the nucleus is determined by the cytoplasmic environment in which it finds itself.

The manner in which the cytoplasm controls selective gene transcription in the nucleus is not fully understood, but there is evidence that proteins may pass from the cytoplasm into the nucleus. However, it is possible that smaller molecules may also be involved (p. 27). It is thus clear that there must be a complex interplay between nucleus and cytoplasm during differentiation, as is well illustrated by studies on the green alga, *Acetabularia,* and the aquatic fungus, *Blastocladiella.*

1.2.2 NUCLEAR-CYTOPLASMIC INTERACTION IN *ACETABULARIA* AND *BLASTOCLADIELLA*

Acetabularia has been extensively used for experimental studies on nuclear-cytoplasmic interaction [2, 10, 19]. A plant of *Acetabularia* consists of a single cell of relatively enormous size, 3–5 cm in height, which has a stalk with basal rhizoids by which it is attached to the substratum, while at the apical end there is a hat-shaped 'cap', in which large numbers of gametes are ultimately formed (Fig. 1.2.2). The various species differ in the morphology of the cap, the details of which need not concern us here.

The large size of the cell makes it possible to remove the nucleus and transfer it to another individual of the same or a different species. When the nucleus of one species is transferred into an enucleate cell of another species, the cap that is formed later is characteristic of the species which supplied the nucleus. Grafts can also be carried out between cells of different species, and if the part from one species is nucleate and from the other enucleate the cap formed resembles more closely the species which supplied the nucleus.

While it is clear that the nucleus ultimately determines the characteristics of the cap and must therefore influence processes

occurring in the cytoplasm, the reverse is also true. Thus, division of the nucleus to give large numbers of daughter nuclei normally occurs just before gamete formation, but if the cap is cut off division of the nucleus does not occur until a new cap is formed.

Although cap formation is ultimately dependent upon influences from the nucleus, the effect of the latter is not immediate and enucleated fragments can continue to function normally for several weeks. Moreover, if the nucleus is removed at a relatively early stage of development, a normal cap can be formed many weeks after removal of the nucleus. Again, if the cell is cut into two to give one part with a nucleus and one without, both parts are capable of regenerating a complete cap. These observations indicate that (1) all the information necessary for the production of the cap passes from nucleus to cytoplasm long before the cap is normally formed; (2) information for the production of the cap may be present in the cytoplasm without being expressed; (3) the information for the production of the cap is very stable. These conclusions strongly suggest that the nucleus produces stable messenger RNA which is released into the cytoplasm where it can continue to exist for relatively long periods. Several types of evidence support this suggestion:

(1) Cytochemical observations have demonstrated that high molecular weight RNA accumulates at the apex of the stalk.
(2) Treatment of the cells with ribonuclease inhibits completely regeneration of both nucleate and enucleate halves, so long as it is present in the sea water surrounding the cells, but the inhibition is reversible in the case of nucleate halves and irreversible with enucleate ones.
(3) Treatment with Actinomycin D, which inhibits the transcription of DNA to give mRNA, does not inhibit regeneration in enucleate fragments (which presumably

Fig. 1.2.2 Plants of *Acetabularia mediterranea.*

already contain stable RNA), but it does inhibit regeneration of the cap by nucleate fragments, which would presumably be dependent upon the synthesis of new mRNA.

This and other indirect evidence suggest that morphogenesis in *Acetabularia* is dependent on the transcription of nuclear DNA into stable, long-lived mRNA. However, no one has yet succeeded in isolating and identifying the cap-forming substances, and therefore the view that they probably represent mRNAs must be regarded as 'unproven' for the present.

Although cap formation depends upon the presence of stable morphogenic substances formed in the nucleus the expression of the genetic information is controlled and effected by regulatory mechanisms in the cytoplasm. The information passed from the nucleus to the cytoplasm can evidently remain unexpressed for several weeks, and the time of cap formation appears to be determined by events that take place in the cytoplasm rather than the nucleus. Control of morphogenesis therefore appears to be at the translational level.

The formation of the cap involves the net synthesis of protein, the synthesis of specific enzymes and the synthesis of specific polysaccharides. The polysaccharides present in the cell wall of the cap are different from those found in other parts of the cell and cap formation therefore requires the synthesis of specific enzymes required for polysaccharide synthesis. Studies on enzyme synthesis indicate that a high proportion of the proteins

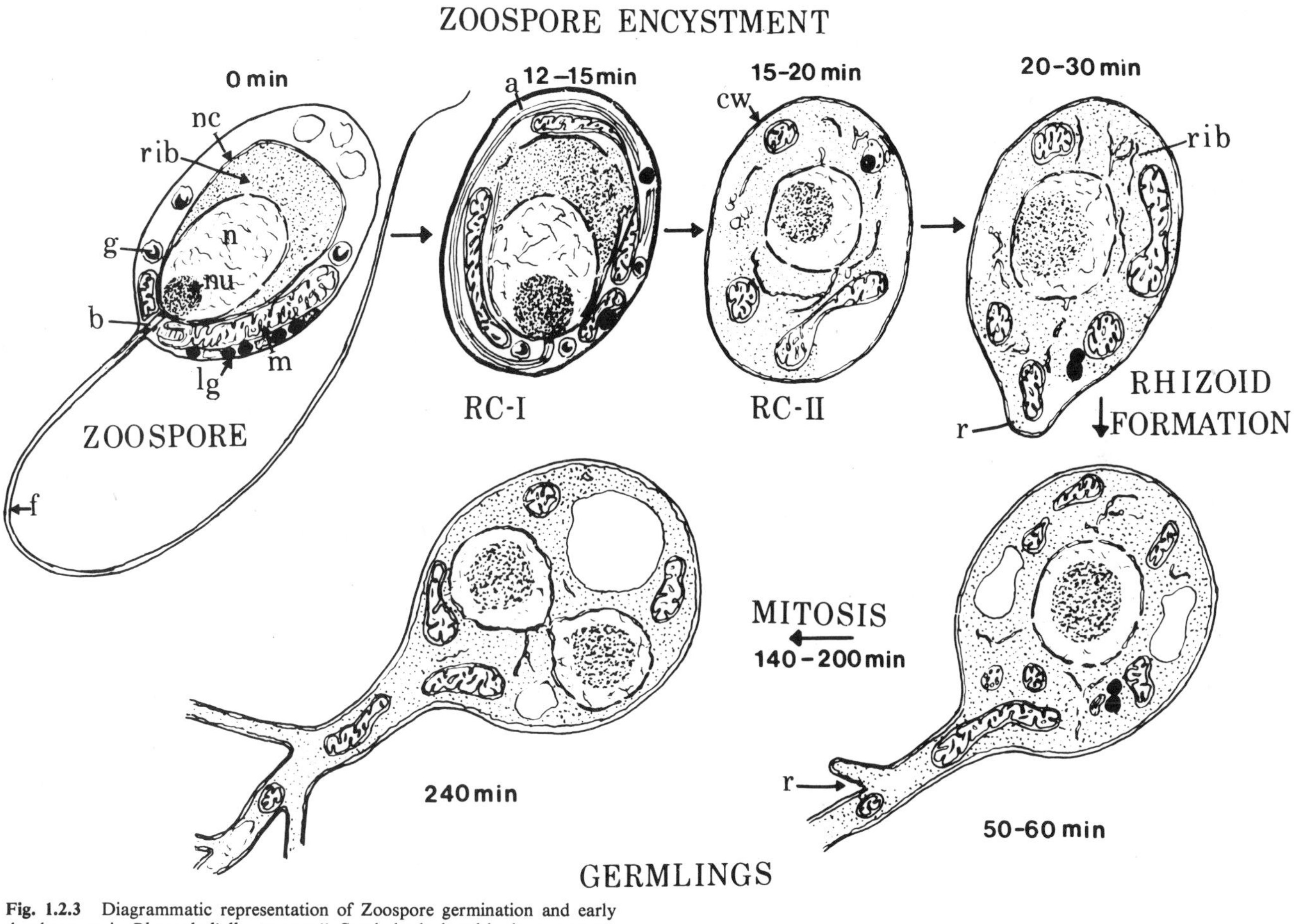

Fig. 1.2.3 Diagrammatic representation of Zoospore germination and early development in *Blastocladiella emersonii*. Symbols: b, basal body; g, gamma particle; m, mitochondrion; rib, ribosomes; n.c., nuclear cap; nu, nucleolus; n, nucleus; lg, lipid granules; a, flagellar axoneme; cw, cell wall; r, rhizoid; f, flagellum; v, vacuole; RC-I, Round Cell I; RC-II, round cell II. Kindly supplied by Dr C.L. Leaver.

of the cell can be formed in enucleate cytoplasm, even though their synthesis is under ultimate nuclear control, as has been shown for lactic dehydrogenase. Moreover not only does the general control of the translation stage in cap formation appear to reside in the cytoplasm, but fine control of the correct sequence of enzyme synthesis appears also to occur at this level. For example, *A. crenulata* contains at least three different phosphatases having different pH optima (5, 8·5, and 12), all of which are synthesized even in the absence of the nucleus, so that the respective mRNAs coding for their proteins have all been released before removal of the nucleus. Nevertheless, the enzymes active at pH 5 and 8·5 are formed before the one active at pH 12, which only appears a very short time before cap initiation, not only in the whole alga but also in enucleate fragments. The implication is that enzyme synthesis does not occur simultaneously, even if the co-ordinating messages are released from the nucleus at the same time; the regulation of the sequence of translation, therefore, probably occurs in the cytoplasm.

The conclusion that morphogenesis in *Acetabularia* is directly regulated at the translation stage of protein synthesis, using pre-formed, long-lived mRNAs is strongly supported by the results of studies on zoospores of the aquatic fungus, *Blastocladiella emersonii* [21]. These zoospores have a single, posterior flagellum and the cell is bounded, not by a wall, but by a single continuous membrane. In addition to the usual organelles, these zoospores contain several special structures, of which the most conspicuous is the *nuclear cap,* which surrounds the anterior two-thirds of the nucleus, and which contains large numbers of pre-formed ribosomes.

The zoospores are released into the water from the parent plant and after a period of free swimming they settle onto a solid substratum and undergo a short period of encystment. During encystment the zoospore undergoes a series of rapid and radical changes in structure (Fig. 1.2.3). About 10 min later a small germ tube appears, marking the commencement of germination. Thus, the period of encystment is very short. One of the most striking and important changes during germination is the rapid disorganization of the nuclear cap membrane and associated release of the enclosed ribosomes, which become dispersed in the cell.

Apart from the morphological changes occurring during encystment and germination the associated changes in nucleic acid and protein synthesis have been studied using various approaches, including the use of inhibitors of nucleic acid and protein synthesis [16]. Not until the spores have germinated and produced germ tubes do measurable amounts of RNA begin to accumulate, at about 40–45 min after encystment. This is followed 30–40 min later by synthesis of protein and about 40 min later still, of DNA. Thus, during the early stages of germination there is no measurable net increase in either RNA or protein.

Spores treated with Actinomycin D encyst, germinate and continue to develop up to the time when the measurable increase in RNA should begin, when they stop growing. Since early protein synthesis is not affected significantly by inhibition of RNA synthesis, it would appear that the ribosomal, transfer and messenger RNAs necessary for it are all present in ungerminated spores.

Polysomes are assembled during the early stages of germination. Actinomycin D fails to reduce significantly the early rise in polysome content, and only a trace of new RNA

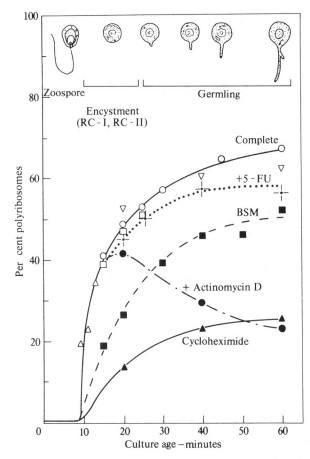

Fig. 1.2.4 The appearance of polysomes during germination and the effects of medium composition and inhibitors. The polysome percentages were determined for cultures germinated under the following conditions: ▽———▽, △———△, ○———○, □———□, complete medium; ■———■, basal salts medium (BSM): +———+, complete medium with 325 µg/ml 5-fluorouracil; ●———●, complete medium with 20 µg/ml Actinomycin D; ▲———▲, complete medium with 5 µg/ml cycloheximide. The control cultures in complete medium were grown at cell densities from $1\cdot69 \times 10^9/1$ to $3 \times 10^9/1$; the BSM and inhibitor treated cultures were grown at $2\cdot61 \times 10^9$–$2\cdot68 \times 10^9$ cells/1. From Leaver & Lovett [16].

synthesis can be detected during the period of maximum polysome formation. These results, illustrated in Fig. 1.2.4, show that the rapid formation of cytoplasmic polysomes occurs by the attachment of 80s ribosomes released from the nuclear cap to the preformed mRNA molecules, and requires no new RNA synthesis.

Cycloheximide inhibits protein synthesis in germinating zoospores. Spores encyst normally, nuclear cap membranes fragment and ribosomes are dispersed normally, but the germ tube does not form. Although germination does not proceed as far as it does in the presence of Actinomycin D, the structural changes associated with encystment apparently require only the protein and RNA found in the non-germinating zoospore. Apparently the spore has all the necessary materials for encystment and early germination. However, the later stages of germination and germ tube emergence do appear to require new protein synthesis.

The activation of protein synthesis using pre-existing ribosomes and stored mRNA observed in the germination of zoospores of *Blastocladiella* presents an interesting parallel with animal embryonic systems in which, as we shall see, the utilization of maternal mRNA in early embryonic development seems well established.

1.2.3 EMBRYONIC DEVELOPMENT AND GERMINATION

In the previous section we have discussed cases in which a single genome is responsible for the characters of the developing system. The situation is different for plant and animal embryos which are produced by proteins coded by three genomes: those of the mother, the father, and the zygote. The diploid maternal genome is active during oogenesis and forms substances which persist into early development. Its products provide nutrition for the embryo, either by depositing substances directly into the cytoplasm (e.g. yolk [22]), or by surrounding the embryo with a feeder layer (e.g. the placenta of mammals).

This genome also controls egg organization and may form substances in the egg which subsequently determine the range of cell types which the embryo can form (e.g. germ plasm of insects, see Part 2). In contrast, substances produced by the diploid paternal genome do not persist far into development. This genome codes for the exine layer of pollen, but the material in this layer is shed soon after pollination and makes no further contribution to the young plant (Chapter 3.5); similarly this genome determines sperm structure but most of the extra-nuclear components of sperm dissipate or degenerate soon after fertilization (e.g. mitochondria, see Chapter 5.2). In practice, we can ignore the effects of the diploid paternal genome once syngamy has occurred. The zygote nucleus

therefore lies in a cytoplasm whose character was determined by the previous activity of the maternal diploid genome; one of the first problems is to discover the extent to which this cytoplasm controls the functions of the zygote nucleus.

There is good evidence that the early development of amphibians and sea urchins depends on proteins synthesized on mRNA templates provided by the diploid maternal genome. Certainly early development depends on protein synthesis, since puromycin (a protein synthesis inhibitor) blocks cleavage. However, if the egg is enucleated or if the nuclear RNA synthesis is inhibited with Actinomycin D, then the egg can still develop up to the early blastula stage. The supply of the mRNA molecules is such that Actinomycin D has no effect on the amount of protein synthesized in early sea urchin embryos. It has recently been possible to identify the mRNA for microtubular protein (a major component of mitotic spindles) and for histones in the sea urchin egg (see Chapter 5.3). Since these proteins are coded by nuclear genes, these studies prove that mRNA of maternal origin is available for use during early development.

The eggs of amphibia and sea urchins are similar to the spores of *Blastocladiella* in possessing long-lived mRNA molecules which can be rapidly activated for protein translation when there is an environmental signal, such as fertilization or good germinating conditions, which require a rapid response from the organism. However, in situations in which the environment is less demanding, eggs may not possess many maternal messenger-RNA molecules. The mammalian egg develops slowly while bathed in the nutrients provided by the mother and it does not appear to possess many maternal mRNA molecules; development can be quickly blocked by RNA synthesis inhibitors and each cleavage division probably depends on the activity of the zygote genome.

The patterns of RNA and protein synthesis found in sea-urchin and amphibian eggs and embryos seem to be paralleled quite closely by the corresponding processes in dormant and germinating seeds. However, whereas in the animal egg we have been considering the changes occurring before and immediately following fertilization, in the case of seeds we are dealing with the changes associated with cessation and renewal of growth of an embryo which, in most cases, has reached a relatively advanced stage of development. Another major point of difference between the plant and animal systems lies, of course, in the fact that the plant embryo normally undergoes considerable dehydration during ripening of the seed and frequently it is the imbibition of water by the tissues when the seed is re-wetted, which triggers off the process of germination, whereas in the majority of animal eggs further development is triggered off by fertilization. In both cases, however, we are dealing with the renewal of activity following a period of quiescence.

Metabolism is effectively absent from dry seeds, but when

they are placed under favourable conditions of moisture, temperature and aeration there is a rapid increase in metabolic activity, as shown by the re-activation of respiratory pathways and increasing respiration rate. There is no doubt that the early stages of renewal of metabolism involve enzymes already present in the dry seed and which become activated by the re-hydration of the tissue. On the other hand, amino acid and protein synthesis can be demonstrated soon after imbibition of the seeds [17, 23], and it has been demonstrated unequivocally that the synthesis of new enzymes occurs [5]. This has been shown for the enzymes isocitratase, α-amylase, protease and peroxidase by demonstrating the incorporation of ^{14}C or ^{18}O labelled (density labelling p. 222) amino acids into the enzyme protein. A parallel increase in polysomes during germination has been reported for the seeds of a number of species [1, 14, 18]. At the same time an increasing ability of ribosome preparations from germinating seeds to incorporate amino acids into protein in cell-free systems has been reported for the early stages of germination [14].

There is considerable evidence indicating that protein synthesis during early germination is dependent upon mRNA already present in the resting seed. Thus, it has been shown for several species of seed that actinomycin D does not inhibit protein synthesis and polysome formation during early germination [3, 20]. It was shown for cotton seeds that actinomycin D inhibits incorporation of ^{32}P into polysomes, but has no effect on the incorporation of ^{14}C amino acids into protein [23]. On the other hand, cycloheximide and puromycin inhibit protein synthesis by germinating seeds [12, 13]. These results suggest that protein synthesis during the early stages of germination does not depend upon parallel RNA synthesis, but is directed by long-lived mRNA present in resting seeds.

If this conclusion is correct, then the question arises as to when this long-lived mRNA is formed. Some evidence on this problem has been obtained from studies on protease in cotton seeds [12, 13]. This enzyme is absent from the resting seed but its activity increases during germination and reaches a maximum after three days. The appearance of the enzyme is completely inhibited by cycloheximide, but is not affected by actinomycin D. It was shown that ^{14}C-amino acids are incorporated into protease, suggesting de novo synthesis of the enzyme.

Developing cotton embryos can be germinated precociously by dissecting them out of the ovules and placing them on a moist substrate, and hence it is possible to determine the time at which transcription of the protease mRNA occurs by precociously germinating embryos at successively younger ages and determining their protease activity. It was found that embryos larger than 85 mg in weight will develop protease in the presence of actinomycin D, whereas smaller embryos will not do so. The results indicated that when embryos reach approximately 60% of their final weight at about 20 days before

embryogenesis is complete, mRNA for protease and presumably for other enzymes necessary for germination is transcribed in the cotyledons. Since these mRNAs are apparently present in the cotyledons during the last 20 days of embryogenesis their translation during this time must be inhibited in some manner. Evidence has been obtained that this inhibitor may be abscisic acid diffusing into the embryo from the ovule wall.

So far, attempts to isolate mRNAs from resting seeds have not been successful, but there have been several reports of the separation of RNA fractions from germinating seeds which show template activity.

1.2.4 CYTOPLASMIC CONTROL OF DNA AND RNA SYNTHESIS

The cytoplasm not only provides for early protein synthesis but also controls the activities of the zygote nucleus. The cytoplasm controls general cell growth and multiplication functions, such as DNA and ribosomal RNA synthesis, and also influences the production of different cell types in the embryo. The effect of egg cytoplasm on DNA and RNA synthesis has been most clearly demonstrated by transplanting nuclei into the large eggs of the South African Clawed Toad (Xenopus laevis). To understand these experiments, it is necessary to digress and to describe the methods which are used to study the synthesis of these molecules.

The synthesis of DNA can be followed by incubating cells with radioactive thymidine. For this precursor to be incorporated into nuclear DNA it must first enter the cell, and then be converted into thymidine triphosphate, and then enter the cell nucleus. The nucleus also contains its own non-radioactive thymidine triphosphate molecules and if these are present in high concentration, then rather few radioactive molecules will actually be included into the newly synthesized DNA. In practice it is rather hard to establish that the radioactive molecules are readily available to label any DNA molecules which are being formed and for this reason it is easier to prove that DNA synthesis is occurring than it is to prove that it is not.

The following experiments have shown that the egg cytoplasm can control DNA synthesis. Sperm nuclei were introduced into eggs at different times after fertilization, and at the same time radioactive thymidine was injected into the egg. It was found that radioactive thymidine was incorporated into the DNA of the male and the female pronuclei between 20 and 40 min post fertilization and that these were only engaged in DNA synthesis during this very limited period of the first cell cycle. The egg cytoplasm was able to induce DNA synthesis in the introduced sperm nuclei only at the times that the egg's own pronuclei were engaged in this activity. Since the egg cytoplasm could also induce DNA synthesis in introduced nuclei when the male and female pronuclei were absent, it seems reasonable to conclude that the induction of DNA synthesis depends only on factors which are present in the cytoplasm or membranes of the unfertilized egg [6, 7].

There are similar reasons for believing that the cytoplasm controls the appearance of newly synthesized ribosomal and transfer RNA molecules. RNA synthesis can be studied by incubating cells in radioactive uridine. As in the case of DNA synthesis, it is necessary to establish that the label is readily available in the cell nucleus and that it is incorporated into RNA. There are however additional problems in the study of the synthesis of this type of molecule. Once DNA is synthesized it is very stable, but RNA molecules may be degraded within a few seconds of their synthesis. For this reason it is particularly difficult to establish that RNA synthesis is not in progress; one might fail to notice a small amount of synthesis because the newly labelled molecules might be broken down before it was possible to look for the presence of radioactivity in RNA.

In practice this difficulty is avoided by studying the appearance of the radioactive precursor in RNA molecules which are relatively stable. It must however be remembered that these will only be a fraction of the total number of RNA molecules which are synthesized and there are good reasons for believing that many of the RNA molecules which are synthesized during early development are rapidly degraded [15]. Another problem is that ribosomal and transfer RNA molecules are processed after or during their synthesis (see Chapter 5.3). They are cut up and altered in conformation soon after transcription and they are not easy to recognize until they have been processed into their mature form. This means that in most experiments one is studying the appearance of the radioactive label in RNA molecules which have been transcribed, escaped degradation, and have been processed into their mature and recognizable form. It is therefore difficult to draw firm conclusions about the control of transcription during development and here we will be concerned with experiments which demonstrate that the cytoplasm can control the rate at which radioactivity enters mature ribosomal and transfer RNA molecules.

In the development of most animal embryos there is a period during which little radioactivity from labelled uridine will enter mature ribosomal RNA. It can be shown, by injection, that this is not because the precursor is unable to penetrate the cell and it is also known that the labelled uridine is rapidly converted into uridine triphosphate and that this enters the nucleus. Newly synthesized ribosomal RNA containing the radioactive label is found in large amounts in the cytoplasm at the late gastrula stage (45,000 cells) in *Xenopus*, at the 32-cell stage in the sea urchin, and at the four-cell stage in the mouse [8]. The following experiment has shown that the cytoplasm controls this process.

Nuclei were transferred from neurula cells into the cytoplasm of enucleate eggs. These donor nuclei are actively engaged in the production of ribosomal RNA and radioactive uridine is rapidly incorporated into mature ribosomal RNA molecules when these nuclei are resident in their own cytoplasm. However, when they are transferred to egg cytoplasm then radioactivity is no longer incorporated into the mature forms of these molecules and this process does not start again until the nuclear transplant embryo has developed up to a stage when it would normally show this activity. It therefore seems that the cytoplasm controls the rate at which these activities occur. The possible molecular mechanisms which effect this control are discussed later (Chapter 5.3). Here we need only note that some type of nuclear-cytoplasmic ratio is involved. Haploid embryos start their rapid increase of incorporation of radioactive uridine into stable RNA about one cell cycle after diploid embryos; this is at the time that nuclear-cytoplasmic ratio in the two types of embryos is similar.

The role of the cytoplasm in controlling DNA and RNA synthesis is also well exemplified by experiments on changes in the cytoplasm during oogenesis. In these experiments, nuclei were obtained from the brain of adult frogs; these nuclei rarely synthesize DNA although they do produce a little RNA. However, when they are injected into unfertilized eggs which had been activated by injection and which were engaged in their own nuclear DNA synthesis, then the brain nuclei started to synthesize DNA as well. The situation was different when the nuclei were injected into eggs at earlier stages of development. In the full grown oocyte, which only synthesizes RNA, the brain nuclei were induced to synthesize RNA only; and in the meiotic oocytes, which are inactive in DNA and RNA synthesis, the brain nuclei do not synthesize any molecules but rather enter division, as their chromosomes condense in the same way as the host cells chromosomes which are in meiosis. The general conclusion from these studies is that states exist in the cytoplasm which can induce both DNA and RNA synthesis and that the brain nuclei respond to these states in much the same way as the resident nucleus of the egg [8].

It is clear from the preceding paragraphs that the activity of brain and neurula nuclei can be controlled by the cytoplasm of early embryonic cells. In these experiments the dominant role of the cytoplasm might be the consequence of the large size of the egg. It is therefore interesting that the cytoplasm of small adult cells can also control nuclear activity. The nuclei of erythrocytes can be introduced into cell cytoplasm using a genetically inactive virus [11]. Erythrocyte nuclei are inactive in both DNA and RNA synthesis but in the cytoplasm of rapidly dividing cells both these functions are induced. It is therefore possible to conclude that there are diffusible factors in most cells which can control the synthesis of DNA and RNA. The nature of the molecules which effect this control is discussed in Chapter 5.3.

1.2.5 CYTOPLASMIC CONTROL OF CELL TYPE

In addition to the control of general nucleic acid metabolism, cell cytoplasm may control cell differentiation. We know very little about the mechanism of this control, although there is considerable information about the strategies which are employed (Chapter 1.3). Here we will only deal with two situations in which there is a little information about the biochemistry of the control.

The polar lobe of Ilyanassa

There is a region of the egg of Ilyanassa which may be involved with the control of RNA metabolism inside the embryo. Ilyanassa is a marine mollusc and its eggs have a clear cytoplasmic region which bulges out during the first three cleavage divisions. It is rather easy to pull the lobe away from the egg when it bulges out, and if it is removed during the first or second cleavage divisions, then the embryo which develops always has a defective mesoderm. The lobe does not contain a nucleus, so that there must be some substance or influence which it normally provides for mesoderm development.

Comparisons have been made between the biochemistry of the lobeless embryos and their normal lobed counterparts [4]. The lobeless embryos incorporate less amino acids into protein and less uridine into total RNA when compared with normal embryos. However the accumulation of uridine into ribosomal RNA is the same in the two types of embryos and presumably the removal of the lobe changes some other RNA species. It is also interesting that the isolated lobes are able to incorporate amino acids into protein, as do the whole enucleate eggs of frogs which were mentioned previously. These results suggest that factors in the lobe may play some part in determining the rates of protein and RNA synthesis. It would be very interesting to know if the effect of the lobe on mesoderm formation depended on its effect on the synthesis of these molecules.

Tissue specific enzymes in *Ciona intestinalis*

Some recent experiments with the embryo of *Ciona* have attempted to unravel the biochemistry involved in the production of enzymes which are characteristic of particular differentiated cells (tissue specific enzymes [24]). *Ciona* and the related *Styela* are marine tunicates; these are in a group of animals which are bottom living and filter feeding as adults. *Styela* has been a favourite animal for early embryological studies because different regions of the embryo can be recognized by differences in colour in some natural cytoplasmic components. These coloured regions are not superficially apparent in the unfertilized egg but they do appear early in development and one early idea was that these cytoplasmic regions controlled cell differentiation.

A study of the appearance of two tissue specific enzymes in the development of *Ciona* has given clearer insight into the importance of regional differences in eggs in the control of specific enzyme synthesis [24]. One of these enzymes is acetylcholinesterase: this enzyme is found at high specific activity in muscle and nervous tissue, and although it is also present in other tissues these large amounts can be taken as indicators of muscle and nerve differentiation. The enzyme hydrolyses acetylcholine and is first detected histochemically in presumptive muscle cells of the early neurula at 8 h after fertilization. The second enzyme is dopa oxidase which catalyses one of the reactions important in melanin synthesis. It is therefore found at high activity in pigment cells and it is first detected in presumptive pigment cells at 9 h after fertilization. First it was decided to find out if new RNA synthesis was required for the appearance of these enzymes. The drug Actinomycin D was applied and this reduced RNA synthesis to 30% of its normal level. Prior to the appearance of the enzymes, there is a sensitive period during which the drug blocks their appearance while later enzyme production is relatively resistant to the drug. This result suggests that new RNA synthesis is required for the appearance of these enzymes; however this drug may have toxic side effects and it is uncertain that its effect on the appearance of enzymes is solely due to its action on RNA synthesis.

Second, the protein synthesis inhibitor puromycin was used to find out if protein synthesis was required for enzyme production. If this drug is applied before enzyme appearance then no enzyme activity is detected but it has no effect on enzyme activity when it is applied for a short time after enzyme appearance. This result suggests that puromycin does not inhibit enzyme activity directly; rather it appears to effect enzyme activity by blocking the synthesis of the enzyme or one of its cofactors. Taking into account the difficulty of interpreting these inhibitor experiments, these results only suggest that both enzymes require new RNA and protein synthesis for their appearance in development. If this conclusion is correct, then it is clear that these enzymes are not formed on messenger RNA molecules which were put down in the cytoplasm by the diploid maternal genome before fertilization. This means that other factors in the cytoplasm must determine the capacity of a cell to form these tissue specific enzymes. The segregation of factors which were responsible for the capacity to synthesize these enzymes was traced with the use of cell division inhibitors (cytochalasin B and colcemid). Cell division was blocked at different stages of development so that the embryos were arrested at the 2, 4, 8, or 16-cell stage and then the embryos were stained to discover which of these cells formed the enzymes. It was found that the capacity to form the two enzymes segregated into different cells very early in development and followed the cell lineages of muscle and pigment cells.

It was next decided to investigate the localization of the

factors which controlled the appearance of these enzymes. Normal development will occur if the cytoplasm is disorganized by centrifugation or if half the volume of the cytoplasm is removed. It is therefore likely that the cell membrane or the stiff cortical layer immediately beneath it may contain the factors which determine the capacity to form these enzymes.

In conclusion, these experiments indicate that regional organization of enzyme forming capacity is probably not based on the storage of enzyme messenger RNA or the storage of inactive enzymes but rather depends on some other component which activates new gene expression at a particular stage of development. Timing of new enzyme synthesis is not related to the number of cell divisions and must be due to a developmental clock which counts neither cell divisions nor cells. In the next chapter we discuss the kinds of factors which may be involved in this control.

1.2.6 REFERENCES

[1] BARKER G.R. & RICHER M. (1967) Formation of polysomes in the seed of *Pisum arvense*. *Biochem. J.*, **105**, 1195–201.

[2] BRACHET J. (1968) Synthesis of macromolecules and morphogenesis in *Acetabularia*. *Current Topics in Developmental Biology*, **3**, 1–36.

[3] CHING T.H. (1972) Metabolism of germinating seeds. In *Seed biology* **2**, pp. 103–218. Academic Press, New York & London.

[4] DAVIDSON E.H. (1968) *Gene activity in early development*. Academic Press, New York & London.

[5] FILNER P., WRAY J.C. & VARNER J.E. (1969) Enzyme induction in higher plants. *Science, N.Y.*, **165**, 358–67.

[6] GRAHAM C.F. (1966) The regulation of DNA synthesis and mitosis in multinucleate frog eggs. *J. Cell. Sci.*, **1**, 363–74.

[7] GRAHAM C.F., ARMS K. & GURDON J.B. (1966) The induction of DNA synthesis by frog egg cytoplasm. *Develop. Biol.*, **14**, 349–81.

[8] GURDON J.B. (1974) *The control of gene expression in animal development*. Oxford University Press.

[9] HALDANE J.B.S. (1932) The time of action of genes and its bearing on some evolutionary problems. *Amer. Nat.*, **66**, 5–24.

[10] HÄMMERLING J. (1963) Nucleo-cytoplasmic interactions in *Acetabularia* and other cells. *Ann. Rev. Plant Physiol.*, **14**, 65–92.

[11] HARRIS H. (1970) *Cell fusion*. Oxford University Press.

[12] IHLE J.N. & DURE L. (1969) Synthesis of a protease in germinating cotton seeds catalyzed by masked mRNA synthesized during embryogenesis. *Biochem. Biophys. Res. Commun.*, **36**, 705–10.

[13] IHLE J.N. & DURE L. (1970) Hormonal regulation of translation inhibition requiring RNA synthesis. *Biochem. Biophys. Res. Commun.*, **38**, 995–1001.

[14] JACHYMCZYK W.K. & CHERRY J.H. (1968) Studies on mRNA from peanut plants: *in vitro* polyribosome formation and protein synthesis. *Biochem. Biophys. Acta*, **157**, 368–77.

[15] KIJIMA S. & WILT F.H. (1969) Rate of nuclear ribonucleic acid turnover in sea urchin development. *J. Mol. Biol.*, **40**, 235–46.

[16] LEAVER C.L. & LOVETT J.S. (1974) An analysis of protein and RNA synthesis during encystment and outgrowth (germination) of *Blastocladiella* zoospores. *Cell Differentiation*, **3**, 165–92.

[17] MARCUS A. & FEELEY J. (1964) Activation of protein synthesis in the imbibition phase of seed germination. *Proc. Nat. Acad. Sci. U.S.*, **51**, 1075–79.

[18] MARCUS A. & FEELEY J. (1965) Protein synthesis in imbibed seeds. II Polysome formation during imbibition. *J. Biol. Chem.*, **240**, 1675–80.

[19] PUISEUX-DAO S. (1970) *Acetabularia and cell biology*. Logos Press, London.

[20] THOMAS H. (1972) Control mechanisms in the resting seed. In *Viability of seeds* (Ed. E.H. Roberts). Chapman & Hall, London.

[21] TRUESDELL L.C. & CANTINO E.C. (1971) The induction and early events of germination in the zoospore of *Blastocladiella emersonii*. *Current Topics in Developmental Biology*, **6**, 1–44.

[22] WALLACE R.A. & BERGINK E.W. (1974) Amphibian vitellogenin: properties, hormonal regulation of hepatic synthesis and ovarian uptake, and conversion to yolk proteins. *Amer. Zool.*, **14**, 1159–75.

[23] WATERS L. & DURE L. (1966) Ribonucleic acid synthesis in germinating cotton seeds. *J. Mol. Biol.*, **19**, 1–27.

[24] WHITTAKER J.R. (1973) Segregation during ascidian embryogenesis of egg cytoplasmic information for tissue-specific enzyme development. *Proc. Nat. Acad. Sci., U.S.*, **70**, 2096–100.

Chapter 1.3
The Formation of Different Cell Types in Animal Embryos

1.3.1 INTRODUCTION

The molecules which control development must determine which cells should synthesize particular proteins by acting either on the transcriptional or post-transcriptional controls of gene expression. At present, the identity of these molecules is unknown (see previous section) and here we will discuss certain experiments with animal embryos which provide information about the properties which these molecules must possess.

The success of nuclear transfer in amphibia makes it certain that the membranes and cytoplasm of the egg can induce nuclei to participate in all the interactions which are required for the differentiation of the cell types of the embryo (Chapter 2.2). We will therefore be concerned with experiments on the one-cell egg before its organization is altered by many cleavage divisions; much of later development consists of cell interactions during morphogenesis and these will be reviewed in Chapter 3.6.

There are at least three kinds of theory which seek to explain how the membranes and cytoplasm of the egg may organize genetic expression in the multiplying nuclei of the cleaving embryo. These theories may be called 'mosaic', 'reference points' and 'non-instructive', and extreme versions of these theories will be described.

1.3.2 THEORIES OF DEVELOPMENT

(a) Mosaic theory

This theory proposes that within the membranes and cytoplasm of the egg there are groups of different instructions. These are thought to be precisely localized in the cytoplasm so that, as cleavage proceeds, these instructions are packaged around the nuclei of different cells and control their behaviour. The instructions are considered to be chemicals which determine what a cell should synthesize and the ways in which the cell should move and interact with other cells. The implication is that there are many different types of instructions, each grouped in different parts of the egg.

(b) Reference points theory

A reference points theory also proposes that the cytoplasm and the membranes of the egg contain instructions. In this case, it is thought that the instructions issue from reference points and that their influence varies with distance from the reference points; as cells are formed so they find themselves in different positions relative to the reference points and as a consequence they become different from each other.

Traditionally embryologists have thought of reference points as areas of the egg which set up embryological fields of influence. They were thought to do this by being sources of gradients and cells were believed to be able to detect the concentrations of substances in these gradients and thus determine their position. In practice, if the reference points provide any signal whose effect changes with distance then they can serve their function in the embryo and there is no compelling reason to believe that they consist of gradients. The reference point theory contrasts with a mosaic theory in that there are thought to be rather few kinds of instructions involved (say four or five different kinds). Both the mosaic and the reference points theories require that the egg cytoplasm and its membranes should contain localized regions which either contain or issue instructions about development. A 'non-instructive' theory does not depend on the localization of instructions in the egg.

(c) Non-instructive theory

This theory of development proposes that the membranes and cytoplasm of the egg contain no instructions for the nuclei which are formed during cleavage. It is proposed that as the egg is cleaved, so nuclei come to lie in different positions relative to each other and as a consequence become different. For instance, as a consequence of post-fertilization metabolism the nuclei on the outside of the embryo may be in cells which contain more nutrients than those on the inside and conversely the inside cells may contain more metabolic waste products than those on the outside. It is proposed that a particular environment is created around each nucleus as a consequence of cleavage. It is also thought that these environments cause the cells to develop different properties so that they may become reference points as development proceeds.

1.3.3 TESTING THE THEORIES

It is important to discover which of these theories best accounts for the development of embryos, because the phenomena of development cannot be explained in molecular terms until we know the class of molecules which may be important in the process. For instance, if a strict mosaic mechanism is operating in the embryo, then one would expect to find the instructions for development contained in large macromolecules which in themselves can carry sufficient information to specify many cell types. Examples of such molecules are nucleic acids, proteins, and glycosoaminoglycans; if the egg is a mosaic then the discovery of the distribution of these molecules becomes very important. In contrast, the action of a reference point mechanism is likely to depend on small molecules which can readily move within a cell and across cell membranes, and in this case it would be necessary to study the distribution of such molecules in the egg.

The following tests may be used to find out whether or not an egg does contain localized instructions, and if it does so then it is possible to find out whether or not these instructions are distributed as a mosaic or issue from reference points.

(a) Test of isolation

For this test, the egg is divided into fragments and the development of each piece is studied in isolation. If the parts of the egg continue to develop as if they were still in place in the egg, then they must contain their own instructions. It is likely that these instructions have a mosaic distribution if the isolated fragments are a small fraction of the total egg volume. In practice, the parts of the egg may develop in two other ways and these are more difficult to interpret.

Parts of the egg often develop fewer structures than they would in the intact egg. The question of whether or not they contain their own instructions cannot always be answered as the expression of these instructions may have been altered by the fragmentation process itself or by the small size of the isolate. In contrast, if parts of the egg form a complete adult then it is possible to conclude that the egg did not contain a fixed mosaic of instructions. It is also tempting to conclude that the egg did not contain reference points either. However this conclusion does depend on the relationship between the fragments and the distribution of reference points in the egg, and if the egg is divided so that each fragment contains a complete set of reference points then each fragment might still be able to form a complete adult. The development of adults from parts of the egg therefore does not exclude the possibility that the egg contains reference points.

(b) Test of transplantation

In this test, a part of the egg is transplanted to a new site in another egg. If the transplanted part develops in the same way as it would before transplantation then clearly it must have contained its own instructions. It is also possible to discover if the transplanted part can act as a reference point by observing if the recipient embryo changes its pattern of development in response to the graft. However, if the graft does not develop as it would without transplantation, then it is impossible to be certain that it lacks instructions for the expression of these instructions may have been altered either by the operation or by mixture with instructions already present in its new site. Once again, this test can demonstrate the presence but not the absence of instructions.

(c) Test of destruction and replacement

This test can be used to distinguish between a mechanism of development based on a mosaic of instructions and one based on reference points. A way of illustrating the application of this test is to compare the egg with a globe of the world. Consider London on the globe. It might influence tourists for two reasons. On a mosaic theory, it would be the kinds of things which are in London which would have the greatest influence on the tourists (Houses of Parliament, Buckingham Palace, etc.), while on a reference points theory the tourists would be influenced by the particular longitude and latitude of London (i.e. on the distance from the reference points of longitude and latitude, Greenwich and the Equator). If London were scooped from the globe with a large spoon then both the Houses of Parliament, etc., and the locality (longitude and latitude) would disappear and it would be impossible to decide what had previously influenced tourists in London. From this discussion it can be seen that if an egg develops abnormally after part has been removed, then either theory could account for the disturbance.

It is only possible to distinguish between the two theories if a part is removed and then another part is put into its place. For instance, if a reference point theory is operating, then the locality of London (and its influence on tourists) can be restored by transplanting Washington in its place because the particular longitude and latitude of London is restored by this operation. In practice the molecules or influence of the reference point would have to move into the Washington graft. While if a fixed mosaic mechanism was operating then Washington would not be able to substitute for London because the Senate has a different influence to the Houses of Parliament.

(d) Test of maternal effect genes

The mosaic and reference points theories of development suppose that instructions are present in particular regions of the egg's cytoplasm and membranes. The instructions and their position may be determined by nuclear genes and if this is the case, then the mother's genotype should affect development by altering the instructions laid down during oogenesis. Take the case where the egg contained a mosaic of instructions; it should be possible to discover genes which in a mother caused the absence or malposition of tissues and organs in her offspring. If such a gene was found then it would be important to show that the same gene did not have a similar effect on development when it was contributed to the embryo by the father and thus it would be possible to show that the gene was primarily affecting the organization of instructions in the egg. Such genes are called maternal effect genes to emphasize that their effect on development stems from their effect on the egg.

The test consists of searching for maternal effect mutations which appear to determine the instructions for development inside the egg. If such are found then the test can provide clear evidence for the localization of instructions in the egg.

However there are many reasons why a failure to detect such mutants does not prove that localized instructions are absent from the egg. First it is difficult to detect and discover the effects of genes which act early in development. Second the instructions and their localization in the egg may primarily depend on cytoplasmic inheritance and processes of self assembly, with the nuclear genes providing no more than the components of the instructions. For instance the pattern on the cell surface of *Paramecium* can be inherited independently of the nuclear genes [48], and it is not yet clear that egg instructions are not inherited in this manner. If they were inherited in this way then the chances of finding tissue specific maternal effect mutations are reduced.

1.3.4 EXPERIMENTAL USE OF THE TESTS

It is now useful to consider what can be learned about the organization of animal eggs by subjecting them to these tests.

We have only included examples in which several tests have been performed on the same kind of egg because a single test does not by itself provide conclusive evidence about egg organization, and may be misleading.

Germ plasm

The eggs of amphibia and insects often contain distinct germ plasm which is localized in one region of the cytoplasm. This cytoplasm can be recognized by its staining properties and it is included into a group of cells during early cleavage; it can be shown that these cells are the primordial germ cells which will eventually form the eggs and sperm (reviewed for amphibia [1], for insects [36, 49]). This is then a case where a visible structure in the egg may have a direct effect on the cells which inherit it.

EXPERIMENTS ON INSECT GERM PLASM

A germ plasm is found in many insect eggs and its effect on development has been thoroughly studied in *Drosophila* embryos. The egg of *Drosophila,* like that of most insects, does not cleave in the usual sense (Fig. 1.3.1). After fertilization, the nuclei divide seven times (about once every 10 min), and migrate out to a position beneath the egg membrane. At this time a small number of nuclei in the posterior pole are surrounded by germ plasm and are cut off into cells; the rest of the nuclei of the embryo divide four more times before cell membranes are formed around them (blastoderm stage, see [19, 49] for *Drosophila;* see articles in [9, 10] for other insects). The late

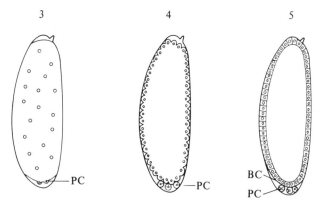

Fig. 1.3.1 Early embryogenesis of *Drosophila.* (1) Fertilized egg with male (MN) and female pronucleus (FN). The egg is surrounded with a vitellin membrane (V), through which the sperm passes by way of a tube, the micropyle (M). The egg is surrounded with a plasma membrane (P). (2) Early cleavage with cleavage nuclei (CN). (3) Pole cell (PC) formation after the seventh cleavage division. (4) Syncytial preblastoderm stage with cell walls starting to form after the twelfth cleavage division. (5) Blastoderm stage with nearly all the nuclei included in a mono-layer of cells beneath the vitellin membrane. From [49].

formation of cell walls in insect embryos is a useful property, for the organization of the embryo up to the blastoderm stage cannot be caused by cell to cell interactions and must directly depend on egg structure. The germ plasm which is incorporated into the primordial germ cells at the posterior pole consists of characteristic polar granules around which are mitochondria and RNA staining material [reviewed 36]. It now appears that the instruction 'form primordial germ cells' is located in this region.

The following experiments demonstrate the special properties of the germ plasm of the egg (Fig. 1.3.2, experiments on *D. melanogaster* [31]). Following other workers, Okada [41] has shown that if the posterior pole is UV-irradiated while nuclei are still at the centre of the egg then no primordial germ cells are formed and the flies which develop into adults are sterile. The irradiation could either destroy a substance or destroy a locality and test (c) was next performed. To distinguish between these two explanations of the results, Okada irradiated eggs at the posterior pole and then injected anterior pole cytoplasm and posterior pole cytoplasm back into this region of the egg. If locality is important, then the injection of any cytoplasm into the irradiated region should restore fertility. However, if a particular instructive substance is destroyed in the irradiated region, then only the reintroduction of this substance into the egg should restore fertility. It is found that only the reintroduction of posterior pole cytoplasm with polar granules can cure the egg's sterility and this result strongly suggests that this cytoplasm contains the instruction 'form primordial germ cells'.

Illmensee then showed that this instruction is antonomous. The posterior egg plasm, which contains the germ plasm, was transferred into the anterior half of a second egg. The nuclei of the second egg moved into this cytoplasm and formed histologically normal primordial germ cells in the anterior part of the embryo. It was possible to show that these were also functional primordial germ cells. To do this, these cells were transferred to the posterior pole of a third egg. This egg was allowed to develop into an adult and it was found that the transferred cells could develop into viable gametes which transmitted their genetic markers to the next generation [31]. This transplantation *tour de force* is the most perfect demonstration of the localization of an autonomous group of instructions in the egg. It should be noted that the destruction of the germ plasm by UV-irradiation does not disturb the development of the rest of the fly and therefore germ plasm does not act as a reference point.

There are maternal mutations of fertility in *Drosophila*, and so it is possible to apply test (d). One maternal effect mutation is *grandchildless (gs)*, which is found in *D. subobscura*. In the homozygous state this mutation causes females to produce sterile sons and daughters. It seems that the mutation has its principal effect on the gonads and does not cause sterility indirectly by effecting other organs in the body: larval gonads

(a) UV irradiation of germ plasm

Result

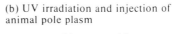

1% flies fertile

(b) UV irradiation and injection of animal pole plasm

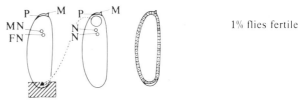

1% flies fertile

(c) UV irradiation and injection of germ plasm

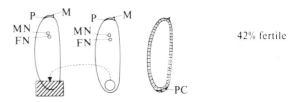

42% fertile

(d) Germ plasm induction of anterior pole cells

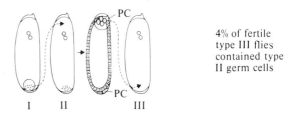

4% of fertile type III flies contained type II germ cells

Fig. 1.3.2 Destruction and replacement of germ plasm in *Drosophila*. (a) UV-irradiation of the posterior germ plasm causes sterility in nearly all the adult flies which develop. This sterility may be cured by injecting germ plasm from an untreated egg (b) and (c). Experiments of Okada *et al.* [41]. (d) Germ plasm from egg I was injected into the anterior pole of egg II. Here it induced the formation of germ cells containing nuclei from egg II. These germ cells were transferred to the posterior pole of egg III. This egg was allowed to develop into an adult and the II germ cells transmitted their genes to the next generation. Experiments of Illmensee [31].

from the offspring of gs/gs females remain sterile when transplanted to wild type hosts, while reciprocally transplanted gonads are fertile. This is then a situation in which one would hope to find an obvious morphological disturbance (such as the absence of polar granules) at the posterior pole of the egg. Unfortunately, all structures in this region appear normal when viewed under the electron microscope; the difference between the embryos of gs/gs mothers and those of wild type mothers seems to be that the nuclei in the affected embryos do not enter the germ plasm directly from the yolk but rather reach it indirectly after remaining beneath the cell membrane in other parts of the egg [21]. Since the late migrating nuclei in the sterile embryos are eventually surrounded by germ plasm but never form primordial germ cells, it seems that this cytoplasm can only impose its instructions at a very short period in embryogenesis. If this analysis is correct, then the effect of the *gs* gene may be on the time that nuclei receive an instruction rather than on the instruction itself. Of course it remains possible that the instruction is altered and that this change cannot be detected by microscopy.

EXPERIMENTS ON AMPHIBIAN GERM PLASM

The germ plasm can also be discerned in sections of some amphibian eggs and this account of its behaviour is based on studies with the anurans *Rana temporaria* and *Xenopus laevis* [1]. The plasm is first observed after fertilization in patches beneath the egg membrane in the vegetal half of the egg; it migrates towards the vegetal pole and concentrates here during the first cleavage division. The patches adhere to the inside of cell membranes and they are found in two to fourteen cells of

Table 1.3.1   Irradiation and replacement of parts of amphibian egg cytoplasm

A. Irradiation at either pole

Pole irradiated	Number of eggs	Number developing into tadpoles	Percentage with primordial germ cells
Vegetal	39	39	0
Animal	20	20	100

Results of experiments with 7752 ergs/mm² of ultraviolet irradiation.

B. Injection of cytoplasm into eggs irradiated at the vegetal pole

Source of cytoplasm	Number of eggs normal after operation	Number developing into normal tadpoles	Percentage of normal tadpoles with primordial germ cells
Vegetal pole from 2-cell	64	51	29
Animal pole from 2-cell	31	28	0

Data from Smith's experiments [47].

the early blastula; in the late blastula they move and surround the nuclear membrane, and the cells containing this material migrate into the genital rudiments and differentiate into the germ cells. The association between this material and the primordial germ cells could be fortuitous or it could be that it carries the instruction 'be a primordial germ cell'.

There are now good reasons for believing that such an instruction is carried either in the germ plasm or in material closely associated with it. Test (c) was applied to this material by Smith several years before the similar studies on insect germ plasm described in the previous section. Smith [47] exposed eggs of *Rana pipiens* to UV irradiation, and showed that if the egg were irradiated at the vegetal pole then tadpoles were formed which lacked germ cells, while if it were irradiated at the animal pole then there was no effect on development. Clearly the irradiation could either destroy a substance or destroy a locality. Employing test (c), it was shown that only the injection of vegetal cytoplasm can cure the sterility of eggs irradiated at that pole and the conclusion must be that this cytoplasm contains instructions for forming primordial germ cells.

The biochemical properties of this instruction are not known in either amphibians or insects. However the germ plasm of amphibians, like that of insects, contains mitochondria and material which stains with RNA reagents. In the amphibian case it has also been shown that the action spectrum of UV sterilization is not too dissimilar from the UV absorption spectrum of nucleic acids.

Taken together, these experiments on germ plasm provide good evidence for the localization of a particular instruction or group of instructions in a small region of the egg. These instructions do not appear to act as reference points because following their inactivation by UV the rest of the embryo appears to develop normally. We will next consider two cases where a region of the egg appears to issue instructions to other regions and thus appears to be acting as a reference point.

Grey crescent region of the amphibian egg

Most of the animal half of the amphibian egg is covered by a black layer of pigment granules which lie beneath the membrane. After fertilization or parthenogenetic activation, these granules migrate upwards on one side of the egg leaving a grey crescent slightly darkened by the granules which have been left behind. The cytoplasm and membranes of the grey crescent are eventually incorporated into the dorsal lip of the blastopore and it has been known for some time that this region of the gastrula organizes the longitudinal axis of the embryo (Spemann's experiments, see Chapters 3.4 and 3.6). The special properties of this region are thought to be the consequence of interactions between the yolky cytoplasm and the material bound to the cell membrane on the animal side of the egg's equator; it is certainly the case that by shifting the relationship between these two structures the position of the blastopore and the number of blastopores in the embryo can be altered [40].

In a series of elegant experiments with *Xenopus laevis,* Curtis has shown that the organizing ability of the dorsal lip of the blastopore derives from instructions close to or in the membrane of the grey crescent region of the egg (Fig. 1.3.3, [12, 13]). The first experiment shows that normal gastrulation can be blocked by cutting out the cell membrane and adhering cytoplasm in the region of the grey crescent of a fertilized one-cell egg. The

embryo divides normally and forms a blastula; cell division continues for some time in this mass of cells but usually gastrulation does not occur, no axis is formed and the cells do not obviously differentiate (Fig. 1.3.3a). This observation suggests that this region of the egg might issue instructions to other parts.

The next experiment provides evidence that particular substances in the grey crescent region are important. Employing test (b), the grey crescent region may be transplanted to the

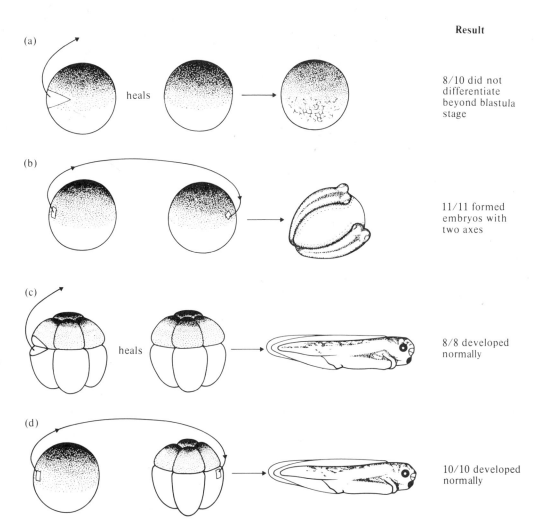

Result

(a)   heals   8/10 did not differentiate beyond blastula stage

(b)   11/11 formed embryos with two axes

(c)   heals   8/8 developed normally

(d)   10/10 developed normally

Fig. 1.3.3 Removal and transplantation of the grey crescent region of *Xenopus laevis.* If the membrane and adhering cytoplasm is removed from the grey crescent region at the one-cell stage, then gastrulation does not occur (a). If this material is removed at the eight-cell stage then development continues normally and this suggests that an instruction has passed from the periphery of the egg towards the centre between these two stages (c). The grey crescent region can be transplanted to a second egg where it causes the formation of a double axis (b). It is unable to impose this effect on an eight-cell embryo (d). See discussion in text; experiments of Curtis [12, 13].

vegetal margin of another egg and in this case two dorsal blastopore lips are formed and the embryo develops a double axis (Fig. 1.3.3b); this result cannot be due to the creation of a novel locality by the operation because animal pole material grafted to the same site does not organize a new axis. A further reason for believing that a unique locality is not involved is that if the grey crescent locality is disturbed by removing 50–100 μm³ of cytoplasm from beneath the membrane, then the grey crescent region can still exert its normal effect and the embryo gastrulates.

It seems most likely that the instructions of this reference point are at first concentrated near the membrane in the grey crescent region and that as development proceeds, so they move away into the cytoplasm. Evidence that this is the case comes from experiments in which the membrane and adhering cytoplasm are removed from the grey crescent region of an eight-cell embryo; in contrast to their removal at the one-cell stage, this operation does not disturb normal development (Fig. 1.3.3c), and it is likely that by this time they are widely distributed in the cytoplasm or have already exerted their effect. It is also the case that once this distribution has occurred, it is impossible for the grey crescent region of a one-cell egg to impose a second axis on the embryo (Fig. 1.3.3d).

These experiments have been interpreted on a gradient model. It is supposed that the membrane and associated material in the grey crescent region is the source of a gradient of instructions

which moves into the cytoplasm as development proceeds. Since the removal of the region at the eight-cell stage does not disturb development it is clear that either these instructions have completed their effect by the eight-cell stage or that regions in the cytoplasm must themselves become local sources of the gradient.

Animal and vegetal regions in the sea urchin

Studies on sea urchin embryos provide a second example of reference points inside an embryo. The sea urchin embryo may be divided at the four-cell stage and each cell is capable of forming a complete larva (pluteus larva [17, 18]). The first two cleavage divisions pass through the animal and vegetal poles of the egg and so each of these four cells contains animal and vegetal pole material. This material is now known to contain reference points and the presence of localized regions in these eggs can only be detected if the egg is divided along the equator lying between the animal and vegetal poles.

A famous series of manipulation studies have been conducted on the eggs of the sea urchin *Paracentrotus (Strongylocentrotus) lividus* by Hörstadius and others [reviewed 27, 29]. The fertilized eggs have a subequatorial pigment band and it is possible to observe the segregation of the pigmented cytoplasm during cleavage and to follow the segregation of other parts of the egg by staining the cell membrane and cytoplasm with vital dyes. Using these

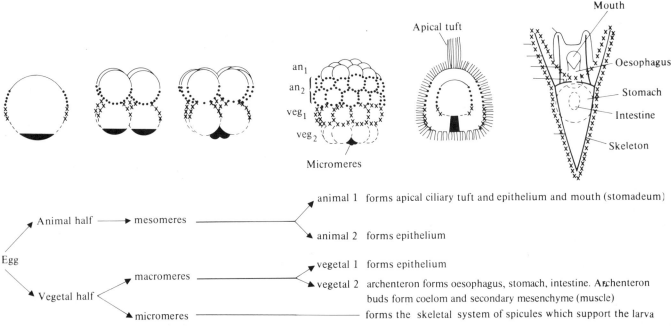

Fig. 1.3.4 Normal development of *Paracentrotus lividus,* diagram of cytoplasmic segregation and cell fate. From Hörstadius [29].

techniques, it can be shown that particular parts of the egg regularly form particular parts of the pluteus larva during normal development. The cell lineage of normal development is shown in Fig. 1.3.4.

The distribution of instructions can be studied by observing the development of isolated parts of the egg (test a). Ideally one would wish to break the egg up into 20 or so fragments, but in practice it is only possible to study the development of isolated animal and vegetal halves. These are broken from the unfertilized egg and subsequently fertilized in isolation; those parts which contained a female nucleus at isolation are diploid while those that lacked this nucleus at isolation are haploid, but the ploidy does not influence their development. The development of these isolates is illustrated in Fig. 1.3.5. Isolated animal halves form apical tufts and ciliated blastulae; in rare cases they contain a mouth but they never form a complete larva. Isolated vegetal halves usually form larvae which lack an apical tuft and mouth; these usually have a reduced epithelium

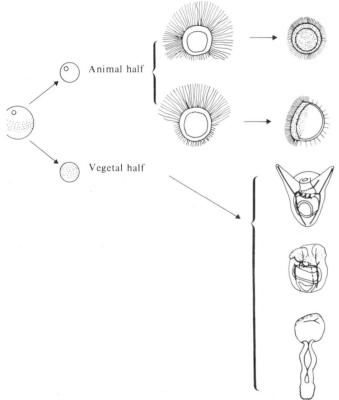

Fig. 1.3.5 Development of animal half and vegetal half of the one-cell *Paracentrotus lividus* egg. The one-cell egg is broken in half and the two halves fertilized. The animal half forms a blastula with an enormously enlarged apical tuft and this never gastrulates. In rare cases the vegetal half will form a complete larva but this never passes through a stage which contains an apical tuft. See discussion in text; experiments of Hörstadius [27, 29].

which restricts the development of the skeleton, and the stomach tends to be large and many exogastrulate.

These observations on half eggs, which show an apparent autonomy of development of some of the structures of the larva, would be consistent with a loose mosaic pattern of instructions in the egg. However further experiments do not fit this view. In these experiments, the embryo is divided into parts at a later stage of development, when it consists of the layers an-1, an-2, veg-1, veg-2, and micromeres (Figs 1.3.4 and 6). It could be argued that the distribution of instructions in the cells of the embryo at this stage of development is not the same as their distribution in the egg; however this seems unlikely because after isolation the cells derived from the animal half of the egg (an-1 and an-2 layers) develop in much the same way as the isolated animal half of the one-cell egg (compare Figs 1.3.5 and 6), and for this reason it will be assumed that little redistribution of instructions occurs.

The idea that there are two sources of instructions in the egg (reference points) comes from the following studies (tests a and b). First there is evidence which supports the theory that there is the instruction 'form the normal derivatives of the vegetal half of the egg' which has its most powerful source at the vegetal pole. The animal half of the egg (an-1 and an-2 layers) is isolated and combined in turn with the layers veg-1, veg-2, and micromeres. The larvae which are formed from these combinations become progressively more normal as cells are added from regions closer to the vegetal pole (Figs 1.3.6 a to d). It should be noticed that cell lineage studies have shown that in these combinations the veg-1 cells and micromeres transmit instructions to the animal cells; for instance, in isolation the animal half forms a huge apical tuft which covers most of the surface of the early embryo (Fig. 1.3.6a), while in combination with veg-1 cells the tuft is restricted to its normal apical position (Fig. 1.3.6b). Similarly, in combinations of animal half cells with micromeres, the an-1 and an-2 cells which normally form only the epithelium and mouth of the larva now forms its oesophagus, stomach, and coelom as well (Fig. 1.3.6d). The development range of veg-2 cells is also extended in these combinations. Normally they never form the skeleton while here they regularly do so (Fig. 1.3.6c).

In addition to this vegetal instruction, there appears to be another instruction with its most powerful source at the animal pole; its effect opposes that of the vegetal instruction. The evidence in favour of this view again comes from studies on isolated cells and again I will assume that the distribution of instructions in these cells is the same as that in the egg. It is possible to assess the strength of the instruction 'form the normal derivatives of the animal half of the egg' by titrating the development of the layers an-1 and an-2 against the addition of different numbers of micromeres (Fig. 1.3.7). The an-2 layer requires the addition of only two micromeres to form a normal larva while the an-1 layer must be combined with four

(a) An$_1$ and An$_2$, alone

(b) An$_1$ and An$_2$, + Veg$_1$

(c) An$_1$ and A$_2$ + Veg$_2$

(d) An$_1$ and An$_2$ + micromeres

Fig. 1.3.6 Development of *Paracentrotus* animal-1 and animal-2 layers with additional vegetal layers. Compare the development of these layers with their normal development in an intact embryo (Fig. 1.3.4). Notice that in combination (b) the vegetal-1 cells restrict the area occupied by the apical tuft.

In combination (c), the vegetal-2 cells (– – –) form the skeleton which they would never do in normal development. See discussion in text; experiments of Hörstadius [27, 29].

micromeres to achieve this result and this observation suggests that the most powerful source of this instruction is at the animal pole. It should be noticed that as in the previous series of experiments, it is the an-1 and an-2 cells which form the oesophagus, stomach, and coelom of the larvae and the range of structures which they form is extended in comparison with normal development.

These results indicate that the egg does not contain specific instructions for every organ (mosaic theory) but rather it appears as if the egg contains two types of instructions whose effects are set up in opposing chemical gradients. In fact, the results need not be explained by a chemical model; any two signals whose effective power changes with distance from the animal and vegetal poles could account for these observations.

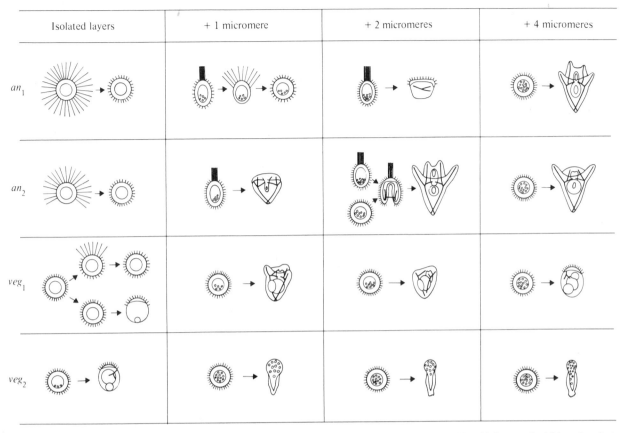

Fig. 1.3.7 Titration of the animal effect of cell layers by addition of micromeres in *Paracentrotus*. The layers animal-1, animal-2, vegetal-1, and vegetal-2 are cultured with different numbers of micromeres. The animal-1 layer requires the addition of four micromeres in order to form a complete pluteus larva while only two micromeres need be added to an animal-2 layer in order to obtain harmonious development. See discussion in text; experiments of Hörstadius [27, 29].

If a chemical gradient model is applied, it remains possible that in the unfertilized egg, the potential sources of each gradient or signal are located at the animal and vegetal poles (reference points) and that they are not fully expressed until the cell layers are formed; it is only at the multicellular stages that the distribution of these instructions has been studied. Also it is clear from the cell combination experiments that cells away from the animal and vegetal poles are themselves capable of issuing animal and vegetal instructions; in other words they are not dependent on a continuous supply of instructions from these poles in order to exert their effect on other cells of the embryo.

The chemical basis of these two gradients of instructions is uncertain. However a gradient of reducing potential can be demonstrated by vital dyes [6] and this gradient does coincide with the instruction gradient in both normal development and in a variety of experimental situations [28]. Further discussion on the possible nature of these instructions may be found in references [30] and [45].

Eggs of eutherian mammals

Some of these tests have also been applied to mammalian eggs to discover the features of their organization. These tests have so far failed to demonstrate the presence of localized instructions.

It has recently become possible to study the development of isolated egg fragments (as in sea urchin experiments), but all published work has been concerned with the use of isolated cells from early cleavage stages (test a). In this case it has been shown that at least one cell of an eight-cell stage rabbit egg can form an adult rabbit [37] and that each cell of a four-cell stage mouse egg can form a complete blastocyst (this is a hollow ball which implants on the uterus [50]. However, isolated cells do not develop very successfully by themselves and it is necessary to combine them with other embryos to follow their full development. In this way the developmental potential of mouse blastomeres has recently been tested in detail [34].

Kelly dissociated four-cell stage embryos and surrounded each isolated cell with cells from another embryo; the isolated cell could be distinguished from those of the carrier embryo both by the coat colour genes and the enzyme marker which it contained (Fig. 1.3.8). Four such composites could be obtained from each dissociated four-cell stage embryo, and they were transferred back to the uterus of a foster mother to continue their subsequent development.

First it was shown that each isolated cell could form parts of the foetus, yolk sac, and placenta of each of the four composites which developed halfway through pregnancy (at this time the embryo has just started to form muscle blocks, the somites). In several cases, three of the four composites were born alive and developed into adult mice; the isolated cell had formed much of the coat colour and germ cell population of these animals. These results are particularly impressive because, unlike sea urchin experiments, the isolated cells form adult and not just larval structures. Similar experiments have been performed by Gardner on later embryos [20]. In this case it could be shown that cells isolated from the inside of the blastocyst were able to colonize a carrier blastocyst to which they had been transferred and that the isolated cell could form part of many of the organs

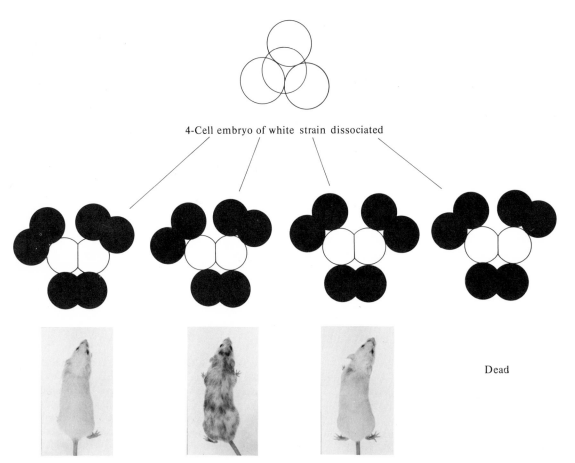

4-Cell embryo of white strain dissociated

Dead

Fig. 1.3.8 Development potential of the blastomeres of a four-cell stage mouse egg. A four-cell egg from a white mouse was dissociated into single cells and each of these was allowed to divide once in isolation. Each of these was then surrounded with blastomeres from a black mouse and the combinations were allowed to develop to the blastocyst stage in culture. The blastocysts were transferred to the uterus of a foster mother and three white mice were born. The coat of each mouse contained cells from the white egg and the cells from this egg had also formed oocytes in each mouse. This shows that at least three cells at the four-cell stage retain the potential to form part of the coat and part of the gonads of an adult mouse. Experiments of Kelly [34].

of the mice which were born. Taken together all these results suggest that there are no localized instructions in the mouse egg, but it should be noticed that the isolation studies on these eggs are not yet as detailed as those performed on sea urchin eggs, and it is still possible that there are localized instructions along the radius of the egg.

The development of different cell types in the mouse embryo appears to depend on the position which the cells occupy during early cleavage. The evidence for this view comes from experiments in which isolated cells were placed inside or outside another embryo [25, 26]. In an outside position the cells tend to form part of the placenta while if they are placed in an inside position then they tend to form part of the foetus and the adult which develops from it. The conclusion must be that if the cells of these early cleavage stages contain any instructions at all then the effect of these instructions is obscured by the effect of cell position. It is interesting to see if it is possible to envisage a mechanism of development which does not depend on any instructions in the egg.

The following is a model which has been applied to the mouse egg. This model proposes that an instructionless egg is divided by cleavage so some cells lie on the outside and others on the inside of a cellular lump. As a consequence of the environment on the outside, the exposed cells are caused to stick to each other and to pump inwards and these cells therefore form a liquid filled vesicle. In contrast, the environment inside the lump causes the enclosed cells to stick to each other and not to pump, and also it causes the small ball of cells which is formed to stick to the sides of the liquid filled vesicle at some point. This simple model accounts for the formation of a blastocyst containing two cell types and an inner cell mass at some point; in other words it can account for the differentiation of cells and the formation of an axis in this cell population, and a series of events of this kind can be used to build up an embryo.

The point of describing such a model is to show that it is possible to conceive of development occurring in the absence of instructions laid down in the egg. It is probable that once the initial differentiation has occurred in such embryos, then the cell groups can act as reference points.

1.3.5 HOW GENERAL IS THE USE OF MOSAIC INSTRUCTIONS AND REFERENCE POINTS?

Despite the wealth of embryological experiments, there are rather few eggs which have been tested as thoroughly as those in the previous section. In addition to the experiments mentioned, there are some beautiful studies on the polar lobes of molluscs [see 5, 7, 16, 23, 52]. Work with these embryos has clearly shown that there are localized instructions in mollusc eggs, but it is not yet clear whether they are distributed as a mosaic or whether they issue from reference points [23]. Because the correct interpretation of these experiments is still uncertain they will not be discussed and for similar reasons many other experiments have been excluded (for review of other experiments, see [15] [43] [51]).

The experiments described in Section 1.3.4 provide examples of parts of eggs which contain but do not issue instructions (germ plasm), and of parts of eggs which issue instructions (grey crescent of amphibia and the animal and vegetal halves of sea urchin eggs). In contrast to these, the mammalian eggs appears to contain no instructions and reference points must be created during cleavage. It would next be useful to discover which of these mechanisms usually operates in eggs.

In this section it will be argued that few, if any, eggs contain a detailed mosaic of instructions laid down during oogenesis. The evidence for this view mainly comes from further studies on the eggs of insects and amphibia.

Mutant and UV studies on insect eggs

Several observations have led some workers to believe that the egg of *Drosophila* contains groups of localized developmental instructions in addition to those for primordial germ cells. One experiment which led to this view involved the UV irradiation of parts of the egg. If the mid-lateral cytoplasm is UV-irradiated during early cleavage, then the flies which develop have missing or crippled legs, distorted body segments, missing head bristles, and bent wings [39]. Unfortunately (without the complete use of test c) it is impossible to decide whether the defects are caused by damaging instructions in this part of the cytoplasm or by some effect on locality (e.g. by changing the time at which nuclei move to the locality, cf. observations on the eggs of gs/gs mothers, see Section 1.3.4: experiments on insect germ plasm). Another approach has been to look for maternal effect mutants. So far several mutants which effect the development of a particular region of the body have been found [41]. For instance, homozygous *almondex* females (amx/amx) produce a few heterozygous daughters when mated to amx/+ males; 80% of these daughters have a distorted abdomen or thorax while the rest of the body appears normal [46], and the abnormalities appear similar to those produced by UV irradiation of mid-lateral cytoplasm (see above). However, as in the case of *grandchildless*, it is not clear that a particular set of instructions is missing from the egg; it is still possible that the time at which the nuclei encounter a particular locality is altered in these eggs and it is also possible that the movement of instructions from a reference point to this region of the egg is confused.

The evidence from studies on insect eggs is more consistent with a reference point explanation of the differentiation of structures other than primordial germ cells. One recent experiment is particularly revealing [32, 33]. Kalthoff irradiated the anterior one eighth of the egg of *Smittia parthenogenetica* before nuclear migration to this region had occurred. The larva which developed consisted of two abdomens which faced each other and no head structures were visible. These bicaudal larvae could be produced either by altering the effect of some reference point which issues the instructions 'form head and thorax structures' or by the mosaic of instructions for abdomen development moving around to the anterior end of the egg and redistributing themselves in the correct arrangement for normal abdomen development. The latter explanation seems unlikely, because the germ plasm and primordial germ cells remain in their correct location when similar double abdomens are produced by UV-irradiation of the anterior pole of another insect *Chironomus dorsalis* [55]. In Kalthoff's experiments the effect of UV irradiation is photo-reversible and this suggests that the UV-induced effect is mediated by nucleic acids. It is also worth noting that the bicaudal condition can be produced by a maternal effect gene in *Drosophila* [4]. In the eggs produced by mutant mothers, the polar granules are located at only one pole of the egg and this observation suggests that while the reference points for arranging the structures of the double abdomen are set up during oogenesis, these do not in addition determine the position of the polar granules.

In addition to maternal effect mutants, there are in *Drosophila,* a large number of homeotic mutants. These are mutants in which a certain cell, tissue, or organ type is transformed into another one, for instance an antenna into a leg [22]. It is important to record that these mutants are never of the maternal effect type and therefore do not obviously involve changes in the instructions inside the one-cell egg. The absence of maternal effect mutants for a single cell, tissue or organ type suggests the absence of general mosaicism in the egg (for further studies on the insect egg see [14]).

Mutant studies in amphibia

It may be said that as in insects, maternal effect genes in amphibia do not obviously disturb the arrangement of particular parts of the embryo. At present the best collection of amphibian mutant genes which affect development is to be found in the axolotl [3]. Maternal effect mutations in this species may prevent development beyond the blastula stage (*o*), may cause the accumulation of large amounts of fluid in the blastocoel (*f*), may lead to the excess production of steroids (*v*), and may cause irregular cleavage (*cl*). These do not appear to block the development of a specific cell, tissue, or organ type and they appear to have more general metabolic effects. Mutations which do have cell, tissue, or organ specific effects (e.g. cardiac non-function (*c*), and eyeless (*e*)) are not of the maternal effect type.

Until maternal effect mutants of specific cell, tissue, or organ types are discovered it seems reasonable to conclude that the eggs of insects and amphibia only contain localized instructions

for the formation of primordial germ cells and it is possible that this conclusion applies to all embryos. It should be noted that several maternal effect mutants which appear to disturb reference points are known: the *bicaudal* mutation is one example and the well known gene which determines the direction of shell coiling in the snail *Limnaea peregra* is another [2].

We can summarize the discussion so far in four statements.

(1) There are several well worked examples of reference points in one-cell eggs.
(2) There are, at present, no reasons for believing that any egg contains many different localized areas of cytoplasm with particular instructions.
(3) Eggs from many phyla can be divided into fragments and in many cases each of the fragments can form complete larvae. The planes in which the egg can be divided to give this result are not random which suggests that many eggs contain reference points.
(4) The primordial germ cells appear to be the only cell, tissue, or organ type of either a larva or an adult which is specified by a localized instruction in the egg.

1.3.6 THE ORIGIN AND ACTION OF REFERENCE POINTS IN THE EGG

The previous discussion suggests that reference points control the location of most gene expression in most early embryos. The molecules involved must be able to convey information to the nuclei of the cleaving embryo and it is pertinent to speculate on the mechanism by which the reference points act.

There is no conceptual problem in setting up the reference points during oogenesis because parts of the egg are in different positions with respect to the cells which surround them. For example, there appears to be direct cytoplasmic continuity between the follicle cells and one end of the insect egg [35], and cytoplasmic deposits in the gastropod mollusc egg correspond to the position of the follicle cells which previously surrounded the egg in the ovary [42]. If the position of the reference points is determined by the cells surrounding the egg, then the structure of the ovary organizes the future development of the embryo. However it is worth mentioning that no abnormality of development has as yet been associated with any abnormal arrangement of the cells surrounding the developing egg, and so the effect of these cells on the organization of the embryo remains uncertain. It is important to discover the extent to which the characters of the egg depend on the characters of the primordial germ cells and the characters of the female body in which they develop. There has been one study of this problem. Blackler [1] grafted primordial germ cells between *Xenopus laevis* and *X. tropicalis* and found that the size and pigment distribution in the eggs which developed was that of the primordial germ cell type and was not controlled by the somatic tissues of the ovary (contrast with sex determination, Chapter 4.1). This approach should be more widely exploited and it would be particularly interesting to find out how the primordial germ cells of gs/gs *Drosophila* females would develop in a wild type host. It is also the case that the mouse egg does not appear to contain reference points; in this case it is necessary to think of ways in which they may be formed during development (Section 1.3.4: eggs of eutherian mammals).

It is known that reference points can exert their effects across cell membranes and the experiments of Hörstadius on sea urchin eggs demonstrate this point particularly clearly. Their action across cell membranes probably restricts the kinds of molecules which can be used. It is known that proteins and large nucleic acids cannot readily pass between cells in contact in tissue culture (see Chapter 3.2) and it is likely that this restriction also applies within the early embryo. If this is correct, then the reference points must act with substances of molecular weights less than 1000.

If reference points do act with small molecules then these molecules are likely to convey information by their mode of presentation to the nuclei of the embryo. At present, detailed studies are being conducted on two models of reference points and their instructions. One possibility is that reference points are sources of chemical gradients and that cells and cytoplasm are able to detect and respond the concentration of the chemical in the gradient [see 11, 38, 53, 54]. Another possibility is that the reference points are sources of pulsatile signals and that the cytoplasm and cells can detect differences in the phase of the signals [see 8, 24]. Both these models are designed to explain pattern formation in comparatively large cell populations but it is also necessary for similar models to explain the very early cell determination which is observed during the early cleavage stages of the embryos discussed in this chapter.

It should also be remembered that cells can interact in other ways (see Part 3), and we still have insufficient evidence to decide which of these methods are used by reference points in the early embryo.

1.3.7 SUMMARY AND CONCLUSIONS

(1) The eggs of many phyla contain regions of membrane or cytoplasm which affect the development of the embryo. These regions can be apparent in the unfertilized egg, but in many cases they only become clearly defined after cytoplasmic movement occurs inside the egg following fertilization or parthenogenetic activation.

(2) The development of primordial germ cells in insects and amphibia seems to depend directly on a localized group of instructions in one region of the egg (germ cell determinants). At present there is no evidence which clearly shows that the development of any other cell, tissue, or organ type solely depends on a localized group of instructions.

(3) The development of the somatic cell, tissue, and organ types appears to depend on a few reference points inside the egg (grey crescent region of amphibia, animal and vegetal poles of sea urchins). These appear to issue instructions, and as development proceeds so the ability to issue instructions is no longer confined to the reference point. The reference points are presumably established by asymmetric processes which occur during oogenesis.

(4) The development of mammals appears to occur from an instructionless egg. In this case the reference points may be set up during cleavage.

(5) Thus the egg or the early embryo contains a few sources of instructions. These are enclosed in different cells and it is probable that the generation of further heterogeneity depends on interactions between these cells (see Chapters 3.4 and 3.6).

(6) Since the development of organization in many embryos appears to depend on reference points, one may speculate that small molecules which can readily move through the cytoplasm and between cells will turn out to be the determinators of cell type heterogeneity.

1.3.8 REFERENCES

[1] BLACKLER A.W. (1966) Embryonic sex cells of amphibia. In *Advances in reproductive physiology*, **I**, 1–28 (Ed. A. McLaren). Logos and Academic Press.

[2] BOYCOTT A.E., DIVER C., GARSTANG S.L. & TURNER F.M. (1930) The inheritance of sinistrality in *Limnaea peregra* (Mollusca, Pulmonata). *Phil. Trans. R. Soc.*, **219**, 51–131.

[3] BRIGGS R. (1973) Developmental genetics of the *axolotl*. In *Genetic mechanisms of development, 31st Symp. Soc. Dev. Biol.*, pp. 169–99 (Ed. F. Ruddle). Academic Press, New York & London.

[4] BULL A.L. (1966) *Bicaudal*, a genetic factor which effects the polarity of the embryo in *Drosophila melanogaster*. *J. Exp. Zool.*, **161**, 221–42.

[5] CATHER J.N. & VERDONK N.H. (1974) The development of *Bithynia tentaculata* (Prosobranchia, Gastropoda) after removal of the polar lobe. *J. Embryol. exp. Morph.*, **31**, 415–22.

[6] CHILD C.M. (1936) Differential reduction of vital dyes in the early development of echinoderms. *Wilhelm. Roux. Archiv. EntwMech. Org.*, **135**, 426–51.

[7] CLEMENT A.C. (1962) Development of *Ilyanassa* following the removal of the D macromere at successive cleavage stages. *J. Exp. Zool.*, **149**, 193–215.

[8] COHEN M.H. (1971) Models for the control of development. In *Control mechanisms of growth and differentiation. Soc. Exp. Biol. Symp.*, **25**, 455–76 (Eds. D.D. Davies & M. Balls). Cambridge University Press.

[9] COUNCE S.J. & WADDINGTON C.H. (1972) (Eds.) *Developmental systems: Insects*, **1**. Academic Press, New York & London.

[10] COUNCE S.J. & WADDINGTON C.M. (1973) (Eds.) *Developmental systems: Insects*, **2**. Academic Press, New York & London.

[11] CRICK F.H.C. (1971) The scale of pattern formation. In *Control mechanisms of growth and differentiation. Soc. Exp. Biol. Symp.*, **25**, 429–38 (Eds. D.D. Davies & M. Balls). Cambridge University Press.

[12] CURTIS A.S.G. (1960) Cortical grafting in *Xenopus laevis*. *J. Embryol. Exp. Morph.*, **8**, 167–73.

[13] CURTIS A.S.G. (1962) Morphogenetic interactions before gastrulation in the amphibian *Xenopus laevis*—the cortical field. *J. Embryol. Exp. Morph.*, **10**, 410–22.

[14] COUNCE S.J. (1973) The causal analysis of insect embryogenesis. In *Developmental systems: Insects* (Eds. S.J. Counce & C.H. Waddington). Academic Press, New York & London.

[15] DAVIDSON E.H. (1968) *Gene activity in early development.* Academic Press, New York & London.

[16] DOHMEN M.R. & VERDONK N.H. (1974) The structure of a morphogenetic cytoplasm, present in the polar lobe of *Bithnyia tentaculata* (Gastropoda, Prosobranchia). *J. Embryol. exp. Morph.,* **31,** 423–33.

[17] DRIESCH H. (1891) Described in English in Morgan T.H. (1927) *Experimental embryology,* pp. 307–13. Columbia University Press.

[18] DRIESCH H. (1892) English translation in *Foundations of experimental embryology,* pp. 38–50 (Eds. B.H. Willier & J.M. Oppenheimer). Prentice Hall, New Jersey.

[19] FULLILOVE S.L. & JACOBSON A.G. (1971) Nuclear elongation and cytokinesis in *Drosophila montana. Develop. Biol.,* **26,** 560–77.

[20] GARDNER R.L. (1975) Analysis of determination and differentiation in the early mammalian embryo using intra- and inter-specific chimeras. In *Developmental biology of reproduction. 33rd Symp. Soc. Develop. Biol.* (Ed. C.L. Markert). 207–236. Academic Press.

[21] GEHRING W. (1973) Genetic control of determination in the *Drosophila* embryo. In *Genetic mechanisms in development. 31st Symp. Soc. Develop. Biol.,* pp. 103–28 (Ed. F. Ruddle). Academic Press.

[22] GEHRING W.J. & NÖTHIGER R. (1973) The imaginal discs of *Drosophila.* In *Developmental systems: Insects,* **2,** 211–90 (Eds. S.J. Counce & C.H. Waddington). Academic Press.

[23] GEILENKIRCHEN V.L.M., VERDONK N.H. & TIMMERMANS L.P.M. (1970) Experimental studies on morphogenetic factors localized in the first and second polar lobe of *Dentalium. J. Embryol. Exp. Morph.,* **23,** 237–43.

[24] GOODWIN B.C. (1963) *Temporal organization in cells. A dynamic theory of cellular control processes.* Academic Press.

[25] HERBERT M.C. & GRAHAM C.F. (1974) Cell determination and biochemical differentiation of the early mammalian embryo. In *Current topics in developmental biology,* **8,** 152–78 (Eds. A. Monroy & A. Moscona). Academic Press.

[26] HILLMAN N., SHERMAN M.I. & GRAHAM C.F. (1972) The effect of spatial arrangement on cell determination during mouse development. *J. Embryol. Exp. Morph.,* **23,** 263–78.

[27] HÖRSTADIUS S. (1939) The mechanics of sea urchin development studied by operative methods. *Biol. Rev. Camb. Phil. Soc.,* **14,** 132–79.

[28] HÖRSTADIUS S. (1952) Induction and inhibition of reduction gradients by micromeres in sea urchin eggs. *J. Exp. Zool.,* **120,** 421–31.

[29] HÖRSTADIUS S. (1973) *Experimental embryology of echinoderms.* Clarendon Press, Oxford.

[30] HÖRSTADIUS S., JOSEFSON L. & RUNNSTROM J. (1967) Morphogenetic agents from unfertilized eggs of the sea urchin *Paracentrotus lividus. Develop. Biol.,* **16,** 189–202.

[31] ILLMENSEE K. & MAHOWALD A.P. (1974) Transplantation of posterior pole plasm in *Drosophila.* Induction of germ cells in the anterior pole of the egg. *Proc. nat. Acad. Sci., U.S.,* **71,** 1016–20.

[32] KALTHOFF K. (1971) Position of targets and period of competence for UV-induction of the malformation 'double abdomen' in the egg of *Smittia* spec. (*Diptera, Chironomidae*). *Wilhelm. Roux. Entwicklungsmech. Organismen.,* **168,** 63–84.

[33] KALTHOFF K. (1971) Photoreversion of the malformation 'double abdomen' in the egg of *Smittia* spec. (*Diptera, Chironomidae*). *Develop. Biol.,* **25,** 119–32.

[34] KELLY S.J. (1975) Potency of early cleavage blastomeres of the mouse. In *The early development of mammals. British Society of Developmental Biology Symposium,* **2,** 97–105 (Eds M. Balls & A.E. Wilde). Cambridge University Press.

[35] KING R.C. (1970) *Ovarian development in Drosophila melanogaster.* Academic Press.

[36] MAHOWALD A.P. (1972) Oögenesis. In *Developmental systems: Insects,* **1,** 1–47 (Eds S.J. Counce & C.H. Waddington). Academic Press.

[37] MOORE N.W., ADAMS C.E. & ROWSON L.E.A. (1968) Developmental potential of single blastomeres of the rabbit egg. *J. Reprod. Fert.,* **17,** 527–31.

[38] MUNRO M. & CRICK F.H.C. (1971) The time needed to set up a gradient: detailed calculations. In *Control mechanisms of growth and differentiation. Soc. Exp. Biol. Symp.,* **25,** 439–53 (Eds D.D. Davies & M. Balls). Cambridge University Press.

[39] NÖTHIGER R. & STRUB S. (1972) Imaginal defects after UV-microbeam irradiation of early cleavage stages of *Drosophila melanogaster. Rev. Suisse. Zool.,* **79,** 267–79.

[40] PASTEELS J.J. (1964) The morphogenetic role of the cortex of the amphibian egg. *Advances in Morphogenesis,* **3,** 363–387 (Eds M. Abercrombie & J. Brachet). Academic Press.

[41] POSTLETHWAIT J.H. & SCHNEIDERMAN H.A. (1973) Developmental genetics of *Drosophila* imaginal discs. *Ann. Rev. Genetics,* **7,** 381–433.

[42] RAVEN C.P. (1964) Mechanism of determination in the development of gastropods. *Advances in Morphogenesis,* **3,** 1–32 (Eds M. Abercrombie & J. Brachet). Academic Press.

[43] REVERBERI G. (1971) Ed. *Experimental embryology of marine and fresh water invertebrates.* North-Holland, Amsterdam.

[44] ROUX W. (1888) English translation in *Foundations of experimental embryology,* pp. 3–37 (Eds B.H. Willier & J. Oppenheimer). Prentice Hall, New Jersey.

[45] RUNNSTROM J. (1966) Considerations on the control of differentiation in early sea urchin development. *Archo. Zool. Ital.,* **51,** 239–72.

[46] SHANNON M.P. (1972) Characterization of the female sterile mutant *almondex* of *Drosophila melanogaster. Genetica,* **43,** 244–56.

[47] SMITH L.D. (1966) Role of 'germinal plasm' in the formation of primordial germ cells in *Rana pipiens. Develop. Biol.,* **14,** 330–47.

[48] SONNENBORN T.M. (1970) Gene action in development. *Proc. Roy. Soc. Ser. B,* **176,** 347–66.

[49] SONNENBLICK B.P. (1950) The early embryology of *Drosophila.* In *The biology of Drosophila,* pp. 62–167 (Ed. M. Demerec). John Wiley & Sons, New York.

[50] TARKOWSKI A.K. & WROBLEWSKA J. (1967) Development of blastomeres of mouse eggs isolated at the 4- and 8-cell stage. *J. Embryol. Exp. Morph.,* **18,** 155–80.

[51] WADDINGTON C.H. (1956) *Principles of embryology.* George Allen and Unwin, London.

[52] WILSON E.B. (1904) Experimental studies on germinal localization, I. The germ-regions in the egg of *Dentalium. J. Exp. Zool.,* **1,** 1–72.

[53] WOLPERT L. (1971) Positional information and pattern formation. In *Current topics in developmental biology,* **6,** 183–224 (Eds A.A. Moscona & A. Monroy). Academic Press.

[54] WOLPERT L., HICKLIN J. & HORNBRUCH A. (1971) Positional information and pattern regulation in regeneration of *Hydra.* In *Control mechanisms of growth and differentiation. Soc. Exp. Biol. Symp.,* **25,** 391–415 (Eds D.D. Davies & M. Balls). Cambridge University Press.

[55] YAJIMA H. (1970) Study of the development of the internal organs of *Chironomus dorsalis* by fixed and sectioned materials. *J. Embryol. Exp. Morph.,* **24,** 287–303.

Chapter 1.4
Origin of Cell Heterogeneity in Plants

1.4.1 UNEQUAL DIVISIONS IN EMBRYO DEVELOPMENT

As we have seen in the preceding section, the early stages of embryogenesis in animals involve the development of a multicellular embryo from a highly polarized egg, in which there are regional differences in the occurrence of macromolecules and other cell components. During the early divisions polarized cytoplasmic components are apparently partitioned into the daughter cells, the nuclei of which find themselves in varying cytoplasmic environments, and this circumstance plays a vital role in the further stages of embryogenesis and differentiation. In the present section we shall see that closely analogous processes and events play an important role in embryo development and cell differentiation in plants.

As in many animal eggs, the polarization of the cytoplasmic components in the eggs of all land plants is already determined before fertilization by influences from the surrounding maternal tissues, but the fertilized eggs of certain algae, such as the brown seaweed, *Fucus,* are initially apolar (unpolarized) and hence provide exceptionally favourable material for studies on the biochemical and cytological processes underlying the establishment of polarity in the early embryo development. The eggs of *Fucus* are released into the surrounding seawater and after fertilization they settle on to a solid substratum, where the development of a localized protuberance is the first sign of the formation of a rhizoid, by which the young plant becomes attached to a rock. The axis of polarity, as indicated by the position of the rhizoid and the orientation of the first mitotic spindle, can be influenced by gradients in various external factors, such as light, heat, pH and osmotic activity [12]. If the fertilized eggs are exposed to unilateral illumination, the side of the cell on the shaded side begins to form a protuberance at about 14 h after fertilization and mitosis takes place so that the axis of the spindle is parallel to the direction of the incident light, and a cell wall is formed at right angles to this, cutting off a larger cell which gives rise to the future thallus and a smaller cell which forms the rhizoid (Fig. 1.4.1). External factors such as light can influence the axis of polarity up to 8 h before the appearance of the rhizoid, but this polarity is labile up to 1–3 h before the latter event, after which the position of emergence of the future rhizoid appears to be irreversibly fixed.

Studies on the ultrastructural changes occurring following the unilateral illumination have shown that 12 h after fertilization (before any visible signs of the emergence of the rhizoidal protuberance) the nuclear surface has become highly polarized with finger-like projections radiating towards the site of rhizoid formation [21]. These projections evidently represent membraneous extensions of the nuclear envelope. Mitochondria, ribosomes and fibrillar vesicles are also concentrated in this region of the rhizoidal half of the zygote, which is thus more densely cytoplasmic. These changes can be observed between 11 and 14 h after fertilization in zygotes which show no rhizoidal protuberances but yet already have the axis of polarity fixed, i.e. at the time when the cell has become determined with respect to the events leading to rhizoid differentiation, but they are not the first signs of polarity in the developing zygote, since it has been shown that if the cells are plasmolyzed, a polarized plasmolysis occurs on the shaded side of unilaterally illuminated zygotes within 15 min of exposure.

In addition to the observable ultrastructural changes described above, there is active conversion of the polysaccharide, fucoidan, into its sulphated form, fucoidin, as shown by the incorporation of ^{35}S into this fraction. Fucoidin appears in a localized area of the cell wall 10–14 h after fertilization, at which time the polarity of the axis has become fixed [16].

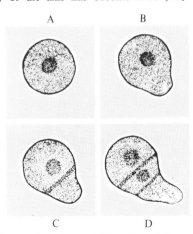

Fig. 1.4.1 Successive stages in the early development of the embryo of *Fucus*. Time after fertilization: A, 4 h; B, 16 h; C, 18 h; D, 26 h. In C a wall has been formed between the rhizoid and thallus cells. In D, further division has occurred. Modified from Jaffé [12].

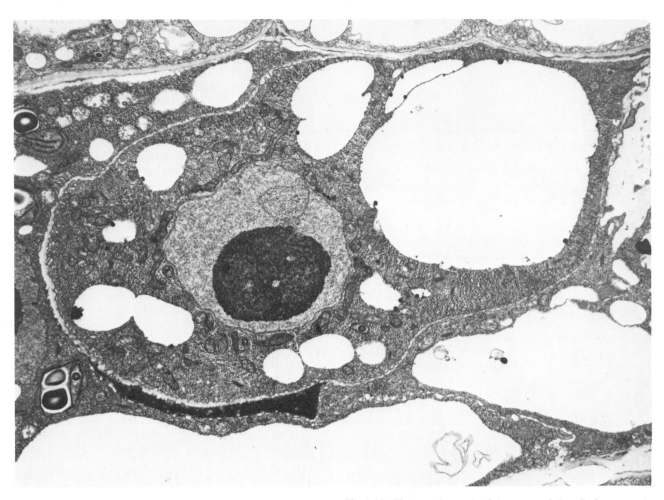

Fig. 1.4.2 Electron micrograph of the zygote of *Capsella bursa-pastoris*, showing marked polarization of the cell, as manifested by the occurrence of the nucleus lying in dense cytoplasm at the micropylar end (left) and a large vacuole at the chalazal end (right). From Sister Richardis Schulz & W.A. Jensen [26].

Germinated embryos show fucoidin localized at the rhizoidal end of the cell. It would appear that fixation of the cell axis is closely associated with the incorporation of fucoidin into the cell wall. The inhibitor, cytochalasin B, which may act on the cell surface or plasma membrane, inhibits the determination of axis polarity and the localization of fucoidin in the cell wall [22], thus further indicating the close connection between these two processes.

Jaffé [12] has shown that the zygotes of *Fucus* generate electrical currents at about the time of axis fixation. If about 100 eggs are placed in a capillary tube and illuminated so as to produce rhizoids towards one end of the tube, then this end becomes markedly electronegative. Apparently each egg drives current through itself, and there is evidence that sodium and calcium ions enter at the rhizoidal end and chloride ions at the thallus end. The amount of ions so moving is sufficient to account for most of the current measured [13]. Two observations suggest that the current feeds back to help fix the polarity. Firstly, the current apparently begins to flow while the axes are being fixed. Secondly, the polarity of *Fucus* eggs can be determined by an imposed voltage [2]. Jaffé suggests that the current may help to bring about differentiation, as well as axis determination, by acting electrophoretically i.e. by generating a field which may localize negatively-charged macromolecules or particles towards a growth point [17]. It is possible that the current is responsible for the transport of fucoidin-containing vesicles towards the wall, where the deposition of the polysaccharide occurs.

A situation similar to that occurring in *Fucus* zygotes, in which the orientation of the first division is not predetermined and is influenced by environmental factors such as light, is found in the spores of many lower plants, such as mosses, ferns, and the horsetail (*Equisetum*), where the spore germinates to give rise to a larger 'thalloid' cell and a smaller 'rhizoid' cell.

The early development of the zygotes of flowering plants shows distinct analogies with that of the lower plants, such as *Fucus,* in that the first division of the zygote is an unequal one [14], but the orientation of the spindle of the first division is evidently determined by the surrounding maternal tissues of the ovule, so that the root end is always towards the micropyle and the shoot end away from it. For example, the egg of Shepherd's purse (*Capsella bursa-pastoris*) is highly polarized, one third to one half of the micropylar end being filled with a large vacuole, while the chalazal end contains the nucleus and much of the cytoplasm of the cell [26] (Fig. 1.4.2). Ribosomes fill the cytoplasm and show little or no aggregation into polysomes. Following fertilization, the cell increases in size, dictyosomes become active and polysomes are formed. The first division is unequal, giving rise to a large vacuolate basal cell and a small densely cytoplasmic terminal cell. The future embryo is derived mainly from the terminal cell, while the basal cell forms the long suspensor. The basal cell undergoes a further unequal division, to form a small cell which, by further transverse divisions, gives rise to the suspensor, while the large end cell enlarges further and becomes sac-like. Thus, it would seem that the division of the whole future plant body into shoot and root is established by the first, unequal division of the zygote.

Not only are unequal divisions important in the early development of the embryos of plants, but they are also a regular feature of the apical cells of the vegetative body of many lower plants. Thus, many algae grow by the activity of a single apical cell, which undergoes unequal division to give a distal daughter cell which continues to constitute the apical cell, and a proximal one which is capable of only a limited number of further divisions and which gives rise to the differentiated, mature tissue of the thallus. The apical meristems of mosses and many ferns likewise consist of single large tetrahedral cells, which divide by unequal divisions, giving rise to daughter cells from the three proximal faces in succession (Fig. 1.4.3). The larger distal daughter cell of each division retains its capacity for further division, whereas the smaller, inner cell is capable of only a small number of further divisions, after which it differentiates.

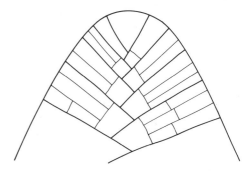

Fig. 1.4.3 Segmentation of tetrahedral apical cell as seen in median section. From F.A.L. Clowes [5].

1.4.2 UNEQUAL DIVISIONS IN CELL DIFFERENTIATION

Unequal divisions appear to play an important role in cell differentiation throughout the plant kingdom, from the procaryotic blue-green algae to the flowering plants [3, 4]. Some clear examples of unequal division in higher plants are provided by developing stomata in the epidermis of the leaves of grasses and other monocotyledons [15, 28]. The adult stomatal complex of grasses consist of narrow, elongated cells, flanked by two subsidiary cells. As a result of more rapid divisions than in the surrounding epidermal cells, rows of small cells are formed. The cytoplasm of a proportion of these cells becomes

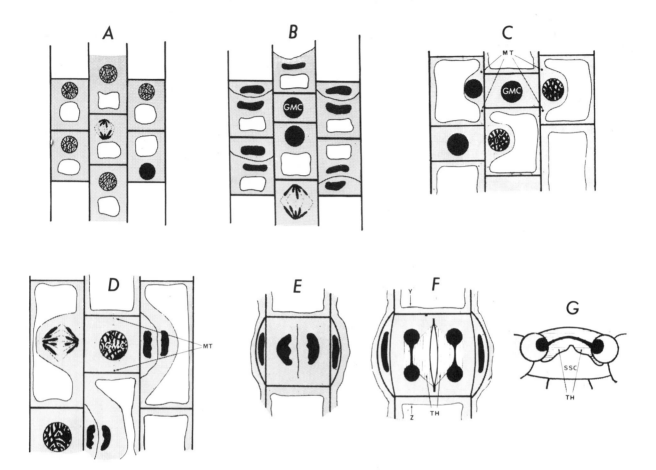

Fig. 1.4.4 Development of a stoma in a grass leaf. Two unequal cell divisions followed by one equal but incomplete cell division give rise to the guard cells and the subsidiary cells of a stoma. In A and B unequal divisions of three cells give rise in C to two large vacuolate cells with a nucleus displaced to one side flanking a smaller non-vacuolate cell, the future guard mother cell (GMC). Unequal divisions in the vacuolated cells give rise to the subsidiary cells of the stoma and finally (D-E) an equal but incomplete division in the small non-vacuolate cell (GMC) gives rise to the two guard cells. The position of the preprophase band of microtubules (MT) which appears to guide the alignment of the cell plate is shown in C and D. After the division of the guard mother cell in D-E, the two nuclei partially separate into two dumb-bell shaped bodies (F) and areas of wall thickening (TH) are laid down. A section through F (Y-Z) is shown in G. SSC is the substomatal cavity. From Clowes and Juniper [6].

polarized so that the cytoplasm is denser near the distal ends of these cells. Before they divide, the nucleus becomes displaced towards the denser end of the cell and the cytoplasmic vacuoles occupy the other end. The division which follows produces a distal small cell with a smaller, heavily staining nucleus, and a large, more weakly staining proximal cell, the cytoplasm of which rapidly becomes vacuolate (Figs 1.4.4 & 5). The distal cell becomes the guard mother cell and the proximal one becomes a normal epidermal cell. The guard mother cell later divides transversely to form two equal guard cells. In this latter division the new cell wall divides the cytoplasm into two similar halves and so no difference between the daughter cell ensues.

Unequal divisions also occur in the epidermal cells immediately adjacent to the guard mother cell (GMC), to form two subsidiary cells, one on each side of the GMC. Before division, the nucleus of the epidermal cell occupies a position close to the GMC. The ensuing mitosis leads to the formation of a small daughter cell laterally adjacent to one side of the GMC, and a larger one which forms a normal epidermal cell. Electron microscopic studies show that the migration of the nucleus of the epidermal cell to the side adjacent to the GMC is accompanied by a characteristic distribution of microtubules. A

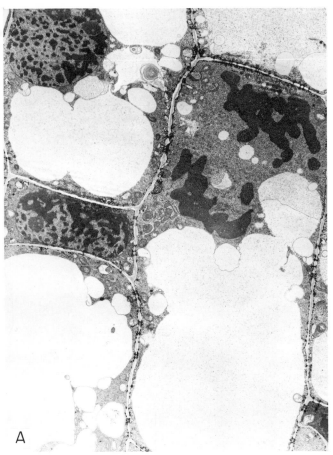

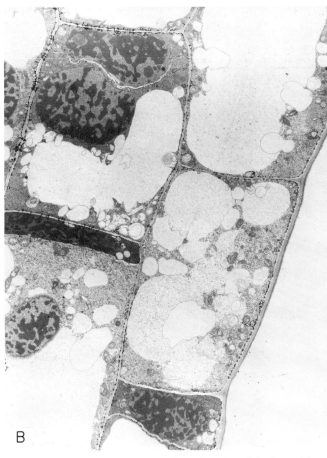

Fig. 1.4.5 A. Unequal division of an epidermal cell in the developing internode of oat (*Avena sativa*). Prior to mitosis the nucleus and much of the cytoplasm has moved to the end of the cell on the right. Mitosis is in progress. On the left a small densely cytoplasmic cell and a large, highly vacuolated cell, the products of an unequal division. The small cell will probably develop as a guard mother cell. B. Short and long cells in the meristematic region of the internodal epidermis of oat. Note the dense cytoplasm of the short cells, presumptive guard mother cells, and the much greater degree of vacuolation in the long cells. From E.G. Cutter [8]. Photograph by Dr P.B. Kaufman.

band of microtubules appears to run in a circular manner around the cytoplasm near the wall adjacent to the GMC (Fig. 1.4.4). This suggests that a specific synthesis or organization of the microtubules occurs at a position dictated by the adjacent GMC, and it has been postulated that substances produced by the GMC diffuse to the epidermal cell through the plasmodesmata [19].

A number of analogous examples of unequal division leading to differentiation occur in higher plants. Thus, in the formation of root hairs on the roots of grasses, such as *Phleum pratense*, certain elongated cells in the epidermis undergo an unequal division, giving a smaller cell with denser protoplasm at the distal end and a longer cell at the proximal end [3] [Figs 1.4.6 &

1.4.7). The smaller cell forms a protuberance which grows into a root hair, whereas the larger cell remains a normal epidermal cell. Other examples of unequal division which lead to differentiation are found in the development of the pollen grain, where the pollen mother cell undergoes an unequal division giving rise to the generative nucleus which lies in a region of dense cytoplasm and the vegetative nucleus lying in the remaining volume of the cell.

It is clear that unequal division plays an important role in the origin of cell heterogeneity in plants. The essential features of such divisions are that the parent cell first becomes polarized by the establishment of a cytoplasmic gradient in respect of macromolecules or other components, followed by a division in which

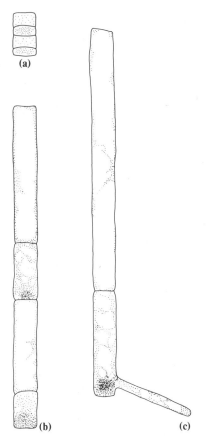

Fig. 1.4.6 Development of root hair initials in root of the grass *Phleum pratense*, at successive stages of development (a-c). The smaller of the cells formed by unequal division (a) gives rise to the root hair cell (c). From Sinnott & Bloch [28].

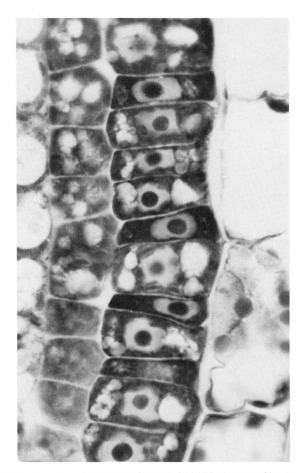

Fig. 1.4.7 Formation of trichoblasts (root hair initials) in the root epidermis of *Hydrocharis morsus-ranae*. Unequal division has just given rise in each case to a small cell which is the trichoblast and a larger cell which is the epidermal cell (e) x. From Cutter & Feldman [9].

the mitotic spindle is orientated along the gradient, so that the daughter nuclei are located in different cytoplasmic environments.

1.4.3 POLARITY

Polarization of the cytoplasm and orientation of the axis of the mitotic spindle are manifestations of a far-reaching polarity of plant cells and tissues, which is an essential feature of many aspects of differentiation and morphogenesis in plants [4, 38]. There are many ways in which this polarity is manifested—for example, in the pattern of regeneration of buds and roots from segments of stem or root, in which buds regenerate from the morphologically upper end and roots from the basal end. This polarity is not simply due to gradients of metabolites since, once established, it is normally persistent throughout the life of a given piece of tissue or organ. Moreover, it cannot be changed by reversing the orientation of a stem cutting, by planting it upside-down.

The polarity of a multicellular tissue appears to reflect the polarity of its constituent cells, as is shown by the fact that after they have been subjected to plasmolysis, each constituent cell of the filamentous alga, *Cladophora,* can regenerate into a new plant, with a rhizoid forming from the basal end of each cell. The structural basis of this polarity of each cell is not known, although it has been assumed that there must be some 'cyto-skeleton' within the cytoplasm, based upon the orientation of macromolecules or other sub-cellular structures such as microfilaments, but this hypothetical cytoskeleton has not yet been identified. Alternatively the cell polarity may be based upon polarization of the plasma membranes at the end walls of the cell [10]. One of the most important manifestations of cell polarity is seen in the polar movement of the hormone, indol-3yl acetic acid ('auxin'), which moves in the plant in a strictly basipetal manner, by cell-to-cell transport (p. 121).

1.4.4 APICAL MERISTEMS OF SEED PLANTS

As we have seen, many lower plants, including various algae, mosses and many ferns, grow by means of a single apical cell. Other lower plants, including the club-mosses (*Lycopodium*) and some 'eusporangiate' ferns have several initial cells, and all seed plants (gymnosperms and angiosperms) grow by apical meristems comprising many cells.

Shoot apices*

In the seed plants two main layers of cells can be distinguished at the shoot apex: (1) the *tunica* consisting of one or several layers of cells which divide predominantly by anticlinal walls (i.e. perpendicular to the surface), and (2) the central *corpus,* in which divisions are both anticlinal and periclinal (i.e. parallel to the surface) (Fig. 1.4.8). Apart from these broad zones, which can be distinguished solely by the orientation of the planes of division, other zones may be distinguished in the apices of many species, although in some species not all the zones are present. Those other zones include: (1) the *central mother cells,* (2) *peripheral* or *flank meristems,* and (3) a *central rib meristem* (Fig. 1.4.9). The rate of division in the central mother cell zone is low, but active divisions at the boundary of this zone give rise to the other two zones. There is considerable variation between species with respect to the general shape of the apex and to the number of zones which are recognizable and yet the apices of all these species produce similar organs viz. stems, leaves and buds. Thus, it would appear that no special morphogenetic significance can be attached to the various zones recognizable in the shoot apex.

Leaves are initiated in the shoot apical region in a regular pattern which is described below. The leaf primordium is initiated by divisions which commence in a group of cells on the flanks of the apex. These divisions may originate in the outermost layers of the tunica, as in many grasses, or they may occur in deeper layers and may involve both the tunica and the corpus. Thus, the extent to which the tunica and the corpus are involved in the initiation of leaf primordia varies greatly from species to species, and even within a single species, depending

* For detailed accounts of plant apical meristems see Clowes [5], Cutter [8], Wardlaw [36].

Fig. 1.4.8   Median section through the shoot apical meristem of *Alternanthera philoxeroides,* showing the two-layered tunica overlying the central corpus. From E.G. Cutter [7].

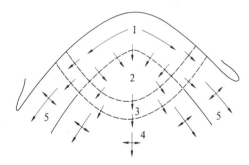

Fig. 1.4.9   Diagram of a median section of the shoot apex of *Chrysanthemum morifolium,* 1, the tunica; 2 central mother cell zone; 3, zone of cambial-like cells; 4, the rib meristem; 5, the peripheral zone. Arrows indicate tissues to which meristematic cells contribute. From Popham [20].

upon the nutritional and other conditions under which the plant is growing.

Lateral buds normally originate somewhat later than leaf primordia, again by divisions which originate in a group of cells on the flanks of the apex. As with the leaf initiation, the extent to which these divisions occur in the tunica and the corpus varies greatly from one species to another. As a result of these divisions the bud primordium first develops as a protuberance which later develops an apical structure similar to that of the main shoot apex of that species.

The siting of leaf primordia

The arrangement of leaf primordia at the shoot apex is normally highly regular and characteristic of the species. A spiral arrangement is very common [Figs 1.4.10 & 11], but other

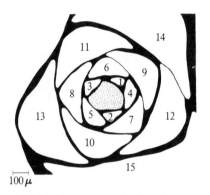

Fig. 1.4.10   Transverse section through the shoot apex of *Saxifraga* showing spiral phyllotaxis. Numbers refer to successively older leaves. From Clowes [5].

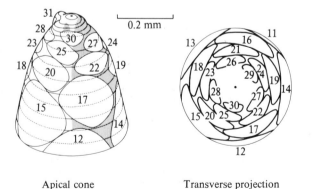

Apical cone   Transverse projection

Fig. 1.4.11   Three-dimensional and transverse projections of the apex of a 15-day-old flax seedling. Dotted lines indicate serial transverse sections. Numbered areas on apical cone represent bases of successive leaf primordia. From Williams [39].

arrangements are frequently found, as when the leaves are arranged (1) alternately on opposite sides of the apex; (2) in opposite pairs, with successive pairs at right angles to each other; or (3) in whorls, when a number of leaves arise around the apex at the same level. These arrangements of leaf primordia are, of course, reflected in the arrangement of the mature leaves on the stem (referred to as phyllotaxis).*

In the spiral leaf arrangement, a line drawn from the youngest through successively older primordia constitutes the '*genetic*' or *developmental* spiral. It is a logarithmic spiral given by the equation $r = a\theta$, where r is the radius, θ the angle of divergence between two points on the spiral and a is a constant. This implies that the radius will increase in geometrical progression as it sweeps through successive angles from the starting point.

The angle of divergence between successive primordia varies from species to species and depends upon the density of packing of leaf primordia at the apex, and where this is high it approaches the limiting value of 137·5°. Thus, spiral arrangement at a shoot apex can be reproduced by marking points consecutively around a centre at a constant divergence of 137·5° and radially at a distance which increased in geometrical progression. The successive points represent the positions of the leaf primordia, and a line through them coincides with the genetic spiral.

Clearly, this regular siting of leaf primordia at the shoot apex is an intriguing phenomenon and earlier botanists were fascinated by the mathematical properties of the system. However, here we are concerned primarily with the developmental aspects of phyllotaxis and, in particular, with the problem of what influences operate to regulate the very precise manner in which leaf primordia are sited. Earlier workers were of the opinion that the apex possessed rather mysterious mathematical properties which controlled phyllotaxis. More recent theories, however, are based on the assumption that the regularity of phyllotaxis is partly a problem of geometrical packing of leaf primordia within the shoot apical region, in conjunction with influences from existing primordia on the emergence of new leaf initials.

The first of such theories is the Available Space Theory, put forward by van Iterson [11], in which it is postulated that a certain minimum space between existing primordia is necessary before a new primordium can form. If successively older primordia are referred to as P_1, P_2, P_3 etc. and the next primordia to emerge as I_1, I_2, I_3 etc. (in that order), then the regularity of the pattern indicates that I_1 will arise between P_3 and P_5 (Fig. 1.4.12), at an angular divergence from P_1 of approximately 137·5°. It will be seen from Fig. 1.4.11 that, as the apex grows, the primordia 'recede' from the centre of the apex and concurrently both the size of the primordia and the

* For general accounts of phyllotaxis see Clowes [5], Sinnott [27], Richards [23, 24], Williams [39].

space between them increase. Thus, on the Available Space Theory, I_1 appears where it does because a certain minimum free area of the surface of the apex is required before a new primordium can arise, and the region between P_3 and P_5 is the first such area to arise as the growth of the apex proceeds. The theory was tested by M. & R. Snow at Oxford, using elegant surgical techniques on the shoot apex of lupin (*Lupinus albus*) [29]. In one experiment they made cuts in the area in which I_2 would be expected to arise, thereby reducing the free space available, and it was found that no primordium developed in the space, presumably because it was reduced below the minimum area necessary.

An alternative theory was put forward by Schoute [25] who postulated that the centre of a leaf primordium is first determined and that as it grows it produces a substance which inhibits the formation of other primordia within a certain radius, so that new primordia cannot be formed until the distance between neighbouring primordia has increased sufficiently to produce an area which is outside the inhibitory fields of either (Fig. 1.4.12). Thus, according to this theory, the factor controlling the initiation of a primordium is not simply a minimum superficial area between adjacent primordia, but a space which is free from the inhibitory influence of neighbouring primordia.

Evidence tending to support this theory was provided by the experiments of Wardlaw [35]. Using fern apices, in one experiment he isolated the site of I_1, from the neighbouring existing primordia by two radial cuts and found that it then grew *more rapidly* than normally (Fig. 1.4.13), suggesting that it

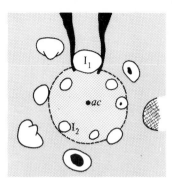

Fig. 1.4.13 The effect of isolating a leaf initial (I_1) by deep radial incisions. The isolated initial has become much larger than the surrounding primordia. After C.W. Wardlaw [35].

had thereby been released from the inhibitory effects of neighbouring primordia. In further experiments, Wardlaw destroyed the site at which I_1 was due to arise, and found that although this treatment had no effect on the position of I_2 and I_3, it did affect the positioning of I_4 (which would normally be sited between I_1 and P_2), so that it arose nearer the expected I_1 position than is normally the case, i.e. it was apparently shifted towards the I_1 position, in the absence of any inhibitory effect from this primordium.

It can be shown from the geometry of the shoot apex, that I_1 will arise at a divergence of 137·5° from P_1 if it occurs at a point which divides the angle between P_3 and P_5 in the inverse ratio of their respective ages, i.e. the new primordium is displaced towards the older of the two primordia. The developmental significance of this observation is not clear, but it may indicate that the inhibitory effect of a growing primordium decreases as it increases in age.

For details of the further development of leaves and of differentiation in the stem, the reader is referred to Cutter [8]. A detailed discussion on the role of hormones in the differentiation of vascular tissue is given in Chapter 4.3.

1.4.5 ROOT APICES

In certain respects the root apex is a simpler system than the shoot apex, due to the fact that it does not produce lateral organs corresponding to leaves and lateral buds, and secondary roots arise adventitiously at some distance from the apex itself. On the other hand, the root apex produces a cap, whereas no equivalent structure is produced by the shoot apex.

It is possible, by studying the planes of division, to trace the cell-lineages within the shoot apical region (Fig. 1.4.14). In this way it ought to be possible to identify the initial cells of the promeristem itself, from which the observed files of cells are

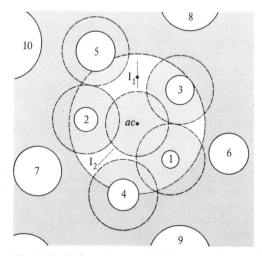

Fig. 1.4.12 The growth centre and field concept as it may apply to the apex of *Dryopteris*. The apex is seen from above. ac, apical cell; 1–9, leaf primordia in order of increasing age; I_1, I_2, position of next primordia to arise. The hypothetical inhibitory fields around the growing primordia are indicated by stippling. From C.W. Wardlaw [35].

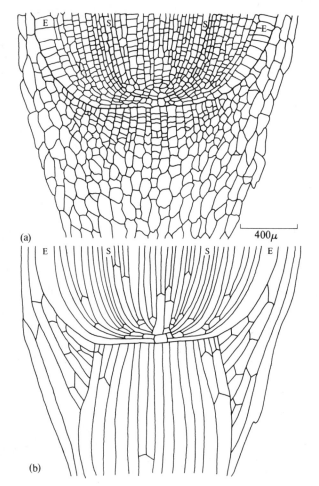

(a)

400μ

(b)

Fig. 1.4.14 (a) Median section of root tip of *Zea mays*. (b) Outline of cell lineages of the same root apex. E, epidermis; S, outer layer of stele. From Clowes [5].

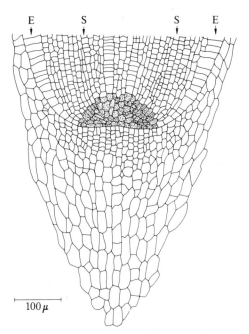

100 μ

Fig. 1.4.15 Median section of root apex of *Zea mays* with quiescent centre shaded. E, epidermis; S, outer layer of stele. From Clowes [5].

derived. Earlier workers interpreted the cell patterns to indicate that the promeristem consisted of two or three layers of initials, from which the main tissue layers of the root (vascular stele, cortex, epidermis and root cap) were held to be derived. However, more recent studies point to a different structure of the promeristem.

Various approaches, including feeding the roots with tritiated thymidine (the incorporation of which can be used as a measure of DNA synthesis—see p. 10), have indicated that DNA synthesis and mitosis proceed at a much slower rate in a group of cells in the meristem region than in the surrounding cells, and hence this zone has been called the *quiescent centre* [5] (Fig. 1.4.15). The nuclei of the cells in the quiescent centre remain for long periods in the G.1 (pre-synthesis) phase of mitosis, but they appear to be capable of undergoing active mitosis under certain conditions.

Active cell division occurs at the boundary of the quiescent centre and the derivative cells give rise to the files of cells seen in the older parts of the root. Hence these actively-dividing cells are held to be the promeristem, which thus takes the form of an inverted 'cup'. The derived cells give rise to the tissues of the root itself on the proximal side and to the root cap on the distal side.

The quiescent centre varies considerably in size both between species and within the same species, and is often absent in young roots. These fluctuations in size involve corresponding changes in the initial cells, and raise the question as to whether there is a permanent promeristem. But as Clowes [5] states, 'if the promeristem is regarded as a collection of initials lying over the surface of the quiescent centre then the promeristem is not permanent. But there still must be a group of cells from which these initials are derived. At some times they are quiescent and at other times they are meristematic. In this sense then, there is a permanent promeristem'.

The function of the quiescent centre is not known. It is possible that its existence is simply the consequence of the geometry of the root i.e. in order to maintain the shape and organization of the apex there must be regional differences in the rates of cell division.

The root presents a particularly favourable object for the study of the progress of cell and tissue differentiation, since the successive stages are set out in a linear sequence in passing from the meristem to the older, differentiated parts (Fig. 1.4.16) (see review by Torrey [33]). The boundary between the central vascular cylinder and the cortex becomes visible immediately behind the region of the promeristem, being recognizable by the difference in size of the cells, which are radially enlarged in the surrounding cylinder of cortical cells. Thus, it is clear that histogenesis may occur at a very short distance from the promeristem itself. The endodermis is recognizable at an early stage, as the innermost layer of the cortex. Similarly, the outermost layer of the central cylinder, the pericycle, is delimited within 100 μ or less of the apical initial region in some roots. The first observable changes to occur in the central cylinder is the blocking out of the future xylem groups, by the radial enlargement of certain cells. On the other hand, the first cells to differentiate into mature cells are the sieve elements of the protophloem, which may occur within a distance of 230 μ from the promeristem, in slow growing roots (Fig. 1.4.16).

It is evident that certain steps in the process of differentiation in the root are established very near the promeristem itself. This consideration, together with the fact that in some roots cell lineages can apparently be traced back to specific layers of cells in the meristem region itself, led Hanstein in 1868 to postulate the existence of three *histogens* in the root, viz. *plerome, periblem* and *dermatogen,* which were held to give rise to the vascular cylinder, cortex and epidermis respectively. Thus, it was held that the cells derived from the different layers of initial cells were in some way *predetermined* to give rise to the various types of tissue. The hypothesis was also extended to shoot apices.

When only roots are considered this is still a tenable hypothesis, but the formation of leaf traces (vascular tissue) from cortical cells in the shoot (p. 218) seem to invalidate its applicability to the shoot. Moreover, if the promeristem consists of a broad cup-shaped region of cells surrounding the quiescent centre, instead of distinct layers of initial cells, part of the supporting evidence for the Histogen Theory is removed. However, the fact remains that distinct layers of cells can be seen even within the zone of the quiescent centre (Figs 1.4.14 & 15), and these give rise to promeristem cells at the boundary of the quiescent centre, which may already be programmed to form one of the major tissue types, as postulated in the Histogen Theory; that is to say, we cannot exclude the possibility that the quiescent centre contains cells of different developmental potential.

In a transverse section through the vascular cylinder of a mature root the xylem is seen to consist of a variable number of triangular arms between which lie arcs of phloem tissue and various hypotheses have been put forward to account for this pattern of tissues. It is shown in Chapter 4.3 that hormones play an important role in the development of vascular tissue, but in the shoot the differentiation of cambial derivatives into xylem and phloem does not appear to be determined primarily by hormones. Similarly the pattern of arrangement of vascular tissues appears to be determined and controlled in the meristem itself, as will be shown below.

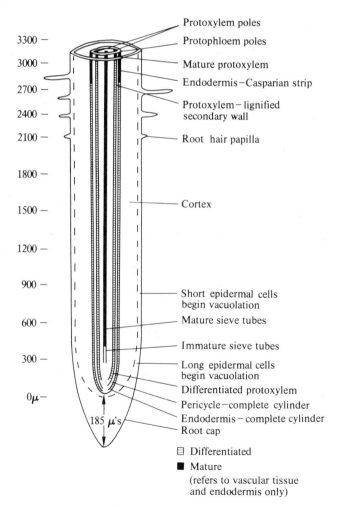

Fig. 1.4.16 Diagram of the root tip of *Sinapis alba,* to illustrate stages of differentiation of various tissues. From Peterson [18].

1.4.6 THE SELF-ORGANIZING PROPERTIES OF APICAL MERISTEMS

Several lines of evidence indicate that shoot and root apices are self-organizing systems. This is almost self-evident from the fact that adventitious buds and roots may arise within parenchymatous tissues of roots and shoots and, in many species, in unorganized callus tissue (p. 214). Moreover, we shall see that adventive 'embryoids' may be developed from liquid suspension cultures of plant cells (Chapter 2.3), and that these embryoids are formed from single cells or small groups of a few cells.

Apart from these various examples of organization of apical meristems *de novo*, existing meristems also have the capacity for regeneration if they are cut or damaged. Thus, if the apical region of lupin is divided into four segments by two vertical cuts at right angles, each of these segments is capable of regenerating a complete, normal apex [1]. It has been suggested that the pattern of differentiation at the apex is determined primarily by the pattern of the pre-existing mature tissue, since the procambial strands (primordial vascular tissue) are continuous with the existing vascular tissue and they extend well into the region of the apical meristem in both shoots and roots. However, several types of experiment indicate clearly that the pattern of differentiation in the apical region of both shoots and roots is not controlled by the pre-existing tissues. Thus, when the apical meristem of the fern, *Dryopteris,* was isolated on a plug by four vertical cuts so that any influence of pre-existing vascular tissue was excluded, the normal phyllotactic sequence of leaf primordia was maintained [34]. A cross section through

the plug of pith tissue on which the apex was borne showed that the stele had a quadrangular form, reflecting the shape of the plug (Fig. 1.4.17). In spite of this, the new vascular tissue formed above the plug followed a normal pattern, so that the abnormal form of the vascular tissue in the plug did not affect the normal pattern of vascular differentiation above it, indicating that the fern apex is a self-determining region in this respect.

Experiments on root apices have led to similar conclusions. Thus, if the terminal 0·5 mm of pea root is cut off and cultured on a nutrient medium, instead of the triarch (i.e. a three-stranded) xylem formed in normal pea roots, a monarch or diarch condition was found until the root had grown further, when it reverted to the normal triarch condition [31]. Again, if pea roots are decapitated and allowed to regenerate in the presence of the hormone indole-3yl acetic acid (IAA), the new vascular tissue is hexarch, but if transferred to a medium lacking IAA they ultimately become triarch again [32]. Thus, the pattern of the vascular system of the root appears to be determined by the conditions prevailing with the apical meristem itself and not by the pre-existing vascular tissue.

Similar conclusions are indicated by experiments in which the terminal 2 mm of root tips was cut off and the tip rotated and replaced on the stump of older tissue; it was found that the vascular tissue which later differentiates at the tip is out of line with that in the stump. Finally, it has been shown that if root tips are bisected, new apices will regenerate from the two halves.

These various experimental results clearly indicate that apical meristems of higher plants are self-organizing and self-determining systems [37]. It would seem that the apical meristems represent dynamic systems in a state of equilibrium and that interference with the system, by surgical or other treatments, is counteracted by compensating changes within the system until it reaches the equilibrium configuration.

1.4.7 CONCLUSIONS

We have seen that unequal division marks the first step in the establishment of cell heterogeneity at various stages in the plant life-cycle. Where such division occurs in the zygote it results in the establishment of the two main subdivisions of the plant body, viz. thallus and rhizoid in algae and other lower plants, and shoot and root in higher plants. Other examples of unequal division occur in the final stages of development of organs, as in the formation of stomata and trichoblasts.

Unequal division is preceded by a visible polarization of cytoplasmic components within the parent cell, and it seems very probable that there is also a polarization of cytoplasmic determinants which become differentially partitioned between the two daughter cells, so that their subsequent divergent patterns of differentiation are controlled by these factors. If this

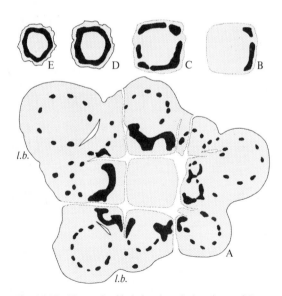

Fig. 1.4.17 The result of isolating the apical meristem of *Dryopteris* by four vertical cuts, as seen in serial transverse sections. The isolated meristem has grown on and has given rise to a short shoot. A, Section taken near the base of the experimental region, showing the incisions whereby the apex was isolated on a plug of pith tissue. B–E, progressively higher sections of the central plug, showing the vascular tissue (black) which has differentiated below the growing apex. After C.W. Wardlaw [34].

interpretation is correct, then we seem to have a clear parallel with the 'mosaic' theory of animal embryo development.

This type of developmental control in plants is closely connected with the phenomenon of polarity (p. 35), since the initial polarization of cytoplasmic components is established either by external factors, such as light or gravity, as in *Fucus* eggs, or by the surrounding maternal tissue in the zygote and stomatal initials in higher plants. Organized development in plants does, indeed, depend entirely upon close control of the orientation of cell division, and where this does not occur we obtain only a disorganized callus. Bünning has epitomized this principle in the statement, 'no differentiation without polarity'.

Many lower plants, including some algae, bryophytes, most ferns and other vascular cryptogams, have single apical cells which show unequal division, giving rise to a distal (outer) cell which retains its meristematic activity and a proximal (inner) cell whose derivative cells have the capacity for only a limited number of divisions before they differentiate and mature. However, unequal division does not seem to be the primary means of establishing cell heterogeneity in the subsequent stages of differentiation of these lower plants and in higher plants, in which there is not a single but several apical initial cells, there is no clear evidence for the occurrence of unequal divisions or for their role in the establishment of cell heterogeneity. It would thus appear that we must seek some other origins of cell heterogeneity in the shoot apices of higher plants. The self-organizing properties of the apical meristems of higher plants appear to indicate a more advanced type of organization in which the fates of particular groups of cells is not predetermined by their own intrinsic properties but by the properties of the system as a whole, which controls the pattern of differentiation of its various cellular constituents.

When we consider the types of cell heterogeneity which originate in the shoot apical region, it becomes apparent that it constitutes a zone of *organogenesis*, in which the primordia of stem, leaves and buds are established, and that only at a later stage of development of these organs does heterogeneity arise at the cell and tissue level. We know little about the manner in which different groups of cells become committed along the 'leaf', 'bud' or 'stem' pathways of development, but it would appear that the distinction between corpus and tunica layers plays little or no role in organogenesis. On the other hand, some other zones within the apex (Fig. 1.4.9) may have some morphogenetic significance. Thus, the development of the internode depends upon polarized cell divisions in which the axes of the mitotic spindles are orientated parallel to the main axis of the future stem and such divisions become apparent earliest in the rib meristem (Fig. 1.4.8).

We have seen that the arrangement of leaf primordia (phyllotaxis) can be related to the geometry of the shoot apex and its pattern of growth. The siting of the new primordium is conditioned by the nearest neighbouring primordia, which appear to be surrounded by inhibitory fields, and the new primordium apparently arises in an area in which these inhibitory effects are absent or minimal.

In the shoot apex, procambial tissue is blocked out in the boundary zone between the rib meristem and the peripheral meristems. However, the further differentiation of vascular tissue in the stem is profoundly influenced by the developing leaf primordia, and there is strong evidence that hormones produced by the young leaves play an essential role in the differentiation of vascular tissue (Chapter 4.3). Thus, several different types of process appear to be involved in the establishment of cell heterogeneity in shoot apices of higher plants.

The situation in root apices appears to be rather simpler, since it is not complicated by the problems associated with the formation of lateral organs, such as leaves and buds (lateral roots are initiated at some distance from the apex in the mature regions of the root). The various tissue zones of the mature root are blocked out at an early stage and differentiating phloem can be recognized at only a short distance from the meristem itself. Moreover, cell lineages giving rise to these various tissue zones can be traced back to initial cells at the periphery of the quiescent centre (Fig. 1.4.14). Thus, the old concept of specific histogens, destined to give rise to the main tissue zones of the mature root, is consistent with the histological evidence and may have some validity for roots, but not for shoots. However, we remain completely ignorant as to the nature of the processes by which cell heterogeneity originates in root apices.

1.4.8 REFERENCES

[1] BALL E. (1948) Differentiation in the primary shoots of *Lupinus albus* L. and of *Tropaeolum majus* L. *Symp. Soc. Exp. Biol.*, **2**, 246–62.
[2] BENTRUP F.W., SANDON T. & JAFFÉ L.F. (1967) Induction of polarity in *Fucus* eggs by potassium ion gradients. *Protoplasma*, **64**, 254–66.
[3] BLOCH R. (1965) Histological foundations of differentiation and development in plants. *Encycl. Plant Physiol.*, **15**(1), 146–88.
[4] BÜNNING E. (1952) Morphogenesis in plants. *Survey of Biological Progress*, **2**, 105–40. Academic Press, New York.
[5] CLOWES F.A.L. (1961) *Apical meristems*. Blackwell Scientific Publications, Oxford.
[6] CLOWES F.A.L. & JUNIPER B.E. (1968) *Plant cells*. Blackwell Scientific Publications, Oxford.
[7] CUTTER E.G. (1967) Morphogenesis and developmental potentialities of unequal buds. *Phytomorphology*, **17**, 437–45.
[8] CUTTER E.G. (1971) *Plant anatomy, part 2 organs*. Edward Arnold, London.
[9] CUTTER E.G. & FELDMAN L.J. (1970) Trichoblasts in Hydrocharis. I, Origin, differentiation, dimensions and growth. *Am. J. Bot.*, **57**, 190–201.
[10] GOLDSMITH M.H.M. & RAY P.M. (1973) Intracellular localization of the active process in polar transport of auxin. *Planta, Berl.*, **111**, 297–314.
[11] van ITERSON G. (1907) *Mathematische und mikroscopisch-anatomische studien über blattstellungen*. Jena.
[12] JAFFÉ L.F. (1968) Localization in the developing *Fucus* egg and general role of localizing currents. *Adv. Morph.*, **7**, 295–328.

[13] JAFFÉ L.F., ROBINSON K. & NUCCITELLI R. (1974) Local cation entry and self-electrophoresis as an intracellular localization mechanism. *Ann. N.Y. Acad. Sci.*, **238**, 372–83.

[14] JENSEN W.A. (1964) Cell development during embryogenesis. *Brookhaven Symp. Biol.*, **16**, 179–202.

[15] KAUFMAN P.B., PETERING L.B., YOCUM C.S. & BAIC D. (1970) Ultrastructural studies on stomata development in internodes of *Avena sativa. Am. J. Bot.*, **57**, 33–49.

[16] NOVOTNY A.M. & FORMAN N. (1974) The relationship between changes in cell wall composition and the establishment of polarity in *Fucus* embryos. *Devel. Biol.*, **40**, 162–73.

[17] NUCCITELLI R. & JAFFÉ L.F. (1974) Spontaneous current pulses through developing fucoid eggs. *Proc. Nat. Acad. Sci., U.S.A.*, **71**, 4855–59.

[18] PETERSON R.L. (1967) Differentiation and maturation of primary tissues in white mustard root tips. *Canadian J. Bot.*, **45**, 319–31.

[19] PICKETT-HEAPS J.D. & NORTHCOTE D.H. (1966) Cell division in the formation of the stomatal complex of the young leaves of wheat. *J. Cell Sci.*, **1**, 121–28.

[20] POPHAM R.A. (1958) Cytogenesis and zonation in the shoot apex of *Chrysanthemum morifolium. Am. J. Bot.*, **45**, 198–206.

[21] QUATRANO R.S. (1972) An ultrastructural study of the determined site of rhizoid formation in *Fucus* zygotes. *Exp. Cell. Res.*, **70**, 1–12.

[22] QUATRANO R.S. (1973) Separation of processes associated with differentiation of two-celled *Fucus* embryos. *Devel. Biol.*, **30**, 209–13.

[23] RICHARDS F.J. (1948) The geometry of phyllotaxis and its origin. *Soc. Exp. Biol. Symp.*, **2**, 217–45.

[24] RICHARDS F.J. (1951) Phyllotaxis: its quantitative expression and relation to growth in the apex. *Phil. Trans. Royal. Soc. B*, **235**, 509–64.

[25] SCHOUTE (1913) Beiträge zur Blattstellungslehre. *Rec. Trav. Bot. Neerl.*, **10**, 153–325.

[26] SCHULZ R. & JENSEN W.A. (1968) Capsella embryogenesis: The egg, zygote and young embryo. *Amer. J. Bot.*, **55**, 807–19.

[27] SINNOTT E.W. (1960) *Plant morphogenesis.* McGraw Hill, New York.

[28] SINNOTT E.W. & BLOCH R. (1939) Cell polarity and differentiation in root hairs. *Proc. Nat. Acad. Sci. U.S.*, **26**, 223–27.

[29] SNOW M. & SNOW R. (1933) Experiments on phyllotaxis II. The effect of displacing a primordium. *Phil. Trans. Ry. Soc. B*, **222**, 363–400.

[30] STEBBINS G.L. & SHAH S.S. (1960) Developmental studies of cell differentiation in the epidermis of monocotyledons II, Cytological features of stomatal development in the Gramineae. *Develop. Biol.*, **2**, 477–500.

[31] TORREY J.G. (1955) On determination of vascular patterns during tissue differentiation in excised pea roots. *Amer. J. Bot.*, **42**, 183.

[32] TORREY J.G. (1957) Auxin control of vascular pattern formation in regenerating pea root meristems grown *in vitro. Amer. J. Bot.*, **44**, 859.

[33] TORREY J.G. (1965) Physiological bases of organization and development in the root. *Encycl. Plant Physiol.*, **15**(1), 1256–1327.

[34] WARDLAW C.W. (1947) Experimental investigation of the shoot apex of *Dryopteris aristata.* Druce. *Phil. Trans. Roy. Soc. B*, **233**, 415–51.

[35] WARDLAW C.W. (1949) Experiments on organogenesis in ferns. *Growth* (suppl.), **13**, 93–131.

[36] WARDLAW C.W. (1965a) The organization of the shoot apex. *Encycl. Plant Physiol.*, **15**(1), 966–1076.

[37] WARDLAW C.W. (1965b) *Organization and evolution in plants.* Longmans, Green & Co. Ltd., London.

[38] WAREING P.F. & PHILLIPS I.D.J. (1969) *The control of growth and differentiation in plants.* Pergamon Press, Oxford.

[39] WILLIAMS R.F. (1974) *The shoot apex and leaf growth—a study of quantitative biology.* Cambridge Univ. Press (in press).

Conclusions to Part 1

This part of the book is concerned with the ways in which different cell types originate in early development. From the studies which are discussed it is possible to make several general statements about development.

1. DEVELOPMENT INVOLVES THE ORDERLY EXPRESSION OF GENES

Current approaches to the problems of development are based upon the premise that development involves the orderly expression of genes. To a large extent this tenet is self-evident, since it is clear that many genes are expressed only at specific stages in the life-cycle. There is still very little direct experimental evidence to support the hypothesis although it is consistent with certain observations such as the finding that different classes of RNA are detected at different stages of the development of the frog *Xenopus* (p. 11) and observations on the sequential appearance of different phosphatase activities during cap formation in the alga *Acetabularia* (p. 8).

2. GENE EXPRESSION MAY BE CONTROLLED AT DIFFERENT LEVELS

This subject is treated in detail in Chapter 5.3, but we have already seen examples of control of gene expression at different levels. Nuclear transplant experiments with *Xenopus* demonstrate control of transcription of RNA by nuclear genes, while studies with enucleate *Acetabularia* and with the fungus *Blastocladiella* show that gene expression may also be controlled at the level of protein translation from messenger RNA. There may, indeed, be a considerable time interval between the synthesis of mRNA and its translation in protein synthesis, as shown from studies on long-lived mRNA in amphibian and sea-urchin embryos and in seeds of higher plants (p. 11). These examples seem to indicate that although selective gene activation may be controlled at the transcription level (i.e. in the nucleus), the *time* of gene expression in enzyme synthesis is, in some instances at least, controlled at the translation level (i.e. in the cytoplasm). These various studies have also emphasized the vital role played by the cytoplasm in controlling gene expression and we next consider how the cytoplasm and membranes of developing systems appear to determine the origin of different cell types.

3. ONE CAUSE OF CELL HETEROGENEITY IN EARLY DEVELOPMENT IS THE SEGREGATION OF CYTOPLASMIC DETERMINANTS

The over-riding importance of cytoplasmic factors in establishing cell heterogeneity is demonstrated by several examples from studies on both animals and plants. Thus, the studies on the polar lobe of *Ilyanassa* (p. 12) indicate the presence of a cytoplasmic factor which affects the development of mesoderm, and those on *Ciona intestinalis* point to the existence of cytoplasmic determinants which segregate into different cells and lead to cell lineages with different capacities to synthesize tissue-specific enzymes. The segregation of cytoplasmic determinants is also strongly indicated by the various examples of differentiation resulting from unequal division in plant cells. The consequences of such unequal divisions strongly point to the polarized distribution of determinants in the cytoplasm of the parent cell, which become differentially partitioned between the two daughter cells in mitosis. Thus, in such cases we appear to have examples of the establishment of cell heterogeneity as a result of intrinsic differences within the cytoplasm.

These examples from plants appear to show clear analogies with the studies in various animal eggs which appear to show the 'mosaic' type of developmental control.

4. PARTICULAR PARTS OF ADULT ANIMALS ARE NORMALLY FORMED FROM PARTICULAR PARTS OF THE EGG

This phenomenon is demonstrated by the formation of the skeleton from the vegetal region of the sea urchin egg (*Paracentrotus*). This regularity of development leads to the view that the egg may contain a mosaic of developmental instructions. However, the only clear evidence that such instructions exist are provided by experiments on the germ plasm of insects (*Drosophila*) and amphibia (*Rana*).

5. MOST PARTS OF THE ANIMAL EGG ARE DEVELOPMENTALLY LABILE

The developmental fate of the cells of early embryos can be altered by changing their position. Thus the fate of four and eight cell blastomeres of the mouse can be readily manipulated by altering their position with respect to other cells. Similarly there are few limitations on the lability of parts of other animal eggs (discussed further in Chapter 3.6).

Isolation and transplantation have identified some regions of animal eggs which can control the development of other regions (e.g. grey crescent of amphibia and animal and vegetal poles of sea urchins). The effect of these controlling regions (reference points) is to produce regularity in development, but such regions are probably not required for the development of many plant and animal embryos.

There is little doubt that cells in growing plant apices are also developmentally labile because their fate can be changed by surgical experiments.

6. DEVELOPING SYSTEMS ARE ABLE TO SELF ORGANISE

This statement is supported by the evidence that parts of animal eggs are able to form a whole embryo (sea urchin and mouse), that combined animal eggs can form a single individual (mouse), that single teratocarcinoma cells can develop in a regular way (Chapter 2.3), and that whole plants can be regenerated from unorganized cell aggregates and pollen grains (Chapter 2.4).

In this part of the book we have been principally concerned with the origins of cell heterogeneity in the early development of plant and animal embryos and in the apical meristems of plants. In many cases cell heterogeneity has been shown to originate from differential partition of cytoplasmic factors between the two daughter cells in mitosis i.e. the cell heterogeneity originates from *intrinsic* differences within a mother cell. However, in the later stages of development, differences may be induced by factors originating from outside the affected cells i.e. by factors *extrinsic* to these cells. The nature of these extrinsic factors, which may be either short-range or long-range (i.e. hormonal), is discussed in Parts 3 and 4. Before we can consider these problems, however, we need to discuss some of the consequences of the establishment of cell heterogeneity in early development and the nature of the resulting restriction of developmental potential (Part 2).

Part 2
Determination and Pluripotentiality

Chapter 2.1
Determination and Stability of the Differentiated State

2.1.1 INTRODUCTION

Because it is a sequential process, development in a multicellular organism consisting of many different organs and tissue types must, of necessity, involve the divergence of cells or groups of cells of common lineage into contrasting paths of development at successive stages. Since the spatial relations between the various regions of a developing plant or animal embryo must follow a very regular pattern if a normal organism is to result, a particular organ or tissue of the mature organism is likely to be derived from groups of cells occupying a specific part of the embryo, in different individuals. In this sense one may say that specific regions of the embryo are destined to form a particular part of the mature organism. This idea was embraced in the statement of Hans Driesch that 'the fate of a cell is a function of its position'. As it stands, this statement might be taken to imply that the pattern of differentiation of any given cell is determined by its position in the organism and that it retains its full development potential, so that if it were transferred to another part of the organism it would differentiate in a manner normal for that part. While it does, indeed, appear that many plant cells retain their full developmental potential (Chapter 2.4), development in animals appears to involve progressive restriction of developmental potential, so that most animal cells are no longer capable of forming a whole organism.

Such cells are described as 'determined' if they retain this restriction through many cell generations and 'determination' is the process by which their developmental potential becomes limited. However, the expression of this restricted developmental potential is profoundly influenced by the environment in which the cell occurs, and can be modified by transferring it to another part of the organism. For instance, the range of cell types formed from the belly ectoderm of the frog gastrula may be extended by transferring it to a new site. Thus, the actual pattern of development is the result of an interaction between the intrinsic potentialities of the cell and influences from its environment. However, as is shown in Chapter 2.2, the restriction of developmental potential in determination is not accompanied by any irreversible changes in the genome of the cell. Moreover, experiments on plants and animal embryos, which are described later in this chapter, show that some cells can develop in a specific manner without loss in their developmental potential, so that differentiation is not the inevitable precondition or consequence of cellular differentiation.

2.1.2 EXTRINSIC AND INTRINSIC FACTORS IN THE DEVELOPMENT OF ANIMAL CELLS

Several processes affect the pattern of development in cells in animal embryos, some of which are discussed in detail in other sections of the book. Here we are concerned to identify some of the factors, both extrinsic and intrinsic, which affect the direction of cell differentiation.

(a) Going clonal (cell allocation)

It is now clear that many animal embryos have a rather definite pattern of cell positions during normal development, and that the pattern of differentiation is determined by the position which a cell occupies in the embryo. Fate maps can be constructed and this would be impossible if the cells of the embryo intermingle in an irregular way.

Fate maps are often made by staining the embryo early in development with vital dyes and subsequently tracing the position of cells which contain the dye in later stages. The fate of the blastomeres of sea urchins has been studied in this way (Fig. 1.3.4) and the fate map of the amphibian Bombinator is illustrated here (Fig. 2.1.1). Both these maps show that a particular tissue or organ in the adult is formed from small localized groups of cells in the early embryo. In the amphibian, the cells which will form the eye are positioned in a pair of groups on either side of the blastula; one group will form the lens cells and the other group will form the retina. During normal embryogenesis, the lens will not be derived from other cells of the embryo and the lens has 'gone clonal', and cells have been allocated to lens development at this stage [13]. Strictly, the onset of clonal development of a tissue is the time at which the mixing between the cell lineages which will form this tissue and those which will form others becomes minimal.

Maps of development may be constructed with less direct methods. The map of *Drosophila* development was made with the help of genetic mosaics because it is difficult to apply vital dyes to the cells of the early embryo, which are surrounded by a hard egg case. It is, however, possible to mark the nuclei of the embryo with particular genes which are expressed in the exterior structures of the adult. This procedure depends on the use of

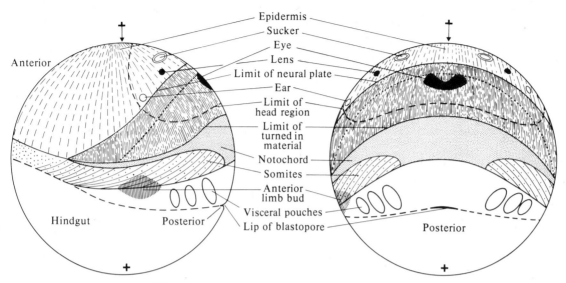

Fig. 2.1.1 Location of the organ forming areas on the early gastrula of the amphibian *Bombinator*. This is a composite of the results obtained by Vogt and his associates by the use of vital dyes. The diagram on the left is the left side of the gastrula and that on the right is the posterior view of the gastrula.

gynandromorphs; these are animals that are mosaics of male and female tissue. There are several stocks of *Drosophila* in which female embryos (XX) lose one X chromosome from the nucleus at the first nuclear division; following this loss the embryo then contains two nuclear populations, one remains

female (XX) and the other is male (XO). These stocks can be arranged so that the X chromosome which is retained in the male cell carries a recessive gene, such as *yellow* or *singed* which affects the colour or morphology of the bristles on the surface of the adult. Since these recessive genes will only be

Fig. 2.1.2 A sketch of cell allocation in the *Drosophila* embryo representing the right half of the fly. This map of cell allocation was obtained by Hotta & Benzer with the use of gynandromorph mosaic mapping (see text).

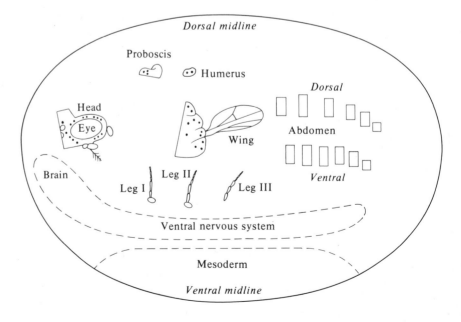

expressed in XO (male) cells it is possible to discover which parts of the adult are formed from them (recently discussed by Hotta & Benzer, [12]).

Following the formation of a male and female nucleus at the first division it appears that their daughter nuclei move away from the centre of the egg in coherent groups and come to lie in a monolayer beneath the egg membrane, dividing the surface of the blastoderm into male and female halves (blastoderm stage, see Fig. 1.3.1). It also appears that the orientation of the spindle of the first nuclear division is random so that the borderline between the male and female cells may be in any orientation with respect to the egg cytoplasm. It seems that the male and female nuclei have an equal chance of multiplying and forming tissues so that the borderline between the two nuclear populations may fall between or in the middle of the cell groups which are allocated to form the tissues of the adult fly. If these assumptions about the formation of mosaics are correct, then the probability of the borderline falling between the cells allocated to different tissues increases as the distance between them increases, and the distance between the allocated groups may be measured by scoring the frequency with which the male-female borderline separates them (Fig. 2.1.2). This type of map is similar to the fate maps of sea urchins and amphibia obtained with vital dyes, but unlike these it does not describe the fate of a group of cells at any particular stage of development but rather indicates the distance between groups of cells allocated to particular tissues. The fact that such maps can be constructed clearly demonstrates that the allocation of cells to particular structures of the adult is a regular process and confirms the orderliness of development which is observed in amphibian and sea urchin embryos with vital dyes.

Recent experiments in *Drosophila* also allow a detailed study of the time and geometry of cell allocation and provide fascinating information about the control of cell fate inside the embryo. These experiments depend on the use of X-rays to induce mosaics at different times in embryogenesis. As in the previous experiments, it is important that the genetically distinct cells produced by X-irradiation should be recognized by genes expressed in the cuticle of the adult insect. It is usual to use stocks of *Drosophila* which are heterozygous for cuticle genes; the X-rays induce somatic recombination so that after division the heterozygous cell forms two cells, each homozygous for a different cuticle gene. In this method of mapping it is assumed that the X-rays induce somatic recombination at the moment at which they are applied, that somatic recombination occurs only in a single cell, and that the fate of cells is not changed by the treatment (reviewed Gehring [8]). If irradiation is conducted at the blastoderm stage, then the homozygous clones are always found in very restricted parts of the adult fly which develops and this shows that in the intact embryo developmental restriction has already started; in fact the clones are always restricted to adult structures formed from one imaginal disc (Fig. 2.1.6). As

irradiation is conducted at progressively later stages of development, so the size of the clones becomes smaller and the range of parts of the adult fly which they form becomes progressively restricted. By itself, this is not a surprising observation because cells irradiated late in development have less time to divide and form large clones before the adult fly emerges. What is more surprising is that the clones are not simply restricted by their growth potential but they are also limited to definite regions of the adult fly. For instance, clones which are produced by X-irradiation during the first two instars of larval life tend to be restricted to the upper or lower surface of the wing and within these surfaces to either the front or back part (reviewed Garcia-Bellido [7], Bryant [1]). Such studies show that cell allocation may be a very accurate process and that it occurs early in development. It is now known that cell allocation in *Drosophila* embryos can be affected by genes and it should soon be possible to study the genetics of the process [7, 23].

The conclusion from these fate maps and cell lineage studies in animals is that inside the intact organism, groups of cells are allocated in a regular way to the formation of particular tissues. If it is the case that cell allocation is controlled by genes, which may be the situation in *Drosophila*, then these studies may become particularly informative about the mechanism of normal development.

(b) Intrinsic factors in the egg

In Part 1 we have provided many examples of embryologies in which the cleavage products of eggs can form fewer cell and organ types than the egg itself. For instance, the eggs of anuran amphibians and insects possess germ plasm and cells which contain it are subsequently able to form germ cells. When the anuran amphibian egg divides equatorially at the eight-cell stage, the four cells in the animal half lack germ plasm and can never form germ cells; they have suffered a developmental restriction.

(c) Embryonic induction

The development potential of cells may also be restricted by interactions with other tissues. This subject is fully discussed in Part 3 and here we may only note one example. The optic cup induces lens formation in the epidermis of amphibians and this interaction prevents the epidermal cells from subsequently forming other structures, such as skin and the lining of the ear.

So far we have only discussed the restrictions of developmental potential which occur and appear to remain stable within the environment of the intact organism. However, the analysis of developmental restriction has mainly been performed on cells or groups of cells isolated from the organism. The reason for using isolation tests for this purpose is that they distinguish situations in which developmental restriction is imposed on cells or groups of cells by their environment from those in which the restriction is inherent to the cell or cell group.

2.1.3 THE OCCURRENCE AND STABILITY OF DETERMINATION IN PLANTS

The phenomenon of determination has received much less attention in plants than in animals. This is probably because botanists have been fascinated with the 'totipotency' of plant cells as illustrated by the ease with which whole plants can be regenerated from tissues which have ceased cell division and have undergone a considerable degree of differentiation. Thus, in many plant species new plants can be regenerated from leaf cells, and we shall see in Chapter 2.4 that embryos capable of developing into mature plants can be produced from almost any part of the carrot plant and of other species. Thus, all cell differentiation is apparently not irreversible in plants, and to this extent it can be argued that restriction in developmental potential is not a necessary consequence of differentiation although the two processes are often linked in animal development.

However, determination of *organs* appears to occur during normal development in plants, as shown in leaf development [16]. Thus, in the fern apex leaf primordia arise first, followed later by the axillary bud primordia. The early stages of development are very similar in both leaves and buds (p. 36), but whereas the leaf primordium soon becomes flattened and shows dorsiventrality, the bud remains radially symmetrical. Moreover, the leaf becomes an organ of 'determinate' growth (in the sense that it does not maintain growth indefinitely), whereas the bud is an organ of indeterminate growth in that it remains potentially capable of unlimited growth. However, at a very early stage of its development the leaf primordium of the fern, *Dryopteris,* can be induced to develop into a bud by making a deep tangential cut between the shoot apical cell and a very young presumptive leaf primordium (Fig. 2.1.3) [19, 3]. With slightly older leaf primordia this surgical treatment is ineffective and the primordium continues to develop into a leaf.

Determination of leaf primordia has also been studied by removing young leaf primordia of the fern, *Osmunda,* at different stages of development and transferring them to a sterile culture medium. When the youngest primordia, P_1–P_5 were treated in this way they developed not as leaves, but as shoots which regenerated roots and so formed complete plants [16]. However, progressively older primordia showed an increasing tendency to develop as leaves, and P_{10} always did so [18]. Thus, it appears that very young leaf primordia are not yet irreversibly committed to develop as leaves, but beyond a certain stage they do become determined and when transferred to a culture medium at this stage, they develop into complete, miniature leaves. Thus, although there is good evidence that most living cells of the mature leaf retain the capacity to undergo 'de-differentiation' and to regenerate whole plants, determination of *organs,* such as leaves (and probably also flowers), appears to occur during normal development.

It may reasonably be asked, how can one have determination of organs without determination of cells? It is possible that the individual cells of leaf primordia suffer no intrinsic restriction of developmental potential and that once the primordium has been channelled into the 'leaf' pathway then the orderly sequence of stages occurring during leaf development represent a 'chain reaction' in which one developmental stage triggers off the next, as has been suggested by Heslop-Harrison for flower development [10]. Whether or not this hypothesis is correct, it is clear that it is possible to construct a model for leaf 'determination' which does not involve a restriction of developmental potential. However, it would appear that other instances of true determination, comparable with the phenomena in animals, do occur in the apical meristems of plants.

At first sight it might appear paradoxical to suggest that meristematic cells exhibit determination, since they are usually regarded as the archetype of cells which are totally uncommitted and undifferentiated. However, there are several well-known phenomena in plants which strongly indicate that determination can occur in shoot apical meristems [21]. One of the best-known examples is seen in the phenomenon of *vernalization.* Many plant species are able to initiate flowers only after they have been exposed to chilling temperatures (0–5°C) for several weeks. Biennial plants which remain

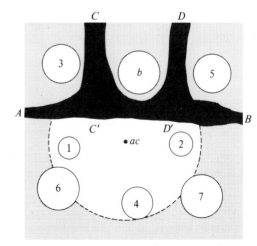

Fig. 2.1.3 *Dryopteris dilatata.* An apex in which the I_1 position has been isolated from the apical cell region (ac) by a deep tangential incision (AB) and from leaf primordia 3 and 5 by radial incisions (CC'', DD''). A bud, b, was formed in what is normally a leaf position (×30). After C.W. Wardlaw [19].

vegetative during the first growing season and flower in the following spring are common examples of species with a chilling requirement for flowering. This induction of flowering by chilling is known as vernalization (reviewed by Purvis [15]; Wareing & Phillips [22]).

The 'winter' varieties of cereals, such as rye and wheat show very similar behaviour to that of biennial plants, since they will flower in the following summer if sown in the autumn, so that they are exposed to a period of chilling during the winter, but they will not do so if they are maintained under warm conditions during the winter. Now, the chilling requirement of winter rye can be met even at the seed stage, since if the seeds are imbibed with water, and maintained at 0–5°C for five to six weeks, they will flower in due course if they are planted and maintained under favourable conditions for growth (Fig. 2.1.4).

Flower initiation involves a transformation of the shoot apex from producing leaf primordia to producing flower primordia and this involves a radical reorganization of the structure of the apex. Now, this transformation of the shoot apex to the flowering condition does not occur in rye embryos during the chilling treatment itself, but is only manifested several months and many cell generations later, after the embryo of the seed has grown into a mature plant.

Once the rye plant has become fully vernalized, it appears

Fig. 2.1.4 Effect of vernalization in winter rye. Left: plants from seed chilled at 1°C for six weeks before sowing. Right: plants from seed germinated at 18°C. From O.N. Purvis (1934) *Ann. Bot.*, **48**, 919.

that the condition is transmitted to all new tissues formed subsequently. Thus, if the main shoot apex is removed, so that the lateral shoots are stimulated to grow and then these are decapitated to stimulate secondary order laterals and so on, it is found that even fourth order laterals are still fully vernalized although the apices of these laterals were not present at the time of the chilling treatment. Evidently the vernalized condition is transmitted from a parent cell to its daughters in cell division and it does not appear to be 'diluted' in the process.

In some species, the effects of chilling can be reversed if the embryos or plants are exposed to high temperature (45°C) for several hours immediately after the completion of the chilling treatment, but if the high temperature treatment is delayed for two days, the reversal of the vernalized state is no longer possible.

The nature of the biochemical changes occurring during vernalization are still completely unknown, but they evidently occur in the cells of the shoot apex itself, since it is possible to vernalize the excised shoot apices of rye embryos if they are placed on a sterile nutrient medium and exposed to chilling temperatures, and they will regenerate into whole plants and ultimately flower. Whatever the nature of the processes occurring during vernalization, they evidently involve metabolism, since the rye embryos require sugars and an adequate oxygen supply for the chilling treatment to be effective.

Although, in rye, the vernalized state becomes irreversible at a certain stage, in some perennial species, such as the garden Chrysanthemum (*C. morifolium*) and Michaelmas Daisy (*Aster novi-belgii*) which have a vernalization requirement, the new shoots which emerge from the base of the plant at the end of the growing season require to be vernalized anew before they can initiate flowers in the following summer. Moreover, even with rye the embryos which the vernalized parent plants produce require fresh vernalization, indicating that the changes involved in vernalization must be obliterated during the development of the pollen grains and embryo-sacs. Evidently there is a 'slate-cleaning' process during gametogenesis, which wipes out the effects of past vernalization treatment.

The familiar phenomenon of an alternation of free-living gametophyte and sporophyte generations in ferns provides another example of a difference in behaviour of apical meristems which is stable and can be transmitted through many cell generations. Normally the gametophyte is haploid and the sporophyte diploid, but the marked differences in morphology and life-cycle between the two generations cannot be attributed to the difference in chromosome number, since it is possible to produce experimentally diploid, or even tetraploid, gametophytes and haploid sporophytes.

A third example of stable differences in the shoot apical meristems of an individual plant is seen in the phenomenon of phase-change in woody plants (reviewed by Doorenbos [5]; Wareing [20]). The majority of woody plant species show a distinct *juvenile phase,* lasting from 1 to 30 years, during which they are unable to initiate flowers, but ultimately they attain the *adult* condition which is marked by the ability to form flowers if other conditions are favourable. There are other differences between the two phases affecting a number of vegetative characters, including leaf shape, phyllotaxis, thorniness etc. A good example of this phenomenon is provided by ivy (*Hedera helix*), in which the base of the vine shows a number of juvenile characters, including palmate leaf shape, whereas the upper part of the vine shows adult characters, including ovate leaves and the capacity to form flowers (Table 2.1.1).

Table 2.1.1   Distinguishing characters of juvenile and adult ivy (*Hedera helix*)

Juvenile characters	Adult characters
Three or five-lobed, palmate leaves	Entire, ovate leaves
Alternate phyllotaxy	Spiral phyllotaxy
Anthocyanin pigmentation of young leaves and stem	No anthocyanin pigmentation
Stems pubescent	Stems glabrous
Climbing and plagiotropic growth habit	Orthotropic growth habit
Shoots show unlimited growth and lack terminal buds	Shoots show limited growth terminated by buds with scales
Absence of flowering	Presence of flowers

One of the most striking features of this phenomenon is the stability of the two phases. If cuttings are taken from juvenile and adult parts of an ivy vine, they continue to show the characteristic features of the shoots from which they were derived, for many years (Fig. 2.1.5). Under natural conditions the juvenile stage ultimately shows the transition to the adult condition, and reversion from the adult to the juvenile phase can be induced in rooted cuttings by various treatments, including growing under fairly high temperatures and treatment with gibberellic acid.

It is clear that phase change does not involve a permanent genetic change, since ultimately it is the adult part of the vine which produces the seeds from which juvenile seedlings are formed. If sterile callus cultures, derived from juvenile and adult ivy tissue, are grown on the same nutrient medium, differences in the two cultures may be seen, with the 'juvenile' cultures maintaining a higher rate of growth and more abundant root regeneration than the 'adult' cultures. Thus, there must be intrinsic differences in the meristematic cells themselves, and the differences between juvenile and adult shoots cannot depend only on differences in the organization of the two types of apex [21]. Evidently we are dealing with two relatively stable states of differentiation which do not involve genetic change, but which can be transmitted through cell division without loss of differentiation.

The phenomena of vernalization, alternation of free living gametophytes and sporophytes in ferns, and phase change in

Fig. 2.1.5 Left: plant of juvenile ivy. Right: plant from cutting taken from adult part of parent vine. Courtesy of Dr V.M. Frydman.

woody plants all involve meta-stable changes affecting shoot apices, and must depend upon activity of different parts of the genome in the two alternative states. This conclusion raises the question whether the difference between shoot and root apices is analogous with the other phenomena we have been considering. Once shoot and root apices have been established in the embryo they are remarkably stable structures which may retain their identity over many cell generations. It is very rare for a root apex to be converted directly into a shoot apex or vice versa, although adventitious buds may arise from mature root cells and adventitious roots in stem tissue. Thus, it would seem that the cells of either shoot or root apices, or both, may have already entered the 'shoot' or 'root' pathways at the embryo stage and in this respect are determined.

However, this suggestion is in conflict with the current opinion that cells in the meristematic regions of shoots and root apices are intrinsically uncommitted with respect to their future differentiation, and it is assumed that the behaviour of the meristematic cells in the shoot and root apices is the result of the organization and biochemical activities of the apex as a whole. We have seen that there is good evidence that some aspects of differentiation at the shoot apex are controlled by the organization of the apex as a whole, as in the siting of leaf primordia which appears to be regulated by inhibitory influences coming from existing primordia (p. 37). However, the self-organizing properties of the shoot apex are not incompatible with some degree of determination within the apex as a whole, as illustrated by the fact that in ivy two different but stable states (juvenile and adult) of the shoot apex may exist, indicating that the cells in the apical meristems can be determined with respect to their general pathway of development and yet remain undetermined with respect to their future differentiation at the cell and tissue level. It is this general type of determination which is postulated as being involved in differences in overall organization and behaviour of root and shoot apices. However, the question of whether or not apical meristems show some degree of determination must remain open until direct experimental evidence is available.

2.1.4 THE STABILITY OF DETERMINATION IN ANIMALS

At present research workers are paying particular attention to two situations in which the stability of determination may be investigated away from the limitations of embryogenesis: these studies have been conducted on transdetermination in the imaginal discs of *Drosophila* and on the expression of differentiated functions by cells in culture.

(a) Transdetermination in imaginal discs

The development of higher insects such as *Drosophila,* is divided into two quite separate phases. A larva emerges from the egg and after a period of growth transforms during metamorphosis into an adult. This metamorphosis is unlike that of amphibians which only involves the growth of a few new organs and the resorbtion of unnecessary structures such as the tail (see Chapter 4.4). In contrast, the adult insect is very different from the larva and its progenitor cells lie in small groups (the imaginal discs) inside the larva. At the early blastoderm stage, the cells of the imaginal discs appear to segregate from those which will form the larva, and the disc cells appear to play no part in larval life. During metamorphosis, the larval organs are resorbed and the disc cells start to proliferate and generate the adult structures. Exceptionally the Malpighian tabules of the larva are retained throughout metamorphosis.

Particular structures of the adult are formed from specific discs (Fig. 2.1.6) in a regular manner. For instance, the external part of the head of the fly is formed from the labial discs, the imaginal cells of the clypeolabrum and the eye antennal discs;

these discs occur in pairs as do those which form most of the other parts of the body and it is only the genital apparatus which is formed from a single disc. The cells of different discs look identical under the electron microscope despite the fact that they will subsequently form an astonishing variety of structures including precisely organized chemoreceptors, mechanoreceptors, jointed limbs, wings, and the regular array of ommatidia in the compound eye.

It is now known that the morphological uniformity of the disc cells conceals the fact that each disc is determined to form particular adult structures during metamorphosis. The evidence for this view comes from elegant transplantation experiments in which the discs are isolated from one larva and transplanted to the body cavity of another larva. The second larva is allowed to develop into an adult and the transplanted disc will form an adult structure in the body cavity of the fly. It is found that one disc will always form one set of adult structures and that its developmental fate is not controlled by the position which it occupies in the body cavity of the host larva.

Imaginal discs can also be used to study the stability of determination over several years. Such tests are possible

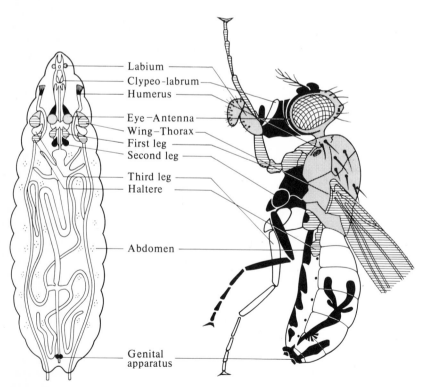

Fig. 2.1.6 The imaginal discs of *Drosophila*. This diagram shows the position of the discs in the larva and their corresponding adult derivatives. After Nöthiger [14].

because the discs can be propagated without the formation of adult structures. This is achieved by transferring the discs to the body cavity of an adult; here they will never experience the conditions which promote metamorphosis and the development of adult structures. The disc cells normally stop dividing just before the emergence of the adult fly, but in the abdomen of a fly they continue to grow and duplicate and at the end of adult life (two to three weeks), they are dissected out and transferred to a new adult. With this technique it is possible to propagate the discs for many years, and at any time the state of determination of the disc may be checked by transferring part of it back to a larva when it is forced to develop adult structures as the host proceeds through metamorphosis (Fig. 2.1.7).

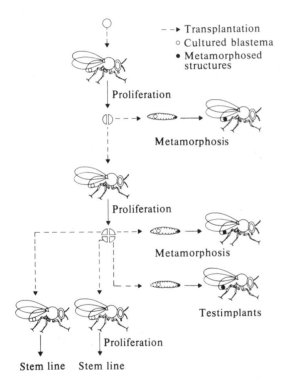

Fig. 2.1.7 The method of culturing and transplanting the imaginal discs of *Drosophila*. The diagram illustrates the method of propagating the discs and testing their state of determination by passing the discs through metamorphosis. After Gehring [8].

These studies have shown that determination may be stable through many cell generations but that on occasion the cells of the disc may be transdetermined to form a new structure. The transplantation of the male genital disc can serve as an example [9]. The genital disc can be recognized by the structures associated with the anal plate and it was found that these

continued to be found in some of the discs for up to 70 transfer generations, that is about three years after the original disc was transferred to the abdomen of a fly. This observation demonstrates that this determination can be inherited for many cell generations. However at the 6–10th transfer generations new structures began to appear and antennal and leg structures were found in some of the discs. At about the same time discs started to form parts of the wing in the 15th transfer generation discs started to regularly form part of the thorax. The observation of transdetermination clearly demonstrates that determination does not involve an irreversible loss of the ability to form other cell types.

Taken together with the observations on vernalization and phase growth in plants we can conclude that cell populations can propagate a state of development over long periods.

(b) The expression of differentiated functions by cells in culture

Many cell types can be cultural *in vitro* and in these situations they are able to continue to synthesize their cell type associated biochemical properties. These cells may be cloned, that is grown up in isolation from each other, and they still continue to express their differentiated functions, which demonstrates that the state of determination is a cell autonomous property. The following is a list of some of the differentiated properties which may be retained by cells in culture for over two years: liver cells and albumin synthesis, pigment cells and the synthesis of melanin, nerve cells and the production of acetylcholine, lymphoid cells and the production of immunoglobulins. Cells with these functions have all been cloned and many of these clones continue to express the differentiated function (reviewed by Davidson [4]).

It is also important to discover how often a single differentiated cell will form daughter cells which continue or are able to synthesize the same differentiated product. This has been investigated with cartilage cells, which synthesize chondroitin sulphate, and with pigment cells, which synthesize melanin. Coon dissociated chick cartilage into single cells and found that more than 98% of these formed colonies which continued to synthesize chondroitin sulphate [2]. Nine of the single cells were grown until they had divided 36 times. During their growth, they were dissociated into single cells on three occasions, and after each dissociation it was found that each of the 200 or more colonies which grew up from these single cells formed chondroitin sulphate. This observation clearly demonstrates that differentiated single cells may inherit their state of determination through many cell generations. He next investigated whether the state of determination was unstable if the expression of the differentiated function was suppressed (chondroitin sulphate not detected). In this case, the cells were grown in a medium with particular protein components which prevented cartilage production and it was shown that after 20 generations the cells could still synthesize chondroitin sulphate if they were returned to normal medium. This observation shows that although a differentiated function may be undetected, nevertheless the determined state may remain stable. However if cartilage cells are grown under different conditions to those employed by Coon, then they may form several different morphological cell types and it is clear that the stability of cell type depends on the conditions of cell growth [11].

Experiments on pigment cells show that in some cases cells may switch their cell type in culture and that determination is not necessarily stable [6]. A population of pigmented cells (99·9% pigmented) was obtained by culturing the pigmented retina of a chick. After a period in culture, a group of pigmented cells was dissociated and the single cells which grew always formed pigmented colonies. However after the colonies had grown for a period and had been subcultured to new bottles, then two of the eight colonies studied started to form groups of organized cells which looked like small lenses. These groups of cells were shown to be synthesizing lens proteins because antibodies prepared against lens fibres reacted with them. These experiments demonstrate that a cell synthesizing melanin can give rise to cells synthesizing lens proteins and show that determination may be unstable.

2.1.5 CONCLUSIONS

The conclusion from this section on cell determination is that cells may undergo certain changes in their developmental potential during differentiation, and that such changes may be relatively stable. In later chapters it will be shown that this stability does not usually involve irreversible changes in the genome. Thus nuclei from differentiated cells can code for the normal development of amphibia (Chapter 2.2), and many cells in an adult plant retain the capacity to form a whole plant (Chapter 2.4). It is also the case that unusual animal cells, called teratocarcinoma cells, are able to form many tissue types after growth in culture for several years (Chapter 2.3). These observations imply that development and differentiation in plants do not require that cells should lose the capacity to repeat the life cycle of the organism, and it is possible that only technical difficulties have prevented similar experiments with animals.

2.1.6 REFERENCES

[1] BRYANT P.J. (1974) Determination and pattern formation in the imaginal discs of *Drosophila*. In *Current topics in developmental biology*, **8**, pp. 41–80 (Eds A.A. Moscona & A. Monroy). Academic Press.

[2] COON H.G. (1966) Clonal stability and phenotypic expression of chick cartilage cells *in vitro*. *Proc. nat. Acad. Sci. U.S.*, **55**, 66–73.

[3] CUTTER E.G. (1956) Experimental and analytical studies of pteridophytes XXXIII. The experimental induction of buds from leaf primordia in *Dryopteris aristata* Druce. *Ann. Bot.*, **20**, 143–65.

[4] DAVIDSON R.L. (1974) Control of expression of differentiated functions in somatic cell hybrids. In *Somatic cell hybridization*, pp. 131–50. (Eds R.L. Davidson & F. de la Cruz). Raven Press, New York.

[5] DOORENBOS J. (1965) Juvenile and adult phases in woody plants. *Encycl. Plant Physiol.*, **15**(1), 1222–35.

[6] EGUCHI G. & OKADA T.S. (1973) Differentiation of lens tissue from the progeny of chick retinal pigment cells cultured *in vitro*: a demonstration of a switch of cell type in clonal cell culture. *Proc. nat. Acad. Sci. U.S.*, **70**, 1495–99.

[7] GARCÌA-BELLIDO A. (1972) Pattern formation in imaginal disks. In *The biology of imaginal disks*, pp. 59–91. (Eds H. Ursprung & R. Nöthiger). Springer-Verlag, Berlin.

[8] GEHRING W. (1972) The stability of the determined state in cultures of imaginal disks in *Drosophila*. In *The biology of imaginal disks*, pp. 35–57. (Eds H. Ursprung & R. Nöthiger). Springer-Verlag, Berlin.

[9] HADORN E. (1967) Dynamics of determination. In *Major problems in developmental biology*, pp. 85–104. (Ed. M. Locke). Academic Press.

[10] HESLOP-HARRISON J. (1967) Differentiation. *Ann. Rev. Plant Physiol.*, **18**, 325.

[11] HOLTZER H. & ABBOTT J. (1968) Oscillations of the chondrogenic phenotype *in vitro*. In *The stability of the differentiated state*, pp. 1–16. (Ed. H. Ursprung). Springer-Verlag, Berlin.

[12] HOTTA Y. & BENZER S. (1973) Mapping of behaviour in *Drosophila* mosaics. In *Genetic mechanisms in development*, 31st Symposium of the Society for Developmental Biology, pp. 129–67. (Ed. F.H. Ruddle). Academic Press.

[13] McLAREN A. (1972) Numerology of development. *Nature, Lond.*, **239**, 274–76.

[14] NÖTHIGER R. (1972) The larval development of imaginal disks. In *The biology of imaginal disks*, pp. 1–34 (Eds H. Ursprung & R. Nöthiger). Springer-Verlag, Berlin.

[15] PURVIS O. (1961) The physiological analysis of 'vernalisation'. *Encycl. Plant Physiol.*, **16**, 76–122.

[16] STEEVES T.A. (1961) A study of the developmental potentialities of excised leaf primordia in sterile culture. *Phytomorphology*, **11**, 346–59.

[17] STEEVES T.A. (1966) On the determination of leaf primordia in ferns. In *Trends in plant morphogenesis* (Ed. E.G. Cutter). Longmans, London.

[18] STEEVES T.A. & SUSSEX G.M. (1957) Studies on the development of excised leaves in sterile culture. *Am. J. Bot.*, **44**, 665–73.

[19] WARDLAW C.W. (1949) Experiments on organogenesis in ferns. *Growth* (Suppl), **13**, 93–131.

[20] WAREING P.F. (1959) Problems of juvenility and flowering in trees. *Journ. Linn. Soc. London*, **56**, 282–89.

[21] WAREING P.F. (1971) Some aspects of differentiation in plants. *Symp. Soc. Exp. Biol.*, **25**, 323–44.

[22] WAREING P.F. & PHILLIPS I.D.J. (1970) *The control of growth and differentiation in plants*. Pergamon Press, Oxford.

[23] LAWRENCE P.A. & MORATA G. (1976) The compartment hypothesis. In *Insect development*, 8th Symposium of the Royal Entomological Society, pp. 135–51. (Ed. P.A. Lawrence). Blackwell Scientific Publications, Oxford.

Chapter 2.2
Pluripotentiality of
Cell Nuclei

2.2.1 INTRODUCTION

The pluripotentiality of a nucleus is its ability, under appropriate conditions, to direct the differentiation of all the various cell-types which make up an individual animal or plant, even though most nuclei present in an adult organism will have already promoted the differentiation of one particular kind of specialized cell. One possibility is that all animal and plant cell nuclei are truly pluripotential. The principal alternative is that, as somatic cells specialize, their nuclei gradually lose their pluripotential state and only the nuclei of gametes retain the capacity to form an adult. It is important to know whether nuclear pluripotentiality is characteristic of specialized cells; if it is not, this would at once suggest that gene expression in specialized cells might depend on, or be caused by, the loss or permanent inactivation of genes which would never normally be expressed during the further specialization of such cells or their daughters. Conversely, to demonstrate that somatic cell nuclei are pluripotential would be important since this would eliminate one class of mechanisms capable of controlling gene expression.

Any cell which, like a fertilized egg, can by itself give rise to a complete individual containing normal cells of all types must necessarily contain a pluripotential nucleus. As explained in Chapter 2.4 botanists have been remarkably successful in growing whole plants (which bear seed) from somatic cells. The work of Steward [25] is a classical example of this kind. The conclusion from such experiments is that callus cells derived from plant leaves, stems and roots (and therefore also their nuclei) are pluripotent. Almost certainly the cells which have given rise to whole plants are parenchymal cells, and evidently these must have had totipotent nuclei. It has not, however, been shown beyond doubt that other more obviously specialized plant cells, such as sieve cells and sclereids of the phloem can give rise to complete plants when cultured singly, and these are not therefore known to be totipotent.

In animals, it has proved impossible to grow a complete individual from a single somatic cell. The main contribution from cell growth experiments to the problem of nuclear pluripotentiality comes from cases in which a specialized cell gives rise to daughter cells which specialize in a different way. Even this happens very seldom, and the only fully substantiated example in vertebrates is the differentiation of lens cells from other eye cells, such as those of the pigmented iris, the cornea,

and the retina [5] (see Chapter 2.3). On the basis of these results, no generalization can be made about the pluripotentiality of nuclei. The most that can be said is that, in both animals and plants, *some* kinds of somatic cells can change through several intervening cell divisions, into certain other cell-types. A total lack of nuclear pluripotentiality is not therefore an essential feature of cell differentiation.

The most direct test of nuclear potentiality in animals is provided by nuclear transplantation experiments. These have been carried out successfully on Amphibia for 20 years, and have yielded clear conclusions on the developmental capacity of nuclei from several different types of specialized somatic cells. Nuclear transfer experiments are discussed in detail below. Quite a different kind of experiment which gives complementary information involves nucleic acid hybridization with DNA from different kinds of somatic cells. The precision of this technique is only now reaching the point at which genes present in one copy per genome can be recognized. The technique makes it possible in some cases to say whether a known gene is present in a population of cells, and if so by how many copies it is represented in a somatic cell genome. Since this technique is still being improved in sensitivity and since current conclusions are tentative, the results of nucleic acid hybridization are discussed only briefly.

2.2.2 NUCLEAR TRANSPLANTATION EXPERIMENTS

Design

The aim of a nuclear transplant experiment is to transfer the nucleus from a specialized cell into an unfertilized egg whose own nucleus has been removed. If the experiment is successful the enucleated egg cytoplasm and the transferred nucleus will develop into a normal individual. In this case it is clear that the transplanted nucleus contained enough information to promote the differentiation of all specialized cell-types in the normal individual. An enucleated egg is not capable, on its own, of development or of any kind of cell differentiation. Furthermore the characteristics of nuclear-transplant embryos are always those of the nuclear, and not cytoplasmic strain or species [9]. Therefore the development of a nuclear-transplant egg depends

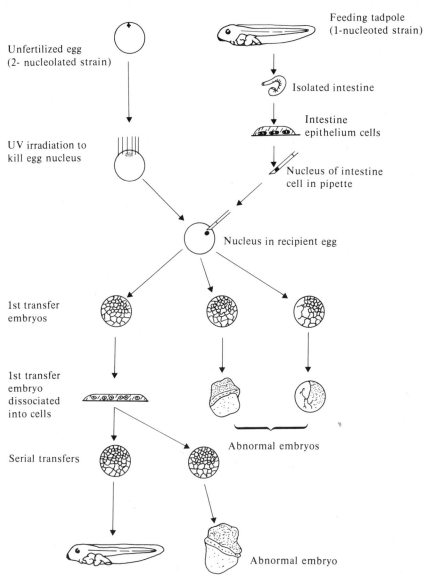

Fig. 2.2.1 The transplantation of nuclei from intestine cells of *Xenopus*. Ultraviolet irradiation kills the nucleus of unfertilized eggs. Donor cells are obtained by dissecting out the intestine, and then dissociating its component cells in versene-containing medium. A single cell is sucked up into a pipette, so as to break the cell but not its nucleus. The whole broken cell is injected into the recipient egg cytoplasm; the donor cell cytoplasm is 10^{-5} times the volume of the egg, and has no effect on development, as tested by injecting cytoplasm without a nucleus. When a first-transfer embryo has reached the blastula stage, its cells may be dissociated and used to make further transfers. Some of these will usually develop into normal tadpoles, and others will form abnormal embryos, for reasons explained on page 59.

on information provided by the transplanted nucleus and its daughter nuclei. Thus a nucleus which supports normal development after transplantation to an enucleated egg is said to be pluripotent.

Methods

The technique by which nuclei can be transplanted successfully from specialized cells is based on that of Briggs and King [1]. Amphibia are the only group of animals in which it has so far been possible to transplant nuclei from differentiated cells [12]. Attempts to transplant nuclei in insects have recently been extended with success to blastoderm nuclei of *Drosophila* [16], though these cells are not cytologically specialized.

The methods which have been used to transplant nuclei in Amphibia differ in detail according to species, but all have the following characteristics in common (Fig. 2.2.1). The nucleus of an unfertilized egg is conveniently located just under the surface of the animal pole. It can therefore be removed surgically by a needle [18], or can be eliminated by ultra-violet irradiation [8]. Donor cells can be separated by various treatments; trypsinization or incubation in EDTA achieves separation of cells from most tissues. After the recipient egg and donor cells have been prepared, the third step in the technique is to suck a single cell into a pipette. This must be sufficiently narrow to break the cell membrane, but large enough not to distort the nucleus, which therefore enters the pipette, and later the egg, intact.

An important aspect of well-designed nuclear-transplant experiments is the use of a nuclear marker. This is a means by which it can be proved that nuclei of the resulting embryo have been derived from the transplanted nucleus, and not from an incomplete removal of the egg nucleus. In *Xenopus*, a very convenient nuclear marker is provided by a deletion of ribosomal genes, a mutation associated with the absence of a nucleolus [7]. In heterozygous form, this mutant is entirely viable, and all its cells contain a single nucleolus; wild-type cells usually have two nucleoli since they have two chromosomes with ribosomal genes able to form a nucleolus at the site of the nucleolus organizer.

The last characteristic of nuclear transplantation experiments that should be mentioned is the use of serial nuclear transplantation. This involves the same procedure as has been described for first transfers except that instead of donor nuclei being taken from the cells of an embryo which originated from a fertilized egg, they are taken from a young embryo that is itself a result of a nuclear transplant experiment (Fig. 2.2.1). The effect of this procedure is similar to that of the vegetative propagation of plants. It enables a large number of genetically identical copies to be made of the originally transplanted nucleus. Adult frogs can be obtained by nuclear transplantation, and these are similar to frogs reared from fertilized eggs (Fig. 2.2.2). There are certain additional reasons why it is particularly helpful to make use of serial transplantation when analysing the nuclei of specialized cells, as explained on page 60.

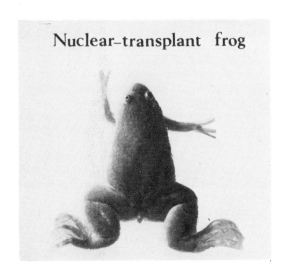

Nuclear–transplant frog

Control frog from fertilized egg

Fig. 2.2.2 A nuclear-transplant frog of *Xenopus laevis*, prepared by serial transplantation of nuclei from cultured tadpole epidermis cells. The frog is similar to one grown from a fertilized egg. Both frogs are she same age. From [13].

Results demonstrating the pluripotentiality of somatic cell nuclei

The first decisive evidence that a nucleus of a specialized cell is totipotent came from experiments conducted on *Xenopus* using intestinal epithelium cell nuclei [10]. These cells were chosen as a source of donor nuclei because their differentiated state is clearly evident from their possession of a striated border, a specialization related to absorptive function. The results of transplanting nuclei from intestinal epithelium cells of feeding larvae are shown in Table 2.2.1. It is seen that 7% of all explant of adult frog skin under these conditions react with anti-keratin antibody three days after the explant had been made. The results obtained with skin nuclei are summarized in Table 2.2.1. It is seen that normal or nearly normal tadpoles can be formed by the serial transplantation of nuclei from cells which grow out from skin explants, and that this happens in over 3% of all cases. This figure is enormously higher than the percentage of cells which grow out from skin explants and which do *not* react with keratin antibodies (<0·1%). As in the experiments with intestinal epithelium cell nuclei, a nuclear marker was used in the adult skin cell experiments. The results with adult skin cells

Table 2.2.1 The development of embryos prepared by transplanting nuclei from tadpole intestine and frog skin cells into unfertilized eggs of *Xenopus laevis*.

Donor cells	Total number of transfers	Percentage of total transfers reaching	
		tadpoles with muscle and nerve cells	tadpoles with muscle, nerve, lens, heart, blood cells etc.
Intestinal epithelium cells of feeding tadpoles [10]	726	20%	7%
Cells grown in culture from adult frog skin [28]	129	4.6%	3.1%
Blastula or gastrula endoderm cells [13]	279	65%	57%

Note: the percentage survival includes the results of first and serial transfers. Details of experiments are given in the references cited in the first column.

transplanted nuclei could be shown to be capable of promoting the differentiation of muscle, nerve, lens, heart and blood cells even though such nuclei had already participated in the differentiation of an intestinal epithelium cell. Nearly three times that proportion of transplanted nuclei promoted development as far as larvae containing functional muscle and nerve cells. A small number of intestinal epithelium cell nuclei yielded adult frogs which became sexually mature as males and females. In these cases the nuclei were evidently pluripotential in the complete sense. The reasons why most nuclear transplants did not lead to development of normal mature adults is not fully understood, but is discussed on pages 59–61. These results with intestine nuclei were validated by the use of the nuclear marker in all experiments, and clearly show that some of the nuclei of specialized cells can promote development and differentiation of a kind totally unrelated to their own.

More recently these experiments have been extended by transplanting nuclei from skin cells of adult frogs [13, 28]. It is not self-evident that cells which grow out from a piece of adult skin will themselves be specialized skin cells, since most adult tissues contain a certain proportion of fibroblasts which are particularly likely to grow out of an explant in culture. The most direct evidence that the cells used for nuclear transplantation experiments were in fact skin cells comes from the use of fluorescent antibodies prepared against frog keratin. It was found [24] that over 99% of all cells that grew out from an

are in complete agreement with those obtained from larval intestinal epithelium cells. In the case of the adult skin cell nuclei, it has not yet been possible to prepare a normal adult frog capable of reproduction. Although we cannot say with certainty that adult skin cell nuclei have retained the capacity to promote *gamete* differentiation, the tadpoles formed from skin cell nuclei contained all other major cell-types in an apparently normal and functioning condition.

These two series of experiments therefore provide a strong argument in favour of the generalization that amphibian cells can specialize without any loss of developmental capacity by their nuclei. It is quite certain that such nuclei have not lost the capacity to promote the differentiation of many different cell-types quite unrelated to that from which a nucleus is taken. Therefore it is certainly correct to say that these vertebrate cells contain pluripotent nuclei; it is very likely that their nuclei are totipotent, though it appears to be hard to demonstrate this, probably for technical reasons.

The results reported so far have been described in detail, but are by no means unique. Kobel *et al.* [19] have obtained tadpoles with myotomes by transplanting the nuclei from cultured melanophore cells. Although the proportion of transplanted nuclei which support the formation of tadpoles was less than 1%, each donor cell was identified under the microscope as one which contained melanin granules; therefore there is no doubt about the specialized nature of those donor cells from which successful results were obtained. Even in *Rana pipiens*, in which species nuclear transplantation results are generally less good than those obtained with *Xenopus*, it has been possible to obtain some nearly normal tadpoles from cells of a newly-metamorphosed frog. McKinnell *et al.* [21] were able to obtain some tadpoles from the nuclei of adenocarcinoma tissue. In the same species, Muggleton-Harris and Pezzella [22] obtained tadpoles after the serial transplantation of nuclei from lens epithelium. Evidently tadpoles containing a range of different cell-types can be obtained from the nuclei of specialized cells not only in *Xenopus* but also in other amphibian species.

The significance of abnormal nuclear transplant embryos

As has been indicated above, the majority of transplanted nuclei from specialized cells do not support normal development. We want to know whether this is because most nuclei in a tissue are not pluripotent. Such a view would be consistent with the results so far mentioned. An alternative point of view is that all nuclei in specialized cells are pluripotent, but that it has not been possible to demonstrate this by means of nuclear transplantation. This would be easily understandable, since the technique of nuclear transplantation is difficult, and since the division cycles of donor nuclei and recipient eggs are very different. This last interpretation is strongly favoured by the detailed analysis of abnormal nuclear transplant-embryo development. Four generally agreed conclusions regarding abnormal transplant-embryo development are as follows.

First it is clear that the developmental capacity of a transplanted nucleus decreases with increasing age of the donor embryo from which it was taken. All those who have compared the ability of nuclei to support development when taken from early as opposed to late stages of development find that blastula nuclei give much more normal results than the nuclei of, for example, neurula and tail-bud embryos. This point was first established by Briggs and King [2] and has subsequently been observed in *Xenopus* and in all other amphibian species so far used for nuclear transplantation [11]. This finding raised the possibility that development might be accompanied by a loss of developmental capacity in nuclei. However this result is also consistent with the view that the decreasing division rate of donor nuclei impairs their capacity to divide normally after transplantation to eggs. The decreasing success of nuclear transplants roughly coincides with the decreasing rate of mitotic division in the donor tissue.

The second commonly agreed generalisation about abnormal nuclear transplant-embryo development is that the decline in developmental capacity of transplanted nuclei is caused by a stable nuclear condition. This was first established by the serial nuclear-transplant experiments of King and Briggs [17], and subsequently confirmed in other species [11]. This conclusion is based on a comparison of the developmental abnormalities observed among serial transplant-embryos derived from a single originally transplanted nucleus. There is considerable variation in the way in which embryos within one clone develop, but consistent differences are usually observed between the various clones derived from different nuclei of the same original donor embryo (Fig. 2.2.3). It is evident that the nuclei derived by transplantation from one original somatic nucleus differ in a stable, and possibly genetic, way from the nuclei derived from another somatic nucleus.

The third general point about abnormal nuclear-transplant embryos is that chromosome abnormalities are the cause of the stable differences between transplanted nuclei, which result in their reduced developmental capacity. It is now agreed that the great majority and probably all developmental abnormalities observed in transplant embryos are caused by chromosome abnormalities. This relationship was first observed by Briggs *et al.* [3] and has now been substantiated in detail by several authors, e.g. Hennen [15]. Chromosome abnormalities of the kinds observed did not exist in the donor cells used, and they must therefore have arisen as a result of nuclear transplantation. It is clear that most, if not all, of the serially transmitted abnormalities in nuclear-transplant clones can be attributed to abnormal chromosomes. Such a basis for developmental abnormalities would explain why the differences between clones are serially propagated. There is, therefore, no reason to think that the stable differences observed between serial clones may represent any kind of normal genetic difference between somatic nuclei.

The chromosome breakage just referred to occurs soon after nuclear transplantation. It is not known why this should be, but it seems very likely to be due to the fact that most somatic cells divide very infrequently (about every one to two days), and take about 7 h to replicate their chromosomes; yet any transplanted nucleus that is to participate in normal development must start and finish replication of its chromosomes within one hour, since eggs always undergo cytoplasmic cleavage within 1 to 2 h of activation by micropipette penetration. Chromosome abnormalities do indeed occur more often with nuclei taken from slow dividing (differentiated) cells than with fast dividing blastula nuclei. Furthermore, DNA synthesis in newly transplanted nuclei often extends beyond the normal time, and sometimes continues into the first mitosis, e.g. Gurdon and Laskey [13].

The chromosome breakage which occurs after nuclear transplantation suggests a reason for the very beneficial effect of carrying out serial nuclear transfers. All the results which argue for the pluripotent state of somatic cell nuclei have been based on serial nuclear transfer experiments. It is known that the chromosome breakages occur not only at the first division but also at subsequent divisions of transplanted nuclei [4]. A detailed study of the cleavage pattern promoted by nuclei from slow-dividing cells [13] has shown that incomplete chromosome

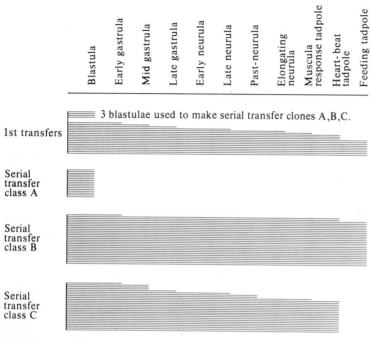

Fig. 2.2.3 Variation in development between serial-transfer clones. In this experiment endoderm cells from an elongating neurula were used to make first transfers. When three of these had reached the blastula stage, each was dissociated into its component cells to provide nuclei for serial transfers. Each horizontal line in the figure represents the development of an individual embryo. It can be seen that all embryos in clone A died as late blastulae, whereas nearly all embryos in clone B developed into entirely normal tadpoles. Each clone represents the development promoted by mitotic products of one endoderm cell nucleus in the original donor embryo. The interpretation of this experiment is explained on page 00, and further details are given in reference [11].

replication and disjunction often results in only one of the first blastomeres receiving a nucleus with a complete set of chromosomes, the other blastomere often lacking a nucleus altogether. The complete nucleus continues to divide, but some of its daughter nuclei are themselves defective. Consequently an abnormal 'partial' blastula is formed which is a mosaic of normal and abnormal cells. By making serial transfers from such blastulae, some of the normal nuclei, which are now dividing fast and are no longer subject to chromosome loss at division, can be selected and permitted to form a normal embryo. While many serial transplant embryos develop abnormally because the transplanted nuclei have abnormal chromosomes, a few receive normal nuclei. In these last cases a complete embryo with normal chromosomes should develop, and this would indicate the true developmental capacity of the original somatic nucleus. This is the most plausible explanation for the beneficial effect of carrying out serial nuclear transfers.

The fourth general statement about abnormal transplant-embryos concerns specificity of the chromosomal and developmental abnormalities revealed by serial transplantation. It is conceivable that the chromosome abnormalities observed in transplanted nuclei might be of different kinds according to the tissue from which the nuclei were taken. In this event the abnormalities concerned might tell us something about the condition of the nuclei in different cell-types. In fact, there is no evidence that the chromosome abnormalities can be related to any particular cell-type, since a wide range of abnormalities arise when nuclei are taken from any one of several different cell-types. The cell-type specificity of developmental abnormalities of embryos has been investigated by comparing the embryos derived from transplanted endoderm nuclei with embryos grown from transplanted mesoderm or ectoderm nuclei. Such an analysis has been made by DiBerardino and King [6]. In this case, the great majority of embryos studied showed developmental abnormalities which were in no way related to the donor cell-type. If attention was limited to the very small number of embryos with completely normal chromosome sets, a slight tendency was detected for nuclei from one cell-type to be associated with developmental abnormalities of a particular kind. Nevertheless, it is clear that over 95% of the embryos obtained in this experiment showed no cell-type specificity in the kinds of developmental abnormalities observed (see Gurdon [14] for a fuller discussion of these results). In conclusion we can say that there is no strong case for believing that the chromosomal or developmental abnormalities in nuclear-transplant embryos are donor cell-type specific.

All these points just summarized clearly favour the view that those transplanted nuclei which fail to promote normal development may do so on account of an incompatibility in their division rate and that of an activated egg. It is understandable that a newly transplanted nucleus is often unable to adjust rapidly to the very fast division rate set by an egg, when it has come from cells which were dividing, and replicating their chromosomes, at a relatively low rate. It seems that this is an entirely sufficient explanation for the chromosomal and developmental abnormalities observed. There is therefore no reason to think that these abnormalities indicate any lack of developmental capacity in the nuclei of somatic cells before the time of their transplantation. The results of nuclear-transplant experiments are therefore entirely consistent with the view that all somatic cells have pluripotent nuclei, the point which has been clearly proved for *some* nuclei of *some* specialized cells.

2.2.3 MOLECULAR HYBRIDIZATION EXPERIMENTS

By demonstrating the presence of unexpressed DNA sequences in a nucleus, molecular hybridization experiments can contribute to the question of nuclear pluripotentiality. The recognition of genes present as a single copy in a genome has been greatly facilitated by the use of reverse transcriptase, an enzyme which promotes the synthesis of DNA from RNA, and which is obtained from various RNA tumour viruses, such as avian myeloblastosis virus. Messenger RNA which codes for a known protein can be purified from various kinds of specialized cells which synthesize large amounts of this protein and its messenger RNA. Using DNA precursors of very high specific activity, reverse transcriptase can be employed to make DNA complementary to messenger RNA; complementary DNA of immensely high specific activity is then hybridized with DNA from a particular tissue, according to procedures described, for example, in references [23] and [26]. If the gene which codes for the messenger RNA is present in cells of that tissue, the complementary DNA will hybridize to the strand of DNA which coded for the messenger RNA. By this means it is possible to

estimate the number of times that the gene which codes for that RNA is present in a particular tissue.

Using such techniques, evidence has recently been obtained which indicates that certain genes are present as only one copy in a haploid genome. An example of this work is that of Packman *et al.* [23] who have made radioactive DNA complementary to globin messenger RNA. They have concluded from the kinetics of hybrid formation that liver cells and erythrocytes in ducks have the same small number (no more than five) of globin genes. A similar design of experiment has been carried out in respect of chick ovalbumin genes which are expressed fully in oviduct cells. Sullivan *et al.* [26] have concluded that ovalbumin genes are present as only one gene per haploid genome in oviduct and liver tissue (Fig. 2.2.4). These experiments make it possible to identify the presence of a gene which was unexpressed in a specialized tissue. The technology involved in such experiments is being improved all the time, and it should not be long before definite information is available about the presence, in several different cell-types, of genes which are primarily or exclusively expressed in another cell-type. Experiments of this kind which have so far been

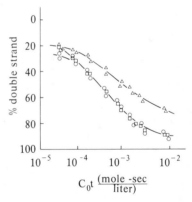

Fig. 2.2.4 The use of molecular hybridization experiments to test for the amplification of genes in different somatic tissues. Highly radioactive 2-stranded DNA is prepared by reverse transcriptase from chick ovalbumin messenger RNA which is present in large amounts in oviduct but not liver tissue. The labelled DNA strands are separated by heat and allowed to reassociate in the presence of unlabelled DNA extracted from oviduct or liver tissue. The rate at which the labelled DNA strands reassociate is accelerated in proportion to the number of ovalbumin gene sequences contributed by the liver or oviduct DNA. The amount of the labelled DNA which has reassociated is shown by the proportion of double-stranded DNA in samples taken at different times after the reaction has begun. The abscissa (C_0t) is a measure of the duration of the reaction in relation to a known concentration of DNA. $\triangle$ represents the rate at which the labelled DNA molecules reassociate in the absence of any added ovalbumin gene sequences. The figure shows that liver DNA ($\bigcirc\bigcirc\bigcirc$) and ovalbumin DNA ($\square\square\square$) accelerate the reassociation of the labelled DNA molecules to the same extent. Hence they must contain about the same numbers of copies of ovalbumin genes, which are therefore not amplified in the tissue where these genes are particularly active (oviduct). This figure is taken from [26].

published make it very likely that genes coding for cell-type specific products are present in one or very few copies per genome, and that they are present in all cells.

2.2.4 CONCLUSIONS

The information derived from nuclear transplantation and molecular hybridization experiments is complementary. Nuclear transplant experiments are able to provide evidence for the existence in cells of a complete range of unexpressed genes. It is assumed that the formation of a normal individual requires the presence of a set of genes which are indistinguishable from those present in a fertilized egg. Clearly genes which have been lost or permanently inactivated cannot be restored by propagating a nucleus mitotically.

The great advantage of determining pluripotentiality by nuclear transplantation experiments is that it is possible by this means to test for parts of a genome which code for unknown proteins or which contain DNA of unknown function. If, for example, regulatory genes exist in animal cells, it would be impossible to demonstrate their presence by current biochemical procedures. But normal development from a transplanted nucleus would argue very strongly for the existence of such genes in a specialized cell.

Molecular hybridization experiments are able to recognize only those genes for which a purified transcribed RNA is available, except in certain cases where it has been possible to isolate genes directly [27]. Such information does not therefore apply to the great majority of the genome. On the other hand hybridization experiments can complement nuclear transplant experiments by providing an estimate of the number of copies of a gene which are present. This method can show, for example, whether a gene which is present in one copy per haploid genome in one tissue has been amplified in another tissue. A phenomenon of this kind takes place in the oocytes of some animals [20], but clearly does not take place in somatic tissues at least for those genes so far studied (p. 61, Chapter 5.3). So long as the number of genes in a genome is more than the minimum necessary, nuclear transplantation experiments cannot distinguish the minimum from many times more than that number. This information can be provided by molecular hybridization experiments, though such information contributes to the exact composition, rather than pluripotentiality, of a nucleus.

In summary, nuclear transplant experiments have clearly established that some specialized cells contain pluripotential nuclei. An analysis of abnormal nuclear-transplant embryos indicates that the failure of most nuclei of a specialized tissue to promote normal development after nuclear transplantation is not due to their lack of pluripotentiality. The clear indication from nuclear transfer experiments is that all nuclei of all cells are pluripotential. Molecular hybridization experiments have confirmed this conclusion at the biochemical level by demonstrating the presence of genes which code for globin or ovalbumin in specialized tissues where these genes are not expressed. In addition, these experiments have shown that gene amplification, which may not necessarily affect the pluripotentiality of nuclei, is not generally involved in cell differentiation. If, as seems likely, the nuclei of all different cells have the same genetic composition, the development of different cell-types from a single fertilized egg must depend upon control of gene expression at the levels of transcription and/or translation.

2.2.5 REFERENCES

[1] Briggs R. & King T.J. (1952) Transplantation of living nuclei from blastula cells into enucleated frogs' eggs. *Proc. Nat. Acad. Sci.*, **38**, 455–63.

[2] Briggs R. & King T.J. (1957) Changes in the nuclei of differentiating endoderm cells as revealed by nuclear transplantation. *J. Morphol.*, **100**, 269–312.

[3] Briggs R., King T.J. & DiBerardino M.A. (1960) Development of nuclear-transplant embryos of known chromosome complement following parabiosis with normal embryos. In *Symposium on germ-cells and development*, pp. 441–77. Inst. Intern. d'Embryol. Fondaz. Baselli, Milano.

[4] Briggs R., Signoret J. & Humphrey R.R. (1964) Transplantation of nuclei of various cell types from neurulae of the Mexican Axolotl. *Devel. Biol.*, **10**, 233–46.

[5] Clayton R.M. (1970) Problems of differentiation in the vertebrate lens. *Curr. Top. Devel. Biol.*, **5**, 115–80.

[6] DiBerardino M.A. & King T.J. (1967) Development and cellular differentiation of neural nuclear-transplants of known karyotype. *Devel. Biol.*, **15**, 102–28.

[7] Elsdale T.R., Fischberg M. & Smith S. (1958) A mutation that reduces nucleolar number in *Xenopus laevis*. *Exp. Cell Res.*, **14**, 642–43.

[8] Gurdon J.B. (1960) The effects of ultraviolet irradiation on the uncleaved irradiation on the uncleaved eggs of *Xenopus laevis*. *Quart. J. Microscop. Sci.*, **101**, 299–312.

[9] Gurdon J.B. (1961) The transplantation of nuclei between two subspecies of *Xenopus laevis*. *Heredity*, **16**, 305–15.

[10] Gurdon J.B. (1962) The developmental capacity of nuclei taken from intestinal epithelium cells of feeding tadpoles. *J. Embryol. exp. Morph.*, **10**, 622–40.

[11] Gurdon J.B. (1963) Nuclear transplantation in Amphibia and the importance of stable nuclear changes in promoting cellular differentiation. *Quart. Rev. Biol.*, **38**, 54–78.

[12] Gurdon J.B. (1964) The transplantation of living cell nuclei. *Adv. Morphogen.*, **4**, 1–43.

[13] Gurdon J.B. & Laskey R.A. (1970) The transplantation of nuclei from single cultured cells into enucleate frogs' eggs. *J. Embryol. exp. Morph.*, **24**, 227–48.

[14] Gurdon J.B. (1974) *The control of gene expression in animal development*. Oxford University Press.

[15] Hennen S. (1963) Chromosomal and embryological analyses of nuclear changes occurring in embryos derived from transfers of nuclei between *Rana pipiens* and *Rana sylvatica*. *Devel. Biol.*, **6**, 133–83.

[16] Illmensee K. (1972) Developmental potencies of nuclei from cleavage, preblastoderm and syncytial blastoderm transplanted into unfertilised eggs of *Drosophila melanogaster*. *Wilhelm Roux' Archiv.*, **170**, 267–98.

[17] King T.J. & Briggs R. (1956) Serial transplantation of embryonic nuclei. *Cold Spring Harb. Symp.*, **21**, 271–90.

[18] King T.J. (1966). In *Methods in Cell Physiology* (Ed. D.M. Prescott). **2**, 1–26. Academic Press, New York.

[19] KOBEL H.R., BRUN R.B. & FISCHBERG M. (1973) Nuclear transplantation with melanophores, ciliated epidermal cells, and the established cell-line A-8 in *Xenopus laevis. J. Embryol. exp. Morph.,* **29,** 539–47.

[20] MACGREGOR H.C. (1972) The nucleolus and its genes in amphibian oogenesis. *Biol. Rev.,* **47,** 177–210.

[21] MCKINNELL R.G., DEGGINS B.A. & LABAT D.D. (1969) Transplantation of pluripotential nuclei from triploid frog tumors. *Science,* **165,** 394–96.

[22] MUGGLETON-HARRIS A.L. & PEZZELLA K. (1972) The ability of the lens cell nucleus to promote complete embryonic development and its applications to ophthalmic gerontology. *Exp. Gerontology,* **7,** 427–31.

[23] PACKMAN S., AVIV H., ROSS J. & LEDER P. (1972) A comparison of globin genes in duck reticulocytes and liver cells. *Biochem. Biophys. Res. Commun.,* **49,** 813–19.

[24] REEVES O.R. & LASKEY R.A. (1975) *In vitro* differentiation of a homogeneous cell population — the epidermis of *Xenopus laevis. J. Embryol. exp. Morph,* **34,** 75–92.

[25] STEWARD F.C. (1970) From cultured cells to whole plants: the induction and control of their growth and differentiation. *Proc. R. Soc. B,* **175,** 1–30.

[26] SULLIVAN D., PALACIOS R., STAVNEZER J., TAYLOR J.M., FARAS A.J., KIELY M.L., SUMMERS N.M., BISHOP J.M. & SCHIMKE R.T. (1973) Synthesis of a deoxyribonucleic sequence complementary to ovalbumin messenger ribonucleic acid and quantification of ovalbumin genes. *J. Biol. Chem.,* **248,** 7530–39.

[27] SUZUKI Y., GAGE LP. & BROWN D.D. (1972) The genes for silk fibroin in *Bombyx mori. J. Mol. Biol.,* **70,** 637–49

[28] GURDON J.B., LASKEY R.A. & REEVES O.R. (1975) The developmental capacity of nuclei transplanted from keratinized skin cells of adult frogs. *J. Embryol exp. Morph.,* **34,** 93–112.

Chapter 2.3
Pluripotentiality of Animal Cells

2.3.1 POTENTIALITY FOR DIFFERENTIATION

Practically all cells in an adult mammal are terminally differentiated or else have a restricted range of differentiative capacity (a determined stem cell). The potentiality of differentiation of a cell should be regarded as the range of types of differentiation which it or its progeny can undergo both in normal and in experimental circumstances. In other words it is the maximum achievable range.

Although there is evidence (see Chapter 2.2) that in general the nuclei of differentiated cells still contain the complete genome and may, if transferred to a suitable cytoplasmic environment, become the nucleus of a cell whose progeny are able to differentiate into all the various tissues of the animal; a differentiated cell itself, as a unit, is restricted—progeny of that cell are in general not able to differentiate into other cell types even when the cells are transferred to a suitable environment. The process of restriction of cell developmental fate is termed cell determination and cells restricted in this way are determined cells.

The determined state of a cell is extremely stable in most cases but there are particular examples of cells which may switch from one type to another. These are the exceptions that prove the rule and they serve to emphasize that cell determination is an epigenetic and not a genetic phenomenon. For example in Wolffian regeneration of the lens in Amphibia the iris gives rise to the new lens [1], and clonally isolated chick pigmented retina cells have been shown to produce lens *in vitro* [2]. Tumours may undergo progression or start to produce hormones which were not produced by the tissue of origin [3]. In each of these cases the cells have been stimulated to divide repeatedly and this may lead to the occasional breakdown of whatever mechanisms maintain the epigenetic state through cell division. In cases where there is a prolonged period of proliferation of cells which are determined but not terminally differentiated the state of cell determination may also undergo a change in the same way. This has been described by Hadorn in the system of imaginal disc transfer in Drosophila and termed transdetermination [4].

As embryonic development proceeds the possible paths of differentiation of particular cell lineages become progressively restricted. The progeny of the zygote itself differentiate into all the cells of the embryo and, in amniotes, the extraembryonic membranes. The zygote, therefore, has complete potentiality for differentiation—it is totipotent. Sooner or later in early development (depending upon the type of embryo) cells arise which do not have this complete range of possible fate, they have become determined to some extent. Cells which have a wide although not necessarily complete range of differentiative capacity are pluripotent cells.

Pluripotent cells exist by definition in all early embryos but whether or not a proliferating population of pluripotent cells continues depends on the type of early embryogenesis. In embryos showing a mosaic type of development the progenitor cells for each type of tissue are very few at the time of their determination and much cell determination takes place during cleavage. In the more regulative type of development a much larger population of cells is built up before determinative events take place in at least some of the cells. The pattern of vertebrate development with induction of the main axial tissues at the time of mesoderm formation leaves a population of ecto-mesodermal cells pluripotent until this time. For instance, although the endoderm in Amphibian embryos becomes determined during cleavage stages and in particular the primordial germ cells are determined by possession of germ plasm at least by the blastula stage, the mesodermal and ectodermal cells are able to give rise to many endodermal, mesodermal and ectodermal derivatives [5]. Indeed, Kocher-Becker & Tiedemann [6] have reported a factor that induces ectoderm to produce mesodermal and endodermal structures including primordial germ cells. (In Urodeles it has recently been shown that the primordial germ cells arise from the animal half of the blastula [7].) With the known necessity of a specific germ plasm for the development of germ cells in this and in other classes of animals this is a surprising result but it would indicate that the determination of germ cells may be only one clear example of the general process and that truly pluripotent cells have this pathway as well as others open to them.

In mammalian embryos, Gardner [8] has demonstrated that the inner cell mass cells of $3\frac{1}{2}$- and $4\frac{1}{2}$-day mouse blastocysts are able to give rise to all tissues excepting the trophoblast, and Levak-Svajger & Svajger [9] have shown that the ectoderm of the embryonic shield of rat embryos is able to give rise to a wide variety of types of tissue when explanted ectopically.

In order that we may understand the control of cell differentiation in embryonic development it is essential to study

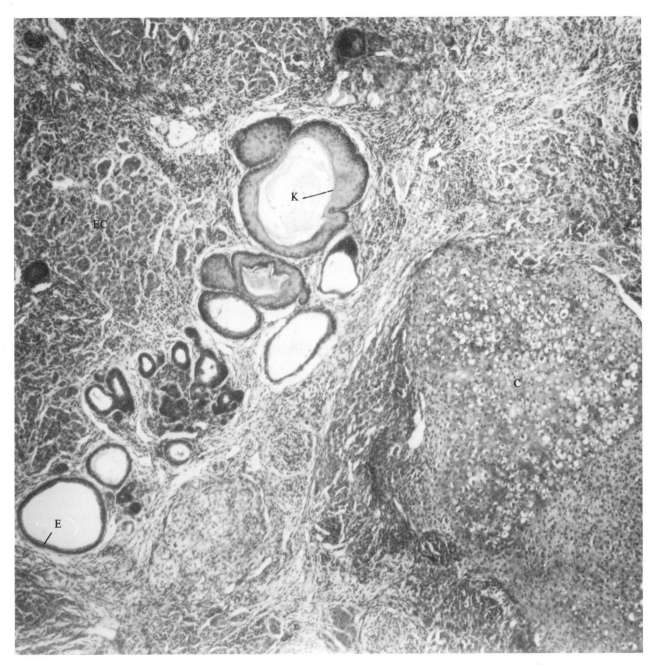

C = Cartilage.   K = Keratinising epithelium.   E = Epithelium surrounding cystic cavity.   EC = Embryonal carcinoma.

Fig. 2.3.1   Section through a transplantable mouse teratoma.

the point at which the cells become determined; that is, we must study the processes by which a choice of potential fate becomes fixed. During embryonic development cell determination takes place in small groups of cells when the embryo is itself small. The process may be traced by the methods of experimental embryology of extirpation, transplantation, isolation and genetic labelling, but biochemical investigation is difficult with such small masses of tissue which are not easily isolated. One possible solution to this practical difficulty may be to make use of tumours of pluripotential cells and it is the use of these terato-carcinomas as a model system of early cellular development which will be discussed here.

2.3.2 TERATOMAS

Teratomas are tumours which arise in the ovary or testis and comprise a disorganized mixture of immature and mature tissues of great variety. In man they are usually benign in the ovary as 'dermoid cysts' but malignant in the testis. Progressively growing teratomas contain an actively growing population of embryonal carcinoma cells and those which have lost these pluripotent cells become a mixture of benign differentiated tissues. A particular strain of mice has been developed by Stevens & Little [10] in which the males have a high spontaneous incidence of testicular teratomas from which transplantable teratocarcinoma tumour lines have been developed, and these have provided an experimental tool for study of the tumours. Stevens [11] and Pierce [12] have reviewed the biology, occurrence and induction, and the experimental investigations with these tumours up until 1967, and there are additional recent reviews [48, 49, 50, 51].

2.3.3 PLURIPOTENCY OF TERATOMA STEM CELLS

The variety of differentiated tissues in the teratomas (Fig. 2.3.1) could result from the parallel growth and differentiation of a variety of determined stem cell lines or from the growth, determination and differentiation of a single pluripotent stem line of cells. That there is indeed a single line of pluripotential stem cells was shown very elegantly by Kleinsmith & Pierce by passaging single cells to produce differentiating teratomas [13]; that is by *in vivo* cloning.

A spontaneously arising testicular teratoma which had been passaged subcutaneously at first was passaged intra-peritoneally and converted to an ascites form. When these tumours grow as an ascites they grow not as single cells in the peritoneal fluid but as small vesicles of cells of various degrees of complexity. Because they bear a close resemblance to early embryos in their organization these vesicles are called embryoid

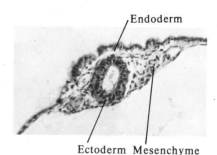

Fig. 2.3.2 Embryoid body. Note the similarity to a cross section of an embryo.

bodies (Fig. 2.3.2). The smaller embryoid bodies are composed of only two layers of cells, of which the outer is parietal yolk sac and the inner is embryonal carcinoma. Kleinsmith & Pierce dis-aggregated these simple embryoid bodies and transferred single embryonal carcinoma cells intraperitoneally into recipient mice. Out of 1700 transfers they obtained 42 multiple differentiating teratomas, thus proving that the embryonal carcinoma cells in teratomas are indeed pluripotent stem cells (Fig. 2.3.3).

2.3.4 ORIGIN OF TERATOMAS

That teratomas are tumours of pluripotent cells is borne out by their mode of origin. The only cells in the adult body whose descendants will include all types of cell are the germ cells and so these cells are, formally at least, totipotent. They will, however, undergo the very particular specializations of gameto-genesis before their totipotency can be realized and so it would not be surprising if they are only effectively totipotent at particular stages in their ontogeny. Stevens has shown that the spontaneous testicular teratomas of the 129 mice arise within the germinal epithelium of the testis, and that the time of origin may be traced back to the 11-12th day of embryo development when the primordial germ cells have migrated to the germinal ridge [14]. When embryo germinal ridges are grafted into an adult host testis they grow into teratomas but only if primordial germ cells are present [15]. So in this case the cell which gives rise to the teratoma is indeed a pluripotent cell in the embryo—the primordial germ cell. Germinal ridges from embryos later than 13 days do not produce teratomas. Although in theory pluripotent, the germ cells are now undergoing specific cytodifferentiation and this may be the cause of their inability to provide a source of pluripotent cells to form a teratoma. Early embryo cells are also pluripotent and teratomas can be induced by implanting these in ectopic sites [16, 17, 18]. In this case the ability to form a teratoma is lost on the eighth day of development. Solter *et al.* [17] have suggested that it is the cells of the ectomesoderm that are

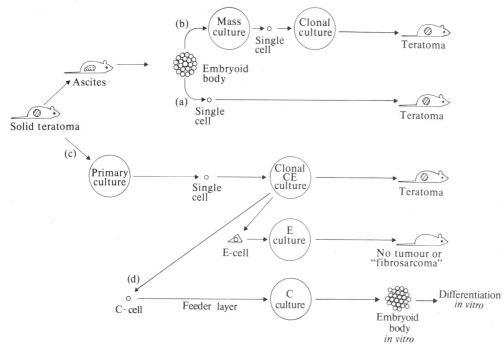

Fig. 2.3.3 Diagram of *in vivo* and *in vitro* cloning of teratoma cells. A teratoma containing many types of tissue is formed from a single cell. References: a–(13), b–(30, 31), c–(32), d–(52).

pluripotent up to this point when primary induction of the embryo causes their determination.

Another type of origin of teratomas has recently been described by Stevens & Varnum in a strain of mice (LT) in which there is a regular parthenogenetic development of ovarian oocytes. When these develop *in situ* in the ovary teratomas are formed [19]. Stevens has compared the early events of each type of origin and finds that a small group of cells seem to start an embryonic development which becomes disorganized. The spontaneous teratomas of 129 (testicular) and LT (ovarian) mice are thus seen to be the result of the ability peculiar to these strains for the germ cells to start spontaneous development *in situ*. In most other strains of mice spontaneous activation of the germ cells is extremely rare and only ectopic embryos are able to give rise to teratomas.

2.3.5 COMPARISON WITH NORMAL DEVELOPMENT

Teratomas have an advantage over embryos for the study of processes by which pluripotent cells become determined and differentiate because they offer a source of large quantities of tissue. It is, however, necessary to compare these processes in teratomas closely with those in normal embryos to ensure that one is not studying an artefactual situation arising out of the disorganized nature of the tumour or out of its being a tumour which may be progressing away from normality over its long passage history.

Histologically the tissues in teratomas are similar to those in the adult or to developing tissues in the embryo. Moreover, embryoid bodies can have a structure closely resembling early embryos. An ultrastructural study of the nervous tissue in teratomas containing a wide range of differentiations and in teratomas in which the predominant differentiated tissue is neural has shown that although individual nerve cells are quite well differentiated, and that synapses are formed between them, there is an overall lack of organization and many of the cell contacts are poorly developed (Fig. 2.3.4). Glial cells, though present, are sparse and myelination is noticeably lacking. The overall picture is one of disorganization which may lead to a lack of function which may in turn be the cause of the relative immaturity of the tissue [20]. Although very many tissues are found in teratomas, and relatively well co-ordinated structures

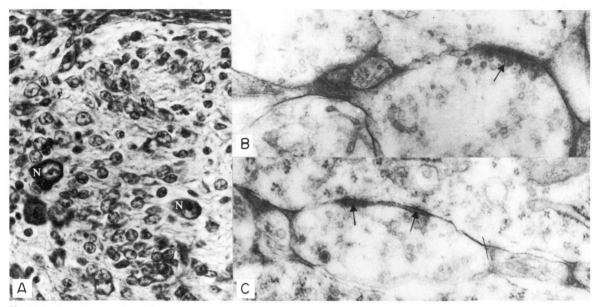

Fig. 2.3.4   Area of nervous tissue (A) and electron micrographs demonstrating synapses in such areas (B) and (C). N = Neuronal cell body and arrows indicate synaptic areas. By kind permission of Dr R.L. Tresman.

such as hair follicles may commonly be found, tissues such as liver which probably have a complex sequence of tissue interactions in their formation [21] are noticeably absent.

That tissue interactions are taking place in the morphogenesis of teratomas is borne out not only by the local organization but also by the experiments of Auerbach, who showed that neural teratoma tissue could provide the inductive stimulus for kidney mesenchyme differentiation trans-filter (as in other induction experiments—see Section 3.4) and that this is a property of foetal but of neither adult brain nor C1300 neuroblastoma tissue. As plentiful immature neural tissue is clearly histologically identifiable in these tumours it is not unexpected that this would have the properties of foetal brain; this does, however, indicate that a search for inductive factors in these teratomas might be rewarding.

Another foetal function found in teratomas is the production of α-foeto-protein, the presence of which may be used as a clinical diagnostic indication for the presence of either a hepatoma or a teratoma [22]. This appears to be synthesized by the embryonal carcinoma cells themselves as its production has been demonstrated in cultures of embryonal carcinoma cells [23]. However immunofluorescence studies on sections of embryoid bodies and tumours indicate that it is mainly found in the endodermal cells [38].

Edidin et al. [24] reported that antibodies raised in rabbits against mouse teratoma embryoid bodies cross-reacted with early embryo cells. Subsequently Artzt et al. [25] have reported

that embryonal carcinoma cells are strongly antigenic in syngenetic mice (i.e. mice of the same genetic strain), and an antibody is raised to a cell surface embryonic antigen. This antibody reacts specifically only against the cell surface of teratoma stem cells, early cleavage embryo cells and sperm. These observations strongly support the homology between the teratoma stem cells and early embryo cells, as do their similarities of ultrastructure [26]. It is known that the recessive alleles of the T-locus in the mouse, which have an early lethal effect in the homozygous condition, affect the cell surface and are expressed on sperm [27, 28], and it has been demonstrated by comparison of adsorption with normal and $t^{12}/+$ sperm that the teratoma antigen is the product of a normal allele of the t locus [29]. As it is believed that the genes of the t series may act at the cell surface and hence affect the normal cellular interaction in embryogenesis, it is extremely interesting to find these expressed on the surface of pluripotent teratoma cells. It will be interesting to discover whether there is any causal relationship between the cell surface expression and cellular determination or whether the former will prove to be only one phenotypic expression of the cell's epigenetic state.

2.3.6 ESTABLISHMENT OF *IN VITRO* CELL LINES OF PLURIPOTENT CELLS

In order that many of the possible experimental manipulations which might be useful in a study of pluripotential cells may be carried out, it is desirable to isolate these cells from the complex of the tumour. One simplification is the availability of simple embryoid bodies in the ascites form of the tumour (see above), and it was from such embryoid bodies that the first two successful tissue culture lines were established [30, 31]. These were both from passaged spontaneous testicular teratomas. Not all teratomas can be obtained in an ascites form, and spontaneous testicular teratomas are only available from particular strains of mice. The next line to be isolated, however, was obtained from the disaggregation of a solid tumour which had been derived from the implantation of a 3-day blastocyst into an adult testis [32]. Subsequent lines have been derived from both the solid tumour [33] and the ascites conversion of an embryo-derived teratoma [34, 35]. This suggests that it may be possible to derive a pluripotent cell line from any genetic strain of mouse, but this may be dependent on the production of a transplantable, progressively growing teratoma (a terato-carcinoma) and many induced teratomas are not transplantable and this property may be strain related [36]. These *in vitro* cell cultures can be cloned and upon re-injection into a mouse give rise to a teratoma containing the usual wide variety of tissues. This reconfirms the result of Kleinsmith and Pierce that the various tissues are formed from a pluripotent stem cell line.

2.3.7 EXPERIMENTS WITH CULTURES OF PLURIPOTENT TERATOMA CELLS

Sub cloning

Pluripotent cells are, by definition, able to become determined and differentiate in a variety of ways. Pluripotent cells *in vitro*, therefore, provide an excellent system for investigating the processes of the choice and establishment of the direction of a cell's differentiation, and those factors which influence this. That a culture of cells gives rise on re-injection to a tumour containing many types of tissue does not prove that there are pluripotent cells in the culture any more than simple tumour passage *in vivo* demonstrates a pluripotent stem line. Sub-cloning from the culture to produce teratoma-forming sub-clones is needed and in all cases reported such sub-clones *are* found, showing that pluripotent cells are present in the culture.

One of the defining features of cell determination is its irreversibility, and thus the loss of pluripotency is necessarily associated with cell determination. This loss may provide an assay for determination of a pluripotent cell. But pluripotency is a positive property of a cell in that it implies the ability to develop further. Changes which block this ability may not necessarily represent cell determination as further development could be blocked for a variety of non-specific reasons. Although loss of pluripotency may be an indication of the cell's becoming determined, gain of the ability to differentiate in a specific direction must also be demonstrated. Histological examination of the tumours formed *in vivo* gives ample evidence of the variety of differentiation to which the pluripotent cell population may give rise, and the cells will also form a variety of cell types *in vitro*.

In vitro differentiation

Small aggregates of teratoma cells from cloned cultures have been found to show areas of histologically recognisable tissue when grown as a floating mass [37] or when cultured on agar or in methyl cellulose. Until recently, however, few results were gained with these techniques and a better assay for pluripotency was re-injection *in vivo*.

Although some of the clonal pluripotent teratoma cultures which have been described have been homogeneous in cell morphology, others [31, 32] are heterogeneous and cells with a

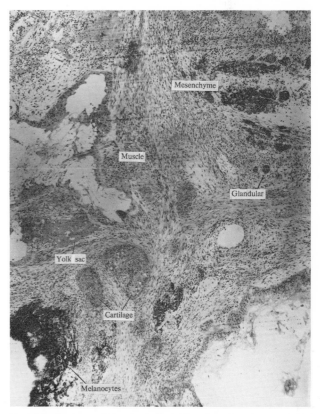

Fig. 2.3.5 Section parallel to the surface of the culture dish of a culture which had been allowed to differentiate for five weeks. The result is a 'teratoma' *in vitro*. In this section there are areas of cartilage, mesenchyme, muscle, pigmented melanocytes, a glandular tissue and parietal yolk sac.

variety of morphologies may be seen. Sub-cloning provides a means of demonstrating the presence of pluripotent cells in the culture and will also demonstrate the presence of any other cell type which will grow progressively *in vitro*.

It has been demonstrated that two classes of clone may be derived from one such heterogeneous clonal teratoma cell line SIKR [32]. Sub-clones in which the characteristic small round piling teratoma stem cells (C cells) are present, are tumourigenetic and form teratomas on re-injection into mice. The other class of sub-clones which contain only monolayering cells of a variety of morphologies (E cells) are only very poorly tumourigenetic and when they do form tumours on re-injection into mice these are not teratomas but contain only a single type of tissue, most usually a fibroblastic type [32, 40]. Similar tumours are found when cultures from a variety of mouse tissues become transformed *in vitro* and are then re-inoculated into a syngeneic mouse [41].

It has now been possible to isolate sub-clones of SIKR which are homogeneously C cells, but the maintenance of these pure C cell cultures is critically dependent upon the continual provision of a feeder layer of killed fibroblastic cells. In the previous heterogeneous sub-clones of SIKR the E cells fulfilled this feeder

requirement and sub-clones in which this heterogeneity was not established at an early stage did not survive [52].

In the presence of feeder cells (E cells or added killed feeder cells) the C cells spread on the surface of the culture and grow actively, but there is little or no differentiation. In the absence of feeder cells the C cells will undergo extensive differentiation in culture. A very complex piled up culture is formed in which a variety of types of differentiated cell may be distinguished. Histological sections cut through the multilayered cell clumps reveal tissues of most of the types listed in the original teratoma from which the cultures were isolated. Cartilage, fibroblast, smooth and striated muscle, neural tissue, pigmented epithelia, keratinizing squamous epithelium and parietal yolk sac endoderm are found [43] (Fig. 2.3.5).

Without feeders the C cells are less firmly adherent to the petri dish and as they form a rounded pile the first stage of

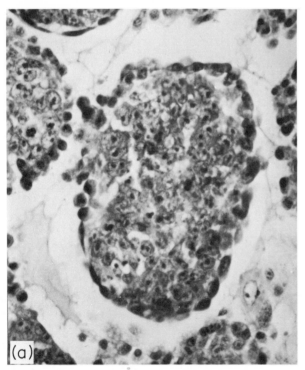

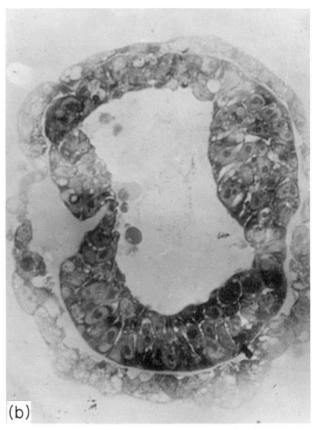

Fig. 2.3.6(a) A simple embryoid body formed *in vitro* from a clonal line of pure embryonal carcinoma cells. Note the outer layer of endodermal cells surrounding an inner core of pluripotent stem cells. ∼ ×200.

(b) A cystic embryoid body three days after the aggregate of stem cells was removed from the surface of the culture dish. A layer of endodermal cells surround cells which have formed themselves into a palisaded layer similar to embryonic ectoderm. At one point there is a cluster of more rounded cells which are similar to the mesodermal cells in an egg cylinder.

differentiation takes place. Cells on the outside of the spherical aggregate of C cells delaminate to form a layer of endodermal cells. The resulting bilayered structure of a layer of endodermal cells surrounding an inner core of pluripotential embryonal carcinoma cells is very closely similar to the arrangement in an early mouse embryo where endodermal cells surround the pluripotent ectodermal cells (Fig. 2.3.6a). This structure is also found in teratomas, in particular in the ascites form of the tumour and is called a simple embryoid body. More complex embryoid bodies—cystic embryoid bodies—are also found *in vitro* by some lines of cells by a progression from the single embryoid bodies. In these a fluid filled cavity has formed and embryo-like development progresses further with the formation of mesenchyme and neural-tube structures (Fig. 2.3.6). Both simple and cystic embryoid bodies attach and spread on tissue-culture treated plastic surfaces and produce multilayered mixtures of variously differentiated tissues [52].

Cell hybridization

Several workers have investigated the role of soluble cytoplasmic or nucleoplasmic factors in establishment and maintenance of a cell's determined state by making cell hybrids between pluripotent teratoma cells and cells of a permanent tissue culture line. Finch and Ephrussi [44] reported that pluripotency was lost after hybridization with fibroblastic cells and this has been confirmed by Jami [45] and in the present author's laboratory by Sit [46]. It is probable that the hybrid cell displays the differentiated characteristics of its fibroblast parent, since fibroblastomas are formed on re-injection of the cultures into a mouse. Hybrids of pluripotent teratoma cells with neuroblastoma cells seem to retain some neural characteristics *in vitro* although not *in vivo* (where they form seemingly fibroblastic tumours) [46]. This observation strengthens the hypothesis that the hybrid follows the differentiation of its determined parent; but pluripotency may be lost as a result of the hybridization procedure *per se*. Sit made hybrids of teratoma cells with each other and all the hybrid (tetraploid) clones which he picked were no longer pluripotent. This observation must lead one to caution in interpretation of hybridization results with these cells.

2.3.8 CONCLUSIONS

It is probably a fortunate feature of mammalian development which allows pluripotent cells to persist as a dividing population. This provides a system for the analysis of early development *in vitro*. As teratomas may be readily produced from normal early embryos, this stem cell does not represent a highly abnormal proliferation akin to that of neuroblastoma cells. That such a population of pluripotential cells can exist argues against a rigid countdown of 'quantal' mitoses as the main organization of

cellular determination as suggested by Holtzer [47]. At the very least it indicates that a 'hold' in the countdown must be possible at a pluripotent condition and that proliferation can proceed indefinitely in the same way as in other stem cell populations. The transitions from a proliferating pluripotent stem cell population to a restricted determined stem cell population may or may not be via a set number of obligate mitoses but it is questions such as these and also questions of the extrinsic and intrinsic factors operating on and in the cell which may be elucidated with cultures of pluripotent cells.

2.3.9 REFERENCES

[1] REYER R.W. (1948) *J. Exp. Zool.*, **107**, 217.
[2] EGUCHI G. & OKADA T.S. (1973) *Proc. nat. Acad. Sci. U.S.*, **70**, 1495.
[3] BOROCHOW I.B. (1965). *Arch. Surg.*, **90**, 101.
[4] HADORN E. (1965) In *Genetic control of differentiation*, Brookhaven Symp., **18**, 148.
[5] NIEUWKOOP P.P. (1973) *Adv. in Morphogenesis*, **10**, 1.
[6] KOCHER-BECKER U. & TIEDEMANN H. (1971) *Nature, Lond.*, **223**, 65.
[7] SUTASURJA L. & NIEUWKOOP P. Demonstrated at XI International Embryological Conference, Sorrento, 1974.
[8] GARDNER R.L. & LYON M.F. (1971) *Nature, Lond.*, **231**, 385.
]9] LEVAK-SVAJGER B. & SVAJGER A. (1971) *Experientia*, **27**, 683.
[10] STEVENS L.C. & LITTLE C.C. (1954) *Proc. Natl. Acad. Sci., U.S.A.* **40**, 1080.
[11] STEVENS L.C. (1967) *Advances in Morphogenesis*, **6**, 1.
[12] PIERCE G.B. (1967) *Current Topics in Development Biology*, **2**, 223.
[13] KLEINSMITH L.J. & PIERCE G.B. (1964) *Cancer Res.* **24**, 1544.
[14] STEVENS L.C. (1970) *J. Nat. Cancer Inst.*, **44**, 923.
[15] STEVENS L.C. (1967) *J. Nat. Cancer Inst.*, **38**, 549.
[16] STEVENS L.C. (1968) *J. Embryol. Exp. Morph.*, **20**, 329.
[17] SOLTER D., SKREB N., DAMJANOV I. (1970) *Nature, Lond.*, **227**, 503.
[18] BILLINGTON W.D., GRAHAM C.F. & McLAREN A. (1968) *J. Embryol. Exp. Morph.*, **20**, 391.
[19] STEVENS L.C. & VARNUM D.S. (1974) *Develop. Biol.*, **37**, 369
[20] TRESMAN R.L. & EVANS M.J. (1975) *J. Neurocytol.*, **4**, 301.
[21] LE DOURIN N. (1962) *J. Embryol. exp. Morph.*, **12**, 769.
[22] SMITH J.B. (1970) *Medical Clinics of North America*, **54**, 797.
[23] KAHAN B. & LEVINE L. (1971) *Cancer Research*, **31**, 930.
[24] EDIDIN M., PATTHEY H.L., McGUIRE E.J., SHEFFIELD W.D. (1971) In *Proceedings of 1st Conference and Workshop on Embryonic and Foetal Antigens in Cancer* (*Eds* N.G. Anderson & J.H. Coggin). USAEC Oak Ridge, pp. 239–248.
[25] ARTZT K., DUBOIS P., BENNET D., CONDAMINE H., BABINET C. & JACOB F. (1973). *Proc. Natl. Acad. Sci., USA*, **70**, 2988.
[26] DAMJANOV I., SOLTER D., BELICZA M., SKREB N. (1971) *J. Nat. Cancer Inst.*, **46**, 471.
[27] BENNETT D., GOLDBERG E., DUNN L.C. & BOYSE E.A. (1972) *Proc. Natl. Acad. Sci., USA*, **69**, 2076.
[28] YANAGISAWA K., BENNETT D., BOYSE E.A., DUNN L.C. & DiMEO A. (1974) *Immunogenetics*, **1**, 57.
[29] ARTZT K., BENNETT D. & JACOB F. (1974) *Proc. Natl. Acad. Sci., USA*, **11**, 811.
[30] KAHAN B.W. & EPHRUSSI B. (1970) *J. Natl. Cancer Inst.*, **44**, 1015.
[31] ROSENTHAL M.D., WISHNOW R.M. & SATO G.H. *J. Natl. Cancer Inst.*, **44**, 1001.
[32] EVANS M.J. (1972) *J. Embryol. Exp. Morphol.*, **28**, 163.
[33] JAMI J. & RITZ E. (1974) *J. Natl. Cancer. Inst.*, **52**, 1547.
[34] BERNSTINE E.G.H., HOOPER M.L., GRANDCHAMP S. & EPHRUSSI B. (1973) *Proc. Natl. Acad. Sci., USA*, **70**, 3899.
[35] JAKOB H., BOON T., GAILLARD J., NICHOLAS J.F. & JACOB F. (1973) *Ann. Microbiol. (Inst. Pasteur)*, **124b**, 269.

[36] SOLTER D. (1972) In *Cell Differentiation* (Eds R. Harris, P. Allin & D. Viza). Munksgaard. Copenhagen.

[37] FINCH B.W. (1970) Thesis Brandeis University.

[38] ENGELHARDT N.V., POLTORANINA V.S. & YAZOVA A.K. (1975) *Int. J. Cancer,* **11,** 448.

[39] PIERCE G.B., DIXON F.J. & VERNEY E.L. (1960) *Lab. Invest.,* **9,** 583.

[40] MARTIN G.R. & EVANS M.J. (1974) *Cell,* **2,** 163–72.

[41] FRANKS L.M., CHESTERMAN F.C. & ROWLAT C. (1970) *Brit. J. Cancer,* **24,** 843.

[42] DAMJANOV I., SOLTER D. & SKREB N. (1971) *Z. Krebsforsch.,* **76,** 249.

[43] EVANS M.J. & MARTIN G.R. (1975) In *Teratomas and Differentiation* (Eds M.S. Sherman & D. Solter). Academic Press, New York.

[44] FINCH B.W. & EPHRUSSI B. (1967) *Proc. Natl. Acad. Sci.,* **57,** 615.

[45] JAMI J., FAILLY C., RITZ E. (1973) *Exp. Cell Res.,* **76,** 191.

[46] SIT K.H. (1973) Thesis. University of London.

[47] HOLTZER H., WEINTRAUB H., MAYNE R. & MOCHAN B. (1972) *Current Topics in Developmental Biology,* **7,** 229.

[48] DAMJANOV I. & SOLTER D. (1974) *Current Topics in Pathology,* **59,** 69.

[49] EVANS M.J. (1975) In *The Early Development of Mammals,* 2nd Symposium of the British Society of Development Biology (Eds M. Balls & A.E. Wild).

[50] STEVENS L.C. (1975) In *The Developmental Biology of Reproduction.* (Eds C.L. Markert & J. Papaconstantinou). pp. 93–106. Academic Press, New York.

[51] MARTIN G.R. (1975) Cell. **5,** 229.

[52] MARTIN G.R. & EVANS M.J. (1975) *Proc. nat. Acad. Sci., U.S.,* **72,** 1441–45.

Chapter 2.4
Experimental Embryogenesis— The Totipotency of Cultured Plant Cells

'The phenomenon of embryogenesis is not necessarily confined to the reproductive cycle. It can be fairly inferred from the recent spectacular development in the culture of single cells of higher plants that any diploid cell, in which irreversible differentiation has not proceeded too far, will, if placed in an appropriate medium, develop in an embryo-like way and produce a complete plant. The whole complex sexual apparatus is not therefore an essential prerequisite for the removal of the effects of ageing and the re-establishment of embryonic properties. The events occurring in the ovule after fertilization thus provide only a special case of embryogeny.'

P.R. Bell [1] reviewing *Recent Advances in the Embryology of Angiosperms* (Edited by P. Maheshwari [2]).

2.4.1 INTRODUCTION

Quite apart from the natural methods of vegetative multiplication seen in higher plants, it is possible to generate new plants from existing plants by a number of manipulative procedures. These involve such processes as the initiation of roots on stem and bud cuttings, of shoot buds on detached roots or root segments, of shoot buds and adventitious roots from leaves and of shoot buds and roots from cultured wound callus. Regeneration via such organogenesis can be achieved over a wide range of plants and animals and is to be distinguished from the origin of new organisms via embryogenesis occurring from individual germ cells or somatic cells. Tokin [3], in 1960, in a survey of somatic regeneration processes in plants and animals advanced the view that developmental processes resulting in the formation of whole organisms from a single somatic cell or even a small group of such cells should be designated somatic embryogenesis, and emphasized that this process is normally preceded by disintegrating processes altering the normal correlations existing between the cells of tissues and organs within the organism.

The appearance of plantlets in plant tissue (callus) cultures and cell (suspension) cultures was accepted at first rather uncritically as evidence of somatic embryogenesis. Against this background, Haccius [4] emphasized that the regenerative process can only be regarded as embryogenesis if the primary structures, from their early development, have a bipolar axis terminating in a shoot and a root pole and if they are quite unconnected with vascular elements in the explant or culture aggregate. In a study [5] of the various forms of morphogenesis exhibited by suspension cultures of *Atropa belladonna* it was possible to distinguish: (i) root initials which later initiate shoot buds; (ii) shoot buds which develop adventitious rootlets; (iii) plantlet structures which pass through a stage when they consist of a shoot bud and a rootlet separated by intervening callus; and (iv) true somatic embryos (often referred to as embryoids) which were, at a very early stage, organized bipolar structures. Only detailed anatomical study of the cellular aggregates of these cultures enabled the somatic embryos to be distinguished ontogenetically from 'embryo-like' structures initiated as endogenous root primordia but subsequently although quickly developing the bipolar organization stressed by Haccius. The extension of these studies [6, 7] now enables more rigorous criteria of embryogenesis in plant cultures to be formulated: embryogenesis involves the formation from single cells of globular embryos which in completing their development pass through stages equivalent to those earlier recognized from studies of the ontogeny of zygotic embryos. This definition embraces embryogenesis as observed both in cultures of somatic origin and in immature pollen grains (microspores).

The concept of pluripotency [8] arose from studies in animal embryology and was used to describe the multiple potency of the parts of the early egg cell, i.e. the concept that the actual fate of these parts was one selected from a number of possible fates. The concept of pluripotency was linked to that of determination: groups of cells were pluripotent in early embryology but at a later stage, due to the action of determining agents or organizers, had their destinies determined. Pluripotency was a property *lost* during the progress of development. The historical context of the term pluripotency led botanists to adopt the term totipotency to express the concept, clearly formulated in the quotation at the head of this chapter, that any diploid (somatic) plant tissue cell, in which irreversible differentiation had not proceeded, could, under the influence of appropriate stimuli, regain the ability to develop like the zygote and produce a new embryo plant. Further, it was clearly implied by authors introducing this term [9] that, in general, differentiation in plant cells *was* a reversible process provided it did not proceed to the point of nuclear breakdown and hence loss of the genome. This concept of the totipotency of plant cells will here be submitted to critical assessment.

Although we are concerned in this chapter with embryogenesis from somatic cells and from germ cells (microspores) in culture, it is important to view this process in context, that is in the light of our knowledge of the normal origins of flowering plant embryos and of the environment in which they normally come to maturity. This leads us to ask whether the zygote is in any way exceptional except in so far as it is the agent of genetic variation and whether the endosperm brought into being by the act of double fertilization is a tissue critical to the initiation of embryogeny and the early nutrition of the embryo. We must also enquire as to whether embryogeny from cells other than the zygote occurs naturally in flowering plants and if so how far such embryogenesis parallels that achieved in culture.

2.4.2 THE EGG CELL AND THE ZYGOTE

The egg cells and zygotes of several species have been the subjects of recent studies by light and electron microscopy [10–15]. These have revealed certain common features of egg cells and zygotes but also have clearly indicated that their structure varies significantly between species.

The egg cell is normally polarized in so far as a large vacuole is present towards its micropylar end and the nucleus and most of the cytoplasm are towards the chalazal end (p. 31). The nucleus is large, the nucleolus prominent but variable in its staining. There is considerable variation in the number of organelles, in the extent of the ER profiles, in the numbers and degree of aggregation of ribosomes and in the extent of storage materials (of which starch is the most consistently present). The variations in fine structure between the egg cells of different species is such as to suggest considerable variation in metabolic activity. Although changes in egg cell structure occur throughout embryo sac development and particularly after anthesis, it is clear that the differences observed cannot be explained as due to differences in the relative maturity of the egg cells studied.

The interval between fertilization and division of the zygote is of variable duration between species and the zygotes of different species show no uniformity in their fine structure at the time of the initiation of their division. The only generalizations that can be made are that there is usually an increase in storage material (starch and lipid), in cellular membranes and in the protein-synthesizing machinery as the zygote prepares to divide. Egg cells and zygotes are neither unique nor uniform in structure.

2.4.3 POLYEMBRYONY

The known instances of polyembryony have continued to increase steadily since the first description of this phenomenon

by Strasburger in 1878 [16]. Polyembryony (the occurrence of more than one embryo in the ovule) can arise by cleavage in the early embryo (cleavage polyembryony), by the development of embryos from other cells of the embryo sac and by the formation of somatic embryos (adventive embryos) from cells of the nucellus or inner integument of the ovule (Fig. 2.4.1). In

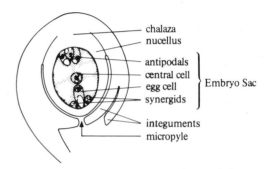

Fig. 2.4.1 The structures of the mature ovule (megasporangium). Adapted from Foster A.S. and Gifford Jr. E.M. *Comparative morphology of vascular plants* which presents a clear account of the development of the ovule and anther, of double fertilization and of embryogeny in the flowering plants.

cleavage embryony a small group of embryo cells proliferate into a number of embryos which then develop independently. Here, as in animals, embryo cells are showing their pluripotency (i.e. they are not determined as to their fate during early embryology). The phenomenon of embryo initiation from synergids, or in rare cases from the antipodals of the embryo sac serves to illustrate further that embryogenesis can occur from cells highly contrasted in fine structure. Where embryo initiation takes place from these cells, the embryos may be diploid (a male nucleus has fused with the mother cell) or haploid. Fertilization is not essential to the induction of embryogenesis within the embryo sac. There are, incidentally, no well authenticated cases of embryo initiation from the central cell or from the triploid endosperm nucleus.

The formation of adventive embryos from the sporophytic (maternal) tissues of the ovule is of particular relevance to the theme of this chapter. The development of adventive embryos from the nucellus is widespread, occurring in individual species over a wide taxonomic range. Nucellar cells adjacent to the embryo sac are most frequently involved and in this way a number of embryos may arise within each ovule instead of or in addition to the zygotic embryo. The adventive embryos of *Citrus* spp. and of *Nothoscordium fragrans* [17] closely resemble in morphology the zygotic embryo but lack a suspensor. Although in some cases initiation of adventive embryos may only occur following pollination or fertilization, in at least six species they arise autonomously; in two genera nucellar embryos can be detected at the megaspore mother cell stage of ovule development.

Polyembryony can be interpreted as resulting from a breakdown of the normal regulatory mechanisms within the ovule which occurs in particular genotypes just as the spontaneous formation of certain somatic tumours (genetic tumours) is characteristic of particular hybrid genotypes in *Nicotiana* [18]. Nevertheless the occurrence of polyembryony does point strongly to the ovule environment as one particularly conducive to embryogenesis if certain restraints on cell function are disrupted.

2.4.4 THE OVULE ENVIRONMENT AND THE NUTRITION OF THE EMBRYO

One objective in attempting to culture *in vitro* immature embryos removed from the ovule has been to identify the special nutritional requirements essential to the progress of embryology. Ideally one would hope to achieve normal embryology *in vitro* starting with the isolated uninjured zygote. In practice, such studies have had to start with embryos at the globular or a later stage in their development [19–21]. Such studies have shown the importance of mineral ions, sucrose, organic N and an appropriate (and often high) osmotic potential. Additionally, for small embryos it has been essential to supply vitamins (particularly B vitamins) and an appropriately balanced mixture of growth hormones. However, satisfactory development in culture of very small embryos of most species cannot be achieved in defined media. In some cases addition of coconut milk (the liquid endosperm of *Cocos nucifera*) or an extract of immature maize kernels to the medium or gelatin-sealing of the embryo to the surface of its own endosperm (or that of a related species) has enabled smaller embryos to be cultured than is possible with defined media. Clearly globular embryos have exacting nutritional requirements and the zygote may well be even more strongly heterotrophic.

2.4.5 THE ROLE OF THE ENDOSPERM

The role of the endosperm in supplying essential metabolites to the embryo during its early development is in considerable doubt. There are cases where twin embryos are formed by fusion of the male nuclei with the egg cell and a synergid and continue their early embryogeny in the complete absence of endosperm. In *Helianthus*, endosperm nuclear divisions, wall formation and increase in endosperm cytoplasm occur during the filamentous and early globular stages of embryogeny. In *Capsella* the endosperm continues its development at least until the embryo is heart-shaped. In *Podolepis jaceoides* and *Minuria*

denticulata there is no evidence of any digestion of the endosperm during the early stages of embryogeny. If then the endosperm is mainly of significance as a source of reserve food material during the *later* stages of embryogeny it would not be expected to be a rich source of factors essential to the initiation of embryogenesis and early segmentation of embryology; it would primarily meet the less exacting requirements of the post-globular embryo. This would not conflict with the value of coconut milk in the nutrition of excised immature embryos; in this particular case the liquid endosperm has to meet the nutritional needs of the coconut embryo which is still very immature when the nut is shed from the parent plant. In most cases, therefore, the special nutrients for early embryogeny are probably routed directly via the nucellus and integument into the embryo sac first over its whole surface and later via the suspensor of the developing embryo. The endosperm is in general *not* the tissue which makes the ovule an environment conducive to embryogenesis. Whether the nucellus acts simply as a reservoir for the essential nutrients and growth regulators entering the ovule from the parent plant or whether it is the centre of their synthesis from such basic precursors as sucrose, amino-acids and mineral ions is uncertain.

Although, in general, the fine structure of the cells of the suspensor suggests that they lack the synthetic capacity of the embryo cells (suspensor cells are often highly vacuolated and contain much fewer ribosomes) they do show certain features indicating that they may be active in absorption and translocation of nutrients to the embryo. In *Capsella* for instance, they show an extensive network of wall projections suggestive of 'transfer' cells [22] and they have numerous plasmodesmata on their transverse walls and on the wall marking the junction of the suspensor with the embryo proper.

This survey emphasizes that embryogenesis occurs naturally from various kinds of cell including somatic cells, that the cells of young embryos can be highly embryogenic, that the suspensor is a dispensable component of the proembryo, that neither fertilization nor endosperm are essential to the initiation of embryogenesis and that, at least beyond a certain stage of embryogeny, it is possible to nurture successfully the zygotic embryo in synthetic culture media.

2.4.6 THE FORMATION OF EMBRYOIDS IN SOMATIC TISSUE AND CELL CULTURES

Curtis and Nichol in 1948 [23] reported that green calluses of embryo origin obtained from the orchids, *Vanda tricolor* and a *Cymbidium* hybrid, developed, after several months in culture, crests on the upper surface of which were carinate projections developing from superficial meristems and that from the lower surface of these projections arose numerous unicellular rhizoids. These structures proliferated by bifurcation and closely

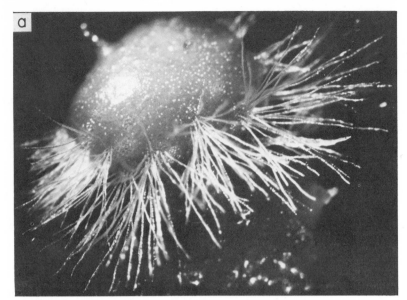

Fig. 2.4.2   Protocorm development and proliferation in orchid cultures. (a) A protocorm of *Cymbidium* developed from a meristem explant after 1 month in culture. (b) Group of protocorms of *Cattleya* developed in culture from a single meristem explant. (Both photographs supplied by the late G.M. Morel, Station Central de Physiologie végétale, C.N.R.A., Versailles.)

resembled in morphology and anatomy the normal protocorm stage in orchid embryo development. Some of these structures developed shoots and roots and came to resemble normal orchid seedlings. Later work [24] was to show that protocorms of tropical orchids could be readily generated by the culture of the apical dome of orchid shoots on simple media, that these protocorms could be propagated in culture by division into fragments (Fig. 2.4.2) and that they would generate orchid seedlings. This technique has now been exploited for the commercial clonal propagation of orchids from several genera (particularly species of *Cymbidium* and *Cattleya*).

Callus cultures from carrot (*Daucus carota*) root tissue were established on solid medium in 1937 by Nobécourt and in 1939 he reported that the cultures spontaneously gave rise to roots [25]. In 1952 Steward adopted a different approach to induce active proliferation from small explants of the storage roots of carrot, submitting them alternately to immersion in a liquid coconut milk-containing medium and exposure to air. Then in 1958 [28, 29] he reported that when the cellular aggregates released into the medium from the explants were transferred to the surface of an agar-solidified medium they gave rise to plantlets with an integrated root–shoot axis and cotyledons of normal morphology (Fig. 2.4.3). In 1959 Reinert [30] noted the formation at the surface of carrot callus of 'smaller shoots' which could be readily released and which

'except for a more or less marked fission or splitting of the cotyledons' corresponded to 'normal bipolar carrot embryos'. Further work quickly established that carrot cultures did give rise, under appropriate conditions, to globular, heart-shaped and torpedo-shaped structures strikingly similar to those observed in zygotic embryogeny. This parallelism was emphasized by describing the structures as embryoids or adventive embryos and the process involved in their origin as one of somatic embryogenesis. This carrot system has become the classical example of embryogenesis in somatic cultures and has been the subject of intensive study and considerable controversy.

Steward [9] has stressed the importance for somatic embryogenesis of: (i) the escape of cells from the restrictions imposed upon them by their location within the plant or within a multicellular aggregate; and (ii) the provision of the special nutrients which nurture the zygote within the ovule and which are competent to support the full capacity of the freed cells to divide and grow. In this latter connection he has attached special importance to the role of coconut milk as a fluid designed by nature for the support of the immature embryo. Steward further considered that the embryos had their origin in free-floating cells which by division generated a cellular clump in which formative influences developed leading to the emergence of growth regions and the development of typical stages of embryology. Experimentally this view was supported by free-floating globular

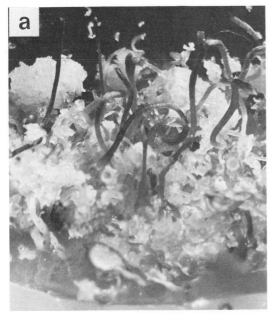

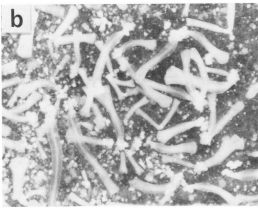

Fig. 2.4.3 Plantlets with integrated root–shoot axis and cotyledons developed in culture from carrot embryoids developing (a) in a callus and (b) in a suspension culture. Note in (b) the presence of plantlets in various stages of development and scattered amongst them the small white cell aggregates in which the embryoids arise. (Photographs by Dr Susan M. Smith.)

and more advanced embryoids in the suspensions and also by the presence of twin cells and aggregates of very few cells which *might* be precursors of the globular embryoids [31].

The interpretation of the few celled aggregates, as figured by Steward and his co-workers, as intermediate stages between free-floating single cells and obvious embryoids was criticized [32] on the grounds that very young carrot embryos are composed of small, densely cytoplasmic cells containing starch whereas the clusters figured by Steward consisted of larger vacuolated cells. Also at this time Halperin and his co-workers [31–33] advanced the alternative view that the embryoids arose from the surface layers of the larger cellular aggregates which always occur in embryogenic carrot suspension cultures. These aggregates (embryogenic clumps) were considered to persist by active cell proliferation and fragmentation due to cell enlargement and separation of cells in their interior. The superficially initiated embryoids were considered to be released, at various stages in their development, as free-floating structures which, if they had already reached the globular stage of embryogeny, were capable of completing their development in isolation from the parent clump. This interpretation has been supported by more recent work [7], and additionally very strong evidence obtained that the individual embryoids have their origin in single superficial cells of the embryogenic clumps [35, 7] (Figs 2.4.4 & 2.4.5).

The use of coconut milk as a supplement to the culture medium was undoubtedly a critical factor in first exposing the occurrence of embryogenesis in carrot cultures and the chemical investigation of the active components of coconut milk has contributed significantly to the development of the new synthetic media used in studies on somatic embryogenesis. However, it is now clear that while coconut milk may be very active in promoting growth through the later stages of embryogeny, it is certainly not essential and may even be inhibitory to embryoid initiation [36]. A very high embryoid yield from carrot suspension cultures can be obtained by maintaining the cultures in a synthetic medium containing a relatively high salt concentration (the inorganic salt solution of Murashige & Skoog [37]), *meso*-inositol, ammonium nitrogen, cytokinin and auxin and then transferring a suitable fraction of the stock culture to the same medium with the auxin omitted. An auxin level essential for continued growth and probably also for embryoid initiation must be lowered to permit the progress of embryoid development. Evidence for the exacting requirements of very young isolated embryos (embryo culture) and the relative simplicity of the culture medium which supports prolific embryogenesis in carrot cultures suggests that the embryogenic cell aggregate plays a special role in the promotion of embryogenesis in its superficial cells, and that the embryoids must remain attached to the aggregate during their early development if they are to survive and mature when released.

Work with two other embryogenic cultures, callus cultures of *Ranunculus sceleratus* [38, 6] (Fig. 2.4.6) and suspension cultures of *Atropa belladonna* [5], also points to the origin of the embryoids from single cells in embryogenic clumps and the work with *Ranunculus* further indicates that the individual cells about to embark on embryogeny and even the two or three-celled proembryos remain in protoplasmic continuity via plasmodesmata with the adjacent cells of the aggregate in which they are embedded. Similarly, epidermal cells of the hypocotyl

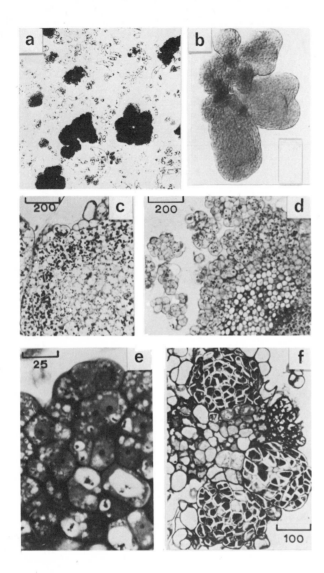

Fig. 2.4.4 Embryogenic aggregates in suspension culture of carrot. (a) A carrot suspension culture showing embryogenic aggregates (dark) and large free cells apparently not dividing. (b) An embryogenic aggregate bearing embryoids at various stages up to young torpedo stage. (c) Section of part of an embryogenic aggregate showing contrast in cell size and in size of starch grains between the superficial and central cells of the aggregate. (d) Section of an embryogenic aggregate showing how the aggregates proliferate by the release of small aggregates of superficial cells and showing that the central cells have thicker cell walls. (e) High power view of superficial cells of an embryogenic aggregate showing their large nuclei, dark nucleoli, abundant cytoplasm and numerous vacuoles (the outer central cells seen in the lower part of the photograph have a single, large central vacuole). Sections (c), (d) and (e) from suspensions cultured in presence of auxin. (f) Section through an embryogenic clump in a suspension cultured in auxin-omitted medium showing oblique sections through three embryoids and the abundant large empty cells characteristic of aggregates which are actually developing embryoids. Sections (c) and (d) stained with periodic acid-Schiff (PAS); sections (e) and (f) stained with toluidine blue (TB). All sections fixed in glutaraldehyde/cacodylate, embedded in glycol methacrylate and sectioned as described by Konar, Thomas & Street [39]. (Photographs (a) and (b) by Susan M. Smith, Botanical Laboratories, University of Leicester; photographs (c), (d), (e) and (f) from McWilliam, Smith & Street [7].)

However, there is good evidence that plantlets have arisen by an embryogenic pathway in cultures from individual species covering a wide taxonomic range: *Citrus microcarpa* [40], *Cichorium endiva* [41], *Macleaya cordata* [42], *Solanum melongema* [43], *Asparagus officinalis* [44], *Sium suave* and *Coriandrum sativum* [45], *Tylophora indica* [46], *Coffea canephora* [47], *Petroselinum hortense* [48], *Bromus inermis* [49] and various Cycads [50] (Figs 2.4.7 & 2.4.8). Further, these embryogenic cultures have variously been derived from embryo tissue, hypocotyls and stems, storage roots, leaf mesophyll, floral buds and nucellar tissue. The ability of cells derived from different plant tissues and organs and from a range of species to produce embryoids in culture can be taken as indicating that cytodifferentiation of plant cells does not involve any loss or permanent inactivation of genetic material; that all living cells of the plant body are capable under the appropriate environmental stimuli of demonstrating their retained totipotency. This may be so but cannot be regarded as established since many unsuccessful attempts have been made to obtain embryogenesis (and/or organogenesis) with cultures derived from many species and since the explants used to initiate cultures always contain parenchymatous cells and usually potentially meristematic cells (organ primordia, cambial cells, pericycle cells and so on). In this connection Gautheret [51] has contended that cells in culture which are derived from young lignified cells or immature sieve cells have their subsequent morphogenetic potential restricted to histogenesis (vascular tissue formation). It should also be borne in mind that cytodifferentiation is often associated with departure from the initial diploid chromosome number and that embryogenesis usually selectively involves diploid cells in cultures containing cells at several higher ploidy levels.

of *Ranunculus* seedlings of culture origin which are about to embark on embryogeny [39] (Fig. 2.4.6) and nucellar cells of *Nothoscordium fragrans* [17] embarking upon adventive embryogeny retain their cytoplasmic connections with adjacent cells. Although, therefore, the young embryoid is certainly clearly delineated from adjacent cells (by severing of cytoplasmic continuity and cutinization of the boundary cell walls) such isolation does not seem to be a precondition for the initiation of embryogeny (perhaps quite the reverse).

The three cases cited above remain the most carefully investigated cases of embryogenesis in somatic cultures.

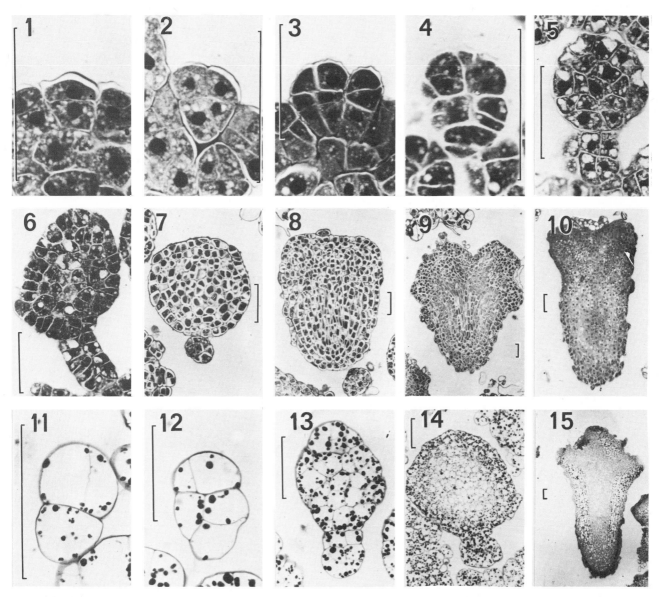

Fig. 2.4.5 Stages in embryo development observed in suspension cultures of a horticultural variety of carrot (*Daucus carota* L.) 1, two-celled stage; 2, four-celled stage; 3, two proembryos, one three-celled and one six-celled; 4, 10–12-celled; 5, globular embryo showing epidermal layer and two-tiered suspensor; 6, later globular stage; 7, fully-developed globular embryo (suspensor still attached) at stage where normally embryos are released from the embryogenic cell aggregates of the culture; 8, young heart-shaped embryo; 9, later heart-shaped embryo showing cotyledonary lobes and beginning of organization of the procambial cylinder; 10, torpedo-shaped embryo. All the above stained with toluidine blue. 11, three-celled stage; 12, five-celled stage; 13, globular embryo at stage when epidermal layer being defined; 14, fully developed globular embryo at similar stage to that shown in photograph 7; 15, torpedo-shaped embryo showing epidermis, cortex and procambial cylinder of the embryo-axis. Photographs 11–15 stained with periodic acid-Schiff (starch grains black). All scales = 50 μm. Material embedded in glycol methacrylate, following fixation in glutaraldehyde/cacodylate, and section at 1–3 μm thick as described by Konar, Thomas & Street [37]. (Photographs by A.A. McWilliam, Lyndsey, A. Withers & H.E. Street, Botanical Laboratories, University of Leicester.)

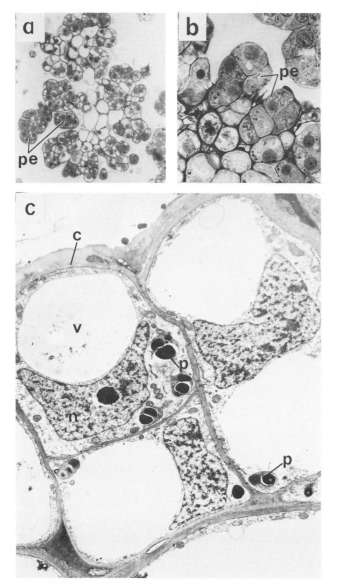

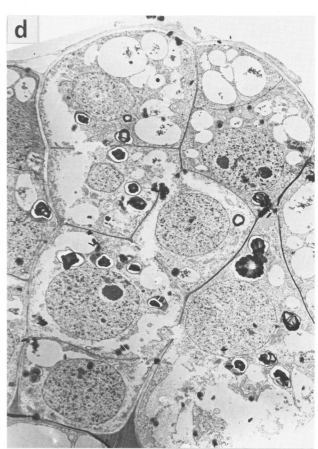

Fig. 2.4.6 Somatic embryogenesis in *Ranunculus scleratus*. (a) and (b) Sections of a callus cultured in auxin-omitted medium showing early stages of pro-embryoid (pe) development. (c) Two-celled proembryoid and an adjacent cytoplasmic-rich epidermal cell of the stem epidermis of a plantlet of culture origin. Plasmodesmata visible in anticlinal cell walls. c, cuticle of epidermis; n, nucleus; p, plastid; v, vacuole. (d) Later stage in development of an embryoid from an epidermal cell. (Photographs from [6] and [38].)

Opposite page

Fig. 2.4.7 Somatic embryogenesis in various species. A, A callus of *Macleaya cordata* bearing numerous globular embryoids. Photographs supplied by H. Lang & H.W. Kohlenbach of the Botanical Institute, Frankfurt-on-Main. B, C, Somatic embryoids of *Coffea canephora*. B, (a) Small plantlet with cotyledons, hypocotyl and primary root; (b) embryoid with developing primary root; (c) globular embryoids. From Staritsky [47]. C, Torpedo embryoids. Photograph supplied by Dr G. Staritsky afd. Tropische Plantenteelt, Landbouwhogeschool, Wageningen. D, Callus bearing globular embryoids of *Zamia*. From Norstog & Rhamstine [49]. E, a callus derived from the nucellus of *Citrus microcarpa* bearing embryoids. F, Stages in development of *Citrus* embryoids. G, Longitudinal section of a fully formed embryo from the *Citrus* nucellar callus. E, F and G, all from Rangaswamy [40].

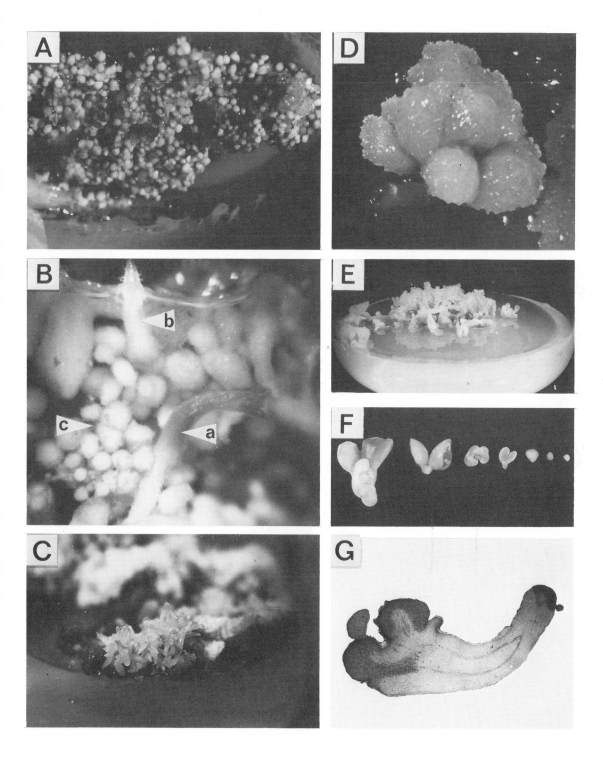

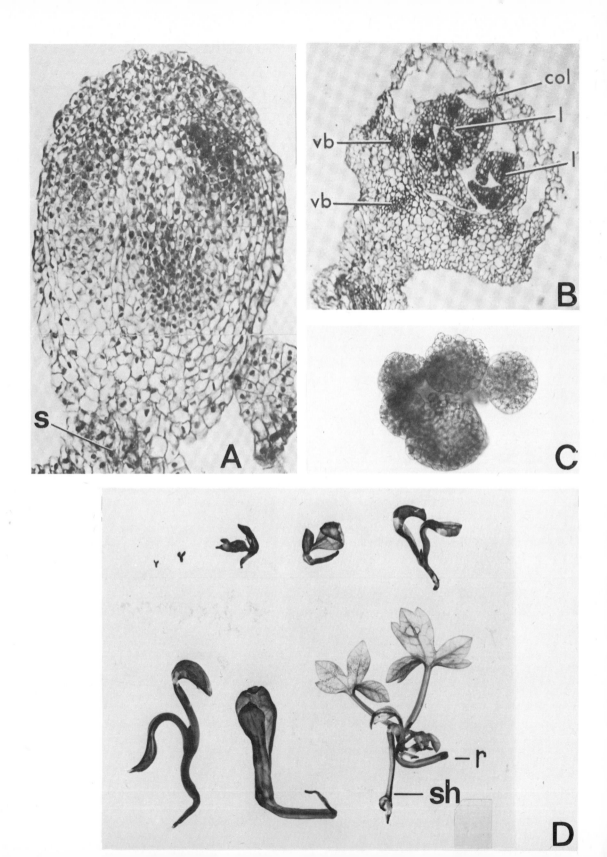

2.4.7 THE INDUCTION OF EMBRYOGENIC COMPETENCE IN SOMATIC CELLS

Halperin [34] has advanced the view that the achievement of embryogenic competence (capacity to express the inherent potential) occurs during the initiation of the culture from the primary explant and that the embryoids develop exclusively from cell clumps derived from such induced cells. The primary culture may, therefore, contain induced and non-induced cells. Active proliferation of these induced cells will be essential for retention of a capacity for embryogenesis; conditions preferentially favouring growth of the non-induced cells will lead to a decline in embryogenic competence. The level of embryogenic competence in the culture can, at any time, be assessed by transferring the culture to an appropriate medium (in the case of carrot, to a medium with the auxin omitted). On this hypothesis failure to obtain embryogenesis in culture would have as its primary cause inappropriate conditions of callus initiation; the conditions of callus initiation would have effectively activated the mechanisms involved in cell division and growth but have failed to achieve the necessary 'dedifferentiation' (induction) to obtain cells with the competence of the zygote.

A number of experimental observations can be considered as supporting the concept developed above, that achievement of growth activation and embryogenic competence are separate steps. Cultures derived from adult and juvenile stem segments of the same ivy plant remained, over many subcultures, consistently different; the tissue cultures from the juvenile phase had a higher rate of proliferation and larger cells [52]. The juvenile-adult difference was retained through the process of callus induction and subsequent growth in culture. In some cases successful initiation of morphogenesis in culture has involved transfer of the cultures through a sequence of media suggesting that the development of the capacity for *organized* growth itself involved more than a single metabolic shift [53]. Again in studies on cytodifferentiation it has been shown that the cultural conditions operating during the induction of cell division can exert a determining influence on the nature of the subsequent differentiation [54].

The most prolific formation of carrot embryoids can be obtained by culturing a suspension of cells derived from mature carrot embryos [31]. Callus derived from small excised embryos of *Cuscuta reflexa* readily generated numerous adventive embryos [55]. Embryoids of *Atropa belladonna*, carrot, *Citrus microcarpa*, *Datura innoxia* and particularly of *Ranunculus sceleratus* frequently develop adventive embryoids over their hypocotyl surface or from the suspensor [5, 7, 38, 56, 57] (Fig. 2.4.9). The embryoids developing from the hypocotyl arose

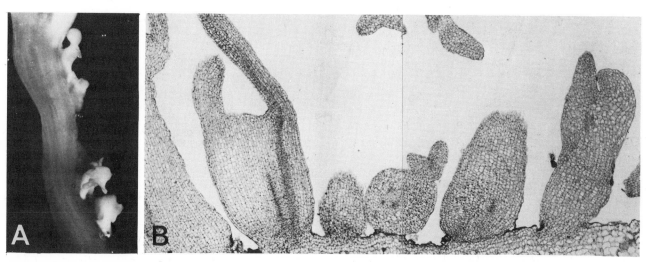

Opposite page

Fig. 2.4.8 Somatic embryogenesis in various species. A, Globular embryoid arising at the surface of an irregular cell aggregate in a suspension culture of *Bromus inermis*. B, Section through a plantlet developing in a *Bromus inermis* culture. C, Group of globular embryos in a culture of *Asparagus officianalis*. D, Stages in the development of embryoids of *Petroselinum hortense* into plantlets. s, Stalk (suspensor); col, coleoptile; l, young leaves; vb, vascular bundles; sh, shoot; r, root. A and B from Constabel, Miller & Gamborg [49]; C supplied by Mrs M. Hellendoorn, Unilever Research, Vlaardingen/Duiven; D from Vasil & Hildebrandt [48].

Fig. 2.4.9 Adventive embryoids developing on young plantlets of tissue culture origin via embryogenesis. Adventive embryoids on haploid plantlets of *Atropa belladonna* (A) and *Datura innoxia* (B) of anther origin. (A) from Rashid & Street [56]; (B) from Geier & Kohlenbach [57].

from individual cells already progressing along the pathway to mature epidermal cells; thus, embryogenic competence can be retained in differentiating cells. A similar phenomenon is clearly involved when embryos develop from synergids and antipodal cells in the embryo sac.

These observations not only suggest that it may be possible to obtain embryogenic tissue and cell cultures from new species by starting with embryo tissue (particularly immature embryo tissue) but indicate that embryogenic competence is a property which can be exhibited by cells in different states of structural differentiation. Hence it can be expected that to induce embryogenic competence in mature tissue cells may call for additional factors over and above those which can induce division and loss of specialized features of cell structure.

2.4.8 THE DIRECT OR INDIRECT (VIA CALLUS) FORMATION OF HAPLOID EMBRYOIDS FROM MICROSPORES

Following the pioneer studies of Guha & Maheshwari [58] with *Datura innoxia* and of Bourgin & Nitsch [59] with *Nicotiana tabacum* and *N. sylvestris*, haploid embryoids (limited at present to some members of the Solanaceae) or haploid callus have been obtained from an increasing number of species by plating their immature excised anthers on solidified culture medium and incubation in light at an appropriate temperature (Fig. 2.4.10). Such anther culture is only successful if the anthers are excised after meiosis in the spore mother cells and release of the microspores from the tetrad sac. Normally the microspores must still be uninucleate; in *N. tabacum* the most responsive stage appears to be just before or during the first mitotic division in the majority of the microspores [60]. Once this mitosis has occurred and starch accumulation commenced the microspores appear to be irreversibly committed to mature into pollen grains. Response beyond the uninucleate stage is reported in *Brassica oleracea* and *Festuca-Lolium* hybrids and here the microspores in each case generate callus. In these species the mature pollen grains are trinucleate and starch accumulation is very delayed [61].

The question of how excision and culture of immature anthers diverts a proportion of the microspores from expressing their gametophytic potential and leads them to develop a competence for embryogenesis or callus development cannot at present be satisfactorily answered. The meiosis leading to the formation of the microspores purges their cytoplasm of the effects of the diploid nucleus and thus apparently makes possible alternative pathways for their further development; the

Fig. 2.4.10 Embryoids of microspore origin in an anther of *Nicotiana tabacum*. From Thesis, Univ. Paris by Brigitte Norreel entitled 'La neoformation d'embryons *in vitro* chez *Daucus carota, Nicotiana tabacum* et *Datura innoxia*', 1971.

environment within the *attached* maturing anther determines that the pathway followed shall be the production of the male gametophyte consisting in the typical case of a vegetative cell and a generative nucleus. Usually a considerable interval separates meiosis from the microspore mitosis and this mitosis is not highly synchronized (in contrast to the preceding meiosis). The anther containing uninucleate microspores contains these at various stages of activation and even under the most favourable conditions of anther culture only a very small proportion embark upon embryogenesis or callus formation. The induction of embryogenic competence may therefore be restricted to microspores which are, at the time of anther excision, at a particular stage in the interphase before the mitosis.

Some and usually many of the microspores in excised anthers become empty. Sunderland [60] has distinguished two developmental sequences which may be followed by the surviving uninucleate spores in *Nicotiana tabacum* anthers. The first and most frequent sequence (Type A) involves an asymmetric division to give two unequal cells, the larger vegetative cell and the smaller generative cell (p. 35). Some of these grains mature into pollen grains and in such cases the vegetative cell will give rise to the pollen tube and the generative nucleus divide once to give the two male nuclei. These pollen grains usually become packed with starch. The other and less frequent sequence (Type B) involves the nucleus entering mitosis nearer to the centre of the microspore and giving rise to two equal cells. In the Type B grains the nuclei are similar in size and staining reaction (resembling the vegetative nucleus of Type A). In Type A the vegetative nucleus is large, diffuse and lightly staining and has a well defined nucleolus, whereas the generative nucleus is smaller, condensed and densely staining.

From detailed microscopic observations it has been concluded that in *Nicotiana tabacum* [62] and *Datura metel* [63] embryoid development takes place from the Type A grains by divisions in the vegetative cell; the generative nucleus in such grains may undergo a limited number of divisions but its products are not incorporated into the embryoid and ultimately break down. In these studies [62, 63] it was considered that Type B grains may play a subsidiary role in embryoid formation.

In excised *Atropa belladonna* anthers embryogenesis only occurs if the microspores at excision are predominantly uninucleate and the pathway of embryogenesis is initiated by an equal division [64]. In *Nicotiana sylvestris* embryogenesis is certainly predominantly, and possibly exclusively, initiated by an equal division in uninucleate microspores [65] (Fig. 2.4.11). A re-examination of the situation in *Nicotiana tabacum* [65] has shown that microspores containing two equal cells (Type B grains) appear in sufficient numbers to yield all the embryoids formed and that grains containing young globular embryoids rarely show any evidence of a generative cell or products of its division. Thus, although divisions do occur in the Type A grains

it is much less clear that they normally continue to give a viable embryoid. Recent work by Nitsch & Norreel [66] has shown that a temperature shock applied to the anthers of *Datura innoxia* enhances embryogenesis and that this is correlated with an increase in the number of microspores showing a shift in the axis of division to produce two equal cells. If it should prove that embryoids normally arise from Type B grains and that division in Type A grains is either limited or goes on to produce only callus, this would be further evidence that the induction of embryogenic competence may have special requirements additional to those required for the induction of meristematic activity.

Embryogenesis from microspores can only at present be achieved for a limited number of species; these species are by no means coincident with those where embryogenesis can be achieved from somatic cultures. Haploid callus, but not embryoids, can be obtained from some excised anthers and such callus may, under appropriate conditions, show organogenesis and/or embryogenesis. A frequent complicating factor in anther culture is that callus arises from the diploid anther tissue although in some instances this can be prevented by use of an appropriate culture medium [64].

Since the anther may be inhibitory to microspore embryogenesis and a source of contaminating diploid somatic callus, attempts are now being made to induce embryogenesis in isolated microspores (pollen culture). Thus it is reported that haploid clones of callus have been obtained from tomato microspores by plating out a suspension of the spores on to filter-paper discs mounted over intact anthers of the same species dispersed in a layer of solidified culture medium [67]. Embryoids of tomato have also been initiated from a microspore suspension by plating this in a culture medium enriched with an aqueous extract of embryogenic anthers of *Datura innoxia* [68]. This new technique of pollen culture may enable microspore embryogenesis to be achieved in a wider range of species and may, in future, enable embryogenesis to be synchronously initiated at a high frequency in the microspores.

2.4.9 EMBRYOGENIC CELLS AND THE ORIGIN OF POLARITY IN THE YOUNG EMBRYOID

Microscopic and ultrastructural studies of the superficial cells of embryogenic cell clumps in somatic cultures have so far failed to identify by their structural features those cells which are immediately about to embark upon embryogenesis. The structure of these superficial cells corresponds closely with that observed in those egg cells and zygotes which are regarded as of high metabolic activity and contain starch and lipid reserves [6, 7, 35]. Many, and possibly all, of these superficial cells (or their division products) are sooner or later involved in embryo-

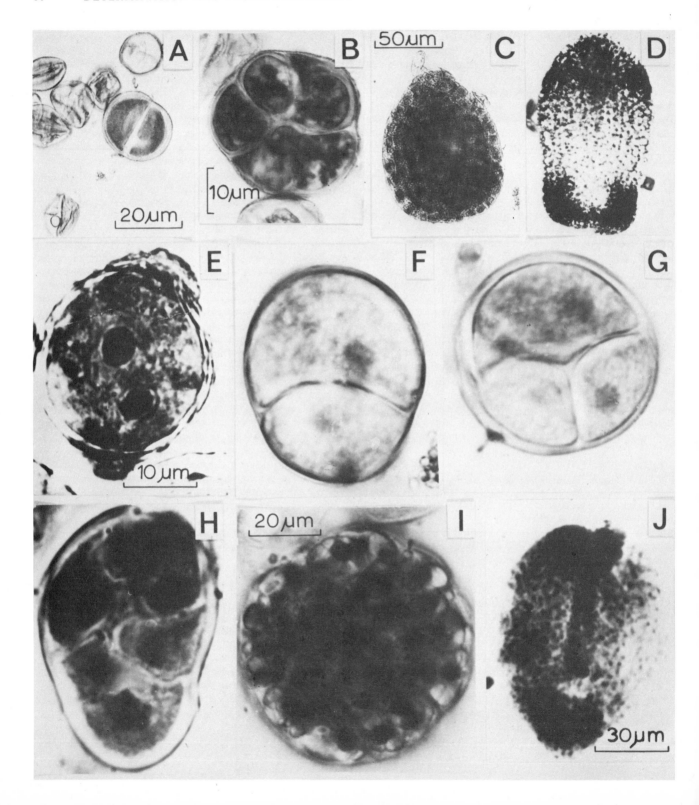

genesis; cultures continue to initiate embryoids over a considerable period and after 2–3 weeks in the minus–auxin medium contain embryoids in all stages of development. It seems, therefore, that identification of the unique features of embryogenic cells is going to require their intensive study at the level of molecular biology. This clearly demands embryogenic cells, uncontaminated by non-embryogenic cells, and synchronized in their progress towards initiation of embryogeny. The best hope of such experimental material seems at the moment to be offered by the new technique of isolated microspore (pollen) culture coupled to an appropriate synchronizing stimulus (temperature shock?).

Embryoids developed from somatic embryogenic clumps are uniformly polarized [5, 7, 33]; the root pole is directed towards the centre of the cell clump and the shoot towards the outside. It is, however, not clear how soon this polarity is established although studies of early segmentation particularly in *Daucus carota* [7] indicate that the first division is uniformly parallel to the surface of the clump and gives rise to a basal and a terminal cell and that the suspensor is developed from the former and the embryoid from the latter. Occasionally a suspensor-like cell can be recognized during microspore embryogenesis (Fig. 2.4.11) and when present is probably indicative of the axis of polarity (root pole adjacent to the suspensor). However, the most frequent pattern of microspore embryogenesis does not involve formation of a 'suspensor' (the inessentiality of a suspensor has already been mentioned with reference to natural adventive embryogeny). There is no evidence that the asymmetric division in Type A microspores has any role in polarization of the incipient embryoid and the evidence that successful embryoids probably have their origin from Type B microspores makes this unlikely. Therefore, it must be concluded that polarity in the globular microspore embryoid arises spontaneously (i.e. in the sense that there is no apparent gradient such as can be postulated for an embedded superficial cell in a somatic embryogenic clump). Here again, pollen culture should provide experimental material for study of the factors which can determine the axis of embryoid polarity and for attempts to recognize the emergence of polarity at the fine structural level.

Opposite page

Fig. 2.4.11 Embryogenesis in anther cultures of *Atropa belladonna* (A–D) and *Nicotiana sylvestris* (E–J). A, First division into two equal cells in an embryogenic microspore. B, Five-celled embryoid. C, Released embryoid with root pole. D, Released embryoid with root pole and two cotyledons. E, Microspore with two similarly staining nuclei. F, Bicellular microspore. G, Four-celled microspore. H, Elongated microspore containing a proembryoid and at the base a 'suspensor' cell. I, Late globular embryoid within microspore wall. J, Released embryoid showing root pole and beginning of cotyledon development. Scale on C applies also to D; Scale on E applies also to F, G and H. A–D from Rashid & Street [64]; E–J from Rashid & Street [65].

2.4.10 THE CLEAVAGE PATTERNS INVOLVED IN ZYGOTE, SOMATIC AND MICROSPORE EMBRYOGENESIS

Embryogenesis in culture has opened up a new avenue for the study of angiosperm embryology and, although the relevant data gleaned so far from this approach is very limited, it clearly calls into question orthodox plant embryology which has been dominated by the so-called laws of embryonomy derived from the works of Souèges and his disciples, particularly Crété. This classical approach has been spelled out in its most authoritarian form by Johansen [69] in such statements as: 'In order to elucidate the laws of embryonomy with the utmost precision and clarity and in as complete a manner as possible, the employment of formulae has become necessary. These formulae indicate precisely the relationships, which the individual cells present to one another during the course of their segmentations, with respect to their origin, to their number, to their disposition and to their eventual destination or their histogenic function.' 'Each and every cell has a reason for its existence, its origin can be demonstrated, its destination determined and its position is invariably the same. A superfluous cell would seriously upset the harmonious balance' (Occam's Razor applied to embryology!). This concept that great significance is to be attached to the order and planes of division in the young embryo, that during these divisions the individual cells of the embryo inherit different cytoplasmic potentialities from different regions of the zygote and that these differences determine from the beginning the exact role they and their daughter cells shall play in constructing the embryo and its parts, is a theory of precise mosaic organization. On this basis there has been erected a key for embryo classification via a list of principal (fundamental) segmentation patterns and variants and since the segmentation pattern is regarded as species specific it has been established as a taxonomic character and as having relevance to evolutionary relationships.

The typical early segmentation sequence reported for zygotic embryonomy in *Daucus carota* rests on the work of Borthwick [70] and is depicted in Fig. 2.4.12. This sequence enables the embryonomy to be classified as a 'minor variant of the Sherardian variation of the Solanad type'. While the later stages of embryogeny of somatic carrot embryoids correspond very closely with those of the zygotic embryo, a comparison of Fig. 2.4.5 and Fig. 2.4.12 shows that there is no correspondence in the early segmentation sequences. However, although the classical approach considers that at least the first four cell generations (up to the 16-celled proembryo) of embryonomy are always highly regular, it is admitted by Borthwick that from the eight-celled stage the planes of division in the carrot embryo become less regular (predictable) and, more important, that whilst usually both the distal cells of the four-celled stage (cells S and D of Fig. 2.4.12) are involved in the formation of the

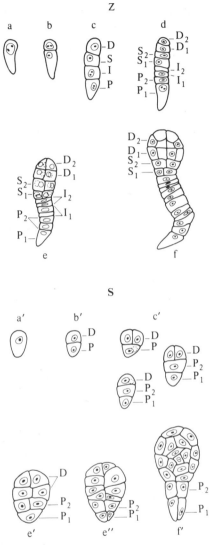

Fig. 2.4.12 Embryonomy of *Daucus carota*. Z, As observed from the zygote by Borthwick [70]. S, As observed in culture (see Fig. 2.4.5). Comparable stages labelled a–f and a′–f′. Cells of the linear four-celled stage of zygotic embryonomy labelled after Borthwick, D, S, I and P. Derivatives of these labelled accordingly, e.g. P → P_1 + P_2. Note the linear eight-celled stage of zygotic embryonomy and the consequent derivation of the pro-embryo and the many-celled suspensor derived from cells I and P.

Turning to microspore embryogenesis we see that normally there is no suspensor although this is a feature of zygotic embryos of the species concerned, and there normally involves derivatives of both of the basal cells of an initial linear tetrad. This is surely incompatible with a mosaic theory. Since the dispensability of a suspensor has long been known from natural adventive polyembryony, how is this accommodated within the so-called laws of embryonomy? Johansen [69] deals with this as follows: 'This only flaw in the rigid application of the law (Occam's Razor) is that we are not yet reasonably clear concerning the role and function of the suspensor... in such instances apparently superfluous cells of the suspensor may be disregarded'.

Cleavage polyembryony and the frequent formation of adventive embryos from individual cells of embryoid plantlets demonstrate the long retained pluripotency (lack of determination) of the cells of angiosperm embryos and this again is incompatible with mosaic theory.

Zygotic embryos develop under the physical restrictions and polarized chemical environment of the ovule of the species whereas somatic embryoid development occurs in the more uniform environment of the shaken culture vessel and the microspore embryoid under the severe restrictions initially imposed by the microspore wall. Although the number of species in which early embryoid segmentation has been followed in somatic and anther cultures is at present very limited, it indicates that in both situations there is more uniformity between species than is observed in ovule embryogenesis. Embryoids follow the tenet of Haldane [71] that the genes which determine interspecific differences come into action late in individual development. Berthold [72] and later D'Arcy Thompson [73] considered that surface tensions are important in determining the form of embryos and their characteristic segmentation, the initial phase of embryo development consisting in enlargement and regular segmentation of a subspherical or club-shaped body by walls of minimum surface. The patterns of segmentation observed during early embryogenesis in culture in no way conflict with this interpretation. D'Arcy Thompson further pointed out that it is at a later time that differential growth sets in to modify the initial regularity and establish localized growth centres so that embryos of quite different systematic affinity initially show closely comparable cellular patterns. The segmentations during early embryonomy are thus here considered to be controlled by physical factors and not to have the phylogenetic significance accorded to them by classical plant embryology. Further, during the early development of the embryo, cells are not considered to inherit different cytoplasmic potentialities but to remain undetermined. This 'regulative theory of organization' is, in contrast to the mosaic theory, supported by studies on embryogenesis in culture and by many observations on whole plant embryology. Jensen [10] in his description of

embryo, there occurs an 'aberrant type' in which all of the embryo is derived from one of the cells (D). Variation incompatible with the mosaic theory is therefore to be observed in normal zygotic embryonomy; a more profound deviation is to be observed in the embryonomy of somatic embryoids.

embryonomy in *Gossypium* states that the globular embryo is produced by apparently random divisions in the terminal (distal) cell of the two-celled proembryo and that no two embryos are exactly alike, thereby failing to obey the laws of Souèges. Similarly Brown & Mogensen [74], in their study of *Quercus gambelii* embryonomy, stress that great variability in cleavage patterns is to be observed.

2.4.11 REFERENCES

[1] BELL P.R. (1965) Angiosperm embryology. *Nature, Lond.*, **205**, 1044.

[2] MAHESHWARI P. (1963) (Ed.) *Recent advances in the embryology of angiosperms*. Int. Soc. Plant Morph. Delhi.

[3] TOKIN B.P. (1964) Regeneration and somatic embryogenesis. In *Regeneration and wound healing*, pp. 9–46. Akadémiai Kiadó, Budapest.

[4] HACCIUS B. (1971) Zur derzeitigen Situation der Angiospermen—Embryologie. *Bot. Jb.*, **91**, 309.

[5] KONAR R.N., THOMAS E. & STREET H.E. (1972) The diversity of morphogenesis in suspension cultures of *Atropa belladonna* L. *Ann. Bot.*, **36**, 249–58.

[6] THOMAS E., KONAR R.N. & STREET H.E. (1972) The fine structure of the embryogenic callus of *Ranunculus sceleratus* L. *J. Cell Sci.*, **11**, 95–109.

[7] McWILLIAM A.A., SMITH S.M. & STREET H.E. (1974) The origin and development of embryoids in suspension cultures of carrot (*Daucus carota*). *Ann. Bot.*, **38**, 243–50.

[8] NEEDHAM J. (1950) *Biochemistry and morphogenesis*. Cambridge Univ. Press.

[9] STEWARD F.C. & MOHAN RAM H.Y. (1961) Determining factors on cell growth: some implications for morphogenesis in plants. *Adv. in Morphogenesis*, **1**, 189–266.

[10] JENSEN W.A. (1965) The ultrastructure and composition of the egg and central cell of *Gossypium*. *Am. J. Bot.*, **52**, 781–97.

[11] DIBOLL A.G. & LARSON D.A. (1966) An electron microscope study of the mature gametophyte of *Zea mays*. *Am. J. Bot.*, **53**, 391–402.

[12] SCHULZ SISTER R. & JENSEN W.A. (1968) *Capsella* embryogenesis: the egg, zygote and young embryo. *Am. J. Bot.*, **55**, 807–19.

[13] VAN WENT J.F. (1970) The ultrastructure of the egg and central cell of *Petunia*. *Acta bot. Neerl.*, **19**, 313–22.

[14] MOGENSEN H.L. (1972) Fine structure and composition of the egg apparatus before and after fertilisation in *Quercus gambelii*: the functional ovule. *Am. J. Bot.*, **59**, 931–41.

[15] NEWCOMB W. (1973) The development of the embryo sac of sunflower, *Helianthus annuus*: before fertilisation. *Can. J. Bot.*, **51**, 863–78.

[16] MAHESHWARI P. & SACHAR R.C. (1963) Polyembryony. In *Recent advances in the embryology of angiosperms*, pp. 265–296 (Ed. P. Maheshwari). Intern. Soc. Plant Morph. Delhi.

[17] DYER A.F. (1967) The maintenance of structural heterozygosity in *Nothoscordium fragrans*, Kunth. *Caryologia*, **20**, 287–308.

[18] BUTCHER D.N. (1973) The origins, characteristics and culture of plant tumour cells. In *Plant tissue and cell culture*, pp. 356–91 (Ed. H.E. Street). Blackwell Scientific Publications, Oxford.

[19] RAGHAVEN V. & TORREY J.G. (1963) Growth and morphogenesis of globular and older embryos of *Capsella* in culture. *Am. J. Bot.*, **50**, 540–551.

[20] RAGHAVEN V. & TORREY J.G. (1964) Effect of certain growth substances on the growth and morphogenesis of immature embryos of *Capsella* in culture. *Pl. Physiol.*, Lancaster, **39**, 691–9.

[21] NORSTOG K.J. & SMOTH J.E. (1963) Growth of very small barley embryos in defined media. *Science*, **142**, 1655–6.

[22] GUNNING B.E.S. & PATE J.S. (1969) 'Transfer cells.' Plant cells with wall invaginations in relation to short distance transport of solutes. Their occurrence, structure and development. *Protoplasma*, **68**, 107–33.

[23] CURTIS J.T. & NICHOL M.A. (1948) Culture of proliferating orchid embryos *in vitro*. *Bull. Torrey Bot. Club*, **75**, 358–373.

[24] MOREL G. (1964) La culture *in vitro* du méristème apical. *Rev. cytol. cytophysiol. végétales*, **27**, 307–14.

[25] STREET H.E. (1973) In *Plant tissue and cell culture*, pp. 1–10 (Ed. H.E. Street). Blackwell Scientific Publications, Oxford.

[26] STEWARD F.C., CAPLIN S.M. & MILLAR F.K. (1952) Investigations of growth and metabolism of plant cells. I. New techniques for the investigation of metabolism, nutrition and growth in undifferentiated cells. *Ann. Bot.*, **16**, 58–77.

[27] STEWARD F.C. & SHANTZ E.M. (1956) The chemical induction of growth in plant tissue cultures. In *The chemistry and mode of action of plant growth substances*, pp. 165–186 (Eds R.L. Wain & F. Wightman). Butterworths Ltd., London.

[28] STEWARD F.C. (1958) Growth and organized development of cultured cells. III. Interpretation of the growth from free cell to carrot plant. *Am. J. Bot.*, **45**, 709–13.

[29] STEWARD F.C., MAPES M.O. & MEARS K. (1958) Growth and organised development of cultured cells. II. Organization in cultures grown from freely suspended cells. *Am. J. Bot.*, **45**, 705–8.

[30] REINERT J. (1959) Über die Kontrolle der Morphogenese und die Induktion von Adventivembryonen an Gewebekulturen aus Karotten. *Planta*, **53**, 318–33.

[31] STEWARD F.C., MAPES M.O., KENT A.E. & HOLSTEN R.D. (1964) Growth and development of cultured plant cells. *Science*, **143**, 20–7.

[32] HOMÉS J.L.A. & GUILLAUME M. (1967) Phenomenes d'organogenese dans des cultures *in vitro* de tissus de carotte (*Daucus carota* L.). *Bull. Soc. Roy. Botan. Belg.*, **100**, 239–58.

[33] HALPERIN W. & WETHERELL D.F. (1964) Adventive embryony in tissue cultures of wild carrot, *Daucus carota*. *Am. J. Bot.*, **51**, 274–83.

[34] HALPERIN W. (1967) Population density effects on embryogenesis in carrot cell cultures. *Expl Cell Res.*, **48**, 170–3.

[35] HALPERIN W. & JENSEN W.A. (1967) Ultrastructural changes during growth and embryogenesis in carrot cell cultures. *J. Ultrastruct. Res.*, **18**, 428–43.

[36] REINERT J. (1973) Aspects of organisation: organogenesis and embryogenesis. In *Plant tissue and cell culture*, pp. 338–55 (Ed. H.E. Street). Blackwell Scientific Publications, Oxford.

[37] MURASHIGE T. & SKOOG F. (1962) A revised medium for rapid growth and bioassays with tobacco tissue cultures. *Physiologia Pl.*, **15**, 473–97.

[38] KONAR R.N. & NATARAYA K. (1965) Experimental studies in *Ranunculus sceleratus* L. Development of embryos from the stem epidermis. *Phytomorphology*, **15**, 132–7.

[39] KONAR R.N., THOMAS E. & STREET H.E. (1972) Origin and structure of embryoids arising from the epidermal cells of *Ranunculus sceleratus* L. *J. Cell Sci.*, **11**, 77–93.

[40] RANGA SWAMY N.S. (1961) Experimental studies on female reproductive structure of *Citrus microcarpa* Bunge. *Phytomorphology*, **11**, 109–27.

[41] VASIL I.K., HILDEBRANDT A.C. & RIKER A.J. (1964) Plantlets from free suspended cells and cell groups *in vitro*. *Science*, **146**, 76–7.

[42] KOHLENBACH H.W. (1965) Über organisierte Bildungen aus *Macleaya cordata* Kallus. *Planta*, **64**, 37–40.

[43] YAMADA T., NAKAGAWA H. & SINOTO Y. (1967) Studies in the differentiation in cultured cells. I. Embryogenesis in three strains of *Solanum* callus. *Bot. Mag. Tokyo*, **80**, 68–74.

[44] WILMAR C. & HELLENDOORN M. (1968) Growth and morphogenesis of asparagus cells cultured *in vitro*. *Nature, Lond.*, **217**, 369–70.

[45] STEWARD F.C., AMMIRATO P.V. & MAPES M.O. (1970) Growth and development of totipotent cells. Some problems, procedures and perspectives. *Ann. Bot.*, **34**, 761–87.

[46] RAO P.S., NARAYANASWAMY S. & BENJAMIN B.D. (1970) Differentiation ex ovulo of embryos and plantlets in stem tissue cultures of *Tylophora indica*. *Physiologia Pl.*, **23**, 140–4.

[47] STARITSKY G. (1970) Embryoid formation in callus tissues of coffee. *Acta Bot. Neerl.*, **19**, 509–14.

[48] VASIL I.K. & HILDEBRANDT A.C. (1966) Variations of morphogenetic

behaviour in plant tissue cultures. II. *Petroselinum hortense*. *Am. J. Bot.*, **53**, 869–74.

[49] CONSTABEL F., MILLER R.A. & GAMBORG O.L. (1971) Histological studies on embryos produced from cell cultures of *Bromus inermis*. *Can. J. Bot.*, **49**, 1415–7.

[50] NORSTOG K. & RHAMSTINE E. (1967) Isolation and culture of haploid and diploid Cycad tissue. *Phytomorphology*, **17**, 374–81.

[51] GAUTHERET R.J. (1966) Factors affecting differentiation in plant tissues grown *in vitro*. In *Cell differentiation and morphogenesis*, pp. 55–71 (Ed. W. Beerman). North Holland Publ. Co.

[52] STONTEMEYER V.T. & BRITT O.K. (1965) The behaviour of tissue cultures from English and Algerian ivy in different growth phases. *Am. J. Bot.*, **52**, 805–10.

[53] STEWARD F.C., KENT A.E. & MAPES M.O. (1967) Growth and organisation in cultured cells: sequential and synergistic effects of growth-regulating substances. *Ann. N.Y. Acad. Sci.*, **144**, 326–34.

[54] FOSKET D.E. & TORREY J.G. (1969) Hormonal control of cell proliferation and xylem differentiation in cultured tissues of *Glycine max* var. Biloxi. *Pl. Physiol., Lancaster*, **44**, 871–80.

[55] MAHESHWARI P. & BALDEV B. (1961) Artificial production of buds from embryos of *Cuscata reflexa*. *Nature, Lond.*, **191**, 197–8.

[56] RASHID A. & STREET H.E. (1974) Growth, embryogenic potential and stability of a haploid cell culture of *Atropa belladonna* L. *Plant Science Letters* 2. 84–94.

[57] GEIER T. & KOHLENBACH H.W. (1973) Entwicklung von Embryonen und embryogenem Kallus aus Pollenkörnern von *Datura meteloides* und *Datura innoxia*. *Protoplasma*, **78**, 381–96.

[58] GUHA S. & MAHESHWARI S.C. (1964) *In vitro* production of embryos from anthers of *Datura*. *Nature, Lond.*, **204**, 497.

[59] BOURGIN J.P. & NITSCH J.P. (1967) Obtention de *Nicotiana* haploids à partir d'étamines cultivées *in vitro*. *Annls Physiol. vég. Paris*, **9**, 377–82.

[60] SUNDERLAND N. (1973) Pollen and anther culture. In *Plant tissue and cell culture*, pp. 205–239 (Ed. H.E. Street). Blackwell Scientific Publications, Oxford.

[61] PANDEY K.K. (1973) Theory and practice of induced androgenesis. *New Phytol.*, **72**, 1129–40.

[62] SUNDERLAND N. & WICKS F.M. (1971) Embryoid formation in pollen grains of *Nicotiana tabacum*. *J. exp. Bot.*, **22**, 213–26.

[63] IYER R.N. & RAINA S.K. (1972) The early ontogeny of embryoids and callus and subsequent organogenesis in anther cultures of *Datura metel* and rice. *Planta*, **104**, 146–56.

[64] RASHID A. & STREET H.E. (1973) The development of haploid embryoids from anther cultures of *Atropa belladonna* L. *Planta*, **113**, 263–70.

[65] RASHID A. & STREET H.E. (1974) Segmentations in microspores of *Nicotiana sylvestris* and *Nicotiana tabacum* which lead to embryoid formation in anther cultures. *Protoplasma* **8**, 323–34.

[66] NITSCH C. & NORREEL B. (1973) Effet d'un choc thermique sur le pouvoir embryogene du pollen de *Datura innoxia* cultivé dans l'anthere ou isolé de l'anthere. *C. r. hebd. Séanc. Acad. Sci., Paris*, **276**, 303–6.

[67] SHARP W.R., RASKIN R.S. & SUMMER H.E. (1972) The use of nurse culture in the development of haploid clones in tomato. *Planta*, **104**, 357–361.

[68] DEBERGH P. & NITSCH C. (1973) Premiers résultats sur la culture *in vitro* de grains de pollen isolés chez la Tomate. *C. r. hebd. Séanc. Acad. Sci., Paris*, **276**, 1281–4.

[69] JOHANSEN D.L. (1950) *Plant embryology: embryogeny of the spermaphyta*. Chronica Botanica Co., Waltham, Mass.

[70] BORTHWICK H.A. (1931) Development of the macrogametophyte and embryo of *Daucus carota*. *Bot. Gaz.*, **92**, 24–44.

[71] HALDANE J.B.S. (1932) The time of action of genes and its bearing on some evolutionary problems. *Amer. Nat.*, **66**, 5.

[72] BERTHOLD G.D.W. (1886) *Studien über protoplasma mechanik*. A. Felix, Leipzig.

[73] THOMPSON W.D'ARCY (1942) *On growth and form*. 2nd Ed. Cambridge Univ. Press.

[74] BROWN R.C. & MOGENSEN H.L. (1972) Late ovule and early embryo development in *Quercus gambelii*. *Am. J. Bot.*, **59**, 311–6.

Conclusions to Part 2

Although the experimental data are diverse and complex, there are clearly close analogies between the phenomena of determination and pluripotentiality in plants and animals which enable us to draw a number of general conclusions.

1. THE DEVELOPMENTAL POTENTIAL OF CELLS MAY BE RESTRICTED BY BOTH EXTRINSIC AND INTRINSIC CONTROLS

The process by which the developmental potential of cells becomes restricted is referred to as *determination*. However, it is important to distinguish clearly between factors influencing cell development arising from (1) the *position* which the cells occupy within the embryo where influences come from the surrounding cells and hence are *extrinsic* to the cells under study, from (2) changes which are *intrinsic* to, or inherent in,

the cells in question and which are still retained even when the cells are removed from their normal position in the organism.

'Fate maps' constructed for animal embryos show that a particular tissue or organ is formed from small localized groups of cells in the early embryo, so that evidently many animal embryos have a rather definite pattern of cell positions during normal development. If the cells of the embryo of *Drosophila* are marked with X-ray induced chromosomal re-arrangements at progressively later stages of development, then the number of parts of the adult which are formed by the marked cells is reduced. Such studies show that cell allocation may be a very accurate process and is under genetic control.

A similar situation is seen in the root tips of plants, where it is possible to trace cell lineages back to localized regions at the surface of the quiescent centre (Fig. 1.4.14). Indeed, the process of cell allocation is almost axiomatic in plants, since the cells are non-motile and their position is determined by the regular patterns of cell division during development, which means that the various parts of a mature organ can often be traced back to certain specific cell locations in the primordial organ.

However, observations on the pattern of cell allocation do not of themselves, enable us to say whether any given groups of cells has become intrinsically committed to a particular path of differentiation. Thus, the establishment of well-defined cell lineages in the root tip which give rise to the various tissue regions of the mature root does not necessarily imply that the cell lineages represent 'histogens' (p. 40) which are intrinsically committed to differentiate into epidermal, cortical or vascular tissue, and are inherently restricted in their developmental potential in this manner. In order to resolve this problem it is necessary to remove cells from their *normal* position and to study their developmental potential in a different cellular environment.

2. THE INTRINSIC CHANGES OCCURRING DURING CELL COMMITMENT ARE USUALLY VERY STABLE

Numerous studies on animal cells indicate that cells can, indeed, undergo an intrinsic restriction of developmental potential which is retained even when such cells are removed from the parent embryo and transferred to another embryo or are cultured *in vitro*. Such intrinsic changes are frequently very stable and are not lost or diminished even after passage through many cell divisions. Thus, the imaginal discs of *Drosophila* can be grown in adult flies for long periods and these discs are usually capable of forming the adult structures to which they were committed by a much earlier embryogenesis. Similarly, the shoot apices of flowering plants can exist in alternative states (as seen in phase change in *Hedera*, in vernalization and in the alternation of gametophyte and sporophyte generations), and these alternative states are usually extremely stable and can be transmitted through many cell generations.

However, although the determined state is very stable in many animal cells there are a number of examples in which cells have been observed to switch from one type to another. Thus, the determined cells of *Drosophila* imaginal discs can sometimes transdetermine and form adult structures which are outside their normal developmental range.

3. RESTRICTION OF DEVELOPMENTAL POTENTIAL IS A PROGRESSIVE PROCESS

As embryonic development proceeds the possible paths of differentiation of particular cell lineages becomes progressively restricted. Thus, although in mammalian embryos the endoderm becomes determined during cleavage stages, the future mesodermal and ectodermal cells are not yet so restricted and can still give rise to endodermal, mesodermal and ectodermal derivatives. A similar progressive restriction of developmental potential is demonstrated by the observation that teratocarcinomas can be formed from primordial germ cells and early embryos but not from mature embryos.

A progressive restriction of developmental potential is also demonstrated by the leaf primordia of the fern, *Osmunda,* which will regenerate whole plants if isolated at a very young stage, but with progressively older primordia there is an increasing tendency to form only leaves. However, it is not known in this instance whether this progressive determination is a function only of the whole organ or whether the constituent cells show an intrinsic restriction of developmental potential.

4. RESTRICTION OF DEVELOPMENTAL POTENTIAL NEED NOT INVOLVE IRREVERSIBLE LOSSES IN THE GENOME

The clearest evidence that there is no irreversible change in the genome as a result of cell commitment comes from nuclear transplant experiments with *Xenopus,* in which it has been shown that nuclei of committed skin cells from adult frogs can code for the development of frog embryos to the stage of metamorphosis.

The regeneration of whole plants from cell cultures derived from various parts of the carrot plant also clearly indicate that the nuclei of differentiated plant tissues are pluripotent (totipotent). However, it cannot yet be stated categorically that *all* living cells of the plant body are capable of showing totipotency, under appropriate conditions, since there have been many unsuccessful attempts to achieve embryogenesis with cultures from many species. Moreover, the tissues used to initiate cultures always contain relatively undifferentiated parenchymatous cells and potentially meristematic cells. Nevertheless, the successful regeneration of whole plants from somatic cells of a number of plant species is in marked contrast to the situation with respect to animal cells, where it has so far proved impossible to grow a complete individual from a somatic cell, although teratomas from mice embryos show a degree of pluripotentiality. Since the changes which occur during biochemical specialization of cells apparently do not depend on irreversible changes in the genome, it is necessary to look elsewhere for the factors which control the determined state.

Part 3
Cell Interactions
in Development

Chapter 3.1
Introduction

It is usual to classify cell interactions into two groups: there are those short range interactions which are thought to depend on the proximity of cells and there are long range interactions which are mediated by diffusible substances which can pass between distant cells. This chapter is mainly concerned with short range interactions and distant interactions are discussed in Part 4.

Short range interactions are important because they provide the means by which cells can change each other's development in a small space without altering distant cells. Since it is now known that cell allocation and cell determination occur in small cell populations (see Part 2), this type of interaction may be involved in setting up populations of heterogeneous cells in developing systems. The chapter illustrates two approaches to the problems of short range interactions. First, the authors describe detailed attempts to elucidate the molecular mechanisms involved and they describe experiments on model systems which are simpler than developing organisms; these models include cells in culture which do not develop at all (Chapter 3.2), parts of amphibian embryos in culture (Chapter 3.4), and pollen-stigma interactions (Chapter 3.5). In all three cases one can expect that a biochemical explanation of the events observed in these model systems will be discovered within the next ten years but it will not be known whether similar mechanisms operate in the intact developing system. Second, the authors discuss how such interactions might be used within the developing organism. In this they are forced to use morphological evidence about cell communication (Chapter 3.3), and to speculate about pattern formation in whole plants and animals (Chapters 3.5 & 3.6). It will be some time before the two approaches meet and provide a complete explanation of the rôle of cell interactions in development.

It is certain that cell interactions are important in development; cells removed from developing systems are able to extend the range of cell types which they would have formed in the whole organism. Thus isolated blastomeres of sea urchin, frog, and rabbit embryos are able to form adults (Part 1), cells derived from plant callus are able to develop into complete plants (Chapter 2.4), and cells from the pre-stalk region of a slime mould slug may form a new slug (Chapter 4.2). One of the roles of cell interactions is to restrict the development of cells (Chapter 2.1). Cell interactions may also extend the developmental range of cells and this is particularly clearly shown in the sea urchin experiments described in Part 1; thus cells of the animal half of the sea urchin embryo can only form a ciliated blastula by themselves, while in combination with micromeres they can develop into many other of the structures of the pluteus larva. At the very least cell interactions are used by developing systems to direct the path of development of embryonic cells.

It is also known that a short range interaction between a pair of cells may exert an effect several cells away. This transmission of effect may be due to actual cell growth and migration; a pollen grain-stigma interaction leads to the growth of the pollen tube and fertilization (Chapter 3.5), and the effect of the grey crescent region on amphibian development may be transmitted by cell migration (Chapter 3.6). The effect of a cell–cell interaction may also be transmitted by a substance; when a pair of cells engage in metabolic cooperation, then the effect of the cooperation can be found in cells connected to the interacting pair and it is thought that a substance is passing between them (Chapter 3.2). Similarly, the micromeres of the sea urchin exert an effect on the formation of large cilia in cells which are several cell layers away and their action is thought to depend on the transmission of a substance between the cells (Part 1). However, it should be noted that there are several ways in which the effect of an interaction between a pair of cells may be transmitted without the passage of molecules away from the interacting pair.

We do not know the mechanism by which cells alter each other's behaviour in developing systems. It is therefore helpful to consider the kinds of mechanism which they may use before coming to the detailed sections of this chapter which describe cell communication in particular experimental situations.

3.1.1 TRANSFER OF LARGE MACROMOLECULES BETWEEN CELLS

One cell might be able to control the behaviour of another by transmitting macromolecules, such as nucleic acids and proteins. These macromolecules are probably involved in the control of protein synthesis within a cell and it would be economical if they were also used in cell to cell interactions. However, large molecules do not readily move between cells which are closely linked (Chapter 3.2), and special mechanisms of secretion and endocytosis would be required if large molecules were passed between the cells of developing systems. Alternatively there might be cytoplasmic connections but such connections are not found between interacting cells in all developing systems. The conclusion is that macromolecular exchange is unlikely to be the general mechanism by which one cell controls the development of another.

In the developing regions of plants, plasmodesmata are regularly found between cells and macromolecular exchange is therefore possible (Chapter 3.3). However, the cells of animal embryos are rarely in direct cytoplasmic contact. In most embryos, cell membranes appear to completely segregate the egg cytoplasm and the early syncytial development of the insect egg is unusual (Fig. 1.3.1 in Part 1). It may be that transitory cytoplasmic connections do occur between cells in many embryos but such cell links have only been observed in sea urchin development [5], and until they have been found in other embryos it is probably fair to conclude that macromolecular exchange does not regularly occur during embryogenesis.

3.1.2 TRANSFER OF IONS AND SMALL MOLECULES BETWEEN CELLS

Cells might change the ionic balance in neighbouring cells and in this way alter their development; in some cases it is known that the ionic composition of the medium around the chromosomes may effect their apparent gene activity [6], and it is therefore exciting that it has recently been found that ions are able to move between cells of developing animals and plants. Ion flow between cells can be demonstrated by introducing a pulsed current into one cell and noting the electronic potential in neighbouring cells and it is found that there are low resistance connections between plant cells joined by plasmodesmata and animal cells linked by gap junctions (Chapters 3.2 & 3.3). The cells of the early embryos of starfish, squid, fish, amphibians and chickens are connected by low resistance junctions and it is therefore conceivable that ions regularly move between the cells of these embryos and are used to mediate cell interactions [7].

In adult animal cells in the body and in culture, molecules as well as ions can pass through these gap junctions; thus the dyes fluorescein (mol. wt. 332) and procion yellow (mol. wt 500) can move between cells connected by low resistance junctions. At present it does however appear that within the embryo, low resistance junctions do not allow the passage of dye molecules and we can only be certain that ions could use these channels of interaction during development [7].

3.1.3 RELEASE OF SUBSTANCES BY ONE CELL WHICH INTERACT WITH THE SURFACE OF ANOTHER

Cells may release chemicals as their cell surface which change the characteristics of adjacent cells. For instance impulses are transmitted between nerves by one cell releasing acetylcholine which changes the permeability of the neighbouring cell membrane. This impulse transmission is a local effect because the acetylcholine is rapidly broken down by the enzyme acetylcholinesterase; it is conceivable that similar types of interaction occur between the cells of developing systems. The only evidence in favour of such a mode of communication are the observations that serotonin, noradrenaline, and acetylcholine are found in developing fish and sea urchin embryos and that analogues and antagonists of these molecules disturb development [2, 3, 4]. It is probably premature to decide whether such molecules are used to influence the development of cells inside the embryo.

3.1.4 INTERACTIONS BETWEEN THE OUTSIDES OF CELLS WITHOUT MOLECULAR EXCHANGE

It is known that substances applied to the outside of cells can rapidly change their behaviour. Plant lectins for instance, can stimulate lymphocytes to divide and form blast cells by acting solely on the cell surface (Chapter 3.5). It seems that moving cells use such interactions to recognize their neighbours and the interactions between pollen and stigma are a particularly clear example of this. Animal cells can recognize each other (Chapter 3.6) and they are presumably using similar mechanisms to those which have been discovered in pollen and stigma.

It seems likely that interactions of this type are of great importance in developing systems. It will, however, be difficult to prove that this is the case. We know that antibodies directed against the surface components of particular cell types will disturb development, but we cannot be certain that the abnormality is caused by a failure of cell recognition rather than the general deleterious effects which might be expected when a cell surface is covered with antibody molecules. It is however impressive that genes which cause abnormal development of mouse embryos also change the cell surface antigens [1], and if it can be shown that the abnormalities of development are solely due to a change in the surface antigens of the cells, then there would be good reasons for believing that much of development involved interactions of this type.

3.1.5 CONCLUSIONS

This completes the list of mechanisms which might be used for short range interactions between cells in developing systems.

It is now necessary to consider whether such interactions are used to compose and regulate the structure of the developing system. This is a difficult problem which has been mentioned in Part 1 and will be discussed again in subsequent chapters (3.2

& 3.3). Briefly, we have to explain the orderly appearance of tissues and organs in different parts of the developing system, and there are two ways of considering this problem (Chapter 3.6). One is to consider that at certain times in development, cells can detect their position relative to reference points in the embryo; that is the cells behave as if they knew their map reference. The other view is that cells never have this information, and that the order of development can be explained by a series of short range interactions which progressively determine cell fate. The proponents of each view have to explain the regulation of development, e.g. the ability of half a developing system to form the whole, and the regeneration of missing parts. In Chapter 3.6, it is argued that a series of short range interactions could explain orderly development; one has to decide whether the explanation is satisfactory.

3.1.6 REFERENCES

[1] ARTZT K., BENNETT D. & JACOB F. (1974) Primitive teratocarcinoma cells express a differentiation antigen specified by a gene at the T-locus in the mouse. *Proc. nat. Acad. Sci., U.S.,* **71,** 811–14.

[2] BUZNIKOV G.A., CHUDAKOVA I.V. & ZVEZDINA N.D. (1964) The role of neurohumours in early embryogenesis. I. Serotonin content of developing embryos of sea urchin and loach. *J. Embryol. Exp. Morph.,* **12,** 563–74.

[3] BUZNIKOV G.A., CHUDAKOVA I.V., BERDYSHEVA L.V. & VYAZMINA N.M. (1968) The role of neurohumours in early embryogenesis. II. Acetylcholine and catecholamine content in developing embryos of sea urchin and loach. *J. Embryol. Exp. Morph.,* **20,** 119–28.

[4] GUSTAFSON T. & TONEBY M.I. (1971) How genes control morphogenesis. *American Scientist,* **59,** 452–62.

[5] HAGSTROM B.E. & LONNING S. (1969) Time-lapse and electron microscopic studies on sea urchin micromeres. *Protoplasma,* **68,** 271–88.

[6] KROEGER H. (1963) Chemical nature of the system controlling gene activities in insect cells. *Nature, Lond.,* **200,** 1234–35.

[7] WARNER A.E. (1975) Pathways for ionic current flow in the early nervous system. In *Simple nervous systems,* pp. 1–25 (Eds D.R. Newth & P.N.R. Usherwood). Edward Arnold, London.

Chapter 3.2
Junctions as Channels of Direct Communication Between Cells

3.2.1 INTRODUCTION

In any multicellular system where the growth and function of the individual cells are interdependent, as they are during the development of an animal from a single fertilized egg, each cell within the system must be in communication with the other cells to permit the required coordination and control. Clearly, such intercellular communication can be of two basic types: direct communication through contacts or junctions formed between contiguous cells, or indirect communication by signal molecules which pass through the extracellular fluids. Indirect communication in the form of humoral, or hormonal mechanisms of functional integration have been recognized for a long time, but direct communication by intercellular junction formation has only recently been discovered, and accordingly it is less well understood and less well documented. It is already clear, however, that intercellular junction formation is a common feature of many cell types in many, if not all, multicellular species.

There are basic differences between the types of intercellular junctions formed in plants and those formed in animals. In plants, the cells are encased in a rigid cell wall which separates the membranes of adjacent cells, but during cell division, when new cell wall material is being laid down, thin areas are left through which cytoplasmic connections are maintained between the daughter cells. These cytoplasmic bridges, or plasmodesmata, allow intercellular movement of many cytoplasmic components (Chapter 3.3).

Animal cells are quite different. They do not have a cell wall and after cell division the membrane between the two daughter cells is constricted to a narrow cytoplasmic connection which eventually breaks, completely separating the newly formed cells which are then free to move their individual ways. Cells, which in the course of their movement come into contact with other cells may, and in many cases do, form specialized junctions through which only certain cytoplasmic components are free to move.

The nature and functions of these junctional channels of direct communication between animal cells will be discussed in this chapter. In particular the following aspects will be covered: the structure of the junctions formed between animal cells, the mechanisms by which they are formed, the nature of the cytoplasmic components which can pass through them, and the possible roles of this form of direct communication in the control of development.

It must be remembered, however, that the subject is a relatively new one. Differences of opinion exist about some of the experimental findings and about some of the interpretations of the available data. The point of view in this chapter is that of the author, but on topics where disagreement exists in the literature reference will be made, where possible, to other views.

3.2.2 DETECTION OF JUNCTIONS BY ELECTRON MICROSCOPY

The earliest descriptions of the morphology of animal cells were, of course, made from observations with the light microscope. With such low resolution, it was impossible to distinguish the individual cell membranes, but points of close apposition between adjacent cells were noticed. In stratified squamous epithelia, densely staining cytoplasmic regions were seen to be associated with regions of cell–cell contact. These regions were called desmosomes or intercellular bridges and it was thought for some time that they represented points of cytoplasmic continuity between the cells.

However, the cell appeared to regain its status as the individual and isolated unit of animal tissues with advent of the electron microscope. Each cell is seen to be surrounded by a continuous three-layered structure, about $7 \cdot 5$ nm in width, called the unit membrane. There is no cytoplasmic continuity between neighbouring cells but junctional and non-junctional regions can be readily identified. In the typical non-junctional regions, the unit membranes are separated by an interspace of 20–30 nm. The junctional regions can be identified either by corresponding cytoplasmic differentiation (plaques) along the inner surfaces of the membranes of adjacent cells, or, by the narrowing of the interspace as the unit membranes come closer together.

The most common types of junction found between animal cells are the desmosome, the tight junction and the gap junction [1]. The septate desmosome appears to be restricted to invertebrates. Much confusion has arisen in the past as a result of the lack of a standard terminology for these junctions and it is unlikely that a definitive classification will be made until more is known about the structures and functions of the various types.

The following is a list of equivalent terms: desmosome or macula adherens; tight junction or zonula occludens; gap junction or nexus (plural nexuses); septate desmosome or septate junction.

The desmosome is a region where the two apposed membranes lie parallel and are separated by a 25–30 nm interspace. The adjacent cytoplasms of both cells have dense fibrillar plaques lying along the inner surfaces of the membranes. These plaques are the sites of attachment of bundles of cytoplasmic filaments, called tonofilaments, which may link the plaques of several desmosomes within a cell but which do not cross between the cells. It is generally agreed that desmosomes are involved in a form of cell–cell adhesion and it has been suggested that, together with the tonofilaments, they form a cytoskeletal system which limits the extent to which a tissue may be deformed by mechanical stress.

The tight junction occurs as a long thin region of union between the membranes of adjacent cells. At the points of union, the total thickness of the junction (14 nm) is somewhat less than the combined thickness of two non-junctional membranes (i.e. 2×7.5 nm). Each junction may be 50–400 nm in length and in groups they act as intercellular seals, preventing the unrestricted movement of extracellular components between the cells. An obvious example of the role of these junctions is the prevention of by-pass diffusion between the cells of the lumenal epithelia of cavity organs.

The gap junction is also an area of close contact between the membranes of adjacent cells. The gap junction is distinguished from the tight junction by an extensive zone rather than a long thin line of contact and by a gap (or more precisely, an electron-luscent zone) of 2–3 nm between the two unit membranes. This produces the typical seven-layered structure observed with the electron microscope in thin sections. This structure is shown in Fig. 3.2.1a.

Until quite recently, junctions had only been examined in stained thin sections such as that shown in Fig. 3.2.1a, which provided useful but very limited information about their structure, however the new and powerful freeze-fracture techniques which have been applied to the examination of cell membranes, have added a great deal more to the picture.

In the freeze-fracture method, the cells joined by a junction are frozen and then broken apart. It seems that in such frozen preparations, the cells tear apart along the lines of least resistance which are the central lamellae of the unit membranes. This of course gives rise to two types of fracture surface, fracture face A, attached to the cytoplasm but facing away from it, and fracture face B, attached to the frozen interspace fluid and facing the cell from which it has been separated. The origin of these faces is shown schematically in Fig. 3.2.2.

When such fractures pass through a gap junction, the central layers of the junction can separate with either cell (see Fig. 3.2.2). Surface replicas can be prepared by heavy metal

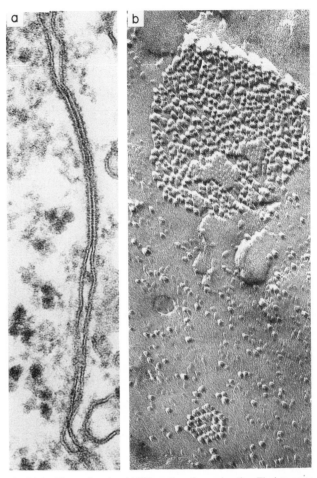

Fig. 3.2.1 The gap junction. (a) Thin section of a gap junction. The interspace between the two-unit membranes narrows to a barely perceptible gap. (b) Replica of a freeze fractured membrane showing gap junctions as aggregates of particles. Note the hexagonal packing of the particles, particularly in the smaller junctional aggregate. Magnification in both photographs, 162,000. Figure taken from Gilula *et al.* [2].

shadowing of the frozen faces. Examination of such replicas with the electron microscope shows the gap junction as an array of particles (on face A or an array of pits on face B). A replica of a fracture face A is shown in Fig. 3.2.1b. The particles are 7–8 nm across and they are packed in a characteristic hexagonal arrangement with a spacing of 9–10 nm between the particle centres [3, 4]. In some freeze fracture replica preparations a small depression, 2–2.5 nm in diameter, can be detected in the centre of each particle. These dimensions compare well with those obtained from negatively stained preparations of membrane fragments containing gap junctions,

(a) Gap junction

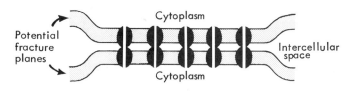

(b) Freeze fractured gap junction

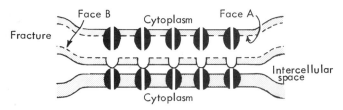

Fig. 3.2.2 Diagrammatic representation of a gap junction. The upper figure (a) is an interpretation of the structural and functional data available showing the interaction of opposing junctional particles or subunits from the two membranes to form hydrophilic channels connecting the cytoplasms of the two cells. The lower figure (b) shows a freeze fractured junction. The membrane of either cell can be split by the freeze fracture technique into two lamellae and when a fracture passes through the region of a gap junction, the junctional subunits separate with the lamella which backs onto the cytoplasm (face A). The particles in face A are matched by corresponding pits in face B. Figure 3.2.1b is a face A replica.

or of isolated gap junctions. Negative staining also shows the presence of a central dot in each junctional particle or subunit.

Gap junctions appear to be important sites of intercellular communication and structural analysis by electron microscopy and X-ray diffraction of isolated junctions suggests a simple explanation for their unusual permeability properties. Aqueous channels (about 1 nm in diameter) penetrate the junctional subunits joining the cytoplasms of coupled cells (Fig. 3.2.2). This junctional permeability has been detected by two quite different approaches, by electrophysiological methods and by biochemical and genetic methods. These approaches, which complement each other rather well, will now be described.

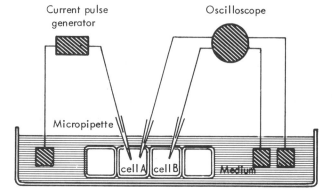

3.2.3 DETECTION OF JUNCTIONS BY INTERCELLULAR ION TRANSFER

Low resistance junctions between non-excitable animal cells were discovered in 1964 independently by Kuffler & Potter [5] who were examining electrical coupling between glia cells in the leach, and by Loewenstein & Kanno [6] who discovered the junctions accidentally while examining the permeability of the nuclear membrane in cells of the salivary glands of *Drosophila* larvae. They noticed that the resistance to electrical current flow between the cytoplasms of two cells in contact was very much lower than that between the cytoplasms and the external medium.

The electrical apparatus which is used for the detection of low resistance junctions is illustrated in Fig. 3.2.3. Electrodes are inserted by micromanipulation into the cytoplasms of adjacent cells and, because of the fluid nature of the membranes, the punctures seal themselves around the glass tips of the

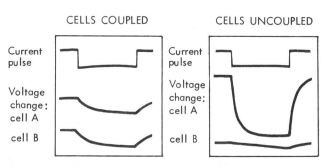

Fig. 3.2.3 The detection of low-resistance junctions. Micropipettes are inserted by micromanipulation, into cells A and B as shown. A current pulse is introduced into cell A and the consequent voltage changes which occur between the cytoplasm of each cell and the surrounding medium are recorded on an oscilloscope. When the cells are coupled by a low resistance junction, similar voltage changes are recorded from the electrodes in both cells. When the cells are not coupled, the voltage change in cell A is large, while that in cell B is barely detectable. Redrawn from Loewenstein [7].

electrodes. Current pulses ($2-4 \times 10^{-8}$ amps; 100–200 msec duration) are passed into cell A (see Fig. 3.2.3) and the consequent voltage changes which occur in cell A and in cell B are displayed on an oscilloscope. If cells A and B are coupled by low resistance junctions, similar voltage changes will occur in both cells. If they are not coupled, the voltage changes occur only in cell A.

Low resistance junctions are formed between many cell types from a wide variety of animal species. Usually measurements are made between large cells but with skill, small cells, such as fibroblasts growing in tissue culture, can be examined [8].

The technique can give quantitatively precise information about junctional resistance between individual cells and demonstrates that, while the movement of ions across the cytoplasmic membrane between the cell interior and the extra-cellular fluid is restricted, intercellular junctions provide low resistance pathways for the transfer of ions between adjacent cells. However, there are clearly practical limitations to the number of cells which can be surveyed in any tissue or organism.

A modification of the approach allows the electro-physiologist to electrically inject charged molecules or ions, directly into individual cells. Fluorescent tracer molecules can be injected through an electrode into one cell and their movement into neighbouring cells followed by fluorescence microscopy. The most commonly used tracer for this kind of work is fluorescein (mol. wt 330). The injected fluorescein appears to diffuse from cell to cell through junctions at points of intercellular contact, but the dye does not diffuse out across the cellular membranes or via the junctional structures into the surrounding medium. Similarly, fluorescein added to the medium does not diffuse into the cells. These observations strengthen the idea that junctions are formed which contain intercellular passages which directly join the cytoplasms of adjacent cells but maintain the insulation of the cytoplasms from the external environment.

A range of different ions and molecules have been injected into cells in an attempt to characterize which cytoplasmic components may pass through intercellular junctions and which may not (see [9]). Some of the evidence is still the subject of controversy but it seems clear that small ions (e.g. Na^+, K^+, Cl^- etc.) and several dyes including fluorescein, neutral red and procion yellow can move freely through the junctions. There are reports that enzymes such as horse radish peroxidase (mol. wt 40,000) and microperoxidase (mol. wt 1800), when injected into a cell can pass into surrounding cells through low-resistance junctions. However, the fixation procedures which are used prior to localization of the enzymes by histochemical methods, may produce artefacts and these reports should be treated with caution (see [9]), particularly in light of evidence from biochemical methods which show that neither specific enzymes, nor cellular proteins in general, are transferred between cells.

3.2.4 DETECTION OF JUNCTIONS BY INTERCELLULAR NUCLEOTIDE TRANSFER

The ability of many types of cells to form intercellular junctions which, unlike the cytoplasmic membrane, are freely permeable to a range of nucleotides, was an accidental discovery made by Subak-Sharpe, Burk & Pitts in 1966 [10].

The techniques used originally to demonstrate nucleotide transfer between cells required the isolation and propagation in tissue culture of mutant cell lines. A mutant hamster cell line (BHK–HGPRT$^-$) which lacks the enzyme HGPRT, hypo-xanthine: guanine phosphoribosyl transferase (E.C. 2.4.2.8) is unable to incorporate ^{3}H-hypoxanthine into cellular RNA and DNA. Such a mutant cell can therefore be distinguished from the wild-type cell after growth in medium containing ^{3}H-hypo-xanthine, by autoradiography of fixed and acid washed preparations. There are no autoradiographic silver grains over mutant cells while wild-type cells are covered with grains. However, if mutant and wild-type hamster cells are grown in mixed culture, grains are found over mutant cells in contact with wild-type cells. This phenotypic change of a mutant cell requires contact with a wild-type cell but can be passed, to a decreasing extent, through chains of mutant cells in contact. Such a gradient of incorporation, extending from small clones of wild-type cells, is shown in Fig. 3.2.4.

It can be shown [11], by counting the grains over individual cells in a 1:1 confluent mixed culture of BHK–HGPRT$^-$ cells and wild-type BHK cells, that all the cells incorporate ^{3}H-hypoxanthine to a similar extent and mutant cells cannot be distinguished from wild-type cells (Fig. 3.2.5). It is clear from the complete absence of non-incorporating cells in such a mixture that all the mutant cells interact with wild-type cells.

This phenotypic modification of mutant cells in contact with wild-type cells could be explained by the cell–cell transfer of any one of a number of cell components. If the enzyme, HGPRT (or the mRNA or the gene for the enzyme) were transferred the mutant cells could gain the ability to incorporate hypoxanthine. If nucleotides, beyond the enzyme block (see Fig. 3.2.6 for relevant metabolic pathways) were transferred then they could be incorporated into the mutant nucleic acid. If labelled RNA or DNA were transferred to the mutant cells, their presence *per se* could account for the observations, or they could be degraded by nucleases and re-incorporated into mutant cell nucleic acid.

Transfer of the enzyme (or its synthesis in the mutant cell as a result of informational nucleic acid transfer) can be ruled out by co-culturing mutant and wild-type cells, separating them, and then examining the mutant cells for their ability to incorporate ^{3}H-hypoxanthine. This has been done using the experimental procedure illustrated in Fig. 3.2.7 and the data (Fig. 3.2.8) show that no enzyme activity can be detected in the separated mutant cells. Reconstruction experiments showed that if the enzyme

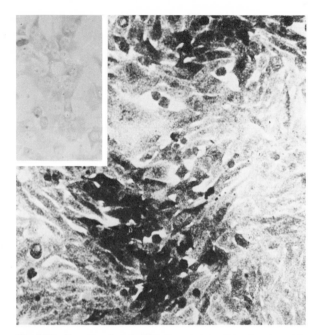

Fig. 3.2.4 Transfer of nucleotides through intercellular junctions. Mutant cultured cells, lacking the enzyme HGPRT, are unable to incorporate ³H-hypoxanthine, and autoradiographs of acid washed preparations show an absence of silver grains over the cells (see insert). Wild-type cells incorporate the labelled base into nucleic acid and appear black in autoradiographs. In the mixed culture shown, labelled nucleotides, synthesized in the few wild-type cells present, have been transferred to the mutant cells through intercellular junctions. It is not possible to distinguish the wild-type cells from the immediately adjacent mutant cells. Even the mutant cells most distant from the wild-type cells show incorporation which is significantly above background. ×96.

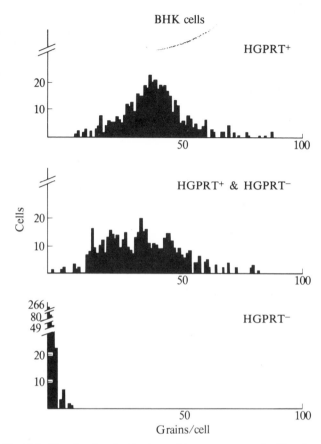

Fig. 3.2.5 Transfer of nucleotides from wild-type to mutant cells. There is another way of showing the phenotypic modification of mutant cells by wild-type cells growing in contact (cf. Fig. 3.2.4). The upper nomogram shows the distribution of autoradiographic grains over wild-type cells cultured in the presence of ³H-hypoxanthine (then fixed and acid washed); all the cells have incorporated the labelled base into nucleic acid. The lower nomogram shows the distribution of grains over mutant cells lacking the enzyme HGPRT; all the cells show little or no incorporation. However, the middle nomogram which shows the distribution of grains over the cells of a 1:1 mixture of wild-type and mutant cells, shows that in the mixed culture both the mutant and wild-type cells incorporate the label. The mutant phenotype has been completely suppressed and it is not possible to distinguish the two cell types.

activity had been present in the mutant cells at the wild-type level for only 5 min, a phenotypic change would have been seen in the separated mutant cells.

This conclusion that enzyme is not transferred between cells was reached independently by Cox *et al.* [12] and by Pitts [11].

Nucleic acid transfer between cells (both DNA and whole chromosomes [13], and RNA and whole ribosomes [14]) has been reported in the literature, but more recent evidence by Pitts & Simms [15] rules this out.

They grew wild-type hamster cells (mutant cells are not required for this method of following intercellular nucleotide transfer) in the presence of ³H-uridine. Such cells will contain labelled uridine, labelled uridine nucleotides and labelled RNA. Uridine, but not nucleotides or RNA can move freely across the cytoplasmic membrane, so the labelled nucleoside (and only the labelled nucleoside) is removed by washing the cells with unlabelled medium. It is a feature of animal cells, labelled and washed in this way, that the uridine nucleotide pool remains

labelled for a long time (reduces to 5% of original activity in 24 h) after washing out the ³H-uridine. This is because a large proportion of the cellular RNA (the high molecular weight nuclear RNA) is unstable and turns over with a short half-life. Eventually, all the label becomes incorporated into the more stable RNA species of the cell, rRNA and tRNA.

When these washed labelled cells are co-cultured with unlabelled cells, labelled components are transferred between

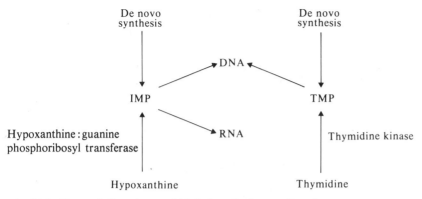

Fig. 3.2.6 The metabolic pathways which lead to the incorporation of hypoxanthine and thymidine into cellular nucleic acid. Mutant cells which lack the enzyme hypoxanthine:guanine phosphoribosyltransferase (HGPRT⁻ cells) are unable to incorporate ³H-hypoxanthine into cellular nucleic acid.

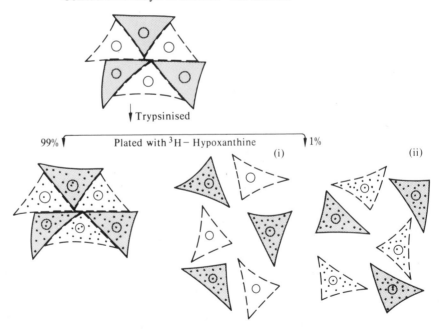

Fig. 3.2.7 The enzyme, HGPRT is not transferred between cells in contact. This diagram illustrates an experiment to test for enzyme transfer; the data from such an experiment are given in Fig. 3.2.8.

Mutant cells, lacking HGPRT are co-cultured under confluent conditions, with wild-type cells. After allowing time for intercellular transfer to take place, the cells are treated with trypsin to form a single cell suspension. Ninety-nine per cent of the cells are dropped into a culture dish containing ³H-hypoxanthine and 1% into another similar dish. The 99% culture forms a confluent layer, while the 1% forms predominantly a culture of single cells. All the cells in the confluent culture should be labelled (cf. Fig. 3.2.5). If enzyme was transferred during the first co-culture, then all the single cells should be labelled too (ii). If enzyme is not transferred, then half the single cells should be labelled (wild-type cells) and half should be unlabelled (mutant cells) (i).

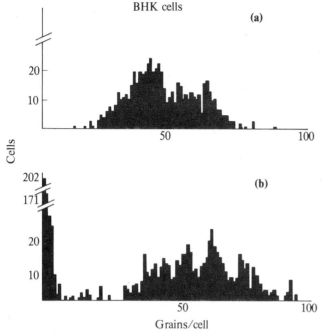

Fig. 3.2.8 For explanation of experimental procedure, see Fig. 3.2.7. In the confluent culture, all the cells are labelled, as expected. In the dilute culture, only half of the cells are labelled. As HGPRT is a stable enzyme, this shows that it is not transferred from cell to cell through intercellular junctions.

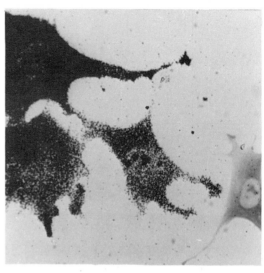

Fig. 3.2.9 Transfer of uridine nucleotides between cells in contact. This autoradiograph of cells grown in tissue culture, shows a donor cell, prelabelled with ^{3}H-uridine (black cell at the top). Labelled uridine nucleotides have passed through intercellular junctions to the originally unlabelled recipient cells and have been incorporated into recipient cell RNA (RNA is synthesized in the nucleoli which are heavily labelled). Note the unlabelled cell, not in contact with the other cells, at the bottom of the photograph. ×450.

cells in contact (Fig. 3.2.9). Contact is required and a gradient of label is formed through a chain of recipient cells in contact with a pre-labelled donor cell (Fig. 3.2.10). The nucleoli of recipient cells are heavily labelled (the nucleoli are the sites of RNA synthesis from nucleotides) and transfer of labelled components through long thin cytoplasmic connections (Fig. 3.2.11 shows a sharp distinction between the extents of labelling in the two cells at the point of contact, rather than a continuous gradient from one to the next).

These features of the system suggest that nucleotides and not RNA are transferred from cell to cell.

Nucleotide transfer can be distinguished from RNA transfer, however, by two simple experiments. When the pre-labelled donor cells are cultured, after washing, in unlabelled medium (chased), the level of activity in the nucleotide pool gradually falls while the level of activity of the cellular RNA gradually increases as the labelled nucleotides are progressively converted to stable RNA species. After 24 h, such chased cells contain about 5% of the nucleotide label present in unchased cells. When chased cells are co-cultured with unlabelled cells very

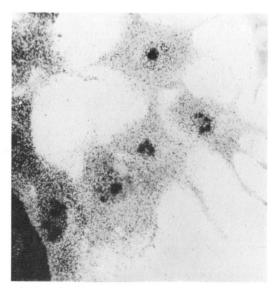

Fig. 3.2.10 Transfer of uridine nucleotides between cells in contact. This autoradiograph shows a gradient of incorporation running from the donor cell (prelabelled with ^{3}H-uridine) through a chain of recipient cells. ×450. Recipient cells which are not in contact (top right) remain unlabelled.

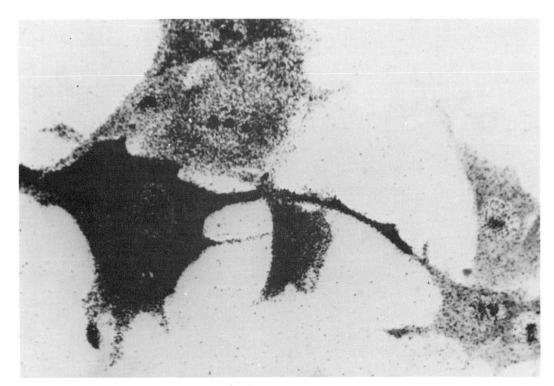

Fig. 3.2.11 Transfer of uridine nucleotides between cells in contact. This autoradiograph shows that transfer can take place through junctions formed at the ends of long cytoplasmic connections. Note the abrupt change of RNA specific activity at the point where the processes from the two cells join. ×560.

little label is transferred (Fig. 3.2.12a). The residual transfer (which shows that intercellular junctions are still formed) has been quantitated and shown to be about 5% of that transferred from unchased cells. This data correlates well with the reduction of nucleotide specific activity in the donor cells but is quite unrelated to the increased specific activity of the donor cell RNA.

This confirms the suggestion that nucleotides and not RNA are transferred between cells in contact. And further evidence can be obtained by using actinomycin D during the co-culture of washed labelled donor cells with unlabelled recipient cells. Actinomycin D blocks RNA synthesis and greatly reduces the amount of label found in the recipient cells (Fig. 3.2.12b). Again the reduction in transfer has been quantitated and shown to be exactly equivalent to the extent of inhibition of RNA synthesis.

The cytoplasmic membrane is impermeable to nucleotides, so, as for ion transfer, it is necessary to propose specific intercellular junctions for nucleotide transfer.

3.2.5 CORRELATION OF GAP JUNCTIONS, ION TRANSFER AND NUCLEOTIDE TRANSFER

The present evidence suggests that ions, dyes and nucleotides all pass through the same junctional structures and that these structures are gap junctions. Several types of experiment support this view, Gilula, Steinbach & Reeves [2] examined cells in tissue culture which could electrically couple and transfer nucleotides, and other cells which failed to couple or transfer nucleotides. The only junctional type that they could detect by electron microscopy, which was present between the interacting cells but could not be found between the cells which failed to interact, was the gap junction.

Brown fat cells in mice are electrically coupled via cellular processes and the only junctional structure which has been detected in these limited regions of contact, is the gap junction [16].

Earlier work suggested that the septate desmosomes found

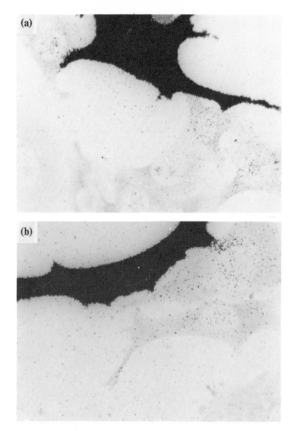

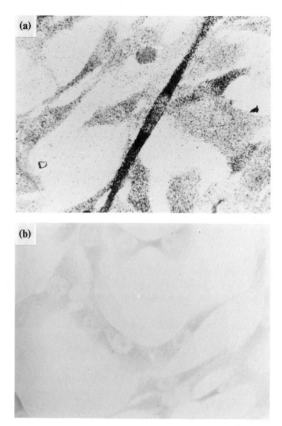

Fig. 3.2.12 RNA is not transferred between cells in contact. (a) If donor cells, prelabelled with ^{3}H-uridine (such as those shown in Figs 3.2.9, 3.2.10 & 3.2.11) are cultured for 24 h in the absence of labelled uridine (chased for 24 h), the specific activity of the cellular RNA changes very little, but the specific activity of the uridine nucleotides falls to a very low level. When such chased cells are used as donor cells, the residual low level of transfer corresponds exactly to the reduced activity of the nucleotide pool and is quite unrelated to the unchanged activity of the donor cell RNA. (b) If unchased donor cells (exactly the same as those in Figs 3.2.9, 3.2.10 & 3.2.11) are co-cultured with unlabelled recipient cells in the presence of actinomycin D, which inhibits RNA synthesis, then the low level of label detected in the recipient cells is directly proportional to the extent of inhibition of RNA synthesis. These observations are explained if nucleotides, but not RNA, are transferred between cells in contact ×500.

Fig. 3.2.13 Transfer of nucleotides from a human skin fibroblast to hamster cells in tissue culture. (a) Autoradiograph of a wild-type human skin fibroblast (the darker, long spindle-shaped cell) cultured in the presence of of ^{3}H-hypoxanthine with mutant hamster cells lacking HGPRT. The mutant cells are all labelled. (b) Autoradiograph of the mutant cells cultured alone under the same conditions. The cells are unlabelled. The mutant cells are unable to incorporate ^{3}H-hypoxanthine, but can incorporate labelled nucleotides obtained via cell junctions, from the wild-type cell. ×210.

between certain invertebrate epithelial cells were the sites of ion transfer. However, gap junctions have now been found between these cells too [17].

3.2.6 LACK OF SPECIES SPECIFICITY IN JUNCTION FORMATION

Junctions are formed readily between chick cells and human cells in tissue culture. No species barriers to junction formation

have so far been detected. Figure 3.2.13a shows a wild-type human skin fibroblast grown with BHK−HGPRT$^-$ cells in the presence of ^{3}H-hypoxanthine. Nucleotides synthesized in the human cell have been distributed to the network of hamster cells. Figure 3.2.13b shows the absence of incorporation when the BHK−HGPRT$^-$ cells are cultured alone in the presence of the labelled base.

It must be pointed out, however, that most of the cross-species investigations have been done in tissue culture, usually with fibroblasts or fibroblast-like cells [18]. Specificity of interaction, both within a species and between species, may exist for other specific differentiated cells.

3.2.7 CELLS WHICH DO NOT FORM JUNCTIONS

Often cells which are known not to be electrically coupled, such as vertebrate muscle fibre cells, are separated by large interstitial spaces.

The inability of cells in tissue culture to form junctions appears to be uncommon, but a few non-interacting cells have been identified of which the most thoroughly characterized is the mouse L cell and its derivatives [11]. L−HGPRT⁻ cells (lacking hypoxanthine:guanine phosphoribosyl transferase) do not incorporate ³H-hypoxanthine either when cultured alone or when co-cultured with wild-type L cells (Fig. 3.2.14). That is there is no phenotypic modification of the mutant cells growing in contact with the wild-type cells. L cells pre-labelled with ³H-uridine, do not transfer labelled uridine nucleotides to unlabelled recipient L cells. Low resistance junctions cannot be detected between L cells and gap junctions have not been found in either thin section or freeze fractured preparations [2].

The inability of L cells to interact and transfer nucleotides is dominant in mixed cultures of L cells with BHK cells or with other cells which interact well. It seems that both cells of an interacting pair must have the ability to form junctions for nucleotide transfer to occur [18]. Two genetically different cells can be fused with inactivated Sendai virus to form a cell containing two different nuclei. After cell division hybrid cells are formed, each with a single nucleus containing both sets of chromosomes [19]. If hybrid L/BHK cells are formed by fusing L cells and BHK cells, they are found to form junctions [20]. L cells must lack some membrane component which is necessary for junction formation. The component is synthesized in the hybrid cells from the BHK genes.

3.2.8 GENETIC ANALYSIS OF JUNCTION FORMATION

The inability of L cells to form junctions with other cell types which can form junctions, and the recessive nature of this inability in hybrid cells, makes a genetic analysis of junction formation possible.

Hybrid cells formed between human fibroblasts (which interact well) and mouse L cells, lose human chromosomes preferentially. This makes it possible to correlate the loss of an activity with the loss of a human chromosome (or part chromosome) and has led to the mapping of a large number of human genes [21].

It has now been shown that loss of human genes from such a hybrid leads to cells with a depleted chromosome number which cannot form junctions. Work is currently in progress to establish which chromosome (or chromosomes) carry genes required for successful junction formation [22].

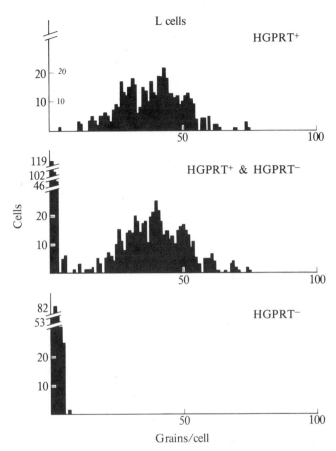

Fig. 3.2.14 Mouse L cells do not form intercellular junctions. These nomograms are directly comparable to those in Fig. 3.2.5. Note the two distinct populations present in the mixed culture, one (about half the cells) which shows wild-type incorporation and one which, like the mutant cells cultured alone, has not incorporated the ³H-hypoxanthine. There is no phenotypic change of the mutant cells in the mixed population.

3.2.9 CHEMICAL CHARACTERIZATION OF GAP JUNCTIONS

The cytoplasmic membranes of animal cells can now be isolated by a number of preparative techniques. Treatment of such membranes with mild detergent (sarkosyl) solubilizes most of the membrane but not the gap junctions. The junctions can then be purified by buoyant density centrifugation from the remaining membrane debris [23, 24].

Examination of purified gap junctions by electron microscopy shows that they are very similar to the structures found in tissues.

They contain 40−50% protein, 5−8% cholesterol and 40% phospholipid. There are only two components in the protein fraction (molecular weights 15,000 and 25,000). This suggests that all gap junctions have the same simple structure although it is possible that the two components represent two classes of closely related proteins which were not resolved by the SDS acrylamide gel technique.

3.2.10 NATURE OF THE CELLULAR COMPONENTS WHICH CAN PASS THROUGH JUNCTIONS

In tissue culture it has been shown that nucleotides, but not RNA or certain enzymes, can pass through intercellular junctions. Other work, using cells with pre-labelled DNA or protein [15], has shown that these macromolecules are not transferred between cells either (see Figs 3.2.15a & b).

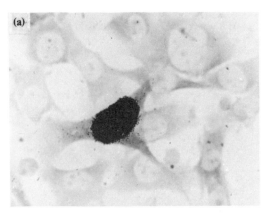

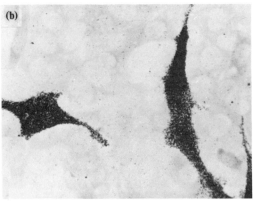

Fig. 3.2.15 DNA and protein are not transferred through intercellular junctions. Autoradiographs showing (a) DNA is not transferred from a donor cell pre-labelled with ³H-thymidine (and chased to deplete thymidine nucleotides), and (b) protein is not transferred from a donor cell prelabelled with ³H-leucine (and washed to remove labelled amino acid). These cells form junctions which are permeable to nucleotides. ×500.

It has been shown recently [22] that other intermediate metabolites including sugar phosphates and choline phosphate and the vitamin derived cofactor tetrahydrofolate are also freely transferred between cells which form junctions.

But how does this work in tissue culture relate to junctions formed between cells *in vivo*?

The available evidence shows that the junctions found *in vivo* are identical to those found between cells in culture. Morphologically, they appear to be the same and three observations *in vivo* suggest that they serve similar functions.

(a) Tumours can be formed in hamsters from BHK−HGPRT⁻ cells and such tumour cells interact with the wild-type host cells resulting in the phenotypic modification of the mutant cells which is typical of tissue culture systems [18].

(b) The Lesch-Nyhan syndrome is an X-linked recessive disease characterized by the absence of the enzyme HGPRT in homozygotes. In heterozygotes, because of random X inactivation, some cells lack the enzyme, while other cells have it, and such heterozygotes are phenotypically normal. This would be expected if purine nucleotides move freely from cell to cell *in vivo*.

(c) Chimeric mice. Adult mice can be made which contain cells derived from two embryos. This is done by mixing the embryos together before the eight-cell stage and then transferring the composite blastocyst back into the uterus of a foster mother, where it develops into a single adult. Mintz & Baker [26] made chimeras between two strains of mice which produced electrophoretically distinct isozymes of isocitrate dehydrogenase. A hybrid enzyme was found in skeletal muscle, but not in cardiac muscle, liver kidney, lung or spleen. Skeletal muscle fibres are multinucleate while in the other tissue the cytoplasms of the individual cells are separated, but often by gap junctions. The experiments demonstrate that neither enzyme monomers nor the mRNA (or the gene) for the enzyme is transferred between cells *in vivo* joined by gap junctions.

Low resistance junctions are formed *in vivo* and in tissue culture and their properties in the two situations are indistinguishable.

It seems reasonable therefore to generalize. Gap junctions provide intercellular channels for the transfer of metabolites such as nucleotides (both purine and pyrimidine), sugar phosphates and choline phosphate, tetrahydrofolate, ions (including Na^+, K^+, Cl^-, I^- and $SO_4^=$) and injected dyes (fluorescein, neutral red, procion yellow). There is no instance yet of a small molecular weight compound which has been shown not to move freely from cell to cell in the experimental systems described. However, these junctions do *not* provide channels for cell−cell transfer of cellular macromolecules such as RNA, DNA and protein.

In those tissues, both *in vivo* and in culture, where all the cells are joined by gap junctions (which is a common situation) the concept of the cell needs reappraisal. In such a situation the cell is no longer the independent and separate unit that it has been thought to be. All the cells in a communicating multicellular population share a common pool of many (possibly all) of the small cellular molecules and ions, but the individuality of each different cell is maintained by its specific macromolecules which it does not share with the surrounding cells.

The number of gap junctions between two cells in a tissue can vary widely. Liver cells are coupled by many large junctions (each with several hundred subunits) while at the other extreme retinal rods and cones are coupled by a few small junctions. In culture cells appear to be joined by a small number of intermediate sized junctions. It seems unlikely that each cellular component (and each dye) is transported through the junctions by a specific transferase system because it would be difficult to fit representatives of each class into small or even intermediate junctions. Furthermore, the fact that junctions contain only proteins of 15,000 and 25,000 molecular weight also suggests that they are not aggregates of different types of transferases. Instead it is easier to think of gap junctions as a group of holes (aqueous channels) without much more than a size specificity.

3.2.11 THE RATE AND EXTENT OF INTERCELLULAR NUCLEOTIDE TRANSFER

It is possible to establish an artificial situation in tissue culture where two different cell mutants depend on mutual nucleotide exchange for growth. Wild-type cells cannot grow in the presence of aminopterin unless the medium is supplemented with hypoxanthine and thymidine (HAT medium; the metabolic block caused by aminopterin and the pathways by which cells can grow in HAT medium are shown in Fig. 3.2.6). Mutant cells lacking either HGPRT or thymidine kinase, cannot grow in HAT medium alone, but in mixed culture (1:1) under conditions where junctions are formed between all the cells, they can grow in HAT medium at the wild-type rate.

Under these conditions, all the purine nucleotides for the HGPRT$^-$ cell's RNA and DNA are supplied by the other mutant. As the growth rate and the nucleic acid content of the cells are known, it is possible to estimate the average flux of purine nucleotides between the cells. It is about 10^6 nucleotides/cell pair/second.

It is also possible to estimate the distance that nucleotides originating from wild-type cells, can diffuse through a mutant cell monolayer. Figure 3.2.5 gives an idea of an experimental situation. A small clone of wild-type cells, growing on a monolayer of mutant cells, causes detectable phenotypic modification of the mutant cells as far as 50 cell diameters from the wild-type cells, 24 h after addition of the label. This is a minimum estimate of the distance and rate of nucleotide movement because there is competition in each cell a nucleotide passes through, for its further transfer or for its incorporation into nucleic acid. A further complication is caused by the continual dilution of the labelled nucleotide pool by *de novo* synthesis of unlabelled nucleotides in each cell. However, it can be seen that nucleotides diffuse quite large distances (in cellular terms) in the course of 24 h.

3.2.12 THE FORMATION AND BREAKDOWN OF JUNCTIONS

Junctions are formed very rapidly, certainly within minutes and probably much faster. They are formed in the absence of protein synthesis and both cells of an interacting pair must be able to form junctions.

A simple model for junction formation is illustrated in Fig. 3.2.16. The model predicts that pre-existing subunits present in the cell membranes, interact and form junctions when cells come into contact. The appearance of the non-junctional membrane prior to junction formation, when examined by the freeze fracture technique, shows the presence of many particulate components which could be gap junction precursor subunits. Such subunits cannot act as channels prior to junction formation (or the membrane permeability would be the same as the junctional permeability) so some structural modification must take place when the junctional components are assembled.

The simplest mechanism of junction formation would involve the direct interaction of individual junctional subunits in the surface of one cell with those in the surface of an adjacent cell brought into contact. The affinity of such subunits for each other could be sufficient to cause junction formation, as the brief maintenance of a point of cell–cell contact would allow other subunits from both cells to diffuse to the site and so form further bridging interactions. The briefest and finest contact between two cells could be sufficient to start such a nucleation event, and the rate of growth of the contact area would be a function of the concentrations of the junctional subunits and the rate at which they can diffuse through the membranes. If the subunits undergo a conformational change, either because of their end–end interactions or because of their lateral interactions as they pack into the typical hexagonal array, which converts a hydrophilic core to a hydrophilic channel permeable to small ions and molecules, then a junction is formed.

It must be stated that the above model is only hypothetical and much more work is required to characterize the structure of the junctional subunits and their arrangement in junctions, before much more can be said about the mechanism of junction formation.

Little is known about the mechanism of junction breakdown. Clearly, cells joined by gap junctions can be physically separated without loss of viability, and it appears that in such situations complete junctions are torn from one cell (where the hole is sealed by the fluid movement of the membrane before

Formation of gap junctions

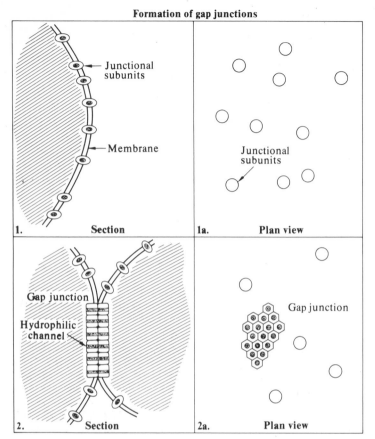

Fig. 3.2.16   A model for gap junction formation. The upper diagrams show a cell membrane in section and in plan view, containing junctional subunits. Each subunit has a hydrophobic exterior and a hydrophilic core (shaded). It is proposed that these subunits have two specific properties. First, they interact, end-on, with similar subunits in an adjacent cell which momentarily comes into contact. The interaction maintains the contact for a brief period giving time for adjacent subunits diffusing through the membrane, to meet and interact. Thus the junction grows by diffusion and interaction of the subunits. A second property of the subunits is that they undergo a conformational change (as a result of end-on interactions, or lateral interactions as they pack in hexagonal arrays) to form a hydrophilic channel leading through the subunits from one cytoplasm to the next.

critical amounts of cell contents leak out) and rapidly removed from the surface membrane of the other cell by ingestion as cytoplasmic vesicles.

The destruction of junctions as a result of artificial separation of the cells is obviously less interesting than the possible *in situ* breakdown of junctions between cells in communication which, for one reason or another, need to isolate themselves. It is possible to imagine a protein molecule ('unzipase') which changes the structure of the junctional subunits by allosteric interaction with their inside surface, thus reducing the affinity between the subunits on adjacent cells.

Such a controlled breakdown could clearly be of value in the organization of a developing system.

3.2.13 JUNCTIONS IN DEVELOPING SYSTEMS

Junctions form an intercellular communication network through which ions and small molecules (including intermediate metabolites and possibly signalling substances such as cAMP) may freely pass.

Crick, in 1970, proposed that a special mechanism which allows the relatively free passage of a developmental signal (morphogen) from one cell to another in a tissue, must be postulated to explain the complex interactions which take place between the cells of a developing system [28]. The junctions described in this chapter offer many of the criteria which must be ascribed to such a special mechanism, if morphogens (whatever they may be) behave like ions and nucleotides. The junctions would allow the ready movement of morphogen from cell to cell but not between the cells and the extracellular fluids. A network of junctions could allow the morphogen to act over distances of perhaps 50 cell diameters or more in a time course of a few hours (see section on distance of nucleotide transfer). Junctions would form rapidly and the pattern of communication would vary if specific junctional breakdown mechanisms operated.

The movement of morphogens directly from cell to cell rather than via the extracellular spaces could clearly produce concentration gradients between adjacent cells which are independent of the geometry of the interacting cells and unaffected by mixing effects caused by convection or other movements of the extracellular fluids (Figs 3.2.17 & 3.2.18).

Is there any evidence that junctions do play a role in development. If so, do they appear to act just as simple holes joining the cytoplasms of all cells, or do they appear to act as elements of developmental organization, by their controlled formation and by their specificity of interaction?

In early amphibian morphogenesis, the outermost embryonic layer, the ectoderm, has a remarkable potential to differentiate into a large number of different tissues. The differentiation of the ectoderm is dependent on the organizer action of the underlying mesoderm. Specific induction phenomena can be observed *in vitro* using mesoderm and ectoderm explants. In one such study, Kelley followed the transfer of labelled components by EM autoradiography from mesoderm, prelabelled with ³H-uridine, to ectoderm [29]. The earliest incorporation in the ectodermal cells was in the nuclei, particularly over the nucleoli. Kelley interpreted the data to mean that either RNA was transferred from mesoderm to ectoderm followed by rapid turnover in the ectodermal cells, or that uridine nucleotides were transferred. In light of the more recent studies outlined in this chapter, it seems most likely that nucleotide exchange was taking place. However, this only shows that morphogens could perhaps move from the mesoderm to the ectoderm during induction through junctions which are permeable to nucleotides. It does not exclude morphogens, or other signals, passing between the two cell types by other unknown routes.

Saxen has shown using explants of mouse embryonic tissue, that an induction process leading to kidney tubule formation can be correlated with gap junction formation between the

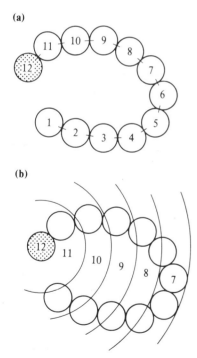

(a)

(b)

Fig. 3.2.17 The movement of a morphogen from cell to cell can produce a clearly defined concentration gradient, whereas a gradient produced by diffusion through the intercellular fluids is more ambiguous and is subject to disturbance by forces which cause fluid movement (e.g. convection, mechanical movement).

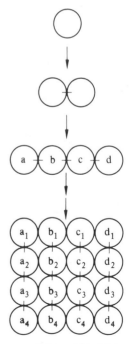

Fig. 3.2.18 Quite complex developmental patterns could be established if the formation and breakdown of intercellular junctions were controlled. For further explanation, see text.

interacting components [30]. Induction and gap junction formation between cellular processes do not take place through filters with a pore size of less than 0·15 μm, but both induction and junction formation can be observed when larger pore sizes are used (Chapter 3.4).

Junctions have been detected by electrophysiological methods in embryos from a wide variety of animal species (see [9]) but as yet there is no evidence of specificity of communication between cells in a developing system (though such specificity has recently been observed in a tissue culture system between fibroblasts and epithelial cells [22]). It may be that junctions provide nothing more than a non-specific communication network between all (or most) cells throughout development and in adult tissues, the specificity of the control processes depending solely on the signalling molecules and the reactions of the recipient cells. Until some observable process of development can be correlated with the formation or the breakdown of specific intercellular junctions, or some morphogenetic signal substance which can pass through intercellular junctions has been isolated and characterized, then the idea that junctions form the basic lines of communication in development must remain no more than an attractive hypothesis.

3.2.14 REFERENCES

[1] McNutt N.S. & Weinstein R.S. (1973) Membrane ultrastructure at mammalian intercellular junctions. *Prog. Biophys. Mol. Biol.*, **26**, 45–101.

[2] Gilula N.B., Reeves O.R. & Steinbach A. (1972) Metabolic coupling, ionic coupling and cell contacts. *Nature, Lond.*, **235**, 262–65.

[3] McNutt N.S. & Weinstein R.S. (1970) The ultrastructure of the nexus. A correlated thin section and freeze cleave study. *J. Cell Biol.*, **47**, 666–88.

[4] Goodenough D.A. & Revel J.P. (1970) A fine structural analysis of intercellular junctions in mouse liver. *J. Cell Biol.*, **45**, 272–90.

[5] Kuffler S.W. & Potter D.D. (1964) Glia in the leech central nervous system: physiological properties and neuron-glia relationships. *J. Neurophysiol.*, **27**, 290–320.

[6] Loewenstein W.R. & Kanno Y. (1964) Studies on an epithelial (gland) cell junction. I. Modification of surface membrane permeability. *J. Cell Biol.*, **22**, 565–86.

[7] Loewenstein W.R. (1970) Intercellular communication. *Scientific American*, (May issue), 79–86.

[8] Furschpan E.J. & Potter D.D. (1968) Low resistance junctions between cells in embryos and in tissue culture. *Curr. Top. Develop. Biol.*, **3**, 95–127.

[9] Bennett M.V.L. (1973) Function of electrotonic junctions in embryonic and adult tissues. *Fed. Proc.*, **32**, 65–75.

[10] Subak-Sharpe J.H., Burk R.R. & Pitts J.D. (1969) Metabolic co-operation between biochemically marked mammalian cells in tissue culture. *J. Cell Sci.*, **4**, 353–67.

[11] Pitts J.D. (1971) Molecular exchange and growth control in tissue culture. In *Ciba Foundation Symposium on growth control in tissue culture*, pp. 89–105 (Eds G.E.W. Wolstenholme & J. Knight). Churchill-Livingstone, London.

[12] Cox R.P., Krauss M.R., Balis M.E. & Dancis J. (1970) Evidence for transfer of enzyme product as the basis of metabolic co-operation between tissue culture fibroblasts of Lesch–Nyhan disease and normal fibroblasts. *Proc. natl. Acad. Sci., U.S.*, **67**, 1573–79.

[13] Bendich A., Vizosa A.D. & Harris R.G. (1967) Intercellular bridges between mammalian cells in culture. *Proc. Natl. Acad. Sci., U.S.*, **57**, 1029–35.

[14] Kolodny G.M. (1971) Evidence for transfer of macromolecular transfer between mammalian cells in culture. *Exp. Cell Res.*, **65**, 313–24.

[15] Pitts J.D. & Sims J. (In press.)

[16] Sheridan J.D. (1971) Electrical coupling between fat cells in newt fat body and mouse brown fat. *J. Cell. Biol.*, **50**, 795–803.

[17] Gilula N.B. & Satir P. (1971) Septate and gap junctions in moluscan gill epithelium. *J. Cell Biol.*, **51**, 869–72.

[18] Pitts J.D. (1972) Direct interactions between animal cells. In *Third Lepetit Colloquium on Cell Interactions*, pp. 277–85 (Ed. L.G. Silvestri). North-Holland, Amsterdam.

[19] Harris H. & Watkins J.F. (1965) Hybrid cells from mouse and man: artificial heterokaryons of mammalian cells from different species. *Nature, Lond.*, **205**, 640–45.

[20] McCargow J. & Pitts J.D. (1971) Interaction properties of cell hybrids formed by fusion of interacting and non-interacting mammalian cells. *Biochem. J.*, **124**, 48P.

[21] Ruddle F.H. (1973) Linkage analysis in man by somatic cell genetics. *Nature, Lond.*, **242**, 165–69.

[22] Simms J., Finbow M.E. & Pitts J.D. (Unpublished results).

[23] Evans W.H. & Gurd J.W. (1972) Preparation and properties of nexuses and lipid enriched vesicles from mouse liver plasma membranes. *Biochem. J.*, **128**, 691–700.

[24] Goodenough D.A. & Stoeckenius W. (1972) The isolation of mouse hepatocyte gap junctions. *J. Cell Biol.*, **54**, 646–56.

[25] Migeon B.R., Kaloustian V.M., Nyhan W.L., Young W.J. & Childs B. (1968) X-linked hypoxanthine-guanine phosphoribosyl transferase: heterozygote has two clonal populations. *Science*, **160**, 425–27.

[26] Mintz B. & Baker W.W. (1967) Normal mammalian muscle differentiation and gene control of isocitrate dehydrogenase synthesis. *Proc. Natl. Acad. Sci., U.S.*, **58**, 592–98.

[27] Goldfarb P.S.G., Slack C., Subak-Sharpe J.H. & Wright E.D. (1974) Metabolic co-operation between cells in tissue culture. (In press.)

[28] Crick F.H.C. (1970) Diffusion in embryogenesis. *Nature, Lond.*, **225**, 420–22.

[29] Kelley R.O. (1968) An electron microscopic study of chordamesoderm-neurectoderm association in gastrulae of a toad, *Xenopus laevis. J. Exp. Zool.*, **172**, 153–80.

[30] Wartiovaara J., Lehtonen E., Nordling S. & Saxen L. (1972) Do membrane filters prevent cell contacts? *Nature, Lond.*, **238**, 407–08.

Chapter 3.3
Junctions between Plant Cells

3.3.1 INTRODUCTION

Although we are beginning to understand some of the inter-actions between the pollen grain and the stigma (Chapter 3.5), we know very little about the interactions between somatic plant cells, and even less about the significance of the connections between them. We are, therefore, forced to adopt the second best approach—to study the anatomy both of the con-nections themselves and the cells they join, and draw what conclusions we can.

Junctions between plant cells, most of which are termed plasmodesmata, have been believed to exist ever since plant cytology became a serious subject in the second half of the 19th century [1]. However, the electron microscope has now revealed that most true cytoplasmic connections usually lie so far below the resolution of the light microscope that nearly all the early work must now be relegated to the division of diffraction-artefacts or fortuitous groupings [2].

The definition of a plasmodesma, the most common of cytoplasmic connections, is discussed later (p. 112). We shall make no distinction between junctions on the basis of size, and we shall include true plasmodesmata, sieve pores, junctions between guard cells, cytomictic channels between angiosperm meiocytes [3, 4] and the cytoplasmic bridges which connect together the cells of certain algae [5, 6].

As we have seen in earlier chapters, animals, being generally soft-bodied, were able to exploit the spaces between cells as a means of transport, upon which has been built the whole vascular system of the higher animal, driven by the heart pump. The plant cell, on the other hand, developing within its own hard 'cardboard box', has been denied the advantages of mechanical pumping systems, but has exploited the cell-to-cell mode of transport.

The plant cell wall is, however, freely permeable both to water and to many soluble materials. Pectins and hemicelluloses, e.g. polymers of galacturonic acid, arabinan, xylan, gold sol particles up to 4 nm in diameter, oil droplets and ions of all kinds can freely pass through an intact cell wall [7, 8, 9].

Passage of soluble substances is restricted where this otherwise permeable wall is impregnated with any waterproof compound, and the casparian strip, a local accumulation on the walls of endodermal cells of what is probably suberin, is shown to prevent the leakage of water or dissolved materials across the inner wall [10, 11] (Fig. 3.3.1).

3.3.2 PLANT/ANIMAL ANALOGIES

In spite of the obvious differences in morphology, there are similarities in the devices used to control cell-to-cell transport in animals and plants and the principal features may be sum-marized in this way:

ANIMAL	PLANT
Desmosomes	*The Plant Cell Wall*
(Intercellular glue connecting to a cytoskeletal framework of microtubules, the whole providing an inter/intracellular framework of mutual support.)	(Complex, biphasic permeable in-tercellular cement, which through the continuity of the adjacent walls, the continuity of those walls with the plasmamembranes and the connections of the plasmamembranes with microtubules create an in-terconnecting inter- and in-traskeletal structure.)
Tight junctions	*Casparian Strips*
(Intercellular seals, preventing the unrestricted movement of substances along cell boundaries.)	(Local impregnations of the cell wall with a waterproof material, probably suberin, preventing the movement of water and dissolved materials between those particular cells.) (Fig. 3.3.1.)
Gap Junctions	*Plasmodesmata and other cytoplasmic continuities*
(Low resistance junctions which may facilitate the passage of materials from cell to cell. Relatively invariable in form.)	(Low resistance junctions, comprising a total cytoplasmic continuity between cells, but sometimes more or less occluded by a fibrous structure, which may facilitate or obstruct the movement of structures and substances from cell to cell. Highly variable in form, dimensions and possible functions.)
Osteocyte Connections	
(In cells of bone (osteocytes). Similarity to plasmodesmata may be brought about by presence of rigid intercellular material (bone) analogous in function to plant cell wall. Cells remain in contact through long processes but cytoplasms do not connect.)	

111

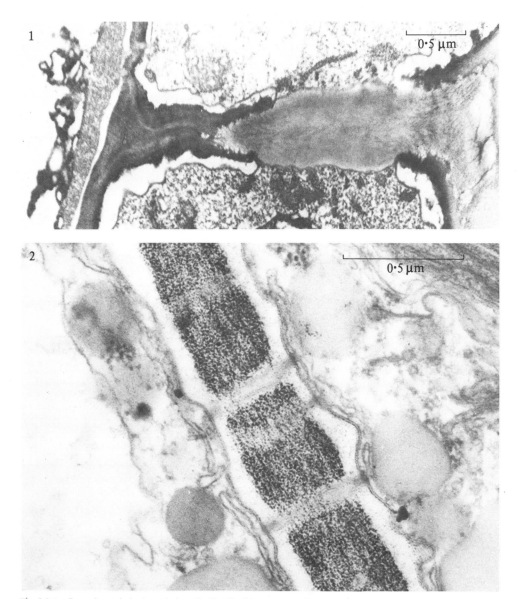

Fig. 3.3.1   Casparian strip in the endodermal cell wall of the tentacular gland of *Drosera capensis*. Electron micrograph kindly supplied by Dr A.J. Gilchrist.

Fig. 3.3.2   Plasmodesmata in the cross walls of the protonema of *Buxbaumia*. Micrograph kindly supplied by Dr D.M.J. Mueller [39].

Plasmodesmata, *sensu stricto,* are fine strands of cytoplasm bounded by a plasma membrane (Figs 3.3.2, 3.3.3) and connect only living cells to one another. Some authors [4] distinguish plasmodesmata from plasma canals on the basis of size, but since both are highly variable and both connect living cells, the distinction does not seem a valid one. Every single living cell in a plant may be connected to all the others through such channels, so that one can think of a single infinitely ramifying plasmamembrane constituting the whole living system of a forest tree.

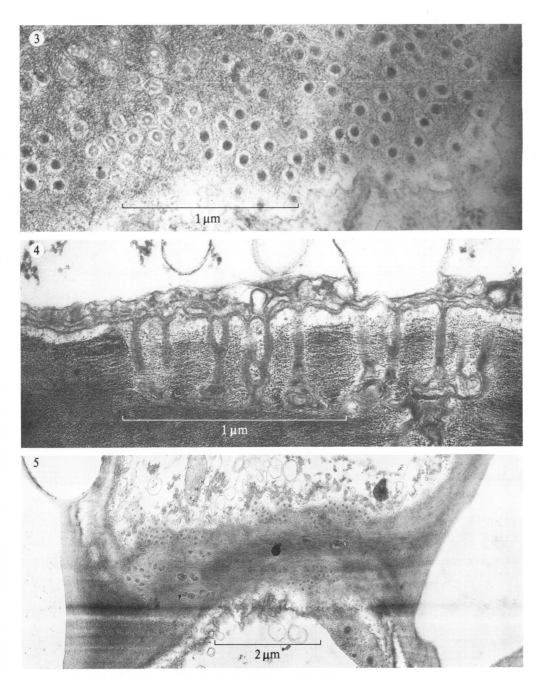

Fig. 3.3.3 T.S. view of plasmodesmata in the bundle sheath of the cotyledon of *Welwitschia mirabilis*. Micrograph kindly supplied by Dr Jean Whatley.

Fig. 3.3.4 L.S. through tissue described in Fig. 3.3.3.

Fig. 3.3.5 Plasmodesmata in the end walls of a young xylem element —otherwise as in Figs 3.3.3 and 3.3.4.

3.3.3 THE ONTOGENY OF PLASMODESMATA

Most plasmodesmata and other cytoplasmic connections are formed on a cell plate; they are, in fact, laid down before the plate has reached the mother cell wall [12, 13, 14]. They are normally simple tubular connections varying in internal diameter from 25 nm up to 80 nm or more. Although at their formation they are usually simple tubes, they can become branched, sometimes on one side as in the companion cell to sieve tube connections [15], sometimes on both sides culminating in complex junctions in the middle termed 'median sinuses' or 'Mittelknoten' [16, 17, 50 and Figs 3.3.4, 3.3.9]. Just how these secondary branches come about is not known, but some speculative ideas will be put forward in Section 3.3.17.

There are also a few well-documented instances where plasmodesmata may be formed between non-sister cells. The best known of these are the plasmodesmata formed between the searching 'hyphae' of parasitic plants, e.g. *Cuscuta* and the cells of the host plant [18, 19, 20]. It is reported that plasmodesmata can be found between the outpushings of ray cell membranes into vessel elements (tyloses) where they meet [21]. Further work on this suggestion at the electron microscope level is necessary.

Rather more difficult to interpret are the partial or possibly complete penetrations from one alien cytoplasm to another seen in the adjacent walls of *Laburnum* and *Cytisus* in the apical tips of the chimaera *Laburnocytisus adami* [22] (Fig. 3.3.6).

3.3.4 THE ESTABLISHMENT OF A PATTERN IN THE DISTRIBUTION OF PLASMODESMATA

An absence of plasmodesmata

If there is a regular and predictable pattern of distribution of plasmodesmata in certain cells then this would imply that a particular pattern of organization has been imposed on the population during development.

Most cell plates at a cell division form plasmodesmata in greater or lesser numbers but there are some instances where few or no plasmodesmata are formed. None are found across the plate between the generative and vegetative cells in the development of the pollen grain in *Dactylorchis fuchsii* and other species [3], nor do they occur at the two cell divisions of the pollen grain meiosis in monocotyledons [31–33]. Plasmodesmata are very few in number or possibly absent across the endosperm walls of *Haemanthus* [23], but it has to be admitted that most of the patterns of cell plate formation, in the examples cited above, have one or more anomalous features. Plasmodesmata are absent from the walls between certain cells, but it is not always known whether this is due to failure to form, or comprehensive occlusion. They are absent between the guard

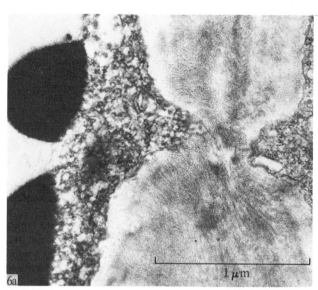

Fig. 3.3.6 **(a)** Part of the interspecific junction in the chimaera *Laburnocytisus*. On the left-hand side of the junction lies *Cytisus purpureus,* and its vacuole contains dark-staining bodies. The cytoplasm of *Laburnum* is on the right and appears to be in connection with that of the *Cytisus*. Micrograph kindly supplied by Dr. J. Burgess [22].

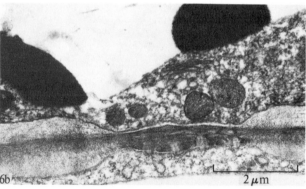

(b) As above. Note the marked concentration of dark-staining fibres at the pit area. From the *Laburnum* side (lower part of the micrograph) this fibrillar layer is partially penetrated by less dense structures.

cells and subsidiary cells of *Opuntia* [24] and of grasses [25], and between the zygote and surrounding cells of *Capsella* [26]. The end walls of the suspensor cells, also of *Capsella,* contain numerous plasmodesmata, but there are no plasmodesmata in the walls separating the suspensor from the embryo sac [26, 27]. There are none between the nucellus and megagametophyte of *Zea mays* [28, 29] and some other higher plants, between the nucellus and embryo sac, and zygote and all other adjacent cells in *Myosurus* [30], between most pollen grains at later stages of their development and between walls separating generative cells from each other and from the surrounding tapetal cells [34]. While these are usually specific examples and it is very dangerous to assume that because nothing is found nothing is present, and there can be little doubt that they are indicative of general categories of plasmodesmatal exclusion at significant times in the differentiation of the cells concerned.

markedly in one direction only, to the establishment of a simple communication pattern (Fig. 3.3.7).

Confirmation of this 'dilution' effect comes from work on the developing cells of the roots of *Hordeum* (barley) [10]. As the endodermal cells extend and mature, the numbers of plasmodesmata through the inner tangential walls fall, but the rate of reduction is consistent with there being neither synthesis nor

Table 3.3.1   Numbers of plasmodesmata in different tissues

Species	Cell Type	Numbers per 100 μm²	References
1. *Dactylorchis-fuchsii* et al.	Cell plate dividing gen. and veg. cell in pollen	Nil	Heslop-Harrison (1968) [3]
2. *Viscum album*	Mature cortical cells	60–240	Krull (1960) [16]
3. *Chara corallina*	Central cell/inter-nodal cell boundary	1,400	Fischer *et al.* (1974) [68]
4. *Zea mays*	T.S. divisions of root cap meristematic cells	1,487	Juniper & Barlow (1969) [12]
5. *Laminaria*	Trumpet cell cross walls	5–6,000	Ziegler & Ruck (1971) [37]
6. *Welwitschia bainesii*	Cotyledon 2·5 cm long	6,000	J. Whatley (pers. comm., 1974)
7. *Osmunda cinna-momea* shoot apex	Anticlinical walls periclinical walls	18,000 (approx.) 8,000 (approx.)	*Hicks & Steeves (1973) [36]
8. *Dryopteris felix-mas*	Root meristem, primary pit fields	14,000	Burgess (1972) [22]

The figure of 100 μm² has been chosen as approximately equal to a small meristematic cell wall.
* Calculated by this author from micrographs published by Hicks & Steeves.

The establishment and distribution of plasmodesmata

The numbers of plasmodesmata formed on a cell plate can vary from none, as we have seen above, up to 140 per μm². A table of some of the measurements made is given (Table 3.3.1), but it should be remembered that the figures are not always strictly comparable since the methods used for calculating them vary.

As the plant matures the numbers of plasmodesmata crossing a given cell wall will usually fall. Most commonly this is due to the fact that no further plasmodesmata will be formed and their numbers will be 'diluted' per unit area, as the primary cell wall expands [12, 38]. The results of a study on the primary root of maize in which the lineage of each cell can be traced, and plasmodesmata are believed never to be formed except on the cell plate, are given below (Table 3.3.2).

Such a situation would lead, as has been shown in the core cells of the root cap of maize, or in all cells which elongate

loss. The frequency of plasmodesmata may fall not only through 'dilution' as demonstrated above, but also by the occlusion of some or all of the plasmodesmata in a given region. One of the best documented examples is the severing of connections between the sporogenous tissue and the archesporial cells in the formation of the pollen grain. All existing plasmodesmata or cytoplasmic connections are sealed off by a massive growth of callose, usually at the end of meiosis II, so that each pollen grain becomes a separate entity [31].

In the sieve tube, at maturity, plasmodesmata of a complex kind exist only between one sieve element and another and between the sieve element and its adjacent companion cell (p. 119). This means that the selective occlusion of many other plasmodesmata must have taken place. Frequently also one sees in older walls of many tissues small bunches of plasmodesmata (Fig. 3.3.8) separated by large areas of unpenetrated thicker wall. This suggests either that extension of the wall has been

Table 3.3.2   Plasmodesmata per 100 μm² in the cap and root tip of maize *Zea mays*.

	T.S. Walls	Longitudinal Walls	
CAP PERIPHERY	513	45	(end of elongation phase)
CENTRAL CAP 200–300 μm from cap junction.	772	75	(period of most rapid elongation of longitudinal walls)
CAP INITIALS	1487	530	(source of plasmodesmata in cap T.S. divisions predominate)
QUIESCENT CENTRE/ CAP JUNCTIONS	576		(the 'oldest' wall in the cap; plasmodesmata crossing this wall have suffered slow dilution since its establishment and are never normally augmented)
QUIESCENT CENTRE (Q.C.)	950	990	(divisions in the quiescent centre, although infrequent, do not show any preferred orientation)
STELE above Q.C. 1 and 2 rows	1415	947	(the root meristem proper; divisions are predominantly transverse but little elongation has yet taken place, hence ratio remains at about 1·5 : 1)
STEELE above Q.C. 3 and 4 rows	1774	889	
STELE above Q.C. 5 and 6 rows	2227	1445	

Direction of elongation (between CAP PERIPHERY and QUIESCENT CENTRE/CAP JUNCTIONS)

Direction of elongation (between QUIESCENT CENTRE and STELE above Q.C. 5 and 6 rows)

Adapted from Juniper & Barlow [12].

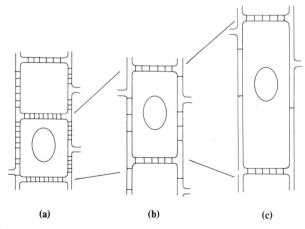

Fig. 3.3.7   Diagram to indicate how plasmodesmata may be 'diluted' by extension.

(a)   (b)   (c)

very uneven, which seems unlikely, or that plasmodesmata have been preserved only at certain points. Occasionally plasmodesmata persist even when a cell has reached the outside of a plant, and remain, presumably plugged, projecting out into space (Fig. 65 in [38]).

3.3.5 STRUCTURAL VARIABILITY

The form of plasmodesmata varies in different cells and can act as a marker of differentiation. Plamodesmata are generally more complex than just a simple, empty, straight-sided canal joining two cells, although such basic structures do exist [39]. Many plasmodesmata, particularly in the young state, are partly filled with a thread-like object first clearly depicted and described in the oat coleoptile [40]. This thread has been variously described as a piece of modified ER trapped at cell division in the new wall, a trapped spindle microtubule or a special structure having some of the characteristics of a microtubule, i.e. having a sub-unit structure generally agreed to be eleven, called a 'desmotubule' [41, 42]. The arguments as to its fine structure and possible origin do not concern us here and

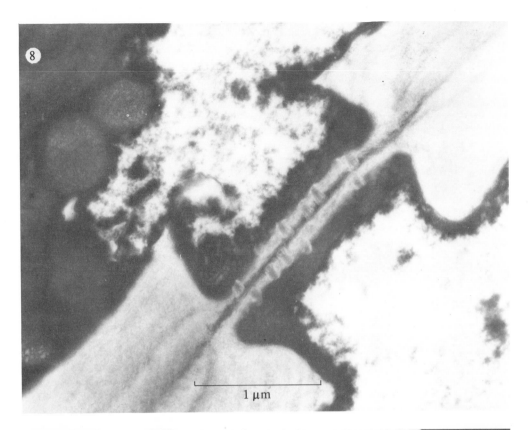

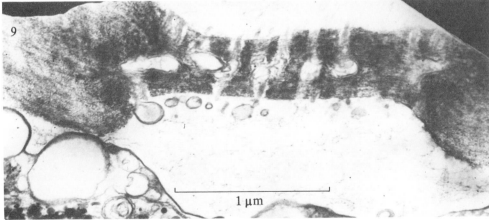

Fig. 3.3.8 A pit field in a *Sarracenia* leaf.

Fig. 3.3.9 A pit field with 'Mittelknoten' in the fruit of *Capsicum annuum*.

are reviewed elsewhere [2, 35]. This thread may make junctions with the ER, but is obviously different from ER itself and may undergo many modifications. Its function is unknown although certain speculations will be considered later. We shall use the word 'desmotubule' in this text without any admission as to its origin.

The more commonly cited deviations from a plasmodesmatal structure with a simple, straight desmotubule are those which have hypertrophied central regions, median nodules or 'Mittelknoten' (Fig. 3.3.9). The median nodule is apparently a secondary development from a 'simple' plasmodesma. This nodule may be important in controlling translocatory fluxes or in completely stopping them; it may also be involved in enzyme-mediated processes initiated within the plasmodesmatal cavity. This idea is supported by the reports of phosphatase activity associated with the plasmodesmata [43, 44] as well as the established hydrolytic activities during sieve plate pore formation [45, 46]. The median nodule may reach a high degree of complexity (Fig. 3.3.9), sometimes coupled with anastomosing desmotubules, sometimes with equal numbers of arms on either side, sometimes assymetrical; sieve tube/companion cell junctions and phloem parenchyma cells provide some of the best examples [47, 48, 49], but there are other locations: *Tamarix* wood fibres [50], secretory cells of *Nepenthes* (Fig. 104 in [38]).

We can speculate that the relatively large, simple, cytoplasmic continuity, which is the normal plasmodesma, is a relic of earlier algal systems; it is excessive in its dimensions and its construction in many instances, and inflexible for the task that it now has to perform in higher plants. In many instances, the plasmodesmatal canal is partially blocked with a desmotubule. It is possible that the desmotubule may serve as some sort of valve system (see below) or an interface along which surface diffusion may proceed and perhaps be controlled. Only in the sieve plate of the phloem, the junctions between grass guard cells, cytoplasmic channels between meiocytes and most algae, in which we can speculate massive and unselective movements occur, is there no obvious constriction. The algae, which lack long distance transport systems such as xylem and phloem, often have simple plasmodesmata, apparently unmodified as in the examples given above, having no central strand or median module (Fig. 3.3.2). As divisions of labour took place in the higher plant, the basic plasmodesma probably diverged and became more flexible; perhaps cells rapidly developed the ability to occlude plasmodesmata over regions where they were no longer required or to enlarge them as in sieve tube companion cell junctions.

3.3.6 DIVERSITY IN A SINGLE CELL

Cells other than phloem

The numbers of plasmodesmata can vary from one wall to another, even in a meristem [12]. In extreme cases there may be no plasmodesmata on some walls and large numbers on others. Sometimes this diversity extends to structure as well. In the phloem the highly modified sieve pores are well described (see below); the sieve tube element/companion cell type is completely different, but presumably has a similar origin and no other walls of the sieve tube possess any plasmodesmata at all, implying that selective occlusion must have taken place. Burgess [35] has also shown that in the root meristem of *Colchicum* the plasmodesmata on the transverse and longitudinal walls differ markedly in structure. Most of these variations are summarized in Fig. 3.3.10. The most extreme of these are, however, in the phloem.

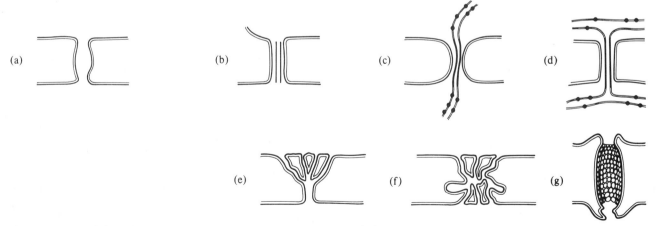

Fig. 3.3.10 Types of plasmodesmata. a: as in *Buxbaumia*. b: as in the lateral walls of *Colchicum* [35]. c: as in the transverse walls of *Colchicum* [35]. d: conventional root cap of higher plant type [12]. e: sieve tube/companion cell type [48]. f: developed 'Mittelknoten' type (Figs 3.3.4 & 3.3.9). g: *Bulbochaete* type [65].

Phloem types

The plasmodesmata of phloem cells represent a separate population of structures, partly because they might be expected to be involved in particularly active fluxes, and partly because in the sieve elements and associated cells the plasmodesmata undergo very extensive differentiation from simple basic types [45–48], in which other organelles or structures may be involved [38].

The general absence of plasmodesmata between parenchyma cells and sieve elements suggests a specific role of companion cells in sieve tube functioning [51, 52, 53, 54], but there are exceptions [55].

A single plasmodesma on the sieve element/companion cell wall develops eight to 15 arms on the companion cell side, but only a single tube, approximately twice the size of a normal plasmodesma, on the sieve element side [15]. The sieve pores finish up as substantial channels, $0.5\,\mu m - 1.5\,\mu m$ across, lined with callose and presumably a very rapid callose-synthesizing system. They are more or less filled with fine protein filaments [56]. This characteristic of a very rapid blocking ability with callose is also shared by the sieve tube/companion cell plasmodesmata, but not apparently by any other plasmodesmata except, possibly, those in *Laminaria* (see below).

3.3.7 CYTOMICTIC CHANNELS

Although not plasmodesmata in the strict sense, cytomictic channels between meiocytes provide cytoplasmic continuity between living cells and are, therefore, relevant. They were discovered in 1908 [57] in *Oenothera* and are clearly seen under the light microscope since they range in dimensions from $0.5\,\mu m$ to $1.5\,\mu m$ [58, 59], i.e. they are of the same order of size as sieve pores. They are probably universal in distribution, but electron microscopical work has so far concentrated on only a limited number of species [3, 31, 32, 33, 60].

These channels are formed in the early leptotene stage of meiosis and whether or not they represent the sites of relic plasmodesmata is unknown. According to Heslop-Harrison [3, 33], all the conventional plasmodesmata linking the host tapetum with the new meiocytes have been eliminated by the time that these inter-meiocyte channels appear. The channels persist throughout the meiotic prophase and, in most species, disappear before meiosis II. After this the young spores become totally isolated from each other within the callose tetrad wall. The channels are relatively large in relation to the dimensions of the walls they transverse and may constitute up to 23.5% of the meiocyte interface surface.

3.3.8 CONNECTIONS BETWEEN GUARD CELLS AND ADJACENT CELLS IN STOMATA

Stomata have an odd and diverse pattern of intercellular connections for which no good explanations have yet been found. In the grasses, the guard cells of each stoma have been found to be inter-connected at each end by a large area [61]. The dimensions are up to $1.5\,\mu m$ across and may allow organelles the size of plastids to pass between them. Plasmodesmata are found between the guard cells and subsidiary cells of a developing stoma of *Avena* [62], but no plasmodesmata at all have been found under the electron microscope between guard and any adjacent cells in a range of other monocotyledons and dicotyledons [63]. They have been seen between guard cells and subsidiary cells of the genera *Nicotiana*, *Datura* and *Phaseolus* with light microscope techniques [64].

3.3.9 ALGAL CELL CONNECTIONS

Diverse forms are found in the algae. In *Bulbochaete* [65] (see Fig. 3.3.10) the inner face of the plasmalemma through the pore is lined with helically arranged particles. In *Chara*, a green alga, some plasmodesmata appear to possess a desmotubule, others appear to lack this structure [66]. In *Volvox aureus* the cytoplasmic bridges more closely resemble cytomictic channels or the connections between stomatal guard cells than normal plasmodesmata of any kind [6].

There are high frequencies of plasmodesmata in the single pit fields of the cross-walls of trumpet cells in *Laminaria* [37]; these plasmodesmata contain some sort of core ('Docht' =wick), although it is not clear what this is and, according to the authors, like sieve tube pores, each plasmodesma is surrounded by callose. The plasmodesmata in the walls between the nodal and internodal cells of *Nitella translucens* [67] vary greatly in morphology, anastomoses and median sinuses; the basic canal diameter, however, is always about 50 nm (internal). Membrane systems are irregularly found within these plasmodesmata. In *Chara corallina* some of the plasmodesmata develop complex 'Mittelknoten' [68] (see Fig. 3.3.10f).

As in fungal cell connections [69, 70] little experimental work seems to have been done to establish their physiological role.

3.3.10 THE CREATION OF MORE COMPLEX PATTERNS IN CELL-TO-CELL COMMUNICATION

Given the diversity in plasmodesmatal construction and distribution already described, further diversity may be achieved, in the service either of physiology or of differentiation, by modifying the pattern of plasmodesmata in a given cell. There are basically six ways to modify cell-to-cell communication, assuming that it is uniform to begin with, to bring about a diverse pattern of distribution.

(1) By determining the plane of division of a cell, in those divisions in which cell plates are known to form plasmodesmata.
(2) By controlling the initial numbers of plasmodesmata on a cell plate.
(3) Where the numbers on the original cell plate formed in any plane are comparable, by bringing about, through differential elongation, a 'dilution' of plasmodesmata on some walls as opposed to others.
(4) By constructing or modifying the plasmodesmata in a different fashion on different walls of the same cell, thereby rendering them permeable to different substances, or at different rates.
(5) By selective occlusion reducing the numbers of plasmodesmata on a given wall, in comparison with other walls.
(6) By operating mechanisms in a given cell whereby plasmodesmata of similar form act as valves in a multi-valved system, some being open and others closed at a given time.

Situations in which assymetric cell divisions are brought about are very well known, some of the best documented being the accessory cells of stomata [71]. All of these walls form plasmodesmata, but their possible functions are unknown.

The numbers of plasmodesmata formed over a given area of cell plate vary enormously (Table 3.3.1). There is, in some individual plants, a slight tendency for most rapidly dividing cells to form more plasmodesmata than slowly dividing cells cv. cap initial cells and quiescent centre cells in maize roots (Table 3.3.2). There is a tendency too for large spindles [72] to produce more plasmodesmata than smaller ones, but there are so many exceptions to both of these observations as to render them of limited value.

The third type of pattern creation shown in Table 3.3.2, is well documented in both maize root tip and endodermal cells, and probably only the difficulty of following complete files of cells in the electron microscope prevents this general principle from being established in other tissues.

The best example of the fourth system, is of course, the sieve tube/companion cell arrangement (see above), but less striking examples are known in, for example, *Colchicum* roots [35]. Burgess has shown that in the transverse walls of the root meristem of *C. speciosum* the dark staining core of the plasmodesma is continuous with the ER, whereas in the longitudinal walls of the same cells, the relationship between the core and the ER is obscure (Fig. 3.3.10).

Selective occlusion is a well established phenomenon and is particularly important in pollen grain development (see below). However, it is not widely appreciated that plasmodesmata of a normal type may persist until the death of the cell (Figs 65 and 123 in [38]).

There is no direct evidence that plasmodesmata act as simple valves, but the presence of statoliths of various types in cells, e.g. amyloplasts or crystalline bodies, or masses of assymetrically placed cytoplasm, could bring about situations depicted in Fig. 3.3.11. Fraser & Gunning [65] have suggested that the bottlenecks at both ends of the complex *Bulbochaete* plasmodesmata (Fig. 3.3.10g) might act as partial valves by restricting or enlarging the orifice.

3.3.11 THE FUNCTIONS OF PLASMODESMATA

The only satisfactory direct observations of the movement of a low molecular weight soluble substance in a plasmodesmatal canal is in the salt glands of *Limonium* and *Tamarix* [73, 74, 75]. Here Ziegler & Lüttge [75], by using silver acetate to precipitate the chloride as electron dense insoluble silver chloride, showed that the plasmodesmata were the paths of movement from mesophyll to gland tissue. These experiments remain the sole unequivocal demonstration of the movement of soluble materials in plasmodesmata channels. Soluble substances often move extremely rapidly and conventional tracking techniques either lack the resolution to deal with canals only 30 nm or a little more in diameter or are rendered ineffectual by the speed of movement.

Where plasmodesmata are obviously present in a transport role it is noticeable that, either they are found in very large numbers as in the end walls of certain conducting cells (see Table 3.3.2), or they become enlarged up to $1 \cdot 5 \, \mu m$ as in the phloem where the only blockage may be protein filaments of very small dimensions [56].

The supposed functions of plasmodesmata are transport of one sort or another, and the control of differentiation. There may, however, be situations in which the one may grade into the other. The former are, obviously, much better understood, if rarely proven, and most of the thoughts concerning the latter role lie well within the realms of speculation.

An indirect approach to the possible function of cytoplasmic connections has been used in the alga *Nitella translucens*, where it was found that the specific electrical resistance of node cells joined by plasmodesmata was 50 times less than those without such connections [67]. Even so the calculated resistance across the node should be 330 times smaller than that actually measured. With the known internal complexity of some plasmodesmata this should cause no surprise. A similar study on a higher plant, *Elodea canadensis*, indicated a resistance of $3 \cdot 1 \, k\Omega cm^2$ for the intact plasma membrane and $0 \cdot 051 \, k\Omega cm^2$

for cell junctions; even the latter is 60 times higher than a theoretical estimate [76].

Tyree [77] concludes, on theoretical grounds, that plasmodesmata constitute the pathway of least resistance for the diffusion of small solutes, and diffusion will be the predominant mechanism of transport across the pores for such solutes.

Plasmodesmata are thought to be responsible, in some way, for polar transport in, for example, *Vallisneria* leaves and *Avena* coleoptiles [78, 79, 80]. Arisz bases his evidence on the observations that inhibitors of membrane passage e.g. potassium cyanide, sodium arsenate and uranyl nitrate, prevent the active movement of salts and asparagine through tissues. Since, for example, chloride cannot diffuse through the plasmamembrane there is no other possibility for this ion to move than through the plasmodesmata. Confirmation of this movement of the chloride ion in plasmodesmata also comes from the work on salt glands (see above).

A most interesting experiment lending general support to the idea of polar plasmodesmatal transport was performed by Tammes [80]. He imposed a permanent twist to the upper part of *Avena* coleoptiles. The phototropic and geotropic stimuli were no longer transmitted longitudinally i.e. to follow the shortest path, but strictly followed the direction of the twisted cells. This is consistent with the view of assymetrically distributed plasmodesmata on the transverse walls, but harder to explain by assuming that the plasmalemma of the transverse walls differs in some significant way from that of the longitudinal walls.

However, before we accept the idea that transport from cell to cell is mediated solely through the plasmodesmata or similar channels, we should consider the ingenious experiments of Goldsmith & Ray [81]. The transport of indole-3-acetic acid along maize coleoptiles is not influenced by cytoplasmic streaming, since this can be inhibited by cytochalasin B. Centrifugation ($1650 \times 1g$ for 10 min) towards the base of the segment promoted the basipetal transport of the auxin whereas apical centrifugation strongly inhibited this transport. Apical centrifugation neither promoted acropetal transport nor reversed the polarity of auxin transport. Basipetal transport seems independent both of the position and pressure of the amyloplasts of the cell, but strongly dependent on the amount of cytoplasm at the basal end of the cell. The authors concluded that the metabolic component of the transport is a polar secretion of auxin localized in the basal plasma membrane of each cell and has nothing whatsoever to do with plasmodesmatal transport.

Apoplastic transport across the endodermis of barley roots is effectively blocked by Casparian strips (Fig. 3.3.1) and other suberin deposits [11]. By implication, the route of movement is symplastic and by further implication the plasmodesmata through the inner tangential walls of the endodermal cells are

functionally important [10]. But again the matter is not settled beyond reasonable doubt [11].

3.3.12 THE MOVEMENT OF VIRUSES FROM CELL TO CELL

Weintraub, Ragetli & Lo [82] showed that particles of the potato virus Y were frequently found in the plasmodesmata of infected *Nicotiana* plants. There seems no doubt that plasmodesmata can act as channels for the transport of infection for some of the smaller spherical viruses; PVY particles are about 12–14 nm across. These authors were convinced that the desmotubule apparently suffered no damage from the passage of the virus and was just pushed to one side (but this point is in dispute for some other viruses and systems).

Although the work of Weintraub *et al.* does not prove that all viruses travel from cell to cell via the plasmodesmata, nor that viruses invariably pass from cell to cell in a complete form, it is known that the seed habit, with few exceptions, 'cleans up' the plant from vegetative and well-disseminated virus infections. The transition from the diploid to haploid state (see p. 123) is characterized either by an absence of plasmodesmatal junctions or by a subsequent occlusion of those that are present. The germinating pollen tube would appear to make doubly sure by forming behind it division walls which are apparently unique in lacking plasmodesmata. Those instances where viruses seem to bridge the seed barrier will repay a fine structural study.

Another way in which we can interpret the evidence for virus movement is by noting that viruses do not move in the phloem unless assisted by photosynthesizing leaves in the light or by artificially applied sucrose [83]. The inference is that they have no power to move through sieve plates unless assisted by a flow of sucrose. We can surmise from the evidence cited above that the potato virus Y would not move through the much smaller channels of normal plasmodesmata unless assisted by some relatively vigorous and viscous flow.

3.3.13 THE TRANSMISSION OF ELECTROTONIC STIMULI

All forms of cell connections, both animal and plant, represent low resistance points between cells. But are they of significance in electrophysiology in plants, as in animals? Williams & Pickard [84] point out that large numbers of plasmodesmata connect the cells most likely to have an electrophysiological role in the tentacles of *Drosera*. It is perfectly possible that they could act as channels for the electrotonic spread of electrical signals, but the suggestion remains to be proved.

3.3.14 THE CONTROL OF DIFFERENTIATION BY INTERCELLULAR CONNECTIONS

Apart from the simple physiological functions proved or suggested in the previous sections, other workers have attempted to relate the existence of cell-to-cell communications with the control of the paths of differentiation.

One of the prime centres for the control of differentiation in the higher plant is the apical meristem and some of the first attempts to correlate plasmodesmatal distribution with mechanisms of the control of differentiation have begun here.

Hicks & Steeves [85, 86] have looked at the shoot apical meristems of the cinnamon fern, *Osmunda cinnamomea*. Their earlier work seemed to indicate that early leaf development in this apex was influenced by the shoot apical meristem; in particular, the normal leaf's dorsiventrality and determinate growth were suppressed when cell continuities between the flank of the promeristem and the leaf site were surgically broken. Plasmodesmata of Type b (Fig. 3.3.10) exist in both anticlinal and periclinal walls between the cells where the interaction is known. The numbers of these plasmodesmata, calculated from the original published micrographs by the present author, are substantial: 18,000 per 100 μm^2 on the anticlinal walls and 8000 per 100 μm^2 on the periclinal walls.

However, the presence of substantial numbers of plasmodesmata of a simple type between donor and receptor sites does not prove their involvement, tempting though this idea may be. Proof of the role of plasmodesmata in such a situation lies in the development of techniques for the spot analysis of very small quantities of soluble substances in very small channels or, alternatively, the discovery of aplasmodesmatal mutants.

3.3.15 PLASMOLYSIS EXPERIMENTS AND THE DISRUPTION OF DIFFERENTIATION

In certain species of fern gametophytes, the normal development from a spore consists of a filamentous stage followed, after a certain interval, by two dimensional growth. Many workers have attempted to use this apparently simple system as a model of cellular control. Attempts have been made to elucidate the role of plasmodesmatal connections between the component cells of the filament [87]. Plasmolysis disturbs the normal synchronous development of the filament and induces abnormal branching [88]. Burgess [89] has taken Nakazawa's work further and has attempted a correlated electron microscope study on the effects of plasmolysis on the filamentous stage of *Asplenium nidus*. However, even using mannitol up to 0·8 M for 1–2 h did not induce all the plasmodesmata to break on both sides of the wall, although the usual disturbances of the pattern of differentiation took place.

After plasmolysis not all the plasmodesmata apparently reform. It is tempting to connect the rupture of plasmodesmatal connections with a divergence in the pattern of cell differentiation, but the electron microscope evidence remains tantalizingly equivocal.

We can speculate that much smaller changes in the osmotic balance of individual cells, with slighter changes in the conformation of the plasmalemma, would act to close or partially close plasmodesmata with constricted throats, particularly those of the type found in *Bulbochaete* (Fig. 3.3.10g).

3.3.16 TISSUE CULTURE EXPERIMENTS

A little work has been done on tissue culture cells [90]. It was possible to show that there were differences between cells at different densities of aggregation, with or without plasmodesmata, but attempts to determine precise correlations as in the fern gametophyte work have not yet been made.

3.3.17 THE SELECTIVE BREAKDOWN OF WALLS OR WALL AREAS

It is a feature of many phases of plant cell development that areas of wall, whole cells or even complete tissues, are broken down to permit a subsequent stage in differentiation. Well-known examples are the end walls of vessels and the cells of the stylar canal just prior to or at fertilization.

When the cells of the aleurone layer of barley (*Hordeum*) seeds begin to break down as a result of the spontaneous production of endogenous hydrolytic enzymes including β-1,3 glucanase [91, 92] the plasmodesmata seem to be regions where substances may be moving out of the cell. As can be seen from the work of Taiz & Jones (Fig. 4d in [91]) and Jones (Figs 4a and 4b in [92]), the enzymes appear to attack not only from the plasmalemma outward, but also from the plasmodesmatal canal outward. As Jones [92] has pointed out there is no evidence, at least for the β-1,3 glucanase, of discharge across the plasma membrane in a vesicular form, the evidence pointing towards soluble enzymes, and therefore the restriction by the size of the plasmodesmatal canal is little barrier to a soluble enzyme. The evidence presented does not prove that the enzymes flow along the plasmodesmata; they could, of course, migrate across the plasma membrane and then flow along the surface of the membrane until they reach the plasmodesmatal channels. However, the existence of groups of plasmodesmata substantially increases the surface area of the plasma membrane in contact with the wall at that particular point. From data published by Laetsch [93], I calculate that across certain areas of the bundle sheath/mesophyll junction in sugar cane

(*Saccharum officinarum*) in a wall 0·5 μm thick a little over 3 μm² of extra membrane is added to 1 μm² of plasma membrane by the plasmodesmatal channel. In *Welwitschia* cotyledon (Fig. 3.3.3) with its more numerous plasmodesmata over 9·0 μm² of membrane surface will be added to 1 μm² of plasma membrane.

If the plasmodesmata are regions where the walls can selectively be eroded away a number of other developmental anomalies may be explained. As we have seen in Section 3.3.5, median nodules or 'mittelknoten' (Fig. 3.3.4) often form in the later stages of plasmodesmatal development. It is noticeable that these cavities generally form in the middle lamella region [94] which is where one would expect the action of hydrolytic enzymes of the types studied by Jones [92] to have the most effect. The flexible membranes of the plasmodesmatal channels may extend slowly into spaces created by the selective degradation of the cell wall. Pits from one cell to another are regions of incomplete deposition of wall material. They are also, at an earlier stage (Figs 3.3.8, 3.3.9), regions of high concentrations of plasmodesmata. An hypothesis based on a slow leakage of hydrolytic enzymes combined with the increase in surface area brought about by the presence of these large numbers of plasmodesmata could account for the failure of substantial wall deposits at these sites. The end walls of vessels, which finally disappear completely are also, through the dilution effect described in Section 3.3.4b, regions of high concentration of plasmodesmata (Fig. 3.3.5). It is also generally observed (Fig. 153 in [38]) that at the breakdown of these end walls traces of fibrous material, probably cellulose microfibrils, persist until they are washed away. This is consistent with the previous observations that none of the enzymes detected as moving across the plasma membrane in this way are cellulolytic. The only difference needed to account for, on the one hand, the total breakdown of aleurone and vessel end walls and the local and partial breakdown at a pit, could be the concentration of the enzymes or the degree of stimulation by gibberellic acid [91].

If the above speculations prove well-founded we can see that plasmodesmata may, in the first stage of their existence, provide a pattern of communication between living cells, and in the second stage, by virtue of the pattern imposed in the primary wall period, determine the existence and extent of the continuities between dead cells.

3.3.18 THE CONTROL OF GENETIC PROCESSES BY CELL COMMUNICATION

Within the anther of the flower are developed the reproductive cells, the meiocytes, which after meiosis will mature into the pollen grain. Sections of ER and certain organelles may, *in vivo*, pass from one meiocyte to another, all the cells together acting as a single coenocyte throughout which streaming may occur. Cytomictic channels assist in this way the exchange of materials and certain organelles between meiocytes during the immediate pre-meiotic period and throughout the meiotic prophase. Heslop-Harrison [3] suggests that this indicates a requirement for the sharing of materials in this early period, which is no doubt useful and economic when the nuclei are identical, but when the new generation of different haploid nuclei emerges, the expression of their multifold new haploid identity, because each meiotic product is different, requires that they should act as isolated protoplasts. The thickening of their independent callose walls renders all the individual meiocytes totally isolated by the onset of meiosis II.

The whole archesporium has a full complement of plasmodesmata, across all of its component walls, just prior to the onset of meiosis. The connections from tapetum to meiocytes are first eliminated (followed by vigorous apparent cooperation between meiocytes); the meiocytes are then separated from one another, the spores separated from each other in the tetrad [31, 32, 33] and finally at a much later stage, the germinating pollen grain separates its generative nucleus from its vegetative nucleus by an aplasmodesmatal wall in the pollen tube [3]. Thus, the new haploid generation achieves both the haploid number and the new situation in which a haploid nucleus is integrated with its cytoplasm by a sequence of controlled and graded independence movements. These movements culminate, like the shedding of a rocket of its final propulsion phase, in the abandonment of its vegetative propulsion mechanism.

There are certain exceptions to these sequences of septation and they may throw some light on the possible functions of cytoplasmic connections. Some of the orchids, e.g. genus *Dactylorchis*, possess pollen in aggregates of several hundred cells and retain the synchrony of their nuclear divisions right through meiosis and the pollen mitosis. Heslop-Harrison [3] suggests that this is a consequence of their sharing, right through the final mitoses, a common cytoplasmic matrix. They are, after the final mitosis, sealed off from each other like most other higher plants. He points out that because, prior to the final sealing off, the genetically different nuclei will mix their products, differences between the haploid nuclei are not expressed at this stage; the cytoplasms of all these cells are homogeneous and their reactions must be homogeneous. No self-incompatibility is known in the genus *Dactylorchis*. It is reasonable that the presence or absence of plasmodesmata or cytomictic channels at these key stages may determine, amongst other things, the genetic reactions of the pollen to the stigma upon which it finally lands (Section 3.3.6). Such connections may also, by the timing of the aplasmodesmatal wall between the vegetative and generative cell in the germinating pollen grain, determine whether or not plastids from the male plastome will pass into the female egg [3].

3.3.19 EXTENDING THE FUNCTIONS OF PLASMODESMATA

Most cell-to-cell connections that we have described apparently do no more than connect cytoplasm to cytoplasm and make it possible for small particles or soluble substances to move from one matrix to another, but other features should now be considered. Components of many plasmodesmata, the desmotubules (Fig. 3.3.10c, d), often but not invariably make contact with the ER. What the functions of these connections are we do not know, but at the simplest level they may operate some sort of valve system (Fig. 3.3.11). Carrying these possibilities a little further, we know that occasional direct membrane connections have been observed in gametophyte cells of *Dryopteris borreri*, of the outer membrane of the chloroplast

envelope with smooth ER, of the plastid envelope with the plasma membrane, and of the nuclear envelope with the ER [95]. In many cells close spatial relationships exist between ER and plastids and microbodies at certain periods of cell differentiation. There is a general tendency for these associations to be more widespread and longer lasting in the lower plants than in the higher plants [96–99]. Although there is some speculation, e.g. on the filling or emptying of plastid storage products at certain times [100], there is no firm evidence as to their function.

In summary, although the experimental evidence for the importance of cell-to-cell communication in plants, as compared with that in animals, is thin, no one would dispute its relevance. Perhaps the best evidence, although indirect, is that at the stage when the ground plan of all future differentiation is being established, i.e. genetic autonomy, the establishment of permanent genetic differences in pollen, the plant takes at least three decisive steps to close off all cytoplasmic continuity (p. 123). These actions would seem to imply not only the importance of cell-to-cell contact but the effectiveness of the individual cell in its control of the intact plasma membrane, the final barrier.

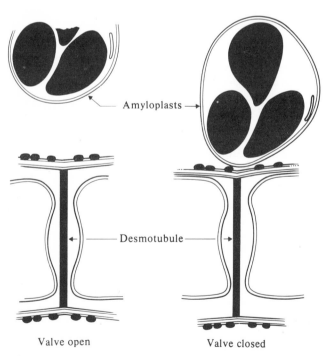

Fig. 3.3.11 Proposed valve mechanism, using amyloplasts to depress the ER valve.

Amyloplasts

Desmotubule

Valve open

Valve closed

3.3.20 REFERENCES

[1] TANGL E. (1879) Ueber offene Communicationen zwischen den Zellen des Endosperms einiger Samen. *Jb. wiss. Bot.*, **12**, 170–90.

[2] ROBARDS A.W. (1975) Plasmodesmata. *Ann. Rev. Plant Physiol.*, **26**, 13–29.

[3] HESLOP-HARRISON J. (1968) Synchronous pollen mitosis and the formation of the generative cell in massulate orchids. *J. Cell Sci.*, **3**, 457–66.

[4] WEILING F. (1965) Zur Feinstruktur der Plasmodesmen und Plasmakanäle bei Pollenmutterzellen. *Planta*, **64**, 97–118.

[5] BISALPUTRA T. (1966) Electron microscopic study the protoplasmic continuity in certain brown algae. *Can. J. Bot.*, **44**, 89–93.

[6] BISALPUTRA T. & STEIN J.R. (1966) The development of cytoplasmic bridges in *Volvox aureus. Can J. Bot.*, **44**, 1697–702.

[7] GAFF D.F., CHAMBERS T.C. & MARKUS K. (1964) Studies of extra-fascicular movements of water in the leaf. *Aust. J. Biol. Sci.*, **17**, 581–86.

[8] KONAR R.N. & LINSKENS H.F. (1966) The morphology and anatomy of the stigma of *Petunia hybrida. Planta (Berl.)*, **71**, 356–71.

[9] COX G.C. & JUNIPER B.E. (1973) Autoradiographic evidence for paramural-body function. *Nature, New Biology*, **243**, 116–17.

[10] ROBARDS A.W., JACKSON S.M., CLARKSON D.T. & SAUNDERSON J. (1973) The structure of barley roots in relation to the transport of ions into the stele. *Protoplasma*, **77**, 291–311.

[11] GILCHRIST A.J. & JUNIPER B.E. (1974) An excitable membrane in the stalked glands of *Drosera capensis. Planta*, **119**, 143–47.

[12] JUNIPER B.E. & BARLOW P.W. (1969) The distribution of plasmodesmata in the root tip of maize. *Planta*, **89**, 352–60.

[13] HEPLER P.K. & NEWCOMB E.H. (1967) Fine structure of cell plate formation in the apical meristem of *Phaseolus* roots. *J. Ultrastruct. Res.*, **19**, 498–513.

[14] EVERT R.F. & DESHPANDE B.P. (1970) An ultrastructural study of cell division in the cambium. *Amer. J. Bot.*, **57**, 942–61.

[15] WOODING F.B.P. & NORTHCOTE D.H. (1965) The fine structure and development of the companion cell of the phloem of *Acer pseudoplatanus. J. Cell Biol.*, **24**, 117–28.

[16] KRULL R. (1960) Untersuchungen über den Bau und die Entwicklung der Plasmodesmen im Rindenparenchym von *Viscum album. Planta,* **55,** 598–629.

[17] COX G.C. (1971) *The structure and development of cells with thickened primary walls.* D.Phil. Thesis, Oxford.

[18] DÖRR I. (1968) Plasmatische Verbindung zwischen artfremden Zellen. *Naturwissenschaften,* **55,** 396.

[19] DÖRR I. (1969) Feinstruktur intrazellulär wachsender *Cuscuta*-Hyphen. *Protoplasma,* **67,** 123–37.

[20] KOLLMAN R. & DÖRR I. (1969) Strukturelle Grundlagen des zwischenzelligen Stoffaustausches. *Ber. dt. bot. Ges.,* **82,** 415–25.

[21] ESAU K. (1965) *Plant Anatomy,* 2nd edn. Wiley, New York.

[22] BURGESS J. (1972) The occurrence of plasmodesmata-like structures in a non-division wall. *Protoplasma,* **74,** 449–58.

[23] HEPLER P.K. Personal communication, July 1974.

[24] THOMPSON W.W. & DE JOURNETT R. (1970) Studies on the ultrastructure of the guard cells of *Opuntia. Amer. J. Bot.,* **57,** 309–16.

[25] BROWN W.V. & JOHNSON S.C. (1962) The fine structure of the grass guard cell. *Am. J. Bot.,* **49,** 110–15.

[26] SCHULZ P. & JENSEN W.A. (1968) *Capsella* embryogenesis: the egg, zygote and young embryo. *Am. J. Bot.,* **55,** 807–19.

[27] SCHULZ P. & JENSEN W.A. (1969) *Capsella* embryogenesis: the suspensor and the basal cell. *Protoplasma,* **67,** 139–63.

[28] DIBOLL A.G. & LARSON D.A. (1966) An electron microscope study of the mature megagametophyte in *Zea mays. Am. J. Bot.,* **53,** 391–402.

[29] DIBOLL A.G. (1968) Fine structural development of the megagametophyte of *Zea mays* following fertilization. *Amer. J. Bot.,* **55,** 787–806.

[30] WOODCOCK C.L.F. & BELL P.R. (1968) Features of the ultrastructure of the female gametophyte of *Myosurus minimus. J. Ultrastruct. Res.,* **22,** 546–63.

[31] HESLOP-HARRISON J. (1964) Cell walls, cell membrane and cytoplasmic connections during meiosis and pollen development. In *Pollen Physiology and Fertilization* (Ed. H.F. Linskens). North Holland, Amsterdam.

[32] HESLOP-HARRISON J. (1966) Cytoplasmic continuities during spore formation in flowering plants. *Endeavour,* **25,** 65–72.

[33] HESLOP-HARRISON J. (1966) Cytoplasmic connexions between angiosperm meiocytes. *Ann. Bot.,* **30,** 221–30.

[34] LEDBETTER M.C. & PORTER K.R. (1970) *Introduction to the Fine Structure of Plant Cells.* Springer-Verlag, Berlin.

[35] BURGESS J. (1971) Observations on structure and differentiation in plasmodesmata. *Protoplasma,* **73,** 83–95.

[36] HICKS G.S. & STEVENS T.A. (1969) *In vitro* morphogenesis in *Osmunda cinnamomea.* The role of the shoot apex in early leaf development. *Can. J. Bot.,* **47,** 575–80.

[37] ZIEGLER H. & RUCK I. (1967) Untersuchungen über die Feinstruktur des Phloems. III. Die 'Trompetenzellen' von *Laminaria*-Arten. *Planta,* **73,** 62–73.

[38] CLOWES F.A.L. & JUNIPER B.E. (1968) *Plant Cells.* Blackwell Scientific Publications, Oxford.

[39] MUELLER D.M.J. (1972) Observations of the ultrastructure of *Buxbaumia* Protonema. Plasmodesmata on the cross walls. *The Bryologist,* **75,** 63–8.

[40] O'BRIEN T.P. & THIMANN K.V. (1967) Observations on the fine structure of the oat coleoptile. II. *Protoplasma,* **63,** 417–42.

[41] ROBARDS A.W. (1968) A new interpretation of plasmodesmatal ultrastructure. *Planta,* **82,** 200–10.

[42] ROBARDS A.W. (1971) The ultrastructure of plasmodesmata. *Protoplasma,* **72,** 315–23.

[43] HALL J.L. (1969) Localization of cell surface adenosine triphosphatase activity in maize roots. *Planta,* **85,** 105–7.

[44] ROBARDS A.W. & KIDWAI P. (1969) Cytochemical localization of phosphatase on differentiating secondary vascular cells. *Planta,* **87,** 227–38.

[45] ESAU K., CHEADLE V.I. & RISLEY E.B. (1962) Development of sieve-plate pores. *Bot. Gaz.,* **123,** 233–43.

[46] NORTHCOTE D.H. & WOODING F.B.P. (1965) Development of sieve tubes in *Acer pseudoplatanus. Proc. Roy. Soc.,* **163,** 524–37.

[47] KOLLMANN R. & SCHUMACHER W. (1962) Uber die Feinstruktur des phloems von Metasequoia glyptostroboides und seine jahreszeitlichen Veranderungen. II. Mitt. Vergleichende Untersuchungen der Plasmatischen Verbindungsbrucken in Phloemparenchymzellen und Siebzellen. *Planta,* **58,** 366–86.

[48] WOODING F.B.P. & NORTHCOTE D.H. (1965) The fine structure and development of the companion cell of the phloem of *Acer pseudoplatanus. J. Cell Biol.,* **24,** 117–28.

[49] MURMANIS L. & EVERT R.F. (1967) Parenchyma cells of secondary phloem in *Pinus strobus. Planta,* **73,** 301–18.

[50] FAHN A. (1967) *Plant Anatomy.* Pergamon Press, Oxford.

[51] SHIH C.Y. & CURRIER H.B. (1969) Fine structure of phloem cells in relation to translocation in the cotton seedling. *Am. J. Bot.,* **56,** 464–72.

[52] EVERT R.F. & MURMANIS L. (1965) Ultrastructure of the secondary phloem of *Tilia americana. Am. J. Bot.,* **52,** 95–106.

[53] NORTHCOTE D.H. & WOODING F.B.P. (1968) The structure and function of phloem tissue. *Sci. Prog. Oxf.,* **56,** 35–58.

[54] WARK M.C. (1965) Fine structure of the phloem of *Pisum sativum.* 2. The companion cell and phloem parenchyma. *Aust. J. Bot.,* **13,** 185–93.

[55] ZEE S.Y. & CHAMBERS T.L. (1968) Fine structure of the primary root phloem of *Pisum. Aust. J. Bot.,* **16,** 37–47.

[56] JOHNSON R.P.C. (1973) Filaments but no membranous transcellular strands in sieve pores in freeze-etched, translocating phloem. *Nature,* **244,** 464–66.

[57] GATES R.R. (1908) A study of reduction in *Oenothera rubrinervis. Bot. Gaz.,* **46,** 1–34.

[58] TAKATS S.T. (1959) Chromatin extrusion and DNA transfer during micro-porogenesis. *Chromosoma,* **10,** 430–53.

[59] KAMRA O.P. (1960) Chromatin extrusion and cytomixis in pollen mother cells of *Hordeum. Hereditas,* **46,** 592–600.

[60] BOPP-HASSENKAMP G. (1959) Cytomixis im elektronmikroskopischen Bild. *Expt. Cell Res.,* **18,** 182–84.

[61] BROWN W.V. & JOHNSON S.C. (1962) The fine structure of the grass guard cell. *Am. J. Bot.,* **49,** 110–15.

[62] KAUFMAN P.B., PETERING L.B., YOCUM C.S. & BAIC D. (1970) Ultrastructural studies on stomatal development in internodes of *Avena sativa. Am. J. Bot.,* **57,** 33–49.

[63] ALLAWAY W.G. & SETTERFIELD G. (1972) Ultrastructural observations on guard cells of *Vicia faba* and *Allium porrum. Can. J. Bot.,* **50,** 1405–13.

[64] LITZ R.E. & KIMMINS W.C. (1968) Plasmodesmata between guard cells and accessory cells. *Can. J. Bot.,* **46,** 1603–5.

[65] FRASER T.W. & GUNNING B.E.S. (1969) The ultrastructure of plasmodesmata in the filamentous green alga, *Bulbochaete hiloensis* (Nordst.) Tiffany. *Planta,* **88,** 244–54.

[66] PICKETT-HEAPS J.D. (1967) Ultrastructure and differentiation in *Chara* sp. I. Vegetative cells. *Aust. J. Biol. Sci.,* **20,** 539–51.

[67] SPANSWICK R. & COSTERTON J. (1967) Plasmodesmata in *Nitella translucens.* Structure and electrical resistance. *J. Cell Sci.,* **2,** 451–64.

[68] FISCHER R.A., DAINTY J. & TYREE M.T. (1974) A quantitative investigation of symplasmic transport in *Chara corallina.* I. Ultrastructure of the nodal complex cell walls. *Can. J. Bot.,* **52,** 1209–14.

[69] HAWKER L.E., GOODAY M.A. & BRACKER C.E. (1966) Plasmodesmata in fungal cell walls. *Nature,* **212,** 635.

[70] HAWKER L.E. & BECKETT A. (1971) Fine structure and development of the zygospore of *Rhizopus sexualis* (Smith) Callen. *Phil. Trans. Roy. Soc. 'B',* **263,** 71–100.

[71] PICKETT-HEAPS J.D. & NORTHCOTE D.H. (1966) Cell formation in the formation of the stomatal complex of the young leaves of wheat. *J. Cell Sci.,* **1,** 121–8.

[72] BARLOW P.W. (1970) Mitotic spindle and mitotic cell volumes in the root meristem of *Zea mays. Planta,* **91,** 169–72.

[73] THOMSON W.W. & LIU L.L. (1967) Ultrastructural features of the salt gland of *Tamarix aphylla*. *Planta*, **73**, 201–20.

[74] ZIEGLER H. & LÜTTGE U. (1966) Die Salzdrüsen von *Limonium vulgare*. I. Mitt. Die Feinstruktur. *Planta*, **70**, 193–206.

[75] ZIEGLER H. & LÜTTGE U. (1967) Die Salzdrüsen von *Limonium vulgare*. II. Mitt Die Lokalisierung des Chlorids. *Planta*, **74**, 1–17.

[76] SPANSWICK R.M. (1972) Electrical coupling between cells of higher plants. A direct demonstration of intercellular communication. *Planta*, **102**, 215–27.

[77] TYREE M.T. (1970) The symplast concept. A general theory of symplastic transport according to the thermodynamics of irreversible processes. *J. Theor. Biol.*, **26**, 181–214.

[78] ARISZ W.H. (1968) Influence of inhibitors on the uptake and transport of chloride ions in the leaves of *Vallisneria spiralis*. *Acta. Bot. Neerl.*, **7**, 1–32.

[79] ARISZ W.H. (1969) Intercellular polar transport and the role of the plasmodesmata in coleoptiles and *Vallisneria* leaves. *Acta. Bot. Neerl.*, **18**, 14–38.

[80] TAMMES P.M.L. (1931) Über den Verlauf der geotropischen Krümmung bei künstlich tordierten Kopeoptilen von *Avena*. *Rec. Trav. Bot. Neerl.*, **28**, 75–81.

[81] GOLDSMITH M.H.M. & RAY P.M. (1973) Intracellular localization of the active process in polar transport of auxin. *Planta*, **111**, 297–314.

[82] WEINTRAUB M., RAGETLI H.W.J. & LO E.J. (1974) Potato virus Y particles in plasmodesmata of tobacco leaf cells. *J. Ultrastructure Res.*, **46**, 131–48.

[83] BENNET C.W. (1937) Correlation between movement of curly top virus and transport of food in tobacco and sugar beet. *J. Agric. Res.*, **54**, 479–502.

[84] WILLIAMS S.E. & PICKARD B.G. (1974) Connections and barriers between cells of *Drosera* tentacles in relation to their electrophysiology. *Planta*, **116**, 1–16.

[85] HICKS G.S. & STEEVES T.A. (1969) *In vitro* morphogenesis in *Osmunda cinnamomea*. The role of the shoot apex in early leaf development. *Can. J. Bot.*, **47**, 575–80.

[86] HICKS G.S. & STEEVES T.A. (1973) Plasmodesmata in the shoot apex of *Osmunda cinnamomea*. *Cytologia*, **38**, 449–53.

[87] MILLER J.H. (1968) Fern gametophytes as experimental material. *Bot. Rev.*, **34**, 361–440.

[88] NAKAZAWA S. (1963) Role of protoplasmic connections in the morphogenesis of fern gametophytes. *Sci. Rev. Tohuko Univ. Ser. IV (Biol.)*, **29**, 247–55.

[89] BURGESS J. Personal Communication, 1974.

[90] SUSSEX I.M. & CLUTTER M.E. (1967) Differentiation in tissue free cells and reaggregated plant cells. *In Vitro*, **3**, 3–12.

[91] TAIZ L. & JONES R.L. (1970) Gibberellic acid, β-1, 3 glucanase and the cell walls of barley aleurone layers. *Planta*, **92**, 73–84.

[92] JONES R.L. (1972) Fractionation of the enzymes of the barley aleurone layer: evidence for a soluble mode of enzyme release. *Planta*, **103**, 95–109.

[93] LAETSCH W.M. (1971) Chloroplast structural relationships in leaves of C_4 plants, pp. 323–349. In *Photosynthesis and Photo Respiration* (Ed. Hatch, M.D., Osmond, C.B. & Slatyer, R.O.). Wiley-Interscience, New York.

[94] EVERT R.F., BORNMAN C.H., BUTLER V. & GILLILAND M.G. (1973) Structure and development of sieve areas in leaf veins of *Welwitschia*. *Protoplasma*, **76**, 23–34.

[95] CRAN D.G. & DYER A.F. (1973) Membrane continuity and associations in the fern *Dryopteris borreri*. *Protoplasma*, **76**, 103–8.

[96] BROWN R. & DYER A.F. (1972) Cell division in higher plants. In *Plant Physiology* (Ed. F.G. Steward). A treatise, Vol. VIC.

[97] LARSON D.A. (1965) Fine structural changes in the cytoplasm of germinating pollen. *Amer. J. Bot.*, **52**, 139–54.

[98] GULLVÅG B.M. (1968) Fine structure of the plastids and possible ways of distribution of the chloroplast products in some spores of archegoniates. *Phytomorph.*, **18**, 520–35.

[99] YOO B.Y. (1970) Ultrastructural changes in cells of pea embryo radicles during germination. *J. Cell Biol.*, **45**, 158–71.

[100] WOODING F.B.P. & NORTHCOTE D.H. (1965) Association of the ER and plastids in *Acer* and *Pinus*. *Amer. J. Bot.*, **52**, 526–31.

Chapter 3.4
Embryonic Induction

3.4.1 DIFFERENTIATION ENTAILS COMMUNICATION BETWEEN CELLS

The development of a mosaic egg, as described in Part 1, implies the presence of cytoplasmic factors unevenly distributed in the egg and consequently segregated into different cells of the developing embryo. Selective loss or amplification of genetic material is another cellular mechanism which might account for the gradual specialization of cells during embryogenesis. Here again, examples of such events are known (Chapter 5.3). However, data of both classic and more recent experiments suggest that such changes in the genetic material of the somatic cells are not regularly involved in cytodifferentiation. A pregastrula stage Amphibian embryo can be cleaved in two with a hair loop, and both halves will develop into complete embryos, provided that the cortical material of the grey crescent is shared between the halves. Nuclear transplantation experiments provide further evidence that the genetic material of the cell is preserved throughout development; a nucleus from an intestinal epithelial cell of an Amphibian tadpole will support the development of a complete embryo when transplanted into the cytoplasm of an enucleated egg (Chapter 2.2). Thus, cytodifferentiation might be regarded as a differential reading of the same genome, and in this chapter we will show that the fate of the individual cell is controlled in part by extrinsic factors. In other words, the 'building plan' of an embryo is to be sought in the organization of tissues and organs rather than at the cellular level, and the guiding forces should be found in the constantly changing microenvironment of the cells and in their intimate associations with other cells. Morphogenetic communication between cells and cell populations would also account for their spatially and temporally synchronized development and the rearrangement of cells required for the creation of a harmoniously built functional organism.

That such theoretically meaningful *inductive interactions* do occur between adjacent cell populations has repeatedly been demonstrated experimentally, either by surgically separating the cells or the organ anlagen or by inserting barriers between them. These measures have affected differentiation, morphogenesis and the maintenance of the differentiated state, and thus a useful definition of embryonic induction is: *communication between cells required for their differentiation, morphogenesis and maintenance.* This definition does *not* imply that we are dealing with interacting processes operating through similar mechanisms or leading to similar consequences. It merely states that cells in higher organisms do not develop independently as preprogrammed units but as targets for various extrinsic stimuli. These latter might be specific determinative signals, less specific permissive conditions or mere nutritional factors required for normogenesis. Much confusion has been created over the last 50 years by searching for a unifying concept or a common denominator for the great variety of known (and unknown) interactive processes.

When considering the numerous experiments relating to embryonic induction, we run into another difficulty. Such experiments deal with single developmental events extracted from the continuous and complicated process of embryogenesis. The cells and tissues studied are at various stages of differentiation and they will contain different cytoplasmic substances and will have experienced several interactions before the experiment starts. The result is that their 'predetermination' and potentialities vary, and the alternative pathways open to them are more or less restricted. To make a synthesis out of such fragmentary observations and to extrapolate it to the normal situation *in vivo* is still premature, and we must content ourselves with considering the examples separately. In doing so, we should be aware that even these simple events, mostly analysed *in vitro,* do not necessarily represent normal conditions and are only to be regarded as model-systems simulating normal developmement as closely as possible.

3.4.2 PRIMARY INDUCTION ILLUSTRATES THE PROBLEMS

One of the first inductive interactions to be recognized was 'primary induction', the process which governs the determination and subsequent morphogenesis of the central nervous system. This in many respects most thoroughly investigated but still only partially understood classic example of induction may be used here to illustrate the basic problems, both solved and unsolved, in the study of inductive tissue interactions [for review, 24].

The experiment performed by Spemann some 60 years ago and illustrated in Fig. 3.4.1 constitutes the basis for studies related to primary induction. When a fragment of the

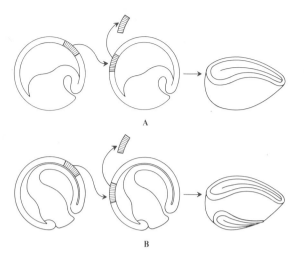

Fig. 3.4.1 Spemann's experiment demonstrating the determination of the presumptive neural plate area. Transplantation at an early gastrula stage (A) leads to normal development of the host whereas a similar experiment performed at late gastrula stage (B) results in the formation of an extra neural plate at the belly side of the donor. After [24].

Fig. 3.4.2 Results of an implantation experiment in which a combined heterogenous inductor was used. The experimentally induced belly ectoderm of the newt gastrula has developed into a disorganized but almost complete embryo.

presumptive neural plate of a young Amphibian gastrula is transplanted to the ventral surface of a host at the same stage of development, the transplant develops in accordance with its new surroundings (A). When the experiment is performed later, at an advanced gastrula stage, the transplant forms an extra neural plate at its new site (B). The inference to be drawn is that the piece of tissue had been *determined* during the short period between the two stages. This determination is apparently caused by the mesoderm; during gastrulation the mesoderm moves beneath and interacts with the ectoderm which is transplanted in experiment B. This postulate is confirmed by two experiments. Removal of the mesoderm or prevention of its invagination blocks development of the central nervous system (CNS), and implantation of the mesoderm into another gastrula induces supernumerary structures in the host. If the tissue implanted is the entire dorsal blastoporal lip of a young gastrula or a combination of certain foreign tissues (e.g. liver and bone marrow from guinea pigs, subsequently called heterogenous inductors), the competent ectoderm can be induced to form a variety of neural, neuroectodermal, mesodermal and endodermal structures (Fig. 3.4.2).

Such observations raise several questions, most of which appear relevant to many inductive interactions:

Do specific substances signal each type of induction?
Is the end-result, seen as completed organogenesis, the consequence of a single determinative event or the product of a train of inductive interactions?

And finally, what are these signal substances and how do they operate?

Let us start by discussing the first two questions posed above, which provide the biological framework for the latter part of this chapter dealing with the complicated and unanswered questions relating to signals, their transmission and nature.

Induction involves more than one signal

Experiments employing various heterogenous inductors exerting different types of effects upon the ectoderm have led to the hypothesis that there are basically two types of primary response, neuralization and mesodermalization [33]. The former

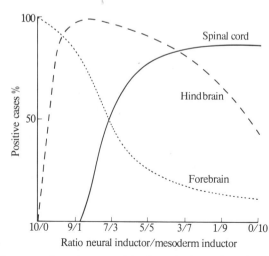

Fig. 3.4.3 Percentage of neural structures belonging to the three major regions induced in a series of experiments where two types of inductors were mixed in different ratios. After [23].

ultimately leads to the formation of cranial neural structures (forebrain derivatives), whereas the latter culminates in the development of purely mesodermal structures (axial mesoderm, limb, etc.). To explain the induction of mid- and caudo-neural structures (hindbrain derivatives and spinal cord), the hypothesis was put forward that these develop in response to a combined action of the primary neuralizing and mesodermalizing principles [34]. Support for this hypothesis was obtained in experiments where two heterogenous inductors were employed simultaneously, one having a neuralizing action and the other regularly inducing mesodermal structures. As expected, such experiments led to the development of secondary hindbrain derivatives and spinal cords and a reduction of forebrain structures. When the ratio of such neural and predominantly mesodermal inductors was altered in semiquantitative experiments, progressive caudalization of the neural structures was associated with increased relative amounts of mesodermalizing/tail inductor (Fig. 3.4.3).

Induction is a multistep process

Since the combined actions needed for the production of regional specialization of the central nervous system could not be ascribed to a single inductive event, the following experiment was performed (Fig. 3.4.4) [25]. Competent ectodermal cells from an amphibian gastrula were exposed to one or other of two heterogenous inductors, one leading to 'pure' neuralization and ultimately to the formation of forebrain structures (guinea-pig liver, A), the other to caudo-mesodermal and spinal cord derivatives but not hindbrain derivatives (guinea-pig bone marrow, B), After allowance of due time for primary induction (24 h later), the inductors were removed and the ectodermal cells disaggregated to single-cell suspensions. When such suspensions were allowed to reaggregate, the expected structures were obtained; in culture A forebrain and eye cups developed while in culture B the spinal cord with associated notocord, muscle blocks, and kidney tubes formed. If the two types of induced suspensions of ectodermal cells, were thoroughly mixed and cultured as a combined aggregate, hindbrain formations were regularly recovered. These hindbrain formations were recognized by their shape and by the presence of ear vesicles. The conclusion was that these had been determined after the initial period of primary induction and before regional differentiation of the CNS was stabilized. To simulate the actual situation *in vivo*, presumptive forebrain cells and mesoderm cells taken from the dorsal mid-line of the embryo were disaggregated and combined. As was to be expected, presumptive forebrain cells from young neurula stages could be 'transformed' to hindbrain and spinal cord structures when cultured in mixed aggregates with mesodermal cells. As a final test, the semiquantitative experiment was repeated, this time with forebrain and mesodermal cells from young neurulas

mixed in different ratios. The results (Fig. 3.4.5) indicated that a gradual increase in the proportion of mesodermal cells resulted in gradual caudalization of the CNS derivatives. From this series of experiments, a rough and still over-simplified scheme for the determination and segregation of the central nervous system can be given. During a short initial phase of interaction with mesoderm cells, the pluripotent ectodermal cells become determined in either a neural or a mesodermal direction. Without any additional directive influence from the mesoderm, neuralized cells form forebrain structures, but continued mesodermal influence alters their further morphogenetic course and they form caudo-neural structures. This secondary type of morphogenetic interaction operates quantitatively and is not strictly specific, as mesenchyme from other sites besides the normal axial region can support regionalization of the CNS. If this observation could be generalized, it would indicate that

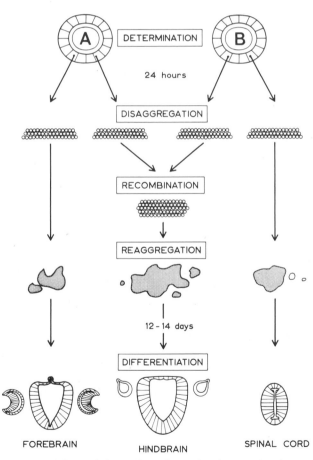

Fig. 3.4.4 Scheme of the experimental procedure demonstrating the two stages, determination and segregation of the central nervous system.

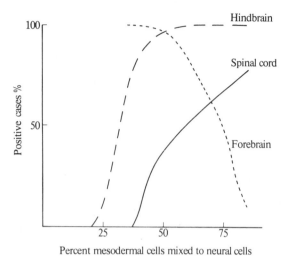

Fig. 3.4.5 Percentage of neural derivatives of various types in an experimental series where cells of the axial mesoderm were mixed in different ratios with neural cells from the prospective forebrain region. After [35].

induction is a multistep process consisting at least of specific, determinative stimuli and less specific, morphogenetic or stabilizing steps. Next we have to test whether this scheme fits what we know of other interactive systems.

3.4.3 EPITHELIO-MESENCHYMAL INTERACTIONS GOVERN ORGANOGENESIS

Most of the recent work related to the specificity problem has been performed on model systems for epithelio-mesenchymal interactions, i.e. morphogenetic interactions between differentiating epithelium and mesenchymal stroma. Although only a few of the many interactive situations explored can be described here, it can be stated that all the experiments show that the differentiation of epithelium and mesenchyme depends on interactions between them. But, as will be seen, such interactions are not necessarily identical in mechanism and the influence of the mesenchyme may vary from determinative ('instructive') actions to entirely non-specific, stabilizing effects.

Dermal mesenchyme determines epidermal differentiation

Differentiation of the vetrebrate epidermis and its derivatives is known to be directed by the underlying mesenchyme, and here we find further examples of an actual 'instructive' influence. Mesenchyme from various sources, when combined experimentally with embryonic chick epidermis, alters epidermal

differentiation over the whole range from the keratinizing, squamous type to ciliated, cuboidal epithelium (Fig. 3.4.6). In such experiments, the regional differences can be demonstrated even within the same embryonic dermis. When dermis from the feather-bearing back is combined with epidermis of the normally featherless ventral skin, feathers form, whereas the reciprocal combination does not support development of feathers [26]. Judging from various transplantation experiments, the feather-forming bias of the epidermis is determined very early in development, but for its expression requires specific, local mesenchymal induction [20].

Such dermo-epidermal interactions are not restricted to the embryonic period, and there is experimental evidence that the maintenance of certain types of epithelium and the continuous differentiation of their basal cells is controlled by the dermis throughout adult life. Billingham & Silvers [4, 5] made recombinations of epidermis and dermis from various sites of adult guinea-pig (sole of the foot, ear and trunk). They showed that the type of epidermal morphology was determined by the dermal mesenchyme and not by a stable phenotype of the epidermis. For example, when epidermis from a pigmented area of the ear was transplanted to an unpigmented area in the sole, the transplant remained pigmented but had the morphological characteristics typical of the sole epidermis.

Perhaps the most convincing example of such specific epithelio-mesenchymal interactions determining the phenotype of the epithelium is that described in tooth morphogenesis [14, 15]. This process is known to be guided by interaction between the enamel epithelium and the mesenchyme of the dental papilla, and the type of tooth formed is ultimately determined by the latter. To test the inductive capacity and specificity of the dental mesenchyme, it was combined with various non-dental epithelia and well-shaped teeth with differentiated enamel and dental matrices were regularly obtained. Figure 3.4.7 illustrates one of their most striking results, a tooth which is a product of dental mesenchyme combined with epidermis from the foot! A reciprocal combination of the presumptive enamel epithelium with various heterologous mesenchymes did not lead to tooth formation, showing that the differentiative capacities of the dental epithelium require, for their expression, a specific mesenchymal influence.

Glandular mesenchyme supports epithelial morphogenesis

Glandular epithelium regularly interacts with its meschymal counterpart and this system has been used in a great variety of experiments. Pancreatic, salivary, mammary, hepatic and many other glandular epithelia have been shown to be morphogenetically dependent on their mesenchymal stroma [for review, 10]. Unlike the readily modified epidermis discussed above, the glandular epithelium, in most instances studied, seems to be strictly limited in its developmental alternatives. A good

Isolated epidermis

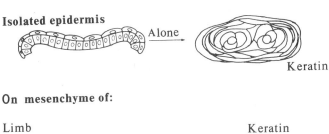

Alone → Keratin

On mesenchyme of:

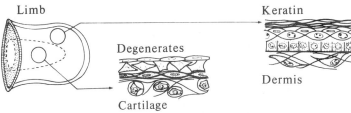

Limb

Degenerates

Cartilage

Keratin

Dermis

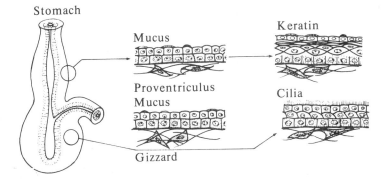

Stomach

Mucus

Proventriculus
Mucus

Gizzard

Keratin

Cilia

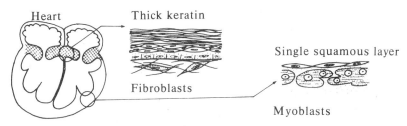

Heart

Thick keratin

Fibroblasts

Single squamous layer

Myoblasts

Fig. 3.4.6 Differentiation of embryonic chick epidermis in combination with various heterologous mesenchymes. After [18].

example is the pancreatic epithelium, which even at a stage when the gland rudiment is no more than a small stalk does not depend on the presence of normal pancreas mesenchyme and can continue its development if cultured with mesenchyme from almost anywhere in the embryo. Morphogenesis and enzyme synthesis proceed after combination of this isolated epithelium with a variety of heterologous mesenchymes [8, 40]. Salivary epithelium at a comparable stage is similarly committed but much more selective in its mesenchymal requirement, as very few of the mesenchymes tested so far can support its normal branching and adenomere formation [12]. The basis for such variations in the specificity of the mesenchymal requirements of these and other glandular epithelia is not known, but most of them respond to the mesenchyme in an all-or-none fashion.

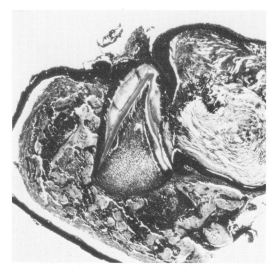

Fig. 3.4.7 Micrograph of a well-differentiated tooth obtained after an experimental combination of dental papillary mesenchyme with plantar epidermis. From [15].

Alteration of the morphogenesis of these epithelia by heterologous mesenchyme is rare. The most striking example is the salivary-like branching pattern of mammary epithelium in combination with embryonic salivary mesenchyme [16].

Parallel with the restriction of their differentiative potentialities, these epithelia gradually acquire independence of specific mesenchymal influences during embryogenesis. This is shown in the following series of experiments [21, 40]. At an early stage an explant of the presumptive pancreatic region of the gut surrounded by its homologous mesenchyme will form pancreatic acini *in vitro*, but fails to do so if the mesenchyme is removed or replaced by a heterologous mesenchyme. Somewhat later stages of the epithelial anlage still requires mesenchymal support, but this becomes less specific, and finally the epithelium reaches a stage at which the mesenchymal influence can be simulated by high concentrations of embryo extract in the culture medium in the total absence of mesenchymal cells. Comparable phases of interactions of varying specificity have been demonstrated in the development of the liver epithelium [17]. At the 5-somite stage of the chick embryo the liver entoderm is determined by the liver-heart mesoderm, which cannot be replaced by any heterologous mesenchyme. Later, this determined entoderm still requires mesenchymal support to become organized into typical liver cords, and now the effect can be simulated by several heterologous mesenchymes.

If such experiments on the varying specificity of the inductive interactions are tentatively generalized and simplified, the scheme might be as follows. During embryogenesis, extrinsic influences guide the differentiation of cells and cell populations through various phases. Initially pluripotent cells become determined and their developmental alternatives are restricted (neuralization, epithelialization, etc.). Subsequently, their organization into polarized, synchronized structures is governed by increasingly less specific, permissive influences (organogenetic tissue interactions). Finally, stages of increasing stability of the organ are reached and stabilizing influences are replaced by maintenance effects which are supportive and probably act through nonspecific channels. Obviously, there are differences in this simplified scheme between different tissues and organs. For example, the constantly renewing epidermis may represent a target cell population which never reaches the stable stage of differentiation and, instead of maintenance effects, requires constant guidance by the underlying dermis. This scheme also emphasizes the great multitude of interactive processes usually lumped under the title 'embryonic induction', and makes it obvious that the various types of interactive mechanisms should be considered separately, as will be done below.

3.4.4 INDUCTIVE COMMUNICATION IS ACHIEVED IN DIFFERENT WAYS

Signal substances are still poorly characterized

If one looks at the different cases of tissue interaction studied, not very much detailed information is available on the transmission of inductive stimuli between cells. Closer chemical characterization of the so-called inducing substances operative in normal development has been largely unsuccessful. Analysis of the substances involved in primary induction points towards their protein nature. A protein fraction, with a molecular weight of 25,000–30,000, that has a mesodermalizing effect when tested upon Amphibian gastrula ectoderm has been isolated from chick embryonic tissues, and substances of similar type have been isolated from other sources [cf. 32]. Whether they are truly responsible for normal development is uncertain and until more information is available, we must content ourselves with discussing the transmission of inductive stimuli at another level. We can discuss the biological systems studied according to what is known about the distance over which the interaction takes place, about the extracellular structures lying between the interacting cells, and about the effects of experimental modifications of the tissue interactions.

Long-range transmission of signals is possible

Primary embryonic induction serves as a typical case of tissue interaction where the need for the physical closeness of the tissues involved has come under careful study. As long ago as the early 1930s Spemann's work on amphibian embryos indicated that certain tissues, even when killed, are capable of

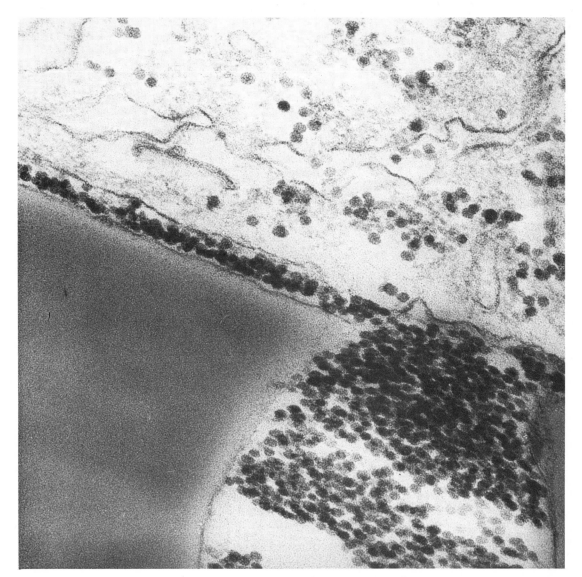

Fig. 3.4.8 Electron micrograph of transfilter induction of *Triturus vulgaris* gastrula ectoderm by axial mesoderm tissue through a Nuclepore filter. Cells of the ectoderm do not penetrate into the filter pores, which contain numerous electron-dense granules. × 135,000. Courtesy of Dr David Tarin.

inducing the development of gastrula ectoderm and that this capacity can be transferred to agar in a cell-free form [1]. These experiments gave rise to the idea of transmission of signals by diffusion which gained further support from the discovery that purified cell-free extracts can mediate both neural and mesodermal induction [cf. 32]. Ultrastructural studies then showed that *in vivo* a wide space separates the ectoderm from the mesoderm at the time of primary induction [13]. Also, it was demonstrated that neuralization can be evoked by the natural inductor tissue across porous filters in the absence of intact cytoplasmic contacts (Fig. 3.4.8). Granular and fibrillar material was found in *Xenopus* gastrulae between the ectoderm and mesoderm during the induction period [13, 31]. This was the case both *in vivo* and *in vitro* induction through filters. The significance of this extracellular material is still obscure.

In several other cases long-range transmission of an inductive influence is also thought to play a role. The formation of pancreatic acini and the initiation of cartilage formation in the somites as the vertebral column develops are two of the events that can be induced through porous filters with the aid of inductors. But the responding tissues in these cases have already started to differentiate and the inducer has merely a supportive function. This was demonstrated by experiments in which the normal inducer was omitted and the extent of cartilage formation was increased by supplementing the culture medium with cell-free tissue extracts or serum of heterologous origin [9, 21, 22]. Such results emphasize the need for caution in interpreting *in vitro* results on induction of differentiation.

Long-range transmission of signals between cells is not restricted to cases of embryonic induction, but is common to many control systems in multicellular organisms, as will be discussed in connection with the effects of hormones on differentiation in Chapter 4.4.

Contact-mediated induction is not excluded

Originally, the idea of the transfilter technique, illustrated in Fig. 3.4.9, was to exclude cytoplasmic contacts between interacting tissues. Recent ultrastructural studies have shown that cells can penetrate into filter with pore sizes of 0·2 micrometres or larger, and that some earlier data may have to be re-evaluated [36]. Induction of kidney tubulogenesis can be taken as an example. It was originally shown that induction took place when the interacting tissues were separated with a Millipore filter 20 to 30 µm thick and with an average pore size of 0·45 µm [11]. An inverse relation was found to exist between the pore size and the thickness of filters across which induction took place. After successful induction, no cytoplasm was found in the 0·1 µm pores of filters less than 18 µm thick, and this was interpreted to mean that transmissible substances were probably involved in the inductive effect. Later on, with Nuclepore filters (which, unlike Millipore filters, have a constant pore size), a minimal pore size of 0·15 µm was established for the transmission of tubulogenic induction. Cytoplasmic processes were found crossing the pores of the Nuclepore filters that allowed

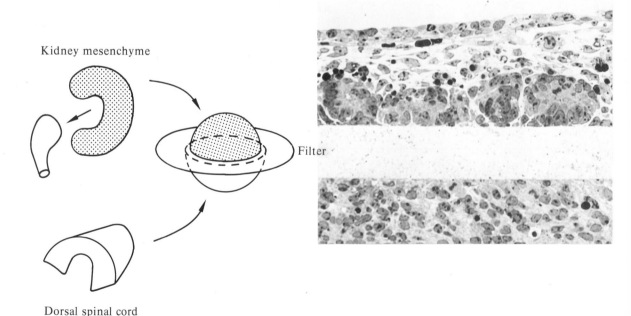

Kidney mesenchyme

Filter

Dorsal spinal cord

Fig. 3.4.9 Illustration of Grobstein's transfilter method of kidney tubule induction [11]. Metanephrogenic mesenchyme is separated from the adjoining epithelial bud and placed on a filter disc in juxtaposition to a piece of spinal cord tissue sealed to the opposite side with agar. Transfilter induction leads to the formation of kidney tubules in mesenchyme.

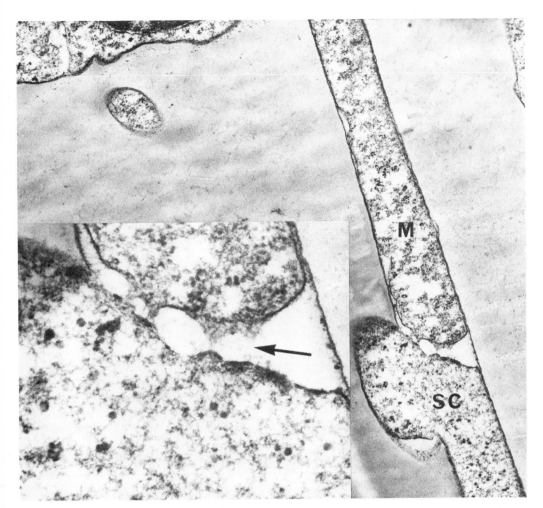

Fig. 3.4.10 Electron micrograph of transfilter kidney tubule induction, showing cytoplasmic processes within the 0·5 μm channels of a Nuclepore filter allowing induction. Insert demonstrates the close apposition of a mesenchymal (M) and spinal cord (SC) process. ×22,000, insert ×57,000. From [37].

induction but not in those that restricted it (Fig. 3.4.10). Hence, it was clear that diffusion could no longer be regarded as the sole explanation for the transmission of kidney tubulogenesis induction. In fact, induction mediated by cell contacts should be seriously considered [37].

Another advantage of the transfilter technique is that contact between the inducing and responding tissues can be broken at any given time. Thus, it was possible to measure the time needed for transmission of the inductive stimulus. To extract the actual transfer time from the total minimum induction time, consisting also of other events like tissue adaptation to *in vitro* conditions,

the following experiment was performed. Kidney mesenchyme was cultured in close proximity to the inducing spinal cord tissue with either one or two Millipore filters interposed. After varying periods of time the mesenchyme was removed from the filter and grown alone. The results showed that with one filter interposed the minimal induction time needed was of the order of 18 h, and that an additional filter prolonged the required induction time by more than 12 h (Fig. 3.4.11).

The diffusion rates of a number of substances, including virus particles, was then tested with these filters. From the diffusion data it was calculated that in order to satisfy the results of the

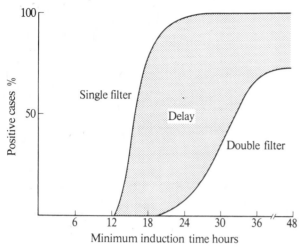

Fig. 3.4.11 Induction of kidney tubules in single and double filter experiments as a function of time of transfilter cultivation with spinal cord. In double filter experiments over 12 h longer transfilter cultivation is required for induction to take place. In the double filter group the induction percentage is also lower. After [19].

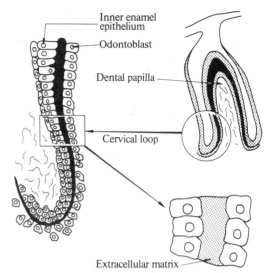

Fig. 3.4.12 Diagram of a developing tooth primordium. Epithelial cells are separated from mesenchymal cells (odontoblasts) by a layer of extracellular material. After [28].

minimum induction time the inducing substance should be irrationally large [19].

In conclusion, concerning transfilter kidney tubulogenesis, transmission of the inductive influence cannot be explained by simple diffusion of molecules across the interspace, but may depend on the cytoplasmic processes extending from the inducing tissue and making contacts with the responding cells. In other systems direct contacts between interacting cells have also been reported *in vivo* as in the case of rat submandibular salivary gland development [7].

Extracellular vesicles may carry inductive information

In all adult organs the epithelial component is separated from the mesenchyme by a layer of extracellular material which is called the basement membrane by light microscopists. Similarly, in developing organs the epithelium is covered by a layer of extracellular material. It is likely that certain components of this material are involved in interactions between neighbouring tissues. One of the most useful models in this respect is the developing embryonic tooth rudiment (p. 132), which offers many of the advantages of the transfilter technique applied to other systems *in vitro*. In the developing tooth epithelio-mesenchymal interaction takes place between the cells of the enamel epithelium and the dentine forming cells of the papilla, which are separated by a 10 to 20 μm thick extracellular matrix

(Fig. 3.4.12). In addition to examination of the situation *in vivo*, the development of the tooth primordium can be studied by isolating the extracellular matrix between the cell layers by microdissection and cultivating on it cells of either epithelial or mesenchymal origin. Under these conditions, cytodifferentiation takes place *in vitro* [29].

What is exceptional in the developing tooth primordium is that electron microscopy has revealed different types of 'matrix vesicles' situated in the fibrillar extracellular material [27]. These vesicles are electron-dense, membrane-limited bodies 0·05 to 0·1 μm in diameter (Fig. 3.4.13). The extracellular matrix also contains cytoplasmic processes, but serial sectioning at the ultrastructural level has shown that the matrix vesicles are not connected to these processes. Some of the vesicles, which contain ribosome-like granules, stain with the indium trichloride staining method for nucleic acids, but not after ribonuclease treatment of the matrix. The affinity for the stain disappears as a result of extraction with phenol, which removes methylated RNAs of relatively low molecular weight. Electron microscopy of the extracted matrix no longer shows vesicles, and the morphogenetic effect of the matrix is also abolished.

In summary, the experiments with the developing tooth primordium indicate that differentiation can be achieved with an extracellular matrix. Both the surface properties of the matrix and the RNA-containing vesicles have been suggested to be involved in the phenomenon. At present it is not known if the vesicles actually participate in the tissue interaction, whether

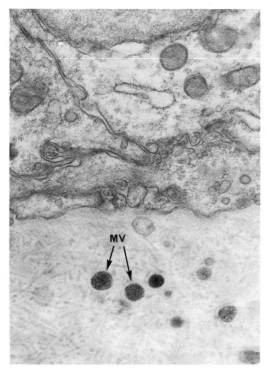

Fig. 3.4.13 Electron micrograph of the extracellular layer between the epithelial and mesenchymal cells in a developing tooth. Besides fibrillar material of collagen nature numerous 'matrix vesicles' (MV) of different types are found embedded in the basement membrane. ×20,500. Courtesy of Dr Harold Slavkin.

they are transferred from one tissue to another and what role the extractable RNA plays in the system.

The intercellular matrix affects morphogenesis

From numerous tissue culture experiments it has become evident that the substratum on which cells are grown affects their growth characteristics. Collagen, widely used as a supporting growth layer, is an integral component of basement membrane structures. Collagenase has been reported to affect the development of certain epithelia (salivary, pulmonary, and ureteral bud), but in these cases impurities in the enzyme preparation might explain the results. More morphogenetic significance is nowadays accorded to the basement membrane glycosaminoglycans (acid mucopolysaccharides) [3].

In the mouse embryonic salivary gland, the participation of extracellular material in epithelial morphogenesis has been the object of special study. During the development of the gland, newly synthesized glycosaminoglycans accumulate on the surface of the salivary epithelium, especially at the sites of epithelial branching within the mesenchyme. If the epithelium is separated from the mesenchyme and treated with hyaluronidase and collagenase, the explant loses its lobes and becomes a round mass. Upon further culture, in direct contact with salivary mesenchyme, it resumes a branching morphogenesis (Fig. 3.4.14). In the rounded mass, the glycosaminoglycans are spread evenly over the surface but when branching resumes they are concentrated at the sites of branch formation. On the other hand, if the epithelium is freed of mesenchyme without enzyme treatment so that most of the surface mucopolysaccharides remain, it retains its shape and continues its branching morphogenesis if grown in direct combination with fresh mesenchyme. Also, when the isolated epithelium is treated only with collagenase in a low concentration, branching continues in a normal fashion [2].

The dependence of normal salivary morphogenesis on an intact cell coat at the epithelial surface seems, in fact, to be attributable to hyaluronidase-sensitive acid-mucopolysaccharide-protein complexes (proteoglycans). Their mode of action is not known. The contractility of the micro-filaments in the epithelial cells has been suggested to require the presence of acid mucopolysaccharides outside. The microfilaments, in turn, have been suggested to control cell shape and initiate the formation of morphogenetic clefts in the developing salivary gland [30].

3.4.5 MOLECULAR MECHANISMS OF INDUCTION ARE STILL UNKNOWN

From this survey of the available data related to the transmission of inductive signals, it appears likely that both specific and less specific substances are involved. In some cases the signals may be of quantitative significance, and may, for instance, control the amount of specific enzyme synthesis by the developing pancreatic rudiment [22]. In other cases the signal might be of a qualitative character, causing the switching on of some genes with the appearance of their specific products (determination). Although the inductive substances involved are not known, it is worth considering three principal categories:

Compounds spatially arranged at the interface between cells leading to reorientation and rearrangement of the adjacent cells.

Molecules at the cell surface interacting with complementary compounds of the other cell type.

Specific, informative molecules transmitted from one cell to another.

The possible importance of spatial arrangement and physical forces is illustrated in certain experiments on embryonic chick

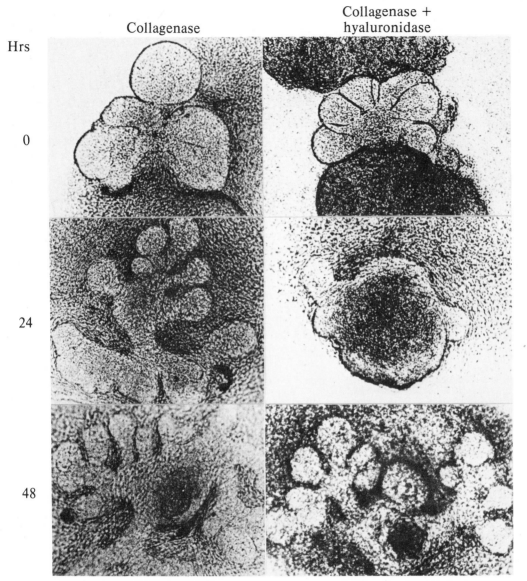

Hrs

Collagenase

Collagenase +
hyaluronidase

0

24

48

Fig. 3.4.14 Mouse submandibular salivary gland epithelium shows normal development after collagenase treatment (0·20 µg/ml) and subsequent cultivation in direct contact with salivary mesenchyme for 24 and 48 h. If hyaluronidase treatment is included, the epithelium rounds up to form a mass which subsequently shows bud formation at 24 h of cultivation with salivary mesenchyme. During further cultivation normal morphogenesis is acquired. From [2].

epidermis, in which differentiation and proliferation are normally controlled by the underlying dermal mesenchyme. This mesenchymal effect can be simulated to a certain extent by a suitable concentration of embryo extract, provided that a physical substrate is available for the basal cells. Such materials as devitalized dermis, collagen gel and Millipore filters have been successfully used [39].

The idea of certain 'template' molecules at the surfaces of interacting cells was put forward long ago by Weiss [38] and has since been shown to operate in various biological interactive processes. Inductive surface interactions could include, for example, antigen–antibody or enzyme-substrate type of reactions, but evidence for their existence is still lacking. Differentiating cells do show surface specialization detectable as specific affinity for and recognition of like cells (Chapters 3.5 & 3.6), and it is conceivable that similar, complementary specialization for inductive interaction between cells of different types can develop.

Finally, the classic hypothesis of transmissible, informative molecules should be mentioned [6]. Recent studies have repeatedly shown that close intercellular contacts create channels for the passage of molecules; most molecules which can pass between cells have molecular weights around 1000 and they might be able to carry specific information (Chapter 3.2).

Again, direct evidence for such a transfer during inductive processes is lacking.

3.4.6 REFERENCES

[1] BAUTZMANN H., HOLTFRETER J., SPEMANN H. & MANGOLD O. (1932) Versuche zur Analyse der Induktionsmittel in der Embryonalentwicklung. *Naturwissenschaften*, **20**, 971–74.

[2] BERNFIELD M.R. & WESSELLS N.K. (1970) Intra- and extracellular control of epithelial morphogenesis. *Devel. Biol. Suppl.*, **4**, 195–249.

[3] BERNFIELD M.R., BANERJEE S.D. & COHN R.H. (1972) Dependence of salivary epithelial morphology and branching morphogenesis upon acid mucopolysaccharide-protein (proteoglycan) at the epithelial surface. *J. Cell Biol.*, **52**, 674–89.

[4] BILLINGHAM R.E. & SILVERS W.K. (1967) Studies on the conservation of epidermal specificities of skin and certain mucosas in adult mammals. *J. Exp. Med.*, **125**, 429–46.

[5] BILLINGHAM R.E. & SILVERS W.K. (1968) Dermoepidermal interactions and epithelial specificity. In *Epithelial-mesenchymal interactions*, pp. 252–66 (Eds R. Fleischmajer & R.E. Billingham). The Williams & Wilkins Company, Baltimore.

[6] BRACHET J. (1950) *Chemical embryology*. Interscience Publishers, New York & London.

[7] CUTLER L.S. & CHAUDHRY A.P. (1973) Intercellular contacts at the epithelial-mesenchymal interface during the prenatal development of the rat sublen-mandibular gland. *Devel. Biol.*, **33**, 229–40.

[8] DIETERLEN-LIÈVRE F. & HADORN H.B. (1972) Développement des enzymes exocrines dans les bourgeons pancréatiques chez l'embryon de poulet en présence de mésenchymes homologues et hétérologues. *Roux. Arch.*, **170**, 175–84.

[9] ELLISON M.L. & LASH J.W. (1971) Environmental enhancement of in vitro chondrogenesis. *Devel. Biol.*, **26**, 486–96.

[10] FLEISCHMAJER R. & BILLINGHAM R.E. (1968) *Epithelial-mesenchymal interactions*, pp. 1–326. The Williams & Wilkins Company, Baltimore.

[11] GROBSTEIN C. (1956) Trans-filter induction of tubules in mouse metanephrogenic mesenchyme. *Exp. Cell Res.*, **10**, 424–40.

[12] GROBSTEIN C. (1967) Mechanisms of organogenetic tissue interaction. *Nat. Cancer Inst. Monogr.*, **26**, 279–99.

[13] KELLEY R.O. (1969) An electron microscopic study of chordamesoderm-neurectoderm association in gastrulae of a toad, *Xenopus laevis. J. Exp. Zool.*, **172**, 153–80.

[14] KOLLAR E.J. & BAIRD G.R. (1970) Tissue interactions in embryonic mouse tooth germs. I. Reorganization of the dental epithelium during tooth-germ reconstruction. *J. Embryol. Exp. Morph.*, **24**, 159–71.

[15] KOLLAR E.J. & BAIRD G.R. (1970) Tissue interaction in embryonic mouse tooth germs. II. The inductive role of the dental papilla. *J. Embryol. Exp. Morph.*, **24**, 173–86.

[16] KRATOCHWIL K. (1969) Organ specificity in mesenchymal induction demonstrated in the embryonic development of the mammary gland of the mouse. *Devel. Biol.*, **20**, 46–71.

[17] LeDOUARIN N. (1964) Induction de l'endoderme préhépatique par le mésoderme de l'aire cardiaque chez l'embryon de poulet. *J. Embryol. Exp. Morph.*, **12**, 651–64.

[18] McLOUGHLIN C.B. (1963) Mesenchymal influences on epithelial differentiation. In *Cell differentiation*, pp. 359–89. Symposia of the Society for Experimental Biology, No. 17. Cambridge University Press, Cambridge.

[19] NORDLING S., MIETTINEN H., WARTIOVAARA J. & SAXÉN L. (1971) Transmission and spread of embryonic induction. I. Temporal relationships in transfilter induction of kidney tubules in vitro. *J. Embryol. Exp. Morph.*, **26**, 231–52.

[20] RAWLES M.E. (1965) Tissue interactions in the morphogenesis of the feather. In *Biology of the skin and hair growth*, pp. 105–28. Angus & Robertson, Sydney.

[21] RUTTER W.J., WESSELLS N.K. & GROBSTEIN C. (1964) Control of specific synthesis in the developing pancreas. *Nat. Cancer Inst. Monogr.*, **13**, 51–65.

[22] RUTTER W.J., KEMP J.D., BRADSHAW W.S., CLARK W.R., RONZIO R.A. & SANDERS T.G. (1968) Regulation of specific protein synthesis in cytodifferentiation. *J. Cell Physiol. Suppl. 1*, **72**, 1–18.

[23] SAXÉN L. & TOIVONEN S. (1961) The two-gradient hypothesis in primary induction. The combined effect of two types of inductors mixed in different ratios. *J. Embryol. Exp. Morph.*, **9**, 514–33.

[24] SAXÉN L. & TOIVONEN S. (1962) *Primary embryonic induction*, pp. 1–271. Academic Press, London.

[25] SAXÉN L., TOIVONEN S. & VAINIO T. (1964) Initial stimulus and subsequent interactions in embryonic induction. *J. Embryol. Exp. Morph.*, **12**, 333–38.

[26] SENGEL P., DHOUAILLY D. & KIENY M. (1969) Aptitude des constituants cutanés de l'aptérie médioventrale du Poulet à former des plumes. *Devel. Biol.*, **19**, 436–46.

[27] SLAVKIN H.C. (1972) Intercellular communication during odontogenesis. In *Developmental aspects of oral biology*, pp. 165–99 (Eds H.C. Slavkin & L.A. Bavetta). Academic Press, New York & London.

[28] SLAVKIN H.C. & BAVETTA L.A. (1968) Odontogenic epithelial-mesenchymal interactions in vitro. *J. Dent. Res.*, **47**, 779–85.

[29] SLAVKIN H.C., BRINGAS P., CAMERON J., LeBARON R. & BAVETTA L.A. (1969) Epithelial and mesenchymal cell interactions with extracellular matrix material in vitro. *J. Embryol. Exp. Morph.*, **22**, 395–405.

[30] SPOONER B.S. & WESSELLS N.K. (1970) Effects of cytochalasin B upon microfilaments involved in morphogenesis of salivary epithelium. *Proc. Nat. Acad. Sci., U.S.A.*, **66**, 360–64.

[31] TARIN D. (1972) Ultrastructural features of neural induction in *Xenopus laevis. J. Anat.*, **111**, 1–28.

[32] TIEDEMANN H. (1971) Extrinsic and intrinsic information transfer in early differentiation of amphibian cells. In *Control mechanisms of growth and differentiation*, pp. 223–34 (Eds D.D. Davies & M. Balls). Cambridge University Press, Cambridge.

[33] TOIVONEN S. (1953) Bone-marrow of the guinea-pig as a mesodermal inductor in implantation experiments with embryos of Triturus. *J. Embryol. Exp. Morph.*, **1**, 97–104.

[34] TOIVONEN S. & SAXÉN L. (1955) The simultaneous inducing action of liver and bone-marrow of the guinea-pig in implantation and explantation experiments with embryos of Triturus. *Exp. Cell Res. Suppl.*, **3**, 346–57.

[35] TOIVONEN S. & SAXÉN L. (1968) Morphogenetic interaction of presumptive neural and mesodermal cells mixed in different ratios. *Science*, **159**, 539–40.

[36] WARTIOVAARA J., LEHTONEN E., NORDLING S. & SAXÉN L. (1972) Do membrane filters prevent cell contacts? *Nature, Lond.*, **238**, 407–08.

[37] WARTIOVAARA J., NORDLING S., LEHTONEN E. & SAXÉN L. (1974) Transfilter induction of kidney tubules. Correlation with cytoplasmic penetration into Nucleopore filters. *J. Embryol. Exp. Morph.*, **31**, 667–82.

[38] WEISS P. (1947) The problem of specificity in growth and development. *Yale J. Biol. Med.*, **19**, 235–78.

[39] WESSELLS N.K. (1964) Substrate and nutrient effect upon epidermal basal cell orientation and proliferation. *Proc. Nat. Acad. Sci., U.S.A.*, **52**, 252–59.

[40] WESSELLS N.K. & COHEN J.H. (1966) The influence of collagen and embryo extract on the development of pancreatic epithelium. *Exp. Cell Res.*, **43**, 680–84.

Chapter 3.5
Cell Recognition and Pattern Formation in Plants

3.5.1 INTRODUCTION

Typical plant cells differ from animal cells in two important respects: (1) they are surrounded by relatively thick cell walls, and (2) consequently, the cells of multicellular plants are non-motile, and their positions within the plant body are determined during growth. The structure of the metazoan animal body depends upon adhesion between the protoplasts of its constituent cells and this, in turn, requires that each cell must be capable of recognizing the neighbouring cells with which it interacts. Hence, in animal tissues, cell recognition and communication occurs across and between cell membranes. In plants, the existence of the cell wall not only prevents the movement of cells within the plant body, but might even seem to remove the need for recognition between adjacent cells, since their relative positions are usually fixed and laid down during growth. However, by analogy with the situation in animals, it is probable that a necessary condition of the multicellular form of life is not only the maintenance of the integrity of the individual tissue, organ or whole plant, but also the need for discrimination or recognition of individual cells within either a mass of cells derived from a common meristem, or a population of gametes of diverse origins. The cell wall has proved to be a physiologically-active component of plant cells, that is strategically sited to play a role in the cell-to-cell transfer of informational molecules.

As we have seen, communication between plant cells is most commonly via the plasmodesmata (Chapter 3.3), but this does not necessarily mean there is communication across cell membranes as in animals, since there is continuity of cytoplasm between adjacent cells through the plasmodesmata, giving rise to the so-called *symplast*. Nevertheless, the necessity for cell recognition can be demonstrated at certain stages in the plant life-cycle, notably in sexual reproduction where motile gametes need to have the capacity for recognizing gametes of the opposite type and especially where outbreeding systems involve compatibility-incompatibility mechanisms. It is not surprising, therefore, that the most striking and clear-cut examples of cell recognition phenomena in higher plants are seen in the interactions between pollen grains and stigma, as has become apparent from recent advances in our understanding of the reproductive physiology of flowering plants.

In this chapter, the term 'cell recognition' is used to define the first event of cell-to-cell communication [1]. All positive recognitions between cells and tissues can be readily interpreted as arising from specific union, reversible or irreversible, between chemical groupings on the surfaces of the interacting cells [7]. Proteins or glycoproteins are the prime candidates, though polynucleotides (RNA) may be involved in amplification of recognition signals [8]. The initial recognition may be visualized as occurring between sterically complementary molecules, rather like enzyme—substrate, and antigen—antibody relationships. Figure 3.5.1 shows a model of the possible

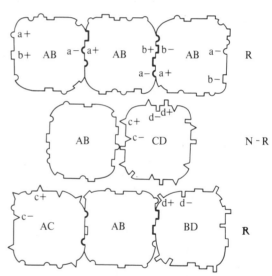

Fig. 3.5.1 Diagram showing possible +/− relationships between recognition units suggested by Burnet [7]. A positive receptor is shown as a projection, the corresponding complementary negative receptor by a depression of the same shape. Where there is a reciprocal relationship (R) allowing receptor union (heavy line), the relation is stabilized. In its absence (N–R), fusion is impossible.

interactions between recognition units, based on the system in the colonial tunicate, *Botryllus* [7]. In experiments, each star-shaped colony can be split into two parts, grown separately, and when rejoined the colonies will fuse. If the colony is rejoined to another unrelated colony of the same species, they reject each other, and a zone of necrosis develops between them. Recognition is based on a single genetic locus with many alleles or recognition genes with each diploid individual possessing two genes. Where the two colonies have at least one common allele, fusion occurs. If the alleles are different, then rejection is initiated. *Recognition* in these terms implies the ability of an organism to positively interact with another organism selected from many alternatives that it may contact.

Recognition sites are distributed on the surfaces of bacterial and mammalian cells. In both gram-positive and gram-negative bacteria, surface antigens are present as protein subunits, separated by several layers of peptidyl glycan polymer wall layers from the plasma membrane [9]. In mammalian cells, the surface antigens may be located partly in the plasma membrane, and also in the glycocalyx, a surface glycoprotein overlay [10], and can show considerable mobility in the membrane lipid matrix [11].

Following *cell recognition,* many types of cell communication can be envisaged. Two major communication systems have been elucidated. Firstly, when two kinds of cell come into *direct* contact, a triggering process may be activated which leads to changes in one of the kinds of cells. This is well demonstrated by the mating interactions of *Chlamydomonas* and yeast. It has proved possible to isolate glycoproteins (proteins with carbohydrate chains attached as an integral part of the molecule) from the cell surface of one mating type, that will neutralize or agglutinate the other [2, 3]. Secondly, when the two cells do not necessarily contact each other, informational molecules can migrate from one cell and bring about differentiation in the other cell. This is demonstrated by the response of sensitized lymphocytes to antigen. Specific substances called lymphokines that may induce cell death, are released by the lymphocyte in response to a signal following antigenic recognition on the cell surface [4]. Host-pathogen relationships in plants may depend on rather similar recognition reactions. The production of inhibitory phytoalexins in soybean plants resistant to the stem rot fungus, *Phytophthora megaspermae,* is stimulated by a protein released from the fungal hyphae on the surface of the soybean leaf [5]. The protein monilicolin A is claimed to have similar functions in red kidney bean [6]. In such cases, cells of one type release informational molecules which specifically interact with receptors on other cells.

In this chapter we will examine two areas that recently have provided evidence for cell recognition mechanisms in plants. Firstly, the pollen wall proteins and their role in pollen-stigma interactions that enables the stigma to recognize whether the pollen is 'self' or 'not-self', and set in train the acceptance or rejection responses. Secondly, we consider the lectins, a group of agglutinating proteins extracted mainly from legume seeds, which induce striking changes in animal cells, but whose role in nature has hardly begun to be understood.

In animals, pattern formation is very dependent upon cell recognition processes as will be shown (Chapter 3.6), but in plants, pattern formation is conditioned by the non-motility of the cells and by the presence of a thick and rigid cell wall. Some examples of the control of pattern formation in the cell walls of pollen grains and xylem cells are considered in the second section of this chapter and show that despite these restrictions pattern is dependent on the precise orientation and polarity of cellular determinants.

3.5.2 GAMETIC COMMUNICATORS: THE POLLEN-WALL PROTEINS

'In preparing the pollen of one of the Amentaceae for the microscope, a considerable quantity was accidentally inhaled before I was aware that it had been thrown off from the catkins so abundantly. A violent attack of sneezing came on in a few minutes. Later on there was a moderate copious discharge of thin serum which kept up for some hours. After the sneezing and coryza had continued for a couple of hours the breathing became very difficult as if from the constriction of the trachea or bronchial tubes, giving me just a slight experience of the misery those have to endure who suffer from the asthmatic form of hay fever. In the course of five or six hours I began to have aching and a sense of weariness over the whole body with pain in the head and spinal column. A very restless night was passed; the pulse rose from its usual (68) to 100. Occasionally there was a slight cough with expectoration of thin frothy sputum, and for twenty four hours I felt as if passing through an unusually severe attack of influenza.'

Charles H. Blackley, 1873.

With these words, written more than a century ago, Blackley [27] first recognized that pollen was a causative agent of hay fever. For pollen to be allergenic, susceptible humans have to be exposed to it, so it has to be present in large quantities in the air. Consequently, wind-dispersed pollens, such as grasses and North American ragweed, *Ambrosia* spp., are the principal sources. Noon, in 1911, showed that saline extracts of pollen contained the active agents now called *allergens.* Since then, work with ragweed and grass pollen has resulted in progress towards the purification and characterization of the allergens responsible for hay fever. These have proved to be proteins or glycoproteins. King and co-workers have shown that Antigens E and K together account for 90% of the allergenicity of ragweed extracts [13] and yet they comprise only 10% of the proteins released from the pollen in saline extracts. We have recently shown that these proteins are located in extracellular wall sites, from which they are rapidly released on moistening [14].

Stratification of pollen walls

The pollen grain wall is an elaborate structure showing great diversity of size and form [15, 16]. The wall is formed from two layers, defined on developmental and chemical criteria (Fig. 3.5.2). The outer sculptured layer forms the *exine*. It is the first wall layer to be deposited (see Section 3.5.6), and is composed of the unique biopolymer sporopollenin [17]. Sporopollenin is remarkable for its strength, plasticity and its resistance to biological degradation. Present evidence suggests it is formed by the oxidative polymerization of carotenoids and carotenoid esters.

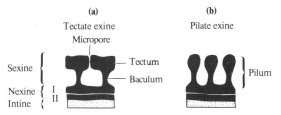

(a) Tectate exine   **(b)** Pilate exine

Micropore

Sexine { — Tectum — Baculum

Nexine { I II Intine {

— Pilum

Fig. 3.5.2 Diagrams showing the stratification of pollen grain walls. (a) shows a wall with a tectate exine and (b) with a pilate exine. Modified from Heslop-Harrison [16].

The exine comprises two distinct sporopollenin layers, named according to Erdtman's terminology [16]. The outermost sculptured layer, the *sexine*, consists of rod-like bacula surmounted either by a head or caput in *pilate* grains (*Lilium*, Fig. 3.5.3). These may be fused together to form the walls of *murate*-pollen (*Raphanus*, Fig. 3.5.3); or covered by a roof-like tectum in *tectate* pollen (*Cosmos*, Fig. 3.5.3). The sexine is endowed with all the patterning attributes remarkable to pollen, so much so that pollen exines preserved in quaternary deposits are identifiable and have provided valuable clues to the flora of the period. Tectate pollen is frequently ornamented by spines, and by micropores which communicate with the crypt-like cavities below (Figs 3.5.3 & 3.5.4). The inner exine layer, the *nexine,* comprises both the foot layer supporting the bacula (called strictly the nexine-1), and also the innermost sporopollenin layer, the nexine-2 (Fig. 3.5.2) which is often fused to the nexine-1 at maturity, though is developmentally distinct. This distinction is clear in *Cosmos bipinnatus*, where the nexine-2 forms a separate inner layer that is differentiated later in pollen development than the outer nexine-1 (Fig. 3.5.3) and forms a space, the cavus, between the two layers. The diversity of exine structure is well-illustrated in Fig. 3.5.3.

The inner wall layer, the *intine,* is similar in composition to the primary cell wall of somatic plant cells (see Section 3.5.5), comprising cellulose, hemicelluloses, pectic polysaccharides and proteins [15]. It is usually directly apposed to, and often interbedded with, the nexine. In many pollens, as in *Cosmos* and *Ambrosia,* the nexine forms an effective seal, permitting intine contact with the exterior surface only at the germinal apertures (Figs 3.5.3 & 3.5.5). In others, the intine at the apertures is

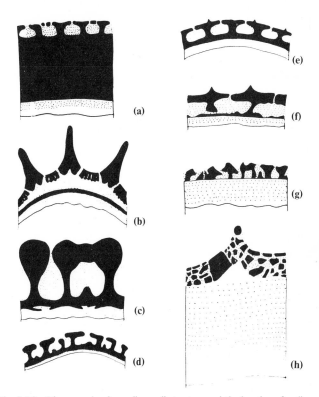

Fig. 3.5.3 Diagrams showing pollen wall structure and the location of wall proteins in different types of grain: (a) *Malvaviscus arboreus*; (b) *Cosmos bipinnatus*; (c) *Lilium longiflorum*; (d) *Raphanus sativus*; (e) *Silene vulgaris*; (f) *Gladiolus gandavensis*; (g) *Populus deltoides*; (h) *Crocus vernus*. Proteins are indicated by stippling. ×350.

covered by a sporopollenin cap of nexine, as in *Silene* and *Malvaviscus* (Fig. 3.5.3). The intine is conspicuously thickened at the aperture regions (Fig. 3.5.4) and shows the most prominent patterning here—with tubules, filaments or ribbons of proteinaceous materials running through the polysaccharide matrix [21], contrasting with the uniformity of the exine.

In many other pollens, the *nexine* layer is virtually absent, and the intine makes direct contact with the exine cavities (Figs 3.5.3 & 3.5.5).

Detection of the pollen-wall proteins

Proteins have been detected diffusing very rapidly from pollen grains in aqueous solutions using biochemical methods [18]. The allergens of grass pollen have been detected after 1 min extraction, though not in the concentrations obtained with prolonged extraction [19]. Tsinger & Petrovskaya-Baranova [20] were first to demonstrate the occurrence of proteins in

POLLEN-WALL PROTEINS

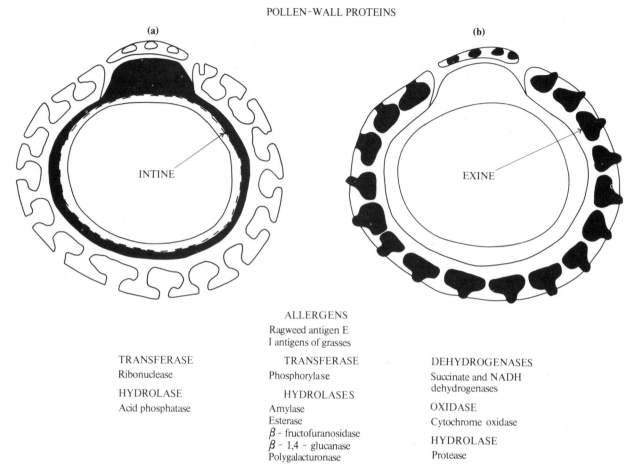

TRANSFERASE	ALLERGENS	DEHYDROGENASES
Ribonuclease	Ragweed antigen E	Succinate and NADH
	I antigens of grasses	dehydrogenases
HYDROLASE	TRANSFERASE	OXIDASE
Acid phosphatase	Phosphorylase	Cytochrome oxidase
	HYDROLASES	HYDROLASE
	Amylase	Protease
	Esterase	
	β - fructofuranosidase	
	β - 1,4 - glucanase	
	Polygalacturonase	

Fig. 3.5.4 Location and nature of the pollen-wall proteins. (a) shows the location of the intine proteins in a grain with a single germinal aperture, such as grass pollen. (b) shows the location of the corresponding exine proteins. The enzymes on the left of the table are exclusively located in the intine, while those on the right are located in exine sites. The proteins in centre have been detected in both sites.

extracellular wall sites by microscopic cytochemical methods in sections of the pollen of *Amaryllis* and *Paeonia*. It is now known, from a survey covering many families and structural types of pollen [21], that a heterogeneous group of proteins occur in pollen wall sites.

Special methods were needed to demonstrate the presence of proteins in pollen-wall sites and were used to confirm that Antigen E, the ragweed allergen, is present in both intine and exine sites in ragweed pollen (Fig. 3.5.5). The principal method used was *freeze-sectioning* [22]. Pieces of anther tissue or dry pollen were simply supported in melted gelatin medium and rapidly frozen. Sectioning was carried out in a cryostat (a refrigerated microtome) at −15°C, and sections as thin as 2 μm could be cut. These were thawed and dried on microscope slides

for staining. Rapid processing before freezing was found essential to prevent diffusion of the proteins from the outer wall sites (see Fig. 3.5.5). *Freeze-substitution* methods have been used to rule out any possibility of diffusion artifacts [23]. Samples were rapidly frozen by dipping into liquid freon held in a container of liquid nitrogen, then transferred to ethanol at −70°C for several days, when the ice crystals in the tissue are sublimed and the specimens are brought to room temperature, embedded in paraffin wax and sectioned under anhydrous conditions.

The presence of proteins in the wall sites was established in several ways. Staining reactions for proteins based on the reaction of dyes with specific groupings in protein molecules provided a visualization of total proteins in the pollen grains.

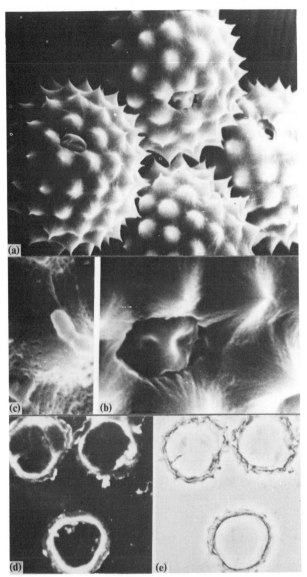

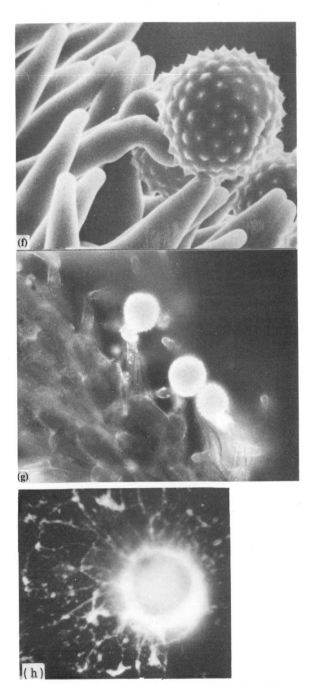

Fig. 3.5.5 (a)–(c) scanning electron micrographs of whole grains of ragweed, *Ambrosia trifida*; (a) ether-washed grains showing densely packed spines rising from the exine, and germinal apertures or colpi (c) ×3400; (b) colpus at higher magnification, showing prominent furrows on flanks of spines. ×11,000; (c) partially-acetolysed grain with some remaining surface lipid (l) showing micropores (m) at base and on flanks of spines. Each micropore is approx. 20 nm in diameter. Acetolysis is an acid treatment—removes all biological materials except exine sporopollenin. ×11,000 (photographs by Dr D.F. Cutler); (d) and (e) pollen of ragweed, *A. trifida*, sectioned to reveal sites of Antigen E within the grain. Preparation by freeze substitution to prevent diffusion of labile proteins during processing. Fluorescence image ×900; (f) scanning electron micrograph of living stigma of ragweed, *Ambrosia tenuifolia*, showing germinating pollen grain, with tube penetrating a stigmatic papillum (t) 15 min after pollination ×1800; (g) pollinated stigma placed in anti-Antigen E serum followed by FITC-labelled anti-rabbit serum, to reveal sites of allergen binding on the stigma surface and around pollen tube. ×375; (h) pollen of A trifida freeze-sectioned in gelatin medium within 30 s. Immunofluorescence method using anti-Antigen E serum showing release of Antigen E from all around the grains.

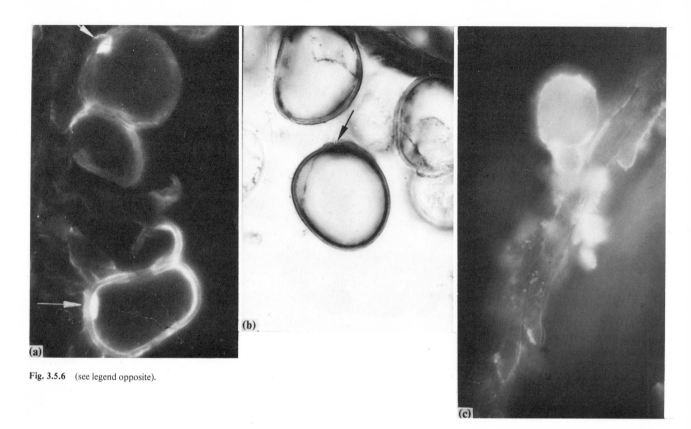

Fig. 3.5.6   (see legend opposite).

Two very sensitive dyes are now available. Coomassie Blue binds to the α-amino-groups in proteins, forming an intense blue-purple reaction product clearly visible by bright field microscopy [38]. The fluorescent protein probe, 1-anilino-8-naphthyl sulphonic acid (ANS) binds to proteins electrostatically. It is detected by its brilliant blue-green fluorescence when viewed by near ultra-violet (blue) light in a fluorescence microscope (see Fig. 3.5.8) [28].

Proteins can be detected by their enzymic activity, and many methods exist depending on the nature of the enzyme and its substrate. Those subject to least diffusion artifacts are simultaneous coupling reactions, which employ a synthetic substrate applied in solution to the section in the presence of a coupling agent, usually a diazonium salt. The tissue enzyme cleaves the substrate, and the released product is coupled to the diazonium salt to produce an insoluble coloured product at the site of the reaction [24]. Acid phosphatase (Fig. 3.5.6) has been used extensively as a marker for hydrolytic enzyme localization in pollen [21]. Alternatively, a natural substrate can be employed in substrate film methods, where sections of pollen are applied to a film of substrate on microscope slide. After a period

for digestion, the reaction is stopped and the substrate stained to reveal the areas of digestion, which appear clear against the coloured substrate background. The demonstration of amylase activity with starch substrate films is shown in Fig. 3.5.7. Many proteins can act as antigens and stimulate specific antibody formation in rabbits immunized with them, and so can be localized in tissue sections by the fluorescent antibody method [25]. These methods make use of the remarkable fidelity with which antibodies, the IgG or γG immunoglobulins of mammalian serum, can recognize and specifically complex with the antigen from which they were prepared. The antibodies can be labelled with a fluorescent dye, e.g. fluorescein isothiocyanate, enabling the sites of the precipitates to be recognized by fluorescence microscopy. Two methods are widely used for antigen localization (Fig. 3.5.9). In the first, the IgG fraction of the serum is fluorescent-labelled for use in the direct method; or secondly, unlabelled specific antiserum can be used first, followed by fluorescent anti-rabbit globulin serum in the indirect method, resulting in enhanced fluorescence. By careful use of such methods, it has proved possible to detect the accumulation of proteins in pollen-wall sites during development.

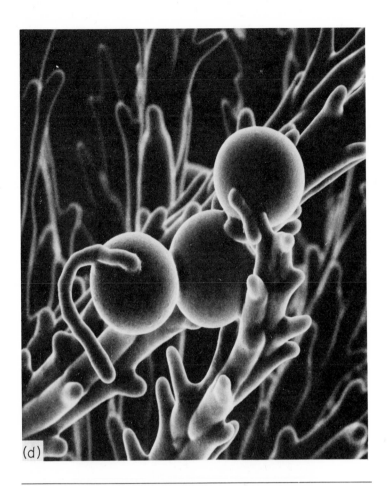

(d)

Fig. 3.5.6 (a) Location of antigenic proteins in the walls of Phalaris pollen using immunofluorescence methods [32]. Activity appears white in the sectioned pollen and is present in intine (arrows) and exine sites; (b) localization of acid phosphatase exclusively in intine, especially where it is thickened at the apertures (arrow); (c) demonstration of the release of antigenic pollen wall proteins from *Phalaris* pollen 5 min after its arrival on the stigma. Pollinated stigmas were placed in specific anti-serum to the pollen protein and, after processing fluorescence from the released antigens, is seen complexed with the stigma surface and around the pollen grains and germinating pollen tube; (d) similar preparation of living pollinated stigma viewed by scanning electron microscopy.

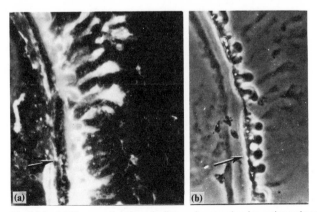

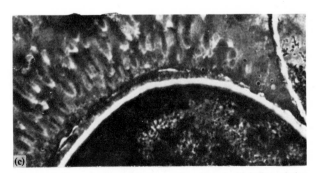

Fig. 3.5.7 (a) Shows section through pilate sexine area showing antigen release from the intine and sexine as wisps of precipitins into the gelatin medium; (b) shows phase contrast image of same field; (c) localization of amylase activity in iodine stained starch substrate films. In (c) frozen-sectioned pollen were placed on the starch film and, after 1 min the reaction stopped with iodine solution. Area of amylase activity appears white in the photographs and shows the enzyme as concentrated in the intine and exine cavities and has diffused into the gelatin medium. ×450.

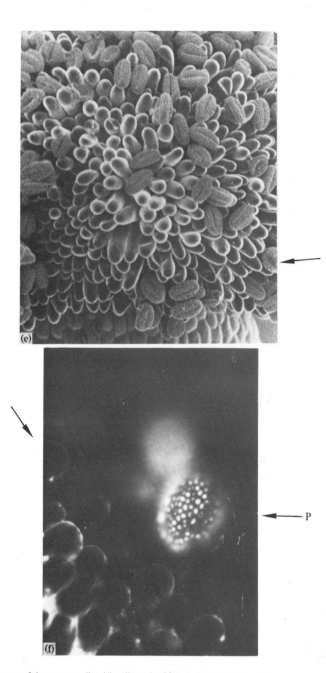

Fig. 3.5.8 (a) Localization of proteins in developing anther of *Iberis sempervirens*, ANS procedure, freeze-sectioned without prior fixation. Proteins present in the tapetal cells show intense fluorescence while only cytoplasmic activity is detectable in the pollen ×350; (b) localization of proteins in mature pollen just before anther dehiscence, ANS procedure as A. The tapetum is absent and intensely fluorescing proteins fill the exine cavities with pollen ×840; (c) scanning electron micrograph of *Iberis* pollen showing the murate sexine and one of the aperture slits; (d) pollen print from similar grain after 3 min exposure to agar gel, Coomassie Blue procedure. The sites of the wall proteins in the exine cavities are clearly demonstrated. The slit running longitudinally down the print is the site of one of the germinal apertures, but release of the intine proteins has not commenced ×1400; (e) scanning electron micrograph of living stigma of *Iberis* showing 1 min after pollination. One grain (arrow) has imbibed and begun to germinate ×210; (f) pollinated stigma exposed to ANS procedure for

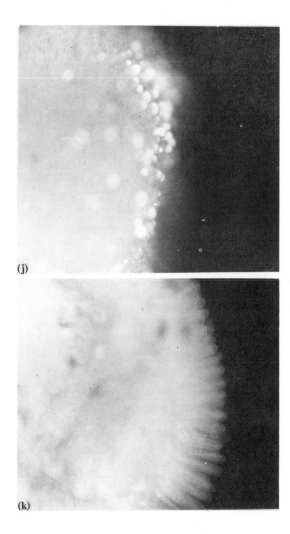

protein localization and viewed by fluorescence microscopy showing pollen grain (P) with fluorescing proteins in exine cavities and release of proteins on the surface of stigma. The pellicle on the surface of the stigmatic papillae (Fig. 3.5.18) is faintly stained (arrow) ×350; (g) self-pollinated stigma of *Iberis* showing location of callose, analine blue technique, fluorescence image. Callose is present in the inhibited pollen tubes and in the stigmatic papillae (arrow) ×140; (h) detail in which the stigma papillae have been teased apart to reveal the characteristic callose plugs ×350; (i) comparable picture after a compatible pollination, no callose plugs are present in the stigma papillae though plugs develop in the pollen tubes ×350; (j) rejection response induced in unpollinated stigma hairs of Brassica oleracea after treatment with self pollen wall protein extracts; callose is accumulated in papillae exposed to the extract ×350; (k) absence of rejection response in similar stigmas to those of K exposed to compatible pollen wall protein extract ×350. From [28].

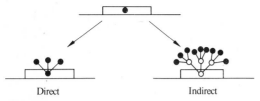

Fig. 3.5.9 Diagram showing immunofluorescence methods for antigen location in tissue sections. In the direct method, the specific antibody is fluorescence labelled and complexes directly with the antigen in the section. In the indirect method, unlabelled specific antibody forms the initial complex and is itself later complexed with fluorescent labelled sheep or goat anti-serum prepared against rabbit IgG.

Origins of the pollen-wall proteins

The surface of pollen grains with its varied and intricate sculpturing, shows two relatively constant features. The first is the presence of cavities of various sizes, from the large muri of *Lilium* and various Cruciferae (Fig. 3.5.4) to the intricate network of cavities connected by micropores to the surface in tectate pollens, as *Malvaviscus* and *Cosmos*. We have found, using very sensitive cytochemical methods, that these *exine* cavities are filled with proteins in mature pollen. A second feature is the presence of *germinal apertures,* usually 1·3 or many (though some pollens have no specific area designated before germination). Through one such site, the pollen tube will emerge. Also at these sites, the inner intine layer usually protrudes to the surface and contains a range of proteins within a polysaccharide matrix—*the intine proteins*. Both these classes of wall proteins are incorporated in the pollen walls at different times during development, and have quite distinct origins.

Exine proteins

SYNTHESIS AND TRANSFER FROM THE TAPETUM

The exine proteins have been detected in mature pollen filling the exine cavities, and diffusing rapidly through surface micropores, if present, when the pollen is moistened.

They have been detected using sensitive protein stains, as ANS (Fig. 3.5.8), by their esterase activity [30] and by fluorescent antibody methods (Fig. 3.5.5).

Evidence from pollens of several flowering plant families including the *Cruciferae, Compositae* and *Malvaceae* has demonstrated that the exine proteins are synthesized in the diploid tapetal cells that surround the anther cavity. During the final period of pollen maturation, the tapetum undergoes dissolution and the proteins are transferred and inserted into the exine cavities. In pollen of *Hibiscus* and *Malvaviscus,* the tapetal cells differentiate around the periphery of the anther cavity but undergo premature dissolution soon after meiosis, followed by breakdown of many of the cellular organelles. The tapetal fluid comes to bathe the developing pollen grains. Transmission electron micrographs at this period of development show the

fluid to contain pigmented lipid droplets and membrane-bounded, proteinaceous, vesicles derived from the endoplasmic reticulum of the original tapetal cells [26]. As the pollen matures the exine cavities are initially empty, but three days before anthesis, proteinaceous materials are injected through the micropores into the exine cavities, while the pigmented lipid materials remain as a surface coating or pollenkitt. In the Cruciferae, the transfer of similar proteinaceous materials from the tapetal cells to the exine cavities has been observed at the fine structural level in the radish, *Raphanus sativus* [27], while in *Iberis,* the events have been observed in the fluorescence microscope using freeze-sectioned anthers stained with the fluorescent protein probe, ANS [28]. The tapetal cells around the anther cavity show intense protein staining until just before pollen maturation (Fig. 3.5.8) when they undergo dissolution. Then fluorescent-stained proteins accumulate in the exine cavities which are murate and open to the surface. There is thus a considerable body of evidence showing that proteins of parental origin made in the diploid tapetal cells are transferred to the exine cavities.

RAPID RELEASE OF EXINE PROTEINS

When tectate pollen is placed in the Coomassie Blue stain-fixing medium, there is an almost immediate release of proteins from all over the surface of the grains. In *Malvaviscus* and *Cosmos,* this occurs within 2–5 s of wetting, and emission is associated with the spines (Figs 3.5.3 & 3.5.5) where release occurs through the micropores at their bases [23, 26]. In Cosmos, dense protein halos surround each spine within a few seconds.

The micropore emission has been confirmed by pollen-print methods, in which the dry pollen is pressed into contact with an agar film for periods as short as 5 s before processing with the Coomassie Blue medium, or by immunofluorescence methods. In *Malvaviscus* and *Cosmos,* release from around the bases of the spines has been detected with contact periods as short as 5 s, while release from the intine at the germinal apertures is delayed for several minutes [29]. In the murate pollen of *Iberis* (Cruciferae), the exine proteins are detected printed off into the gel with remarkable precision in short-term pollen prints (Fig. 3.5.8).

Intine proteins

INCORPORATION INTO WALL SITES

When pollen grains are released from the tetrad, the exine is well-developed, but the intine is not deposited until early in the vacuolate period of pollen development [25]. Intine polysaccharide synthesis is detectable cytochemically using the Periodic Acid-Schiff reaction with freeze-sectioned pollen. Protein synthesis, detected as hydrolytic enzyme activity, is associated with the intine polysaccharides soon after their synthesis has commenced.

Transmission electron microscope observations at this period show the cytoplasm adjacent to the developing intine is rich in polysomes and stratified endoplasmic reticulum in *Cosmos bipinnatus* [21]. The proteins are apparently deposited in layers in the wall between the polysaccharides. By the end of the vacuolate period, the intine proteins are conspicuously present as ribbons or leaflets, especially in the aperture regions where the layer is thickened. Synthesis of the intine and its proteins thus appears to be under the control of the pollen cytoplasm—the gametophytic or haplophase generation.

EMISSION FROM MOISTENED POLLEN

In many pollen grains, the route of release of the intine proteins is via the germinal apertures, where the intine makes contact with the surface (Fig. 3.5.2). The release of the intine proteins at the apertures has been demonstrated in pollen of the Malvaceae and Compositae using a Coomassie Blue stain-fixing medium. Emission of proteins begins after 2–3 min in the medium and continues for 10–15 min. This is especially clear in grass pollen where only a single pore is present (Fig. 3.5.6). In *Gladiolus* and *Iris* pollen (Fig. 3.5.7), where the nexine is virtually absent release of the intine proteins occurs directly to the surface through the exine cavities [25].

Origins of the pollen-wall proteins

From this diversity of evidence, we can conclude there are two quite different classes of proteins in pollen walls. The exine proteins are products of the diploid tapetal tissue—the parental *sporophyte*. In contrast, the intine proteins are synthesized by the pollen cytoplasm—the haploid *gametophyte*.

The time of onset of protein synthesis in the two systems is also different. The exine proteins are transferred from the tapetal cells during the final stages of pollen maturation. However, their synthesis in the tapetal cells appears to have been completed early in meiosis [28]. In contrast, the intine proteins are synthesized after meiosis within the microspores themselves, during the vacuolate period of pollen development [21]. Little is known of the nature of the two classes of proteins. Similar classes of enzymes have been found in both wall sites (esterase and amylase) [30] (Fig. 3.5.4), while others are specific to intine (ribonuclease, phosphatase) or exine sites (various dehydrogenases and oxidases). The ragweed allergen, Antigen E, detected by fluorescent antibody methods, has been found in both sites.

Genetical incompatibility barriers

Many different protein fractions are released from pollen, as many as 18 having been detected in grass pollens [19]. What possible function have these proteins in the biology of the pollen grain? Following the early lead given in 1894 by Green [31], we might expect enzymic proteins to function in the early events of pollen germination on the stigma. However, it now seems that many of these proteins function as *recognition substances* concerned with compatibility reactions on the female stigma, which regulate the breeding system [15, 21]. Charles Darwin described various mechanisms used by plants to regulate cross- and self-fertilization [33, 34] and these included:

(1) incompatibility barriers to prevent hybridization between different species and genera;
(2) self-incompatibility systems, when pollen from one plant fails to set seed on its own flowers, but can cross successfully with other plants of the *same* species.

Discriminations are therefore made at two levels, (1) in the rejection of foreign species and the acceptance of own species, and (2) in the rejection of own pollen and the acceptance of that from another genotype of the same species.

Successful pollination involves acceptance of the pollen grain at the stigma surface, followed by growth of a pollen tube which penetrates the stigma and grows down the style, and so acts as a conduit for transfer of the male gametes to the ovary. Pollen germination involves the uptake of water ('hydration') leading to emergence of the pollen tube. The causes of rejection of foreign pollen are not fully understood. In some instances it is probably due to a number of maladjustments in the relations between the pollen grain and the stigma, such as mismatch in the osmotic relations between the two, which may either prevent initial water uptake or cause bursting of the pollen tube. In other instances, it is thought that the rejection of pollen is controlled by a mechanism similar to that controlling rejection of pollen of the *same* species i.e. it is of the nature of a genetic incompatibility barrier.

Apart from the incompatibility barriers between species and genera which operate to prevent wide crosses, there are two major genetic systems which prevent selfing. *Gametophytic* self-incompatibility is determined genetically by one or two major loci named S and Z, each with multiple alleles, e.g. $S_1, S_2, S_3 \ldots S_n$. The S allele borne in each haploid pollen grain expresses its own phenotype during pollination on the diploid stigma, rejection occurring when the same allele occurs in both pollen and stigma (Fig. 3.5.10). Examples of this self-incompatibility system are to be found in cherry, *Prunus cerasus* (Rosaceae), rye *Secale cereale* (Gramineae) and in tobacco, *Nicotiana* spp. (Solanaceae) [35]. In gametophytic cases, pollen tube growth and penetration of the stigma surface occurs, but is not maintained within the style so that fertilization is prevented.

Secondly, *sporophytic* self-incompatibility is indicated by the identical behaviour of all the pollen from a plant on a given stigma, regardless of segregation of the S alleles during meiosis. This system is found in many Compositae and Cruciferae, and dominance of S alleles is demonstrated, the dominant allele, e.g.

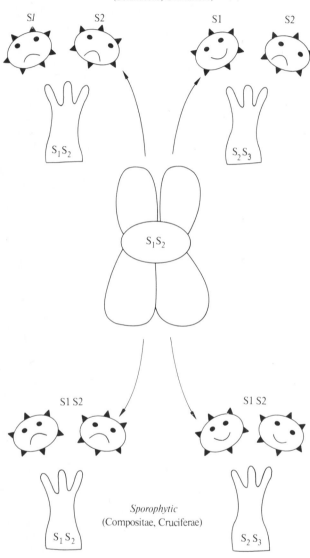

Gametophytic
(Solanaceae, Gramineae)

S_1 S_2 S_1 S_2

$S_1 S_2$ $S_2 S_3$

$S_1 S_2$

$S_1 S_2$ $S_1 S_2$

Sporophytic
(Compositae, Cruciferae)

$S_1 S_2$ $S_2 S_3$

Fig. 3.5.10 Simplified diagram to illustrate gametophytic and sporophytic self-incompatibility in flowering plants. The pollen parent is shown centre, producing pollen grains of haploid genotype either S_1 or S_2. In the gametophytic situation (top), the S allele expresses its own phenotype during pollination on the diploid stigma, whereas in the sporophytic system (bottom) the phenotype of all the pollen grains is determined by the genotype of the parent ($S_1 S_2$). In both the systems illustrated, the left hand pollinations are self-incompatible and the right hand ones compatible for (a) S_1 pollen grains only in the gametophytic system and (b) for all pollen grains of the sporophytic system (assuming that S_1 is dominant). A 'smile' on the pollen grains indicates a compatible pollination and a 'glum' expression an incompatible.

S_1, being expressed by *all* progeny from a heterozygous parent carrying this allele (Fig. 3.5.10). Thus, the phenotype of the pollen grains is determined by that of the parent plant. In this situation, the self pollen germinates, tube growth usually occurs, but the recognition systems enabling pollen tube penetration are incomplete and it is unable to penetrate the stigma surface so that fertilization is again prevented.

Events at the pollen-stigma interface

We have already seen that the pollen wall proteins, especially the exine proteins, are rapidly released in aqueous media. Does this occur on the stigma surface? Cytochemical evidence shows that within minutes of pollen alighting, pollen wall proteins are released and bind to the stigma surface, in the grass, *Phalaris tuberosa* [14]; the daisy, *Cosmos bipinnatus,* and ragweed, *Ambrosia tenuifolia* [38]; the beach cottonwood, *Populus deltoides* [37] and the crucifer, *Iberis sempervirens* [28]. These proteins were detected bound to the stigma surface adjacent to the pollen and around the germinating pollen tubes (Fig. 3.5.5).

Until recently, it was thought that the stigma surface was covered by a waxy cuticle [35]. However, we have demonstrated that the stigma surface is in fact overlaid by a hydrophilic proteinaceous pellicle which provides a sticky surface rather like a fly-paper to trap the pollen grains [39]. The pellicle possesses all the attributes of a recognition receptor site. It is external to the cuticle, providing an initial point of contact with the pollen grain and its wall proteins. High resolution transmission electron micrographs show that pollen wall protein binding occurs at the pellicle [39].

On arrival at the stigma surface, the pollen grain is able to imbibe water from the stigma and to commence the processes of germination within a fluid exudate. In *Silene,* pollen grain volume increases by about 50% within the first ten minutes, but later decreases as pollen tube growth commences [29].

The interaction between the male gametophyte and the stigma begins at the moment of capture of the pollen grain, but in an incompatible combination arrest or retardation may take place at various sites (Fig. 5.3.11). As we have seen, in the *sporophytic* system, the emergent pollen tube either fails to penetrate the stigma surface, or penetrates only a short distance. On the other hand, in most gametophytic systems studied (with certain exceptions, such as the grasses), the incompatible grains germinate and produce tubes which penetrate the stigma and grow down into the style, but growth is arrested at various depths. It is clear that in both systems there must be initial recognition reactions by which there is a discrimination between compatible and incompatible pollen, following which there is acceptance or rejection of the individual grains. We now have to discuss in more detail what is involved in each of these two steps.

'Mentor' pollen and stigma recognition

The role of the pollen wall proteins as recognition substances was first established by experiments in which incompatible pollen was 'disguised' with the pollen wall proteins of the compatible, and made acceptable to the stigmas [37]. In the composite, *Cosmos bipinnatus,* seed set from self-pollinations is very low and a sporophytic system of self-incompatibility has been demonstrated [see 38, 40]. In both crosses between two compatible lines and in selfings, the pollen-wall proteins bind to the stigma surface adjacent to the pollen and around the pollen tubes, providing visual evidence of communication between pollen grain and stigma (Fig. 3.5.5 and [38]). The nature of this interaction can be revealed by pollination experiments on altered stigma surfaces [40]. The stigma surface is first exposed to dead compatible pollen (produced by gamma-irradiation) or to an extract of compatible pollen-wall proteins, followed by self-pollination. Under these conditions, a seed set amounting to 15·9% was obtained, whereas without pre-treatment of the stigma the seed set obtained by self-pollination was less than 1%. Thus, the proteins of the non-self pollen had 'deceived' the stigma so that it allowed germination of normally incompatible self-pollen. It is interesting that the pollen extract remains effective even after heating at 60° for 10 mins, conditions under which only the allergenic components of pollens can survive.

This evidence, together with that from similar experiments with poplar hybrids [37], certainly indicates that the pollen-wall proteins act as recognition factors in determining acceptance or rejection of pollen. However, the nature of the initial recognition reaction and of the processes linking this with acceptance or rejection still remain obscure.

The rejection response of incompatibility

A drawback of these experiments is that definitive results are not available until seed set and often until after progeny testing. In order to investigate the biochemical mechanisms involved in cell-to-cell recognition, it is obvious that a more sensitive and rapid assay is required. Recently three quite independent observations on the cytochemistry of pollen-stigma interactions in the Cruciferae and Compositae [27, 28, 40] have provided just such an assay of self-incompatibility. It was found that both pollen tubes and adjacent stigma papillae become occluded by the β-1,3-linked glucan, callose, within a few hours of an incompatible pollination and that this reaction is absent in compatible matings. Within the family Cruciferae the callose stigma response is produced in all incompatible matings, either within or between species, and even genera [36]. This has led to the development of a direct test for incompatibility based on the production of callose in the stigma cells. The callose is detected cytochemically by its yellow fluorescence after staining with decolorized aniline blue [28]. It occurs as plugs or lenticules of callose formed on the inner face of the stigma cell wall in contact with the incompatible grain (Fig. 3.5.8). In self-incompatible matings in *Iberis* and *Brassica*, where the S alleles of the pollen were present in the stigma (self), the rejection response was induced. Where the alleles were different in pollen and stigma, as in compatible matings, no response was obtained. The rejection is certainly an active developmental response since stigma cells which have been 'narcotized' with ether vapour failed to produce the response. However, ether vapour treated pollen (which rendered it inviable) was still capable of eliciting the callose response in untreated self stigmas. Pollen washed in isotonic buffered mannitol, which removed the wall proteins leaving the pollen viable, also failed to provoke the callose response and was apparently not recognized by the stigmas [36].

The role of the exine fraction of the wall proteins in inducing rejection or acceptance has been demonstrated in several experiments.

(1) Sections of anther tissue, taken prior to transfer of proteins from the tapetum to the exine, are capable of inducing callose formation in the stigmas, suggesting that both synthesis and assembly of the exine proteins occurs in the tapetal cells prior to transfer [28].

(2) Short-term diffusates of the exine materials into cubes of agar gel will induce callose formation in self stigmas but not in compatible stigmas with different S alleles [28, 29].

(3) When intact *Raphanus* pollen was centrifuged over dry Millipore filters, the exine materials were driven into the cavities in the filters. After removing the pollen, squares of filter were placed on unpollinated stigmas, and shown to induce the callose rejection response when self pollen was used [41].

(4) Partly purified short-term extracts, consisting largely of the exine proteins of *Brassica oleracea*, marrow-stem kale, will induce callose formation in self stigmas but not in compatible stigmas [29].

(5) Using extracts partly fractionated by thin layer gel filtration, it has been possible to demonstrate that the protein or glycoprotein fraction is active in promoting rejection [28, 29]. The two major fractions present have molecular weights of ·10,000 and 16,000 daltons (in the range of Cytochrome *c* and ribonuclease).

These experiments have provided evidence that at least for sporophytic self-incompatibility in the Cruciferae, the exine proteins are involved in pollen-stigma recognition events that lead to rejection of self pollen, or acceptance of compatible pollen with different S alleles from those in the stigma cells. Inhibition occurs on the stigma surface and is associated with the cell surface pellicle layer. Callose synthesis is an important and active component of rejection, though its role has not been

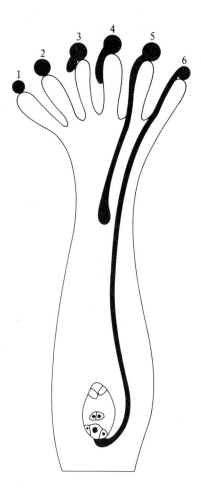

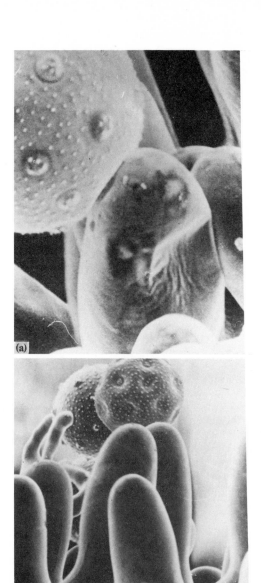

The male gametophyte

In a successful pollination:	In an incompatible pollination:
1 Touchdown on stigma surface (pellicle)	Barrier 1 (foreign pollen not recognised)
2 Hydrates	Barrier 2 (sporophytic SI)
3 Germinates, produces pollentube	
4 Penetrates stigma	Barrier 3 (gametophytic SI)
5 Grows through style	
6 Penetrates female gametophyte	Barrier 4 (sperm/egg interactions)
Discharges gametes	

Fig. 3.5.11   Diagram showing behaviour of pollen grain during germination in successful and unsuccessful pollinations.

Fig. 3.5.12   (a) Scanning electron micrograph of living stigma of *Silene vulgaris* 30 min after pollination showing pollen prints of the germinal apertures on the surface pellicle of the stigma. These were made out by the pollen when it initially adhered to the papillum; (b) as (a) but showing germinating pollen tubes from the imbibed grain at rear of photograph and sterile grain in foreground which has failed to imbibe. From [29].

defined. Callose is produced very rapidly when plant cells are wounded, and it has an important physiological role in the uptake of macromolecules in certain plant cells, acting as a permeability barrier, preventing the entry of large macromolecules [42].

The consequences of pollen-stigma recognition

It is clear that recognition reactions determine the fate of pollen when it has arrived on the stigma. The various reactions involved are summarized in Fig. 3.5.11. The first event after arrival involves an interaction with the sporophyte—the stigma surface. Exine proteins on the surface of the pollen are likely to be involved. The reaction is rapid, and determines whether or not the pollen is accepted by the stigma. If it is not accepted, as in the case of foreign pollen from other families, the pollen remains and fails to germinate (see Fig. 3.5.12). If it is accepted, and in the Cruciferae tests with *Iberis* stigmas show that this occurs with the pollen of all genera tested, the pollen is permitted to take up water and germination proceeds (Fig. 3.5.12). This occurs within 10 min of arrival.

There follows a second recognition event which enables the stigma to accept or reject the germinating pollen, enabling discrimination between pollen from other species and self-pollen from not-self pollen within species. This is achieved through an interaction of the pollen-wall proteins with the stigma surface pellicle, which activates the stigma cell, and sets in train the processes of rejection or acceptance. In sporophytic self-incompatibility in Cruciferae and Compositae, the inhibition occurs on the stigma surface, while in gametophytic self-incompatibility, inhibition of pollen tube growth is manifested at characteristic distances through the stylar tissue. In the Gramineae, pollen tube arrest occurs within 1–2 cells from the point of stigma penetration; in *Petunia* and *Lycopersicum*, arrest occurs deep in the stylar tissue. In *Theobroma cacao* [35] it occurs within the embryosac at the surface of the egg cell itself (Fig. 3.5.11), perhaps suggesting that a further recognition barrier is operative within the female gametophyte itself. The sperm apparently must be acceptable to the egg, just as in mammalian systems.

Gametic cell-to-cell communication thus appears to involve similar general mechanisms to those in somatic cells. Increasingly, we are aware of the role of the cell wall and cell membranes in the reception and transmission of the recognition signals. It is significant that Dickinson & Lewis [27] in their elegant fine-structural study of the events of sporophytic self-incompatibility in the crucifer *Raphanus*, record that the first detectable change in the stigma papillae after arrival of incompatible pollen involved changes in the plasmalemma and the adjacent cytoplasm, demonstrating that these are critically involved in the rejection response.

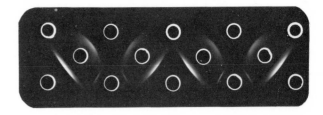

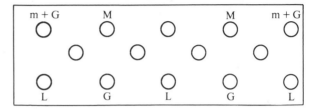

Fig. 3.5.13 Diffusion in agar gel of Con A (in centre wells) against various carbohydrates and glycoproteins. Precipitin lines have formed with glycogen (G), yeast mannan (M) and lima bean lectin (L) but not with a mixture of glycogen and mannose (m + G).

3.5.3 PLANT LECTINS: SOMATIC CELL COMMUNICATORS?

Lectins are cell agglutinating globular proteins or glycoproteins which are widely distributed in the seeds of legumes, but have also been found in tissue extracts of sponges, snails, crabs and fish [43]. These lectins have the ability to agglutinate a wide

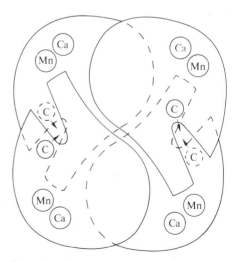

Fig. 3.5.14 Proposed three-dimensional structure of the tetrameric form of Concanavalin A [44] showing binding sites of metal ions manganese (Mn) and calcium (Ca) and saccharides (C). Sugar binding site also occur on the surface of the molecule.

range of mammalian cells. They may also be mitogenic, that is able to stimulate the transformation of small resting lymphocytes into large blast cells which may undergo mitosis. Some are known to restore the property of contact inhibition to certain malignant cells. The specificity of these lectins has, in some cases, been shown to be directed to carbohydrate units of a specific configuration, so that many of the reactions of the lectins can be inhibited by simple sugars. The effects of lectins on cells is dependent on their binding with the carbohydrates of cell surface glycoproteins.

Distribution and properties of plant lectins

As early as 1888, Stillmark found that saline extracts of castor bean seeds specifically agglutinated human erythrocytes. Since then, lectins have been extracted from the seeds of more than 800 species of flowering plants from many families but especially the legumes [43]. In some plants, the lectin is not confined to the seeds but has also been detected in the leaves, roots and bark [43].

Less than twenty plant lectins have been partially purified and biochemically characterized (Table 3.5.1). They have a wide range of molecular weights and polymeric structure. All show agglutinating activity, and some are mitogens.

The best characterized lectin, Concanavalin A (Con A) from jack bean seeds, *Canavalia ensiformis,* is one of three known protein lectins (Table 3.5.1), and is active either as a dimer or a tetramer. Con A comprises as much as $2 \cdot 5 - 3 \cdot 0\%$ of the seed protein. Like other lectins, it is a metalloprotein with specific binding sites for the metal ions Mn^{2+} and Ca^{2+}. The three-dimensional structure determined by Edelman and co-workers in 1972 [44] indicates that the saccharide binding sites are present in deep inpocketings of the molecules (Fig. 3.5.2). Con A binds sugars having the D-arabino configuration at C3, C4 and C6, and an α-configuration at C1 [45], as glucose or mannose units and sterically related structures. Where these

Table 3.5.1 List of some well characterized plant lectins, showing their known properties and specificities. Adapted from [12]

Lectin	Source	Mol. Wt.*	Mitogenic Activity	Sugar†	Specificity: Human Blood
PROTEINS					
Concanavalin A (Con A)	*Canavalia ensiformis*	55,000 (2)	+	D-Man D-Glc D-GlcNac	None
Garden Pea Lectin	*Pisum sativum*	53,000	–	D-Man	None
Wheat Germ Agglutinin (WGA) GLYCOPROTEINS	*Triticum aestivum*	26,000	–	$(GlcNac)_n$	None
Soybean Agglutinin (SBA)	Glycine *max*	110,000 (4)	–	D-Man D-GlcNAc	None
Phytohaem agglutinin (PHA)	*Phaseolus vulgaris* Red Kidney	I 138,000 (8) cv.II 98,000– 138,000 (4)	+	D-GalNAc	None
Lentil Lectin (LcH)	Lens *culinaris*	42,000– 69,000	+	D-Man D-GlcNAc	None
Lima Bean Lectin	*Phaseolus lunatus*	I 269,000 II 138,000	–		A
Dolichos Lectin	*Dolichos biflorus*	140,000	–	D-GalNAc	A
Lotus Lectin	*Lotus tetragonolobus*	I 120,000 II 58,000 III 117,000	–	L-Fuc	H(O)

* Figures in parentheses show number of subunits.
† Abbreviations: D-Man = D-Mannose; D-Glc = D-Glucose; D-GlcNAc = N-Acetyl D-Glucosamine; D-GalNAc = N-Acetyl D-Galactosamine; L-Fuc = L-Fucose.

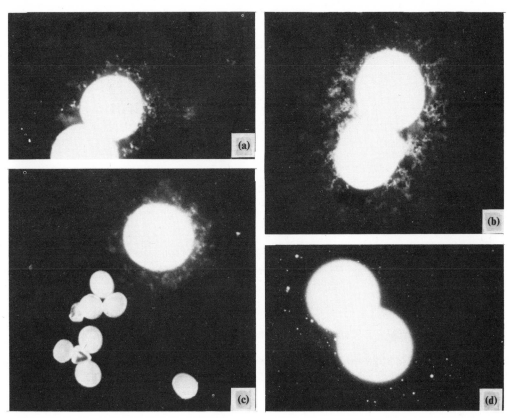

Fig. 3.5.15 Reaction of pollen of *Zea mays* (corn) sprinkled on agar gels containing Con A [60]. Precipitate forms around the grains within 2 min (a) and is very dense after 3 h (b). When pollen of rye grass was included with the corn pollen, precipitate formed only around the corn grains (c). No precipitate formed around grains of corn pollen in gels without Con A (d).

units occur as non-reducing end groups on polysaccharides or glycoproteins, interaction with Con A may cause them to precipitate [45]. Con A can mimic antibody activity in agar gel double diffusion tests [46] forming precipitin lines with several polysaccharides and glycoproteins (Fig. 3.5.3). Insoluble complexes form between Con A and certain human blood group substance glycoproteins, probably because of their N-acetyl-D-glucosamine content [47]. Specific binding of Con A to mammalian membrane glycoproteins induces a range of effects which are described below.

Garden pea lectin from *Pisum sativum* seeds, and wheat germ agglutinin (WGA) from embryos of *Triticum aestivum* are also proteins, but possess different specificities from Con A (Table 3.5.1). WGA binds N-acetyl-D-glucosamine and its oligosaccharides. Its binding site appears to consist of three or four adjacent subsites with differing specificities, located in a cleft in the molecule [48].

The other lectins have proved to be glycoproteins (Table 3.5.1). Of these soybean agglutinin (SBA) was the first plant glycoprotein to be characterized [49]. It has four sub-units and a molecular weight of 30,000 daltons. Each sub-unit contains 4·5% D-mannose and 1·2% N-acetyl-D-glucosamine. Haemagglutinating activity of SBA is inhibited by very low concentrations of N-acetyl-D-galactosamine (0·05−0·1 mM). It amounts to 1·5% of the seed protein, and is present in soybean seeds with three other minor haemagglutinating proteins, called *isolectins*. This multiplicity of lectins has been found in other seeds, including the lentil, *Lens culinaris* and the lima bean, *Phaseolus lunatus*. Lima bean lectin was the first agglutinin to show blood group specific agglutination, and was isolated by Boyd & Reguera in 1949 [50]. There are two isolectins, one with twice the molecular weight of the other, but both have the same blood group specificity. Other lectins have since been found with specificities for the other known blood groups.

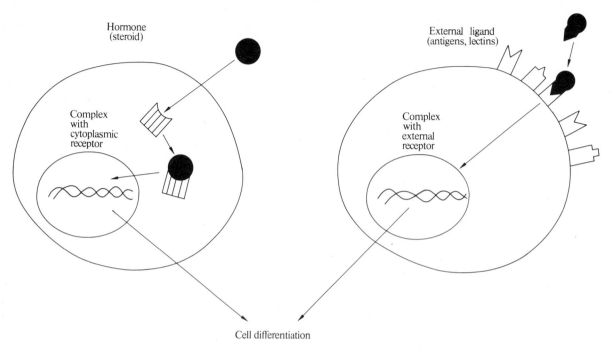

Hormone
(steroid)

Complex
with
cytoplasmic
receptor

External ligand
(antigens, lectins)

Complex
with
external
receptor

Cell differentiation

Fig. 3.5.16 Schematic diagram showing the possible binding interactions between informational molecules and mammalian cells and their developmental consequences. In (a) steroid hormones are shown complexing with specific cytoplasmic receptors; in (b) the binding of external ligands, as antigens, lectins and peptide hormones, to cell surface receptors may lead to cell activation and differentiation. (Courtesy of Dr J. Marchaloris.)

Lectins as inducers of differentiation in animal cells

It is the interaction between specific lectin and surface receptors that induces transformation of lymphocytes, and the events may be summarized as follows (for review, [51]):

(1) *Binding and cross-linking* of specific glycoprotein receptors on cell surface (at least 10 different glycoproteins have been detected in lymphocyte membranes);
(2) followed very rapidly by a *redistribution* of membrane components (patching and capping);
(3) increased cell permeability to various ions and metabolites; increased phospholipid metabolism;
(4) *enzyme activation* (adenyl or guanyl cyclase) catalysing the conversion of ATP → cAMP, promoting blast formation and mitosis.

Such a response bears many similarities to hormone action (Fig. 3.5.16), and this view is supported by two recent reports. Firstly, binding of the red kidney bean lectin, phytohaemagglutinin (PHA), to human platelet cells has been found to mimic the *aggregation and release reaction* characteristically induced by the hormone thrombin [52]. Secondly, binding of the hormone insulin to fat cells and liver membranes is enhanced by low concentrations of WGA [53]. At higher concentrations, both WGA and Con A will competitively displace insulin from the fat cell receptor sites, and mimic its cellular effects. Certainly evidence is accumulating that lectins, by binding specifically to cell membrane receptor glycoproteins, can initiate many of the biochemical events of cell differentiation.

Role of lectins in nature

The lectins occur as a readily solubilized protein (or glycoprotein) fraction of seeds and other parts of plants. Various suggestions have been put forward for their function in the plant cell: as *plant antibodies* produced in response to soil bacteria; as agents of *plant protection* from microbial attack through inhibition of fungal cell-wall-hydrolysing enzymes; involvement in sugar transport and storage; binding of glycoprotein enzymes to form batteries of *multi-enzyme systems*; and for the mitogens, as controlling agents of cell differentiation and mitosis, especially during seed germination [43].

Implicit, in most of these suggestions, is the assumption that lectins are located at appropriate sites throughout the plant body. Early work with the lectin from the pokeweed, *Phytolacca americana* [54] showed that extracts from roots and leaves were more active, both in terms of agglutination and mitogenicity, than those from seeds and fruit. However, work with other lectins has not confirmed this picture. LcH and PHA have been found only in the cotyledons and embryo of the developing and germinating seeds [55, 57], though a protein immunologically related to PHA was detected in the leaves. PHA has been localized in cytoplasmic sites in the cotyledon cells of red kidney beans [55] by the use of specific antibodies labelled with peroxidase. Cytochemical staining methods were used to visualize the sites of enzyme activity. Con A has been located in similar sites in cotyledons of jack beans [56] using fluorescent-labelled serum which binds specifically to the lectins. Activity was associated particularly with the storage protein bodies and starch grains. Another class of lectin-like substances, with a specificity for β-glycosides, however, were located in the intercellular spaces, cell walls, and associated with the cell membrane [56], sites suggestive of a role in cell communication.

However, the principal evidence supporting the recognition role of lectins has come from studies of plant-bacterial symbiosis. PHA has been found to cause agglutination of the bacterium *Rhizobium phaseoli* [58], responsible for root nodulation, and nitrogen fixation. PHA, of course, agglutinates red blood cells, and when sterile-grown roots of red kidney bean were placed in solutions containing red blood cells, the cells agglutinated specifically on the young root primordia. This evidence is preliminary, but nevertheless there is the suggestion that PHA is secreted on the root surface to enable the bacteria to recognize the root sites and initiate their symbiotic relationship with the higher plant cells. Soy bean agglutinin (SBA) has been shown to specifically recognize and complex with 22 out of 25 strains of *Rhizobium japonicum*, the root-nodulating bacteria found only on soybeans [59]. The lectin failed to bind to 23 other *Rhizobium* strains that nodulate on other hosts. Again, there is the suggestion that lectins play a role in determining the specificity of communication between bacteria and their plant cell hosts.

Conclusions

Evidence is accumulating that lectins occur to a greater or lesser extent in plant tissues other than the seeds. This raises the question of whether they play a recognition role in vegetative plant cell differentiation. The plant lectins may have first evolved as a basic recognition system for the maintenance of the integrity of the multicellular plant body, for discrimination between self and not-self. Certainly the phenomenon of scion-root stock compatibility relationships indicates the operation of such a recognition mechanism. Sir Macfarlane Burnet suggests that, in the mammalian system, immunoglobulins may have performed a basic cell recognition role and have been elaborated

during evolution to their present wide range of activities in the immune response [7].

Evidence that lectins can indeed recognize and precipitate plant glycoproteins and carbohydrates has recently been obtained. The jack bean lectin, Con A, is able to differentiate among pollen-wall glycoproteins and carbohydrates of grasses [60]. Con A will precipitate materials diffusing from the pollen of corn, *Zea mays* and its close relatives, while it generally fails to recognize substances released from pollens of the temperate festucoid and chloridoid grasses. Precipitation is rapid, occurring within a few minutes if corn pollen is sprinkled on agar gels containing the lectin (Fig. 3.5.4). The quest for a role for lectins in the plant is currently attracting a lot of interest, and promises as great rewards as have been obtained by investigations in the field of cell recognition in mammalian systems.

The plant cell, enveloped in its thick rigid cell wall, might be considered unavailable to external ligands. This is not so. When cell differentiation is being initiated, the cells are meristematic with thin primary walls carrying conspicuous communication channels—the plasmodesmata and pit fields. Even when fully differentiated, and supported within thickened secondary walls, there are large communication channels within the walls themselves (Chapter 3.3). The plasma membrane of plant cells is no less responsive than its mammalian equivalent to external stimuli—as seen, for example, in the case of the stigma cell in contact with incompatible pollen (Section 3.5.2). The documented cases of cell recognition in plants point to essential similarities with the animal cell model, rather than to differences. Cell recognition in plants can, therefore, be conveniently considered at two operational levels. Firstly, *gametic* recognition functions to enable the selection of a few male gametes out of many for fertilization. Informational molecules, recognition factors, are released by the male gametophytes, the pollen grains of flowering plants, which provide the means of initial communication with the stigma or style of the female reproductive pathway. Secondly, *somatic* recognition, though less well defined, is needed both to maintain the integrity of the individual plant, that is for the recognition of 'self', and also for the selection of specifically-sited cells to proceed along quite different developmental pathways, for example phloem cells and companion cells. The cell communicators that determine such events are under precise genetic control and we can only speculate that the cell membrane and its peripheral proteins, either associated with the membrane or located in the cell wall, are likely to be involved.

In mammalian cells, the internal proteins of the plasma-membrane appear to be involved in determining the spatial organization of adjacent cytoplasmic microfibrillar structures. Their ability to bind to cytoplasmic macromolecules may confer on the membrane proteins a role in the spatial organization of cell growth and differentiation. In this way, it is not difficult to imagine how groups of cells with similar specifications may proceed to initiate new growth centres or specialized tissues in a co-ordinated fashion. This, as we have seen, is the basis for pattern determination.

Recognition mechanisms are currently a central theme of cell biology. Application of the techniques well established in mammalian cell studies to plant cells is proving rewarding. The new information becoming available will have a profound impact on the manipulation of plant breeding systems and somatic cell hybridization, in the control of plant infection, and will also open up new possibilities for the management of pollen allergies.

3.5.4 WALL PATTERN FORMATION IN CELL DIFFERENTIATION

The main structural differences between the various types of cell found in the plant body can almost all be related to differences in the development of either the cell wall or the plastids. Plastid development and the different types of plastid formed are dealt with in Chapter 4. Some of the most striking examples of cell wall patterns are found in the sculpturing of the exine of pollen grains described earlier in this chapter, and this is therefore an appropriate place to discuss some of the factors controlling wall pattern formation in pollen grains, together with that in xylem cells, which provide another striking example.

The young cells at the meristems show pattern in their smooth *primary* walls, largely in terms of their macromolecular architecture, but also through inter-cell cytoplasmic connections, the plasmodesmata, which are often clustered together as pit fields (Chapter 3.3). The major and most characteristic forms of patterning appear during later secondary growth and differentiation of the cells. A variety of macromolecules, together usually with cellulose, provide the wall with its strength, rigidity and lack of extensibility: lignin, suberin, sporopollenin, cutins and waxes. Pattern is seen both in the successive layers of wall materials deposited, but also in the orientation and polarity of thickening. This is strikingly illustrated in Fig. 3.5.19(b), which shows a transverse section of a wheat root, where the endodermal cells show polarized secondary thickening on their inner tangential walls. It is also seen in the formation of pores and callose deposits in sieve tubes, lignification of xylem vessels, collenchyma, the softening of root epidermal cell walls during root hair formation, and perhaps most spectacularly of all in the diversity of pollen grain wall patterns.

The control of pattern formation in plant cell walls involves two discrete processes:

(1) sequential periodicity in wall synthesis: the changeover from secretion of one group of wall polymers to another. These steps are most clearly evident in the transition from primary to secondary cell wall growth and differentiation.

(2) spatial organization in the deposition of wall materials: they are laid down according to a precise and pre-determined pattern, implying the existence of an underlying cytoplasmic organization. This is the aspect that has attracted most work, since it is obviously the basis for pattern control.

At the macromolecular level, pattern is evident in the primary cell walls of sycamore cells [61]. Primary cell walls of this and many other plants are predominantly pectocellulosic in nature, and in sycamore seven wall polymers are involved (Table 3.5.1). The wall is composed principally of parallel chains of cellulose elementary fibrils which provide its strength and flexibility. These are cross-linked by hydrogen bonding to the hemicellulose, xyloglucan. This, in turn, is covalently bound to various pectic polysaccharides, which are cross-linked to the hydroxy-proline rich wall glycoprotein, thought to be important in stabilizing wall structure. From the biochemical data, Albersheim and co-workers were able to construct a model (see Fig. 3.5.17), showing that the primary cell wall of the sycamore cell can be considered as 'a single macromolecule'. Enzymic digestion methods were used to release the wall polymers from cell wall preparations of cultured sycamore cells. These were treated with purified pectinase (endopolygalacturonase), found to be first in a temporal series of enzymes released by the fungus *Colletotrichum* to degrade higher plant cell walls.

We thus see pattern both in molecular terms in the configuration of the wall polymers, and in the deposition of the wall apertures or plasmodesmata. However, it is at the level of secondary wall growth that we are beginning to understand what the control mechanisms of pattern formation may involve. We will examine the experimental evidence from two contrasting systems: the process of lignification in xylem vessels; and the sculpturing of pollen grain walls.

Origin of pattern in xylem vessels

The xylem vessel begins its existence as a large elongate cell with a thin primary cell wall. During differentiation, the remarkable spiral, annular or scalariform anastomosing bands of cellulose and lignin, are deposited as a secondary wall (Fig. 3.5.18). This provides the strength and flexibility for the xylem vessel, which, without nucleus or cytoplasm at maturity, is empty but functional in the transport of water and solutes in the plant.

The differentiation of these secondary thickenings has been the object of many studies, first with the light microscope, later with the transmission electron microscope, in the hope that characteristic changes in the form or distribution of cytoplasmic organelles could be associated with wall pattern formation.

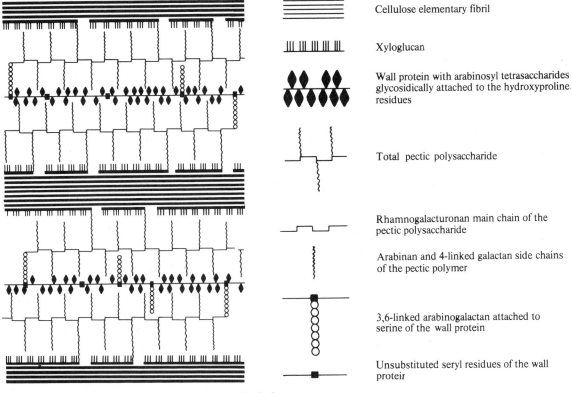

	Cellulose elementary fibril
	Xyloglucan
	Wall protein with arabinosyl tetrasaccharides glycosidically attached to the hydroxyproline residues
	Total pectic polysaccharide
	Rhamnogalacturonan main chain of the pectic polysaccharide
	Arabinan and 4-linked galactan side chains of the pectic polymer
	3,6-linked arabinogalactan attached to serine of the wall protein
	Unsubstituted seryl residues of the wall protein

Fig. 3.5.17 Tentative molecular structure of the primary cell wall of sycamore based on the model of Keegstra, Talmadge, Bauer & Albersheim [61]. The symbols shown to the right are used to represent the various components of the cell wall.

Hepler & Newcomb [62] working with *Coleus* stem segments, observed that the alignment of microtubules, which form a 'cytoskeleton' in the cytoplasm adjacent to the wall, was parallel to the direction of the polysaccharide fibrils. This was confirmed in later work by Wooding & Northcote [63] on *Acer* stems, Cronshaw & Bouck [64] with *Avena* coleoptiles and Pickett-Heaps & Northcote [65] with wheat roots.

The changes observed in microtubule distribution during wall growth have thrown light on their role in the control of pattern formation. In wheat roots the microtubules were initially evenly distributed along the length of the thin primary wall (Fig. 3.5.19, period 1). As secondary thickening began (period 2), the microtubules appeared in clusters between the developing secondary growths. Finally massive deposition of secondary thickening occurred and the microtubules were observed associated with the tops of the thickenings (period 3). In the mature xylem vessel, secondary thickening was complete, and the vessel was empty and functional (period 4). During cell differentiation, the microtubules ran hoop-like round the cell in transverse orientation, and were parallel to each other and the plane of deposition of cell wall polysaccharides.

It has long been known from work on onion root tips [66] that colchicine can not only disrupt the formation of the mitotic spindle, but changes the pattern of growth of differentiating cells, usually elongate cells becoming round or isodiametric. It is now known that colchicine acts as an inhibitor of microtubule formation [67] binding specifically to the tubulin subunits. Green [68] observed that the normally asymmetric cells of *Nitella* were isodiametric in the presence of colchicine, and attributed the effect to the absence of microtubules which conferred stability during wall synthesis.

Pickett-Heaps [69] used the drug to investigate the role of microtubules in secondary thickening patterns in wheat xylem vessels. In treated root tips, microtubules were virtually absent but secondary wall growth proceeded, though with drastically altered organization. Secondary thickenings were irregular in

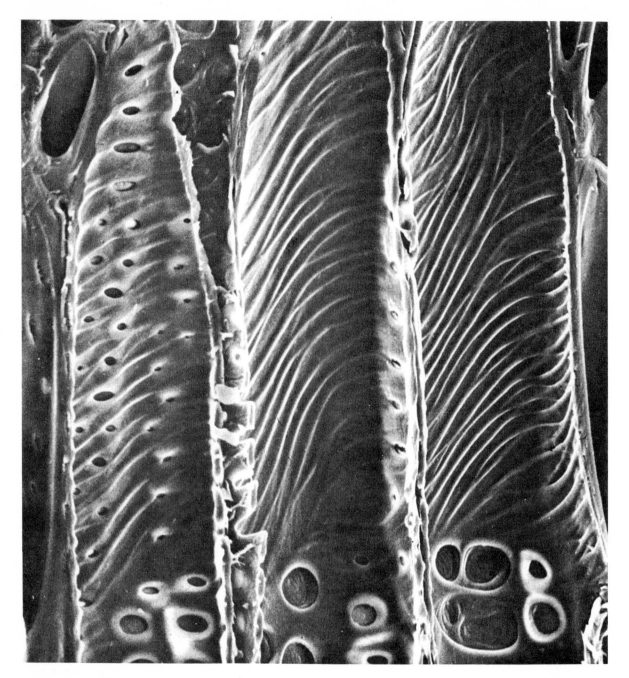

Fig. 3.5.18 Scanning electron micrograph showing helical thickenings on vessel walls of *Pseudopanax arboreum*. The concentration of larger pits at the bottom of the photograph marks the position of a ray passing behind the vessels. (× 1200). (From B.A. Meylan & B.G. Butterfield, *Three dimensional structure of wood*, Chapman & Hall Ltd., London, 1972).

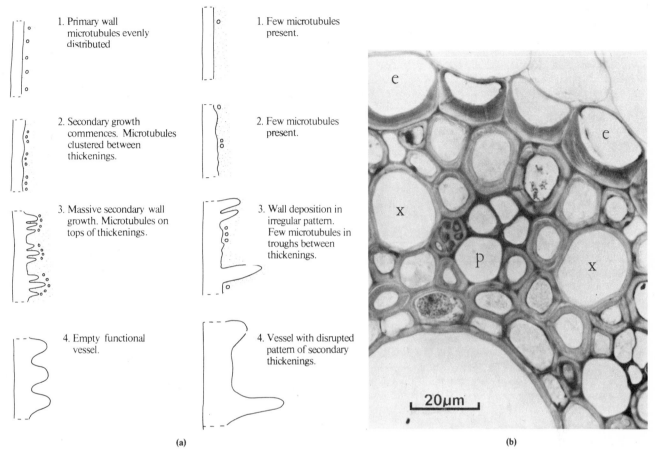

1. Primary wall microtubules evenly distributed

2. Secondary growth commences. Microtubules clustered between thickenings.

3. Massive secondary wall growth. Microtubules on tops of thickenings.

4. Empty functional vessel.

1. Few microtubules present.

2. Few microtubules present.

3. Wall deposition in irregular pattern. Few microtubules in troughs between thickenings.

4. Vessel with disrupted pattern of secondary thickenings.

(a)

(b)

Fig. 3.5.19 (a) Diagrammatic presentation of xylem vessel wall development in wheat roots, adapted from electron micrographs of Pickett-Heaps [69]. The left-hand column shows sequentially the development of a portion of the untreated xylem wall, while the right-hand column shows wall development in tissue treated with colchicine.

(b) Transverse section of the stele of a wheat root showing the secondary walls of the endodermis (e), and xylem (x). Note the polarized thickening of the walls of the endodermis. Photograph by Dr A.E. Ashford.

shape and distribution, and the entire regular pattern was disrupted. We have seen in untreated xylem cells that the microtubule rearrangements precisely reflect the patterns of secondary thickening (see Fig. 3.5.19). But after colchicine treatment, while the thickenings have the usual fibrous texture, there is no regular pattern, deposits varying from almost complete absence to large wart-like outgrowths. These experiments suggest that, while the presence of microtubules is not required for secondary wall growth, they seem to be needed for pattern control. Similar conclusions were reached after a study of the effects of colchicine on secondary thickening patterns of endothecial cells of the anther of *Lilium*, the cells responsible for dehiscence [70]. The evidence in all these cases is by no means conclusive [71].

Control of pattern formation of pollen grain walls

The pollen grain is the male gametophyte of flowering plants. Its function is to provide a protective environment for the male gametes during their development, and also during transport to the stigma. There the pollen germinates, producing a pollen tube which bears the gametes to the embryo sac—the female gametophyte. The processes controlling germination have been considered earlier in this chapter, in relation to proteins residing in sites within the pollen grain walls. We have seen that pollen grains are intricately patterned and sculptured, and that much of this patterning is related to the storage and routes of release of these wall materials during pollen germination [72, 73].

Since the pollen grain is a haploid structure, with a single set of information, we might expect that the exines of any one genotype would show diversity of pattern suggestive of meiotic segregation of the controlling genes. Such differences have been sought, but not found [70, 74]. What has been observed is that each pollen grain shows slight differences in wall formation sufficient to distinguish it from its fellows (Fig. 3.5.20), even though the overall pattern is identical. These modulations arise during deposition of wall materials, in this case, the unique biopolymer, sporopollenin [75]. We have already considered the types of exine patterning and wall formation found in pollen, and it remains here to consider how such patterns emerge, and how they may be controlled.

Two kinds of pattern are displayed in pollen grains: *patterning of the exine layer,* of which there are two kinds—tectate and pilate, from which other patterns can be derived (Fig. 3.5.2); and the *positions of the germinal apertures,* pores or slits (colpi) through one of which the pollen tube may emerge on germination. This is clearly seen in grass pollen, where the position of the single pore is related to the planes of the spindles during the second meiotic divisions (Fig. 3.5.21), and this appears to be a general rule in most pollens [76].

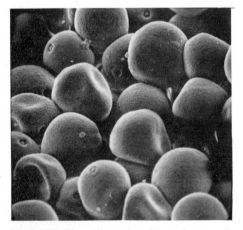

Fig. 3.5.20 Ripe pollen grains of bermuda grass, *Cynodon dactylon*. The single germinal aperture or pore is prominent on each grain, and the exine is finely patterned (× 500).

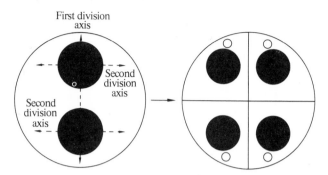

Fig. 3.5.21 Position of the pores of grass pollen in relation to the positions of the first and second meiotic spindles. From Dover [81].

Exine pattern formation

The origins of pattern in pollen exines are indicated in the flow-chart (Fig. 3.5.22). The physiological mechanisms that control polarity and cell-to-cell communication are obviously deeply involved. Synchrony of the cells undergoing meiosis is imposed in the premeiotic interphase, perhaps through onset of the S or DNA synthesis period of the cell cycle. The cells at this time are linked by plasmodesmatal connections through their primary walls. In early prophase, synthesis of a callose (β-1, 3-linked glucan) secondary wall commences, but the cells remain linked together, now by large channels (up to 1·5 µm in diameter) through the callose wall. Effectively, each anther loculus is a syncytium of cells, with constant cytoplasmic communication one with another [77].

At the end of prophase, the channels are sealed by callose, and each meiocyte is completely enclosed within a callose wall of very selective permeability—allowing entry of only simple substrates, such as sugars [78]. The effect of this is to act not only as a restraint to further cell growth but to preserve the independent genetic identity of the cells undergoing the meiotic divisions. During this period, ribosome elimination has occurred during meiotic prophase with associated loss of other cytoplasmic organelles, followed by the period of ribosome restoration, and the re-appearance of a new complement of cytoplasmic organelles after the first meiotic division [79].

Within this callose special wall, during tetrad period, the principal features of exine pattern were detected by Beer in 1911, and other early cytologists. The tetrad period in *Lilium* has been subdivided by Heslop-Harrison [80], on the basis of structural and chemical changes, into four periods:

(1) *pre-pattern period,* when the spore plasmalemma is appressed to the callose wall, with no spore wall present, though membranes of endoplasmic reticulum are associated with future germinal aperture sites;

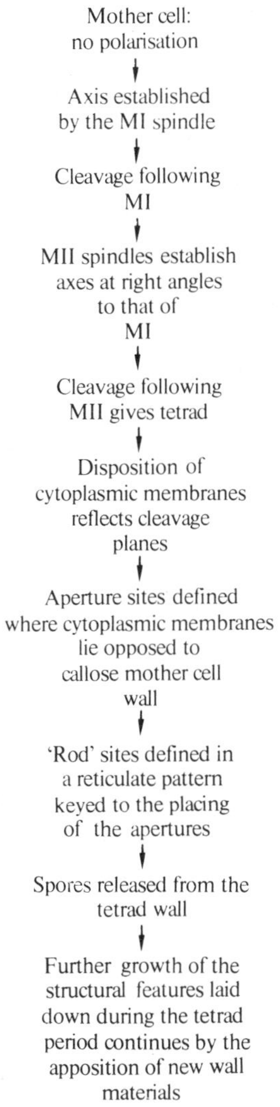

Mother cell:
no polarisation

↓

Axis established
by the MI spindle

↓

Cleavage following
MI

↓

MII spindles establish
axes at right angles
to that of
MI

↓

Cleavage following
MII gives tetrad

↓

Disposition of
cytoplasmic membranes
reflects cleavage
planes

↓

Aperture sites defined
where cytoplasmic membranes
lie opposed to
callose mother cell
wall

↓

'Rod' sites defined in
a reticulate pattern
keyed to the placing
of the apertures

↓

Spores released from the
tetrad wall

↓

Further growth of the
structural features laid
down during the tetrad
period continues by the
apposition of new wall
materials

Fig. 3.5.22 Sequence of events in the development of the patterned exine of *Lilium* pollen grains from the pollen mother cell period through the two meiotic divisions (MI and MII) to release of the spores from the tetrads to pollen maturity. From Heslop-Harrison [82].

(2) *early-patterned period*, when the exine precursor layer, the primexine (largely cellulosic in nature) is laid down forming the first elements of pattern;

(3) *late-patterned period*, when a form of sporopollenin (resistant to acetolysis) is laid down over the primexine, and all the pattern of the mature exine is now present;

(4) *spore-release period*, when the callose special wall is degraded and the microspores released into the anther cavity.

The first evidence of pattern therefore appears at the early-patterned period, with the deposition of the primexine wall by the spore protoplast. This lies between the spore plasmalemma and the callose wall. It is largely cellulosic, since it can be removed from sections by digestion with the enzyme cellulase [80]. On this cellulosic matrix, a form of sporopollenin is deposited, and by late-patterned period, all the intricacies of the final wall patterning can be discerned. The major deposition of sporopollenin occurs after spore release during growth and expansion of the pollen, and this material originates both from the spore protoplast, and from products of the tapetal or nurse cells that surround the anther cavity. Pattern determination is carried out wholly within the meiotic tetrad period.

Experimental control of exine pattern

The control of pattern formation has proved susceptible to experimental manipulation in *Lilium* pollen-mother cells. Heslop-Harrison [70] showed that centrifugation of the developing anthers at meiosis within flower buds caused mechanical displacement of subcellular structure, whose effects on later pollen development could be observed after *in vitro* culture. By sequential treatment, he was able to detect the period when any anomalies were determined. When anthers were centrifuged during prophase, the pollen showed imperfect separation of the grains, producing monads (uninucleate tetraploid grains formed after meiotic failure), dyads (diploid grains formed after failure of the second division) and even enucleate micrograins. Treatment during the second division of meiosis resulted in pollen showing random aperture (colpi) sites, and when delayed until pre-patterned tetrad period, anomalies of exine patterning were produced.

Heslop-Harrison [70] also found that colchicine produced similar effects. When applied during prophase, the grains later showed both imperfect separation and misplaced apertures, while treatments during the period from metaphase 1 until the end of second division, produced striking exine pattern changes. In some grains, the rod-like bacula had fused to form nodules, bosses and even gigantic wart-like structures. In other cases, a roof or tectum had been formed by lateral extension of the heads of the bacula, producing a phenocopy of the tectate exine. The temporal precision of these experimentally-induced modifications to pollen pattern, suggest the operation of a series of diploid parental determinants. Likely candidates include persistent mRNA or informational peptides [77] whose formation would be governed by the pollen-mother cell, for later read-out at the appropriate period during meiosis.

Thus *exine pattern* is determined during the meiotic divisions and laid down precisely during tetrad period with the formation of the primexine. This layer is the first wall layer laid down, and forms a template for later sporopollenin deposition. Exine formation is controlled by parental, sporophytic genes and gene products. This is achieved firstly by determinants that may survive through the meiotic cell lineages, to be activated for primexine deposition; and secondly through deposition on the developing exine of sporopollenin made in the parental tapetal cells surrounding the anther cavity. The experimental modifications induced by chemical and physical treatments, demonstrate conclusively that pattern is determined at these precise periods of development, and that pattern formation requires precisely specified temporal and spatial organization.

Experimental modification of pore pattern

In wheat, *Triticum*, pollen-mother cells, there is a rather extended pre-meiotic interphase, and Dover [81] has taken advantage of this to determine the effects of colchicine on the control of pore pattern, since this grass pollen has a precisely sited single germinal aperture (see Fig. 3.5.20). Treatments applied soon after the final pre-meiotic mitosis produced poreless monads, while treatments at preleptotene and leptotene resulted in four-pored monads. The four pores were randomly distributed and this lack of pattern was correlated with the absence of microtubules in the grains. Dover considered the absence of pores in the earliest treatment, taken together with other cytological effects, indicated the existence of a 'sensitive' period during pre-meiotic interphase when the pore determinants had not been formed or were affected by the presence of colchicine. Thus pore siting is a discrete process, set in train well before exine pattern is effected.

3.5.5 CONCLUSIONS

In both the xylem cells and pollen grains, pattern control has proved susceptible to experimental manipulation by chemical and mechanical treatments. The key to pattern control resides in the developmental changes of the cell wall resulting from (i) the sequential secretion of different groups of polysaccharides and proteins into the wall; and (ii) the spatial organization of deposition. The experimentally-induced pattern changes are the products of disruption in regular pattern control. These have been induced by changes in spatial organization, presumably determined by cellular organelles. Despite the remarkable correlations with microtubule distribution and polarity, the results are inconclusive. The primary controller may yet prove to be the cell membrane, which is strategically sited in relation to the wall, and whose complexities of form and architecture in plant cells have hardly begun to be understood, and have not been detected with the techniques used in these experiments.

3.5.6 REFERENCES

[1] OSEROFF A.R., ROBBINS P.W. & BURGER M.M. (1973) The cell surface membrane: biochemical aspects and biophysical probes. *Ann. Rev. Biochem.*, **42**, 647–82.

[2] WIESE L. (1965) On sexual agglutination and mating G-type substances (Gamones) in isogamous heterothallic Chlamydomonas. *J. Phycol.*, **1**, 46–54.

[3] CRANDALL M.A. & BROCK T.D. (1968) Molecular aspects of specific cell contact. *Science*, **161**, 463–65.

[4] BURNET F.M. (1969 *Self and non-self*. Melbourne and Cambridge.

[5] CRUICKSHANK I.A.M. & PERRIN D.R. (1968) The isolation and partial characterization of monilicolin A, a polypeptide with phaseollin-inducing activity from *Monilinia fructicola*. *Life Sci.*, Oxford, **7**, 449–58.

[6] FRANK J.A. & PAXTON J.D. (1971) An inducer of soybean phytoalexin and its role in the resistance of soybeans to *Phytophthora* rot. *Phytopathol.*, **61**, 954–58.

[7] BURNETT F.M. (1971) 'Self-recognition' in colonial marine forms and flowering plants in relation to the evolution of immunity. *Nature, Lond.*, **232**, 230–35.

[8] BRAUN W. (1973) RNA as amplifiers of specific signals in immunity. *Ann. N.Y. Acad. Sci.*, **207**, 17–28.

[9] BRAUN V. & HANTKE K. (1974) Biochemistry of bacterial cell envelopes. *Ann. Rev. Biochem.*, **43**, 89–121.

[10] BARBER B. & MUSCONA A.A. (1972) Reconstruction of brain tissue from cell suspensions—Part 2. Specific enhancement of aggregation of embryonic cerebral cells by supernatant from homologous cell cultures. *Dev. Biol.*, **27**, 235–43.

[11] NICOLSON G.L. & SINGER S.J. (1971) Ferritin-conjugated plant agglutinins as specific saccharides bound to cell membranes. *Proc. Nat. Acad. Sci.*, USA., **68**, 942–45.

[12] BLACKLEY C.H. (1873) *Experimental researches on the cause and nature of catarrhus aestivus*. Baillier, Tindall & Cox, London.

[13] KING T.P., NORMAN P.S. & LICHTENSTEIN L.M. (1967) Studies on ragweed pollen allergens. *Ann. Allergy*, **25**, 541–53.

[14] KNOX R.B. & HESLOP-HARRISON J. (1971) Pollen-wall proteins: localization of antigenic and allergenic proteins in the pollen grain walls of *Ambrosia* spp. (ragweed). *Cytobiol.*, **4**, 49–54.

[15] HESLOP-HARRISON J. (1971) The pollen-wall: structure and development. In *Pollen development and physiology*, pp. 75–98 (Ed. J. Heslop-Harrison). Butterworths, London.

[16] HESLOP-HARRISON J. (1968) Pollen wall development. *Science, N.Y.*, **161**, 230–37.

[17] BROOKS J. & SHAW G. (1971) Recent developments in the chemistry, biochemistry, geochemistry, and post-tetrad ontogeny of sporopollenins derived from pollen and spore exines. In *Pollen development and physiology*, pp. 99–114 (Ed. J. Heslop-Harrison). Butterworths, London.

[18] STANLEY R.G. & LINSKENS H.F. (1965) Protein diffusion from germinating pollen. *Physiologia Pl.*, **18**, 47–53.

[19] AUGUSTIN R. & HAYWARD B.J. (1962) Grass pollen allergens. IV. The isolation of some of the principal allergens of *Phleum pratense* and *Dactylis glomerata* and their sensitivity spectra in patients. *Immunol.*, **5**, 424–60.

[20] TSINGER N.V. & PETROVSKAYA-BARANOVA T.P. (1961) The pollen grain wall—a living physiologically active structure. *Dokl.. Akad. Nauk SSSR*, **138**, 446–96.

[21] KNOX R.B. & HESLOP-HARRISON J. (1970) Pollen-wall proteins: localization and enzymic activity. *J. Cell Sci.*, **6**, 1–27.

[22] KNOX R.B. (1970) Freeze-sectioning of plant tissues. *Stain Techn.*, **45**, 265–72.

[23] HOWLETT B.J., KNOX R.B. & HESLOP-HARRISON J. (1973) Pollen-wall proteins: release of the allergen Antigen E from intine and exine sites in pollen grains of ragweed and *Cosmos. J. Cell Sci.*, **13**, 603–19.

[24] PEARSE A.G.E. (1968) *Histochemistry, theoretical and applied.* Churchill-Livingstone, London. 2 Vol.

[25] KNOX R.B. (1971) Pollen-wall proteins: localization, enzymic and antigenic activity during development in *Gladiolus* (Iridaceae). *J. Cell Sci.*, **9**, 209–37.

[26] HESLOP-HARRISON J., HESLOP-HARRISON Y., KNOX R.B. & HOWLETT B. (1973) Pollen-wall proteins: 'gametophytic' and 'sporophytic' fractions in the pollen walls of the Malvaceae. *Ann. Bot.*, **37**, 403–12.

[27] DICKINSON H.G. & LEWIS D. (1973) Cytochemical and ultrastructural differences between intraspecific compatible and incompatible pollinations in *Raphanus*. *Proc. R.S. Lond. B.*, **183**, 21–38.

[28] HESLOP-HARRISON J., KNOX R.B. & HESLOP-HARRISON Y. (1974) Pollen-wall proteins: exine-held fractions associated with the incompatibility responses. *Biol. J. Linn. Soc.*, **7**, Suppl. 1, 189–202.

[30] KNOX R.B., HESLOP-HARRISON J. & HESLOP-HARRISON Y. (1975) Pollen-wall proteins: localization and characterization of gametophytic and sporophytic fractions. *Biol. J. Linn. Soc.*, **7**, Suppl. 1, 177–87.

[31] GREEN J.R. (1894) On the germination of the pollen grain and the nutrition of the pollen tube. *Ann. Bot.*, **8**, 225–28.

[32] KNOX R.B. & HESLOP-HARRISON J. (1971) Pollen-wall proteins: the fate of intine-held antigens on the stigma in compatible and incompatible pollinations of *Phalaris tuberosa* L. *J. Cell Sci.*, **9**, 239–51.

[33] DARWIN C. (1876) *The effects of cross- and self-fertilization in the vegetable kingdom.* London.

[34] DARWIN C. (1877) *The different forms of flowers on the same species.* London.

[35] LINSKENS H.F. & KROH M. (1967) Incompatibilität der Phanerogamen. *Encyc. Plant Phys.*, **18**, 506–30.

[36] HESLOP-HARRISON J. (1975) Incompatibility and the pollen-stigma interaction. *Ann. Rev. Plant Phys.*, **26**, 403–25.

[37] KNOX R.B., WILLING R.R. & ASHFORD A.E. (1972) The role of pollen-wall proteins as recognition substances in interspecific hybridization in poplars. *Nature, Lond.*, **237**, 381–83.

[38] KNOX R.B. (1973) Pollen-wall proteins: pollen-stigma interactions in ragweed and *Cosmos* (Compositae). *J. Cell Sci.*, **12**, 421–43.

[39] MATTSSON O., KNOX R.B., HESLOP-HARRISON J. & HESLOP-HARRISON Y. (1973) Protein pellicle of stigmatic papillae is a probable recognition site in incompatibility reactions. *Nature, Lond.*, **247**, 298–300.

[40] HOWLETT B.J., KNOX R.B., PAXTON J.D. & HESLOP-HARRISON J. (1974) Pollen-wall proteins: characterization and role in self-incompatibility in *Cosmos bipinnatus*. *Proc. Roy. Soc. B.*, **188**, 167–82.

[41] DICKINSON H.G. & LEWIS D. (1974) Interaction between the pollen grain coating and the stigmatic surface during compatible and intraspecific pollinations in *Raphonus*. *Biol. J. Linn. Soc.*, **7**, Suppl. 1, 165–75.

[42] KNOX R. B. & HESLOP-HARRISON J. (1970) Direct demonstration of the low permeability of the angiosperm meiotic tetrad using a fluorogenic ester. *Z. Pflanzenphysiol.*, **62**, 451–59.

[43] LIS H. & SHARON N. (1973) The biochemistry of plant lectins (Phyto-haemagglutinins). *Ann. Rev. Biochem.*, **42**, 541–74.

[44] EDELMAN G.M., CUNNINGHAM B.A., REEKE G.N., BECKER J.W., WAXDAL M.J. & WANG J.L. (1972) The covalent and three-dimensional structure of Concanavalin A. *Proc. Nat. Acad. Sci., USA.*, **69**, 2580–84.

[45] SO L.L. & GOLDSTEIN I.J. (1967) Protein and carbohydrate interaction. Application of the quantitative hapten inhibition test to polysaccharide-concanavalin A interaction. *J. Immunol.*, **99**, 158–63.

[46] GOLDSTEIN I.J. & SO L.L. (1965) Protein-carbohydrate interactions. III. Agar gel diffusion studies on the interaction of concanavalin A, a lectin isolated from Jack bean, with polysaccharides. *Arch. Biochem. Biophys.*, **111**, 407–14.

[47] CLARKE A.E. & DENBOROUGH M.A. (1971) The interaction of Concanavalin A with blood-group-substance glycoproteins from human secretions. *Biochem. J.*, **121**, 811–16.

[48] ALLEN A.K., NEUBERGER A. & SHARON N. (1973) The purification, composition and specificity of wheatgerm agglutinin. *Biochem. J.*, **131**, 155–62.

[49] WADA S., PALLANSCH M.H. & LIENER I.E. (1958) Hemagglutinating substances for human cells in various plants. *J. Biol. Chem.*, **233**, 395–400.

[50] BOYD W.C. & REGUERA R.M. (1949) Hemagglutinating substances for human cells in various plants. *J. Immunol.*, **62**, 333–40.

[51] CUATRECASAS A. (1974) Membrane receptors. *Ann. Rev. Biochem.*, **43**, 169–214.

[52] MAJERUS P.W. & BRODIE G.N. (1972) The binding of Phyto-hemagglutinins to human platelet plasma membranes. *J. Bio. Chem.*, **247**, 4253–57.

[53] CUATRECASAS P. & TELL B.P.E. (1973) Insulin like activity of Concanavalin A and wheat germ agglutinin—direct interactions with insulin receptors. *Proc. Nat. Acad. Sci., USA.*, **70**, 485–89.

[54] FARNES P., BARKER B.E., BROWNHILL L.E. & FANGER H. (1964) Mitogenic activity in *Phytolacca americana* (Pokeweed). *Lancet 2*, Nov. 21, 1100–101.

[55] MIALONIER G., PRIVAT J.-P., MONSIGNY M., KAHLEM G. & DURAND R. (1973) Insolement, propriétés physiocochimiques et localisation *in vivo* d'une phytohemagglutine (lectine) de *Phaseolus vulgaris* L. (var. rouge). *Physiol. Veg.*, **11**, 519–37.

[56] CLARKE A.E., KNOX R.B. & JERMYN M. (1975) Localization of lectins in legume cotyledons. *J. Cell Sci.* (in Press).

[57] HOWARD I.K., SAGE H.J. & HORTON C.B. (1972) Studies on the appearance and location of hemagglutinins from common lentil during the life cycle of the plant. *Arch. Biochem. Biophys.*, **149**, 323–26.

[58] HAMBLIN J. & KENT S.P. (1973) Possible role of phytohaemagglutinin in *Phaseolus vulgaris* L. *Nature, New Biol.*, **245**, 28–30.

[59] BOHLOOL B.B. & SCHMIDT E.L. (1974) Lectins: a possible basis for specificity in the *Rhizobium*-legume root nodule symbiosis. *Science*, **185**, 269–71.

[60] WATSON L., KNOX R.B. & CREASER E.H. (1974) Concanavalin A differentiates among grass pollens by binding specifically to wall glycoproteins and carbohydrates. *Nature*, **249**, 574–76.

[61] KEEGSTRA K., TALMADGE K.W., BAUER W.D. & ALBERSHEIM P. (1973) The structure of plant cell walls. III. A model of the walls of suspension-cultured sycamore cells based on the interconnections of the macro-molecular components. *Plant Physiol.*, **51**, 188–96.

[62] HEPLER P.K. & NEWCOMB E.H. (1964) Microtubules and fibrils in the cytoplasm of *Coleus* cells undergoing secondary wall deposition. *J. Cell Biol.*, **20**, 529–33.

[63] WOODING F.B.P. & NORTHCOTE D.H. (1964) The development of the secondary wall of the xylem in *Acer pseudoplatanus*. *J. Cell Biol.*, **23**, 327–35.

[64] CRONSHAW J. & BOUCK G.B. (1965) The fine structure of differentiating xylem elements. *J. Cell Biol.*, **24**, 415–31.

[65] PICKETT-HEAPS J.D. & NORTHCOTE D.H. (1966) Relationship of cellular organelles to the formation and development of the plant cell wall. *J. Exp. Bot.*, **17**, 20–26.

[66] LEVAN A. (1939) Cytological phenomena connected with the root swelling caused by growth substances. *Hereditas*, **25**, 87–96.

[67] OLMSTED J.B. & BORISY G.G. (1973) Microtubules. *Ann. Rev. Biochem.*, **42**, 507–40.

[68] GREEN P.B. (1962) Mechanism for plant cellular morphogenesis. *Science, N.Y.*, **138**, 1404–405.

[69] PICKETT-HEAPS J.D. (1967) The effect of colchicine on the ultrastructure of dividing plant cells, xylem wall differentiation and distribution of cytoplasmic microtubules. *Devel. Biol.*, **15**, 206–36.

[70] HESLOP-HARRISON J. (1971) Wall pattern formation in angiosperm micro-sporogenesis. *Symp. Soc. Exp. Biol.*, **25**, 277–300.

[71] O'BRIEN T.P. (1972) The cytology of cell wall formation in some eukaryotic cells. *Bot. Rev.*, **38**, 87–118.

[72] HESLOP-HARRISON J. (1971) Sporopollenin in the biological context. In *Sporopollenin*, pp. 1–22 (Eds J. Brooks *et al.*). Academic Press, London.

[73] HESLOP-HARRISON J., KNOX R.B. & HESLOP-HARRISON Y. (1974) Pollen-wall proteins: exine-held fractions associated with the incompatibility response in Cruciferae. *Theoret. Appl. Genetics*, **44**, 133–37.

[74] GODWIN H. (1968) The origin of the exine. *New Phytol.*, **67**, 667–76.

[75] BROOKS J. & SHAW G. (1971) Recent developments in the chemistry, biochemistry, geochemistry and post-tetrad ontogeny of sporopollenins derived from pollen and spore exines. *Pollen, development & physiology*, pp. 99–114 (Ed. J. Heslop-Harrison). Butterworths, London.

[76] WODEHOUSE R.P. (1935) *Pollen grains.* McGraw-Hill, New York.

[77] HESLOP-HARRISON J. (1966) Cytoplasmic connections between angiosperm meiocytes. *Ann. Bot.*, **30**, 221–30.

[78] KNOX R.B. & HESLOP-HARRISON J. (1970) Direct demonstration of the low permeability of the angiosperm meiotic tetrad using a fluorogenic ester. *Zeitschr. f. Pflanzephysiol.*, **62**, 451–59.

[79] DICKINSON H.G. & HESLOP-HARRISON J. (1970) The ribosome cycle, nucleoli, and cytoplasmic nucleoloids in the meiocytes of *Lilium*. *Protoplasma*, **69**, 187–200.

[80] HESLOP-HARRISON J. (1968) Wall development within the microspore tetrad of *Lilium longiflorum*. *Canad. J. Bot.*, **46**, 1185–92.

[81] DOVER G.A. (1972) The organization and polarity of pollen mother cells of *Triticum aestivum*. *J. Cell Sci.*, **II**, 699–711.

[82] HESLOP-HARRISON J. (1972) Pattern in plant cell walls: morphogenesis in miniature. *Proc. Roy. Inst. Great Britain*, **45**, 335–52.

Chapter 3.6
Pattern Formation in
Animal Embryos

3.6.1 INTRODUCTION

The outward signs of embryonic development, consisting of the gradual differentiation of specialized cell types, must, when put in terms of present-day cell biology, be the products of the selection and expression, in different cells, of particular components of the complete set of instructions for differentiation provided in the egg nucleus. We know little about the control mechanisms which allow cells to express the particular patterns of gene activity which distinguish liver cells, for example, from kidney cells (Chapter 5.3). However, we may assume that similar principles of control operate in all types of specialized cell in higher organisms and, in view of the remarkable evolutionary constancy of the basic biological mechanisms of genetic coding and protein synthesis [1], it is not unlikely that they resemble situations in bacteria where single regulatory genes (for example, catabolite repression and stringent control genes [2]) mediate complex changes in the functional pattern of the genome causing major changes in the properties of the cell.

This chapter will concentrate on the central issue of how different states of genetic commitment are reached by different cells in the developing embryo in such a way that consequent patterns of differentiation conform to the adult body plan of the species; how, given the familiar assumption that the nucleus given to each newly appearing cell is an exact copy of the original egg nucleus from which it derives ([3] and Chapter 2.2), quite distinct components of the genome become selected for expression in different cells. To what extent then is differential gene expression the result of specialization built into regions of the original egg which modify nuclei assigned to them during cleavage?

That the egg *can* contain regional specializations which directly anticipate the layout of the future differentiated organism is demonstrated in species (Fig. 3.6.1) where the egg, while remaining a single cell with one nucleus, can undertake visible differentiation of its parts in an approximation of at least the earliest stages that would normally emerge during multicellular development. Further evidence for specialization within the egg which anticipates future body parts comes from studies where isolated egg fragments, such as animal and vegetal halves of the sea-urchin egg (Fig. 1.3.5) [5], display distinct patterns of continued development appropriate to the parts of the organism that they would normally have gone on to form.

However, in the case of amphibian eggs, halved eggs give rise to complete and normally proportioned embryos no matter which half is isolated (Fig. 3.6.2A). The sole exception, an isolated vegetal half (Fig. 3.6.2B), fails to form any definitive embryonic parts. So, in marked contrast to the situation in the

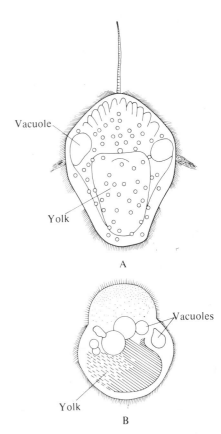

Fig. 3.6.1 Evidence for preformed regional specializations within an egg corresponding to the future layout of the differentiated organism. In the annelid *Chaetopterus* the egg can be induced, by no more than a change in ionic conditions, to undertake differentiation without nuclear division or cellular cleavage in which there is apparent partial fulfilment of larval development including the appearance of vacuoles and yolk-rich regions and the formation of cilia (B). The similar pattern of differentiation of the larva which emerges during the course of normal multicellular development (A) could therefore be the result of organization already built into the single egg cell. From Lillie [4].

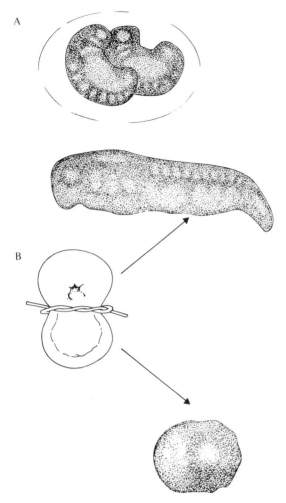

Fig. 3.6.2 A. Evidence that no cells in the amphibian embryo acquire any irreversible commitment to prospective fates on the basis of their derivation from specific parts of the egg. After separation of the cells of the two cell stage embryo complete, normally proportioned embryos develop from both parts of the original egg. This is true regardless of the plane in which cleavage happened to have divided the egg except when it separates animal and vegetal halves (B). In this case the animal half again forms a complete embryo but the vegetal half fails to form any recognizable differentiated body parts. The same results are obtained when parts of embryos are removed up until the early gastrula stage by which time the embryo consists of tens of thousands of cells. From Spemann [6].

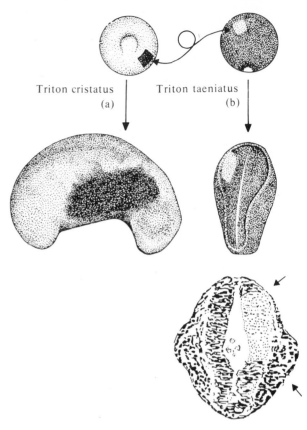

Fig. 3.6.3 Evidence that amphibian embryo cells can fulfil any pattern of differentiation as late as the early gastrula stage of development. When ectodermal cells of the prospective brain region of a *Triton taeniatus* embryo (black) (b) are exchanged early in gastrulation with cells from the prospective body wall of a *Triton cristatus* embryo (white) (a), both transplants formed whatever structures were appropriate to their new contexts. Thus, cells that would have formed body wall now form parts of the nervous system in (b); their perfect integration with their new surroundings (transplanted cells lie between the arrows in the cross-section shown below) indicates that subsequent to transplantation these cells must have received detailed developmental cues from their new neighbours. From Spemann [6].

sea-urchin, there is no evidence whatsoever for any differential predispositions in any axes within the amphibian egg which could be directly responsible for the formation of particular parts of the future tadpole. As late as the early gastrula stage of development the same remains true. Whichever half of the embryo is removed, the remaining cells differentiate into a complete embryo so that no definitive parts of the future tadpole can yet have been finally assigned for differentiation in the cells normally destined to form them. Indeed, as can be shown in detail by transplanting given embryonic cells to a variety of embryonic locations (Fig. 3.6.3), any cell can still fulfil any form of differentiation as required in order to fit its position in the whole emerging embryo. So with regard to achieving specific patterns of differentiation all embryonic cells at this stage seem to be equivalent and open to all possibilities. Although, exceptionally, some cells *do* become committed as definitive germ cells much earlier (Chapter 1.3) they do not at first contribute to

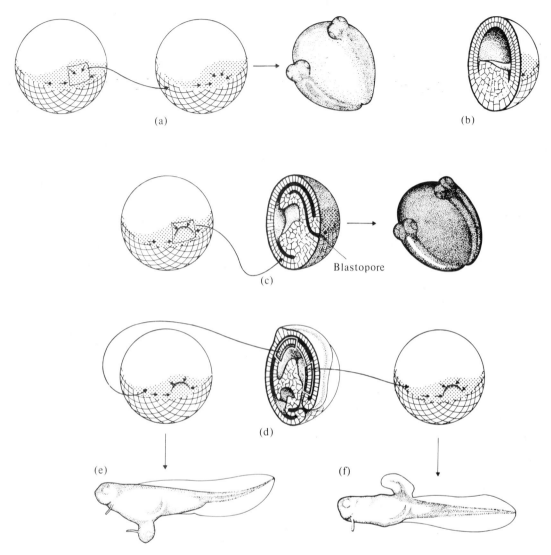

Blastopore

(a)

(b)

(c)

(d)

(e)

(f)

Fig. 3.6.4 Major events in the derivation of the basic layout of the embryo from conditions inherent in the egg. Diagrams (a) to (d) show sequential developmental stages from the egg to the neurula, viewed from similar angles, to illustrate the cell movements involved in gastrulation. Cross-hatching indicates vegetal regions and the dotted area represents the 'grey crescent' of the egg. (a) *Egg*. Transplantation of the grey crescent region of one egg to the prospective belly region of a second fertilized egg promotes the appearance of a second, normally proportioned embryo alongside the primary embryo developing as usual in association with the original grey crescent of the recipient egg [7]. (b) *Blastula stage*. During this stage the egg cleaves to form some tens of thousands of cells. A cavity is formed in the animal hemisphere. (c) *Gastrula stage*. Transplantation (from a white to a similarly staged black embryo) of the immediate derivatives of the grey crescent (the dorsal lip of the blastopore), which are now becoming anatomically distinguished as the focus for invaginative cell movements (shown by arrows), leads to the same results as in (a). The transplanted cells, shown in white, contribute only a small part to the structure of the secondary embryo. The fact that they contribute a variety of differentiated cell types appropriate to the context of their irregular distributions in the embryo, including cells of the neural tube, the notochord and the mesoderm, is evidence that at transplantation they were not already committed to any restricted developmental fates. From Spemann & Mangold [8]. (d) *Neurula stage*. When prechordal plate (e) or notochordal (f) regions of the chordamesoderm are transplanted to identical positions in the belly region of separate gastrula stage embryos, they cause the immediate transformation of the overlying ectoderm into specific regions of the neural plate, whose distinct characters can be readily recognized by the structures which form secondarily around them. The forebrain region (e) leads to a complete set of head structures while the spinal cord (f) leads to tail formation, and thus point to the initial conditions needed for the organization of these structures during normal embryogenesis. The original chordamesodermal transplants fulfil their normal patterns of differentiation despite the foreign environment and can therefore be said to have been already irreversibly committed as specialized cells. From Mangold [10].

the definitive anatomy of the animal; they only do so later by migrating into the gonads when these *are* formed in the later course of embryogenesis.

The central issue of embryology can be restated still more clearly in the light of this evidence from the vertebrate embryo. How might adult patterns of differentiation even begin to emerge if the embryo is initially composed only of cells which are all equally unpredisposed to any of the possible forms of adult cell specialization?

The key to our understanding lies in the following observation. When a particular region of the amphibian gastrula stage embryo (the dorsal lip of the blastopore) is transplanted into the prospective belly region of a second, similarly staged embryo, it causes the formation of a second, complete embryonic axis (Fig. 3.6.4c). As Fig. 3.6.4c demonstrates, the transplanted blastopore lip, shown white, contributes but a small part to the structure of the induced embryo, which is largely formed out of cells which, but for an influence emanating from the transplant, would have formed part of the belly of the host.

By direct analogy with this result we can conclude that *the organization of all the distributed diverse parts of any normally forming embryo is the result of the influence of its associated blastopore lip.* Are we entitled to conclude that this region alone

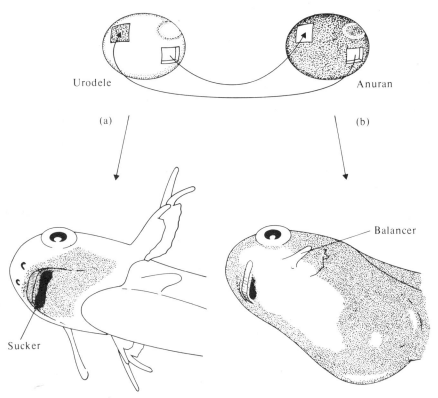

Fig. 3.6.5 The genotype of the species dictates developmental potentialities. Tissue is exchanged between embryos of different amphibian orders whose only major difference in structure is in mouth parts; urodeles (a) possess bilateral 'balancer' appendages below and behind the eyes whereas anurans (b) lack these organs but instead have a sucker in the midline immediately behind the mouth. Prospective belly ectoderm is transplanted (a) from an early anuran gastrula (black) (*Rana esculenta*) to the prospective mouth region of a similarly staged urodele (white) (*Triton taeniatus*) [11] or (b) from a urodele (*Triton taeniatus*) to an anuran (*Bombinator pachypus*) [12]. The graft successfully forms the range of structures appropriate to conform with its surroundings in the host where the structures are in common in both species; for example, the mouth formed in the correct position in (b) in continuity with the mouth of the host. The grafts never formed structures that they would not normally have formed in their own embryos of origin; in (a) the anuran tissue failed to form a balancer where it overlies the appropriate position on the host and in (b) urodele tissue fails to complete the formation of the oral sucker. But grafts did form structures unique to their own species, such as the balancer formed in (b) and the sucker formed in (a) whenever they occupied positions on the host with anatomical relationships corresponding to those of their normal sites of formation. In order for belly ectoderm to realize its potentiality to form a balancer organ at its characteristic location, it must be provided by the host with cues defining the appropriate location. Since the host would never normally form any organ at this particular site, let alone a balancer which is an organ it is incapable of forming, it seems improbable that cues should be available uniquely marking this one location in particular and even less probable that the cues should be specific prerequisites for balancer differentiation.

directly dictates the particular coordinated patterns of differentiation of cells throughout the embryo?

In the experiment depicted in Fig. 3.6.5 tissue taken from the belly region of an embryo of one amphibian species incorporates perfectly into the head structures forming in the embryo of a second species. In so far as the anatomy and patterns of development of the two species are closely similar this merely duplicates the findings of Fig. 3.6.3 and shows that integrating forces in development can act in the normal way between tissues of these different species. And yet it was the invariable finding that, despite this evidence that the transplant was in receipt of organizing cues typical of the host species, transplants from a species lacking a particular appendage, such as a balancer, failed to form that structure where it would normally have appeared in the host (Fig. 3.6.5a). Here we have a clear indication that the response capabilities of the differentiating cells themselves must contribute to embryo-genesis, at least in so far as the genotype of the species sets limits on what the integrating influences in the embryo can do.

Further evidence for the decisive role of the constitution of the developing cells has come from studies of imaginal disks in insects (Chapter 2.1) [13, 14]. These are pockets of embryonic tissue whose differentiation is delayed until metamorphosis of the insect larva. They are then the sources of cells from which specific newly forming adult body parts are derived (Fig. 2.1.6). In place of the surgical techniques used in the above experiments, experimentally induced chromosome recombination can be used to change the constitution of cells genetically *in situ* [15]. Figure 3.6.6 illustrates cases where a 'genetic mosaic' has been formed, composed of the patch of cells derived from a single recombinant cell (shown in white) surrounded by and in continuity with the normally developing disk cells. The

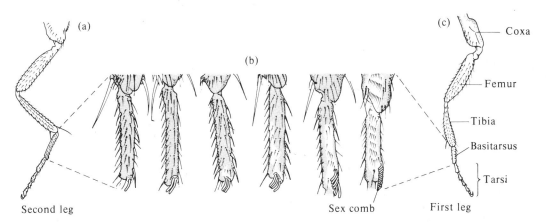

Fig. 3.6.6 Examples of genetic mosaics involving a homeotic mutation which modifies the pattern of development of the second limb imaginal disk in *Drosophila*. This diagram shows various examples (b). taken from separate male flies, of the distal tibia and basitarsus of limbs formed from the second limb imaginal disk which contain patches of mutant cells (shown in white). These patches vary in size according to the number of mitoses that have occurred among the derivatives of an initially affected cell in the time interval since that cell underwent chromosomal recombination. Examples are arranged with, the most nearly complete second limb (a) on the left, and on the right, limbs almost completely transformed into first limbs (c) as a result of the expression of the 'extra sex-comb' homeotic mutation. Regardless of patch size or position, all mutant cells (which can be independently identified by colour as the result of carrying a second, linked mutation affecting their pigmentation [15]) formed only structures typical of corresponding positions in the first limb, despite their context which before recombination had been appropriate to the formation of second limb. From Hannah-Alava [16]. Thus wherever mutant tissue occupied the position corresponding to the sex-comb, which in the male distinguishes the first from the second limb, it formed enlarged sex-comb bristles 'autonomously' regardless of whether it was surrounded by first or second limb structures. However, if the positions of these bristles are examined in detail (comparing cases from left to right in which increasingly large patches of cells are transformed) it can be seen that the definitive longitudinal arrangement of the sex-comb is reached by gradual clockwise rotation of what would otherwise have been the final transverse basitarsal row of bristles in the normal second limb—in the same way as the sex-comb is thought to be formed in the normal male first limb as a transformation of the final transverse row represented in the female first limb [13, 15]. Individual transformed bristles do not move in isolation but depend on the influence of their surroundings: the cases shown here demonstrate that rotation occurs in proportion to the amount of tissue transformed. The entire transformation may be the result of no more than a general augmentation of all aspects of growth in the region. Not only do the bristles become enlarged but there is an enlargement of the surrounding basitarsus and rotation could be due to mechanical forces generated as a result. Such forces would account for the coordination of rearrangements occurring in the affected area—even normal female or second limb bristles lacking the potentiality to transform into sex-comb bristles, which happen to lie in the region, are constrained by their surroundings to move in line with the sex-comb. When the whole region is further enlarged under the influence of the 'eyeless-dominant' mutation any genetically normal male tissue present is caused to transform into surplus sex-comb by its abnormal surroundings [15]. These results show that this prominent component of the structural differences which distinguish first and second limbs may consist of a relatively simple deviation in a pattern of development initially common to both limbs and that this component can occur (due to the action of a single gene which may be identical to that involved in the same difference between male and female first legs) independently of the other components which normally combine to identify the two limbs.

recombinant cells in this example are expressing one of the 'homeotic' class of mutations [14], which, when available for expression in all cells, characteristically cause one imaginal disk to develop in the form of other regions of the body normally derived from entirely different disks. In the cases shown here recombinant cells are differentiating in quite different ways from normal and from their surrounding cells due to the purely intrinsic influence of a single modified gene.

A wide variety of evidence shows that, far from its being the case that influences external to the developing embryonic cell can alone cause it to conform to the overall pattern of differentiation, the cell itself plays a decisive part.

Following sections will present two contrasting models, each of which seems, at the present time, to offer a causal description of embryogenesis and each of which attempts to identify these two factors of cell constitution and external coordinating influences. It must be emphasized that no attempt will be made to give a systematic account of embryonic development or of all possible relevant explanatory models and that much of the discussion which follows is necessarily or deliberately speculative.

3.6.2 THE POSITIONAL INFORMATION MODEL

In considering the wide varieties of differentiating insect tissues which can, by the formation of genetic mosaics, be brought into conjunctions which would never occur during normal development, Stern [15] has emphasized the normality of the patterns of differentiation of the separate parts of the mosaics. Even though, up to the time of genetic recombination, imaginal disk cells are clearly open to modification of their future courses of development, there are few distortions or deficiencies in finally formed adult structures such as might indicate interactions with or dependencies upon the nature of differentiation of nearby tissue. As an example of such 'autonomy' in the development of elements in a pattern, Fig. 3.6.6 shows fully differentiated mosaic limbs, formed out of *second* limb imaginal disk tissue containing patches of cells which, as a result of their expression of a particular homeotic mutation, differentiate atypically in forms characteristic of the *first* limb. Regardless of where a recombinant cell happens to occur, it invariably differentiates in the form in which normal first limb cells at corresponding positions in the first limb would differentiate, even when it is completely surrounded by normal second limb tissue.

On the one hand this result points to the fact that in selecting a pattern of differentiation a developing cell can choose independently of the way in which cells in its immediate surroundings are developing. On the other hand, in order to develop in the one limb as though they were at precisely comparable positions in the other, it would seem that

recombinant cells must be receiving information based on the large scale, general anatomical features that the two limbs have in common and therefore based on distant surrounding tissue as a whole. Pursued to its logical conclusion evidence of this sort means that, for a cell to be able to develop entirely independently of the states of differentiation occurring elsewhere in the tissue, the information it receives must be provided by a means which is itself quite independent of differentiation occurring within the tissue. One way in which this could be done, so that cells would directly and unfailingly be able to develop in accordance with their positions, is that mechanisms are provided within the embryo for the specific purpose of defining positions as such. This is the positional information model [17].

As evidence that position as such is defined in an embryo the experiments shown in Fig. 3.6.5 are important. These show that the potentiality to form the balancer organ characteristic of a particular amphibian species, is inherent throughout the embryonic ectoderm of that species. It can be released in any part of the prospective belly ectoderm which happens to occupy the appropriate highly specific position on the head following transplantation. This can only be explained if the tissues of the host embryo had precisely defined this position and in a unique way so that balancers never form elsewhere. The experiment also shows that the appropriate site is defined even in the embryo of a species in which—since it itself is never normally able to form *any* organ at that site and cannot normally form a balancer of any kind—there is no reason to suspect that this site should be particularly singled out for definition, let alone in a way specific for balancer promotion. These observations can be readily explained if we suppose that any site in the embryo is defined as a matter of course during embryogenesis and in a general way without regard to how this information might be used, so that whatever the potentiality of a cell it can develop in the appropriate way at each and every position.

Having been provided with information about its position in the embryonic tissue as a whole, each cell would then be able to achieve the pattern of differentiation appropriate to its position in the future adult by referring to an inbuilt set of instructions relating differentiation to positional values. At this point the model brings in the parameter of cell constitution and in particular the repository of potentialities which the genome provides. Even when cells carry a mutation which causes a new pattern of differentiation to occur at a particular location, a suitably placed cell could, given positional information, express the pattern of differentiation appropriate to its constitution directly and independently of the constitutions of surrounding cells. So autonomy of differentiation in genetic mosaics is explained. The coordinated development of tissues with different genotypes, including even tissues of different species, is also accounted for provided that the positional information channels of each are of sufficiently general a form to be interchangeable.

Such a requirement could, for example, be met by a simple, perhaps chemical, gradient extending continuously across the whole tissue; changing levels along the gradient could potentially define all positions.

The idea of a gradient has often appealed to embryologists because it offers a ready solution to the problem of generating diversity in a coordinated pattern from the necessarily simple starting conditions which can be provided in the egg. The gradient only has to be thought of as being constantly renewed, perhaps by chemical diffusion from a source, to account for the revision of prospective developmental fates of transplanted cells consistent with new positions in the embryo (Fig. 3.6.3). The formation of a complete tadpole following transection of the embryo (Fig. 3.6.2A) would also be explained if the gradient renews itself in such a way that a complete, normal gradient will be established across the existing extent of the embryo or if, alternatively, cells assess their locations only with respect to their relative proximities to the maximum and minimum of the surviving part of the original gradient. If the dorsal lip of the blastopore is identified as the origin of such a position-defining gradient, the spreading of the influence of the blastopore through the embryo may be related to diffusion of the gradient and the coordination of embryonic development which occurs around this crucial region can be readily accounted for on the positional information model. Graded distributions of developmental capacity have been detected in the amphibian embryo [6, 9] comparable to the gradients of animal and vegetal factors in the sea-urchin (Chapter 1.3, [5]).

However, much important experimental evidence is difficult to reconcile with the positional information theory.

3.6.3 EMBRYONIC INDUCTION

So far we have considered developing systems largely from the point of view of their behaviour as a whole in response to experimental modifications performed at early developmental stages. But when the causal sequence of events leading to the appearance of any given adult structure is examined in detail, it can be shown to be critically dependent on the presence of particular, previously formed embryonic structures.

Such a dependence is implied in the case of the formation of the lens from ectodermal tissue, to take a familiar example, by the finding [6] that lens formation fails after early removal of the eyecup, whose approach towards the ectoderm would normally be associated with the initiation of lens formation (Fig. 3.6.8). However, the eyecup could merely be activating a course of development to which the ectodermal cells had already been committed by other events. As proof that such a developmental response is totally dependent on the approach of an inductor like the eyecup, there is the crucial evidence of the kind shown in Fig. 3.6.7 concerning, in this case, the role of the hindbrain as

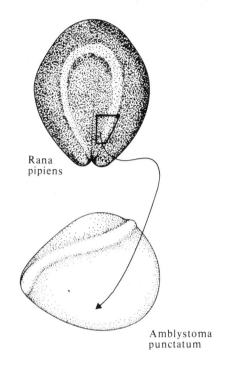

Rana
pipiens

Amblystoma
punctatum

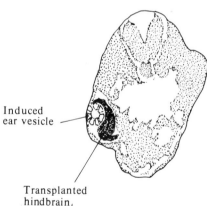

Induced
ear vesicle

Transplanted
hindbrain.

Fig. 3.6.7 Inductive modification of development in one tissue due to the presence of a second transplanted tissue, whose own unmodified development in the new environment proves its own developmental commitment. The prospective hind-brain region of a *Rana pipiens* neurula (black) is transplanted to a position below the flank ectoderm of a similarly staged *Amblystoma punctatum* embryo (white). Although at the time of transplantation the tissue is no more visibly differentiated than the tissue transplanted in Fig. 3.6.3, it has, by this slightly later stage, evidently acquired irreversible developmental commitment because, instead of accommodating to a changed embryonic environment, it fulfils the pattern of differentiation appropriate to its original position. The cross-section of the host embryo at a later stage shows that in the neighbourhood of the transplant host ectoderm, which in the normal course of events would have gone on to form trunk epidermis, has formed a distinct otic vesicle. From Albaum & Nestler [46].

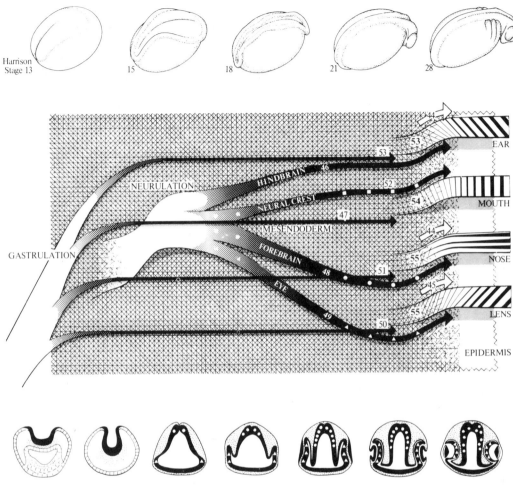

Fig. 3.6.8 Outline of major components in the causal sequences of events leading to the formation of organ rudiments derived from early amphibian ectoderm. This schematic diagram summarizes evidence concerning the derivation of several major organ rudiments from mid-neurula ectoderm (represented as cross-hatched background) and the nature of the responsible inductive agents. Time is represented horizontally and judged against standard stages of embryonic development as shown above. Space is represented vertically and therefore changes of cell position as the result of morphogenetic movements are depicted diagonally. Below are shown, in diagrammatic cross-sections of the head, the sequence of events leading up to lens induction. From the left, gastrulation leads to the formation of mesoderm (stippled) and the invagination and induction of neural plate, whose subsequent rearrangements and specialization into eyecup (shown by triangles) and forebrain (shown by circles) rudiments are traced to the right. (a) *References 46–49* refer to experiments, exactly equivalent to that shown in Fig. 3.6.7, where neural rudiments or the rudiment of the foregut, are transplanted to a variety of foreign sites in similarly staged embryos. In each case the transplant was free of any of the tissues which normally surround it in its original embryonic location which might have had continued inductive actions upon it. Since these tissues' original prospective developmental specializations were not altered by their new circumstances it can be concluded that they were developmentally committed. States of commitment are depicted in the diagram as solid lines. The single

factor of the transplant was sufficient to cause nearby prospective epidermal ectoderm to form discrete and easily distinguishable organ rudiments of specific kinds according to the type of transplant. Either ear (otic vesicle), mouth (stomodeum), nose (nasal placode) or lens were formed in conjunction with hindbrain, foregut, forebrain and eye respectively, as they are during normal embryogenesis. (b) *References 53–55* refer to various experiments which show that the potentiality to form any of these organs is simultaneously present in the same ectodermal cells—arbitrarily taken in all cases from the region near the prospective gills in Stage 22 *Amblystoma punctatum*—presumably in common with ectodermal cells throughout the embryo. We can conclude that *the formation of these induced organs at the particular restricted locations at which they appear during normal embryogenesis can be fully accounted for in terms of selection, from among the multiple potentialities inherent generally in ectoderm, on the basis of the presence of specific inductors. Well before ectoderm begins to lose its multipotency the inductor tissues are discrete and distinctive as a result of having already become developmentally committed to various forms of differentiation.* (c) *References 50–53* document evidence that local mesoderm (or neural crest) have an inductive role in the formation of these ectodermal derivatives. (d) *References 44–45* refer to examples of abnormal inductors having identical inductive effects to those considered here. The normality and specificity of their actions may be due to their abilities to trigger the appropriate inductive actions in local mesoderm.

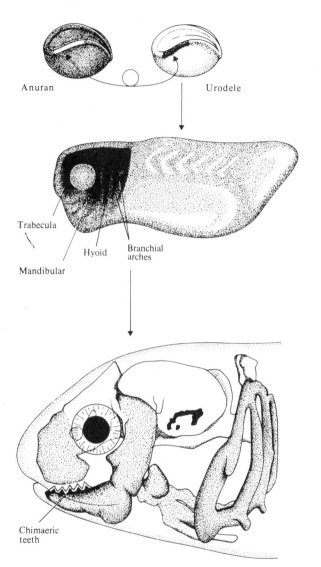

Anuran

Urodele

Trabecula

Hyoid   Branchial
arches

Mandibular

Chimaeric
teeth

Fig. 3.6.9 Regionally specific inductive cues unused in the tissues of one species can be revealed by tissues from another species with the appropriate responsiveness. The neural crest of a urodele neurula (*Triton*, white) is replaced by vitally stained crest from an anuran (*Bombinator*, black) whose subsequent migration through specific routes in the mesoderm into the host's head (to form the named visceral arches) is seen to follow a pattern common to both species. The enamel organ components of the teeth, formed from the stomodeal region of the ectoderm in urodeles, are normally absent in anurans but, as the present experiment shows, this is not because anurans lack the necessary inductors. Anuran neural crest will sustain the formation of enamel organs in urodeles in which the neural crest has been removed and in which they would otherwise fail to appear; chimaeric teeth are thereby formed with enamel from urodele ectoderm and dentine from anuran neural crest. From Wagner [19]. So anuran neural crest which itself normally has no such inductive action can perform the function of a specific inductor. Thus normal induction may not depend directly on a property specifically provided for the purpose in the inductor, but may reflect the responsiveness of uncommitted embryonic cells to forms of specialization which inductive tissues anyway possess as a function of their own independent states of commitment and differentiation. In the closely comparable situation considered in Fig. 3.6.5 the formation of a normal balancer by competent ectoderm transplanted to a species lacking such an organ requires no more than that this species possesses, as part of its own normal anatomy, structures homologous in kind and in location to those which normally induce the balancer. Since there is evidence that the underlying mandible is such an inductor [22] it is clear that the necessary homologous structures are present and there is direct evidence [19] that—as with tooth induction by mandible from a species lacking enamel organs—such a neural-crest derived tissue of a species lacking the balancer can indeed substitute for the normal inductor.

inner ear by ectoderm, far from being 'autonomous', can be entirely accounted for by the single factor of the presence of hindbrain in its immediate vicinity. As Fig. 3.6.7 illustrates, a purely local inductive interaction decisively controls ectodermal development in spite of the fact that the ectoderm remains at an embryonic location which would normally have predisposed it to develop in a different way and in spite of any influences that the ectoderm may continue to receive from more distant parts of the embryo as a whole which would be consistent with its original prospective form of development.

Fig. 3.6.8 summarizes evidence which shows in the same way how the formation and location of many of the major vertebrate organ systems are the result of embryonic induction of ectoderm. The neurula ectoderm can be shown, regardless of location, to possess the potentialities to form a large number of organs. Differentiation in the form of one particular organ can be totally and solely controlled by specific inductor tissues like forebrain or eyecup, which themselves have been formed at earlier stages of development and which can be shown to be irreversibly committed to distinctive forms of differentiation by the time ectoderm potentialities become selected. So in the case of the normal embryogenesis of these induced organs, each of which normally develops in association with the specific tissues known to be capable of their induction, the particular locations at which these initially uniformly available developmental programmes occur can be fully accounted for in terms of embryonic induction alone. When the size of the eyecup is experimentally varied [6], no matter which part remains, it is only the ectoderm lying immediately opposite the eyecup tissue which undergoes a uniform and coherent transformation resulting in the formation of a lens of matching size. Thus

inductor of the inner ear. The formation of a complete and normal looking inner ear vesicle by *belly* ectoderm following the transplantation of hindbrain beneath it establishes the following points. The potentiality to form inner ear is not confined to the particular region of ectoderm which actually differentiates in this way during normal embryogenesis but is present equally in other ectodermal cells which would normally form other quite distinct differentiated structures, in this case flank epidermis. What is more the experiment shows that hindbrain alone is able to evoke this developmental potentiality.

Evidence of this kind stands in marked conflict with the approach discussed in the previous section. The formation of an induction also accounts for the size of embryonic rudiments. This example illustrates the general point that ectoderm only reacts developmentally where a suitable inductor lies in immediate conjunction with it. We can sum up the general features of embryonic induction by saying that single, specific forms of cell differentiation are selected and activated in embryonic cells as the direct result of a pre-programmed, invariant interaction with cells which themselves have already become committed to a single specific form of differentiation, which is quite independent of, and in no other way connected with, the form of differentiation induced.

In attempting in the following sections to account for the way in which one inductive cell might be able to cause selective gene expression in uncommitted cells undergoing induction, we will have to consider the roles of both sets of cells. It cannot be assumed that patterns of differentiation brought about in uncommitted cells by induction are entirely due to the external factor of the inductor, for we have already seen that each developing cell contributes to its own developmental potentialities. On the contrary, the induced cells might achieve their developmental specializations on the basis of their own abilities to detect and respond appropriately to specific features of the inductor.

The simplicity of the influence immediately due to the inductor, in terms both of spatial organization and of informational content, is brought out by studies [9] in which various and bizarre agents have been applied to embryonic tissue in the course of attempts to identify the chemical nature of inductors. No more than a change in pH of the culture medium is sufficient to duplicate the action of the chordamesoderm in causing isolated gastrula ectoderm to undertake neural as opposed to epidermal differentiation [18]. A large number of chemical agents, applied either diffusely or focally to gastrula tissue, initiate the formation of distinct head or tail parts indistinguishable from those normally induced by parts of the chordamesoderm (Fig. 3.6.4d) [9]. The very simplicity and diversity of agents capable of bringing about quite normal and specific patterns of inductive consequences (Fig. 3.6.8d) [43], point to the possibility that the responding embryonic tissues contain all the information needed for their own possible future specific patterns of development, while the inductor provides no more than a basis for a choice among these alternatives.

Sufficiently specific features to account for the choices made by responding cells may be provided in the states of developmental commitment which, as we have noted, are characteristically already possessed by tissues with inductive properties. On this view we would not have to propose any feature of inductors specifically associated with their actions on other cells in addition to properties they must anyway possess (Fig. 3.6.9). The possibility that embryonic cells may in effect be able to recognize the states of specialization of committed cells by reference to some aspect of their emerging adult differentiation and to use these cues to control their own development, is supported by evidence from at least one developmental

situation. During embryogenesis the axonal processes of nerve cells have to grow through long and complex routes in the nervous system in order to reach neurones with which they must connect in order to come into functional association. The specificity of the routes followed by the axons of outwardly identical nerve cells indicates that the developmental history of the cell has conferred on it high degrees of specialization which allow the axon to distinguish and follow its own choice of routes among the vast range of alternatives constantly available to axons (Fig. 3.6.10). An axon must be selecting routes on the basis of distinctive characteristics associated with the tissues it meets while growing. These may be provided by local nerve cells possessing their own specializations derived, like the particular specialization of the cell from which the axon grows, during the original independent embryonic development of the various parts of the nervous system.

In the sections which follow we discuss whether embryonic induction, regardless of the nature of the cellular interaction involved, could alone form the basis for an explanation of all the phenomena of development. In particular it is necessary to consider the nature and derivation of the potentialities of embryonic cells undergoing induction. Only if the potentialities of these cells were entirely unrestricted could the specific character of their subsequent differentiation be solely ascribed to equally specific properties of the inductor; otherwise additional embryological mechanisms may have to be brought in. The induction model requires that the formation of all adult organs is preceded by the existence of an array of inductors with corresponding locations and specificities. How are we to account for the formation of the inductors themselves if, as the evidence for the equivalence of all early embryonic cells implies, they are not built into the egg from the start? How is the dorsal lip of the blastopore related to the formation of the first inductors? If specific inductive interactions are to be the only mechanism for choosing all cell specializations then it is important to know how the necessary confrontations between uncommitted embryonic cells and the appropriate inductor occur in such a way that cells of the correct specialization are formed at the location required in adult anatomy.

The role of cell movements

The way in which cell movements may serve as the means of achieving the confrontations between uncommitted tissues and their inductors which decide where new tissues will be formed, can be illustrated by following the sequence of events preceding lens formation (Fig. 3.6.8) [6, 67]. In the course of their formation during the massive cell movements of gastrulation, the specialized cells of the notochord and prechordal plate come to underlie the uncommitted ectoderm. This confrontation leads to the induction of the forebrain rudiment (Fig. 3.6.4d). As a result of later cell movements within this rudiment outgrowth of the optic vesicle occurs along with its associated commitment of

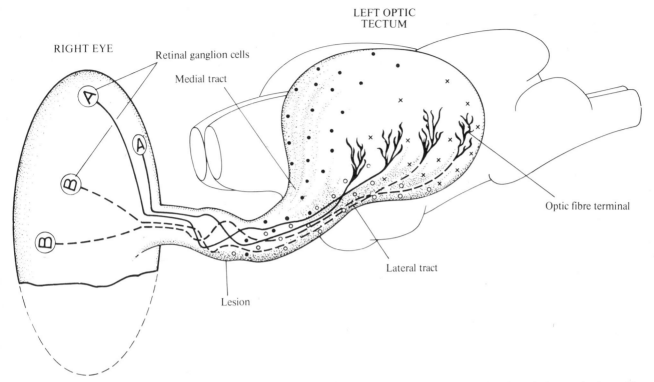

LEFT OPTIC TECTUM

RIGHT EYE

Retinal ganglion cells

Medial tract

Optic fibre terminal

Lateral tract

Lesion

Fig. 3.6.10 Specific growth patterns of nerve fibres during regeneration of the optic nerve show that fibres can discriminate features of independently developed regions of the nervous system. After section of the optic nerve in a goldfish, optic nerve fibres regrow, in exact repetition of events during embryonic development, from retinal ganglion cells into the optic tectum where each fibre terminates at specific locations according to the part of the retina it derives from. In order to reach a specific position in the brain, fibres must *actively* select their growth paths because they initially start out in haphazard array due to the tissue distortions produced at the site of the optic nerve lesion. Each fibre behaves differently in selecting a path unique to itself and probably does so independently of other fibres because, as shown in this diagram, fibres grow normally after removal of surrounding fibres by ablation of ventral parts of the retina. Although fibres originating from retinal ganglion cells in the various regions of the retina are uniform in their structural and physiological characteristics, individual fibres must be specialized with regard to growth potentialities, varying according to the regions of developmental origin of their cell bodies in the retina, in order to explain these observations. Furthermore, despite the structural homogeneity of central nervous tissue, alternative pathways must also have specializations, which must be formed early in embryogenesis prior to and independently of the arrival of optic fibres, if actively selecting fibres are to be guided into appropriate paths. Rather than suppose that the whole of the route to be followed by each type of fibre is individually and continuously marked out by a single cue uniquely matching that fibre type, it would be more economical if, at points where fibres follow similar routes, they were guided by the same local cues. Tissues associated with different pathways would then only need to possess a single and perhaps diffusely arranged distinguishing characteristic. By way of illustration, all fibres from dorsal retina, shown here, initially become segregated in common into the lateral optic tract because they are all attracted to the same open circle cue diffusely associated with this tract and repelled by filled circles associated with the alternative medial tract. Even having behaved similarly so far fibres may later diverge, according to type, to reach different regions in the optic tectum due to A-type fibres avoiding and B-type fibres seeking a region with a distinctive cue shown by crosses. Pathway cues meeting these requirements may well exist in the form of the specializations of neurones composing the nuclei which are characteristic of the various regions of the brain through which optic fibres pass—specializations which each nucleus must have (like optic fibre specializations) to enable it to make the connexions with other brain regions required for its own particular functional role. These nuclei are formed independently and in highly orderly and predictable arrangement (as is the retina) in the embryogenesis of the central nervous system. We can conclude that what optic fibres are specialized for is the recognition of the states of differentiation of cells in various, independently developed parts of the brain. After Attardi and Sperry [20].

cells as eye cells. With the eventual approach of the eyecup to the ectoderm the confrontation specific for lens induction occurs.

During these earlier stages of embryogenesis cells are able to move as coherent sheets and it is easy to see the association between movements at distinct locations and distinct inductive interactions. But as conditions in the embryo change at later stages, embryonic cells tend increasingly to reach their definitive positions by diffuse migratory movements through the embryonic mesenchyme or through the vascular system. Cases in point include the germ cells, the neural crest [21, 22] (Fig. 3.6.9), and angiogenic cells themselves [23]. To the extent that these cells are uncommitted embryonic cells, they can meet a wide range of specific inductive cues at the various locations that they reach, and so give rise to a wide range of differentiated cell types. And to the extent that they are already committed, such migratory cells can act inductively on as yet uncommitted cells throughout the embryo, and so enormously increase the complexity of spatial arrangements and types of cells in tissues.

Differential rates of mitosis and the formation of patterns in static cell populations

Further variations in the final spatial pattern of differentiated cells can be brought about within static populations of differentiating cells by the activity of cells which have retained a primitive embryonic condition. Continued mitosis by these cells can vary the distribution of already differentiated structures (Fig. 3.6.16) or, by providing cells open to later embryonic induction, add to them. The elongation of the developing vertebrate limb is entirely due to addition of tissue at the tip of the limb-bud by an apical reservoir of proliferative ectodermal and mesodermal cells (Fig. 3.6.12A). Additional variation of pattern results from interactions between differentiated cells including those between cells of the same type.

Thus melanocytes may interact to control each other's spatial arrangement or inductively to control pigmentation (Fig. 3.6.11). Inhibitory interactions may contribute to the arrangement of bristles in insect cuticle (Fig. 3.6.16).

At all stages of limb outgrowth mesodermal cells just formed in the proliferative zone possess an initial predisposition to differentiate into cartilage [63]. One could account for the

Fig. 3.6.11 The formation of pigmentation patterns specific to their breed of origin by melanocytes transplanted to a non-pigmented avian breed. Neural crest tissue of a male Barred Rock chicken embryo was transplanted into the early wing-bud of a female White Leghorn embryo. The patterns of barring within each feather and their variations across the wing are characteristic of the breed and the sex of the donor of the melanocytes [24]. Banded pigmentation patterns are thought to be the result, following initial random invasion of the feather rudiment by premelanocytes, of the inhibition of pigmentation in white areas by inhibitory factors formed by pigmenting areas. Variations in banding patterns in different body regions and within each feather may be due to variations in the growth rates of the feather which lead to different rates of accumulation of inhibitory factor [24]. Feathers in the non-pigmented breed are presumably able to promote the normal patterns of inhibition between transplanted premelanocytes because they undergo variations in growth homologous to feathers in the donor breed. In amphibian skin patterns of pigmentation reflect dispersive or associative movements of melanocytes [22].

Fig. 3.6.12 A. The proximo-distal succession of formation of structures during the outgrowth of the chick limb-bud. At three stages of limb outgrowth a comparable apical region of the limb-bud was removed. The limb parts which later differentiate in the remaining limb-bud stumps are shown on the left. From the increasingly complete limbs that are formed it can be concluded that at the earliest stage only the most proximal limb parts have been laid down and that during subsequent stages the apical region has formed successively more distal parts; successively formed generations of cells are marked by dots and circles. Explants of the apical regions of limb-buds to the coelomic cavity (on the right) successfully generate all the limb parts that had not already been laid down at the time of explantation. This confirms that the apical zone of a limb-bud is the source of the cells from which future distal limb parts are to be formed. It is also evident that all the prerequisites for the continued laying down of a succession of limb parts are inherent within this proliferative zone and that influences from the rest of the embryo are unnecessary. From Saunders [25].
B. The proliferative limb tip retains a constant developmental potentiality at all stages of outgrowth. The limb tip of a late stage hindlimb (white) was transplanted onto the proximal stump of a similarly staged forelimb (black). Even though the transplant had already formed most adult limb parts (as shown by parts formed from the donor limb stump shown on the left) it retains an unchanged ability to form all limb parts, including, as shown here, proximal parts. In accordance with its new context the transplant forms successive distal parts starting from the proximal level of the host limb stump. That the new limb parts were formed by the transplant is shown by the marker carbon particles inserted at the original junction between the tissues *and* by the fact that all new

parts were of *hindlimb* type. From Kieny [26]. This experiment demonstrates that the proliferative limb tip selects each successive limb part to be formed on the basis of cues from the immediately neighbouring tissues of the already formed limb proximal to it (see Fig. 3.6.13). The experiment also demonstrates the homology of structure and of development of the vertebrate forelimb and hindlimb. Even though the tissues of the two limbs retain their abilities to eventually generate the minor distinctions of structure which normally characterize them, they can share their basic patterns of initial laying down of a proximo-distal succession of cartilage rudiments.
C. When mesoderm from proximal levels of the hindlimb (prospective thigh tissue, shown white) is transplanted beneath the apical ectoderm of the forelimb bud, it appeared to adopt the proximo-distal level of differentiation of its new surroundings; it formed distal hindlimb parts such as toes and claws [27]. A possible explanation, which is difficult to test, is that under the proliferative stimulus of the surrounding apical growth zone of the host limb-bud, the transplanted mesoderm formed distal parts autonomously (like explanted limb tips in A) without interacting with mesoderm of the host limb. This experiment shows that limb type is carried as a property of the mesoderm; the ectoderm of the host limb is totipotent because it contributes to the formation of the hindlimb structures (scales and claws) under the inductive action of the hindlimb mesoderm. The mosaic limbs obtained in this experiment are remarkably similar to the forelimbs partially transformed into hindlimbs in animals carrying the 'ametapodia' mutation [28], whose effects may therefore be ascribed to a primary mesodermal modification in distal limb parts.

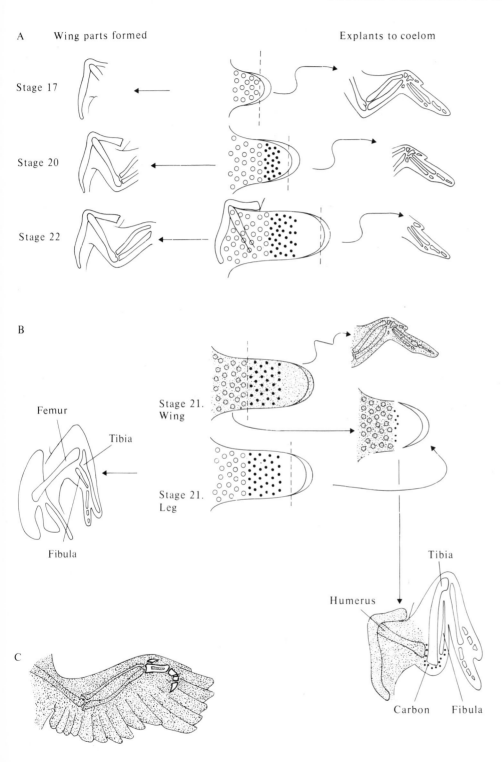

A   Wing parts formed

Explants to coelom

Stage 17

Stage 20

Stage 22

B

Femur

Tibia

Fibula

Stage 21.
Wing

Stage 21.
Leg

Tibia

Humerus

Carbon   Fibula

C

arrangement of the elements of the limb skeleton in terms of perhaps quite simple mechanical forces governing the way in which the proliferative zone varies the spatial distributions of newly added cells and so dictates the sites of future cartilage condensations. In basic outline the skeletal elements of the vertebrate limb are arranged so that at increasingly distal levels there is a duplication of elements for each element already formed more proximally. This pattern of development could be explained if the proliferative zone were morphogenetically disposed in such a way that, as it lays down new cells which will form successively more distal limb parts, it concentrates them towards the forward and rear borders of each cartilage rudiment which is already in process of differentiation proximal to it. The simpler non-duplicating succession of elements formed in the digits might reflect modified morphogenetic circumstances affecting the proliferative zone at later stages of development. In addition to the formation of cartilage the single mesodermal cell type can freely differentiate into a number of adult cell types, some of which may themselves be able to undergo transformation of type [36]. Transformations such as ossification of cartilage occur according to internally generated circumstances, including mechanical circumstances [31], so that mesoderm on its own can generate considerable complexity of arrangement of a number of cell types. As a result of this arrangement the mesoderm can directly control the development of hairs, feathers, scales, claws and nails which are formed later in development by multipotent ectoderm under the inductive influence of specific components of the mesoderm [29]. Additional anatomical complexity is generated by the superimposition of patterns of invading cell types such as muscle, blood vessels, neural crest and nerve fibres. Their distributions and the results of their inductive effects in the limb can also be determined by mechanical conditions created autonomously by the mesoderm.

This example illustrates how an organ of considerable anatomical complexity can be formed simply by the control of the distribution of multipotent embryonic cells without otherwise requiring more than a number of inbuilt, fixed rules for the induction of the contributing cell types. In this case (like many others—for example, the cranio-caudal sequence of formation of the neural plate—where patterns of differentiation may appear to change according to their times of origin), it can be shown that there is no change as time passes in the developmental rules operating. There is direct evidence that apical cells retain constant potentialities at each stage of outgrowth (Fig. 3.6.12B). Any changes in their products in time must therefore be due to progressively changing cues, either inductive or morphogenetic, from more proximal limb parts in process of differentiating.

Clearly the final adult structure of any complex organ which, like the limb, is composed of a number of different specialized cell types will not only reflect the inductive effects of these cells but also interactions between them resulting from the anatomical and physiological properties which characterize the cells once they begin to differentiate. Differentiated cells display a bewildering variety of highly specific and adaptive potentialities to respond to circumstances including, for example, trophic dependence on other cells [30] and responsiveness to mechanical stresses [31], to metabolic conditions [32] or to nerve impulse patterns [33]. As part of their states of developmental specializations, some cells may synthesize hormone receptors whose activation by a hormone can be used to regulate their rates of mitosis and differentiation [34], or even, as in the case of withdrawal of the tail at metamorphosis in amphibian tadpoles, rates of selective cell death (Chapter 4.4). The differences in the reproductive systems in the two sexes in vertebrates, which are due to hormonally induced differential development in initially identical embryonic tissues, illustrates the potency of such mechanisms in bringing about variations in adult anatomy. In this way the details of adult anatomy can become highly adapted to their functions. Anatomical complexity generated in this way as a secondary consequence of properties of the differentiated states of cells, clearly does not add to the complexity of embryological control mechanisms required. These mechanisms have merely to be sufficient to generate the basic types of specialized cells in the first place.

The states of embryonic cells prior to the initiation of differentiation by induction

If we trace the sources of the multipotent cells involved in various inductions it emerges that they are derived from a population of cells which, judging by their unchanging appearance and consistent anatomical location, may serve continuously throughout development as a reservoir of stem cells which remain free of any restriction in their potentialities. Thus organs formed at the neurula stage (Fig. 3.6.8) are all formed from multipotent cells from the ectoderm which is made up of cells derived, without any apparent anatomical modification, directly from the external cells of the blastula. Even organs which are only formed towards the end of development, such as hairs and claws, are induced from multipotent cells of the stratum germinativum of the epidermis which itself is anatomically the direct descendent of the ectoderm. These considerations raise the possibility that this multipotent cell population may simply owe its properties to the fact that its cells are unchanged from the totipotent cells of the blastula.

Further light on the persistence in fully developed tissues of cells with high degrees of multipotency comes from the phenomenon of regeneration. The regeneration of the vertebrate limb (Fig. 3.6.13) restores the spatial layout of specialized structures which was achieved during embryogenesis by means

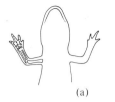  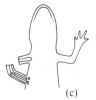 

(a)   (b)   (c)   (d)

Fig. 3.6.13 Regeneration of the amphibian limb. The distal end of the limb is first implanted into the flank to establish a second blood supply, then divided at the elbow (b). Both stumps underwent regeneration (shown in black in (d)). The forward limb stump simply replaced all missing parts to form a complete, normal limb. In the other case a duplication of structures was the result [35]. Taken together these two patterns of regeneration are explained by and combine to demonstrate the operation of the same, single rule which governs the sequence of formation of limb parts during development; the proliferative tip

(paralleling events during embryonic development, the regeneration blastema provides a source of cells from which increasingly distal limb parts are formed) only lays down the next most *distal* limb parts appropriate to the already formed limb parts with which it is in contact. This evidence from regeneration proves that the proliferative tip is influenced in its formation of new structures *only* by formed structures it is in immediate contact with; it forms new structures appropriate to these cues and regardless of overall context or polarity of the rest of the formed limb.

of a complex sequence of morphogenetic and inductive interactions led by a multipotent growth zone as described above. The only reasonable explanation for this remarkable recovery of structure is that the same sequence of events can be repeated including the continuous supply of cells with sufficient multipotency that they can successively form the range of mesodermal cell types appropriate to each proximo-distal limb level. The regeneration blastema, which is the source of these cells, would have to acquire its mesodermal-type cells from the limb stump from which it is formed. There is direct evidence [36] that tissues of the differentiated proximal limb stump such as cartilage can give rise to a range of cell types sufficient to duplicate the range arising from the original mesoderm of the limb-bud—either as a result of an inherent potentiality for multiple forms of specialization of these cells, or possibly as a result of the persistence among them of some unchanged mesodermal cells. As for the replacement of parts originally derived from the limb-bud ectoderm, there is evidence to show that adult epithelial tissues in general (of which the stratum germinativum of the epidermis of the limb stump is an example) have an unexpected multipotency amply sufficient to permit them to differentiate in the variety of ways required. For example, differentiated mammary gland epithelium can undertake cornification [37], and epidermis, when in association with visceral connective tissue, can differentiate into ciliated secretory epithelium [38]. Corneal epithelium can even form feathers when suitably induced [39].

Regeneration of the lens from pigmented cells of the epithelium of the iris [40, 41] provides even more remarkable evidence for the persistence of multipotency in differentiated cells. These cells derive from a proliferative zone, at the margin of the optic cup, which supplies cells which differentiate into later formed parts of the retina and so might be expected, like its counterpart at the tip of the vertebrate limb, to have embryonic properties. The fact that these cells can form lens tissue, which is normally provided directly by neurula stage ectoderm, can

only mean that despite having passed through all the events of neural induction and eye formation (Fig. 3.6.8) these cells have retained a range of potencies equal to that of their original stem cells, late gastrula ectoderm. During the regeneration of most tissues, any totipotent cells present will not be exposed to conditions any different from those in their normal surroundings, which are consistent with the original form of differentiation of the tissue, and so would not be expected to show major departures from regeneration in the same form. That transformation of iris to lens should stand out as a rare exception in this respect may be explicable in terms of unique anatomical features of the eye which permit the inductive factors [41, 67] normally maintaining the lens to act on cells of the iris after lens removal.

It can therefore be concluded that throughout development there are some cells which retain a complete range of potentialities similar to that of blastula cells, and that it is from such cells that most newly differentiating cells are formed by induction. Thus regardless of time or location—whether the cells subject to induction are derived from the stratum germinativum itself or from representatives of the original ectodermal stem cell line like those at the growing margin of the eyecup—the choice of a cell's pattern of differentiation must be solely due to the specific nature of its inductor. The formation of all differentiated cell types may be accounted for in terms of two factors; the specific state of the inductor which may be based on its own previous commitment to a certain form of differentiation and, secondly, a cell with the potentiality to switch on any form of differentiation derived directly as a result of what each cell inherits from the egg. There must be built into the embryo a set of rules dictating the fixed relationship between a particular type of inductor cell and the form of differentiation which results in the induced cell. Of course the available evidence does not exclude the possibility that embryonic cells may reach their final adult states of differentiation by passing through intermediate states of specialization which are reversible like the states of differentiation of some epithelial cells. It may be, for example, that migrating neural crest cells sequentially modulate their differentiated states according to the changing inductive conditions they meet.

Given that the formation of each new cell type depends on confrontation between uncommitted embryonic cells and the appropriate inductor, then the locations at which a given cell type is formed, and indeed whether it is formed at all, is a direct reflexion of the morphogenetic events which bring about confrontations. These are not limited to movements of inductor tissues towards embryonic cells. Simply by determining the distribution in time and space of cells emerging from the reservoir of multipotent stem cells these events will control their access to the potential inductors already formed and in this way control the emergence of adult anatomy. Consistent with this view is the evidence that apparently quite distinct classes of embryonic tissues (such as the ectoderm-derived neural crest and the blastoderm-derived primary mesoderm) give rise to a number of common differentiated cell types including cartilage (Fig. 3.6.9). Neural crest also forms nerve cells indistinguishable from those formed from the earlier derived neural plate and from the later derived cranial placodes which contribute to the formation of the vagal or trigeminal ganglia [42]. The unique circumstances of the iris are suggested again in the fact that muscle cells originate from its ectoderm-derived cells; throughout the rest of the body muscle is derived only from the mesoderm [42]. Evidence of this kind indicates that the traditionally distinguished embryonic germ lines such as ectoderm or neural crest do not represent classes of cells with irreversibly restricted ranges of potentialities. The different ranges of cells formed by these classes of embryonic cell may merely reflect their patterns of access to inductors as dictated by the temporal and spatial patterns of their origin from the reservoir of multipotent stem cells.

The initiation through the blastopore of the earliest developmental commitment of cells by embryonic induction

But is it possible to account for the origin of the first inductors, given the evidence for equivalence of all parts of the vertebrate egg? How then could the first regional specialization like the dorsal lip of the blastopore in the amphibian be accounted for? Or must regional distinctions, whether in the form of egg mosaicism or of positional information, be built in from the start to account for the localization of the first developmental events?

In terms of the inductive model described above any change in cell state is the result of a confrontation between multipotent stem cells and the appropriate inductive conditions. During amphibian development gastrulation is the first morphogenetic event which might serve to bring some of the homogeneous population of multipotent blastula cells into a new and potentially inductive environment within the embryo (Fig.

3.6.4). Gastrulation is known to be associated with the formation, in appropriate spatial arrangements, of parts of the chordamesoderm which, as a result of their inductions of parts of the neural plate, are essential preliminary steps in the laying down of embryonic anatomy as a whole (Fig. 3.6.4d). The whole of embryogenesis may be accounted for in terms of the consequences of gastrulation and there is no need to hypothesize any prior states of specialized commitment among embryonic cells. Up to this stage all blastula cells may retain the state of multipotency of the egg cell from which they are directly derived by cleavage and the first specialization may follow only as a result of the new locations reached by cells through gastrulation. Consistent with this view is evidence (Fig. 3.6.4c) that cells of the dorsal lip of the blastopore give rise to a multiplicity of differentiated cell types. The fact that in each case the form of differentiation is in accord with the location reached by the cell after gastrulation suggests that differentiation is a consequence of final location and that while moving inwards through the blastopore cells retain a wide multipotency.

Embryonic regulation

Given this interpretation we may envisage that the dorsal lip of the blastopore plays no more than the morphogenetic role of providing a mechanical focus for the cellular rearrangements of gastrulation. If this mechanical process is all that has to be provided for in the preformed organization of the amphibian egg, then the formation of complete embryos from half eggs (Fig. 3.6.2A) becomes no more than a reflection of the properties of this preformed organization which allow it to generate a blastopore. A diffuse arrangement of mechanical forces, such as is shown by the arrows in Fig. 3.6.4, could still form a blastopore after ablation of most parts of the blastula, but be unable to do so in isolated vegetal half embryos (Fig. 3.6.2B); hence their failure to regulate into whole embryos like all other halves.

Gradients

The graded distributions of potentialities for the formation of animal and vegetal parts which are built into the sea-urchin egg ([5], Chap. 1.3), and similar gradients in other embryos, can be interpreted as being gradients of morphogenetic dispositions to undergo the cellular rearrangements which, just as with amphibian gastrulation, lead to the formation of definitive larval parts. It may well be possible to quite accurately *describe* the complex consequences resulting from the influence of the blastopore (the spread of the influence of a transplanted blastopore lip over a host embryo (Fig. 3.6.4c), the way in which the transplant and host blastopore fields interact in marking out two embryonic axes and so on) in terms of the formation of embryonic parts according to positions on gradients of diffusing material set up by the blastopores. But we know in fact that the layout of the embryo depends on cell-to-cell inductions consequent on the arrangements taken up by gastrulating cells. It is by promoting these cellular rearrangements that the dorsal lip of the blastopore has an influence on the blastula as a whole. Only detailed knowledge of this kind concerning the control of development of the finally observed differentiated elements in a pattern, can establish whether any graded properties initially built into any particular developmental situation are themselves directly concerned with the control of formation of that pattern.

Mosaic eggs

Thus far we have accounted for amphibian development in a way which allows that, until the first inductive interaction at gastrulation, all cells remain as genetically uncommitted as the egg from which they derive. The one precondition for development, the set of morphogenetic conditions necessary to produce the blastopore, must, of course, exist as preformed regional specializations built into the original amphibian egg cell. This mosaicism needs only to be of a mechanical kind, based perhaps on differential yolk distributions in the egg which, through the process of internal subdivision during cleavage, could be directly passed on to early embryonic cells. Do the discrete, non-regulative patterns of differentiation which occur in fragments of the eggs of many invertebrates signify a different

form of mosaicism in these eggs? Is it necessary to hypothesize that controlling factors, already built into different regions of such non-regulative eggs, directly bring about the specialization of early cleaving nuclei, which in vertebrates seems to be achieved only as a result of interactions by induction between separate, independently developing blastomeres?

A true comparison between non-regulative and regulative eggs depends on being able to test the potentialities of cells at equivalent early stages of development. In large vertebrate embryos it is an easy matter to prove the initial lack of commitment of all embryonic cells (Figs. 3.6.2, 3.6.3). But the small sizes and rapid development which are characteristic of non-regulative invertebrate embryos mean that technically difficult manipulations of small numbers of cells must be performed if similar tests are to be used. Mosaicism has usually been assessed by looking for autonomous differentiation of parts of the future larva within isolated fragments of the egg or early embryo [56]. The formation of reasonably complete half larvae out of half eggs (Fig. 1.3.5) is no indicator of the original degree of anticipation of future larval structure in the egg because much of the detail could have been formed later by organizational processes within the half embryo. For this reason the degree of original mosaicism that can be claimed can only be as great as the number of larval structures that can form autonomously from appropriate isolated egg fragments. In practice relatively few larval parts have ever been so identified [56].

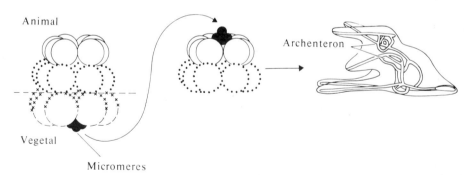

Fig. 3.6.14 Evidence for the initial equipotentiality of cells of the sea-urchin embryo. The micromeres (shown in black) which become distinct at the vegetal pole of the 16-cell stage embryo are normally destined, as the first cells to undergo invagination, to form primary mesoderm. After transplantation at the 32-cell stage to the animal pole of an animal half-embryo, the micromeres lead to the formation of a complete set of vegetal parts. Most of these, including the archenteron, are presumably formed in the normal way, by neighbouring cells (in this case animal cells) which invaginate after the micromeres. It is therefore likely that cells of the animal pole can form all larval cell types. A second gastrulation, leading to the formation of a second set of archenteric structures, occurred at the cut surface of the animal half-embryo and again the local animal cells which would normally have formed epidermis demonstrate their equipotentialities in contributing to vegetal structures. Thus the abilities of isolated animal fragments of sea-urchin embryos to differentiate autonomously

(Fig. 1.3.5) cannot be due to any preformed restrictions in their development potentialities. It could be due to the continued normal occurrence in this part of the embryo of the oral invagination which results in the formation of distinctive mouth parts as a result of the specific inductive conditions met by the invaginating cells; normally the oral invagination would occur in the context of structure already formed by the earlier vegetal invagination and similarly specific conditions may still be generated within an isolated animal half-embryo. The present results show that the cut surface of an animal half-embryo does indeed tend to undergo vegetal development. This finding indicates that even the ability to initiate gastrulation, which is characteristic of micromeres at a very early stage of development, is not a property acquired directly as a result of their derivation from a particular part of the egg because cells with other derivations can, in the appropriate circumstances, also acquire this role. From Horstadius [5].

According to these criteria the sea-urchin egg behaves as a mosaic egg in that its animal and vegetal halves differentiate into distinct larval parts after isolation (Fig. 1.3.5). However, these separate patterns of development could be entirely accounted for in terms of the morphogenetic rearrangements of cells which occur in the animal and vegetal halves of the embryo. The first event in embryogenesis is, as in vertebrates, gastrulation. This occurs at the vegetal pole and leads to the formation of internal structures such as mesoderm, skeleton and archenteron. The distinct pattern of differentiation which occurs in the animal half also involves an invagination, but one which forms mouth parts. Without needing to hypothesize any initial specialization in the potentialities of invaginating cells, these mosaic differences may arise from distinctions in the circumstances of the two invaginations; since, for example, the animal invagination occurs later, the cells involved will meet circumstances already changed as a result of the developmental responses of earlier invaginating cells (Fig. 3.6.14). So perhaps even in the case of the highest degrees of mosaicism that have been described in invertebrate eggs what is actually distinctive about different regions of the egg is differential morphogenetic predispositions stored in the egg in order to initiate and control future embryonic development.

That this is an adequate explanation for mosaic behaviour of egg fragments and that such behaviour does not necessarily imply any special features of non-regulatory invertebrate embryos (such as direct commitment of cells according to sites of origin in the egg) is brought out by the fact that similar behaviour can be demonstrated in regulatory amphibian embryos at a stage when none of its cells are yet committed. When the prospective ectodermal, endodermal and chordamesodermal regions of an early gastrula stage embryo are isolated they each go on to differentiate characteristically in accordance with their prospective fates [57] even though none of these are at the time committed to any particular developmental fates (Figs 3.6.2, 3.6.3). This is to be expected if these components have within them the differential pre-conditions, derived from regions of the egg, needed to ensure that they play their respective roles in gastrulation. Actual distinct patterns of differentiation would then only be achieved later, and in the normal way, as a result of the distinctive inductive conditions met following separate patterns of morphogenesis. In exogastrulation [58] we get an indication of the complexity of pattern that can be generated as a direct consequence of the stored preconditions for gastrulation. Exogastrulation, which occurs in the absence of the ectoderm, shows how the endomesodermal components in gastrulation can fulfil their roles quite autonomously.

In invertebrates we can get a comparable indication of the complexity of organization for future development stored as preconditions in the egg from the case of *Chaetopterus* (Fig. 3.6.1). Since the organization that appears in the egg cell here cannot be due to differential gene expression, it can only be due to the activation of events of which the egg itself is capable, including the rearrangement of existing cell organelles. In other words one can conclude that the comparable degree of organization reached during normal larval development in this species may be controlled directly by the egg, without requiring the differential control of gene expression in larval cells until later. The orderly patterns of egg cleavage, which characterize non-regulative invertebrate eggs [56], may be an indication of similar mechanical constraints operating to control future cellular relations from the earliest stages of development. The polar lobe of some invertebrate embryos (Chapter 1.2) may exert its powerful formative effects by exerting similar morphogenetic constraints.

It is proposed that there is no fundamental difference in the mechanisms for generating embryonic organization in the 'mosaic' eggs of some invertebrates and in vertebrate eggs. Even within the molluscs, which are unlikely to vary greatly in their basic mechanisms of developmental control, species with small eggs are classified as mosaic, whereas, in the case of the cephalopods which have large eggs, there is evidence that the eggs are regulatory like those of vertebrates [56]. Thus the significant variable would appear to be speed of development. Vertebrates, like cephalopods, show considerable delay in the onset of the first formative rearrangements of cells, presumably as a consequence of the larger size of their eggs which is dictated by the need for large-scale yolk storage. It is this which facilitates the demonstration of the uncommitted state of the earliest embryonic cells.

3.6.4 POSITIONAL INFORMATION OR EMBRYONIC INDUCTION?

The most substantial area of contention for these two models of development remains the evidence [17] that in some situations a developing cell can control its own pattern of differentiation under conditions where no local specific inductive cues would be expected to exist (Fig. 3.6.5), or where such cues would be expected to impose a different pattern of differentiation, as in the case of genetic mosaics involving homeotic mutations in insects (Fig. 3.6.6). In the latter case, a purely internal genetic change in a cell which is part of the rudiment of one organ, can cause that cell to differentiate as if it were part of another organ *in spite of* surrounding circumstances which would otherwise have been sufficient to allow it to develop as part of the first organ. Even when the cell is completely surrounded by normally differentiating tissue of the first organ, it can still develop independently of the inductive circumstances which would be expected to be uniquely conducive to development as the first limb.

However, the tissue combinations which are brought about

by the genetic mosaic technique involve tissues with various degrees of structural homology. All the structures concerned are cuticular derivatives of segmentally arranged imaginal disks, which have many features in common in their development. For example, they all undergo similar patterns of eversion at metamorphosis, they are affected by the same mutations specifically blocking disk development [14, 64] and many show the duplicative pattern of regeneration [13, 14] which is indicative of a polarized sequence of development like that of the vertebrate limb (Fig. 3.6.13). The two organs involved in the extra sex-comb mutation (Fig. 3.6.6), the first and second legs, are obviously similar in the arrangement of most of their component parts. Indeed, it is only because of this homology that it is possible to judge the extent to which tissue of one kind develops appropriately according to its position in the context of tissue of the other kind. In circumstances such as these, in which homologous structures in the two parts of the genetic mosaic may provide interchangeable inductive cues, distortions of pattern revealing inductive interdependence would be expected to be, at the most, small.

The embryogenesis of insects is sufficiently similar to that of vertebrates [59] that there is no reason to doubt that it could similarly be based on inductive interactions [60]. Detailed examination of genetic mosaics in which autonomy seemed to apply reveals that the site of formation of cuticular organelles such as bristles *are* influenced on a small scale by surrounding tissue context (Figs 3.6.6, 3.6.15). It is established, from studies of mosaics of disk tissue formed either genetically or by mechanical mixing of cells, that the formation of bracts is solely dependent on the influence of other previously formed organelles [13, 14]. The formation and arrangement of bristles can in general be adequately accounted for in terms of inductive and morphogenetic interactions which are known to operate during the development of insect cuticle (Figs 3.6.6, 3.6.16). Thus the available evidence does not uniquely require an explanation in terms of the positional information model to the

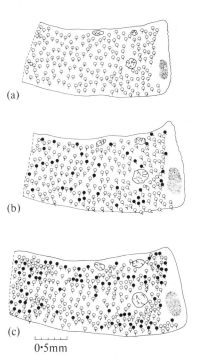

(a)

(b)

(c)

0·5mm

Fig. 3.6.16 Evidence for the dependence of bristle formation in insect cuticle on influences due to already formed cuticular organelles. Bristles added (shown in black) to the cuticle of an abdominal segment of the bug *Rhodnius*, as the larva grows from its 4th (a) to its 5th larval instar (b), occupy spaces over certain minimum distances from already formed bristles. Thus stretching or growth of spaces between bristles may directly control future bristle formation by moving potential bristle cells out of range of an inhibitory influence due to cells which have already so differentiated. In (c) the pattern was modified experimentally by mechanically stretching the cuticle; further bristles were added in the same way [61]. Where various kinds of organelle are formed by the cuticle these may or may not influence one another's development according to similar inhibitory interactions [62]. Bristles formed at different times may vary in size (larger bristles—like sex-comb bristles—may be those which are formed earliest) and where one large bristle is missing several small ones may be able to appear later [62]. Given these variables much of the observed variety of cuticular patterns of differentiation may be referable to patterns of variable rates of cuticular growth, as in the case of the sex-comb (Fig. 3.6.6). Patterns of development in particular locations, such as are selectively affected in mutations like achaete and scute [15], may involve cuticular regions made distinctive by their particular rates of growth and their particular bristle-forming potencies.

Fig. 3.6.15 Small-scale dependence of bristle formation on its surroundings revealed in genetic mosaics involving the achaete mutation. *Drosophila* flies carrying the 'achaete' mutation lack bristles at certain locations: in particular, the anterior and posterior dorsocentral bristles (normally located on the dotted lines in the diagram) are missing on the thorax. In genetic mosaics involving patches of mutant tissue (shown in black) these bristles fail to appear if the normal bristle locations are occupied by mutant tissue but occasionally a single bristle will form at a new location in nearby normal tissue [15]. This finding suggests that the locations of the two normally formed bristles may be dependent on the nature of surrounding tissues—perhaps, on mutual inhibitory interactions.

exclusion of the induction model. The induction model incorporates most of the observations whereas many, such as cases of nonautonomy just referred to, are difficult to account for on the alternative model. The fact that organs can be caused to form far from their normal positions or contexts by homeotic mutations is at first sight incompatible with either model.

On the grounds of a similar homology of the tissues concerned, none of the evidence for the autonomy of differentiation of elements in a pattern (on which the positional information model is founded), can be taken to exclude the possible explanation of the findings in terms of embryonic induction. Thus combinations of forelimb and hindlimb tissue in vertebrates (Fig. 3.6.12C) may become integrated in their patterns of differentiation because the two limbs are broadly homologous in their anatomy and in their patterns of development (Fig. 3.6.12B). The normal development of an organ such as a balancer, in tissue of one species transplanted to a species normally without any organ at the relevant location (Fig. 3.6.5), is hardly surprising if, as seems likely (Fig. 3.6.9), organs homologous to its normal inductor are present in the host species. The same point applies to melanocytes of one avian breed which form their own uniquely characteristic patterns of pigmentation by making use of the tissues of another breed (Fig. 3.6.11). It is clear (Figs 3.6.3, 3.6.4, 3.6.5, 3.6.7, 3.6.9, 3.6.11) that a difference in the species of origin of embryonic tissues is no barrier to their normal involvement in inductive interactions at all stages of development.

3.6.5 ON THE GENETIC CONTROL OF DEVELOPMENT—HOMEOTIC MUTATIONS

By specifically causing cells with prospective developmental fates as part of one organ to differentiate in the form of a completely different organ, homeotic mutations would appear to duplicate exactly the effect of embryonic inductions (as seen in Fig. 3.6.7). Thus they offer the prospect of providing a direct link between the embryonic cell interactions which control differentiation, and the store of information in the form of genes (provided initially by the egg) which must tell the cells how to do it. Nor are homeotic mutations unique to insects or to imaginal disks. A dominant mutation has been described in the chicken in which the forelimb develops substantially in the form of the hindlimb [28] (Fig. 3.6.12C).

As a result of separate homeotic mutations the imaginal disks which would normally form such highly specialized head and tail parts as the eye, the proboscis, the antenna or the genitalia can be caused to form one of the body parts such as wings, legs, halteres or body segments [14] which, since they have a greater number of structural features obviously in common, may be considered less highly divergent forms of what were perhaps initially similar patterns of development in all disks. The disks of these latter organs also commonly transform among themselves (different mutations can bring about transformations in either direction between any of them) but it is notable that they rarely if ever transform into head or tail structures. Two points can be made about the nature of homeotic transformations in general. Firstly, it is clear that, if the expected pattern of differentiation fails in a disk due to the action of the mutation, each disk has the potentiality to express one of a number of other forms of

cuticular development. This shared pattern of multipotency is perhaps not surprising, in view of what has already been said about the multipotency of epithelia in vertebrates, since imaginal disks are primarily composed of cells directly derived, apparently unchanged, from the ectodermal stem cell line as it existed early in development. The fact that future patterns of differentiation can be changed in many way by homeotic mutations, even when they are made operative by chromosomal recombination of mid-larval stages, confirms that all imaginal disks may retain the identical and complete multipotency of ectoderm until they start to differentiate at metamorphosis.

Secondly, it is known that imaginal disks cultured *in vivo* spontaneously transform ('transdetermination', Chapter 2.1) into the patterns of differentiation typical of other disks [14] in the same way as is seen to occur due to homeotic mutations but in this case in the absence of heritable genetic changes. Similar transformations of organ-type occur during regeneration of appendages in adult insects [14]. Thus, in line with the principle that mutations are usually deleterious, the direct and single effect of the mutation might be confined to causing failure in the particular patterns of differentiation needed in a given imaginal disk to allow it to fulfil its characteristic form of development: the substitution of another organ may be quite incidental to the genetic change. It is a possibility that some homeotic mutations are lethal to the cells expressing their effects and that the homeotic substitutions are simply the result of anomalous regeneration from surrounding unaffected ectoderm. However, the fact that homeotic mutations in general appear to result in transformations from more specialized structures to less, and not the reverse, suggests that the particular substitute for a highly specialized disk may represent a degree of tissue organization which that disk would normally pass through in attaining its specialized development, now revealed because development beyond this point is impossible due to the elimination by the mutation of the required additional specialized developmental pathway.

The only homeotic mutation in which the mode of action of the affected gene can be pinpointed is the extra sex-comb mutation in *Drosophila* (Fig. 3.6.6) involving the substitution of the second leg by a first leg. These two organs are obviously homologous in structure. In fact many anatomical regions in the one are identical to those in the other and the major differences between them, such as the presence of a sex-comb only in the first, affect only particular regions. The use of genetic mosaics involving this mutation (Fig. 3.6.6) makes it possible to show that the sex-comb may be no more than a single, regionally circumscribed extension to a pattern of development which is intitially common to both limbs. The difference may consist simply of an enlargement of bristles in a certain region and associated morphogenetic rearrangements. This example illustrates how a single mutation can result in a homeotic substitution where the distinctions between two organs depend on the activation of a component of development controlled initially by one gene.

The way in which homeotic mutations substitute single and complete organs one for another might suggest that the single affected genes are directly and specifically concerned with the initiation and the coordination of the entire unitary and specific pattern of development of given imaginal disks. Mosaics of first limb and second limb tissue (Fig. 3.6.6) show that structural components like the sex-comb are formed by developmental processes which are confined to a limited part of the imaginal disk and are unaffected by the limb-type of surrounding areas. Thus the organs derived from imaginal disks should be regarded as composites formed by the various separate modes of cuticular development necessary to give rise to the patterns of cuticular differentiation which typify them. In the case of the insect limbs there is no need to think in terms of any overall organ-specific developmental characterization, because the two organs are largely the same in structure. Even the most highly specialized organs may only derive their individuality through the accumulation of the effects of separate multiple, diverse modifications to what might initially during development have been a common pattern of imaginal disk development. As we have already noted—and we might expect given their common starting points as initially equivalent ectoderm—there is evidence that imaginal disks share a number of features in their early development, and this may be true even of disks which, when fully differentiated, lack any obvious homology with less highly specialized organs. Granted the possibility that more specialized organs are formed by the cumulative addition of modifications to earlier stages of development shared with other disks, failure to develop beyond each point of modification will automatically result in differentiation of the disk in the form of an organ equivalent to the point of developmental elaboration already reached. The result will appear as a substitution, within the boundaries that would normally have been occupied by the more specialized organ, of a complete organ similar to that which would normally develop elsewhere in the body. The more such accumulated modifications are involved in a disk's normal development, the greater likelihood that separate mutational blocks can result in differentiation in the form of several other less specialized organs. Thus the genes affected by homeotic mutations may simply be structural genes needed for the modified patterns of differentiation characteristic of more specialized disks. We may conclude that homeotic mutations do not necessarily involve a special class of genes; they may be structural genes as various as the component structural specializations which make various disk-derived organs distinctive.

The ametapodia mutation in the fowl [28] may be described as homeotic because it causes the substitution of parts of one organ (e.g. terminal feathers of the wing) by parts of another (e.g. clawed digits of the leg) (Fig. 3.6.12C). That this mutation

should stand as a rare exception in vertebrates, in which homeotic mutations are hard to find, while numerous examples are known in insects, can be readily accounted for in terms of what has already been said about the prerequisites necessary for mutations to have such effects. In vertebrates the two limbs are perhaps the only instances of discrete organs which share much of their early development in common but which may require, for the eventual structural divergence of one from the other, the activation of additional genes in only one limb.

We have already discussed how morphogenetic properties of the mesoderm can generate diversity in anatomical arrangement of single tissue types during the formation of the vertebrate limb. Given that many of the differences between forelimb and hindlimb are no more than differences of size and arrangement of otherwise identical structural elements, we might expect that the mesoderm could be directly responsible for the distinctive patterns of development of the two limbs. There is clear evidence that specific developmental preconditions for forelimb and hindlimb formation are inherent in limb-bud mesoderm from an early developmental stage [63]. Because this limb-type property is retained by mesoderm after its cells have been randomly mixed [63] it is likely to be a simple characteristic shared equally by all the cells, as might be expected of a morphogenetic predisposition; the fact that randomly mixed mesoderm taken from both limbs mechanically separates into its two components [63] shows that this characteristic can indeed manifest as a specific morphogenetic disposition.

The establishment of limb-type by limb-bud mesoderm is further demonstrated in the experiment shown in Fig. 3.6.12C. Here hindlimb mesoderm causes forelimb ectoderm to differentiate into hindlimb parts such as claws presumably, as in the normal development of such structures, as a result of the inductive conditions generated in the mesoderm [29] during the course of its continued development. This outcome is remarkably similar to the effects of the ametapodia mutation. So, by relating these effects to the inductive role of mesoderm, we may be able to describe the mutational defect in terms of embryonic induction. The evidence shows that during *normal* development particular forms of gene expression, leading to the patterns of ectodermal differentiation characteristic of each limb, can be activated specifically in one limb—even though these genes are equally available in the ectoderm of both limb-buds—as a result of unique inductive properties present in the mesoderm of that limb alone. Just as during the normal course of limb outgrowth the mesoderm creates diversity of structure and of inductive conditions out of initially simple starting conditions, the mesoderm is capable of generating such limb-specific inductive conditions despite the general earlier homology of development of both limbs. The effect of the mutation could be accounted for either as a failure in mesoderm which prevents its divergence into forelimb-specific morphogenesis from an otherwise common pattern of limb development or as a failure in the structural genes necessary for the differentiation of forelimb structures.

In the case of imaginal disks we are even further from being able to identify how structural genes may be differentially activated at specific locations to bring about diversity of patterns of differentiation. However, there is evidence that mechanisms are not unlike those we have described for vertebrates; there are correlations between pattern formation and morphogenetic events which could be providing for specific inductive interactions. Thus differential rates of mitosis (Fig. 3.6.16) can and do operate morphogenetically during disk development to expose cells to conditions required for bristle differentiation. Sex-comb differentiation is correlated with morphogenetic rearrangement of the prospective sex-comb bristles (Fig. 3.6.6). Imaginal disks may derive their different patterns of development from their characteristic morphogenetic contexts in larval tissues and in this regard it is interesting that mesoderm is known to be involved in controlling their organ-specific development [64]. Consistent with such a basis for organ-specific disk development is the ease with which transformations can occur under the conditions of in vivo disk culture and their correlation with high rates of cellular proliferation in the disk [14].

The evidence of embryonic induction indicates that any structural genes can be activated anywhere and in any pattern in the embryo because all possible forms of differentiation are available, until induction, in initially totipotent embryonic cells. Identical patterns of gene expression can be induced at widely different embryonic locations provided, as may be the case in homeotic substitutions, identical inductors exist at those sites. As was evident from variations in the timing of the exposure of the extra sex-comb mutation by chromosomal recombination and consequent variations in the populations of cells carrying the mutation (Fig. 3.6.6), the affected gene was only employed in modifying development in cells which happened to occupy the specific region of the sex-comb; the prior spatial availability of the gene is an irrelevant consideration. Patterns of activation of genes may show very orderly and apparently significant spatial arrangements. These will directly reflect proliferative and morphogenetic histories of cells expressing the genes. But correlations between clonal history and differentiation of cells cannot be taken to indicate any causal connexion, other than by way of induction, until operative inductive forces have been clarified. Tentatively we can conclude that the only forms of genes which are required for development are the structural genes which are the immediate objects of selection by induction. At least as regards genes mediating differentiation of major cell types, the evidence of embryonic induction indicates that no other genes contribute to the choice of cell specialization. Even on the evidence of homeotic mutations, which of all known mutations appear most directly to be involved in the fundamental control of patterns of development, there is no need to invoke any new class of genes.

3.6.6 A SECOND LOOK AT EMBRYONIC INDUCTION

In previous sections, we have argued that embryogenesis can be accounted for in terms of inductive interactions which can *directly* select particular states of gene expression from among the *whole* range of alternatives available in the genome inherited from the egg. Central to the argument is the contention that each induction has potential access to the whole range of alternatives because the embryonic cells undergoing induction retain the variety of potentialities of the egg itself. So the selection of both the forms of differentiation of cells and their spatial arrangements are the direct consequences of the morphogenetic movements of embryonic cells into their inductive locations. In the following sections we attempt a closer identification of mechanisms responsible for morphogenetic movements and of inductors themselves.

The nature of the mesoderm

It is difficult to believe that the epithelial tissues such as ectoderm or stratum germinativum, which are the sources of most newly induced cells, are themselves solely responsible for the morphogenetic movements which bring them to the locations where the inductions begin. In practice they move in the context of mesoderm, a tissue which, in its differentiated form as connective tissue, is specialized for generating and adapting to mechanical forces [31]. In the form of the vertebrate notochord or the skeleton of the sea-urchin embryo, the mesoderm is concerned, from a very early stage of development, with the determination of shape. The mechanical characteristics of mesoderm are clearly displayed in the way in which it dictates the paths of migration of neural crest [21] (Fig. 3.6.9). Here the mesoderm, by directly controlling the way in which these uncommitted embryonic cells reach various locations, determines the inductors they will eventually meet. But since neural crest also forms cell types, such as cartilage, which contribute to identical tissues developing directly from mesoderm, this is an example of how mesoderm continually modifies its own development on the basis of the way in which it has already developed. In limb development, as we have seen, the developing mesoderm of the apical proliferative zone is continually adapting its own morphogenetic disposition according to the nature of mesodermal structures already formed and in process of differentiation. These characteristics of mesoderm suggest that it may in general be in a position to play the crucial role of generating, transmitting and integrating the morphogenetic forces acting most directly on embryonic cells prior to induction. Even later in development, as the mesoderm continues to grow, it will contribute, alongside the morphogenetic consequences of the differentiation, growth and movement of already developed organs, to the control of subsequent inductions.

Underlying the distinctions made between the primary embryonic germ layers of classical embryology, one can compare, on the one hand, the continuous surface epithelial cells (including both ectoderm and endoderm) and, on the other, the internal mesoderm. The epithelial layers may be regarded as derivatives of, or representatives of, the original embryonic stem cell line from which all differentiated cells derive in the way discussed above. Concerning the mesoderm, a distinction can be made between the *secondary* mesoderm, with which we are concerned here, and *primary* mesodermal cells [65] which (as cells directly derived from external blastula cells by gastrulation that give rise to a number of distinct, non-transforming cell types including endothelium and muscle) may be considered to be true totipotent stem cells distinguishable only because of the time and location of their origin from the stem cell reservoir. The secondary mesoderm, which is also derived from the primary and can be identified as the connective tissue cell type with its multiple, transformable modes of differentiation [36], should be regarded as a finally committed cell type rather than as a germ layer.

Mesoderm as a specific inductor

In addition to its mechanical role mesoderm has a remarkable range of specific inductive actions. These include its first major induction of neural plate (Fig. 3.6.4), the induction of endodermal derivatives such as the lungs and pancreas and the induction of ectodermal derivatives such as hairs, feathers and of mouth parts like the salivary glands [29]. The mesoderm acts both as a determinant of the organ specificity and as a contributor to the structure of the organ in the case of the limb, and, if its neural crest-derived components are included, also of the dorsal fin of the tadpole and the balancer [22]. Having added the role of mesoderm in the inductions of lens, inner ear, stomodeum and nasal placode (Fig. 3.6.8c), it is hard to think of any induction in which the mesoderm has not been implicated in the determination of the specific character of the resulting differentiation. In view of the relative uniformity, throughout the body, of the cell types composing the connective tissue derived from secondary mesoderm, it is not easy to see how the mesoderm could possess the enormous range of specialized forms which would be required for its many inductive specificities.

The role of the mesoderm in the induction of the inner ear and lens and the further characterization of inductors

In the case of organs such as the inner ear which are derived from the neurula ectoderm (Fig. 3.6.8) there is evidence that local mesoderm is involved in their induction together with a quite separate, specific underlying inductor such as, in this case, the hindbrain. The involvement of two quite dissimilar tissues might be explained if the mesoderm plays only the morphogenetic role of causing the invagination that brings ectoderm into contact with hindbrain which alone might be responsible for the specific inductive outcome. But this can be excluded because local mesoderm can, like the hindbrain (Fig. 3.6.7), induce an inner ear when implanted *alone* under distant ectoderm [66]. In any case the contact between the invaginating ear vesicle and the hindbrain is never marked; the vesicle is embedded at most points in mesoderm. So the hindbrain itself may have no more than a morphogenetic role of mechanically conditioning the mesoderm around it so as to allow that tissue alone to bring about ectodermal invagination and induction.

But it does not follow that the mesoderm is the ultimate single-handed inductor of the inner ear. The formation of the organ which is subsumed under that name is a gradual process involving outgrowth and partitioning of regions of the originally simple inner ear vesicle [53, 66]. When normal surrounding conditions are disturbed, as for example after rotation of the vesicle or transplantation to foreign tissues, major distortions occur in the development of the vesicle (including duplication of outgrowths and swelling) and differentiated components characteristic of the fully formed inner ear, such as hair cells, fail to appear [53, 66]. In any organ formed as a result of morphogenetic movements like invagination, it is hard to distinguish a particular time of commitment to a form of differentiation as opposed to arrival at a morphogenetic state which ensures that commitment to differentiation will occur at some later time. In the case of an organ, such as the inner ear, the idea of a single determinative event of 'induction of the inner ear' is clearly inappropriate anyway. Since the adult inner ear is the result of multiple cell specializations arrived at continuously during a series of morphogenetic modifications of the original vesicle, no single specific inductor is to be expected.

We can begin to redefine the role of the local mesoderm in terms of the way in which it can control the multiple events which eventually go to make up the formation of an inner ear, in much the same way as we have already described the role of mesoderm in limb formation. Clearly mesoderm may play a continuing decisive role in controlling the shape of the developing vesicle but the particular patterns of differentiation of the organ are the result of multiple circumstances such as conditions created within the vesicle itself, new tissues forming around it (particularly the capsular cartilage [53]) and the various conditions reached by parts of the vesicle during their outgrowth. Just as limb mesoderm may dictate later limb structure by controlling the patterns of ingrowth of invasive cells like neural crest, so the mesoderm surrounding the inner ear may exercise a specific influence on later inductions of the epithelial tissue, purely mechanically, by controlling the approach of the invasive tissues which are directly responsible for various inductions of various places.

Lens induction, despite the apparent similarity of its

invaginative origin from the ectoderm to that of the inner ear, illustrates a quite different adaptation of epithelial cells to their circumstances. In this case local mesoderm may well be involved mechanically (as it is in assisting the approach of the optic cup) in transmitting the forces which cause ectoderm to invaginate opposite the eyecup. The evidence of lens regeneration shows quite clearly that there exist within the eye a complex combination of mechanical and biochemical circumstances, some of them to some extent defined [41], which initiate the formation of a lens. These conditions may be of a relatively simple and nonspecific kind in view of the fact that the same transformation can occur due to the conditions of tissue culture alone [40]. So if we accept the comparability of the lenses formed during regeneration and during embryonic development, we can suggest that it is the appearance of these same circumstances within the gradually developing eyecup (Fig. 3.6.8) which is the immediate and specific inductor of normal lens development that has been ascribed to the eyecup by classical experimental embryology [41].

However, the structure of lenses varies considerably according to the conditions under which they are formed, as for example after transplantation of the lens rudiment to foreign sites in the embryo [50]. Clearly lens development is not an all-or-none event but a series of events open to possible disruption by interference with a number of contributing causes. Mesoderm is not only involved in so far as it assists in the formation of the initial vesicular shape of the lens and in exposing the prospective lens to its normal specific inductor, but there is evidence [50] that, together with endoderm, it directly promotes lens differentiation and growth. Even the adult lens is subject to important continuous influences from its environment which mechanically control its size and shape, and which also promote and sustain a mitotic growth zone on its rear surface [67]. It is clear that initial 'induction' of the lens cannot be distinguished from later modelling and growth of the lens; they are continuous processes. Like hairs and feathers which are similarly formed from a single differentiated cell type, many of the features of lens structure may be the result of the way in which newly formed lens cells, due to a responsiveness acquired through induction, differentiate and arrange themselves according to mechanical conditions and especially their relation to already differentiated lens cells. Induction may in part consist of the release of this self-assembling characteristic simply by the initiation of the appropriate mechanical conditions.

The apparent sufficiency of the influence of single tissues like the hindbrain and the eyecup as explanations for the immediate induction of specific organs, like the inner ear and the lens, has recommended these classical instances of embryonic induction for the search for the identity of the relevant specific inductors. However, even though the comparison with lens regeneration may have taken us nearer the identification of factors responsible for the specific commitment of cells as lens cells, the

organs that we identify as normal adult inner ear or lens differ only in degree from, for example, the limb in the extent to which they are only the result of a number of interlocking developmental control mechanisms.

3.6.7 SUMMARY AND CONCLUSION—THE CELL RECOGNITION MODEL

In the foregoing sections an attempt has been made to show that it is possible in principle to account for the spatial arrangements and specialized forms of differentiation achieved by cells during embryogenesis in terms of inbuilt rules for the selection of specific structural gene expression on the basis of the particular inductive conditions of a cell's immediate environment. A complete description of development in these terms would be premature and enormously involved but the examples considered above were chosen to demonstrate that rules of embryonic induction would alone be sufficient to account for the development of complex adult organs and to encompass situations in which the need for quite different mechanisms, such as the availability of positional information, has been argued. While it is difficult to positively exclude the existence of other mechanisms, embryonic induction is the necessary and the sufficient explanation for all the observations made in embryonic tissues described here; all changes in the differentiative states of cells correspond, according to predictable rules applying throughout embryogenesis, to changes in the immediate environment of the cells, regardless of the history or broader context of the cells.

The case that the inductive conditions are the sole determinants of cellular specializations rests on evidence that cells undergoing induction are otherwise unrestricted in their potentialities; they can undertake *any* forms of differentiation specific to each possible inductor presented to them. The validity of this contention is greatly strengthened by further evidence that even after induction, when they have adopted one specific form of differentiation, some cells retain a wide multipotency in that they can adopt new forms of differentiation when placed in new contexts; for example, transformation of iris cells into lens. Although reversibility of the differentiated state may be the norm in primitive organisms like *Hydra* [68] it is apparently exceptional in vertebrates; it is a common observation in experimental embryology that after a certain stage of development many amphibian tissues continue in their original patterns of differentiation despite being exposed to a variety of different conditions by transplantation to new embryonic locations. The irreversibility of the states of certain differentiated cells is no reason to doubt the reversibility of others. Irreversibility may be due to nothing more than the inability of the cell to remove the products of the earlier differentiated state. For an organism like *Hydra* with a simple cellular organization which does not change during the life cycle, definitive states of cellular differentiation can be maintained (despite their reversibility) by the continued operation of the inductive forces responsible for initial differentiation. But this cannot be true for higher organisms in which, in the course of subsequent morphogenetic changes, cells may move from their inductive contexts after initial differentiation. Particularly in continuously growing tissues, in which the early acquired form of differentiation must be passed on to cells added after embryonic induction, the differentiated state may be maintained and rendered irreversible by cytoplasmic factors capable of transmission during mitosis.

The induction model therefore stands in opposition to theories which imply that the particular line of descent of a cell prior to commitment to a definitive form of differentiation—either due to its descent from a particular region of the egg or due to spatial or temporal features of pattern of mitosis of its ancestors—can itself contribute to the future restricted specialization of the cell. Evidence from amphibian embryos strongly suggests that no specialized cells of the adult owe their specializations to their pattern of derivation from the original substance of the egg; even germ cells, though normally derived from an apparently specialized 'germ plasm' in the egg, can be caused to form in other regions according to circumstances [69]. Even in the so-called mosaic eggs of invertebrates the situation appears to be the same. Specializations within the egg and early embryo, including gradients, would appear to be of less direct, perhaps morphogenetic, significance for the control of embryogenesis. In fact the induction model requires no more of the egg than that it carries the potentialities for the possible forms of differentiation of all the adult cells descended from it and that—as we might expect of mitosis as a basic cell function carried out in a relatively standardized way regardless of the specialization of the cell—this is what is passed on to the immediate cellular derivatives of the egg cell which compose the early embryo.

Evidence has been presented here that the specific conditions which serve to induce differentiation in embryonic cells can be closely equated with the states of specialization of cells already undergoing differentiation by the stage of embryogenesis in question or their immediate products. This feature may be explained—and any implication that inductors in any way anticipate their inductive effects or instruct embryonic cells how to respond can be avoided—if we introduce the notion that embryonic cells are provided with receptivities to their potential inductors which they use to control their selection of the structural genes mediating their eventual differentiation. If we suppose that all embryonic cells start out equally provided with a battery of such receptivities, then the differentiated state triggered by any particular receptivity will only be activated after the corresponding inductive cell types have undertaken differentiation. Thus from a starting point of equivalent cells a

temporal sequence of activation of specific differentiated cell types can occur. The first activation of specialized cell states to occur during embryogenesis could be explained if embryonic cells were specifically receptive to conditions generated internally within the embryonic cell mass such as they meet during gastrulation in vertebrates. This model of inductive interactions, which allows the mechanism for selection of specific cell states to operate within the same single cell boundaries as the structural genes which are selected for activation, may be referred to as the cell recognition model since it is based on the discrimination by one cell of the specific, independently derived, differentiated states of other cells.

We have seen that the time and location of specific inductive interactions depends directly and crucially on complex patterns of morphogenesis which bring about confrontations between uncommitted embryonic cells and their inductors. Despite the evidence that the mesoderm is formed by a single differentiated cell type, it is this tissue which is largely responsible for the enormously complex and coordinated pattern of embryonic morphogenesis. By considering the evolution of advanced species we may see how this comes about.

While the types of differentiated cell components remain remarkably constant in the vertebrate series, species differ widely in the arrangements and proportions of their specialized tissues. As we might expect from the evolutionary stability of established cell characteristics [1], the rules of embryonic induction appear, on the evidence of a number of inter-species transplantation studies described above, to remain constant among species—even in the sea-urchin the rules of gastrulation can be compared with those of vertebrates. Thus the most immediate basis for species variation may be morphogenetic and may reflect the potentiality for morphogenetic variation intrinsic to mesoderm together with inevitable differences in the morphogenetic starting conditions dictated in the egg by such species-specific needs as yolk storage. Comparison between organs such as vertebrate forelimbs and hindlimbs or insect imaginal disks, whose homology of structure and development imply common origins in evolution as identical and simpler structures, shows the same basis for evolutionary variation Most of the differences in structure of such organs are in the arrangement of common differentiated cell types. Through the initially random variations of morphogenesis which will inevitably occur during the embryogenesis of species early in evolution not only will new and potentially advantageous arrangements of already differentiated cell types occur but so will new combinations of cells. A particularly favourable spatial combination of cells may be retained during evolution perhaps because the cells can cooperate in the formation of a new differentiative product which is selectively advantageous to the species. Subsequently this advantage may be consolidated by further evolutionary adaptation of the specialized forms of the cells to their functional roles in the new context. Whatever the original basis for the association of the two cell types—it might be morphogenetic mixing of already differentiated cells or it might be due to functional cooperation between multipotent cells in which the differentiative product released from one cell allows the other cell, with some of its metabolic needs thereby provided for, to adopt a complementary and more specialized metabolic state—their forms of differentiation can evolve towards greater adaptive specialization while the original basis for the mutual dependence can remain. The mechanisms which may originally have modified the specialization of one cell on the basis of direct functional dependence on the other can continue to mediate cell interactions even when, following evolutionary divergence in the forms of specialization of the cells, the resultant forms of differentiation appear quite unconnected functionally. Where the same differentiated cell types occur at quite separate locations in the organism (as for instance in two imaginal disks which initially in evolution developed identically) due to the different contexts they will be subject to different adaptive pressures. Gene duplication [1] may provide the opportunity, long term, for a structural gene initially used in the same way at both locations to undergo divergent specializations, but only if cells at the same time acquire distinct receptivities which allow the appropriate specialization to occur at the appropriate site.

Thus we can imagine that the opportunity for evolutionary advance can occur, as a consequence of initially random morphogenetic variation, at any point during the embryogenesis of primitive organisms. The most highly evolved species will be the products of the very gradual accumulation, through the successive modifications occurring at each evolutionary stage to the products of all earlier stages, of multiple, separate adaptations. Vertebrate embryonic cells have, on this view, derived their proposed inductive receptivities to the differentiated states of earlier and quite independently developing inductor cells precisely because it was to these same environments (formed by the same differentiated cell types existing in embryos which were products of earlier evolution) that the receptivities became evolutionary adapted. A new form of cell differentiation arises by induction by an already differentiated cell because it only evolved in the first place as a result of unique features of the context of the same, earlier evolved inductor cell type. During evolution it was the morphogenetic consequences of already evolved differentiated cells including mesoderm that gave rise to embryonic contexts which promoted the evolution of new cell types. So during the embryogenesis of more highly evolved organisms the morphogenetic conditions leading up to the induction of each newly differentiating cell type may be seen as the same direct mechanical consequence of the differentiation of earlier formed tissues as occurred in evolution. The mesoderm, while only a particular form of specialized cell, will contribute to the morphogenetic conditions, as it did progressively during evolution, as a result of its responsiveness to foregoing morphogenetic differentiations. In general, the sequence of appearance of new differentiated cell types and their morphogenetic consequences leading to further inductions is a direct reflexion of how successively evolved species achieved variation from foregoing species.

Thus an element of the recapitulation of evolution is inevitable in the embryogenesis of highly evolved species. This does not mean every component of the development of any one organism intermediate in evolution must be retained in later evolved organisms. When specific components due to earlier evolution come into conflict with each other or with the needs of later evolved species, they can readily be lost. The longest evolved components will be the most resistent to loss if only because they have to be retained for their inductive and morphogenetic consequences in later evolved embryos. Perhaps it is for these reasons that structures like the notochord and pronephros are retained in higher vertebrates even though they do not contribute directly to adult survival advantage; they are lost during embryonic stages. The earliest stages of embryogenesis will, because they evolved earliest in more primitive organisms, tend to have more in common when compared in different species. The same considerations dictate that organs with common evolutionary origins such as imaginal disks or vertebrate limbs will retain greatest primitive homology during their earliest developmental stages. The gradually accumulating divergences in the development of such organs, leading especially in terminal parts to increasing complex and specific forms of differentiation, reflect the sequence of evolutionary accumulation of developmental variation.

We have referred in this chapter to the highly specific receptivities which cells can acquire as components of their specialization following induction. These may allow them to further modify and refine their patterns of differentiation in adaptation to their functional circumstances and in accordance with their own functional roles. An example is the steroid receptor, formed by a gene activated in certain specialized cells, which specifically controls the mode of differentiation of its own cell and its later function when triggered by signals from other cells [34].

Given that cells can express such properties it is not implausible that particular forms of structural gene expression may be triggered in embryonic cells as the result of receptivities specifically tuned to the states of developmental commitment of specialized cells in their environment. *Any* theory of development must include among the preconditions for embryogenesis provided within the egg the means of selectively activating the regulatory genes which must themselves be responsible for picking out and activating alternative states of structural gene expression appropriate for differentiation of the particular cell. It is worth pointing out that the number of receptivity states which would have to be provided to mediate embryonic inductions would not have to exceed the limited number of major classes of differentiated cell types that occur in vertebrates. We have seen that variants of these major classes, including modulatory forms of differentiation of a cell type like the mesodermal cell, may be controlled by receptivity mechanisms acquired secondarily as a result of initial induction. The number of receptivities of any embryonic cell at any one time may be even fewer if an initial receptivity state is itself used to trigger further alternative receptivity states prior to developmental commitment.

By the inclusion, in addition to structural gene states programmed into the genome of the species, of genes mediating inductive receptivities this model for the control of embryogenesis is compatible with the limitations imposed by the properties of cells as we know them. The fully differentiated adult organism is entirely the sum of the differentiated states of single cells and we know that single cells alone provide the means, by way of their protein synthesizing machinery, for the expression of single mutually exclusive alternative structural gene states. Thus instructions for specialization can reach each cell, on the basis of the unique replicating properties of DNA, reliably and without restriction in the copy of the egg nucleus handed down through successive mitoses. With the rules for embryonic induction programmed in the form of genes mediating receptivity states, embryonic induction provides a means whereby one part of the single genome inherited from the egg can control others in such a way as to generate diversity of cell specializations. By reference to neighbouring specialized cells, whose states of differentiation are products of one structural gene component, the receptivity genes of an embryonic cell can directly and predictably activate a second quite independent set of structural genes within its own cell boundaries. Thus the preconditions for development which the egg cell is capable of providing and their means of delivery and exploitation in all the cells of the developing organism can all be contained in no more than a random, linear array of quite independently operating genes.

By explicitly specifying that one differentiated cell type will form next to another, embryonic induction represents a direct and highly predictable way of ensuring that differentiated cells are appropriately related in the adult. Because induction operates between two neighbouring cells it can control development at the single cell level with the maximum possible precision. Given the mechanism of embryonic induction the specialization of embryonic cells is inseparable from their locations. This mechanism alone may appear quite inadequate for the immensely complex task of coordinating the phenomena of embryogenesis in such a way as to ensure a perfectly integrated adult organism—particularly since no direct provision is made, either in the genome or in the egg, for the control of some of the most striking features of embryonic development including the temporal ordering of developmental stages and morphogenesis. And yet we must not forget that the entire process must be achieved by way of specialized states that can be adopted by single cells and must ultimately be the result of properties inherent in the single cell of the egg.

We have presented evidence that many adult vertebrate cells retain unchanged the properties of the egg cell and, as the larval type differentiation undertaken by the *Chaetopterus* egg cell shows (Fig. 3.6.1), the egg cell is no different from later formed cells other than with regard to such specializations as yolk-storage. The complexity and indirectness with which embryonic stages lead up to the formation of an adult organism incline one to think of them as a sequence of specialized events designed specifically for the end in view. But the continuity of cell properties throughout the life cycle indicates that it is misleading to distinguish between the adult state and the processes by which it is attained. On the view taken in this Chapter, embryogenesis is no more than the sum of activations of structural genes which mediate adult cell differentiation; and the adult is no more than the sum of the conditions for these activations generated during embryogenesis.

3.6.8 REFERENCES

[1] OHNO S. (1970) *Evolution by gene duplication.* Springer-Verlag, Berlin.

[2] LEWIN B. (1974) *Gene expression.* Wiley, London.

[3] GURDON J.B. (1974) *The control of gene expression in animal development.* Clarendon Press, Oxford.

[4] LILLIE F.R. (1902) Differentiation without cleavage in the egg of the annelid *Chaetopterus pergamentaceus. Wilhelm Roux Arch. Entwicklungsmech. Organismen,* **14,** 477–99.

[5] HORSTADIUS S. (1935) Uber die Determination im Verlaufe der Eiachse bei Seeigeln. *Pubb. Staz. Zool. Nap.,* **14,** 251–479.

[6] SPEMANN H. (1938) *Embryonic development and induction.* Yale University Press, New Haven.

[7] CURTIS A.S.G. (1960) Cortical grafting in *Xenopis laevis. J. Embryol. exp. Morph.,* **8,** 163–73.

[8] SPEMANN H. & MANGOLD H. (1924) Uber Induktion von Embryonalanlagen durch Implantation artfremder Organisatoren. *Wilhelm Roux Arch. Entwicklungsmech. Organismen,* **100,** 599–638.

[9] SAXEN L. & TOIVONEN S. (1962) *Primary embryonic induction.* Academic Press, London.

[10] MANGOLD O. (1933) Uber die Induktionsfähigkeit der verschiedenen Bezirke der Neurula von Urodelen. *Naturwissenschaften,* **21,** 761–66.

[11] SPEMANN H. & SCHOTTE O. (1932) Uber xenoplastische Transplantation als Mittel zur Analyse der embryonalen Induktion. *Naturwissenschaften,* **25,** 463–67.

[12] ROTMANN E. (1935) Reiz und Reizbeantwortung in der Amphibienentwicklung. *Verh. d. zool. Ges.,* **37,** 76–83.

[13] GARCIA-BELLIDO A. (1972) Pattern formation in imaginal disks. In *The biology of imaginal disks* (Eds H. Ursprung & R. Nothiger), pp. 59–91. Springer-Verlag, Berlin.

[14] GEHRING W.J. & NOTHIGER R. (1973) The imaginal discs of *Drosophila.* In *Developmental systems: insects* (Eds S.J. Counce & C.H. Waddington), vol. II, pp. 211–90. Academic Press, London.

[15] STERN C. (1968) *'Genetic mosaics' and other essays.* Harvard University Press, Cambridge.

[16] HANNAH-ALAVA A. (1958) Developmental genetics of the posterior legs of *Drosophila melanogaster. Genetics, N.Y.,* **43,** 878–905.

[17] WOLPERT L. (1969) Positional information and the spatial pattern of cellular differentiation. *J. theor. Biol.,* **25,** 1–47.

WOLPERT L. (1971) Positional information and pattern formation. *Current Topics in Devel. Biol.*, **6**, 183–224.

[18] HOLTFRETER J. (1945) Neuralization and epidermization of gastrula ectoderm. *J. exp. Zool.*, **98**, 161–209.

[19] WAGNER G. (1949) Die Bedeutung der Neuralleiste für die Kopfgestaltung der Amphibienlarven. *Rev. Suisse Zool.*, **56**, 519–620.

[20] ATTARDI D.G. & SPERRY R.W. (1963) Preferential selection of central pathways by regenerating optic fibres. *Expl. Neurol.*, **7**, 46–64.

[21] WESTON J.A. (1970) The migration and differentiation of neural crest cells. *Adv. in Morphogenesis*, **8**, 41–114.

[22] HORSTADIUS S. (1950) *The Neural Crest*. Oxford University Press, London.

[23] HARRIS P.F. (1974) The development of the cells of the blood. In *Differentiation and growth of cells in vertebrate tissues* (Ed. G. Goldspink). Chapman & Hall, London. 209–262.
COOPER M.D. & LAWTON A.R. III (1974) The development of the immune system. *Sci. Amer.*, **231**, Nov., 58–72.

[24] WILLIER B.H. & RAWLES M.E. (1944) Melanophore control of the sexual dimorphism of feather pigmentation pattern in the barred Plymouth Rock fowl. *Yale J. Biol. Med.*, **17**, 319–40.
NICKERSON M. (1944) An experimental analysis of barred pattern formation in feathers. *J. exp. Zool.*, **95**, 361–98.

[25] SAUNDERS J.W. (1948) The proximo-distal sequence of origin of the parts of the chick wing and the role of the ectoderm. *J. exp. Zool.*, **108**, 363–403.

[26] KIENY M. (1964) Etude du mécanisme de la régulation dans le développement du bourgeon de membre de l'embryon de poulet. II. Régulation des déficiences dans les chimères 'aile-patte' et 'patte-aile'. *J. Embryol. exp. Morph.*, **12**, 357–71.

[27] SAUNDERS J.W., GASSELING M.T. & CAIRNS J.M. (1959) The differentiation of prospective thigh mesoderm grafted beneath the apical ectodermal ridge of the wing bud in the chick embryo. *Devel. Biol.*, **1**, 281–301.

[28] COLE R.K. (1967) Ametapodia, a dominant mutation in the fowl. *J. Hered.*, **58**, 141–46.

[29] DEUCHAR E.M. (1975) *Cellular interactions in animal development*. Chapman & Hall, London.

[30] HARRIS A.J. (1974) Inductive functions of the nervous system. *Ann. Rev. Physiol.*, **36**, 251–305.

[31] MATHEWS M.B. (1975) *Connective tissue: macromolecular structure and evolution*. Springer-Verlag, Berlin.
PRITCHARD J.J. (1972) The control or trigger mechanism induced by mechanical forces which cause responses of mesenchymal cells in general and bone apposition and resorption in particular. *Acta Morph. neerl-scand.*, **10**, 63–9.

[32] NEBERT D.W., ROBINSON J.R., NIWA A., KUMAKI K. & POLAND A.P. (1975) Genetic expression of aryl hydrocarbon hydroxylase activity in the mouse. *J. Cell. Physiol.*, **85**, 393–414.

[33] KEATING M.J. (1974) The role of visual function in the patterning of binocular visual connexions. *Brit. Med. Bull.*, **30**, 145–51.
LOMO T., WESTGAARD R.H. & DAHL H.A. (1974) Contractile properties of muscle: control by pattern of muscle activity in the rat. *Proc. Roy. Soc.*, **B187**, 99–103.

[34] THOMAS P.J. (1973) Steroid hormones and their receptors. *J. Endocrin.*, **57**, 333–59.

[35] BUTLER E.G. (1955) Regeneration of the urodele forelimb after reversal of its proximodistal axis. *J. Morph.*, **96**, 265–81.

[36] NAMENWIRTH M. (1974) The inheritance of cell differentiation during limb regeneration in the axolotl. *Devel. Biol.*, **41**, 42–56.

[37] BILLINGHAM R.E. & SILVERS W.K. (1968) Dermoepidermal interactions and epithelial specificity. In *Epithelial-mesenchymal interactions* (Eds R. Fleischmajer & R.E. Billingham), pp. 252–66. Williams & Wilkins, Baltimore.

[38] McLOUGHLIN C.B. (1968) Interactions of epidermis with various types of foreign mesenchyme. In *Epithelial-mesenchymal interactions* (Eds R. Fleischmajer & R.E. Billingham), pp. 244–51. Williams & Wilkins, Baltimore.

[39] COULOMBRE J.L. & COULOMBRE A.J. (1971) Metaplastic induction of scales and feathers in the corneal anterior epithelium of the chick embryo. *Devel. Biol.*, **25**, 464–78.

[40] YAMADA T. & McDEVITT D.S. (1974) Direct evidence for transformation of differentiated iris epithelial cells into lens cells. *Develop. Biol.*, **38**, 104–18.

[41] REYER R.W. (1962) Regeneration in the amphibian eye. In *Regeneration* (Ed. D. Rudnick), pp. 211–65. Ronald Press, New York.

[42] WARWICK R. & WILLIAMS P.L. (1973) *Gray's Anatomy* (35th Edition). Longman, Edinburgh.

[43] WADDINGTON C.H. (1938) Studies on the nature of the amphibian organization centre. VII. Evocation by some further chemical compounds. *Proc. Roy. Soc. B.*, **125**, 365–72.

[44] WOERDEMAN M.W. (1938) Inducing capacity of the embryonic eye. *Koninkl. Akad. Weten. Amst.*, **41**, 336–43.

[45] OKADA Y.K. (1949) The factors responsible for the so-called free lens. I. Limitations to the capacity of the nose to induce lens. *Proc. Jap. Acad.*, **25**, 45–50.

[46] ALBAUM H.G. & NESTLER H.A. (1937) Xenoplastic ear induction between *Rana pipiens* and *Amblystoma punctatum*. *J. exp. Zool.*, **75**, 1–10.

[47] STROER W.F.H. (1933) Experimentelle Untersuchungen uber die Mundentwicklung bei den Urodelen. *Wilhelm Roux Arch. Encklungsmech. Organismen*, **130**, 131–86.

[48] KUCHEROVA F.N. (1945) Inductive influence of forebrain upon body epithelium in Amphibia. *C. R. Acad. Sci., U.R.S.S.*, **47**, 307–309.

[49] LEWIS W.H. (1907) Lens formation from strange ectoderm in *Rana sylvatica*. *Am. J. Anat.*, **7**, 145–69.

[50] JACOBSON A.G. (1958) The roles of neural and non-neural tissues in lens induction. *J. exp. Zool.*, **139**, 525–58.

[51] YNTEMA C.L. (1955) Ear and nose. In *Analysis of development* (Eds B.H. Willier, P.A. Weiss & V. Hamburger), pp. 415–28. W.B. Saunders, Philadelphia.

[52] CUSIMANO T., FAGONE A. & REVERBERI G. (1962) On the origin of the larval mouth in the anurans. *Acta Emb. morph. Expt.*, **5**, 82–103.

[53] YNTEMA C.L. (1950) An analysis of induction of the ear from foreign ectoderm in the salamander embryo. *J. exp. Zool.*, **113**, 211–44.

[54] ADAMS A.E. (1924) An experimental study of the development of the mouth in the amphibian embryo. *J. exp. Zool.*, **40**, 311–80.

[55] LIEDKE K.B. (1955) Studies on lens induction in *Amblystoma punctatum*. *J. exp. Zool.*, **130**, 353–80.

[56] REVERBERI G. (1971) Experimental embryology of marine and fresh-water invertebrates. North-Holland, Amsterdam.

[57] HOLTFRETER J. (1938) Differenzierungspotenzen isolierter Teile der Anurengastrula. *Wilhelm Roux Arch. Entwicklungsmech. Organismen*, **138**, 657–738.

[58] HOLTFRETER J. (1933) Die totale Exogastrulation, eine Selbstablösung des Ektoderms vom Entomesoderm. *Wilhelm Roux Arch. Entwicklungsmech. Organismen*, **129**, 669–793.

[59] POULSON D.F. (1945) Chromosomal control of embryogenesis in Drosophila. *Amer. Nat.* **79**, 340–63.

[60] COUNCE S.J. (1973) The causal analysis of insect embryogenesis. In *Developmental systems: insects* (Eds S.J. Counce & C.H. Waddington), Vol. II, pp. 1–156. Academic Press, London.

[61] WIGGLESWORTH V.B. (1959) *The control of growth and form: a study of the epidermal cell in an insect*. Cornell University Press, Ithaca.

[62] LAWRENCE P.A. (1973) The development of spatial patterns in the integument of insects. In *Developmental systems: insects* (Eds S.J. Counce & C.H. Waddington), Vol. II, pp. 157–209. Academic Press, London.

[63] ZWILLING E. (1968) Morphogenetic phases in development. In *The emergence of order in developing systems* (Ed. M. Locke), pp. 184–207. Academic Press, New York.

[64] EL SHATOURY H.H. (1955) Lethal no-differentiation and the development of the imaginal discs during the larval stage in *Drosophila*. *Wilhelm Roux Arch. Entcklungsmech. Organismen*, **147**, 523–38.

[65] HAY E.D. (1968) Organization and fine structure of epithelium and mesenchyme in the developing chick embryo. In *Epithelial-mesenchymal interactions* (Eds R. Fleischmajer & R.E. Billingham), pp. 31–55. Williams & Wilkins, Baltimore.

[66] HARRISON R.G. (1969) *Organization and development of the embryo* (Ed. S. Wilens). Yale University Press, New Haven.

[67] COULOMBRE A.J. (1965) The eye. In *Organogenesis* (Eds R.L. DeHaan & H. Ursprung), pp. 219–51. Holt, Rinehart & Winston, New York.

[68] BURNETT A.L. (1973) *Biology of Hydra*. Academic Press, New York.

[69] KOCHER-BECKER U. & TIEDEMANN H. (1971) Induction of mesodermal

and endodermal structures and primordial germ cells in *Triturus* ectoderm by a vegetalizing factor from chick embryos. *Nature,* **233,** 65–6.

The author is indebted to J.E. Brown, A.C. Read and E.F. Wood.

Conclusions to Part 3

1. SHORT RANGE INTERACTIONS IN DEVELOPMENT

Cell interactions are required to coordinate development and in animal embryos short-range interactions are concerned with determination and differentiation in small cell populations. They may restrict the development of cells and in Part 2 we noted that pluripotential cells may be held in particular states by interaction with other cells. Short-range interactions in plants are well established for pollen–stigma recognition reactions, but it is not yet clear whether somatic development involves comparable short-range effects. The number of known examples of short-range interactions between cells in differentiation in plants is very small. Thus, at present the importance of short-range interactions in differentiation is well established only for animal embryos and most of the available information on this subject relates to animal development.

2. DEVELOPMENT COULD BE COORDINATED BY THE EXCHANGE OF MOLECULES BETWEEN CELLS

Many different types of molecules can pass between the cytoplasms of adjacent cells. Animal cells in embryos are connected by low resistance gap junctions and these probably have the same properties as tissue culture cells with similar junctions. Experiments in cell culture show that these junctions allow the passage of a wide variety of small molecules (M.W. below 1,500) between cells in contact, but that proteins and nucleic acids do not move through them. It is therefore likely that if the coordination of development depends on these junctions, then the morphogenetic signals are small molecules (see Chapter 1.3).

In contrast most plant cells are connected by cytoplasmic channels, the plasmodesmata, and the distribution of these connections in development is well mapped. These channels presumably allow the exchange of small molecules, proteins and nucleic acids between neighbouring cells because small viruses pass through them and because they are broken when neighbouring cells are required to synthesize different protein coats (e.g. during meiosis in self-incompatible flowering plants).

There is thus good evidence that the cells in developing systems could communicate by molecular transfer. The problem is to discover whether this mode of communication actually controls development.

3. DEVELOPMENT CAN BE CONTROLLED BY RECOGNITION OF MACROMOLECULES

It is clear from experiments on aggregation of animal cells that they must be capable of recognizing cells of the same and of other types and the capacity for cell recognition must be an essential requirement for the coordination of the complex cell movements involved in morphogenesis. The basis of cell recognition in animal cells is still not clear but there is considerable evidence that macromolecules, such as glycoproteins, at the cell surface may play an important role.

The recognition of pollen by the stigma in plants is more fully understood and serves as a model of how macromolecular contact may be used in cell interactions. Compatible pollen alighting on the stigma immediately releases proteins from the pollen case which bind to the stigma and after the imbibition of water the pollen germinates and the pollen tube grows towards the ovule. In an incompatible cross, either the pollen fails to germinate or the growth of the pollen tube is blocked by the deposition of callose. This blocking reaction can be elicited from the stigma solely by the application of a protein (or glycoprotein) fraction of the pollen wall and it is clear that the response depends solely on contact with this particular macromolecule.

It is possible that comparable cell recognition phenomena are important also in interactions between somatic plant cells in development of the plant body although the immobility of plant cells raises the question as to whether it is essential for somatic cells to be able to recognize their neighbours. However, it is well established from grafting experiments that some plant genotypes will form compatible graft unions, whereas other combinations are incompatible, suggesting that somatic plant cells may, indeed, have the capacity for distinguishing between self and not-self.

4. EMBRYONIC INDUCTION MAY DEPEND ON CLOSE CONTACT BETWEEN INTERACTING CELLS

Embryonic induction is communication between cells which is required for their differentiation, morphogenesis, and maintenance. Such communication is required for the formation of adult structures such as the kidney, liver and lens in animals. Cell contact is required for this communication in kidney induction but in other cases it appears that granular material (nerve tube induction) or the extra-cellular matrix (tooth induction) might be involved. In cases where contact is necessary, then low resistance gap junctions or surface macromolecules could be the channels of communication, but in cases where contact is not required it is likely that macromolecular recognition is an important part of the mechanism.

Note that although we have emphasized two particular modes of communication between developing cells, there are many other methods which they could use (see Introduction to this Part). It is still uncertain which of these is generally employed to elicit specialized gene expression during cell interactions.

5. PATTERN FORMATION MAY ONLY DEPEND ON SHORT-RANGE INDUCTIVE INTERACTIONS

It is important to decide whether the short-range interactions which have been discussed could account for the formation of the complex array of tissues in the adult. It is argued that pattern formation need not and probably does not involve interactions which depend on all the other cells in the embryo. Rather pattern formation is seen as the consequence of specific short-range inductions which select a set of differentiated gene expressions from multipotential stem cells. Embryonic development then consists of a series of confrontations between inducing and responding tissues and any initial heterogeneity in the egg (e.g. Chapter 1.3) is mainly responsible for setting up the morphogenetic movements which permit these confrontations. If this view of development is correct, then the short-range interactions which are discussed in this part of the book must be the method by which regular cell diversity is created in developing systems.

Part 4
Hormonal Control of Development

Chapter 4.1
Introduction

4.1.1 LONG RANGE INTERACTIONS

This section of the book is concerned with studies on substances produced in one part of a developing organism which are transported or move to other parts where they exert their effects. For simplicity these substances will be called hormones, but it should be noted that in addition to well known plant and animal hormones, such as indole-3-acetic acid (IAA) and thyroxine, these substances also include cyclic-AMP (adenosine 3′,5′-monophosphate) and ammonia which are rarely described as hormones. The effects of these substances on development do not depend on a close association between the producing and the responding cells.

Hormones may be produced very early in development. Thus the small embryo in germinating seeds produces gibberellic acid, and serotonin and adrenaline are found in the early cleavage stages of sea urchin and fish embryos [1, 2]. The function of these early plant hormones is well studied but the function of the hormones found in early animal embryos is not understood.

4.1.2 HORMONES AS CELL TYPE DETERMINANTS

Hormones could control the differentiation of plants and animals in several ways. Thus, on the one hand they might act as cell type determinants and on the other hand they might induce differentiation in determined cells. To demonstrate that a hormone is a cell type determinant, the experimenter should show that it does not merely provide permissive conditions for determination but actually instructs the cell to be determined in a particular way. Formally it is necessary to show that a pair of identical cells can be forced to form different mature tissues of an organism by exposure to different hormones or hormone concentrations. Experimentally it is difficult to furnish such evidence. However there is one study in this section which shows that all the cells of a slime mould amoeba can be forced to form microcysts (spores) by exposure to physiological concentrations of ammonia (Chapter 4.2). In the absence of this 'hormone', the amoebae would have formed a fruiting body with stalk and spore cells, and it appears that all the cells are forced to follow one path of development and that ammonia is a cell type determinant.

At first sight, the induction by IAA of highly differentiated vascular tissue in hitherto unspecialized parenchymatous tissue in plants (p. 219) would seem to afford another example of hormonal control of cell type determination. However, although IAA is necessary for normal differentiation of xylem tissue this response can only be elicited in cells arising on the inner side of the cambium, and differentiation into xylem and phloem appears to be predetermined by a radial polarity within the cambial region (p. 224). Thus, the hormone appears to be stimulating differentiation in determined cells, although the *de novo* development of a vascular strand as a whole appears to be hormonally controlled.

Again, in the hormonal induction of buds and roots in un-differentiated callus tissue we do not seem to be dealing with the direct induction of specific cell types, but rather with the initiation of shoot apical meristems and root meristems as organized wholes, from which specific cell types arise later in a normal (but unknown) manner. The fact that each type of plant hormone has a wide spectrum of effects (p. 217), the type of response depending upon the nature of the target tissue, also indicates that the specific pattern of differentiation is already determined in the target cells and that the role of the hormone is to evoke the differentiation process. Certainly many experiments on plants and animals suggest that hormones generally act by eliciting the differentiation of determined cells.

4.1.3 HORMONES AS INDUCERS OF DIFFERENTIATION

In plants experiments have been carried out with dioecious species (which bear male and female flowers on separate plants) such as hemp (*Cannabis sativa*), and with those monoecious species such as cucumber (*Cucumis sativa*) or squash (*Cucurbita pepo*) where the male and female organs, although on the same plant, are borne in separate, distinct flowers. Application of auxins to male hemp plants leads to the formation of female flowers and the same treatment results in an increased ratio of female to male flowers in squash. In organ culture experiments it has been shown that young flower primordia of cucumbers excised from nodes which would normally have formed male flowers can be induced to develop into female flowers if treated with auxin (indole-3-acetic acid).

Ethylene has similar effects to auxin in experiments such as those described above and since application of auxins to plants often leads to enhanced ethylene production it has been suggested that the effects of auxin in these systems are in fact mediated by the gaseous hormone. Gibberellins on the other hand tend to increase the ratio of male to female flowers, and indeed, in certain lines of cucumber which usually produce only female flowers, application of gibberellins leads to the formation of functional male flowers. The few studies so far carried out on endogenous hormone levels in plants showing these effects suggest that sex expression is effected by a balance between endogenous auxins and gibberellins, since in plants showing pre-dominantly male characteristics gibberellin levels are high, whereas wholly female plants tend to have high auxin levels.

Similarly the differentiation of functional ovaries and testes can be controlled by altering the hormone concentration in the water in which amphibian embryos and larvae are reared (reviewed [5]). In these experiments it seems that the hormone has a direct effect on the somatic tissues of the gonad and that it only indirectly influences the sex of the germ cells. The primordial germ cells of *Xenopus laevis* are labile and will form sperm when transplanted to a male host and eggs when transplanted to a female host, whatever their genetic sex [6]. It seems that it is the sexual nature of the gonad in which the germ cells take up residence which determines whether they form sperm or eggs. The somatic tissues of the gonad are similar in male and female embryos during early development, and consist of an outer cortex surrounding an inner medulla. During ovarian development the cortex proliferates and contains the germ cells while the testes develop by proliferation of the medulla which encloses the germ cells. The hormones induce the proliferation and differentiation of either the cortex or the medulla and thus indirectly determine the cell type which the primordial germ cells will form.

There are many other examples of hormones acting as inducers of determined cells. The following is a list of a few other hormones which act in this way: gibberellic acid induction of α-amylase synthesis in the aleurone layer of barley seeds (Chapter 4.3), thyroxine induction of metamorphosis in amphibia (Chapter 4.4), erythropoietin induction of haemo-globin synthesis in the mouse embryo [7], and glucagon induction of liver enzymes in the rat [8]. It seems that hormones rarely act as cell type determinants during the course of normal development.

The response to hormones

Hormones often exert their effects in only a few cells of the developing embryo. In adults, it is known that cells which do respond to hormones possess special mechanisms. Target cells of protein hormones have receptors on the cell surface and a system by which the binding of a hormone to the receptor changes the cyclic-AMP levels inside the cytoplasm and mediates the hormones' action [9]. Target cells of steroid hormones contain receptors in the cytoplasm which bind the hormone, a specialized nuclear membrane which allows the receptor-hormone complex to enter the nucleus, and particular acidic proteins in the chromatin which associates with the receptor hormone complex [10, 11]. If these specialized mechanisms are required for hormone action in the embryo, then the formation of receptors must be an early developmental event in those cells which respond to hormones. It is therefore particularly interesting that a start has been made on the detection of thyroxine receptors in the early amphibian embryo (Chapter 4.4).

It might be thought that the synthesis of hormone receptors would be one of the first signs of determination in a cell. This attractive idea appears to be incorrect in at least one case. The liver enzyme, TAT (tyrosine amino transferase) can be induced in foetal liver by cyclic AMP at least two days in development before it can be induced by the hormones glucagon and cortisone. This suggests that liver cells are predetermined to synthesize TAT before they have the receptors required to respond to these hormones [8]. It remains to be seen whether hormone induced structures are also predetermined to develop in a particular way before they are able to respond to a hormone.

4.1.4 HORMONES AS CO-ORDINATORS WITHIN THE ORGANISM AS A WHOLE

Although hormones do not appear to act as determinants of specific cell types, they do seem to play an important role as co-ordinators of growth and development within the organism as a whole. This is well illustrated in the critical role of thyroxine controlling metamorphosis in amphibian tadpoles, where a very wide range of tissues undergo fundamental changes in direct response to the hormone (p. 232). A similar situation appears to be the case in insect metamorphosis. Here the hormone, ecdysone, triggers all the changes involved in moulting while the levels of juvenile hormone direct the morphological and biochemical results of the moult. High levels of juvenile hormone yield larval moults; low levels, larva to pupa transformation; and in the absence of juvenile hormone hemi-metabolous larvae and holometabolous pupae form adults. Thus these two hormones direct the overall organization of the insect [12].

The transition from the vegetative to the flowering condition in plants also involves a radical change in the structure of the shoot apical meristems and frequently in the plant as a whole, since when an apical meristem changes from the vegetative to the reproductive condition it loses its capacity for unlimited growth and becomes a structure of determinate growth, as a result of which it ceases to produce leaves and produces a flower or inflorescence. In many species this transition is controlled by environmental factors, such as seasonal variations in daylength or winter chilling temperatures (p. 349). There is much experimental evidence suggesting that daylength changes are detected by the leaves, although the response occurs in the shoot apices. Hence some sort of a signal must be transmitted from the leaves to the apices under inductive conditions, which stimulates the apex to undergo the transition from the vegetative to the reproductive condition. Although there is much circumstantial evidence that this signal is of a hormonal nature, so far attempts to isolate the hypothetical 'flower hormone' have proved unsuccessful. Gibberellins have been found to stimulate flowering in a number of species, but in most species flowering does not appear to be regulated by endogenous gibberellins (p. 350).

Apart from the hormonal control of such major switches in the life-cycle as metamorphosis or entry into the flowering phase, hormones play an important co-ordinating role within the organism as a whole, as seen in the development of secondary sex characters in animals and co-ordination of growth of the different regions of the plant body—for example, in the stimulation of growth in the maternal tissue of fruits, following pollination and development of the embryo.

4.1.5 REFERENCES

[1] BUZNIKOV G.A., CHUDAKOVA I.V., BERDYSHEVA L.V. & VYAZMINA N.M. (1968) The role of neurohumours in early embryogenesis. II. Acetylcholine and catecholamine content in developing embryos of sea urchin. *J. Embryol. Exp. Morph.,* **20,** 119–28.

[2] BUZNIKOV G.A., CHUDAKOVA I.V. & ZVEZDINA N.D. (1964) The role of neurohumours in early embryogenesis. I. Serotonin content of developing embryos of sea urchin and loach. *J. Embryol. exp. Morph.,* **12,** 563–73.

[3] GALUN E., JUNG Y. & LANG A. (1962) Culture and sex modification of male cucumber buds *in vitro. Nature, Lond.,* **194,** 596–98.

[4] WAREING P.F. (1971) Some aspects of differentiation in plants. In *Control Mechanisms of Growth and Differentiation. Soc. Exp. Biol. Symp.,* **25,** pp. 323–344 (Eds D.D. Davies & M. Balls). Cambridge University Press.

[5] BURNS R.K. (1961) Role of hormones in the differentiation of sex. In *Sex and Internal Secretions,* **1,** 76–158 (Ed. W.C. Young). The Williams and Wilkins Co., Baltimore.

[6] BLACKLER A.W. (1966) Embryonic sex cells of amphibia. In *Advances in Reproductive Physiology,* **I,** 1–28 (Ed. A. McLaren). Logos and Academic Press.

[7] COLE R.J. & PAUL J. (1966) The effects of erythropoietin on haem synthesis in mouse yolk sac and cultural foetal liver cells. *J. Emb. exp. Morph.,* **15,** 245–60.

[8] GREENGARD O. (1971) Enzymic differentiation in mammalian liver. *Essays in Biochemistry,* **7,** 159–205.

[9] ROBISON G.A., BUTCHER R.W. & SUTHERLAND E.W. (1971) *Cyclic AMP.* Academic Press, New York.

[10] O'MALLEY B.W. (1971) Unified hypothesis for early biochemical sequence of events in steroid hormone action. *Metabolism,* **20,** 981.

[11] O'MALLEY B.W., SPELSBERG T.C., SCHRADAR W.T., CHYTIL F. & STEGGLES A.W. (1972) Mechanism of interaction of a hormone-receptor complex with the genome of a eukaryotic target cell. *Nature, Lond.,* **235,** 141.

[12] WILLIS J.H. (1974) Morphogenetic action of insect hormones. *Ann. Rev. Ent.,* **19,** 97–115.

Chapter 4.2
Signalling Systems in Dictyostelium

4.2.1 INTRODUCTION

It may be helpful to illustrate communication systems in development by picking one example and examining it in some depth. The cellular slime moulds, and especially the species *Dictyostelium discoideum* which was discovered by K.B. Raper in the 1930s, have become popular organisms for the study of development. The reasons for this popularity are not hard to find:

(1) in the first place, they have a remarkably simple life history in which there is differentiation into only two cell types: stalk cells and spores;
(2) they are very easy to grow in the laboratory;
(3) it is even possible, as a result of recent work, to grow certain strains without a bacterial food supply; they can be grown axenically;
(4) part of their life cycle takes place on a surface, and therefore is two-dimensional, making experimental analysis that much easier;
(5) it has recently been discovered that they have sexuality, therefore proper genetical studies are now possible;
(6) finally, they are eukaryotes and therefore presumably any information they divulge in our experiments has a good chance of applying to the development of other higher animals and plants.

The life cycle of *Dictyostelium discoideum*, which takes roughly four days in the laboratory, is in itself curious and unusual. The cellular slime moulds differ from other soil amoeba in this respect; most other soil amoeba (e.g. *Hartmanella* and *Acanthamoeba*) form cysts by encapsulating individually in a resistant cellulose wall, while the slime moulds come together in aggregates and form communal fruiting bodies. In the case of *Dictyostelium* (by which we will mean *Dictyostelium discoideum*) there may be from a few hundred up to hundreds of thousands of amoeba in a large fruiting body. But 'large' is relative and a hundred thousand cells will only produce a fruiting body a few millimetres high that consists of a delicate stalk made up of large vacuole cells (Fig. 4.2.1). At the tip of the stalk there is a spherical lemon-shaped mass of cells, each one of which is an encapsulated spore, similar to the cysts of the solitary amoebae. In nature these small fruiting bodies project into interstices in the soil or in the small cavities between sticks and leaves in humus. It is presumed that gathering the spores in one large mass provides an effective means of dispersal and that

this is its adaptive advantage. But this is a subject which, in itself, is poorly understood and in need of further investigation.

A new generation can easily be started from one spore. Normally, the amoebae are grown on a suitable bacterial food source. (*Escherichia coli* or *Aerobacter aerogenes* are the most commonly used food organisms.) It is possible to grow up the bacteria beforehand and add spores to the bacteria, or the two can grow in a medium that will support the growth of the bacteria which, in turn, supplies the feeding and dividing amoebae. The growth can take place either in liquid culture or on the surface of a solid medium and numerous permutations are possible [3], however development beyond the amoeboid stage requires a surface covered with a water film.

On an agar surface the amoebae begin the morphogenetic phase of their life cycle. The cells double their rate of

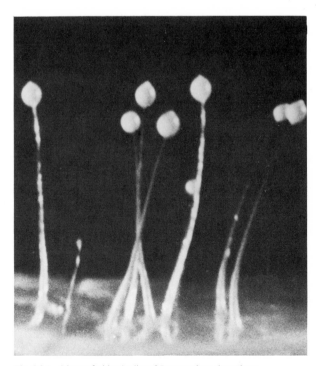

Fig. 4.2.1 Mature fruiting bodies of *Dictyostelium discoideum*.

204

locomotion for a brief period, after which they become very sluggish. Certain groups of cells appear to become focal points and begin to attract other cells. The cells soon pour in and form streams that branch outward (Fig. 4.2.2). In time lapse motion pictures the process is very dramatic, for among other things, the cells move inward to the aggregation centre in pulses spaced about 5 min apart. These pulses radiate outward and are waves of fast inward movement of the amoebae.

Once the cells are collected into a mass, the mass begins to push upwards and then falls sideways on the substratum producing a migrating slug (or 'pseudoplasmodium', which is synonymous with cell mass) (Fig. 4.2.3). This slug has remarkable properties: it is extraordinarily sensitive to environmental conditions and will go toward light of extremely low intensity and move up very weak heat gradients.

The duration of the migration period is also sensitive to environmental conditions; it will be brought to an end by a very slight drop in the ambient humidity and an increase in illumination. The slug rights itself and stalk formation now begins on the anterior end of the cell mass. Large vacuolate cells appear and these are pushed down through the cell mass to form the base of the stalk. Soon the whole mass rises into the air as the stalk cells move onto the tip and become incorporated into the stalk. The stalk itself is made up of a tapering cellulose cylinder which is secreted by the prestalk cells before they enter into the stalk and become vacuolate and eventually dead. As the stalk elongates so the posterior mass of pre-spore cells differentiates into spores and it slowly rises with the stalk so that it ends up at the apex (Fig. 4.2.3).

As in an annual plant, maturity means quiescence; there is no change until one of the spores germinates to start a new generation. In this case, the period of active development includes the entire life cycle.

What I have described is the asexual reproduction. It has been shown that the cellular slime moulds are capable of parasexual development and that by this means, it is possible to obtain genetic recombination [5, 14]. An exciting new development is the possible discovery of true sexuality. Some time ago, Blaskovies & Raper described macrocysts which appear during aggregation in some species of cellular slime moulds. It is now evident from the work of a number of laboratories, that these macrocysts probably contain zygotes [7, 8, 10]. *Dictyostelium discoideum* appears to be heterothallic, that is, it has opposite mating types. If two of the standard laboratory strains of opposite type are mixed, they will produce macrocysts. There is still a problem with the germination of macrocysts, but it is already possible to study development in *Dictyostelium* using genetic analysis as a tool.

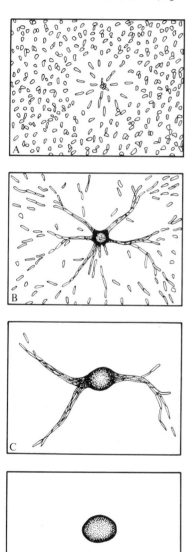

100 μ

Fig. 4.2.2 Four drawings (semi-diagrammatic) of successive stages of aggregation in *Dictyostelium.*

4.2.2 EVIDENCE FOR DIFFERENT SIGNALS

We will be concerned entirely with chemical signals which are the only kind that have been demonstrated in the cellular slime

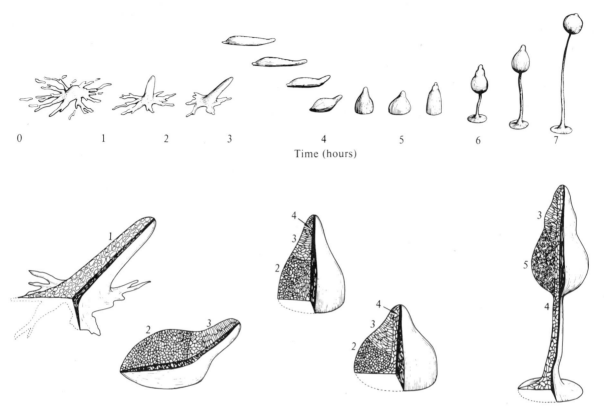

Time (hours)

Fig. 4.2.3   Development in *Dictyostelium discoideum. Above:* The aggregation, migration, and culmination stages shown in an approximate time scale. *Below:* Cutaway diagrams to show the cellular structure of different stages. 1, Undifferentiated cells at the end of aggregation; 2, prespore cells; 3, prestalk cells; 4, mature stalk cells; 5, mature spores. Drawing by J.L. Howard, courtesy of the *Scientific American.*

moulds; it is quite possible that there are other signalling methods which have not yet been discovered. Already a surprising number of different kinds of chemical signals are known. For some, we simply have evidence of their existence, but no clue as to their chemical nature. We know some of the chemical details of others, but we still lack a great deal of important information.

The signals could be classified in a number of ways. For instance, one could separate the signals which control or direct cell movement from those that are responsible for differentiation. My method will be to discuss what is known about signalling for each stage of development; it will be a life-history classification. One virtue of this system is that the life cycle begins with cells which communicate and are separate from one another; then the cells come together in a cell mass and the cells signal one another while in intimate contact; and finally, when the cell masses are developing there is a communication between these masses by what we might think of as a 'social' chemical hormone system. As we shall see, the signalling from cell to cell, when the cells are in contact, gives a more orderly and precise pattern than when signalling occurs at a distance.

In any small organism that develops in the outside world, it is essential that its life cycle be geared to physical changes in the environment. It must respond to changes in temperature, humidity, light, food, and so forth, so that it can maintain itself safely in a dangerous and fluctuating environment. It does this by responding to slight changes in the outside world, for instance, a drop in humidity causes the cessation of migration of the slug, probably because if the cell mass does not fruit quickly and form resistant spores, the naked amoebae will die from desiccation. It also responds to the environment by the cell mass moving towards light and towards heat; presumably these taxes are mechanisms whereby the fruiting body will be located at an optimal site for spore dispersal. There are other examples, but this is not the kind of signal that is our primary concern. Rather, we will examine those signals that are between cells or groups of cells; they are the signalling systems generated by the organism itself.

Signals in spore germination

In *Dictyostelium*, as with many fungi and other plants, a higher percentage of spores germinate when the spores are sparsely sewn rather than in a dense concentration. This apparent inhibition can be removed if the spores are carefully washed, and furthermore, if the washings of spores is concentrated it will greatly inhibit spore germination [3, 14]. Presumably this autoinhibition prevents excess germination in dense spore populations. This serves as a mechanism to prevent germination in the sorus, and since only a few amoebae are needed to produce a new generation it is obviously good strategy not to germinate all one's spores if there is an excess of spores at any one time. In contrast, the spores are stimulated to germinate in the presence of substances associated with food, namely amino acids [14].

Signals in feeding or vegetative amoebae

If amoebae are put in a concentrated drop on the surface of agar, they will spread and move away from the drop. This could be by random motion of the amoebae (similar to the diffusion of ions away from a salt crystal placed in water) or there could be a repellant produced by the amoebae which causes them to move away from one another (negative chemotaxis). There are two lines of evidence to support the latter hypothesis. One is an unpublished experiment of E.W. Samuel in which he placed a concentrated mass of amoebae on a small square of cellophane. He placed this near a few vegetative amoebae on an agar surface and they oriented away from the square; if he then moved the square 90° to one side of the amoebae, they changed their direction 90° and again moved away. Another test was suggested to me by Lee Segal: if the movement is similar to diffusion, then the distance that the amoebae moved away from the drop would fall off as the square root of time, while if there is a repellant it would be linear with time. Actual measurements show a linear relationship, again supporting the idea of a repellant.

Unfortunately we have no idea of the chemical nature of the repellant. We do know that different substances which are given off by bacteria attract amoebae (folic acid, cyclic AMP, and other unknown substances). However, this is not a morphogenetic signalling system but rather a mechanism whereby the amoebae can locate their food. When the food is all gone, or when there is an even distribution of food over the surface, then the amoebae are evenly spaced, presumably by this cell repulsion signal which acts at a distance between cells. Such a mechanism will make for even, efficient grazing in the presence of food, and an effective searching mechanism in the absence of food.

Signals between aggregating amoebae

The mechanism of aggregation has been the primary concern of many investigations for some years. The first clear indication that it was a chemotaxis was a key experiment of Runyon [3]. He showed that if one placed an aggregation centre on one side of a dialysis membrane, and sensitive amoebae on the other side, then the amoebae were attracted to the point opposite the aggregation centre. He suggested that this was chemotaxis effected by a small molecule which could get through the cellophane membrane. This conclusion was supported by other evidence [1], but attraction to acrasin (as the attractant is called) was not proved until the work of Shaffer [3, 18]. He developed a test in which he sandwiched some sensitive preaggregation amoebae under a small block of agar on a glass slide, and then repeatedly added water containing acrasin (by taking drops of water from around aggregating centres in another dish) to the edge of this agar block. In a matter of minutes the amoebae under the block streamed to the edge. He also showed that the substance broke down readily for if he lengthened the interval between additions of acrasin to the edge of the block, no attraction occurred. Subsequently, he showed that this was because there is an enzyme, an acrasinase, that destroys the activity of the attractant.

In the 1950s and early 1960s there was a great wave of activity in various laboratories to discover the chemical nature of acrasin but none of these succeeded. Part of the difficulty was that the Shaffer test was inconvenient as an assay of acrasin in chemical analysis. Independently, Konijn and our laboratory devised assays for acrasin that worked on essentially the same principle [5]. A drop of cells was placed on agar containing a test substance and if it had acrasin activity, the cells worked away from the drop, further and more rapidly than in the control. It is presumed that this happens because the cells produce a large amount of acrasinase which diffuses into the agar around the drop and removes all the acrasin from that area. The outer edge of amoebae are therefore in a gradient of acrasin which is more concentrated as one goes away from the drop. Since the cells are positively chemotactic, they will go up a concentration gradient and therefore spread rapidly away from the drop.

Using this kind of test, in a collaborative effort with Konijn, we were able to show that the acrasin for *Dictyostelium* is cyclic AMP. First it was shown that the cells at the aggregation stage are extraordinarily sensitive to cyclic AMP, the optimum concentration being 10^{-7}M. Next it was shown that the amoebae produced large quantities of an extracellular phosphodiesterase that specifically converted cyclic 3′,5′ AMP into 5′ AMP, a compound devoid of any chemotactic activity. This, then was presumably the acrasinase first demonstrated by Shaffer. Thirdly, it was shown that the amoebae both synthesized and secreted cyclic AMP. Finally, evidence has accumulated that while the phosphodiesterase is secreted continuously by amoebae undergoing normal development on an agar surface, the cyclic AMP secretion rises abruptly at the beginning of aggregation, attaining a level 100 times greater than the level of vegetative amoebae. Furthermore, it had been known for some time that the sensitivity to acrasin also greatly increases (again about 100 times) at the onset of aggregation. These are the bases for the contention that cyclic AMP is an acrasin.

What we do not know is exactly how gradients of cyclic AMP orient cells. This is being approached experimentally on two levels. One is the question of what is occurring on the surface of the responding amoebea, and the other is the overall coordination of the aggregation movements.

It is clear that if an amoeba is in a gradient of cyclic AMP, there must be more cyclic AMP molecules surrounding its anterior end. Recently it has been shown that there are, in fact, at least two phosphodiesterases: one extracellular, as we have seen, and the other membrane bound. One interesting theory put forth is that the membrane bound phosphodiesterase is associated with a receptor site for the cyclic AMP during chemotaxis [16, 17]. The evidence for this suggestion comes from the fact that the membrane bound enzyme, unlike the extracellular phosphodiesterase, increases specifically at the aggregation stage, and then subsequently falls off by the migration stage. There is corroborative evidence that during aggregation the cyclic AMP is broken down at the cell surface and does not enter the cell intact. Whether this hypothesis turns out to be correct or not, there will be a continued interest in the structure and the activities of the cell membrane during aggregation.

Another problem in aggregation concerns the role of the pulses of fast inward movement of the aggregating amoebae. The first to appreciate the possible significance of these waves was Shaffer who suggested that acrasin did not necessarily distribute itself in one big gradient, but that when the cells were still separate, each amoeba produced a puff of acrasin and oriented the next amoebae by producing a small, local gradient. He assumed that the presence of acrasin stimulates a cell to make its own puff of acrasin, and that subsequently each cell goes into an insensitive, refractory period, a model which closely parallels the transmission of impulses in nerves or in the heart of animals. The idea has been subsequently pursued by Robertson & Cohen [5] who argue, on the basis of some interesting theoretical considerations, that not only aggregation, but all aspects of development, including the pattern of differentiation, are governed by these waves. The alternative is that the waves, which indeed do not always seem to be present might be incidental and secondary and not in themselves of major morphogenetic significance.

Another interesting approach to the broader problem of the factors which govern aggregation comes from the theoretical ideas of Keller & Segel [5]. Following the tradition of earlier mathematicians, especially Rashevsky & Turing, they suggest that the overall pattern of aggregation is an instability phenomenon. It can be predicted that a uniform sheet of cells producing both extracellular cyclic AMP and phosphodiesterase will break up into a series of hills and valleys if there is a change in the production and sensitivity of response to cyclic AMP. It is thought that these hills and valleys are ultimately gradients (or perhaps the relay type gradients of Shaffer) that produce the aggregation of the cells into the centres. We shall return to these ideas of Keller & Segel when we discuss the patterns of aggregations over a wide field of amoebae. Here I want merely to emphasize that in one way or another, gradients are produced by a small molecule, acrasin, and an enzyme which destroys the acrasin, and that aggregation is the ultimate result. It should be added that not all species of cellular slime mould respond to cyclic AMP, but, as yet, we do not know the chemical nature of the other chemotactic systems.

Signals in the multicellular stages

Once the cells come together in a mass, it becomes increasingly difficult to analyse the signalling system. This is regrettable, because some of the most interesting features of development take place at this stage. Here we will consider two main aspects but the interpretation of each will necessarily involve a considerable amount of speculation. Those two aspects are: (1) the nature and the method of producing an internal gradient within the migrating slug; and (2) the way in which a regular division of the slug occurs so that the ratio between the number of anterior prestalk cells and posterior prespore cells remains constant. No doubt the two aspects are related and the first leads directly into the second. But before we look for models of signalling mechanisms to account for these phenomena, it is worth considering what is known about development of the migrating slug.

There is some old evidence for an acrasin gradient along a slug, but as evidence it has severe limitation [2]. It was shown that if a slug is placed near aggregating amoebae, the direction of orientation of the amoebae can be used as an index of the concentration gradients of acrasin surrounding the slug. By this method, it was shown that during migration there was more acrasin produced at the anterior end and that there was a gradient moving posteriorly. Only at the very end of migration, when the cell mass is righted to begin fruiting, does the gradient disappear and the acrasin is only secreted at the tip. The difficulty with these observations is that there is no assurance that what is given off to the outside reflects the internal concentration; it may simply reflect the permeability of the slime sheath to cyclic AMP over different regions of the slug. For this reason, a number of laboratories are attempting direct chemical methods to determine differences in cyclic AMP in various regions of the slug.

using histochemical techniques, initially the cells appear uniform, but later the anterior cells become larger and have different staining properties [3]. If cells are stained with a vital dye before aggregation, they will first form uniformly coloured slugs which later, after a period of migration, turn darker at the posterior end and blanch at the anterior end. Furthermore, this difference which can be revealed in living slugs with vital dyes, corresponds to the prespore and prestalk regions shown in the histological work.

Extensive studies have been done examining the proportion of stalk cells to spore cells [3]. These have involved a variety of different methods and they have been done both by measuring the final differentiated state and the prestalk and prespore cell in the slugs. It is, of course, possible to achieve a rather extra-ordinary range in size, since size depends upon the number of amoebae that enter an aggregate. The result of these painstaking experiments is that in all species examined, there is a proportional relation between the number of stalk and spore cells. In *Dictyostelium discoideum* the relation is linear (isoallo-metric rather than positively or negatively allometric) which means simply that the per cent stalk or spore mass is constant whether the slug consists of a few cells or hundreds of thousands of cells. This rule applies under normal growth conditions (but see below).

There is now compelling evidence that cells sort out and rearrange themselves within the slug; they do not keep to the positions which they occupy immediately after aggregation. I will not pursue here all the lines of evidence except to point out that Takeuchi showed in an early study, that if he made antibodies to spores (by injecting spore suspensions into rabbits) and then he conjugated these antispore sera with a fluorescent compound, he could stain amoebae at different stages of development. Different amoeba contained different amounts of spore antigens before aggregation, and shortly after aggre-gation, all the cells with spore macromolecules ended up in the posterior (prespore) region, while the anterior (prestalk) region had nothing but cells without any fluorescence. This was so despite the fact that a short time previously, during aggregation, the cells that fluoresced were randomly distributed [5].

There has been one recent experiment of Garrod & Ashworth [12] that is of special interest. They showed, using a mutant of *Dictyostelium* that can be grown axenically, that if it is grown with excess glucose in the medium, there is a larger per cent of spores; they can shift the normal proportions. They then mixed glucose grown cells (which were marked with a radioactive label) and unmarked cells which were grown without glucose, and they were able to show that the glucose grown cells became spores and the cells grown in minimal medium became stalk cells. To put the matter in crude terms, it would appear that the well fed cells are tipped in the spore direction, a reasonable enough strategy since they will have to start a new generation on their stored reserves.

These are the facts; now we want to speculate on what kinds of signals could be operating in the cell mass to account for the proportional development of the spore and stalk cell regions. It is a basic premise that there must be some kind of a gradient inside the slug and that this gradient somehow ends up in two discontinuous regions, so regulated that their ratio is the same regardless of the size of the slug. One very obvious possibility is that the gradient is of some morphogen such as cyclic AMP.

A gradient is a method of signalling, of imparting spatial information, but before we examine that proposition, we must ask how a gradient could have arisen in the first place. There are at least three types of hypotheses (which are not mutually exclusive), and let us examine each briefly.

The simplest notion would be that the tip is the former centre of the aggregate; it initially is the prime source of cyclic AMP and it continues to maintain the dominant position. This hypothesis then is one of a pure concentration gradient of a particular morphogen. The second possibility is a variation of the first which considers the tip of the slug not so much the point of highest concentration of a morphogen, but a pacemaker region. Its dominance is due to its pacemaking ability, and the waves of morphogen production and destruction pass down along the slug. Robertson & Cohen [5] favour such a model and believe that it can account for all the observed phenomena including the division line between prestalk and prespore cells.

Finally one could speculate that the sorting out of the cells might itself be of key importance in setting up the gradient. One could assume that the preaggregation cells have a range of adhesive properties and they, in some way, sort out by 'differential adhesion' as proposed by Steinberg [19] for the mechanism of sorting out in vertebrate embryos. The cells form a gradient and this gradient is not only one of adhesion between cells, but there could be numerous other gradients including spore proteins, cyclic AMP production, energy storage, and so forth, which could exist along the axis of the slug.

But in any one of these hypotheses, we have not said enough; we have not said precisely how the gradient turns into a fixed proportion of prestalk and prespore zones regardless of the size of the cell mass. To emphasize the importance of this, let me give one more fact. Raper [5] showed many years ago that if a slug was cut into pieces, then each piece, could eventually produce a normally proportioned, miniature fruiting body. This means that any portion of the slug and of the gradient within the slug is capable of giving the correct proportions. The spore macromolecules of Takeuchi, then, are not fixed, but reversible; one can isolate a part of a slug that normally would give all spores, and cause the prespore cells in the anterior end to turn into prestalk cells. The inescapable conclusion is that there must be some sort of polar signal that is sent forward or backward in the slug (or both) that must carry information.

Unfortunately, we do not know what the nature of the signal might be. Again, we do have hypotheses. One is the waves or pulses could, making various assumptions, manage to be such a signal. A new and intriguing model has been devised by McMahon [15]. He suggests that each cell, which is known to be arranged in a polar fashion during aggregation, and therefore presumably in the slug, has a different set of enzymes at opposite ends. He assumed that at the anterior end it has a concentration of adenyl cyclase which produces the cyclic AMP and at the posterior end it has the membrane bound phosphodiesterase which converts the cyclic AMP to $5'$ AMP. Again, making some further reasonable assumptions, he can show that the distribution of cyclic AMP would soon become discontinuous in such a system; high in the anterior end and low in the posterior end, and that the boundary between the two would be very sharp. Furthermore, the position of the boundary in the cell mass would depend on the size and shape of the whole mass, hence accounting for regulation. It will be of enormous importance to the problem of development in general to find out what the actual mechanism might be, for proportional development is the rule rather than the exception in the development of all organisms (see Chapter 3.6).

Not all the signals within the multicellular mass are merely concerned with the division between spore and stalk cell. There are innumerable other details of the shape of the final fruiting body that need signals in order for them to be perfectly achieved in the same fashion in successive generations. But unfortunately, we know nothing of the mechanisms or the signals; they simply have not been investigated. Here are a few examples of the kind of controlled phenomena that need analysis. All the control mechanisms for the movements of the cells during stalk formation and the rising of the sorus into the air. The control mechanism for the deposit of the tapering stalk; not only its secretion by the prestalk cells, but the control of its taper. The control of the formation of the basal disc. One could easily add to this list.

Signals between multicellular masses

If one examines the locations of centres of aggregation on a culture dish, it is obvious that their distribution is non-random (Fig. 4.2.4). This can, furthermore, be verified by simple measurements of the distances to the nearest neighbours and by application of appropriate statistical tests. It turns out there are a number of reasons why there is a non-random distribution and all of the reasons involve signals in one form or another.

The simplest reason is that as aggregation chemotaxis begins so certain areas will soon have fewer amoebae than others, and in these sparse regions new centres are less likely to form. Here the signal is simply acrasin chemotaxis. This is undoubtedly related to the previously discussed instability phenomena which have been emphasized by Keller & Segel [5]. Because of changes in (1) the amount of acrasin secreted, (2) the sensitivity to acrasin; and (3) the change in rate of the random motility of

the cells, a uniform distribution of acrasin producing cells will break up into a series of hills and valleys and these will appear in some regular, non-random pattern.

Another factor which inhibits secondary centres from forming near the primary ones is the fact that if a centre is in a steep gradient of acrasin it will disintegrate [5]. In other words, even if a centre did arise by accident in a zone close to another centre, it would be overruled.

It would seem that these mechanisms would be sufficient to assure non-random spacing of centres, but, in fact there appears to be an additional special mechanism to assure non-random spacing. There is a considerable amount of evidence that a centre gives off a substance which inhibits other centres from forming in the near neighbourhood. In *Dictyostelium*, there is evidence that the substance is volatile, for if two layers of cells are opposed on agar surfaces separated by an air space, the centres from one surface will inhibit the centres on the opposite surface. Our first suspicion was that the volatile signal might be CO_2, but Feit [9] repeated the experiment with the layers of cells at different ages and put the culture dishes both upside down and right side up. That is, the advanced cells with their centres already established were either on the 'floor' of the culture dishes, or the 'ceiling'. The centre inhibition effect was very much greater when the older centres were on the 'floor'; clearly the volatile substance was lighter than air and therefore could not be CO_2. Feit suggested it might be ammonia, and Lonski

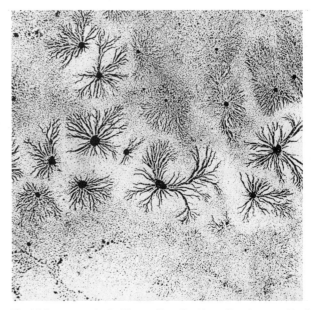

Fig. 4.2.4 Aggregation in *Dictyostelium discoideum*. Very low magnification to show many aggregates. Photographs of K.B. Raper.

[13] has demonstrated in some detail that NH_3 will produce this effect in concentrations equivalent to those which the organism is known to produce.

There is the interesting related question of why such great pains should be taken to insure a non-random distribution of centres. One presumes that it is again closely related to the effectiveness of spore dispersal. Previously I suggested that this might be the adaptive advantage of collective masses of spores borne on stalks, and now we are adding the suggestion that it is important to have the spore masses effectively spaced. It is possible to show some further signal-mediated controls which support this contention.

In very sparse populations of amoebae of *Dictyostelium* where minute fruiting bodies are produced, there is no migration and, therefore, the location of a centre is also the location of a fruiting body. But in dense populations, the aggregation cell mass goes through a period of migration which can totally alter the pattern of distribution of cell masses and therefore the distribution of the fruiting bodies. The point has been made that the migrating slugs go towards light and heat, mechanisms which again must help in producing more effective dispersal by placing the cell masses in optimal positions. There is still another spacing mechanism operating by signals between cell masses that further ensures the optimal spacing of the spore masses (or sori) in the air. Its existence was first appreciated when the cell masses about to culminate (that is, rise into the air to form fruiting bodies) were pushed close together and it was noted that they tended to veer away from one another as they rose into the air. It was subsequently shown that the rising cell masses produce a volatile substance which repels any neighbouring cell mass. The evidence for this came from a series of observations, the most convincing of which was to place a small piece of activated charcoal near the cell mass and the stalk would curve right over towards the charcoal. It is presumed that the charcoal actively absorbs the volatile repellant and therefore, there is less of it on the charcoal side, which results in its movement toward the charcoal (Fig. 4.2.5). One result of this volatile signal is that the sori are kept far apart as they are raised above the substratum.

Besides the signalling between rising cell masses, this volatile repellant serves another important function. It is responsible for the orientation of the fruiting body with respect to the substratum. For many years it was a puzzle why very small fruiting bodies should always grow at right angles from the substratum, regardless of its orientation. The correct explanation appears to be that the highest concentration of the repellant is retained near the substratum. The repellant is diluted away from the substratum and the fruiting body therefore grows in this direction.

This idea can be tested a number of ways. If the cell mass is covered with mineral oil, fruiting will occur up into the oil, but the shape of the stalk will be utterly convoluted and disoriented.

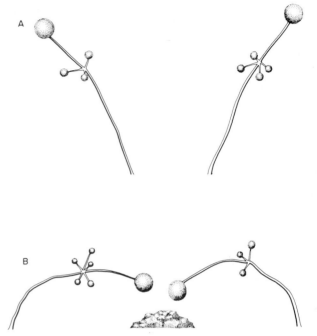

Fig. 4.2.5   (A) Two fruiting bodies of *Polysphondylium pallidum* bending away from each other. (B) Two similar fruiting bodies bending towards some activated charcoal. Drawing by R. Gillmor.

A cell mass can be placed at the bottom of a crack of agar, or at the base of an agar cliff, and no matter what the location, the sorus will eventually lie mid-way between the walls. One can even put the cell mass on the edge of a right angle agar cliff, and the stalk will jut out so that it is exactly 135° from the top and the side of the cliff (Fig. 4.2.6).

Nothing is known concerning the chemical nature of the volatile signal. Lonski (unpublished) made some analyses using gas chromatography of what substances are given off by

Fig. 4.2.6   A fruiting body of *Dictyostelium* rising on the edge of an agar block. Drawing by K. Zachariah.

culminating *Dictyostelium*, and he found, besides CO_2 and NH_3, ethane, ethanol, acetaldehyde, and ethylene. Any one of these could be active; this is an important project for future experiments.

Signals which control major developmental pathways

In *Dictyostelium* a cell can differentiate either into a spore or into a stalk cell. These are radically different pathways, for a spore is a condensed resistant stockpile of reserves, while a stalk cell is a dead cell which used virtually all its energy in the manufacture of thick cellulose walls. A spore also has cellulose walls, but their construction must use only a small part of the spore's energy. All the rest is put into reserves. There is, for instance, an accumulation of the trisaccharide trehalose which is used as a quick source of energy upon germination [14].

Certain aspects of this difference has already been stressed when we discussed spore-stalk proportions; here I want to begin a more molecular analysis of the process. It is unfortunately a subject that is not even in its infancy but at what would more

Dictyostelium, or each isolated amoeba can feed and then bypass all these stages and immediately encyst (Fig. 4.2.7). These 'microcysts' appear similar in structure to spores, except that their cellulose envelope is spherical rather than elliptical.

Recently Lonski [13] has undertaken to discover the normal signal in nature which is responsible for this developmental switch. He noticed that if a culture of *Polysphondylium* was very crowded, the frequency of microcyst formation increased but that the effect of crowding was totally eliminated by adding activated charcoal to the dish. In a series of experiments, he demonstrated that the crowding effect could be achieved on sparse populations of amoebae by adding physiological concentrations of NH_3 which suggests that this may be the normal signal in nature for the microcyst pathway.

The reason that this microcyst formation may be of such importance to the study of the biochemistry of developmental pathways is that it is possible to obtain 100% microcysts, and superficially, at least, this would appear to be equivalent of having all spores. On the other hand, with cyclic AMP and *Dictyostelium* (especially in Chia's mutant) it is possible to obtain 100% stalk cells. It should, therefore, be possible to study

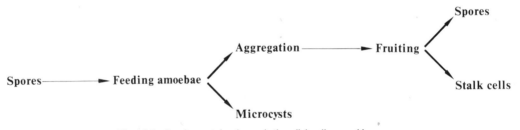

Fig. 4.2.7 Developmental pathways in the cellular slime moulds.

aptly be called a foetal stage. We know that well-fed cells turn into spores as previously discussed, and that high concentrations of cyclic AMP will cause cells to turn into stalk cells [4]. The latter has been studied recently in some detail by Chia [6] on a mutant of *Dictyostelium* that is especially sensitive to cyclic AMP. He has been able to show that there is indeed a direct connection between cyclic AMP and stalk cell formation. Furthermore it appears to be quite a different effect from the chemotaxis effect for only cyclic AMP produces stalk cells, while various other cyclic nucleotides are effective as chemotactic agents. From these experiments one might conclude that cyclic AMP has two functions in *Dictyostelium*: a chemotactic signal and a stalk cell differentiation signal.

It has been known for some time that some species of cellular slime moulds, such as *Polysphondylium pallidum,* have two alternate pathways of development. The amoebae can either aggregate and form fruiting bodies as we have described for

the biochemical differences between these two pathways, for they can, in this fashion, be separated.

4.2.3 THE BIOCHEMISTRY OF SIGNALLING

Besides the substances which are exchanged between cells and groups of cells, there are key messages within cells. This is a whole new level of signalling which is as essential as the external signalling we have been discussing. I am referring to the subject of gene expression or gene action. It is our basic assumption that all the various features of development are ultimately controlled by genes. This applies to the production of stalk cells, or spores, the chemotaxis of aggregation, the migration and culmination movements, and so forth. Let us look briefly at what is known of signals inside the cell and then finally see what can be done to bring together the external level of signals with these inner ones.

Transcription and translation

In order for development to occur it is necessary to synthesize new substances. The initiation of this process largely comes from transcription: *messenger* RNA coding for a particular protein is constructed directly on the DNA of the genome. This RNA then goes out into the cytoplasm to direct the manufacture of a specific protein by translation (Chapters 1.1, 1.2, 5.3).

Nothing could fit the category of a developmental signal better than *messenger* RNA. For this reason, a large amount of work has been done on examining the nature of such messages from *Dictyostelium*. The pioneer experiments were those of Sussman and his co-workers [14] who showed that if RNA synthesis inhibitors were added to the *Dictyostelium* amoebae at different stages in their morphogenesis, then specific proteins and specific enzymes were not synthesized. On this basis, they suggested that at a particular time in the life cycle, the *messenger* RNA for a certain enzyme was synthesized. By adding protein synthesis inhibitors in parallel experiments they could also find the moment that the message was translated. Using these methods, they suggested that the *messenger* RNA is not immediately used up, but can remain stable for varying periods of time, the length of time being different for different messages.

The difficulty with these experiments lies in the inhibitors. More recently, Firtel and his associates [11] have developed methods for isolating the *messenger* RNAs of *Dictyostelium*. The messenger molecules are relatively small and, as one would expect, different *messenger* RNAs appear at different times during the course of development. Apparently the standard use of actinomycin D as an inhibitor does not completely stop messenger synthesis; this can be done by also adding another inhibitor. When this is done it is clear that the stability of all the *messenger* RNAs is about the same. It is in the order of $1\frac{1}{2}$ h, and never extends to a period as long as 10 h, which had been previously suggested for the messenger of some enzymes. From this it would appear that there is little in the way of translational control, at least as far as timing is concerned. This does not rule out the possibility that the amount of protein is regulated to some degree at the translational level.

Enzyme activity

All other substances besides proteins are synthesized by the enzymes which have been produced by transcription and translation. An enzyme may promote a certain reaction, and in this case the quantity of the substance may be controlled by the amount of substrate, the amount of enzyme and the conditions which influence the activity of the enzyme. As Wright [20] has emphasized, there are many ways in which a series of interlocking enzyme steps could control the amounts of all sorts of key substances. To put the matter another way, there are

numerous enzyme controlled reactions which can, and undoubtedly do play an essential role in the internal signalling system.

One might consider these totally cytoplasmic signals rather than the nucleus-to-cytoplasm signals we have discussed before. But, of course, it is well known that the cytoplasm may affect the nucleus; under different cytoplasmic conditions different *messenger* RNAs will be synthesized. So all the signals within the cell are related and interdependent. At any point in the life cycle, even at the spore stage, there is a combination of nuclear and cytoplasmic information, all of which is necessary for further development.

The timing of signals

One of the most important problems in developmental biology is to understand the mechanism whereby the events of development are timed. There are two ways in which we use the concept of timing: one is to imply a strict sequence of events, and the other is to literally mean what time a particular event occurs.

Since developmental stages usually occur in a particular sequence, one assumes that there is some kind of strict order. This has been looked at on the molecular level in a most interesting way by Newell & Sussman [14] who examined the synthesis of three enzymes that formed sequentially during the migration stage. After the three syntheses were complete, they took the slugs and disrupted them into a rubble of single cells. Soon thereafter the cells came together to form reconstituted slugs, and during this process a new round of synthesis occurred for the three enzymes. The sequence of synthesis was identical to that of normal development; the only difference was that they occurred in abnormally rapid succession. As a result the slugs now had twice the amount of these three enzymes, yet the subsequent development was normal. By the use of inhibitors, they were able to show that this sequence of resynthesis involved a new set of transcriptions as well as translations. One presumes, therefore, that the sequence is somehow engraved in the genome, but that the rate at which the sequence is transcribed and translated is under external control of some unknown sort.

Clearly not all sequences are controlled by the genome. There must be others that are controlled entirely at the level of the cytoplasm. For instance, one could imagine all the enzymes necessary to produce a key substance being present, but the substance can only be produced in a strict sequence if the substrate needed for one reaction is the product of a previous one.

Sequences can also be determined by signals between cells. Consider the crudest case where the question of whether *Polysphondylium pallidum* produces microcysts or undergoes normal development depends upon the amount of exogenous ammonia.

Or consider the fact that aggregation in *Dictyostelium* cannot occur without extracellular cyclic AMP. In this latter example, some biochemical consequence of starvation results in the vastly increased production of, and sensitivity to cyclic AMP.

If one looks at signals at the three levels, nuclear, cytoplasmic and intercellular (and even between cell masses), it is clear that the signals from one level can either act on its own level or on any of the other levels.

A large share of the signals which determine exactly when a particular chemical reaction is to occur come from the external environment. As we have already pointed out, a decrease in humidity stimulates stalk cell and spore differentiation; light hastens the initiation of aggregation; and there are other examples. This is reasonable and expected, for in this way such a vulnerable and exposed micro-organism can exist in harmony in a changing physical environment.

Signals which involve spacing

This is a subject we have already covered in some detail. We talked about the stalk–spore cell ratio and how it was kept constant in different sized cell masses, and even readjusted in each portion of a large cell mass cut into fractions. We also discussed the pattern of aggregations to show how there was some signalling between cell masses.

In the former case, all the hypotheses which bear on possible mechanisms, assume that there must be some form of signal which travels either up or down (or both) the major axis of the slug. This signal might be a substance such as cyclic AMP, and various suggestions were made as to how the morphogen could come to be distributed in a gradient. It is unfortunate that we do not know which of the hypotheses is correct. We gave some specific models. One was that preaggregation cells differ and that these sort out by differential adhesions producing a gradient of cells with different properties. Another model was that each cell had cyclic AMP synthesis enzymes at its anterior end and cyclic AMP elimination enzymes at the posterior end, and by having all the cells aligned in the same direction, a two-zoned distribution will result. There is no reason why both mechanisms could not operate together, and there are obviously other possibilities.

These signals which we discussed previously all occur at the most peripheral level; we now ask the question of how the genome could control such signals. The answer is very simple. All the cell properties which lead to the communication necessary to produce spacing are engraved in the genome. The substances on the cell surface which are responsible for adhesion between cells are synthesized ultimately through the transcription–translation route. If they are substances other than proteins, for example, polysaccharides, this statement is still true for enzymes are responsible for their synthesis and the structure of enzymes is, of course, entirely determined by the

structure of the DNA. In McMahon's model there is a slight further complication. Not only is there the synthesis of adenyl cyclase and phosphodiesterase, but they themselves are spaced in each individual cell. One might postulate that this polarized distribution of the enzymes was set up during aggregation and is the direct result of the cells orienting in an acrasin gradient.

In the second case, that is the pattern of centre distribution, one assumes that initially any cell or any group of cells has an equal chance of becoming a centre. But new centres which might emerge near established centres will be obliterated for two reasons: they will not survive in steep acrasin gradients or in the presence of an inhibitor (possibly NH_3) secreted by the older centre. Again we can account for the production of cyclic AMP and NH_3 by the normal routes of genome related biosynthesis. It is also easy to understand how an inhibitor might suppress centre formation for there must be a whole host of chemical reactions, any one of which could be stopped by a specific chemical signal. It is less easy to understand how a steep cyclic AMP gradient can destroy a centre unless it is simply that it overrules the gradient of the weaker new centre, and each individual cell is beckoned to the older centre.

In general, therefore, the production of substances is either a direct result of the activity of the DNA in protein synthesis, or an indirect result through the action of enzymes whose structure is coded by DNA. But the spacing of substances to form a pattern has a far more indirect connection with the genome. It is the result of secondary consequences involving time of production, the particular combination of substances produced at one time, and initial local differences in the amount of these substances. All of these circumstances (and others as well) are the ingredients of pattern formation. They all involve signals which, for the most part lie well outside the nucleus of the cell, even though the nucleus is the ultimate source of all the chemical ingredients that go into the signalling that makes the pattern.

4.2.4 CONCLUSIONS

The cellular slime moulds have many of the characteristics of the development of higher animals, and at the same time they have some very conspicuous advantages as experimental organisms. Development in general consists of a signalling system out of which comes a consistent pattern from one generation to the next. All the substances of the organism are ultimately produced by the DNA of the nucleus, but even in a spore which begins a generation, many of the proteins, and in fact, all the molecules and internal structure of the spore cell were manufactured in one of the preceding cycles. Each generation begins with at least a cell. It has an active nucleus and an active cytoplasmic metabolic machinery which is ready and capable of the synthesis of all manner of substances as development unfolds. Development involves more than just the synthesis of the proteins and the other substances, but the timing of the synthesis and the spacing of the products of synthesis. Since the chemical reactions are mutually interdependent, that is, the product of one may affect the outcome of another, the production, the timing, and the spacing of the substances to produce an adult, rely utterly upon the exchange of signals. In fact, as we have seen, the entire development of *Dictyostelium* can be analysed in terms of the signalling systems. These signals pass between parts of a nucleus, between parts of the cytoplasm of a cell, between cells, and even between cell masses. Furthermore, signals pass from any one of these levels to another so that all the levels are in constant communication. This means a remarkable variety of signals in the development of even so simple an organism as *Dictyostelium*. But this intricate network of multilevel signals is precisely what produces consistent form from one generation to the next in a multicellular organism.

4.2.5 REFERENCES

* The reviews are indicated by an asterisk and give full references to particular experiments discussed in the text.

[1] BONNER J.T. (1947) Evidence for the formation of cell aggregates by chemotaxis in the development of the slime mold *Dictyostelium discoideum. J. Exptl. Zool.,* **106,** 1–26.

[2] BONNER J.T. (1949) The demonstration of acrasin in the later stages of the development of the slime mold t2Dictyostelium discoideum. J. Expl. Zool.,* **110,** 259–71.

[3] *BONNER J.T. (1967) *The Cellular Slime Molds,* 2nd Ed. Princeton University Press, Princeton, New Jersey.

[4] BONNER J.T. (1970) Induction of stalk cell differentiation by cyclic AMP in the cellular slime mold *Dictyostelium discoideum. Proc. Nat. Acad. Sci., USA.,* **65,** 110–13.

[5] *BONNER J.T. (1971) Aggregation and differentiation in the cellular slime molds. *Ann. Rev. Microbiology,* **25,** 75–92.

[6] CHIA W.K. (1974) Ph.D. Thesis, Princeton University.

[7] CLARKE M.A., FRANCIS D. & EISENBERG R. (1973) Mating types in cellular slime molds. *Biochem. Biophys. Res. Com.,* **52,** 672–8.

[8] ERDOS G.W., RAPER K.B. & VOGEN L.K. (1973) Mating types and macrocyst formation in *Dictyostelium discoideum. Proc. Nat. Acad. Sci., USA.,* **70,** 1828–30.

[9] FEIT I.N. (1969) Ph.D. Thesis, Princeton University.

[10] FILOSA M.F. & DENGLER R.E. (1972) Ultrastructure of macrocyst formation in the cellular slime mold, *Dictyostelium mucoroides:* extensive phagocytosis of amoebae by a specialized cell. *Develop. Biol.,* **29,** 1–16.

[11] FIRTEL R.A., BAXTER L. & LODISH H.F. (1973) Actinomycin D and the resolution of enzyme biosynthesis during development of *Dictyostelium discoideum. J. Mol. Biol.,* **79,** 315–27.

[12] GARROD D. & ASHWORTH J.M. (1973) Development of the cellular slime mould *Dictyostelium discoideum.* In *Microbial Differentiation* (Eds J.M. Ashworth & J.E. Smith), *Symposium Soc. Gen. Microbiol.,* **23,** 407–35.

[13] LONSKI J. (1973) Ph.D. Thesis, Princeton University.

[14] *NEWELL P.E. (1971) The development of the cellular slime mould *Dictyostelium discoideum:* a model system for the study of cellular differentiation. In *Essays in Biochemistry* (Eds P.N. Campbell & F. Dickens), **7,** 87–126. Academic Press, New York & London.

[15] MCMAHON D. (1973) A cell-contact model for cellular position determination in development. *Proc. Nat. Acad. Sci., USA.,* **70,** 2396–400.

[16] MALCHOW D. & GERISCH G. (1973) Cyclic AMP binding to living cells of *Dictyostelium discoideum* in presence of excess cyclic GMP. *Biochem. Biophys. Res. Com.,* **55,** 200–4.

[17] MALCHOW D., NÄGELE B., SCHWARZ H. & GERISCH G. (1972) Membrane-bound cyclic AMP phosphodiesterase in chemotactically responding cells of *Dictyostelium discoideum. Eur. J. Biochem.,* **28,** 136–42.

[18] *SHAFFER B.M. (1962 & 1964) The Acrasina. *Adv. Morphogenesis,* **2,** 109–82; **3,** 301–22.

[19] STEINBERG M.S. (1970) Does differential adhesion govern self-assembly processes in histogenesis? Equilibrium configurations and the emergence of a hierarchy among populations of embryonic cells. *J. Expl. Zool.,* **173,** 395–434.

[20] *WRIGHT B.E. (1973) *Critical Variables in Differentiation.* Prentice-Hall, Englewood Cliffs, New Jersey.

Chapter 4.3
Hormones and
Differentiation in Plants

4.3.1 INTRODUCTION

It is a characteristic of higher plants that there is a division of labour between the various parts, expressed as a difference either in form or function, or both. Such specialization has, as a prerequisite, mechanisms for ensuring that the growth and development of the constituent parts occur in an orderly and co-ordinated fashion. The very fact that growth and development *are* highly integrated processes implies complex interactions between parts of the organism. Further, while the pattern of a plant's development and the final form which it will assume must in the last analysis be a reflection of the genetic material encoded in its cells, equally, it is highly improbable that the organizational processes involved can be achieved by each cell acting as an individual in isolation.

At the simplest level it is clear that because of specialization certain essential nutrients will be synthesized or taken up in particular areas and translocated to others. Such differences extend not only to the obvious cases of sugars from the leaves and mineral nutrients from the roots but also to such factors as vitamins, whose synthesis appears to be restricted to the leaves, and certain amino acids which may only be produced in the roots or shoots respectively. Nevertheless, whereas the continuance of normal functions depends on such interchanges as these, in the main they do not appear to determine how or where growth and development will proceed.

The demonstration by Darwin of the dependence of phototropic curvature in coleoptiles on the transmission of a stimulus from the tip to the growing zone at the base was the first evidence for the directional movement of a substance with growth regulating properties at a specific locus. This and later work led to the use of the term *growth hormone* in plant physiology, implying a substance active in small quantities and having its effect at a point distant from its site of production. In the years that have followed it has become clear that most if not all processes in plants are regulated to some extent by hormones, from the day-to-day control of growth and development to changes induced by both seasonal and non-seasonal environmental fluctuations (see Chapter 6.1).

This account is concerned with the effects of substances produced internal to the plant but extrinsic to the cells or tissues whose development they may affect, as distinct from factors intrinsic to the cell whose role is discussed in Chapter 1.4. It goes without saying that hormones are the principal factors relevant to this study although we should not lose sight of the fact that other substances, such as those referred to earlier, fall into the category of extrinsic factors as we have defined them and as we shall see, they may be of considerable significance in development, especially when they interact with hormones.

As mentioned above, hormones appear to be involved in most developmental processes in plants. Representatives of the five main groups of plant growth hormones are shown in Table 4.3.1 along with an abridged list of their effects. A cursory examination of the diversity of developmental phenomena in which hormones are involved shows that it would be impracticable and largely unnecessary to survey the whole range of hormonal effects in this chapter, especially since several excellent accounts are to be found in the literature [1–4]. For the same reasons it is clear that in most cases hormones cannot possibly be determining the way in which a cell or tissue may respond but rather that in most cases their role is more likely to be that of inducing certain responses, the nature of which are predetermined by factors intrinsic to the cell. This latter 'pre-programming' which would occur prior to the reception of the hormonal stimulus might, of course, itself be a consequence of hormonal control. This account therefore seeks to explore to what extent hormones determine developmental processes and how such responses may be modified in particular cases.

Another aspect which is apparent from an examination of Table 4.3.1 is that in most cases more than one hormone is involved in the mediation of particular developmental processes. Indeed in a large number of instances it appears that the levels of hormones in a tissue *relative* to one another are a more important consideration that their absolute concentrations. Examples of such interactions are described in several of the succeeding sections.

In approaching such a problem it seems appropriate to consider, first, hormonal effects on systems whose future pattern of development is least determined. The systems fitting such criteria would appear to be meristematic tissue or dedifferentiated callus.

4.3.2 ROOT AND SHOOT INITIATION

Shortly after the recognition of kinetin as a growth regulating substance involved in the control of cell division in tobacco stem

Table 4.3.1 Involvement of hormones in plant growth and development. A tick indicates that the hormone in question has been shown to affect the process in some way although not necessarily in all instances.

Type of hormone (showing a representative example of each class)	Cell elongation	Cell division	Induction of primary vascular tissue	Induction of secondary vascular tissue	Root and shoot initiation	Breaking of seed dormancy	Senescence	Abscission of flowers fruits and leaves	Fruit growth	Sex expression	Control of stomatal aperture
Auxins Indole-3-acetic acid (IAA)	✓	✓	✓	✓	✓	✓	✓	✓	✓	✓	
Gibberellins Gibberellic acid (GA$_3$)	✓	✓	✓	✓	✓	✓	✓	✓	✓	✓	
Cytokinins Zeatin		✓	✓	✓	✓	✓	✓	✓	✓		✓
Inhibitors Abscisic acid (ABA)	✓	✓			✓	✓	✓	✓	✓		✓
Ethylene	✓	✓			✓	✓	✓	✓	✓	✓	

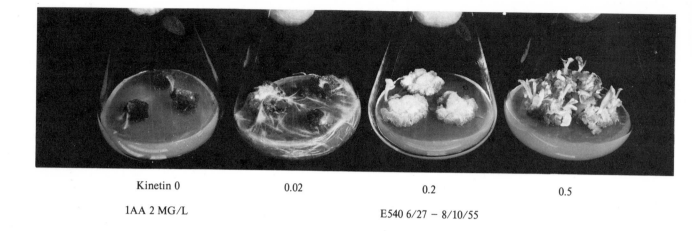

Kinetin 0   0.02   0.2   0.5

1AA 2 MG/L   E540 6/27 − 8/10/55

1.0   2.0   4.0   10.0

Fig. 4.3.1 Effect of indole-3-acetic acid (IAA) and kinetin on growth and organ formation in 44-day-old tobacco callus cultures grown on nutrient agar. A range of concentrations of kinetin were used but all cultures contained IAA at a concentration of 2 mg/l. Reprinted from *Symp. Soc. Expl. Biol.*, **11** (1957). Photograph kindly supplied by Professor F. Skoog.

callus cultures, Skoog & Miller [5] discovered that, in the same type of culture, manipulation of the ratio of auxin to kinetin in the growth medium resulted in different patterns of development. If both auxin and kinetin levels were high or low the cultures grew as amorphous, undifferentiated masses of callus. On the other hand, a high auxin to kinetin ratio led to the induction of roots in the callus and a low auxin to kinetin ratio caused shoots to be produced (Fig. 4.3.1). Since the cells of such cultures are apparently identical and may be diverted into either one of two different pathways of development merely by altering the ratio of one hormone to another it seems likely that in this case at least the hormones themselves alone determine the pattern. Of course, Skoog & Miller also found that root and shoot initiation could be modified by other substances such as sugars and amino acids in association with particular hormonal

concentrations, but these effects were quantitative rather than qualitative.

A somewhat analogous situation has been observed in the regeneration of roots and shoots from segments of *Taraxacum* and *Cichorium* rhizomes. In these experiments, wherever the segments were cut, meristematic areas were initiated and these differentiated predominantly into roots or shoots depending on whether they appeared on the morphologically lower or upper ends respectively. Here again is an example of morphologically identical tissues entering different developmental pathways depending on their position in relation to the cut.

This early work also showed that endogenous auxins tended to accumulate at the morphologically lower ends of the tissue and further that the treatment of the segments with indole-butyric acid (a synthetic auxin) led to initiation of roots at both

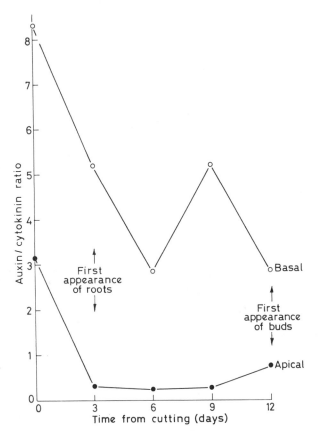

Fig. 4.3.2 Quantitative variations in auxin/cytokinin ratio in the apical and basal zones of *Cichorium* root cuttings during regeneration of roots and shoots. Adapted from J.P. Nitsch (1967) In *Wachstums regulatoren bei Pflanzen. Rostock.*

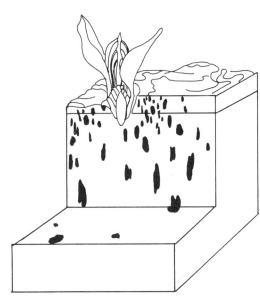

Fig. 4.3.3 Induction of vascular tissue in *Syringa* callus. An apex of *Syringa* bearing two to three leaf primordia was grafted into a block of callus. After 54 days the block was sectioned. Vascular regeneration is shown by dark bands. From R.H. Wetmore & S. Sorokin (1955) *J. Arnold Arboretum,* **36.**

4.3.3 INDUCTION OF VASCULAR TISSUE

In callus

It has been known for some time that hormones are involved in the induction of vascular tissue. Thus, both Camus and Wetmore and Sorokin showed that grafting a bud into *Syringa* callus tissue led to the formation of xylem in the callus (Fig. 4.3.3). Later, Wetmore & Rier [9] demonstrated that the same effect could be achieved by application of auxin and sucrose at discrete points in the callus, suggesting that these were the factors transmitted by the bud which are responsible for vascularization. In both systems the pattern of development was similar, that is, nodules of actively dividing cells were initiated at fairly precise but different distances from the bud or point of auxin/sucrose application; in addition, a few nodules always flanked the cuts themselves. It is significant that variations in the absolute amounts of auxin and sucrose and their relative concentrations markedly affected the induction of nodules and their subsequent pattern of development. A summary of the main findings of this work in given in Table 4.3.2.

Variation of IAA concentration above $2 \cdot 9 \times 10^{-7}$M did not appear to affect the type of differentiation but whereas at all concentrations a ring of nodules appeared, the diameter of this ring increased with increasing concentration. Since *Syringa* callus does not show polar transport of auxins the point at

ends. Later, Vardjan & Nitsch [6] investigated the same system and showed that although, after cutting, the changes in auxins and cytokinins were complex, nevertheless the alterations in their levels were such that a much higher auxin/cytokinin ratio occurred in the lower ends than in the upper ends of the cuttings (Fig. 4.3.2). The observed redistribution of auxins and cytokinins in this system reflect what we know about the transport of these substances about the plant. Thus, plant tissue usually shows a basipetal polarity with respect to auxins (that is, movement is predominantly from the apex towards the base), whereas cytokinins—which appear to be principally synthesized in root tissue—move in the opposite direction.

Thus, it appears that Skoog & Miller's findings can be applied to a natural situation of root and bud regeneration where differing auxin to cytokinin ratios are achieved chiefly by redistribution of these substances.

Table 4.3.2   Induction of vascular tissue in *Syringa vulgaris* cambial callus tissue by application of IAA and sucrose (adapted from [9]).

Concentration of substances supplied at top of callus (Molar)*		Type of development observed†
IAA	Sucrose	
0	0	Normal callus growth
$2 \cdot 9 \times 10^{-6}$	0	As above plus a very small amount of vascular tissue
0	$11 \cdot 6 \times 10^{-2}$	
$2 \cdot 9 \times 10^{-6}$	$2 \cdot 9 \times 10^{-2}$	Circle of nodules of dividing cells, a few mature xylem elements
$2 \cdot 9 \times 10^{-6}$	$5 \cdot 8 \times 10^{-2}$	Circle of nodules of dividing cells, strong xylem differentiation
$2 \cdot 9 \times 10^{-6}$	$8 \cdot 7 \times 10^{-2}$	Circle of nodules of dividing cells, strong xylem and phloem differentiation
$2 \cdot 9 \times 10^{-6}$	$11 \cdot 6 \times 10^{-2}$	Circle of nodules of dividing cells, strong phloem differentiation

* All calluses were supplied in the growth medium with a mixture of mineral ions and adenine sulphate, vitamins and nitrogen source as well as IAA at $2 \cdot 9 \times 10^{-7}$ M.
† This refers to development *above* the level of the medium only.

which differentiation occurs is presumably related to a specific concentration of auxin; indeed, similar, subsequent work by Jeffs & Northcote [7] showed that both auxin and sugar moved along diffusion gradients in callus and that the concentrations of auxin and sucrose at which differentiation occurred in *Phaseolus vulgaris* callus were $1 \cdot 4 \times 10^{-7}$ M and $2 \cdot 2 \times 10^{-2}$ M respectively. In all cases phloem forms to the outside of the xylem which many workers have taken to mean that a lower concentration of IAA favours phloem induction. In addition, examination of Table 4.3.2 shows that sugar concentration appears to determine the pattern of development. Where, instead of an agar block, micropipettes containing sugar and auxin were inserted into *Syringa* callus blocks a continuous ring of xylem and phloem separated by an organized meristem was produced (Fig. 4.3.4).

Thus, like root and shoot initiation, here is a situation where the pattern of a cell's development appears to be controlled solely by the effect of two extrinsic factors (but see below, p. 220). Also, since the same cells will either remain as undifferentiated callus or become meristematic and later develop into phloem or xylem, it seems unlikely that they are predetermined to respond in a specific way.

It should be noted that in the type of experiment just described, the xylem tissue which is formed is abnormal in that the cells are isodiametric. It has been suggested that this is due to the fact that in the growing plant the apical region (probably the main source of auxin for the uppermost internode) is constantly moving away from any one point in the stem so that any given cell behind it will receive a gradually decreasing amount of auxin. Since the optimal auxin concentration for elongation is usually much higher than that for vascularization we would expect the former process to occur nearer to the apex than the latter. Indeed, in experiments with callus where micropipettes containing auxin are inserted in the tissue some cell elongation does occur near the point of insertion. Such

findings demonstrate that in the intact plant there is a carefully defined developmental sequence depending on auxin concentration, namely elongation followed by vascularization, which is achieved, at least in part, by having a moving source of auxin production.

Because of the important role played by sugar in vascular induction Jeffs & Northcote have investigated the effect of different sugars upon this process in *Phaseolus* callus. They were able to show that of the fifteen sugars tested, only sucrose,

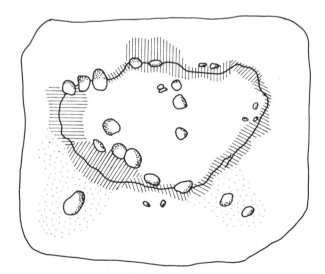

Fig. 4.3.4   Diagrammatic interpretation of a transverse section through a block of *Syringa* callus taken 450 µm below the point of application of an agar block containing 4% sucrose and 0·5 mg/l NAA (naphthalene acetic acid) after 54 days. Nodules are represented by irregular small circles, the cambium by a black line and cellular regions laid down by cambial activity by radially placed parallel lines. Areas of haphazard mitotic activity are shown by stippling. From R.H. Wetmore & J.P. Rier (1963) *Amer. J. Bot.*, **50.**

trehalose and maltose induced organized nodules and xylem, and of the remainder only cellobiose, lactose, glucose and fructose and the last two combined produced xylem but not nodules. Sucrose does not therefore appear to be necessary merely as an energy source—glucose would be just as effective in this respect and certainly it can replace sucrose as the energy source in maintaining callus cultures. Further, there appears to be a requirement for a specific glycosidic configuration in that sucrose, trehalose and maltose have an α-glucosyl residue at the non-reducing end. Such specificity would imply an interaction with a specific binding site. Torrey has noted a similar dependence on sucrose for the induction of xylem strands in roots.

In intact plants

It seems logical to continue this discussion with a consideration of vascular regeneration or induction in an intact plant, since it does not follow that these processes show the same characteristics there as in undifferentiated callus. The pattern of vascular development in any one higher plant may differ very substantially from that in another. Can this development of pattern be controlled solely by the hormone system, or do other systems of organization take part? It is inappropriate to deal extensively with this question here except insofar as it relates to the question of to what extent hormones are determining

development, and we must take this to include not only the qualitative nature of the tissue produced but also the pattern in which it becomes arranged.

It has been known for some time that IAA is an important factor in the regeneration of vascular bundles after wounding. Thus, Jacobs showed that removal of young leaves above a wound inhibited xylem regeneration but that supplying IAA to the petiole of such an excised leaf overcame this effect. LaMotte & Jacobs later showed that similar systems were operative for phloem regeneration (the pattern of this type of regeneration is shown in Fig. 4.3.5).

Further studies by Jacobs and his collaborators [10, 11], using mainly *Coleus* tissue, showed that there was also an excellent correlation between the rate of normal xylem and phloem differentiation and the rate of auxin production by the leaf which that vascular strand would serve. These workers also demonstrated that in addition to the normal basipetal

Fig. 4.3.5 Phloem regeneration around a wound made seven days earlier in one side of an internode of *Coleus blumei*. The wound (centre triangular-shaped, light area) severed three phloem bundles. Photograph kindly supplied by Dr C.E. LaMotte.

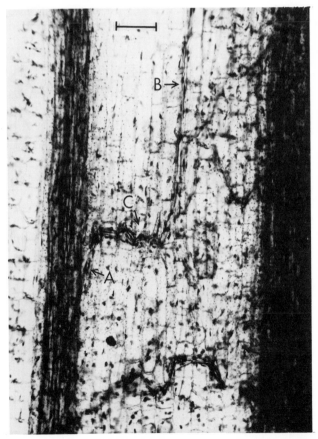

Fig. 4.3.6 Longitudinal section through a stem of *Coleus blumei* showing acropetally (A) and basipetally (B) developing vascular strands joined by vascularized parenchyma tissue (C). From C.E. LaMotte & W.P. Jacobs. *Stain Technol.*, **37**, 63–73. Photograph kindly supplied by Dr C.E. LaMotte.

development, some development of xylem strands occurred acropetally (that is, towards the apex) from the next lower internode and this is in agreement with their demonstration of a small acropetal movement of auxin in *Coleus*. It was also observed that the developing basipetal and acropetal strands did not always meet directly but usually ceased development as they reached the same level, at which point the intervening parenchyma cells differentiated into xylem elements thus joining the two strands (Fig. 4.3.6). In addition, it was found that parenchyma cells required approximately ten times more auxin in order to differentiate into xylem elements during regeneration experiments than did procambial tissue in normal development. It is suggested that this provides a mechanism for the selection of which cells will vascularize in that only where a localized increase in auxin concentration occurs—for example, due to the interruption of normal translocation by wounding—will cortical tissue differentiate into xylem. The formation of bridges between basipetally and acropetally developing strands can be explained in the same way if these strands are transporting auxin as Jacobs suggests.

It is interesting to compare this work with that of Sachs [12, 13] who has studied the effect of IAA upon the induction of vascular tissue in pea plants. Using a variety of elegant surgical techniques he was able to show that vascular strands from buds connected preferentially to other xylem strands from buds which were previously removed. He demonstrated a similar effect in roots with differential application of auxin apically and laterally. Thus, where vascular strands were induced to form by a lateral application of auxin these joined the vascular cylinder at a wide angle if the latter was not supplied with the auxin and at an acute angle if it was so supplied (Fig. 4.3.7). In other words, vascular strands well supplied with auxin tend to repel other strands whilst those deficient in auxin tend to attract them.

This observation would account, at least in part, for the patterns of vascular arrangement which may be observed. For example, it would explain the presence of leaf gaps in the primary body as being due to inhibition of a basipetally-developing vascular strand within the stem by the leaf strands which would be well supplied with auxin during leaf expansion (see Fig. 4.3.8). Sachs also showed that repulsion of an induced strand by another strand well supplied with auxin only occurred if the latter was supplied with a higher concentration of auxin than the former, suggesting that repulsion was not due to a supra-optimal concentration of auxin in the vicinity of the main strand. He explains repulsion by supposing that vascular elements well supplied with auxin induce longitudinal polarity in the tissue adjacent to them so that auxin in a neighbouring induced strand cannot pass at right angles through such cells and cause xylem formation by so doing. 'Attraction' is explained in an analogous fashion, by vascular tissue deficient in auxin acting as a sink for auxin from an attached strand.

Sachs summarizes his findings in a general hypothesis suggesting that, once formed, vascular tissue offers a preferred pathway for the movement of the inducing stimulus; which is another way of saying that the tissue becomes polarized—at least with respect to auxin transport. He further suggests that this polarity be imposed on adjacent tissues.

We can summarize the work described in this section by observing that auxin may be a limiting factor in the induction of vascular tissue in both callus tissue and in higher plants. However, whereas in callus auxin clearly determines the site of vascular induction and appears to influence (in association with sucrose) the type of tissue formed, in the intact plant constraints are imposed on the locus of induction. These constraints appear to derive partly from the nature of the cells themselves and partly from their polarity and the polarity of adjacent tissues.

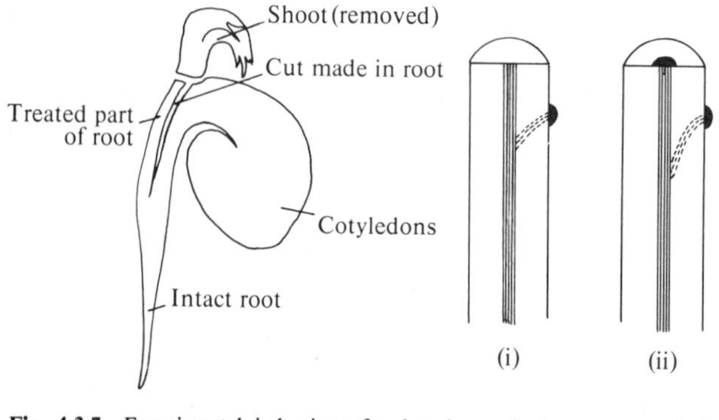

Fig. 4.3.7 Experimental induction of xylem by auxin in pea roots. Left. Diagram of three-day-old pea seedlings showing the cuts made to separate the part of the root used for experiments. Right. Treated roots in face view. (i) Auxin applied laterally only. (ii) Auxin applied laterally and at the apex. The dotted line shows newly-formed xylem. From T. Sachs (1968) *Ann. Bot.*, **32**.

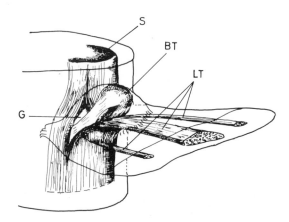

Fig. 4.3.8 Diagram of nodal region of *Phlox* showing stele (S), bud trace (BT), leaf traces (LT), and leaf gap (G). After Miller & Wetmore from Earres & McDaniels, *An Introduction to Plant Anatomy*, McGraw-Hill.

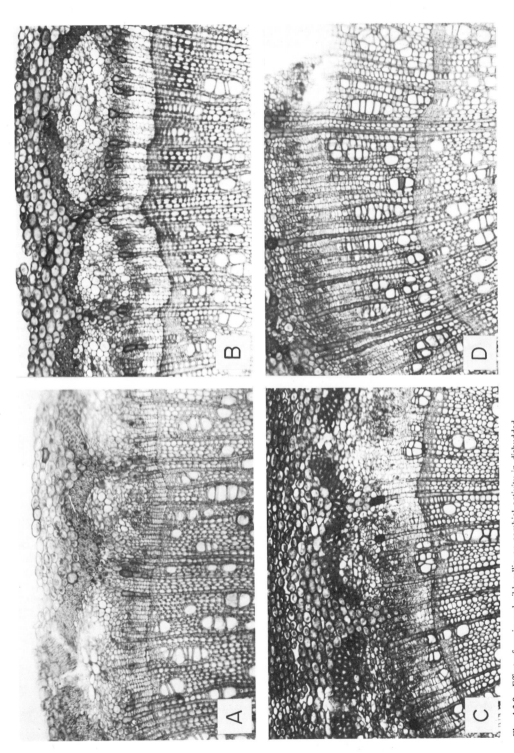

Fig. 4.3.9 Effect of auxin and gibberellin on cambial activity in disbudded poplar (*Populus robusta*) twigs. A. Untreated control; B. Treated with gibberellin (GA₃) only; C. Treated with indole-3-acetic acid (IAA) only; D. Treated with IAA + GA₃ in combination. Photographs supplied by Professor P.F. Wareing.

4.3.4 INDUCTION OF SECONDARY THICKENING

We have so far only considered the role of hormones in the control of primary tissue formation, but it is now well established that these substances are involved in the formation of secondary tissues, in particular secondary vascular tissue.

It has long been known that the appearance of developing buds on deciduous trees in the spring is accompanied by the onset of cambial division in the trunk and differentiation of the cells produced by this cambial activity. This observation together with work indicating that developing buds are rich sources of hormones led a number of workers to suggest that these substances were involved in the initiation of cambial activity. The work of Wareing *et al.* [16] on this subject showed that apical application of hormone mixtures to disbudded shoots of *Acer pseudoplatanus* led to cambial division and the differentiation of xylem and phloem. When IAA was applied alone, isolated areas of xylem differentiation were observed. With gibberellic acid alone, there was extensive division internal to the cambium but the newly formed tissue did not differentiate. The effect of the two hormones applied together was to cause extensive cambial division and differentiation into xylem elements. Some division also occurred to the outside of the cambium with this treatment (Fig. 4.3.9). Further studies with IAA and GA_3 alone or in different combinations indicated that both hormones were required for normal development of xylem and phloem but that little or no phloem differentiation occurred with IAA alone and vice versa for xylem and GA_3. Similar observations were made with herbaceous plants and woody conifers.

Clearly these results resemble many of the findings on the induction of primary vascular tissue, and it seems that, in the main, hormonal application leads both to cambial division and determination of the cambial derivatives, whether we are considering primary or secondary vascular tissue formation. Nevertheless, whereas in callus tissue hormone treatment not only stimulates cambial division but also *initiates* the cambium itself—in secondary thickening we are often dealing with a pre-existent cambial zone, and the different aspects introduced by this additional factor are well illustrated by the work of Siebers [14] with hypocotyl tissue of *Ricinus communis*.

These studies showed that when blocks of interfascicular tissue were removed from the hypocotyls and then replaced in the inverse direction that not only did the interfascicular cambium form in the same layer of cells but that, furthermore, its derivatives differentiated into xylem on the outside and phloem on the inside (Fig. 4.3.10). Plugs of pith parenchyma inserted in place of interfascicular tissue only showed differentiation, and then to a very slight extent, when the outer cells of the graft had formed wound tissue, i.e. when they had dedifferentiated to some extent. Siebers' conclusion, which

seems inescapable, is that the cells in which the interfascicular cambium will arise are already determined at the stage of formation of the primary meristem ring, not only in their position but also in the fate of the products of their division—that is, they exhibit a radial polarity.

Although space does not permit discussion of other work on this subject it is important to realize that normal development appears to depend on a very complex interaction of factors and that in addition to auxin and sucrose other hormones and sugars appear to play significant determinative roles (cf. Table 4.3.1). It is not yet clear whether the wide range of responses observed represent real differences in the requirements of specific plants or organs or whether they relate to differing levels of particular hormones within such tissue so that the determinative factor depends on that hormone in shortest supply.

4.3.5 MOBILIZATION OF RESERVES IN CEREAL ENDOSPERM

While the work described so far has been useful in establishing that hormones can induce differentiation, it tells us nothing about the nature of the controls exercised by hormones at the subcellular or molecular level which underlie determination. On the other hand there is a wealth of information about the biochemical changes induced by hormones in a variety of systems where cell division and/or differentiation are not involved. By studying such systems it may be possible to identify the level of cellular organization at which hormones have their primary effects and perhaps in so doing to suggest possible mechanisms for hormone action in differentiation itself. The discussion will be limited to what are probably the two best-documented systems.

Early workers on the control of mobilization of reserve materials in the endosperm in cereals tended to stress the metabolic inertness of the endosperm itself and suggested that the necessary enzymes were secreted by the germinating embryo. However, in 1958 Yomo showed that starch in isolated barley endosperm would undergo hydrolysis if the latter was incubated in the same culture flask as isolated embryos. These latter did not themselves produce significant amounts of amylases and hence they must have been secreting some other factor which affected the endosperm, causing it to produce enzymes. Yomo purified this factor inducing amylolytic activity and also showed that comparable amounts of a GA_1/GA_3 mixture could produce the same effect. At the same time, but quite independently, Paleg demonstrated the same phenomenon [18].

By various surgical treatments MacLeod & Millar have shown that the aleurone layer of the endosperm is essential for the appearance of amylolytic activity. This layer, present in the seeds of all cereals, surrounds the endosperm and has a very

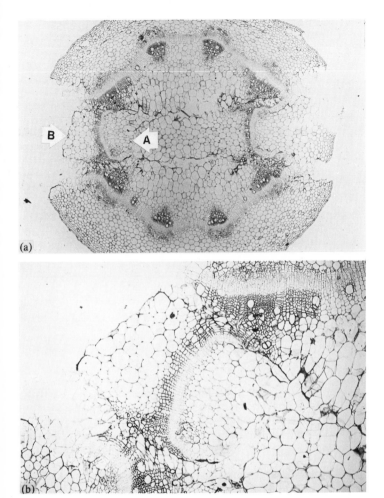

Fig. 4.3.10 (a) Transverse section through 17-day-old hypocotyl of *Ricinus communis* with an inverted plug of interfascicular tissue (BA) 10 days after inversion. At the time of inversion the seedling showed no signs of secondary growth. (b) Detail of (a) showing inverted tissue block and surrounding tissue. Note the inverted position of the xylem. Photographs kindly supplied by Dr A.M. Siebers.

characteristic appearance. The cells have very thick walls and contain prominent structures termed 'aleurone grains' (see Fig. 4.3.11). Later work showed clearly that isolated aleurone layers release α-amylase in response to treatment with GA_3. Furthermore, the ultrastructural changes induced by GA_3 treatment in isolated aleurone layers are identical to those occurring in intact tissue during germination (see below). Several workers have also shown that gibberellins are produced by barley seeds in the early stages of germination. The timing and precise site of synthesis are not entirely clear but it seems

likely that the scutellum produces gibberellin in the first two days after imbibition and the axes one to three days later [19].

Thus, the work on enzyme formation and gibberellin production taken together suggest that during germination gibberellins are transported from the embryo to the aleurone layer where they induce far-reaching physiological and morphological changes. Obviously this is a classical case of hormonal effect. What then can we say about the nature of the changes induced by gibberellins?

The time course of the development of α-amylase activity shows a lag period of some 8–10 h (see Fig. 4.3.12), after which a steady rate of enzyme production is maintained for some 30 h. Most of the enzyme is released into the incubation medium. In addition to α-amylase a number of other enzyme activities

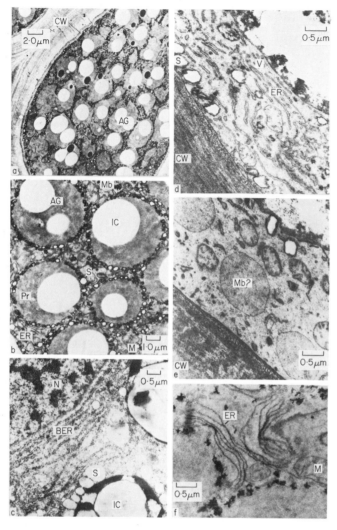

Fig. 4.3.11 (a, b) Section through imbibed wheat aleurone tissue fixed in KMnO₄. CW, wall; AG, aleurone grains; pr, protein; Mb, microbody (glyoxysome); IC, internal cavity (electron transparent zone probably left by loss of protein-carbohydrate complex during fixing/sectioning); S, spherosome; M, mitochondria; ER, short segments of endoplasmic reticulum. (c, d, e) Section through wheat aleurone tissue four days after imbibition, fixed in KMnO₄. ER, stacks of endoplasmic reticulum; V, vesicles budding from ER; N, nucleus; Mb, microbody (glyoxysome), note much larger size than in b. Other lettering as for a, b. (f) Section through wheat aleurone tissue from embryoless half seed four days after imbibition. ER, stacks of endoplasmic reticulum, but note that vesiculation does not occur (cf. d). (b, c, d and f) from D.L. Laidman, A.J. Colborne, R.I. Doig & K. Varty (1974) In *Mechanisms of Regulation of Plant Growth* (Eds R.L. Bieleski, A.R. Ferguson & M.M. Creswell), *Bulletin* **12**. The Royal Society of New Zealand, Wellington. (e) from R.I. Doig, A.J. Colborne, G. Morris & D.I. Laidman (1975) The induction of glyoxysomal enzymes activities in the aleurone cells of germinating wheat. *J. Expl. Bot.* Photographs kindly supplied by Dr D.I. Laidman.

appear in response to treatment, namely protease, ribonuclease, endo-β-glucanase, pentosanase, acid phosphatase, isocitratase and malate synthetase.

Much early work attempted to determine whether these various enzyme activities arose by release from bound forms or as a consequence of *de novo* synthesis. Many inhibitors of protein and nucleic acid synthesis, provided they were added early enough and under the right conditions, were shown to inhibit α-amylase production and it was also demonstrated that ¹⁴C amino acids are incorporated into α-amylase in the presence but not the absence of GA₃. However, doubts still remained that perhaps the effect of inhibitors was an indirect one. The problem was finally solved by using the technique of 'density labelling' developed by Filner. In this method seeds are imbibed in H₂¹⁸O and treated with GA₃. During breakdown of reserve protein the amino acids which are produced become labelled with ¹⁸O and are utilized for protein synthesis; thus any α-amylase produced under these conditions is denser than that which might be pre-existent in the tissue which would only contain ¹⁶O (see equation below). The two forms may then be separated by prolonged centrifugation.

$$\text{Reserve Protein} \xrightarrow[\text{proteolysis}]{H_2{}^{18}O} \begin{array}{c} R.CH.NH_2C^{16}O^{18}OH \\ \Updownarrow \\ R.CH.NH_2C^{18}O^{16}OH \end{array} \longrightarrow \begin{array}{c} \text{amylase} \\ {}^{16}O^{18}O \end{array}$$

Using this technique it was shown that at least the α-amylase and protease components arise as a consequence of *de novo* synthesis and further work demonstrated that there are in fact at least four α-amylase isozymes in the barley system which all arise by *de novo* synthesis. This work also showed that the β-amylase components which had also been detected are already present as zymogens.

However, proof that GA₃ induces enzyme synthesis in aleurone layers takes us only one step further forward since there are numerous possible points at which control over enzyme synthesis may be exercised [20]. Subsequent attempts to define the primary site of the GA₃ effect have taken two basic approaches, namely (1) studies on the kinetics of enzyme production and release, especially in relation to the effects of specific inhibitors of protein and nucleic acid synthesis, or (2) investigations on changes in various macromolecular components in the aleurone layer, in particular nucleic acids.

A very significant observation in this work was that the system has a continuous requirement for GA₃ and a rapid decrease in enzyme production results from washing the aleurone layers free of the hormone. On the other hand if returned to GA₃ such tissues recommence enzyme production without a significant lag (Fig. 4.3.13).

Given the right conditions, inhibitors of protein and nucleic acid biosynthesis will inhibit α-amylase production.

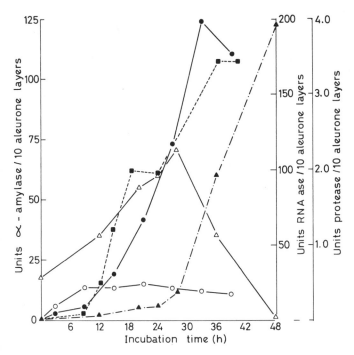

Fig. 4.3.12 Time course of synthesis and release of α-amylase, protease and ribonuclease from barley aleurone layers during incubation in gibberellic acid, α-amylase: in aleurone layers (○), in medium (●); protease: in medium (■); ribonuclease: in aleurone layers (△), in medium (▲). Adapted from J.E. Varner & G.R. Chandra (1964) *Proc. Nat. Acad. Sci. (U.S.)*, **52**; J.V. Jacobsen & J.E. Varner (1967) *Plant Physiol.*, **42**; M.J. Chrispeels & J.E. Varner (1967) *Plant Physiol.*, **42**.

Nevertheless, whereas cycloheximide (an inhibitor of protein turnover) 6-methyl purine, azaadenine and bromouracil (inhibitors of RNA synthesis) have such effects even if added some considerable time after the addition of GA₃, actinomycin D does not affect α-amylase production if added 8 h after GA₃, even though it does inhibit RNA synthesis if supplied at this time. Since actinomycin D specifically inhibits DNA-dependent RNA synthesis this seems to indicate that such processes (i.e. transcription) are only important in the early phases of GA₃ action, although the work with the other inhibitors suggests that there is a continuous requirement for other processes involving nucleic acid and protein synthesis.

What is happening to the cell as a whole while these dramatic biochemical changes are occurring? Laidman and his co-workers [17a, b] have followed the ultrastructural changes in wheat and attempted to relate these to biochemical events (Fig. 4.3.11). These results are similar whether the aleurone tissue is derived from whole germinating seeds or from embryoless half seeds treated with GA₃. The only difference observed is one of

time scale, being much more rapid in the latter case. The aleurone grains are surrounded by a unit membrane and contain three distinct areas; an electron transparent zone and two electron dense zones one of which contains protein and the other phytin (the mixed potassium, magnesium and calcium salt of inositol hexaphosphate). On the outside of the unit membrane are many lipid-containing spherosomes. The endoplasmic reticulum is present as short segments (Fig. 4.3.11b). As germination proceeds the spherical appearance of the aleurone grains is lost and there is an increase in their volume. This is followed by a proliferation of rough endoplasmic reticulum (RER) (Fig. 4.3.11c), followed by distension of cisternae of the same and a marked vesiculation of the RER and dictyosomes (Fig. 4.3.11d). The microbodies (glyoxysomes) appear to increase in number and size (Fig. 4.3.11e). This would appear to be borne out by the observation that the appearance of glyoxysomal enzyme activities such as isocitratase and malate synthetase are dependent on the presence of the embryo. The volume of the aleurone grains as a whole decreases and the number of spherosomes is also reduced. Eventually the fibrillar nature of the cell wall is lost.

There seems to be good agreement between the biochemical and ultrastructural studies. Thus the observed decrease in

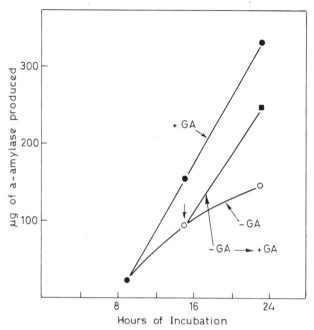

Fig. 4.3.13 Effect of gibberellic acid (GA₃) added and removed at different times upon production of α-amylase by barley aleurone layers. All aleurone layers incubated for 7 h in 0·5 μm GA₃, then either in GA₃ for a further 16 h (●), or GA₃ removed by four consecutive 30 min rinses (O), or GA₃ added back at 15 h (■). From M.J. Chrispeels & J.E. Varner (1967) *Plant Physiology*, **42**.

protein nitrogen is reflected in a breakdown of the protein fraction of the grains and the changes in endoplasmic reticulum are in accord with the occurrence of active protein synthesis. It should be noted however, that the formation of endoplasmic reticulum itself does not appear to be gibberellin dependent (Fig. 4.3.11f). Similarly, by analogy with other systems the behaviour of the RER and dictyosomes seems to infer a secretory mechanism with the sort of selective properties which are exhibited by the aleurone system.

Clearly GA₃ plays a multifaceted role in this system since not only does it initiate complex cellular reorganization but subsequently controls the production of selected enzymes, the biosynthesis of at least one of which requires its continued presence. Nevertheless other hormonal factors appear to be important in the full expression of the developmental response, i.e. the mobilization and utilization of the endosperm reserves. Thus Laidman and his co-workers [17a] have shown that cytokinins, probably derived from the endosperm itself, initiate mobilization of at least part of the triglyceride reserves and promote retention of macronutrient mineral ions in the aleurone tissue. IAA in association with glutamine appears to control mobilization of another part of the triglyceride reserves, induces

neutral lipase activity and activates phytase (which degrades phytin). Indeed, imbibition itself, irrespective of hormonal factors leads to some changes in the tissue—for example the *initiation* of phytin hydrolysis and the formation of endoplasmic reticulum. These findings together with those of other workers showing that ethylene may promote the release but not the synthesis of α-amylase and that absciscic acid specifically inhibits GA₃-induced phenomena in barley aleurone suggests that the whole range of processes observed represent a complex and highly co-ordinated system whose controls emanate from different parts of the organism in a defined sequence.

Despite the extensive work on this system we still cannot say with any confidence precisely how GA₃ exerts its effects although it is at least possible to define the processes which are involved. The results with inhibitors of nucleic acid and protein synthesis are consistent with the hypothesis that the expression of the GA₃ effect requires the synthesis of enzyme-specific RNA molecules, but the search for such RNA species has not proved very rewarding. Zwar & Jacobsen have shown a correlation between incorporation of uridine or adenosine into a specific RNA fraction (constituting less than 1% of total RNA) and the appearance of α-amylase activity both in the presence and absence of inhibitors [20a]. Whether this RNA species does represent m-RNA for hydrolase formation, as the authors suggest, must await considerable advances in our understanding of nucleic acid biochemistry in higher plants. Even so, the fact that DNA-dependent RNA synthesis is only necessary in the early stages (see above p. 227) whereas there is a continuous requirement for GA₃ suggests that the latter plays other roles.

In conclusion it is clear that endogenous gibberellins produced by the barley embryo are an essential part of the control mechanism whereby the breakdown of reserves in the endosperm is synchronized and co-ordinated with the initiation of growth in the embryo. However, the special type of response elicited by the hormone in the aleurone layer is evidently determined by the 'preprogramming' of the target tissue, since the same hormone will produce quite different effects in other tissues (p. 217). Thus, in this system the hormone does not appear to be determining the pattern of development, as appears to be the case in the induction of buds and roots and of vascularization described in the preceding sections.

4.3.6 STEM AND COLEOPTILE EXTENSION

The final system which it is appropriate to consider is, in some respects, the best known—namely elongation of stem or coleoptile segments in response to added hormones. When sections of coleoptiles are cut from an intact organ and placed in solutions of auxins, such as IAA, growth is greatly stimulated, and the system has long been used for assaying auxin activity in plant extracts. The growth in such sections is entirely by cell enlargement, and involves extensive water uptake and cell wall growth. However, Wright has shown that the sensitivity of coleoptile sections to various types of hormone varies with their age, and they are most sensitive first to cytokinins, then gibberellins and finally auxins. It is well known, however, that if sections are excised during the latter stage then their rate of elongation falls dramatically compared to that obtaining in the intact plant. Incubation of such sections in an auxin solution of appropriate concentration leads to an increased rate of elongation after a lag of some 10–15 min [21]. This lag period presumably represents the time taken for auxin to penetrate the tissue, reach its site of action, initiate whatever metabolic processes are necessary for the acceleration of growth and also for these processes to begin to have an effect. The system requires a constant supply of auxin and the growth rate decreases rapidly if this is removed.

Using a variety of inhibitors, a number of workers have shown a requirement for continued nucleic acid and protein synthesis during growth. However, inhibitors of nucleic acid and protein synthesis either do not affect the length of the time lag before growth accelerates or may even shorten it.

The shortness of the lag time for growth induction makes it unlikely that an effect on transcription is involved here and the effects of inhibitors suggest that while nucleic acid and protein biosynthesis are necessary for continued extension they are not involved in the processes operative in re-initiating it in the first place. Further, no one has yet convincingly demonstrated rapid increases in biosynthetic or hydrolytic enzymes—which are necessary for continued growth—in response to auxin treatment. In fact, if *Avena* coleoptiles are incubated in auxin but prevented from extending by the presence of a high concentration of mannitol then the increase in the activity of cellulose synthetase usually observed with auxin treatment does not occur. This seems to indicate that many of the effects of auxin on this system are indirect, that is, they are induced by the increased growth rate itself, which in turn requires a constant supply of auxin.

Recent work has shown that a reduction in the pH of the incubation medium leads to an almost instantaneous, albeit transitory, increase in the rate of coleoptile elongation. More recently still, Rayle [22] has shown that very soon after its application, IAA increases the activity of a proton pump in *Avena* coleoptiles. It is proposed that the resultant fall in pH activates an enzyme or enzymes cleaving linkages within the cell wall, resulting in a change in mechanical properties thus allowing additional extension. These observations indicate that this may be the mode of action of auxin in this system. Precisely how such rapid changes might be brought about by hormones is not known, although they might be caused by effects on membrane permeability. Indeed, such effects appear to be involved in some other instances. For example, in the control of the closure of stomata by abscisic acid the hormone appears to act by causing a very rapid efflux of potassium and organic acids from the stomatal guard cells.

The analogies between the coleoptile system and the results on sequential addition and removal of GA_3 in the barley aleurone system are obvious (see p. 226). Presumably, in the case of the coleoptile, the process inducing the proton pump, that is, the determinative step, occurs before the phase of auxin sensitivity, so that by that time the cellular apparatus mediating the first step in the developmental process—growth—is already organized and merely awaits the hormonal signal.

4.3.7 CONCLUSIONS

Is it possible to come to any conclusions as to the role of hormones in either the channelling of plant tissues into particular pathways of development or in the initiation of pre-determined events? It seems appropriate to consider this question on two levels. In the first instance it seems likely that hormones, often in association with other substances are capable of initiating meristematic areas. It is also clear that the subsequent differentiation of cells produced by such meristems may also be controlled by hormonal stimuli, although in this case the hormones may only represent the signal for cells to develop in a way for which they are already programmed. Thus the observed distribution of phloem to the outside of the xylem would be a consequence of the presence of a radial polarity in the cambium and not directly of hormone concentration on the cells derived from it, although part of the determinative process in cambium formation might involve induction of a radial polarity with respect to the transport of hormones or other substances, thus affecting the supply of these factors to the cambial derivatives on either side. Since single cells in suspension culture can be induced to differentiate into xylem elements by judicious application of cytokinin and auxin it would appear that cells do not require to be programmed to respond in this way to hormones but rather that their position may determine what balance they will receive.

Similar hypotheses could account for the induction of primary vascular tissue in the work of both Sachs & Jacobs. Thus, both workers observe that promeristematic tissues or those adjacent to the vascular cylinder vascularize preferentially, and it is immaterial whether we attribute this to inherent polarity or sensitivity. Experimentally induced vascularization in cortical tissue probably represents an abnormal response although it is significant that here higher concentrations of auxin are required, whether applied artificially or derived from an expanding leaf. In any case regeneration after wounding presumably represents a homeostatic mechanism such that tissue only vascularizes when normal conditions are disturbed and we cannot exclude the possibility that cortical tissue is programmed to respond to auxin in this way when a certain concentration is exceeded. This would apply equally to vascular strands from expanding leaves.

Hypotheses like those outlined above do not of course explain how the observed changes may be brought about. The barley aleurone work clearly implicates nucleic acid metabolism at some level, at any rate in the initiation of the process, but are all hormonal effects necessarily at the transcriptional level? The results with cell elongation would suggest not. Any hypothesis has to account for the enormous variety of hormonal responses, but clearly the relative simplicity and paucity in the numbers of growth hormones together with the low concentrations of these which are effective, seem to require mechanisms not only for amplification but also for translating a message composed of relatively few pieces of information into a signal initiating a complex developmental sequence. It seems to the author that there are two principal alternatives, although these are not mutually exclusive. One hypothesis which has some experimental support is that hormones bind to specific proteins. These could be enzymic or not and through them control could be exercised at any of the possible stages from RNA transcription through to selection or regulation of particular metabolic pathways. Such a hypothesis would also accommodate the frequently observed requirements for specific concentrations or combinations of hormones, since such effects could be explained by supposing that particular proteins have different binding characteristics. Alternatively, the sequential pattern of gene activation observed during development could be achieved by the triggering of one stage by that preceding it so that any given process would only need the activation of relatively few sections of the genome in the first place—perhaps by a hormone.

Either of these alternatives would imply rather different modes of action of hormones in controlling differentiation of cells or tissues. In the first case presumably determination would involve the presence of specific proteins in particular cells. In the other, a restriction of template activity would be imposed to ensure that only specific areas could be transcribed in response to a hormonal stimulus. A combination of both possibilities would be equally if not more attractive in that it would accommodate both the type of result described above and such hypotheses as those of Bonner regarding the role of histones—basic proteins which are known to prevent transcription in some cases by masking portions of the genome (p. 338).

These two possibilities are by no means exhaustive and indeed only provide partial answers since if, for example, specific binding sites *are* required and if these *are* proteins then how is the production of these controlled by hormones in those instances where the latter appear to determine patterns of development? Such a system would require that hormones can not only react with binding sites and initiate developmental responses but that in some cases these very binding sites are themselves initiated by hormones in the first place.

While it is true that we do not have an *experimentally proven* explanation for the mode of action of hormones in the control of plant development, nevertheless the picture is in many ways much clearer than it was 25 years ago, largely because we have come to understand more about control mechanisms in general. Similarly, we now know that hormonal effects range from cases where these substances clearly determine the developmental response through a whole spectrum of situations where a complex interaction occurs between the hormone(s) and factors intrinsic to the plant tissue itself. Further advances are dependent on developments both technical and conceptual not only in the hormone field but also in other aspects of plant physiology and biochemistry.

4.3.8 REFERENCES

[1] AUDUS L.J. (1973) *Plant Growth Substances*, 3rd edition. Leonard Hill Ltd., London.

[2] LEOPOLD A.C. (1964) *Plant Growth and Development*. McGraw-Hill, New York.

[3] WAREING P.F. & PHILLIPS I.D.J. (1970) *The Control of Growth and Differentiation in Plants*. Pergamon Press.

[4] WILKINS M.B. (Ed.) (1969) *Physiology of Plant Growth and Development*. McGraw-Hill, London.

[5] SKOOG F. & MILLER C.O. (1957) Chemical regulation of growth and organ formation in plant tissues cultured *in vitro*. *Symp. Soc. exp. Biol.*, **11**, 118–31.

[6] VARDJAN M. & NITSCH J.P. (1961) La regénération chez *Cichorium endiva* L. étude des auxines et des 'kinines' endogènes. *Bull. Soc. Bot., France*, **108**, 363–74.

[7] JEFFS R.A. & NORTHCOTE D.M. (1967) The influence of indole 3-yl acetic acid and sugar on the pattern of induced differentiation in plant tissue culture. *J. Cell Sci.*, **2**, 77–88.

[8] WETMORE R.M., DEMAGGIO M. & RIER J.P. (1964) Contemporary outlook on the differentiation of vascular tissues. *Phytomorphology*, **14**, 203–17.

[9] WETMORE R.H. & RIER J.P. (1963) Experimental induction of vascular tissues in callus of angiosperms. *Am. J. Bot.*, **50**, 418–30.

[10] JACOBS' W.P. (1952) The role of auxin in differentiation of xylem around a wound. *Am. J. Bot.*, **39**, 301–9.

[11] JACOBS W.P. & MORROW I.B. (1957) A quantitative study of xylem development in the vegetative shoot apex of *Coleus*. *Am. J. Bot.*, **44**, 823–42.

[12] SACHS T. (1968) The role of the root in the induction of xylem differentiation in peas. *Ann. Bot.*, **32**, 97–117.

[13] SACHS T. (1968) On the determination of vascular tissue in peas. *Ann. Bot.*, **32**, 781–90.

[14] SIEBERS A.M. (1971) Initiation of radial polarity in the interfascicular cambium of *Ricinus communis* L. *Acta. Bot. Neerl.*, **20**, 211–20.

[15] TORREY J.G. (1963) Cellular patterns in developing roots. *Symp. Soc. Expl. Biol.*, **17**, 285–314.

[16] WAREING P.F., HANNEY C.E.A. & DIGBY J. (1964) The role of endogenous hormones in cambial activity and xylem differentiation. In *The Formation of Wood in Forest Trees* (Ed. M.H. Zimmerman), pp. 323–44. Academic Press, New York.

[17a] LAIDMAN D.L., COLBORNE A.J., DOIG R.I. & VARTY K. (1974) In *Mechanisms of Regulation of Plant Growth* (Eds R.L. Bieleski, A.R. Ferguson & M.M. Creswell) *Bulletin* **12**. The Royal Society of New Zealand, Wellington.

[17b] DOIG R.I., COLBORNE A.J., MORRIS G. & LAIDMAN D.I. (1975) The induction of glyoxysomal enzyme activities in the aleurone cells of germinating wheat. *J. Exp. Bot.*, **26**, 387–97.

[18] PALEG L.G. (1961) Physiological effects of gibberellic acid. III. Observations on its mode of action on barley endosperm. *Plant Physiology,* **36,** 829–37.

[19] RADLEY M. (1967) Site of production of gibberellin-like substances in germinating barley embryos. *Planta,* **75,** 164–71.

[20] VARNER J. (1971) The control of enzyme formation in plants. *Symp. Soc. Expl. Biol.,* **25,** 197–206.

[20a] ZWAR J.A. & JACOBSEN J.V. (1972) A correlation between an RNA fraction selectively labelled in the presence of GA and amylase synthesis in barley aleurone layers. *Plant Physiol.,* **49,** 1000–4.

[21] EVANS H.L. & RAY P.M. (1969) Timing of the auxin response in coleoptiles and its implications regarding auxin action. *J. Gen. Physiol.,* **53,** 1–20.

[22] RAYLE D.L. (1973) Auxin-induced hydrogen ion secretion in *Avena* coleoptiles and its implications. *Planta,* **114,** 63–73.

Chapter 4.4
The Hormonal Control of Amphibian Metamorphosis

4.4.1 INTRODUCTION

Amphibian metamorphosis is the process by which a more or less aquatic larval form is transformed into an adult adapted to a more or less terrestrial existence. It is seen at its most dramatic in the anuran amphibia (frogs and toads) where the larvae (tadpoles) differ from the adults not only in their manner of respiration but also in their locomotory and feeding habits. In these species, metamorphosis represents an intensive period of growth and developmental change during which the structure and function of almost every organ in the body is radically modified for adaptation to a new way of life. Unlike other adaptational changes, however, this process is begun and usually completed in anticipation of the change of environment.

Some of the changes seen during metamorphosis involve growth and maturation of the tissues concerned; limbs are formed, lungs develop, new enzyme activities appear within the liver. Other changes result in the regression and death of organs; the gills and the tail, which will not be needed in adult life, are completely resorbed. A list of the most salient morphological changes of metamorphosis is given in Table 4.4.1. The biochemical changes which have so far been characterized are listed in Table 4.4.2.

Table 4.4.1 Anatomical changes during anuran metamorphosis.

1. *External Features*

Tissue/organ	Changes	Function
Head	Loss of horny beaks to mouth. Widening of mouth.	Adaptation to new diet.
	Development of tympanum. Repositioning of eyes.	Accommodation to aerial sensory input.
Limbs	Complete growth of fore- and hind-limbs.	Locomotion on land.
Tail	Complete resorption.	Loss of swimming as major mode of locomotion.
Skin	Pigmentation changes. Hardening.	Protective colouration. Protection against water loss on land.

2. *Internal Features*

Organ/system	Changes	Function
Digestive system	Development of muscular tongue. Major shortening of gut. Repositioning of anus.	Change from vegetarian to car- nivorous diet.
Respiratory system	Resorption of gills. Development of lungs. Development of hyoid cartilages and muscles for respiration.	Change from water- to air-based respiratory system.
Reproductive system	Development of gonads.	Sexual maturity only in adult form.
Nervous system	Degeneration of Mauthner cells. Growth of new neurones and nerves.	Denervation of de- generating tissue. Innervation of new structures.

Table 4.4.2 Biochemical changes during anuran metamorphosis.

Tissue/organ	Change induced	Function
Eye	Change in visual pigment from porphyropsin to rhodopsin.	—
Liver	Synthesis of urea cycle enzymes.	Excretion of urea.
	Synthesis of serum albumin.	Maintenance of homeostasis.
	Synthesis of cerulo- plasmin.	Connected with changed iron utili- zation?
Erythropoietic tissue	Change from synthesis of larval haemoglobin to adult haemoglobin.	Lower affinity oxygen carrier for air-based respiration.
Gut	Synthesis of hydrolytic enzymes.	Resorption of tissue.
	Appearance of peptic activity in foregut.	Change to digestion of animal tissue.
Skin	Melanin synthesis. Induction of Na^+-K^+- ATPase.	Protective colouring. Maintenance of electrolyte balance.
	Serotonin synthesis.	—
	Changes in collagen de- position and breakdown.	Changes in mechanical properties of skin for terrestrial life.
Tail	Synthesis of hydrolytic enzymes	Resorption of tissue

unse blood contains (hypused.

phases of metamorphosis

The exciting discovery was made earlier this century that this entire, coordinated developmental process is under the obligatory control of the thyroid hormones. It is induced only by a spontaneous or artificial increase in the thyroxine (T_4) or tri-iodothyronine (T_3) levels in the blood and will only be continued if thyroid hormone is constantly present (reviewed [1]).

The nature of the hormonal control of metamorphosis

Morphological studies of the events of spontaneous metamorphosis [2, 3, 4] in various anuran species have led

Etkin [1] to define three distinct phases of metamorphic change (see Fig. 4.4.1). Pre-metamorphosis is defined as an initial period of extensive growth but little developmental change. This is followed by prometamorphosis during which growth continues but conspicuous developmental changes such as the growth of hind legs takes place. Prometamorphosis ends with the emergence of the fore-limbs and the dramatic third and final stage of development, the metamorphic climax sets in. This is a comparatively brief period in which profound morphological changes occur very rapidly, the most conspicuous being the complete resorption of the large muscular tail. The tadpole is transformed into an immature froglet and metamorphosis is

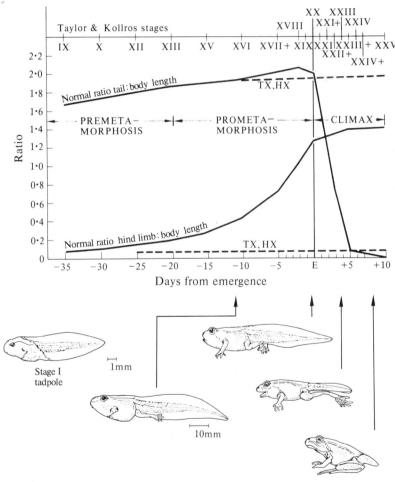

Fig. 4.4.1 The pattern of typical anuran metamorphosis (*Rana pipiens*). Data for animals grown at 23°C ± 1. Time on the lower abscissa is shown in days before or after the emergence of the second fore-leg (E) which is taken as the beginning of metamorphic climax. The upper abscissa indicates the Taylor-Kollros stages of the animals. The ratios of hind leg length (H.L.L.) to body length (B.L.) and tail length to body length are shown for normal tadpoles by the solid lines. Comparable data for thyroidectomized (TX) or hypophysectomized (HX) tadpoles are shown by the broken lines. The drawings show the morphological appearance of the tadpoles at the indicated stages of metamorphosis. After Etkin [1] and Weber [48].

completed. As a generalization it may be said that most of the growth and maturational changes of metamorphosis occur during the slower prometamorphic period, whereas the faster, resorptive phenomena are activated during metamorphic climax.

The following experiments demonstrate that the coordinated development of each of these stages can only be reproduced if the thyroid hormones are presented to a thyroidectomized larva in a specific manner. Thyroidectomized tadpoles will not metamorphose. They can be induced to do so by a single large dose of thyroid hormone (T_3 or T_4), but this results in an unco-ordinated pattern of metamorphic changes which appears to be the consequence of all the changes starting simultaneously. Etkin has shown that thyroidectomized tadpoles can be taken through coordinated metamorphosis if they are exposed to the following sequence of hormones. Levels of thyroxine of less than 1 µg/l ($1 \cdot 3 \times 10^{-9}$M) are sufficient to maintain the normal changes of premetamorphosis. Prometamorphosis is initiated and will proceed more or less normally if the hormone con-centration is increased from 4×10^{-9} to 26×10^{-9}M over a 21-day period. The onset of climax requires the abrupt raising of the hormone level to approximately 325×10^{-9}M (or 250 µg/l) but beyond climax only very low maintenance levels of the hormone are needed.

From these and other morphological studies has come the concept that metamorphosis is achieved by a crescendo of thyroid hormone secretion into the tadpole blood stream, reaching peak levels at metamorphic climax. After this peak of activity, the thyroid becomes essentially inactive and seems to play a very subsidiary role in the adult life of these animals (see [5] for review of thyroid function in adult amphibia).

Kollros has suggested that the tissues of the tadpole are set to start their metamorphic changes when the concentration of thyroxine in the blood reaches different levels. If this theory is correct, then it should be possible to halt metamorphosis at particular stages by, for instance, exposing a thyroidectomized tadpole to the concentration of thyroxine required for pro-metamorphosis but not to that required for metamorphic climax.

However, it is also possible, as Etkin [1] has suggested, that all of the tissues respond to all concentrations of the hormone, but that each tissue requires a different total amount of hormone to complete its metamorphic response. The idea underlying this theory is that the hormone is required to react stoichio-metrically in various chemical cellular processes. Since the rate of accumulation of the hormone by the different tissues would be determined by the blood concentration, it would follow that tissues requiring a large amount of hormone would undergo their metamorphic response very slowly at low blood hormone levels. Nevertheless, if this theory is correct, a thyroidectomized tadpole exposed to prometamorphic levels of thyroxine should eventually be able to undergo complete metamorphosis. It has

not yet been possible to distinguish between these two possibilities because of the technical problems of maintaining thyroidectomized tadpoles on low doses of thyroxine for very long periods of time (years) to assess their responsiveness.

The endocrinology of metamorphosis

In order to account for the orderly sequence of changes during metamorphosis it is necessary to find out how the tadpole manages to increase its blood thyroxine levels in a controlled manner. In amphibia and mammals the secretion of thyroid hormone is under the control of the pituitary and hypothalamus [7].

In adults, thyroxine inhibits both the hypothalamic secretion of TRH (TSH-releasing hormone) and hypophyseal secretion of TSH (thyroid stimulating hormone) and the circulating levels of T_3 and T_4 are maintained at a low level (Fig. 4.4.2). If this system operated in tadpoles then they would never have sufficient levels of thyroxine for metamorphosis. There is circumstantial evidence to suggest that environmental factors (e.g. light and temperature) influence the immature hypothalamus of the tadpole so that, for a brief period of its development, thyroxine stimulates rather than inhibits the release of TRH [8]. A self-accelerating system would thus be

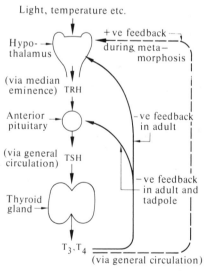

Fig. 4.4.2 Schematic representation of the regulation of thyroid hormone secretion in adult and larval amphibia. TRH-TSH releasing hormone, TSH-thyroid stimulating hormone, T_3-tri-iodothyronine, T_4-thyroxine. In adults and young tadpoles, TRH and TSH release are suppressed by thyroxine (solid lines). This control must change to allow thyroxine levels to increase during metamorphosis. Perhaps light changes the hypothalamus so that thyroxine stimulates TRH release (dotted lines).

generated giving rise to a crescendo of thyroid hormone secretion. It will be necessary to accurately measure the amounts of TRH, TSH and thyroxine in the tadpole before the details of the process can be resolved [9].

Although metamorphosis is entirely dependent upon the thyroid hormones, its onset may be influenced by other hormones, the best documented effects being produced by mammalian prolactin [10, 11] which has been shown to promote tadpole growth and inhibit metamorphosis. This action is in some ways similar to that of the well-known 'juvenile hormone' of insect metamorphosis [12]. However, it is not known if prolactin is utilized by the tadpole to achieve a regulated metamorphosis.

4.4.2 THE ACTION OF THYROID HORMONE IN METAMORPHOSIS

Characteristics of the thyroid hormone induced changes

Before considering the biochemistry of thyroid hormone action, it will be useful to consider certain general features common to all of the hormone-induced changes of metamorphosis.

THE LOCAL NATURE OF HORMONE ACTION

Virtually every cell and tissue type within the tadpole is affected by metamorphosis and all of these changes appear to result from the direct local action of the hormone. Thus Kaltenbach [13] has shown that if a thyroxine-cholesterol pellet is implanted into a section of the tadpole tail fin, the rest of the tail will remain intact whilst the region around the implant alone undergoes resorption. Similarly, Wilt [14, 15] has shown that if thyroxine is applied to one of the tadpole eyes, only that eye will develop the adult visual pigment rhodopsin, whilst the other eye retains the larval pigment porphyropsin. It can, therefore, be concluded that there are no systemic intermediates between thyroxine and its action on a tissue.

THE CONTINUAL REQUIREMENT FOR HORMONE

Thyroid hormone must be continually present, at a concentration or range of concentrations specific to that particular tissue, for any one of the developmental changes associated with metamorphosis to be carried through to completion. This was well demonstrated by Etkin [1] who thyroidectomized tadpoles at the very onset of metamorphic climax so that tail regression began but ceased abruptly as the hormone disappeared rapidly from the circulation. Strange creatures with half-resorbed tails were thus produced which remained so for the rest of their lives.

THE PREDETERMINED NATURE OF THE METAMORPHIC RESPONSES

It is important to recognize that during metamorphosis the thyroid hormones elicit tissue specific responses which were predetermined earlier in development. Thus, a hind-limb bud transplanted onto the tail will survive and grow during metamorphosis whilst the tail itself regresses. If, conversely, the tail is transplanted onto the back, it will still be resorbed whilst the tissue around it matures and differentiates (organ transplants reviewed [16]). Cells in the skin which appear histologically identical have been shown to react differently to thyroid hormone, and implantation of thyroxine pellets into the central nervous system will produce regression of the giant Mauthner neurones, whilst adjacent nerve cells are seen to grow [17, 18].

It is clear that thyroid hormones do not induce a select class of genes in the tissues on which they act; rather they modify the biosynthetic pathways of the different tadpole tissues in a manner peculiar to each tissue.

Biochemical systems studied

We have chosen biochemical changes in the liver as our main example of tissue growth and maturation during metamorphosis and biochemical changes in the tail as the main example of tissue regression during metamorphosis.

CHANGES IN THE TADPOLE LIVER DURING SPONTANEOUS AND INDUCED METAMORPHOSIS

Most of the studies into the liver changes during metamorphosis have involved the tadpoles of *Rana catesbeiana* (the American bull-frog) since this species is considerably larger, and therefore provides more experimental material, than the other readily available anuran larvae. These changes have been studied during both induced and spontaneous metamorphosis.

New or preferential protein synthesis

During metamorphosis extensive changes occur in the proteins synthesized by the liver (Table 4.4.2, Fig. 4.4.4 & Fig. 4.4.5). The synthesis of serum albumin increases at least 20-fold, and production of ceruloplasmin, the copper-containing serum protein which is also synthesized by the liver parenchymal cells increases abruptly, blood levels rising from 50 to 100-fold during metamorphosis. A transition is also seen in the type of haemoglobin synthesized by the liver erythropoietic tissue, larval haemoglobin being replaced by an adult molecule with lower affinity for oxygen.

The best documented changes, however, are those concerning the synthesis of the urea cycle enzymes of the hepatocytes. During the tadpole stages, nitrogenous waste is excreted almost

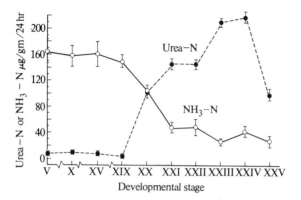

Fig. 4.4.3 Changes in nitrogen excretion during spontaneous metamorphosis of *Rana catesbeiana* at 25°C. Urea-N and ammonia-N were determined for larvae at various stages as defined by Taylor & Kollros [2]. This staging system was devised for *Rana pipiens* (see Fig. 4.4.1). It is also used to stage the development of the closely-related *Rana catesbeiana* although the actual process of metamorphosis takes much longer in this species. After Frieden & Just [9].

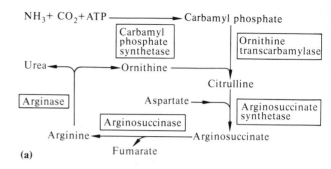

(a)

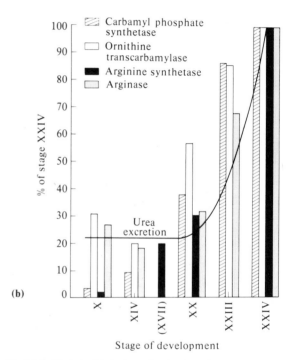

(b)

Fig. 4.4.4 (a) A simplified scheme showing the main substrates, products and enzymes of the urea cycle in amphibia. This accounts for the fixation of ammonia and carbon dioxide and leads to the elimination of nitrogen as urea. (b) The correlation of changes in urea cycle enzyme activities with urea excretion during spontaneous metamorphosis of *Rana catesbeiana*. The larvae were staged using the Taylor and Kollros system (see Legend to Fig. 4.4.3). Note that arginine synthetase activity represents the sum of argino-succinate synthetase and argino-succinase activities. After Cohen [20].

exclusively as ammonia but the change to terrestrial life, with much more limited possibilities for free diffusion, necessitates a shift to excretion of a much less toxic substance. There is a striking shift away from ammonia excretion so that 90% of nitrogenous waste is excreted as urea at metamorphic climax (Fig. 4.4.3).

Studies largely from Cohen's laboratory (see [19] for review) have shown that this enhanced urea production correlates with marked increases in the activities of the five urea cycle enzymes (see Fig. 4.4.4). The most dramatic change appears early in metamorphosis, before any of the gross morphological effects are detectable. The five enzymes show different relative increases in activity during metamorphosis and are found in different parts of the cell (carbamyl phosphate synthetase and ornithine transcarbamylase being mitochondrial and the other enzymes extra-mitochondrial). However, recent studies from Cohen's group [20] have shown that the increases in the activities of all five enzymes both during induced and spontaneous metamorphosis, occur simultaneously and in a highly correlated manner demonstrating that thyroid hormone induces a concerted and coordinated response of all the enzymes of the cycle.

It is important to recognize that so far it has not been technically possible to decide whether or not the urea cycle enzymes and other new proteins of the adult liver are present in very low amounts before metamorphosis. Their increased production will be described as 'new protein synthesis' but there may be low levels of synthesis prior to metamorphosis.

DNA synthesis

In other developmental systems, the *de novo* synthesis of proteins as a result of hormone action has been found to be associated with an obligatory initial phase of DNA synthesis and cell replication. Often the combined action of two or more hormones is involved, with one hormone being primarily responsible for stimulating cell proliferation before the synthesis of new proteins can be induced by further hormone(s). Two such, well-documented, examples are the insulin/hydrocortisone/prolactin-induction of milk protein synthesis in cultures of mouse mammary gland [21] and the oestrogen/progesterone-induced synthesis of avidin by the chick oviduct [22]. In some cases, however, it seems that one hormone may be responsible for both the proliferative phase and for specific product synthesis, as in the oestrogen-induced synthesis of ovalbumin by the chick oviduct [22].

Early histological studies indicated that the tadpole liver does not undergo cell division during metamorphosis [23] and it has been therefore generally considered that thyroid hormone related changes do not involve an obligatory proliferation phase. Atkinson *et al.* [24], studying DNA synthesis in *Rana catesbeiana* during spontaneous and induced metamorphosis, were able to show a transitory increase in incorporation of ^{3}H-thymidine and ^{14}C-lysine into liver nuclear DNA and protein, 2–4 h after administration of thyroxine. This was followed by a larger, more sustained increase in ^{3}H-thymidine incorporation beginning four days after the onset of hormone administration. This second increase coincides with the earliest detectable increases in serum albumin and urea cycle enzyme synthesis and might suggest that the synthesis of these new proteins is a consequence of the proliferation of new hepatocytes under the influence of thyroid hormone. However, it has not been possible to detect an increased rate of cell division in the hepatocytes proper during metamorphosis [25, 26]. The enhanced DNA synthesis probably reflects changes in other cell populations present in the liver, a likely possibility being the generation of a new erythroid cell line for the transition from larval to adult haemoglobin [27].

Although the role of nuclear DNA synthesis in the response of the hepatocytes to thyroid hormone may not prove obligatory, obviously in those tissues which are rapidly formed during metamorphosis, such as the limbs and the lungs, a greatly enhanced rate of DNA synthesis and cell division is one of the earliest and more important effects of hormone action. A clear role for cell division as a pre-requisite for the maturation of the serous skin glands in response to thyroxine has also been shown in organ cultures of the skin of the South African clawed toad, *Xenopus laevis* [28].

As well as changes in nuclear DNA synthesis, Atkinson *et al.* [24] also reported enhanced mitochondrial ^{3}H-thymidine incorporation in response to thyroid hormone, beginning one hour after treatment and persisting for three days. Enhanced mitochondrial DNA polymerase activity has also been reported as one of the earliest detectable effects on the liver after thyroid hormone administration [29]. These findings probably relate to the increase in size and number of the hepatocyte mitochondria

which is known to occur as part of the response to thyroid hormone [25, 26].

RNA synthesis

A cell may change the range of proteins it synthesizes by two main devices. Firstly new m-RNA molecules may move into the cytoplasm either as a consequence of the synthesis of new m-RNA or as a consequence of a change in the processing or transport of m-RNA. Secondly, there may be a change in the cytoplasm so that a new range of m-RNA molecules is translated into protein (control mechanisms of protein synthesis reviewed [30, 31] and Chapter 5.3).

There is some evidence that the induced synthesis of liver proteins involves a change in transcription. One of the first effects of the hormones is to change the uptake and pool size of RNA precursors so that measurements of rates of synthesis have to be corrected for this effect. Taking this into account there is an increased incorporation of precursors into nuclear and subsequently cytoplasmic RNA [32, 33]. But investigations by base analysis, sucrose gradients, polyacrylamide gels, and DNA : RNA hybridization have not revealed the appearance of large quantities of new types of m-RNA [33, 34, 35]; rather these studies show that the bulk of the newly synthesized RNA is r-RNA and that, as in other developmental processes [36], the increase in cytoplasmic RNA labelling is mainly due to the appearance of newly synthesized ribosomes which are found in polysomes. However, these techniques might not be sensitive enough to detect new m-RNA synthesis and the observation that the induction of carbamyl phosphate synthesis can be blocked by actinomycin D [37] and that there is an increase in the template activity of isolated liver chromatin [38] argues that the hormone acts by affecting transcription. It remains for someone to unequivocally demonstrate new m-RNA synthesis.

One surprising observation is that the euchromatic nuclei of the hepatocytes during the early tadpole stages become heterochromatic at the start of metamorphosis and remain heterochromatic throughout the process. Heterochromatin is often associated with genetic inactivity and it is therefore possible that the increase in RNA synthesis during metamorphosis is the consequence of a more rapid transcription of a select class of genes with the concomitant inactivation of a larger number of genes which code for features of larval life.

As well as being implicated in the regulation of m-RNA synthesis, there is some evidence to suggest the involvement of thyroid hormone in controlling the availability of m-RNA within the tadpole liver. Studying the *in vitro* synthesis of carbamyl phosphate synthetase by cubes of tadpole liver in response to thyroxine, Cohen's group [19, 39] found that:

(i) the lag phase in induction of enzyme synthesis which is normally seen upon hormone administration *in vivo* was not found *in vitro*.

(ii) Actinomycin D is not immediately effective in preventing completely the induction of carbamyl phosphate synthetase and only becomes so over a period of 48 h.

This suggests that some m-RNA for the enzyme is already present with the liver cells and, *in vitro* at least, thyroxine is able to enhance its use for translation into active enzyme.

Changes in translation and protein turnover

Four days after thyroid hormone treatment there is an increase in the rate of protein synthesis in the liver (see Fig. 4.4.5). This increased rate of translation is associated with the appearance of new hormone-induced ribosomes in the cytoplasm, enhanced amino acyl-tRNA transferase activity [40] and a preceding increase in amino acid uptake. It seems unlikely, however, that any of these processes directly controls the rate of protein synthesis in this system.

There are also changes in the relative amounts of the two major liver leucine tRNAs during metamorphosis [41]. Relative changes in the amounts of iso-accepting tRNAs have been reported in other systems in association with quantitative changes in protein synthesis, but the significance of these changes is not understood at present.

The key urea cycle enzyme carbamyl phosphate synthetase which increases 30-fold in activity in liver extracts during metamorphosis has been shown by immunological techniques to be synthesized initially as an inactive precursor molecule [39]. This was also found to be true for glutamate dehydrogenase, another mitochondrial enzyme whose activity is enhanced during metamorphosis. Using cubes of tadpole liver subjected to thyroxine treatment *in vitro* or *in vivo*, it was found that the increase in enzyme activity seen in response to the hormone for both these enzymes results from a multiplicity of hormone-induced effects. As well as enhanced synthesis of enzyme precursors as a result of enhanced m-RNA synthesis or usage (see section on RNA synthesis, above), a faster rate of conversion of the inactive precursors to active enzyme is seen in response to thyroxine. The level of active enzyme was shown to result from the balance of precursor conversion and enzyme degradation. Thyroid hormone was shown to stimulate both of

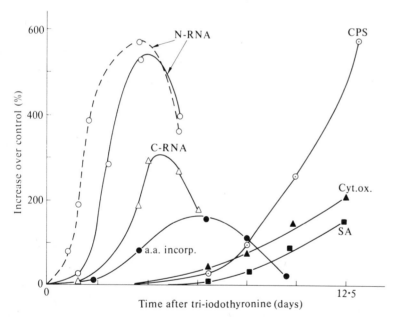

Fig. 4.4.5 Schematic representation of the stimulation of nuclear RNA synthesis, its turnover into cytoplasm of liver of *Rana catesbeiana* tadpoles, stimulation of amino acid incorporation and the induction of *de novo* protein synthesis following treatment of tadpoles with tri-iodothyronine. O---O, specific activity of nuclear RNA labelled with ³H-uridine not corrected for changes in distribution of radioactivity in the acid-soluble fraction; O——O, specific activity of nuclear RNA after correction for changes in acid-soluble radioactivity; △——△, specific activity of RNA recovered in cytoplasmic ribosomes and polysomes; ●——●, specific activity of protein recovered in the labelled microsomal fraction 40 min after administration of a mixture of ¹⁴C-labelled amino acids; O——O, carbamyl phosphate synthetase activity; ▲——▲, specific activity of cytochrome oxidase in mitochondrial fraction; ■——■, serum albumin accumulation in blood. The earlier stimulation of labelling of nuclear RNA when not corrected for acid-soluble radioactivity is presumably due to a more rapid action of the hormone on the uptake or pool size of uridine. The abrupt downward trend of the curves is due to dilution with nucleotides released by the regression of tail, gut, gills etc. Data compiled from Tata [32, 33].

these processes but at 24°C, the maintenance temperature for the tadpoles and cultures used in these experiments, the stimulation of precursor conversion was greater than that of enzyme degradation.

Reorganizational changes in the liver

Although thyroid hormone stimulates the synthesis of new ribosomes there is no increased accumulation of ribosomes in the liver during the first 6–10 days of hormone treatment. Instead, it has been shown by double-labelling experiments, that not only is ribosomal synthesis stimulated but so also is ribosomal degradation so that although the total ribosomal number remains fairly constant there is a considerable ribosomal turnover, the old ribosomes present before metamorphosis being replaced by a new population [35].

Simultaneously with this enhanced ribosomal synthesis there is a proliferation of the membranes of the endoplasmic reticulum as judged by membrane phospholipid synthesis, especially of those membranes to which ribosomes are bound, the rough endoplasmic reticulum (RER) (see [42] for further details). The newly-formed ribosomes are largely present in this RER fraction and appear to be more tightly bound to these membranes than ribosomes of the control untreated liver tissue. It is also in this rough endoplasmic reticulum fraction that the increased protein synthesizing activity noted 4–8 days after thyroid hormone induction is largely located (see Fig. 4.4.6).

These biochemical findings correlate well with ultrastructural studies of the liver during induced metamorphosis which show that the hormone produces a striking structural reorganization of the rough endoplasmic reticulum (Fig. 4.4.7). In early pre-metamorphic tadpoles these membrane-ribosome complexes are largely present as simple, sometimes vesicular, structures. Under the influence of thyroid hormone there is a proliferation of complex double lamellar arrays of membranes which then spread throughout the cytoplasm, though often appearing to originate in the peri-nuclear region [42].

These observations suggest two exciting possibilities:

(i) that the increased protein synthesizing activity within the liver giving rise ultimately to the new metamorphic proteins may be restricted largely to the ribosomes which are newly formed under the stimulus of thyroid hormone; and
(ii) that there may be a topographical segregation within the cell of these new ribosomes engaged in the synthesis of a specific group of proteins, by means of their attachment to particular structures of the endoplasmic reticulum.

This second suggestion involves a novel interpretation of the role of ribosomal attachment to the membranes of the endoplasmic reticulum. The function characteristically assigned to the RER is that of the synthesis of secretory proteins, the

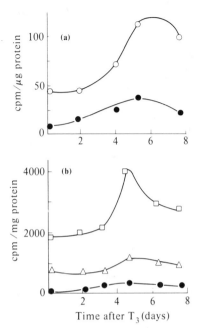

Fig. 4.4.6 Co-ordinated proliferation of endoplasmic reticulum and the increase in hepatic protein synthesis *in vivo* in the heavy rough membranes (densely packed ribosomes attached to membranes) during tri-iodothyronine-induced metamorphosis of young bullfrog tadpoles. (a) The appearance of newly synthesized (^{32}P-labelled) heavy rough membrane phospholipids (O) and RNA (●); (b) the recovery of labelled nascent proteins formed *in vivo* after a short pulse of radioactive amino acids in heavy rough membranes (□), light rough membranes (△), and free polysomes (●). For details, see Tata [35].

export of which is known to occur via the cisternae of the endoplasmic reticulum. It seems possible in view of the observations made on tadpole liver, that this attachment may play a wider role and might be used much more generally in cells to achieve effective compartmentalization of the production of particular proteins within the cell. In this respect it is interesting to note that a preferential shift from free to membrane-bound ribosomes is characteristic of growth and development even in non-secretory tissues and that the inducible intracellular enzyme, serine hydratase, is almost exclusively synthesized on membrane-bound and not free ribosomes in rat liver [43].

Extensive ultrastructural changes are seen in other organelles of the hepatocyte during thyroid hormone action (Fig. 4.4.7). One of the most dramatic and striking effects is seen in the mitochondria, which increase in size and number so that they occupy a much greater proportion of the cytoplasm. The broad lamellar cristae seen in the early tadpole stages are replaced by smaller more tubular structures and the overall shape of the

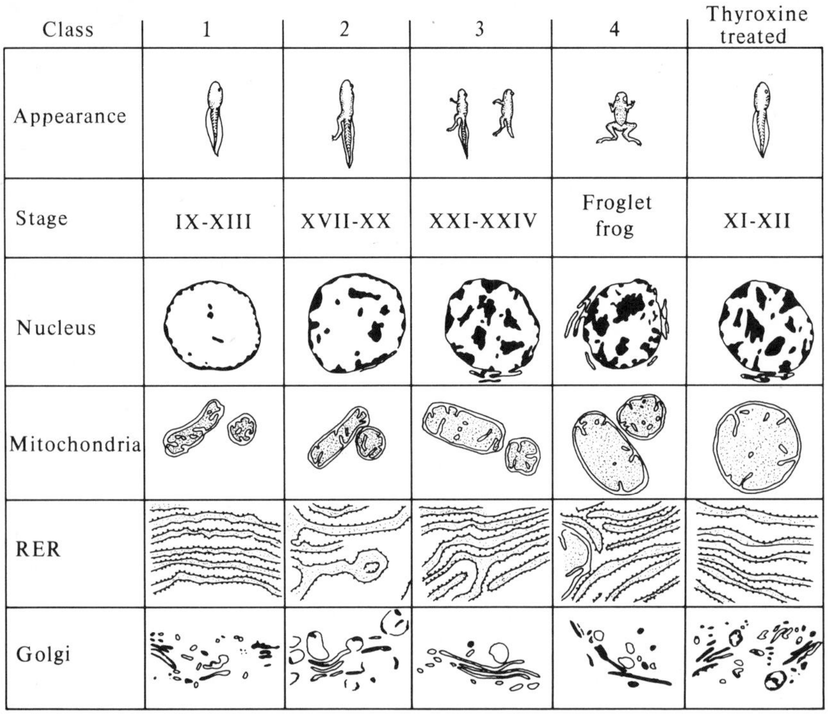

Class	1	2	3	4	Thyroxine treated
Appearance					
Stage	IX-XIII	XVII-XX	XXI-XXIV	Froglet frog	XI-XII
Nucleus					
Mitochondria					
RER					
Golgi					

Fig. 4.4.7 Schematic diagram summarizing the salient ultrastructural changes in the liver parenchymal cells during spontaneous and thyroxine-induced metamorphosis of *Rana catesbeiana*. After Bennet *et al.* [28].

mitochondrion changes from an elongate to a more spherical structure. Frequently newly-proliferated RER may be seen intimately associated with these new mitochondria and it seems likely that this is the site of synthesis of the two mitochondrial urea cycle enzymes (carbamyl phosphate synthetase and ornithine transcarbamylase) and the various mitochondrial dehydrogenases whose synthesis in enhanced during metamorphosis. In the case of carbamyl phosphate synthetase which is a large protein structure (mol. wt. 315,000) composed of various subunits, Cohen has proposed [20] that precursor forms of the subunits are made by the RER and are transported into the mitochondria. Final processing and assembly of the subunits into the completed protein then occurs in the mitochondrial inter-cristal space. Mitochondrial DNA synthesis is one of the earliest changes noted in response to thyroid hormone (see section on DNA synthesis, above) and it seems possible that a phenomenon similar to that seen in the ribosomal population also occurs amongst the mitochondria—a combination of proliferation and turnover so that finally a new generation of mitochondria with new functions and structure is set up within the cell.

Summary of changes in the liver

As a result of the action of thyroid hormone upon the liver, the synthesis of a select group of proteins (such as albumin, ceruloplasmin and urea cycle enzymes) is dramatically increased. Studies on the enzyme carbamyl phosphate synthetase indicate that the increased synthesis in this case results from effects of the hormone on mRNA transcription, mRNA usage and conversion of an inactive precursor form of the enzyme into active enzyme.

Extensive morphological changes accompany this increased synthesis of specialized proteins. There is a co-ordinated proliferation of ribosomes and membranes of the endoplasmic reticulum giving rise to large amounts of RER perhaps specifically engaged in making the new proteins. New mitochondria are generated, probably as new intracellular structures to house some of the new functions of the liver.

STUDIES OF TAIL RESORPTION

Unlike the gut or skin where metamorphosis produces a subtle interplay of tissue destruction and tissue proliferation, the response of the tail to thyroid hormone is entirely one of tissue involution, leading eventually to the complete disappearance of this organ. It is therefore ideally suited to the study of the mechanism of thyroid-hormone induced tissue destruction within the tadpole.

The tail tissues are the most insensitive of the tadpole tissues to thyroid hormone, regression only beginning with the onset of metamorphic climax (see Fig. 4.4.1), when blood thyroid hormone levels are thought to be highest. In tadpoles of *Xenopus laevis* which have been used for many of the studies on tail regression, resorption is very rapid. The tail length begins to decrease at approximately day three of climax and within three or four days the tail is reduced to a tiny cone of residual tissue. During climax, the blood levels of thyroid hormone begin to decline due to the activation of the adult feed-back inhibition of the hypothalamus (see Fig. 4.4.2). It is interesting in this respect that, unlike the tissues developing during prometamorphosis which require continuously increasing levels of thyroid hormone for continued change, the tail actually increases in sensitivity to the hormone with progressive regression and thus involution may be completed in a declining blood hormone concentration [9].

Morphologically, tail regression is characterized by early degenerative changes in the epidermal and connective tissues so that the tail fins are quickly resorbed (see Fig. 4.4.8). Muscle degeneration giving rise to bundles of degenerating myofibrils occurs almost simultaneously, but the nervous structures appear to be most resistant to histiolysis and disintegrate only during the later stages of regression. It is well established that there is enhanced macrophage activity within the regressing tail.

Studies on the mechanism of tail regression have been greatly assisted by the discovery that tail tissue involution, very similar in timing and magnitude to that seen *in vivo* can be induced *in vitro* in organ cultures of isolated tadpole tails treated with thyroid hormone [44]. Successful culture of tail tips in a variety of tissue culture media has been achieved although sensitivity to the thyroid hormones appears to be greatest when a simple, nutrient-free saline solution is used (see [45] for review).

Mechanism of tail regression

Clearly the total disintegration of the comparatively large tadpole tail must involve intensive enzymic digestion of some kind and from numerous studies it has become established that a large number of hydrolytic enzymes such as cathepsins and acid phosphatase increase 5 to 50-fold in specific activity during tail regression (see Table 4.4.3). These enzymes are known to be

Table 4.4.3 Changes in tadpole tail enzymes during spontaneous metamorphosis

	Ratio of enzyme activities Metamorphic climax*	
	---	---
	Premetamorphic tadpoles	
Enzyme†	Specific activity	Total activity
β-glucuronidase	34‡	3–4
	12–50‡	2–3
Deoxyribonuclease (acid)	50‡	2–3
	3‡	—
Ribonuclease (acid)	20	1
Cathepsin	22‡	2
	50‡	2–3
Phosphatase (acid)	5–20	1–1·5
Phosphatase (alkaline)	1·5–2·8	—
Collagenase	10	—
Proteinases (acid)	9·2	—
Dipeptidases, tripeptidases	3–7	—
Catalase	6	—
ATPase (Mg++)	0·5	—
Aldolase	0·2	—

* Where possible, data are based on enzyme measurements performed on tadpole tails that were reduced to 10% of their original weight. The activity of enzymes per remaining protein in tail (specific activity) increases dramatically while the total activity does not; this is because there is very little of the tail left at this point in metamorphic climax.
† Enzymes that have been reported not to change, or to decrease in activity ratio include alkaline proteinases, alkaline ribonuclease, amylase, lipase, glutamic and succinic dehydrogenases.
‡ Data from different studies (see Ref. [9]). After Frieden & Just [9].

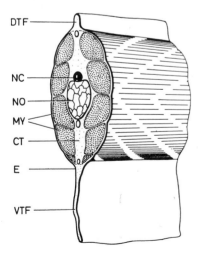

Fig. 4.4.8 Tail tissues of a *Xenopus laevis* larva. CT, Connective tissue; E, epidermis; MY, myomeres; NC, nerve cord; NO, notochord; DTF, dorsal tail fin; VTF, ventral tail fin. After Weber [48].

lysosomal in other species and tissues. The exceptional hydrolase is Mg^{++}-activated ATPase which is a non-lysosomal enzyme associated with the muscle proteins of the tail, and which declines in activity during tail regression. Similarly, enzymes associated with the energy metabolism of the tissue such as aldolase, also decrease in activity during resorption.

The association of these hydrolytic enzymes with lysosomes in other tissues led to the proposition, based on an original idea of de Duve [46] that the controlled release of these enzymes from lysosomes within the tail might be responsible for the tissue degradation produced in metamorphosis. The disintegration of the lysosomes was presumed to result from some direct interaction of the hormone with these organelles and the fact that lysosomolytic agents such as vitamin A could promote tail resorption [47] lent support to this concept.

There are, however, several reasons for believing that thyroxine does not act directly on the lysosomes. First of all, the controlled morphogenetic pattern of tail regression and the latent period of several days seen before T_4-induced regression is initiated, argue against a non-specific lysosomal activation. Secondly, it has become obvious that both RNA and protein synthesis are necessary for the induction of tail regression by thyroid hormone. The first indication that protein synthesis might be important for tail regression came from the studies of increased enzyme activity already described. It was found that not only the specific activity but the total activity of the hydrolases involved in regression increased as resorption continued (see Table 4.4.3). Thus increased synthesis of these enzymes must be occurring during the resorption process.

That this new protein synthesis is obligatory for the resorption process has been shown by the use of inhibitors of RNA and protein synthesis. Weber [48] observed that injection of *Xenopus laevis* tadpoles with low doses of actinomycin D almost completely inhibited tail resorption and prevented the increase in cathepsin activity seen in the tails of uninjected control animals. Using organ cultures of tail tips from *Xenopus laevis*, Eeckhout [49] found that T_4-induced tissue resorption could be inhibited by puromycin, and Tata [50] has shown that T_3-induced regression of tail tips from *Rana temporaria* may be completely abolished by treatment with either Actinomycin D, puromycin or cycloheximide (see Fig. 4.4.9). Tata was further able to demonstrate that there is actually a burst of RNA and protein synthesis prior to, or coincident with, the onset of regression *in vitro* (Fig. 4.4.10). Polyacrylamide gel analysis [35] of the RNA synthesized during tail regression *in vivo* has shown that, as in the liver, this stimulation of RNA synthesis is largely as a result of enhanced ribosomal RNA synthesis.

Ultrastructural studies of tail regression

Histochemical location of the increased acid phosphatase and cathepsin-like esterase activities during the actual process of tail

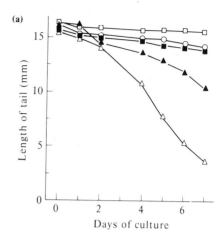

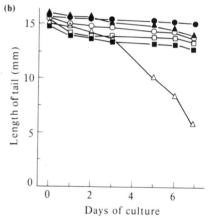

Fig. 4.4.9 (a) Inhibition by actinomycin D of the regression of isolated tadpole tails induced by tri-iodothyronine. Both the hormone and inhibitor were present in the culture from the beginning. O———O, Control; △———△, T_3 1 μg/ml; □———□, actinomycin D 5 μg/ml; ■———■, T_3 1 μg/ml and actinomycin D 5 μg/ml; ▲———▲, T_3 2·5 μg/ml and actinomycin D 5 μg/ml. (b) Inhibition by puromycin and cycloheximide of T_3-induced regression of the isolated tadpole tail. The hormone and inhibitors were added 18 h after the culture was begun. O———O, Control; △———△, T_3 0·8 μg/ml; ●———●, puromycin 80 μg/ml; ▲———▲, T_3 0·8 μg/ml and puromycin 80 μg/ml; □———□, cycloheximide 40 μg/ml; ■———■ T_3 0·8 μg/ml and cycloheximide 40 μg/ml. From Tata [50].

regression was studied by Weber [51] who found that these enzymes are largely located within the phagocytic vacuoles of certain macrophage-type cells present in the sub-epidermal connective tissues. The macrophages must be of local origin, since resorption occurs *in vitro* in the absence of a blood circulation and thus this histochemical evidence suggests that the increased synthesis of lysosomal acid hydrolase noted in tail regression is related to local stimulation of certain mesenchymatous cells into phagocytic activity.

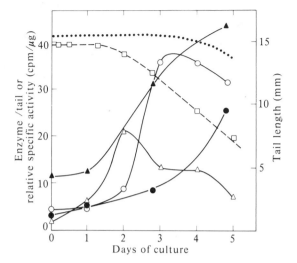

Fig. 4.4.10 Accumulation of cathepsin (●) and deoxyribonuclease (▲) and burst of additional RNA and protein synthesis during regression of tadpole tails induced in organ culture with tri-iodothyronine added to the medium. ○———○, incorporation of ^{3}H-uridine into RNA (12·5 h after the addition of 10 μCi); △———△, incorporation of ^{14}C-labelled amino acids into protein (8·8 h after the addition of 1·6 μCi of a mixture of labelled algal amino acids). Incorporation on day 0 is the average value for controls; all other points refer to samples to which T_3 was added. ....., tail length of controls over the duration of the experiment; □———□, length of tri-iodothyronine-treated tails showing a marked onset of regression between the second and third day after culture. Curves compiled from data of Tata [50].

However, ultrastructural studies of tail degeneration [51, 52, 53] show that except in the case of the collagen fibres and the matrix of the connective tissues of the tail, tissue disintegration is not caused directly by macrophage invasion or by the secretion of degradative enzymes from the macrophages. In the case of the outer epidermal cells, degeneration appears to result mainly from a form of severe dehydration, leading to shrivelling, loss of structural integrity and finally in the sloughing of these cells. In the muscle tissue degradative changes typical of hydrolytic enzyme action do occur but these are always well advanced before macrophage ingestion occurs, and the spatial progression of these changes would not correlate with any possible diffusion of secreted enzymes from the phagocytic cells.

These findings indicate that the macrophages play only a secondary role in tail resorption, as has been seen in other tissues [46] and are involved mainly in the removal of tissue debris after cell death and disintegration have been effected by some other means. How this initial cell death is induced by the thyroid hormone is an intriguing question, particularly since the ultrastructural evidence suggests different ways of death for the different tissue types, such as the epidermis and the muscle tissue, within the tail. In the case of muscle tissue, death by

auto-digestion would seem to be a real possibility and it may prove that in this tissue, new non-lysosomal enzymes, quite distinct from the acid hydrolases of the macrophages are induced by thyroid hormone as 'death' enzymes within this tissue.

Summary of tail resorption studies

The regression of the tadpole tail is associated with the enhanced local activity of numerous acid hydrolase enzymes. An early suggestion that this represented a controlled disintegration of the cellular lysosomes as a result of thyroid hormone action now seems to be disproved. It has been shown instead that, as in the tissues which undergo maturation during metamorphosis, tail regression results from hormone induced changes in RNA and protein synthesis within the tissue. The newly synthesized hydrolase enzymes appear to be largely associated with certain mesenchymal cells which are activated into phagocytic activity during regression, however, and therefore are probably not primarily responsible for tissue death and disintegration within the tail. How the thyroid hormone might trigger off this initial cell death is a compelling question which is yet to be further investigated.

The initial site of hormone action; the thyroid hormone receptors

It can be seen from the experiments described above on the tail and liver during metamorphosis that many detailed studies into the reprogramming of cellular biochemistry as a result of thyroid hormone action have been performed. The initial step in the chain of events leading to hormone-induced biochemical changes, is probably the binding of the hormone to cellular receptors molecules. It is difficult to identify the receptors for thyroid hormones. This is not because of a lack of thyroid hormone binding molecules; rather there is a wide range of cellular proteins which bind thyroxine and it is not known which of these is essential for hormone action.

Griswold et al. [54] have reported that ^{14}C-labelled thyroxine administered in vivo to tadpoles of Rana catesbeiana was concentrated into the nuclear fraction of the liver and appeared to be tightly bound to the chromatin. Transport of the hormone into the nucleus was almost completely inhibited if the injected tadpoles were maintained at 5° instead of 25°, which leads the authors to believe that this phenomenon is related to the well-known low temperature inhibition of T_4/T_3-induced metamorphosis. This preliminary work might suggest that the thyroid hormone action on the target tissues is mediated via a receptor system similar to that known for steroid hormones. The action of these hormones on their target tissues involves the binding of the hormone to a cytosol receptor protein. This is believed to induce some conformational change in the receptor such that this protein may now enter the nucleus. The hormone-receptor complex thus becomes strongly bound to components of the chromatin which results in changes in the pattern of RNA transcription (see [55] & [56]).

Since metamorphosis may be induced in premetamorphic tadpoles much earlier than the natural timing of events it is clear that the hormone receptors are present in the tissues well ahead of thyroid gland activation. Tata [57] has shown that in *Xenopus laevis* larvae competence to respond to thyroid hormone, as judged by biochemical parameters such as synthesis of DNA, RNA, phospholipids and protein, and changes in anion permeability, is acquired at a very early stage in development (between Nieuwkoop-Faber stages 38–45; that is at a developmental stage earlier than stage 1 in the Taylor & Kollros system for *Rana pipiens*). Studies of thyroid hormone binding to these larvae [58] showed that an excellent correlation existed between the appearance of high affinity hormone binding and the acquisition of biochemical responsiveness to the hormone (see Fig. 4.4.11). This suggests that there is a defined point early in development at which the first group of previously undetermined cells within the tadpole, become determined with respect to their response to thyroid hormone.

An excellent demonstration of the acquisition of competence as a defined stage in development is shown by the studies of Weber [59] on *Xenopus* tail regression. If the tail tip is amputated a few days before metamorphic climax a new tip will begin to regenerate but this will only be resorbed as a detectably later event than resorption of the bulk of the previously differentiated tail.

Originally [60] it was reported that all the tadpole tissues acquire the capacity to respond to thyroid hormone at the same time, but a careful study by Moser [61] has shown in *Rana temporia*, that the various organs acquire sensitivity over a definite period of time. Reference has already been made to the differences in sensitivity to thyroid hormone which exists between the different tadpole tissues (see p. 234) and it seems clear that these differences in sensitivity and the acquisition of sensitivity are intimately connected with the production of a co-ordinated pattern of response to the thyroid hormone.

As yet no studies to distinguish the characteristics of hormone binding by the different tadpole tissues have been reported. It would be of great interest to know how these tissue specific differences in hormone sensitivity relate to the physical properties, concentration and distribution of any specific thyroid hormone binding proteins demonstrable within the various tissues.

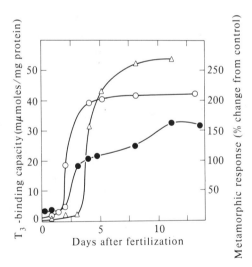

Fig. 4.4.11 Correlation between the appearance of strong temperature-sensitive binding capacity for tri-iodothyronine (●) and the acquisition of a metamorphic response to thyroid hormone by developing *Xenopus* larvae. The temperature-sensitive binding component(s) was calculated by subtracting the binding capacity for tri-iodothyronine at 25°C from that expressed at 5°. The metamorphic response to tri-iodothyronine is illustrated for the increase in the rate of RNA synthesis (△) or the diminution of uptake of PO_4^{3-} ions (○) when *Xenopus* larvae at different stages of development are exposed to 10^{-9} M tri-iodothyronine. Data from Tata [57, 58].

4.4.3 REFERENCES

[1] ETKIN W. (1964) Metamorphosis. In *Physiology of the Amphibia* (Ed. J.A. Moore), pp. 427–68. Academic Press, New York & London.

[2] TAYLOR A.C. & KOLLROS J. (1946) Stages in the normal development of *Rana pipiens* larvae. *Anat. Rec.*, **94**, 7–23.

[3] ETKIN W. (1932) Growth and resorption phenomena in anuran metamorphosis. I. *Physiol. Zool.*, **5**, 275–300.

[4] NIEUWKOOP P.D. & FABER J. (1956) Normal table of *Xenopus laevis* (*Daudin*). North Holland Publishing Co., Amsterdam, Holland.

[5] GORBMAN A. (1964) Endocrinology of the amphibia. In *Physiology of the Amphibia* (Ed. J.A. Moore), pp. 371–425. Academic Press, New York & London.

[6] KOLLROS J.J. (1961) Mechanisms of amphibian metamorphosis: hormones. *Am. Zoologist*, **1**, 107–14.

[7] ETKIN W. (1968) Hormonal control of amphibian metamorphosis. In *Metamorphosis* (Eds W. Etkin & L.I. Gilbert), pp. 313–48. Appleton-Century-Crofts, New York.

[8] ETKIN W. (1963) Metamorphosis-activating system of the frog. *Science*, **139**, 810–14.

[9] FRIEDEN E. & JUST J.J. (1970) Hormonal responses in amphibian metamorphosis. In *Biochemical Actions of Hormones*, Vol. 1 (Ed. G. Litwack), pp. 2–52. Academic Press, New York & London.

[10] BERMAN R., BERN H.A., NICOLL C.S. & STROHMAN R.C. (1964) Growth promoting effects of mammalian prolactin and growth hormone in tadpoles of *Rana catesbeiana*. *J. Expl. Zool.*, **156**, 353–60.

[11] BERN H.A., NICOLL C.S. & STROHMAN R.C. (1967) Prolactin and tadpole growth. *Proc. Soc. Exp. Biol. Med.*, **126**, 518–20.

[12] WYATT G.R. (1971) Insect hormones. In *Biochemical Actions of Hormones*. Vol. II (Ed. G. Litwack), pp. 386–490. Academic Press, New York & London.

[13] KALTENBACH J.C. (1959) Local action of thyroxin on amphibian metamorphosis. IV. Resorption of the tail fin in anuran larvae affected by thyroxin–cholesterol implants. *J. Expl. Zool.*, **140**, 1–18.

[14] WILT F.H. (1959) The organ specific action of thyroxin in visual pigment differentiation. *J. Embryol. Exp. Morphol.*, **7**, 556–63.

[15] OHTSU K., NAITO K. & WILT F.H. (1964) Metabolic basis of visual

pigment conversion in metamorphosing *Rana catesbeiana*. *Develop. Biol.*, **10**, 216–32.

[16] NEEDHAM J. (1942) Metamorphosis, competence and the last inductors. *Biochemistry and Morphogenesis*, pp. 447–55. Cambridge Univ. Press, London & New York.

[17] WEISS P. & ROSSETTI R. (1951) Growth responses of opposite sign among different neuron types exposed to thyroid hormone. *Proc. Nat. Acad. Sci.*, **37**, 540–56.

[18] PESETSKY I. & KOLLROS J.J. (1956) A comparison of the influence of locally applied thyroxine upon Mauthner's cell and adjacent neurones. *Expl. Cell Res.*, **11**, 477–82.

[19] COHEN P.P. (1970) Biochemical differentiation during amphibian metamorphosis. *Science*, **168**, 533–43.

[20] WIXOM R.L., REDDY M.K. & COHEN P.P. (1972) A concerted response of the enzymes of urea biosynthesis during thyroxine-induced metamorphosis of *Rana catesbeiana*. *J. Biol. Chem.*, **247**, 3684–92.

[21] TOPPER Y.J. (1970) Multiple hormone interactions in the development of mammary gland *in vitro*. *Rec. Prog. Hormone Res.*, **26**, 287–308.

[22] O'MALLEY B.W., McGUIRE W.L., KOHLER P.O. & KORENMAN S.G. (1969) Studies on the mechanism of steroid hormone regulation of synthesis of specific proteins. *Rec. Prog. Hormone Res.*, **25**, 105–60.

[23] KAYWIN L. (1936) A cytological study of the digestive system of anuran larvae during accelerated metamorphosis. *Anat. Rec.*, **64**, 413–41.

[24] ATKINSON B.G., ATKINSON K.H,, JUST J.J. & FRIEDEN E. (1972) DNA synthesis in *Rana catesbeiana* tadpole liver during thyroxine and tri-iodo-thyronine-induced metamorphosis. *Develop. Biol.*, **29**, 162–75.

[25] BENNETT T.P., GLENN J.S. & SHELDON H. (1970) Changes in the fine structure of tadpole (*Rana catesbeiana*) liver during thyroxine-induced metamorphosis. *Develop. Biol.*, **22**, 232–48.

[26] BENNETT T.P. & GLENN J.S. (1970) Fine structural changes in liver cells of *Rana catesbeiana* during natural metamorphosis. *Develop. Biol.*, **22**, 535–60.

[27] BRUNS G.A.P. & INGRAM V.M. (1973) The erythroid cells and haemoglobins of the chick embryo. (C) Erythropoiesis during metamorphosis of the bullfrog tadpole, *Rana catesbeiana*. *Philos. Transact. Royal Soc., London*, **266**, 22t–305.

[28] McGARRY M.P. & VANABLE J.W. (1969) The role of cell division in *Xenopus laevis* skin gland development. *Develop. Biol.*, **20**, 291–303.

[29] CAMPBELL A.M., CORRANCE M.H., DAVIDSON J.N. & KEIR H.M. (1969) The metabolism of DNA in the liver during precocious induction of metamorphosis in *Rana catesbeiana*. *Proc. Roy. Soc. Edinburgh B.*, **LXX**, 295–310.

[30] The Mechanism of Protein Synthesis (1969) *Cold Spring Harbor Symposia in Quantitative Biology*, Vol. XXXIV (Numerous articles on eucaryotic protein synthesis) Cold Spring Harbor, L.I., New York.

[31] Transcription of Genetic Material (1970) *Cold Spring Harbor Symposia in Quantitative Biology*, Vol. XXXV (Articles on eucaryotic RNA synthesis). Cold Spring Harbor, L.I., New York.

[32] TATA J.R. (1965) Turnover of nuclear and cytoplasmic ribonucleic acid at the onset of induced amphibian metamorphosis. *Nature*, **207**, 378–81.

[33] TATA J.R. (1967) The formation, distribution and function of ribosomes and microsomal membranes during induced amphibian metamorphosis. *Biochem. J.*, **105**, 783–801.

[34] WYATT G.R. & TATA J.R. (1968) The hybridization capacity of ribonucleic acid produced during hormone action. *Biochem. J.*, **109**, 253–58.

[35] RYFFEL G. & WEBER R. (1973) Changes in the pattern of RNA synthesis in different tissues of *Xenopus* larvae during induced metamorphosis. *Expl. Cell Res.*, **77**, 79–88.

[36] TATA J.R. (1970) Regulation of protein synthesis by growth and developmental hormones. In *Biochemical Actions of Hormones*, Vol. I (Ed. G. Litwack), pp. 89–133. Academic Press, New York & London.

[37] KIM K.H. & COHEN P.P. (1968) Actinomycin D inhibition of thyroxine-induced synthesis of carbamyl phosphate synthetase. *Biochim. Biophys. Acta.*, **166**, 574–77.

[38] KIM K.H. & COHEN P.P. (1966) Modification of tadpole liver chromatin by thyroxine treatment. *Proc. Nat. Acad. Sci.*, **55**, 1251–55.

[39] SHAMBAUGH G.E., BALINSKY J.B. & COHEN P.P. (1969) Synthesis of carbamyl phosphate synthetase in amphibian liver *in vitro*. Effect of thyroxine. *J. Biol. Chem.*, **244**, 5295–308.

[40] DE GROOT N. & COHEN P.P. (1962) Amino acid activating enzymes and pyrophosphatase in frog and tadpole liver. *Biochim. Biophys. Acta.*, **59**, 595–604.

[41] TONOUE T., EATON J.E. & FRIEDEN E. (1969) Changes in leucyl-tRNA during spontaneous and induced metamorphosis of bullfrog tadpoles. *Biochim. Biophys. Res. Commun.*, **37**, 81–8.

[42] TATA J.R. (1971) Protein synthesis during amphibian metamorphosis. In *Current Topics in Developmental Biology*, **6**, 79–110. Academic Press, New York & London.

[43] IKEHARA Y. & PITOT H.C. (1973) Localisation of polysome-bound albumin and serine dehydratase in rat liver cell fractions. *J. Cell Biol.*, **59**, 28–44.

[44] SCHAFFER B.M. (1963) The isolated *Xenopus laevis* tail: a preparation for studying the central nervous system and metamorphosis in culture. *J. Embryol. Exp. Morphol.*, **11**, 77–90.

[45] WEBER R. (1969) Tissue involution and lysosomal enzymes during anuran metamorphosis. In *Lysosomes in Biology and Pathology*, Vol. I (Eds J.T. Dingle & H.B. Fell), pp. 437–61. North-Holland Publishing Co., Amsterdam & London.

[46] DE DUVE C. (1963) The lysosomal concept. In *Lysosomes*. Ciba Foundation Symposium (Eds A.V.S. de Reuck & M.P. Cameron), pp. 1–31. J. & A. Churchill, London.

[47] WEISSMAN G. (1961) Changes in connective tissue and intestine caused by vitamin A in amphibia, and their acceleration by hydrocortisone. *J. Expl. Med.*, **114**, 581–92.

[48] WEBER R. (1965) Inhibitory effect of Actinomycin D on tail atrophy in *Xenopus* larvae at metamorphosis. *Experientia*, **21**, 665–66.

[49] EECKHOUT Y. (1966) Aspects biochimiques de la metamorphose. *Rev. Questions Sci.*, **27**, 377–410.

[50] TATA J.R. (1966) Requirement for RNA and protein synthesis for induced regression of the tadpole tail in organ culture. *Develop. Biol.*, **13**, 77–94.

[51] WEBER R. (1964) Ultrastructural changes in regressing tail muscles of *Xenopus* larvae at metamorphosis. *J. Cell Biol.*, **22**, 481–87.

[52] FOX H. (1972) Tissue degeneration: an electron microscopic study of the tail skin of *Rana temporaria* during metamorphosis. *Arch. Biol. (Liège)*, **83**, 373–94.

[53] FOX H. (1972) Muscle degeneration in the tail of *Rana temporaria* larvae at metamorphic climax. *Arch. Biol. (Liège)*, **83**, 407–17.

[54] GRISWOLD M.D., FISCHER M.S. & COHEN P.P. (1972) Temperature-dependent intracellular distribution of thyroxine in amphibian liver. *Proc. Nat. Acad. Sci.*, **69**, 1486–89.

[55] JENSEN E.V., NUMATA M., BRECHER P.I. & DESOMBRE E.R. (1971) Hormone-receptor interaction as a guide to biochemical mechanism. In *The Biochemistry of Steroid Hormone Action*. Biochem. Soc. Symp. 32 (Ed. R.M.S. Smellie), pp. 133–60. Academic Press, New York & London.

[56] SAMUELS H.H. & TOMKINS G.M. (1970) Relation of steroid structure to enzyme induction in hepatoma tissue culture cells. *J. Mol. Biol.*, **52**, 57–74.

[57] TATA J.R. (1968) Early metamorphic competence of *Xenopus* larvae. *Develop. Biol.*, **18**, 415–40.

[58] TATA J.R. (1970) Simultaneous acquisition of metamorphic response and hormone binding in *Xenopus* larvae. *Nature, Lond.*, **227**, 686–89.

[59] WEBER R. (1967) Biochemical and cellular aspects of tissue involution in development. In *Morphological and Biochemical aspects of Cytodifferentiation*. *Exp. Biol. Med., Vol.I*, (Eds E. Hagen, W. Wechsler & P. Zilliken), pp. 63–76. S. Karger, Basel & New York.

[60] ETKIN W. (1950) The acquisition of thyroxine—sensitivity by tadpole tissues. *Anat. Record*, **108**, 541.

[61] MOSER H. (1950) Ein Beitrag zur Analyse der Thyroxin wirkung im Kaulquappenversuch und zur Frage nach dem Zustandekommen der Fruhbereitschaft des Metamorphosereaktions—systems. *Rev. Suisse Zool.*, **57**, (Suppl. 2) 1–144.

Conclusions to Part 4

1. HORMONES COORDINATE DEVELOPMENTAL CHANGES

As well as the short-range interactions between cells and tissues discussed in Part 3, development also requires the coordination of changes occurring in different parts of the organism. In animals such long-range interactions generally depend on the movement of substances ('hormones') through the organism which initiate changes in different tissues, as in the wide range of changes occurring during amphibian metamorphosis in response to thyroxine. The situation in plants is modified by the constraints imposed by the presence of cell walls, so that hormones can generally affect development only during the growth phase in the meristematic regions of the plant. Nevertheless, hormones are of overriding importance as co-ordinators of development between different parts of the plant, at the cell, tissue and organ levels. Thus, correlation between the development of a leaf and its vascular connections is ensured by the circumstance that development of the vascular tissue is stimulated by IAA produced in the growing leaf itself.

Hormones also appear to play an essential role in developmental responses to environmental factors, as in the induction of metamorphosis in insects and of flowering in plants by seasonal changes in daylength and temperature.

2. HORMONES USUALLY DO NOT DETERMINE CELL TYPE BUT INDUCE DEVELOPMENTAL CHANGES IN ALREADY DETERMINED CELLS

Hormones often elicit growth and differentiation in pre-existing tissues without inducing new cell types. These hormones can control whether flowering plants and amphibians have male or female reproductive organs by stimulating the growth of either male or female primordia in ambisexual organisms. The multiple effects of each type of plant hormone (p. 217) clearly indicates that the specificity of the response must lie in the 'programming' of the target tissue. Thus, gibberellic acid may cause flowering, germination or α-amylase production in different organs and tissues. Again the normal production of xylem on the inner side of the cambium in response to IAA can be changed if a sector of cambial tissue is reversed in its orientation, suggesting that the derivative cells on the normally inner side are pre-determined to produce xylem (p. 224).

There are other cases when hormones set up the necessary conditions for the formation of new cell types. Cyclic AMP is required for the aggregation of some slime moulds and without aggregation the cells will not differentiate unless they are exposed to special conditions; the hormone is therefore a necessary condition for the subsequent stalk and spore cell development.

3. IN RARE CASES HORMONES MAY BE DIRECT CELL TYPE DETERMINANTS

In a few cases it has been shown that cells may become determined and differentiate by the direct action of a hormone. Thus, physiological concentrations of ammonia will induce 100% spore formation in sparse cultures of the slime mould, *Polysphondylium,* and there is also evidence that this may be the normal signal for spore formation in this species. The production of vascular tissue in parenchymatous cortical tissue by applied IAA would also seem to provide an example of induction of differentiation by a hormone in cells not pre-determined to follow this pattern. Another example of determination induced by hormones is the regeneration of organized bud and root primordia in undifferentiated callus tissue by appropriate concentrations of IAA and kinetin; this is apparently an example of organogenesis induced by hormones, but the specific cell types formed by the meristems is probably not determined by hormones.

4. MECHANISM OF HORMONE ACTION

In many cases hormones induce changes in RNA metabolism and in some instances the hormone is thought to exert this effect either by directly bonding to chromatin or by binding to the chromatin in association with a hormone receptor (e.g. some of the changes in amphibian metamorphosis). This change in RNA metabolism is followed by the massive synthesis of particular proteins (e.g. α-amylase in barley seeds and carbamyl phosphate synthetase in frog liver), and it is likely that the synthesis of messenger RNAs for these proteins are induced by the hormones.

It should be noted that hormones may also have a direct effect outside the cell nucleus. There is evidence that the action of IAA on cell wall enzymes during coleoptile extension and the cyclic AMP induction of movement in aggregating slime moulds do not immediately depend on the action of the nucleus.

In several examples, the pattern of development appears to be controlled by changes in hormone concentration; thus, in amphibians, the different stages of metamorphosis appear to be controlled by the levels of thyroxine. There are several examples from plants, in which the pattern of differentiation is markedly affected by the relative concentrations of two different types of hormone, as in bud and root regeneration in callus tissue, and the production of xylem and phloem by the cambium.

Part 5
The Molecular Biology
of Development

Chapter 5.1
Self-Assembly

5.1.1 INTRODUCTION

In this chapter we shall be discussing the extent to which the development of an organism occurs by a process of self-assembly. It is immediately necessary to clarify the meaning of self-assembly for we shall not be concerned here to talk about reductionism nor about the philosophical understanding of the properties of things, however valid or even relevant these subjects may be [1]. The aim is a restricted one; the question at issue is whether the component parts of an organism have, by themselves, the potential to spontaneously aggregate with the relationships that result in the organism. To what extent is the final form of an enzyme, cell organelle or larger structure, a consequence of its component parts? Is it a property of the parts, a part of their regular behaviour, to aggregate in this way? What are the conditions required to enable them to aggregate thus? Is it possible to understand the aggregation of the parts in terms of physical forces? To what extent may the demand of biological usefulness supplement the minimization of free energy as a factor in understanding biological structures? By posing these restricted questions we neither pre-suppose answers to the broader philosophical questions nor prejudice their urgency; rather the aim is to see to what extent self-assembly is a useful concept in discussing biological structures.

The question of self-assembly may be raised at various levels in the organism. Does molecular folding occur by self-assembly and do molecules then self-assemble to form higher levels of aggregation such as filaments, membranes, and organelles? We may then ask if these structures self-assemble to yield cells and finally whether cells self-assemble to make tissues. I shall start with the long polymer molecules, proteins, nucleic acids and lipids. This is not quite arbitrary though there is a sense in which self-assembly occurs before this level. Atoms may be regarded as the product of the self-assembly of more elementary particles which exist in highly symmetrical states of low energy. Small molecules are similarly stable states of the aggregation of atoms. Experiments on the prebiotic synthesis of amino acids and sugars show that when energy in a variety of forms is introduced into certain gaseous mixtures of small molecules [2], common monomers of biology are formed. We will exclude discussion of these from self-assembly because of the high energies involved and in particular because of the breaking and making of many covalent bonds. We shall also exclude, on grounds of high energy, the self-assembly of coacervates and microspheres [3] which occurs when mixtures of polypeptides are heated. The self-assembly under discussion is the kind that could occur in living animals. Thus extremes of energy are not permitted. A start will be made at the molecular level and the aim will be to see how far up the hierarchy of levels of organization the concept of self-assembly is appropriate.

A further point about self-assembly that requires clarification is the interpretation of 'self'. Once more we deliberately side-step the philosophical questions of the extent to which the concepts of physics such as force, energy, length and mass which are estimated by operations involving the behaviour of matter, can be described as properties of the parts of organisms. The only aspect considered here will be the experimental question 'Can the parts alone, aggregate to form the organism or a more complex part of it?' Even at this stage it is necessary to further delimit the self-assembly under consideration. Kellenberger (in [13]) has suggested that the general process of whereby a biological structure takes its shape be called morphopoesis. He then defines morphopoesis of the first level as that in which the only subunits which take part in the assembly process are those which will themselves form part of the completed structure. *This will be the meaning of self-assembly in this chapter.* It rules out cases where spatial information is imposed externally by a template which is not itself part of the final structure. It also excludes systems where other subunits such as enzymes are required to effect some of the steps in assembly. If thoroughly followed, this definition of self-assembly would also exclude the participation of solvent molecules and so small molecules and ions such as water, small anions and cations are arbitrarily discounted as subunits when the assembly of macromolecules is being considered.

We are thus left with a very elementary definition of self-assembly as a process in which macromolecules and larger units, either fold or aggregate into specific structures by themselves in aqueous solutions of small molecules. Our basic question is 'How far does this kind of self-assembly occur in the development of organisms?'

A model for the type of self-assembly we shall discuss is crystallization. When the environmental conditions are right, crystallization occurs spontaneously and the resultant microscopic symmetry of the crystal is a direct consequence of the intermolecular interactions. To deal with biological materials however we must have a broad understanding of crystallization which led Bernal [4] to introduce the term 'generalized crystallography'. We are not only interested in the regular three-dimensional lattices but also in the linear arrays of identical interaction which form helices and the two-dimensional arrays comprising the closed surfaces of the regular solids. It was pointed out by Crane [5] and Caspar & Klug [6] that the high symmetry of helical polymers and icosahedral viruses makes it likely that these structures occur by self-assembly. In self-assembly two subunits may interact with an energetically favourable interaction. If a third subunit is added and can adopt the same relation to the second as the second to the first, then a highly symmetrical structure will result. A symmetry operation is one which usually involves rotation, translation or reflection of an object and the result of the operation is to leave the object indistinguishable from its initial state [7]. Symmetry therefore, depends on equivalence within the object and self-assembly depends on equivalent intersubunit interaction. Equivalence results from there being a favoured, low-energy interaction between the assembling or assembled units. At the molecular level a distinguishing feature is that subunits are *precisely* identical so symmetrical relationships can be precisely obeyed. In this chapter we shall survey some examples of self-assembly at the molecular level and it will now be obvious that we may expect the symmetrical filaments and spheres to form by self-assembly.

Self-assembly has already been discussed by Bernal [4], Calvin [8] and Kushner [9] and in symposia edited by Wolstenholme & O'Connor [10], Allen [11], Hayashi & Szent-Györgyi [12], Engström & Strandberg [13] and Timasheff & Fasman [14]. In this chapter we will try to give an up-to-date account of the available evidence. Since a large number of topics has to be covered references will only be to key papers and to recent reviews which could be consulted for more detailed information.

5.1.2 MOLECULAR FOLDING

Protein molecules crystallize and this implies that each molecule is folded into the same three-dimensional conformation. It is possible to ask whether this conformation can arise spontaneously or whether it requires some external information to fold correctly. At the level of molecular folding it is now widely accepted that the three-dimensional structure of the molecule is determined solely by the sequence of amino acid residues. At present the basis of this view is almost entirely experimental with only partial indication of the physical principles behind the protein folding, and we shall now review some of this evidence.

Helix–coil transitions

Many homo-polypeptides (polymers of a single type of amino acid) wilt take up the regular α-helical conformation in aqueous solution. There is a conformation which results in the main chain of the polypeptide following the locus of a helix of pitch 5·4 Å with a 1·5 Å axial translation between adjacent amino-acids (Fig. 5.1.1). Clearly the relation between neighbours in

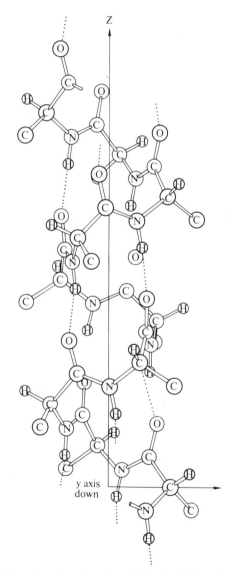

Fig. 5.1.1 The α-helix conformation of a polypeptide chain. This is a right-handed helix of pitch 5·4 Å with an axial shift of 1·5 Å between adjacent amino acid residues.

this helix is energetically favourable. That this is so has been demonstrated in a number of ways. Ramachandran *et al.* [15] pointed out that the relation between neighbouring amino acids in a polypeptide chain with planar peptide groups, could be represented by two angles ϕ and ψ (see Fig. 5.1.2). The definitions of ϕ and ψ recommended by the IUPAC-IUB Commission [16] is used here. Wooden space-filling molecular models may be used to show that, if one assumed that atoms could not approach each other more closely than the van der Waals distance*, the possible (ϕ, ψ) values were severely limited and indeed could be shown to occur in about five patches in the (ϕ, ψ) plot. This exercise was repeated by Liquori [17] who used Lennard-Jones potential functions to represent the molecular interactions and showed quantitatively that there were five minima in the energy contours for poly-L-alanine. The lowest minimum from Liquori's calculations occurred at the (ϕ, ψ) values in the α-helix. Hence the common occurrence of the α-helix can be accounted for and deviations from it have to be explained.

In studies on polyglutamic acid, Doty *et al.* [19] had shown the α-helix arrangement in solution existed at pH values when the carboxylic groups were uncharged, but the helix gave way to a random coil at pH values where the groups were charged. Mutual repulsion between the similarly charged acid groups is obviously causing the helix to unfold. Similarly poly-L-lysine was examined [20] and it was found that the optical rotation at neutral pH was 100° more negative than that at pH 12. Now the extent of optical rotation has been used as a measure of helix content. The change in optical rotation occurred over a pH range of 1 at pH 10, the pKa value of the amine group. The values of the optical rotation indicate that polylysine is helical when the amino group is uncharged but when it is charged by the uptake of a proton at low pH the molecule unfolds. Raising the pH brings about α-helical formation again. In 1M salt solution plus buffer, this transition is moved to a lower pH presumably because the ionic solvent decreases the charge–charge repulsions which unfold the helix.

Another factor which can cause destabilization of the α-helix is bulky side-chains. For this reason valine and isoleucine do not form α-helices. Work on the conformations adopted by

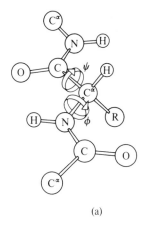

(a)

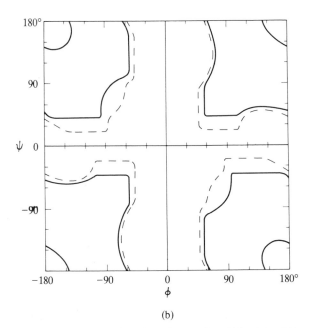

(b)

Fig. 5.1.2 Diagram to show the angles ϕ and ψ which relate neighbouring amino acids in a polypeptide chain.

* For an uncharged atom it is possible to estimate a so-called van der Waals radius. The sum of the van der Waals radii of two atoms is frequently used to evaluate the contact distance between the atoms. Two physical interactions are involved. One is the force of *repulsion* between the negatively charged electron clouds of the atoms. This force, frequently termed the Born repulsion force, varies as $1/r^{12}$ where r is the distance between the atomic centres. The other has a more complex origin and is a force of *attraction*. Two permanent dipoles have a residual force of attraction between them. Atoms with equal numbers of positive and negative charges in nucleus and electron cloud respectively have no permanent dipole when the electron positions are averaged over time. However, a 'snapshot' of an atom would reveal an instantaneous dipole and London calculated that the average effect of interactions of these instantaneous atomic dipoles was an *attractive* force proportional to $1/r^6$. The force arises because one instantaneous dipole induces a dipole in the other atom. Experimentally the force may be estimated by measurement of the dispersive effect they induce into rays of light. This is because they depend on *induced* dipoles and hence the polarisability of atoms. This, in turn, determines their effect on the refractive index of light. The combined effect of the Born *repulsive* force and the London dispersion *attractive* forces was calculated by Lennard-Jones to give an expression for the energy of interaction as $(a/r^{12} - b/r^6)$. The value of this expression is very large when r is small (less than 1 Å) indicating that a large amount of energy would be required to bring atoms as close as this. However an energy minimum does exist around r values of 3–4 Å and this corresponds to the van der Waals distance. The van der Waals radius of an atom denotes the atomic volume which was assumed by van der Waals in his modification of the perfect gas equation. For details, see a standard text such as [18].

sequential polypeptides [21] has shown the increasing destabilization of the α-helix by progressively increasing the fraction of valine in poly(glu-val).

Hence it is reasonable to conclude that the nature of the side-chains, which is often determined by the solution medium, does enable polypeptide chains to self-assemble into a specific conformation.

Coiled-coils

In addition to determining the helical parameters of a polypeptide chain, the particular sequence of amino acids residues has now been clearly associated with the way in which the polypeptide chain may interact with other chains. This is evident in the relatively simple case of fibrous proteins where there is a regularity of sorts along the polypeptide chain.

In collagen, glycine occurs as every third residue for the whole of the tropocollagen molecule (1011 residues plus two teleopeptides without this regularity accounting for an additional 41 residues—see summary in [22]). There is also a high fraction of proline and hydroxyproline residues in collagen and the effect of proline in a polypeptide chain is to make $\phi = -70°$. Thus a single chain folds up as a helix with approximately three residues per turn. This results in all the glycine residues occurring approximately above each other on one side of the helix. The tropocollagen molecule is then formed when three such single chains come together parallel with their glycine edges in contact; the individual helices must be twisted about each other so as to bring the glycines exactly above each other to form the core of a three-strand rope. Sequential polypeptides (gly-pro-hypro) and (gly-pro-pro)$_n$ to give (gly-pro-pro)$_n$ adopt a collagen-like conformation [23, 24].

The three-strand collagen molecule can be broken down by heat treatment to form gelatin. This collagen−gelatin transition is accompanied by a loss of order and the original collagen molecule of molecular weight 300,000 breaks down to monomers of molecular weight 100,000, the so-called α-chains. The complete amino-acid sequence of a single α-chain has been determined (see summary in [22]) and for 1011 residues glycine occurs in every third position, an essential factor for the adoption of the collagen fold. Gelatin may be renatured by cooling and triple helical molecules are reformed. This has been demonstrated by the fact that stretched gelatin yields an X-ray diffraction pattern similar to native collagen [25] and infra-red studies show the reappearance of the N−H stretching frequency characteristic of collagen when gelatin gels are cooled [26]. The primary sequence thus determines the molecular interaction between the three α-chains to form a tropocollagen molecule.

A similar situation has been shown to exist in the α-proteins of muscle. In 1953, Crick [27] suggested that the X-ray diffraction pattern of the α-proteins could be accounted for by a structure in which two α-helices coiled around each other to

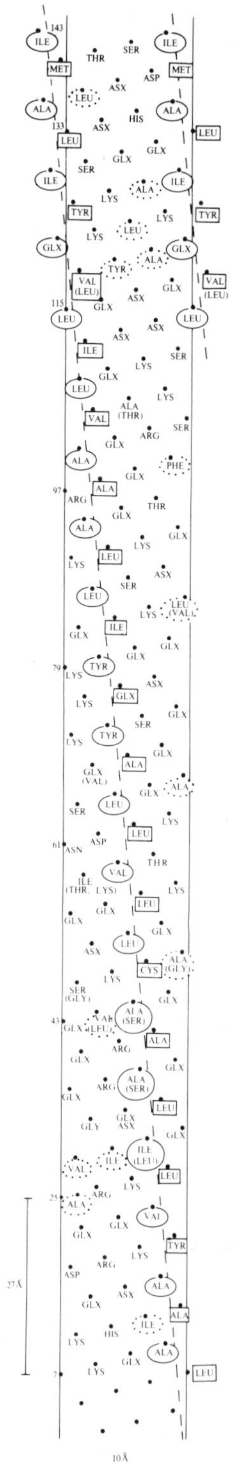

Fig. 5.1.3 A radial projection of the amino-acids in tropomyosin in the α-helical conformation. This projection is made by drawing radii through each amino acid residue and making their intersection on a cylindrical sheet concentric with and surrounding the α-helix. The cylindrical sheet is then cut along a line parallel to the helix axis and opened into a flat sheet.

form a two-strand coiled-coil. He further suggested that this supercoiling might be brought about if the apolar amino acid residues were arranged in a regular way such that they lay on a long-pitch helix round the outside of an α-helix. Two such α-helices could then twist around each other so as to bring the apolar helices into contact in the centre of the two-strand rope. Coiling the α-helices would straighten out the apolar helices so that they met in a line along the core of the two-strand rope. X-ray evidence for the two-strand structure was provided for paramyosin [28] and α-keratin [29], and Woods [30] showed that a tropomyosin molecule could be dissociated in 8M urea into two chains of equal molecular weight. Elegant confirmation of Crick's suggestion has now come from studies on the amino-acid sequence of tropomyosin [31]. The sequence of amino acids for about one half of the tropomyosin molecule has been determined and it has been shown that apolar residues occur regularly along the chain. When the chain is plotted in the α-helical conformation, the apolar residues trace out a helix of about the pitch of the coiled-coil required by Crick (see Fig. 5.1.3). Once again the sequence of the amino acids clearly determines the molecular interactions and thus directs the construction of the tropomyosin molecule.

Globular proteins

The globular proteins are more complex than the fibrous proteins in that they are not mainly regular in structure and the geometric relations between nearest neighbours in the polypeptide chain are not identical so the chian does not adopt a regular helical conformation for the whole of its length like collagen or the α-proteins. However the globular proteins do have the common feature that though the main polypeptide chain follows an irregular course, it usually winds up in a compact manner with roughly spherical or ellipsoidal shape. The fact that globular proteins crystallize implies that the molecules have adopted precisely identical three-dimensional structure and they also appear to do so in solution (Fig. 5.1.4).

The self-assembly of a molecule into its three-dimensional structure from an unfolded state was first clearly demonstrated with ribonuclease [32]. The ribonuclease molecule contains four disulphide bonds. When these bonds are broken by reducing the sulphydryl groups, the molecule will unfold with accompanying loss of enzyme activity. There are over one hundred theoretical ways in which the eight reduced half cysteine residues could recombine, but in an oxidizing environment the molecule refolds to form the original structure in high yield with full enzyme activity.

Other globular proteins have since been denatured by various means and renatured to restore structure and activity. Sometimes heating followed by cooling can effect the transitions. Thus the primary structure is sufficient to determine the tertiary structure of globular as well as fibrous proteins.

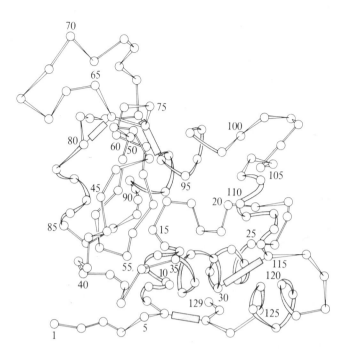

Fig. 5.1.4   Schematic drawing of the main chain of lysozyme. (By W.L. Bragg, from Blake *et al.* (1965) *Nature*, **206**, 757–63.)

Because of the increased complexity of the situation, it is not so easy to see why globular proteins adopt the three-dimensional structure that they do. However the same general feature of apolar residues occupying the interior of the molecule applies here as in fibrous proteins. In some cases it is possible to see why certain amino acids residues are essential for the preservation of a given tertiary structure. For example only glycine in cytochrome *c* could permit the shortish chain to encompass the prosthetic group. In general, however, it is more difficult to understand the folding mechanism. The most recent experiments and ideas on how globular proteins fold have been reviewed [33]. A basic question is whether or not the molecule folds up as it comes off the ribosome. This is made unlikely by the observation that a peptide consisting of the first twenty amino acids of ribonuclease appears to take up no definite tertiary conformation. Presumably then, the whole molecule is required to stabilize the native structure. This idea is supported by the work of Anfinsen and colleagues on staphylococcal nuclease. The bond following amino acid 126 can be split to give a polypeptide containing amino acids 1–126 from the amino terminal end of the molecule. This is 85% of the molecule and it is found to possess no definite structure and to lack catalytic activity. Other experiments reveal that fragment 6–48 will react with 49–149 to yield an active enzyme. Thus nearly all of the protein is required for it to take up the 'native' conformation.

Experiments on staphylococcal nuclease by Jardetzky and co-workers have provided information about the steps by which the protein folds into its native conformation. Proton magnetic resonance is used to probe the conformation of recognizable residues in the polypeptide chain. Jardetzky finds that when the molecule is converted from a 'random coil' to a specific tertiary structure by varying the pH, the change does not take place in one step. Fragments of the molecule 'click' into the native conformation one by one and it has been possible to suggest a sequence of steps by which the random coil adopts the native conformation. Anfinsen believes that the kinetic folding and unfolding of the molecule will result in transitory 'native' fragments which may interact with each other to form a nucleus for the folding of the rest of the molecule.

We may conclude then that the mechanism of the folding of globular proteins certainly conforms to our requirements for self-assembly. The primary sequence alone, in the appropriate solvent conditions, is sufficient to determine the unique and complex manner in which the polypeptide chain will fold. However, it also appears that nearly all of the molecule is required for the folding process and the stability of the unique tertiary structure depends on the interaction of virtually the whole molecule. At present it seems unlikely that the molecule folds up simply from one end to the other and we are far from understanding the physics of protein folding. This is not surprising in view of the fact that the physics of folding of polyethylene is still a topic of active research. For our purpose we note that the form adopted by the molecule is determined by the amino-acid sequence and hence, by the sequence of the bases in the DNA coding for the protein.

5.1.3 MOLECULAR AGGREGATION—(1) LIKE MOLECULES

Once macromolecules such as proteins have folded into their native conformation it is possible to ask if this tertiary structure is sufficient to determine the arrangement which these molecules will adopt in the next level of the structural hierarchy, that is, in structures such as fibres, organelles, membranes and viruses. This is frequently the main point at which self-assembly of biological materials is discussed (see e.g. [9]) but it is obviously one of a series of levels.

Many biological structures are in fact composed of identical subunits assembled, like bricks, to form a larger structure. The genetic advantage of this was early recognized [34]. The DNA need only be long enough to specify one subunit and a defect in

a subunit involves the loss of only one subunit rather than one member of the larger structures.

Well established examples are available of self-assembly of long molecules to form fibrous structures.

Collagen

Collagen molecules may be separated from connective tissue by extraction with salt at low temperatures or at acid pH at room temperatures. The so-called tropocollagen molecules obtained are 3000 Å long and 15 Å in diameter with the triple helical structure described in the section on coiled-coils above. A solution of tropocollagen may be reprecipitated and the precipitate examined in the electron microscope. This reveals fibrils with a 640 Å periodic banding pattern indistinguishable from the fibrils isolated from native tendons. It is clear that the structure of the molecule is sufficient to define the mode of molecular packing in the fibril.

In native tendons the molecules are held together by covalent bands formed between modified lysine residues (see [35] for review). The lysine is oxidized enzymatically to allysyl residues which have aldehyde groups. These then crosslink covalently. Covalent crosslinking is not essential for fibril formation since banded fibrils may be reconstituted from tropocollagen in which the conversion of lysine to aldehyde is prevented and the molecular arrangement in latharytic rat tail tendon is the same by X-ray diffraction (Miller & Parry, unpublished) as that in native tendon.

Some understanding of the kind of interactions that might be responsible for specifying the molecular packing in collagen comes from studies of the amino-acid sequence [22].

The primary sequence of one α-chain is now known from the pooled results of various workers. If two such linear sequences are laid alongside each other, it is possible to estimate by rather crude criteria how the intermolecular interaction would vary as the sequences are moved past each other. The interaction is scored by assigning +1 when either, two oppositely charged residues or two polar residues are within range of each other. These estimates reveal a periodicity of 670 Å (or 234 residues) in the intermolecular interactions which tallies well with structural information available by X-ray diffraction and electron microscopy (see Fig. 5.1.5). The main implication for our purposes is that this periodicity emerges from the amino-acid sequence alone, thus it is possible to give a chemical account of the origins of the self-assembly of the molecules. We also note that the most clearly defined periodicity is followed by the large apolar residues. The hydrophobic interactions are shown to be important in molecular aggregation just as they were in molecular folding.

The tropocollagen molecule can be precipitated from solution to yield other regular polymorphic structures which are not usually observed in native collagen. One of these, so-called SLS,

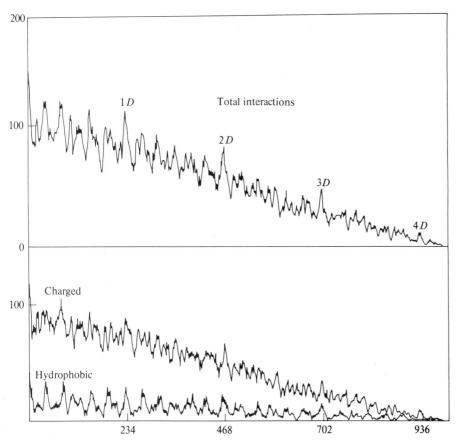

Fig. 5.1.5 A computer plot of the number of hydrophobic and charge interactions and their total between two collagen molecules (on the ordinate) as a function of the stagger between them. The stagger (on the abscissa) is measured in residues. See [22].

in which neighbouring molecules are lined up in parallel register with each other, is readily understood since it occurs in the presence of ATP. The diphosphate could be thought of as bridging similarly charged residues. Other polymorphic forms have symmetric periodicities and this indicates that the molecules are assembling in anti-parallel array. The relative stagger between the molecules in these symmetric structures has also been shown to involve regular interaction between apolar residues [36]. Thus in collagen we begin to understand the origins of molecular specificity which determines the different ways in which the molecules may self-assemble.

There is evidence that in native tendon, the tropocollagen molecules interact by staggers of 670 Å between nearest neighbours so as to form a microfibril with a 5_1 topology [37, 38]. If this microfibril interpretation is correct it will illustrate

interesting design features. Helical symmetry occurs when amino acids and nucleotides polymerize and it is again seen to emerge when identical subunits engage in identical interactions. The 670 Å stagger is long enough to ensure that the addition of one molecule leads to an increase in the length of the microfibril while at the same time allowing a large degree of molecular interaction. The helical arrangement suggests that in addition to specific axial shifts between molecules there will also be specific azimuthal relations and it may be possible to trace this specificity to the molecular structure [58] coupled with the amino-acid sequence [22]. The helical arrangement also suggests the microfibril as a 'unit tendon' in which the continuity of molecular contacts is parallel to the axis along which the stress is applied to the tendon. We can now appreciate this design feature as a direct outcome of the primary sequence mediated through the molecules first folding and then aggregating together in a specific manner.

Muscle

Muscle affords an excellent example of a series of levels of order all of which are essential for the biological function of contractility which involves transduction of the chemical energy from the hydrolysis of ATP into mechanical work.

In striated muscle, single fibres of diameter 70 μ have a regular banded appearance of period 2–3 μ. In electron micrographs of longitudinal sections through muscle fibres, this banding is seen as due to the in-register arrangement of myofibrils of about 1 μ diameter; each longitudinal unit of length 2–3 μ is termed a sarcomere. When muscle shortens, the sarcomeres shorten by the same fraction and this has been accounted for by two sets of filaments sliding between each other (Fig. 5.1.6) [39, 40]. The thick filaments are in the centre of the sarcomere and the thin filaments originate from the Z-disc and interpenetrate between the thick filaments. Thus muscle shortening occurs by the filaments sliding between each other without themselves changing length. Crossbridges have been visualized in the electron microscope and shown by this method and X-ray diffraction to originate from the thick filaments (see review in [41]). A suggested mechanism for muscle contraction is that in which the crossbridges interact cyclically with the thin filaments and pull them into the centre of the sarcomere thus shortening the sarcomere and hence the whole myofibril.

A satisfying illustration of self-assembly is provided by studies on the molecular packing within the thick filaments. The thick filaments are composed largely of the protein myosin. This molecule is about 1500 Å long and consists of a α-helical rod of

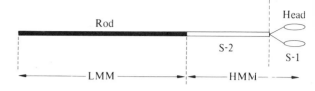

Fig. 5.1.7   Diagram of a myosin molecule.

length 1350 Å to which is attached a globular, enzymatically active head [42, 43] (Fig. 5.1.7).

Myosin molecules may be dissolved from thick filaments by solutions of high ionic strength (0·6M). When the ionic strength of such myosin solutions is lowered, it was observed with the electron microscope that the molecules reaggregated to form rodlets [44]. A detailed study of this reaggregation process and the structures of the aggregates was made by H.E. Huxley [42]. The spindle-shaped aggregates were closely similar in appearance to the thick filaments of muscle though they exhibited a wide variation in length in contrast with the precisely equivalent filaments in muscle. However the synthetic filaments were always polarized about their centre. The centre of each synthetic filament differed from the rest of the length in that it did not have rough projections on the surface but was relatively smooth. This so-called 'bare zone' in the centre of the synthetic filaments was observed to be of a fairly constant length 1500–2000 Å. The projects from the rest of the filament surface were presumed to be the globular heads of myosin molecules.

Huxley's [42] results suggest that the assembly of thick filaments takes place in two stages (Fig. 5.1.8). First myosin molecules came together pointing in opposite directions with the

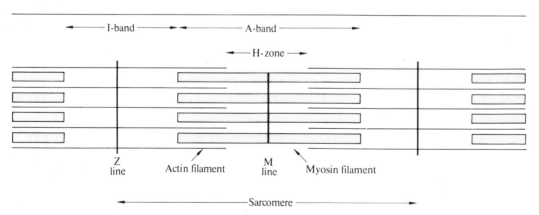

Fig. 5.1.6   A diagram of the arrangement of thick and thin filaments in a sarcomere of striated muscle.

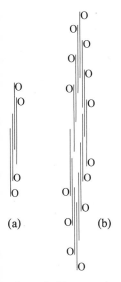

Fig. 5.1.8 Two stages in self-assembly of myosin molecules to form synthetic thick filaments. (a) Molecules assemble rod-to-rod pointing in opposite directions to form a nucleus. (b) The nucleus grows in both directions by the aggregation of myosin molecules pointing in the same direction to form a bipolar thick filament.

rod portions overlapping. This accounted for the constant length 'bare zone' in the centre of the filament. The ability of myosin to form anti-parallel dimers has been supported by Harrington *et al.* [45] on the basis of high speed sedimentation equilibrium studies of myosin association. More detailed study of the anti-parallel interaction of myosin by electron microscopy has revealed that myosin molecules and rods are able to form anti-parallel segments *in vitro* with overlap lengths of 1300, 900 and 430 Å (see [46] and for a review [47]). These studies reveal that *in vitro* distinct modes of anti-parallel packing of myosin rods is possible.

The second stage of thick filament assembly follows the first when the nucleating anti-parallel segments (the bare zones) are elongated by the parallel interaction of rods on opposite sides of the segment (Fig. 5.1.8(b)). Myosin molecules in opposite halves of the filament point in opposite directions but molecules in the same half point in the same direction. Compound segments of myosin rods have also been shown *in vitro* to form both polar and bipolar patterns [48]. As Huxley [42] pointed out, this bipolar symmetry of synthetic thick filaments is a design feature essential for the functioning of the sliding filament model of muscle contraction where thin filaments are moved in opposite directions in opposite halves of a sarcomere. Huxley [42] further demonstrated that the thin filaments are polar when interacted with globular portions of myosin and that they too point in opposite direction in opposite halves of a sarcomere.

Huxley's work shows that the thick filament design features are a consequence of the self-assembling properties of the myosin molecule.

Once again we note that, as with the tropocollagen molecule, the myosin molecule is capable of polymorphism in which the modes of molecular interaction are distinct. The question arises as to whether all of these modes are utilized *in vivo* and Harrison *et al.* [46] have proposed a scheme for the molecular arrangement in the thick filament which does involve a number of the distinct molecular contacts. It may also be possible to generate varieties of filament structures by modulation of the same set of modes of molecular contacts. Sobieszek [49] has shown that myosin from vertebrate smooth muscle may be reprecipitated to yield long filaments that have no central 'bare zone' but rather a 'bare zone' is located on one side at the ends of the filament. A possible interpretation is that these filaments are formed by association of small bipolar units such as myosin dimers. These filaments have a clear 140 Å axial periodicity which is taken as a visualization of myosin heads.

In summary, the thick filaments of muscles are known to have different symmetries but it is likely that they are all based on similar interactions of myosin molecules. A general model for thick filament structure has been suggested by Squire [50, 51, 52] in which the myosin contacts are closely similar. It is worth noting however, that other factors are likely to be involved in determining thick filament architecture. Though the shapes of myosin molecules isolated from different kinds of muscle are closely similar, there are known variations in other properties such as solubility, ATP-ase activity and presence of low molecular weight components. Furthermore it has been established that there is another major protein constituent of the thick filaments the so-called C-protein [53] which appears to be regularly arranged with the periodicity of the myosin heads.

A special type of thick filament occurs in certain non-striated muscles of molluscs; this is the paramyosin rich 'catch' muscles of the oyster aductor muscle and the anterior byssus retractor muscle of mussels. Filaments are of diameter 100–1500 Å in other words about 4–6 times that of the thick filaments of vertebrate striated muscle. Myosin occurs as a macromolecular veneer on the surface of these filaments while the rest of the filament is composed of the protein paramyosin. This is rod-shaped, some 1300 Å long having the two-strand coiled α-helix structure of the α-proteins and is devoid of enzymic activity. The detailed molecular arrangements of the paramyosin within these filaments has been investigated [54, 55]. For our purpose we note that electron micrographs of 'catch' muscle thick filaments from which myosin has been removed, shows a characteristic regular pattern. This so-called checker-board pattern has a true axial period of 725 Å which reduces to 145 Å when the structure is projected on to the fibre axis. The existence of polar triangles in the checkerboard reveals that the filaments are bipolar about their centre and electron microscopy of filaments which have been 'shadowed' by stain on both sides show the filament has a helical structure [56].

Paramyosin may be precipitated *in vitro* and reconstituted filaments with various banding patterns are observed by electron microscopy indicating the ability of paramyosin to make different polymorphic structures. As in the case of myosin, these polymorphic forms and the bipolar filament structure point to different modes of packing of the paramyosin molecule. One of the polymorphic patterns has the symmetry of the native checkerboard pattern and closely similar periodicities; it may be constructed by appropriate arrangement of other observed polymorphic aggregates and Cohen *et al.* [55] conclude that it is likely that the thick filaments of catch muscle are formed by self-assembly of paramyosin molecules. They also suggest that polymorphism indicates that the different modes of molecular bonding are likely to have closely similar energies and it is not unlikely that dynamic transitions between some of these states are important in certain contractile functions such as 'catch' in which the muscle can develop tension for prolonged periods. Paramyosin is now known to occur quite generally. It has been isolated from insect flight muscle [57] and it has been suggested that it is a core for the myosin containing filaments in certain muscles.

It is clear then that the muscle proteins self-assemble in the same way as the proteins of connective tissue. The primary sequence (the distribution of apolar residues) determines how the two molecular chains fold together to produce a two-strand rope. These ropes then self-assemble to produce higher level structures namely the myofilaments which have design properties appropriate to the biological function of contractility. Though the complete primary structure of myosin is not yet determined, tropomyosin once more provides evidence that this is the mechanism. The apolar amino acids that do not participate in the two-strand bonding of the molecule, occur at axial intervals of about 28 Å in the half of the molecule for which the sequence is known. This is the axial spacing between subunits in the thin filaments of muscle where tropomyosin occurs. Thus the primary sequence may determine the intermolecular interactions.

Microtubules

Considerable interest now centres on the structure and function of microtubules since it has now been shown that they occur widely in eukaryotic cells. The term microtubule was first introduced by Slautterback [59] with reference to the tubules in cytoplasm but the term took on a wider significance when tubules with apparently identical structures were found in a wide range of situations. It is well known that the cilia and flagella of eukaryotic cells are more complex than the flagella of bacteria; they possess a 9 + 2 arrangement of fibre 'doublets' and it is now

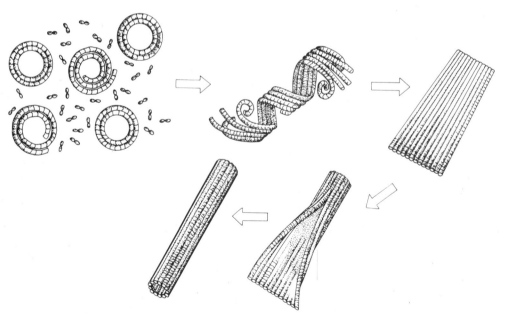

Fig. 5.1.9   Proposed pathway of microtubule assembly. After Kirschner M.W., Lawrence S.H. & Williams R.C. (1975) *J. Mol. Biol.*, **99**, 263–76.

well-established that each fibre has a microtubular structure. Microtubules have been found in plant cells [60], the tails of spermatozoa, mitotic spindles, axopodial fibres which project from the surface of various protozoa and as the long neurotubules of nerve fibres. They have been associated with cytoplasmic streaming, with directing the flow of materials such as synaptic vesicles within cells and with imparting asymmetry on cell shape. Reviews have been provided by Porter [61] and Stephens [62].

The accepted relationship between these microtubules is based on a common structure (Fig. 5.1.9). The cylindrical appearance is about 240 Å diameter. In electron micrographs of tranverse cross sections through microtubules they are observed to consist of a ring of 12–13 subunits evenly distributed round the circumference suggesting a bundle of 121–13 protofilaments. The sensitivity of this appearance to the tilt of the specimen from normal to the electron beam precluded the possibility that the protofilaments followed a helical path and indicated that they were oriented parallel to the microtubular axis. Electron micrographs of longitudinal sections through the structures revealed that the protofilaments have a 40 Å axial periodicity and are spaced about 50 Å apart round the microtubule circumference. X-ray diffraction studies on wet and dry preparations of oriented gels of sea urchin flagellar outer fibre doublets indicate a half-staggered surface lattice of 40 Å × 53 Å in the wet state. This is consistent with a 12- or 13-fold rotational symmetry for the microtubule. On drying the lateral register of the surface lattice is lost and the structure diffracts as a ring of protofilaments each with a 40 Å axial periodicity. This suggests the importance of protofilaments as stable assemblies; support of this comes from electron microscopy reported below. In the electron microscope an 80 Å axial periodicity is sometimes observed and this may be related to dimerization of 40 Å subunits, also discussed below.

A microtubular protein was characterized by Renaud et al. [63, 64] from ciliary outer fibre doublets. The protein migrated as a single band on acrylamide gel electrophoresis. The molecular weight of the monomer is about 60,000 and it can readily exist as a dimer. Mohri [65] compared the amino-acid composition of the microtubule protein with that of the muscle proteins actin, myosin and tropomyosin and also with that of flagellin, the protein from bacterial flagella. In agreement with previous findings, Mohri found the amino-acid composition of the microtubular protein resembled that of actin, but he considered that they were sufficiently different to justify the name 'tubullin' for the former. This suggestion has proved appropriate. The molecular weights of actin (45,000) and tubullin (60,000) are now recognisably different, and the tubullins resemble each other more than they do the actin molecules from the same species. There are less than 40 tryptic peptides from actin while tubullin provides at least 50 and there is no obvious relation between the peptides. There is thus strong evidence for the distinctiveness of tubullin, but also it is clear that there are distinct tubullins. Peptide mapping shows that at least 80% of the peptides from the A- and B-outer fibres of the doublets in cilia are identical but up to 20% can differ. Other differences in properties of tubullins are well known and had led to a suggestion that there were four kinds of tubullin from A-outer fibres, B-outer fibres, central fibres and cytoplasmic microtubules [66], but at present a complete, consistent classification has not been achieved.

Self-assembly of microtubules has been studied at various levels. Before microtubules were recognized in mitotic spindles, Inoue [67, 68] had noted the dependence of birefringence of the spindles on temperature. On lowering the temperature, birefringence disappears but reappears on raising the temperature again. Similar behaviour has been observed in the axopods of the heliozoon *Actinosphaerium* [61]. Axopods are spikes about 500 μ in length that project from the spherical surface of the protozoan and they possess a highly birefringent axoneme running along the central core. The axonemes consist of microtubules [69]. When *Actinosphaerium* is exposed to low temperature [70] the birefringence of the axoneme disappears. Electron micrographs of sectioned axonemes in this condition revealed that the microtubules had disassembled leaving amorphous material. On raising the temperature, the axopods resumed their normal appearance and the birefringence returned. These observations clearly suggest that microtubules are capable of self-assembly though the steps involved are not obvious. Furthermore, the temperature dependence of the polymerized structure is similar to that of T.M.V. where polymerization is also an endothermic reaction. All microtubules, however, do not show the same temperature dependence as mitotic spindles and heliozoan axonemes. The outer fibres of cilia and flagella do not dissociate in the cold.

Behaviour suggestive of self-assembly was more difficult to observe with the outer fibres. Some degree of reconstitution of parts of microtubules was achieved by Gibbons [71, 72]. He was able to remove the 'projections' from the outer fibres of a cilium and show that this was accompanied by disappearance of the ATP-ase activity. The 'projections' could be added back on to the outer fibres with some restoration of response to ATP.

Attempts to reconstitute microtubules from tubullin were not at first successful. Tubullin dimers from cilia were constituted to yield aggregates of 40 Å wide protofilaments. The significance of these protofilaments as stable units which could then assemble into microtubules is supported by the X-ray observations on drying microtubule gels where the protofilaments eventually scatter independently. It is also suggested by Behnke's [73] observation that lateral aggregation of protofilaments to yield microtubules occurred on warming cold-dissociated microtubules of blood platelets.

A more satisfying demonstration of self-assembly of microtubules was finally provided by Stephens [74]. He obtained the 60,000 molecular weight tubullin monomer from sperm tail of sea urchin by treatment with the detergent Sarkosyl. This mild treatment was followed by simple dilution and addition of salt and resulted in the reassociation of tubullin into several fibrous forms some closely similar in appearance to the original microtubules. Just as fibrous actin contains ADP, microtubules contain GDP though whether the polymerization of microtubules, like that of actin is dependent on dephosphorylation is not yet clear. Stephens found that small protein aggregates of about 1000 Å in size were essential for nucleation of polymerization; if they were centrifuged off only random aggregates were found. In the absence of magnesium the polymerized forms were ribbons which could aggregate into clusters. When diluted in the presence of 0·1 mM magnesium chloride [62, 75] Sarkosyl solutions of tubullin yielded microtubules so demonstrating that magnesium is required for protofilaments to aggregate into microtubules.

Tubullin and its association properties once again resemble the muscle proteins myosin, paramyosin and tropomyosin in exhibiting polymorphism which can be controlled by the species of the medium. It is particularly important for our purpose of enquiring into the role of self-assembly in the development of organisms, to note its occurrence in microtubules since these structures appear to play a part in shaping the cell, in determining the asymmetry of an organism and thus in influencing the possible geometric congruences or incongruences with neighbouring bodies. The ability of microtubules to dissociate and associate rapidly and to respond to environmental changes with modifications of form obviously provide a mechanism for translating minute alterations in concentrations of fluid media into dynamic and novel macroscopic effects. The extent to which this mechanism is used *in vivo* remains to be investigated, but the evidence described in this section leaves no doubt that the tubullin molecules are sufficient to determine their own self-assembly into biologically functional microtubules.

To summarize the discussion so far, we can say that aggregation of collagen molecules to form fibrils is beginning to be understood in terms of amino acids along each molecule. This is not yet the case for the aggregation of myosin or tubullin, but both of these molecules clearly demonstrate self-assembly and also show a range of different types of aggregation can occur *in vivo*. We can thus ask how the different modes of aggregation are controlled *in vivo*. Does the solvent play a part or is a different mechanism operative? Different modes of aggregation do occur in the same solvent conditions. An observation on muscle that may be significant for the assembly of molecules *in vivo* is that the muscle cell proteins are synthesized by polyribosomes lined up in the growing muscle [76]. Thus the final structure might be governed both by the self-assembly properties of the molecules and their mode of synthesis.

Bacterial Flagella

Flagella are fine fibres that adhere to the outside of many cell surfaces and by actively beating are probably the source of a primitive cell motility. Bacterial flagella (Fig. 5.1.10) are simpler than the flagella of eukaryotic cells. Bacterial flagella are frequently about 140 Å in diameter and they may be dissociated by heat or acid pH into monomers of the protein flagellin of molecular weight about 40,000 [77]. Flagellin lacks ATP-ase activity [78]. The structure of bacterial flagella has been determined by electron microscopy and X-ray diffraction [79, 80, 81, 82]. The subunits are arranged on a set of co-axial helices or on rows parallel to the flagellar axis. The periodicity of the structure projected on to the flagellar axis is 52 Å. The appearance in the electron microscope is of about eight to ten rows of subunits. X-ray diffraction studies on reconstituted filaments formed from flagellin show that they have a similar structure to the original flagella.

Reconstitution of the flagella from flagellin requires 'seeds' of parts of the flagella to initiate assembly. A particular model of flagellar motion [83] involves waves of structural perturbation moving along the flagellum converting it from a straight to a helical form. Movement of the region of perturbation along the flagellum brings about rotation of the helix which could propel the bacterium. This model predicts a high energy of activation for self-assembly of flagellin, such as is observed.

Asakura and co-workers [84] have studied the reconstitution of the flagella of the bacterium *Salmonella*. The flagella have a

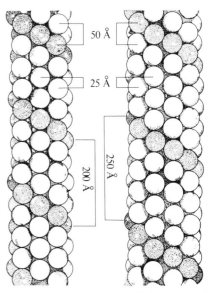

Fig. 5.1.10   Diagram of proposed models for bacterial flagella [79].

wavy appearance. A mutant form is also wavy but with a wavelength about half that of the normal. Seeds of normal flagella will stimulate flagella assembly in a solution of mutant flagellin but the wavelength is of the mutant type, while the new flagella has a normal wavelength when a solution of normal flagellin is seeded with mutant flagella. Either form can seed either type of growth, but the wavelength of the resulting flagella is determined by the monomer rather than the initiating seed.

We conclude that not only do we have another example of self-assembly in bacterial flagella, but also a particularly clear indication that the form of the flagellum is determined by the flagellin molecule. Seeding may be useful in starting self-assembly but it is not able to alter the normal mode of interaction of the molecules.

5.1.4 MOLECULAR AGGREGATION—(2) UNLIKE MOLECULES

Viruses

Self-assembly of biological structures at the molecular level was recognized early in experiments on the reconstitution of virus particles [85]. Since then the viruses have continued to provide examples of many aspects of this subject and many of the guiding principles which inform our understanding of self-assembly were derived from studies on the viruses.

Crick & Watson [34] suggested that small viruses were composed of identical subunits packed together in a regular way to form an outer coat around the infective nucleic acid. The subunit pattern could provide genetic economy and at the same time efficiency in the face of the inevitable 'mistakes' leading to defective structures. Crick & Watson [34] also suggested that the small viruses would have cubic symmetry and this was shown to be the case for bushy stunt virus. The symmetry is a consequence of packing identical subunits in as equivalent ways (and hence as lowest energy states) as possible. Linear arrays of subunits would be expected to yield helical structures. The general types of virus structure found—rod-shaped, helical viruses and spherical viruses—confirm this view of the principles involved in virus assembly which was expounded by Caspar & Klug [87]. We shall examine the virus types in turn and discuss the information they have continued to provide even recently, on the mechanism of self-assembly. A recent review is by Eisenberg & Dickson [88].

TOBACCO MOSAIC VIRUS (T.M.V.)

T.M.V. is the most thoroughly studied virus and is the paradigm of the rod-shaped, helical viruses. The virus particle is 3000 Å long with an outer diameter of 180 Å; it has a hollow core, 40 Å

Fig. 5.1.11   Diagram of the subunit arrangement in tobacco mosaic virus.

wide (Fig. 5.1.11). The virus is composed of over 2000 identical protein subunits of approximately ellipsoidal shape with one axis 70 Å and the other two 20–25 Å. In the normal virus the subunits are arranged on a shallow helix of pitch 23 Å with just over 16 subunits per turn of the helix. The RNA is continuous and arranged on a helix also of pitch 23 Å but at a radius of about 40 Å. There are three nucleotides per protein subunit. In the normal T.M.V. helix the protein subunits are in equivalent positions but there does exist a Dahlemenshe strain in which this structure is subject to a periodic perturbation which results in the subunits being in slightly different packing environments. The mode of inter-subunit bonding in the Dahlemenshe structure is constant but the bonds are slightly deformed to give quasi-equivalent packing of the subunits. Quasi-equivalence is a concept introduced by Caspar & Klug to describe the principles involved in the packing of subunits in the spherical viruses (see section below on complex bacteriophages) and the generality of its application is obvious from its occurrence here.

The general features of the self-assembly of T.M.V. are well established. The virus is dissociated by one of a variety of techniques into protein subunits and nucleic acid which can then be recombined to form the virus particle. Experiments show that the protein subunits can be assembled by themselves to form

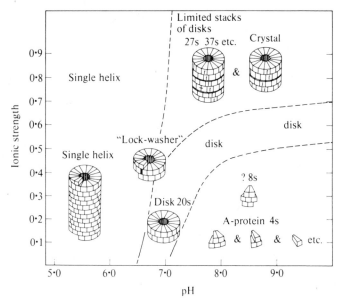

Fig. 5.1.12 This 'phase diagram' shows the conditions of ionic strength and pH under which the various polymorphic aggregates of tobacco mosaic virus exist.

virus-like particles. The preferred structure of the protein subunits alone is a particle composed of stacked discs (see summary in Fig. 5.1.12). Each disc is composed of 17 subunits arranged in a circle and two such circles then aggregate together to form the disc. Discs stack on top of each other making a cylindrical particle. Conditions have been derived however, in which the protein subunits alone will aggregate to form a particle with the same helical parameters as normal T.M.V. though of variable length. Thus the information for the virus structure must be present in the structure of the protein subunits. Protein subunits will readily recombine with RNA to form normal, helical virus particles so the RNA both facilitates the formation of the helical as against the stacked disc structure, and acts as a length determinant.

One attraction of the idea of equivalence or at least quasi-equivalence of subunits in viruses was that in the growing virus it would enable each new subunit to add on in a manner identical (because, presumably, energetically favourable) to each other subunit. Recent work on the *in vitro* association of the protein subunits has shown however, that this is not the mechanism of self-assembly [80, 91, 92, 93]. Figure 5.1.12 shows that under different conditions of pH and ionic strength, a variety of types of stable aggregates of protein monomers exist. The smallest aggregates are those sedimenting around 4S of which an important kind is the trimer. These 4S species will aggregate to form a disc. The units add on to the growing disc

with a constant free energy indicating that disc formation is not a cooperative phenomenon. A detailed analysis of the structure of the disc shows that two rings of subunits make a polar disc, that is, each ring is polar; it can be regarded as having a top and a bottom. When they come together the bottom of one comes in contact with the top of the other.

The importance of the disc is clear from the observations of Butler & Klug [92] that RNA of TMV will recombine with 34 subunit discs to form T.M.V. virus particles within 5 min. When RNA is added to 4S subunits it takes 6 h before virus particles are produced. The virus helix therefore, grows by the addition of discs and this is a much faster process than the growth of discs from 4S particles. An advantage of virus construction from discs is that it would overcome the initially high activation energy involved in forming the first turn of a helix from monomers.

T.M.V. in the normal helical form contains two groups which titrate abnormally with a pK near 7. Caspar [94] suggested that this could be due to two carboxylic groups forced close enough to each other to take up a proton and form a carboxyl-carboxylate pair to avoid mutual repulsion. Durham & Klug [91] found that the stacked discs titrate normally which suggests that in this structure the carboxylic groups are not forced close to each other. However this disc may be converted to the helical form by the addition of RNA or protons. The disc is the stable aggregate of protein subunits because of the repulsion of the carboxylic groups, but interaction with RNA converts discs to the helical form.

The lesson to be derived from these studies on T.M.V. is that while there is here a classic example of molecular self-assembly involving both like and unlike molecules, the ideas of equivalence and symmetry which are so well illustrated by the structure of the final aggregate may not allow straight-forward prediction of the actual mechanism of self-assembly.

SPHERICAL VIRUSES

The spherical viruses, so called because of their shape, differ in size and complexity but all appear to be composed of protein subunits symmetrically arranged on a spherical surface to form a coat which surrounds the nucleic acid. Bushy stunt virus was the first shown by X-ray diffraction to have the subunits arranged so as to give icosahedral symmetry [86]. The principles of subunit packing were described by Caspar & Klug [87]. The first concept is that equivalent subunits may be expected to assemble with equivalent bonding systems. The definition of a symmetrical array is one in which equivalent units are related by some geometrical operation such as translation, mirror reflection or rotation. In proteins which are enantiomorphic with L-amino acids, mirror symmetry is not possible. Helical structures arise from a combination of the operations of rotation and translation parallel to a line and the various line group symmetries result. If a closed surface is to be formed about a point, then the symmetrical operations are likely to be rotation axes through the point and the groups of self-consistent rotation axes are represented by the well-known point groups of symmetry. If the subunits are to be uniformly distributed in three dimensions about a point, then the cubic point groups appear as likely ways in which the virus protein coat might be constructed. This was recognized by Crick & Watson [34]. The kinds of polyhedra produced by the cubic point groups have been known for some time (see e.g. Plato [95]); they are the cube, tetrahedron, octahedron, icosahedron and dodecahedron. The vertices of these regular polyhedra lie on the surface of a sphere and it is now well established that the symmetry of many of the spherical viruses is icosahedral.

The relation between the cubic symmetry of the spherical viruses and the equivalent of subunit and subunit bonding has also been described by Caspar & Klug [87]. Plane lattices in which there is exact equivalence can be folded so as to form closed surfaces. Icosahedral symmetry can be constructed from a net composed of equilateral triangles by converting some of the six-fold axes into five-fold axes [96]. When the number of subunits in the virus is larger than the number of vertices in the icosahedron (60) then the bonding between the identical subunits becomes quasi-equivalent; i.e. the overall icosahedral symmetry is preserved and to allow this, the precise equivalence of all bonds is relaxed. This quasi-equivalence however involves only a small and regular deviation from exact equivalence. In tomato yellow mosaic virus there are 180 subunits, so each vertex of the icosahedron is associated with three subunits. Each trimer is identical and each trimer has itself a local three-fold axis. The manner in which the subunits group determines the morphology of the virus particle. Icosahedral symmetry has been confirmed by electron microscopy of negatively stained viruses [97]. Presumably the icosahedral symmetry is preferred over that of the other Platonic solids since, for a sphere of given diameter, it has the largest number of subunits (Fig. 5.1.13).

The point of this somewhat abstract geometrical discussion to our concern with self-assembly is as follows. The suggestions of Caspar & Klug [87] about icosahedral symmetry for spherical virus protein coat structure have been confirmed by observations. These suggestions were based on the assumption of subunit equivalence and equivalent inter-subunit bonding and a further prediction which follows from these assumptions is that such structures should self-assemble. This has also been confirmed experimentally for the spherical viruses [98, 99, 100]; tobacco mosaic virus, cucumber mosaic virus, chlorotic mottle virus. Viruses with similar structure are sometimes stabilized by different kinds of intermolecular interactions, protein-protein or protein-nucleic acid and the success of reconstitution of the virus naturally depends on the conditions of dissociation (see [88]).

Another feature of the spherical viruses which they have in common with other self-assembling systems is polymorphism. The papilloma-polyoma family of viruses are spherical viruses but unfractionated preparations sometimes show tubular forms in which the protein subunits are in a hexagonal array on the surface of a cylinder. Frequently the cylinders have rounded, hemi-spherical ends. Thus the subunits can aggregate into either the closed spherical surface or the cylindrical surface. Presumably the free energy and modes of bonding in these polymorphic forms are closely similar.

We conclude from these studies that the close connection between equivalence, symmetry and self-assembly is well established. Furthermore, the spherical viruses show that symmetry can be conserved at the expense of quasi-equivalence in the inter-subunit bonding.

COMPLEX BACTERIOPHAGES

Viruses under this classification are more complex than the rod and spherical viruses in two ways. First their shape—they are not simple but contain, in the case of the T-even bacteriophages, a polyhedral head, a tail comprised of core and sheath and tail fibres. Secondly they are not composed by the assembly of subunits of the same type, but of a number of different types. Because of this complexity it has not yet been demonstrated that a T-even bacteriophage can self-assemble *in vitro* like the rod and spherical viruses. However these systems are invaluable in revealing the likely extent of self-assembly in biological systems and the sort of situations in which it might not be sufficient.

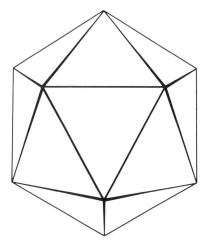

Fig. 5.1.13 An icosahedron.

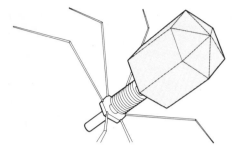

Fig. 5.1.14 Diagram of the structure of a T-even bacteriophage virus.

The structure of the T-even bacteriophages is well established (Fig. 5.1.14) and a substantial understanding has been gained about the sequence of events by which an intact virus particle is assembled. This has been possible because the virus is a bacteriophage; this has allowed the methods of bacterial genetics to be applied to the problem. A large number of mutants have been analysed and it is now possible both to provide a genetic map of the virus and to demonstrate which genes are responsible for the manufacture of the proteins for various parts of the bacteriophage. This work has been described in reviews and articles [101, 102, 103]. A recent detailed review of the T-even bacteriophages is given by Eiserling & Dickson [82]. Here we shall only summarize the results and relate them to our discussion of self-assembly. Figure 5.1.15 shows the sequence that has been worked out for the steps in the assembly of T4 phage. The phage contains at least 15 proteins and probably many more. It is evident from Fig. 5.1.15 that major parts of the phage are assembled separately and then these completed parts themselves assembled to make the phage. Since all of the protein components appear to be present simultaneously, the assembly must occur by specific molecular interactions. The tail of the phage resembles a rod-shaped virus and the head a spherical virus; this makes it easy to think that these parts could form by self-assembly. The sheath of the tail for example is made up of helically arranged identical protein subunits. Parts of the tail have been self-assembled *in vitro* [104]. The head is significantly different from icosahedral symmetry however, to ensure that the constituent subunits could not occupy equivalent positions; this suggests that self-assembly may be too simple a mechanism. The genetic evidence shows that most of the steps in the assembly of a phage particle are under the control of gene products. This is so to the extent that mutant gene products produce polymorphic forms of parts of the phage. The mechanism whereby these gene products, presumably proteins, exert their control over the phage assembly is still largely obscure but subject to intensive investigation [82]. For our purpose the main question is whether the effect is catalytic, whether the proteins are enzymes. We wish to know if the role

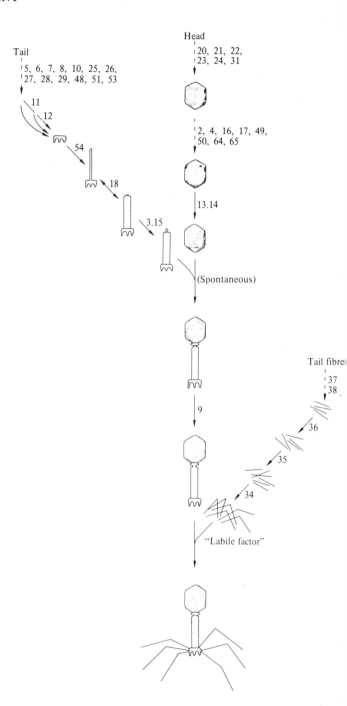

Fig. 5.1.15 Sequence of steps in the assembly of subunits to form a T-even bacteriophage.

of a controlling gene product is to bring about a step in the assembly process without the gene product itself becoming incorporated into the final structure. It seems very likely that this is what happens. Certainly the role of the gene product P63 appears to be to catalyse the attachment of tail fibres and Kellenberger [103] has speculated on the possible allosteric effects of gene products in causing changes in molecular structure so as to bring about assembly. This is revealing for us because it suggests that self-assembly in the simplest form where only those subunits involved in the final structure are required for the assembly, is not sufficient for structures of the complexity of T4 phage.

Multi-enzyme complexes

There are systems where function is related to the specific assembly of enzyme molecules into complexes. These complexes are said to have quaternary structure and they can range from dimers upwards. The topic is an extensive one and so I will merely summarize the present understanding in general terms which bear on our subject and refer to recent reviews which may be consulted for details of the primary sources.

Studies on the structure of haemoglobin by Perutz [105] represent the history of much of structural molecular biology in miniature. Haemoglobin was the subject for the first information about the atomic arrangement in proteins and since the functional haemoglobin unit is a tetramer of two α- and two β-polypeptide chains, it also provided our first information about how protein subunits interact at the atomic level, gave an introductory example of the molecular elucidation of a long known physiological phenomenon, the Bohr effect and yielded the initial insights into molecular diseases. Haemoglobin may be dissociated in two $\alpha\beta$ dimers. The contacts between $\alpha_1\beta_1$ dimers are weaker than the contacts between $\alpha_1\beta_1$. This latter contact undergoes considerable change on deoxygenation while the $\alpha_1\beta_1$ contacts are only minimally altered.

The respiratory pigments in many invertebrates are either haemocyanin or haemerythrin. These molecules can exist in elaborate quaternary structures. The monomers aggregate to form closed rings and the rings then stack to form cylinders. In both cases the state of aggregation is closely related to and may control the state of oxygenation of the molecules [106, 107]. The state of aggregation is in turn controllable by the environmental conditions of pH, ionic strength and temperature. Though much remains to be learned about these molecules, they clearly provide the possibility of a system where the self-assembly of molecules is controlled by the environment and by this means a regulation of the molecular function is obtained. Given the appropriate conditions the molecules can self-assemble to form large aggregates; specific molecules will selectively aggregate from solutions containing haemocyanin from different animals.

There are many examples of enzymes which are only functional in oligomeric form [108] and the environmental dependence of the extent of oligomerization indicates that the monomers self-assemble. This is even the case when the functional oligomer is composed of different types of monomer [109].

Ribosomes

Ribosomes occur in virtually all types of cell where they comprise part of the machinery of protein synthesis. The general structure of ribosomes is well known—they are made up of two subunits, a large one that sediments in the ultracentrifuge at 50S and another smaller one at 30S in animal cells. The complete ribosome sediments at 70S. The molecular weights of the 50S and 30S subunits are $1\cdot8 \times 10^6$ and $0\cdot7 \times 10^6$ respectively and they are readily dissociated from the 70S particle by lowering the concentration of magnesium ions from 10^{-2} M to 10^{-4} M. The ribosome is about 60% RNA and 40% protein. The 50S subunit contains two RNA molecules, one 23S and a smaller 5S as well as 43 different proteins; the 30S subunit contains only one RNA molecule sedimenting at 16S and 23 different proteins. Here we shall describe some of the evidence concerning the extent to which the 70S ribosome may be formed spontaneously by self-assembly from RNA and proteins.

It has been suggested [110, 111] that the 30S and 50S subunits do not come together spontaneously but require an RNA and tRNA. Thus they probably come together for protein synthesis but dissociate again when this is completed.

While the 50S and 30S particles are stable they will dissociate on centrifugation in caesium chloride into proteins and 'core particles' which are ribonucleoproteins. These 'core particles' can reassemble with the split proteins to yield 50S and 30S subunits which in turn may be joined to yield functional 70S ribosomes. The 'core particles' were inactive and were only reactivated if reassembled with their own split proteins.

More detailed analyses of these subunits and their constituents has been carried out by Nomura and his colleagues. They have shown that some of the split proteins are essential for certain functions such as binding messenger or transfer RNA or amino-acid incorporation; others are not essential but stimulatory.

Traub & Nomura [112] have shown that a 23S core particle (derived from a 30S ribosomal particle) may be dissociated into free 16S ribosomal RNA plus core protein (CP30S). A functionally-active 30S subunit can then be assembled from the 16S r-RNA and the mixture of ribosomal proteins. This reassembly is a specific reconstitution. The above reconstitution was done using ribosomes from *E. coli*; neither yeast 16S ribosomal RNA nor rat liver 18S ribosomal RNA can replace *E. coli* ribosomal RNA in the reconstitution, nor is degraded *E. coli* ribosomal RNA effective. Core protein from 50S subunits

(CP50S) cannot effectively replace the CP30S protein used above. Thus the specificity of the rRNA and the core protein is demonstrated. It was also shown that streptomycin resistance lies in CP30S.

A similar reconstitution of the 50S subunit proved more difficult. Nomura & Erdmann [113] noted that the high temperature apparently required to drive the assembly, caused denaturation of the proteins. They therefore selected a thermophilic bacterium *Bacillus stearothermophilus* and achieved a successful reconstitution of a 50S subunit from their r-RNA and protein. Maruta *et al.* [114] investigated the conditions for reassembly of *E. coli* 50S subunits and showed that previous attempts at reconstitution had probably failed because of denaturation of the proteins. They designed a method to avoid denaturation and obtained successful reconstitution of 50S subunits of *E. coli*. The confirmed that though reconstitution of 30S subunit is not dependent on 50S, the reverse is not true, and they established the necessity for the 5S r-RNA in the 50S reassembly.

These studies were taken a stage further in an ambitious project [115] to map the sequence of steps by which the 30S subunit of *E. coli* was assembled. Their method was to take each of some twenty proteins and test them separately for binding to the 16S r-RNA.

The binding tested was rather specific since coulombic forces were decreased by using high ionic strength solutions. While this has the advantage of preventing non-specific electrostatic binding of basic proteins to RNA, it may also inhibit specific electrostatic interactions which are important *in vivo*. Only three proteins (termed 4a and 4b & 14) bind strongly and most showed no binding, so those binding were taken as the first to be added to r-RNA when the 30S subunit is assembling. The 18 non-binding proteins were then tested for binding to complexes of r-RNA and 4a, 4b and 14 protein. Three showed up as strong binders (5, 9 and 10a) while the other fifteen did not bind. This technique was continued to reveal the sequence of binding of all the proteins to 16S r-RNA and their interdependence of binding. The method is based on the consideration that if protein B will not bind until protein A does, then protein B must follow protein A into the complex.

Mizushima & Nomura [115] conclude that they have established the topology of inter-protein contacts in the complete ribosome. Experiments are underway in an attempt to determine the three-dimensional molecular arrangement in ribosomes (see e.g. [116]) and it will be of interest to observe developments in this exciting field.

For the purpose of our argument we must summarize. There appears good evidence that the final form of a 30S or 50S subunit is a consequence of the constituent r-RNA and protein molecules. Reconstitution takes place by self-assembly and the temperature dependence of some reactions suggest that hydrophobic interactions may be involved [9]. A ribosome is a fairly complex organelle and takes part in the highly integrated process of protein synthesis. It is interesting that self-assembly can lead to structures with this demanding complexity and that this involves self-assembly of a large number of different protein components. The importance of a proper sequence of addition is also evident. It implies that for the rapid assembly of ribosomes which occurs *in vivo* in a few seconds, the protein must be presented to the assembling ribosome in the correct order. In other words temporal presentation of the parts assists self-assembly.

Membranes

Membranes form the outer coat of cells and of parts of cells. Their functions can be complex and range from simple physical support to a selectively permeable barrier. Their proper function is central to many biological systems. The basic membrane structure is formed by a combination of phospholipid with protein. Phospholipids form bilayers. In these the phospholipid molecules are arranged on a two-dimensional lattice with the molecular axis normal to the plane of the layer. In one layer the molecules are parallel and in register so that all the phosphate groups are on one side of the layer; on the other side are the fatty acid ends of the lipid molecules. Thus a single layer has one side apolar and the other highly charged. Two such layers adhere with their apolar faces in contact to form a bilayer. The existence of the bilayer structure in phospholipids and membranes has been well established by X-ray diffraction. Lecithin layers can be stacked together and the X-ray diffraction pattern from the stacks has a series of reflections which are orders of approximately 48 Å. The intensities of these orders are consistent with a bilayer structure. Stacks of lecithin bilayers have been studied by varying the charge on the layer surface and observing the effects on the inter-bilayer spacing in various solvents. This suggests that they are stabilized by a sensitive force balance.

Membranes contain varying ratios of lipid to protein and as yet, the exact nature of the interaction of the protein with the lipid bilayer is not known. However, there is good evidence from X-ray diffraction and electron microscopy that the protein molecules in some membranes are regularly arranged on a lattice in the plane of the membrane [117, 118]. When membranes are delipidated with phospholipase or ethanol: ether, the basic structure, frequently remains suggesting that the proteins are important in membrane stabilization. When the proteins are digested by enzymes frequently the membrane collapses [119]. There are many experiments which suggest that complex lipo-protein systems such as mitochondria will self-assemble from smaller units. Certainly the lipid bilayer readily forms by self-assembly. In many biological tissues such as keratin [120]; muscle (Miller, unpublished) and collagen [38] the X-ray diffraction patterns show reflections which index on

48 Å. These are of variable character and appear to become more intense as the tissue keratinizes or gets older. Pollard *et al.* [119] noted that preparations of chromaffin granules showed a similar set of reflections of variable intensity. This may indicate that as membranes break down, the phospholipids may readily recrystallize as bilayers by themselves.

Many reasonable suggestions have been made to the effect that cells may adhere by relatively simple physical forces (electrostatic and van der Waals forces) which may be estimated from some familiarity with the membrane structure [121, 122, 123, 124]. It has also been suggested that the different compositions of cells in terms of protein and sugar molecules may eventually account for specificity of cell recognition [125]. This topic is at present under development since regular progress is being made in our understanding of membrane structure, the physical forces themselves and the phenomena of cellular recognition. If these speculations are confirmed then the concept of self-assembly even in its simplest form may be sufficient to account for interactions between cells and open the way for formalisation of theories of development at this level [126].

5.1.5 CONCLUSIONS

In this survey we have considered examples of self-assembly of various biological structures. Self-assembly is defined here as a mechanism in which only the constituents of the final structure take part in the assembly process. By constituents we mean mainly macromolecular subunits and not water molecules or other small solvent ions. This type of self-assembly occurs at several levels and is well exemplified by the proteins. The order of amino acids in a protein (its primary structure) is determined by the order of nucleotides in the DNA of the gene coding for the protein. The first level of self-assembly is the adoption of a secondary structure by the protein. This is a regular coiling of the molecule to form a helix such as the α-helix or collagen triple helix. Both of these helices are determined by the primary sequence of the protein but the actual structure can also be affected by the solvent conditions. A second level of self-assembly is when a protein molecule folds into an irregular three-dimensional tertiary structure. While the actual course of the main polypeptide chain is irregular in such structures, the tertiary structures of all molecules of a given protein are identical under the same conditions. Yet another level of self-assembly is when two fibrous molecules such as in myosin, coil about each other to form a two-strand coiled α-helix. In tropomyosin this is now seen to be due to a regularity in the positioning of apolar amino acid side-chains in the single α-helices, so once more the self-assembly is determined by the primary sequence of the protein. In the case of the collagen molecules of connective tissue, self-assembly of the molecules

into native-like fibrils is well known. The origin of this is now known to lie in the amino-acid sequence. Where the recognition process between these long molecules is due to like-like interactions, it seems possible to state a general principle which relates the symmetry of the molecule to the symmetry of the aggregate. If two molecules are related by a given symmetry operation in the aggregate, then the molecule itself will contain that same symmetry element in at least quasi-form. Thus the collagen molecules which are staggered axially by a distance D in the fibril, themselves have a quasi-D-period in their apolar residues. It is likely that the role of the solvent is to control the symmetry of the interacting residues on the molecule and thus control the symmetry of the aggregate. In collagen the three-dimensional structure of the fibril has also been related to the helical symmetry of the molecule. Thus the fibrous proteins have proved valuable in providing examples of a situation where biological specificity and hence self-assembly may be understood in terms of physico-chemical interactions between molecules of known structure. Collagen may be the first example where all of the processes can be understood from the molecular level through molecular folding and molecular aggregation to aggregation of fibrils into tissues of known biological function.

Self-assembly of globular subunits also takes place to yield specific aggregates. Globular protein subunits comprise flagella, microtubules, viruses and ribosomes and while it is not as easy at present in these cases to recognize the origins of these interactions in the amino-acid sequence, it is likely that similar principles will operate as in the fibrous systems and in haemoglobin which was the first to be understood at this level.

We see then, that self-assembly according to our simple definition certainly occurs up to the level of cell organelles and some animal tissues, and this despite the fact that some of these are intracellular such as muscle, and others extracellular as collagen. However we also noted that in the case of a larger virus, assembly may be enzyme assisted and thus not meet our severe definition. It is also difficult in the case of intracellular systems, to detect the effect of the surrounding organization of the cell, but *in vitro* experiments help to clarify which assembly mechanisms are, at least in principle, capable of self-assembly. We may then inquire whether the particular organization within the cell serves to enhance assembly rates or not. Frequently a complex system involving templates is, or may be converted to, one which meets our simple definition by considering a larger 'aggregate' which includes previous aggregates plus templates etc., so there are examples where the question of self-assembly can become purely semantic and unhelpful. However, the value of the simple definition is that where a system does conform to it, then it can be possible to trace an unbroken line of causal steps from the gene to the form of the aggregate. This is obviously of considerable significance for our understanding of development and something close to that situation may soon exist for the extracellular connective tissues of metazoans. For interactions between cells it may be that the architecture of molecular packing on the cell surface which is possibly a self-assembling system, will in turn provide the platform for a rationalization of cellular recognition processes. But this has not been demonstrated yet. It is necessary to have a grasp of the properties of the elementary parts in order to interpret completely an experiment in terms of self-assembly.

5.1.6 REFERENCES

[1] KOESTLER A. & SMYTHIES J.R. (eds) (1969) *Beyond Reductionism.* Hutchinson, London.
[2] See e.g. ORGEL L. (1973) *The Origins of Life.* Chapman & Hall, London.
[3] See e.g. FOX S.J. & DOSE K. (1972) *Molecular Evolution and the Origin of Life.* Freeman, Reading.
[4] BERNAL J.D. (1967) *The Origin of Life.* Weidenfeld and Nicholson, London.
[5] CRANE H.R. (1950) *Sci. Monthly,* **70,** 376.
[6] CASPAR D.L.D. & KLUG A. (1962) *Cold Spring Harbour Symp. Quant. Biol.,* **27,** 1.
[7] WEYL H. (1952) *Symmetry.* Princeton University Press, Princeton, New Jersey.
[8] CALVIN M. (1969) *Chemical Evolution.* Oxford.
[9] KUSHNER D.J. (1969) *Bacteriological Reviews,* **33,** 302.
[10] WOLSTENHOLME G.E.W. & O'CONNOR M. (eds) (1966) *Principles of Biomolecular Organisation.* Churchill, London.
[11] ALLEN J.M. (ed.) (1967) *Molecular Organisation and Biological Function.* Harper & Row, New York.
[12] HAYASHI T. & SZENT-TYORGYI A.G. (eds) (1966) *Molecular Architecture in Cell Physiology.* Prentice-Hall, Englewood Cliffs, New Jersey.
[13] ENGSTRÖM A. & STRANSBERG B. (eds) (1969) *Nobel Symposium No. 11* on *Symmetry and Function of Biological Systems at the Macromolecular Level.* Wiley, New York.
[14] TIMASHEFF S.N. & FASMAN G.D. (eds) (1971) *Biological Macromolecules. Vol. 5A.* Dekker, New York.
[15] RAMACHANDRAN G.N., RAMAKRISHNAN C. & SASISEKHARAN V. (1963) *J. Mol. Biol.,* **7,** 95.
[16] IUPAC-IUB, Commission on Biochemical Nomenclature (1970) *J. Mol. Biol.,* **52,** 1.
[17] LIQUORI A.M. (1966) In ref. [10], page 40.
[18] SEBERA D.K. (1964) *Electronic Structure and Chemical Bonding,* Chap. 11. Blaisdell Publishing Company, New York.
[19] DOTY P., WADA A., YANG J.T. & BLOUT E.R. (1957) *J. Polymer Sci.,* **23,** 851.
[20] APPLEQUIST J. & DOTY P. (1962) In *Polyamino Acids, Polypeptides and Proteins* (ed. M.A. Stahlman), p. 161. Univ. of Wisconsin Press, Madison, Wisconsin.
[21] FRASER R.D.B., HARRAP B.S., MACRAE T.P., STEWART F.H.C. & SUZUKI E. (1965) *J. Mol. Biol.,* **12,** 482.
[22] HULMES D.J.S., MILLER A., PARRY D.A.D., PIEZ K.A. & WOODHEAD-GALLOWAY J. (1973) *J. Mol. Biol.,* **79,** 137.
[23] ANDREEVA N.S., MILLIONOVA M.I. & CHIRGADZI Y.N. (1963) In *Aspects of Protein Structure* (ed. G.N. Ramachandran), p. 137. Academic Press, New York.
[24] TRAUB W., YONATH A. & SEGAL D.M. (1969) *Nature,* **221,** 914.
[25] GERNGROSS O. & KATZ J.R. (1926) *Kolloid-Z.,* **39,** 181.
[26] ROBINSON C. & BOTT M.J. (1951) *Nature,* **168,** 325.
[27] CRICK F.H.C. (1953) *Acta Crystallographica,* **6,** 685, 689.
[28] COHEN C. & HOLMES K.C. (1963) *J. Mol. Biol.,* **6,** 423.
[29] FRASER R.D.B., MACRAE T.P. & MILLER A. (1965) *J. Mol. Biol.,* **14,** 432.
[30] WOODS E.F. (1967) *J. Biol. Chem.,* **242,** 2859.
[31] HODGES R.S., SODEK J., SMILLIE L.B. & JURASEK L. (1972) *Cold Spring Harbour Symp. Quant. Biol.,* **37,** 299.
[32] EPSTEIN C.J., GOLDBERGER R.F. & ANFINSEN C.B. (1963) *Cold Spring Harbour Symp. Quant. Biol.,* **28,** 439.
[33] FREEDMAN R. (1973) *New Scientist,* **58,** 560.
[34] CRICK F.H.C. & WATSON J.D. (1956) *Nature,* **177,** 473.
[35] TRAUB W. & PIEZ K.A. (1971) *Adv. in Protein Chemistry,* **25,** 243.
[36] DOYLE B.B., HULMES D.J.S., MILLER A., PARRY D.A.D., PIEZ K.A. & WOODHEAD-GALLOWAY J. (1974) *Proc. Roy. Soc. B.* (in press).
[37] SMITH J.W. (1968) *Nature,* **219,** 157.
[38] MILLER A. & WRAY J.S. (1971) *Nature,* **230,** 437.
[39] HUXLEY A.F. & NIEDERGERKE R. (1954) *Nature,* **173,** 971.
[40] HUXLEY H.E. & HANSON J. (1954) *Nature,* **173,** 973.
[41] HUXLEY H.E. (1969) *Science,* **164,** 1356.
[42] HUXLEY H.E. (1963) *J. Mol. Biol.,* **7,** 281.
[43] LOWEY S., SLATER H.S., WEEDS A.G. & BAKER H. (1969) *J. Mol. Biol.,* **42,** 1.
[44] JAKUS M.A. & HALL C.E. (1947) *J. Biol. Chem.,* **167,** 705.
[45] HARRINGTON W.F., BURKE M. & BARTON J.C. (1972) *Cold Spring Harbour Symp. Quant. Biol.,* **37,** 77.
[46] HARRISON R.G., LOWEY S. & COHEN C. (1971) *J. Mol. Biol.,* **59,** 531.
[47] COHEN C. & SZENT-GYÖRGYI A.G. (1971) In *Contractility of Muscle Cells and Related Processes* (ed. R.J. Podolsky), Prentice-Hall, Englewood Cliffs, New Jersey.
[48] KENDRICK-JONES J., SZENT-GYÖRGYI A.G. & COHEN C. (1971) *J. Mol. Biol.,* **59,** 527.
[49] SOBIESZEK A. (1972) *J. Mol. Biol.,* **70,** 741.
[50] SQUIRE J.M. (1971) *Nature,* **233,** 475.
[51] SQUIRE J.M. (1972) *J. Mol. Biol.,* **72,** 125.
[52] SQUIRE J.M. (1973) *J. Mol. Biol.,* **77,** 291.
[53] OFFER G. (1972) *Cold Spring Harbour Symp. Quant. Biol.,* **37,** 87.
[54] ELLIOTT A. & LOWY J. (1970) *J. Mol. Biol.,* **53,** 181.
[55] COHEN C., SZENT-GYÖRGYI A.G. & KENDRICK-JONES J. (1971) *J. Mol. Biol.,* **56,** 223.
[56] ELLIOTT A. (1971) *Phil. Trans. Roy. Soc. London Ser. B,* **261,** 197.
[57] BULLARD B., LUKE B. & WINKELMAN L. (1973) *J. Mol. Biol.,* **75,** 359.
[58] SEGREST J.P. & CUNNINGHAM L.W. (1971) *Nature New Biol.,* **234,** 26.
[59] SLAUTTERBACK D.B. (1963) *J. Cell Biol.,* **18,** 367.
[60] LEDBETTER M.C. & PORTER K.R. (1963) *J. Cell Biol.,* **19,** 239.
[61] PORTER K.R. (1966) In Ref. [10], p. 308.
[62] STEPHENS R.E. (1971) In Ref. [14], p. 355.
[63] RENAUD F.L., ROWE A.J. & GIBBONS I.R. (1966) *J. Cell Biol.,* **31,** 92A.
[64] RENAUD F.L., ROWE A.J. & GIBBONS I.R. (1968) *J. Cell Biol.,* **36,** 79.
[65] MOHRI H. (1968) *Nature,* **217,** 1053.
[66] BEHNKE O. & FORER A. (1967) *J. Cell. Sci.,* **2,** 169.
[67] INOUE S. (1952) *Bull. Biol.,* **103,** 316.
[68] INOUE S. (1952) *Expl. Cell Res. Suppl.,* **2,** 305.
[69] KITCHING J.A. (1964) In *Primitive Motile Systems in Cell Biology* (eds R.D. Allen & N. Kamiya), p. 445. Academic Press, New York.
[70] TILNEY L.G. & PORTER K.R. (1967) *J. Cell. Biol.,* **34,** 327.
[71] GIBBONS I.R. (1965) *Arch. Biol.,* **76,** 317.
[72] GIBBONS I.R. (1967) In *Formation and Fate of Cell Organelles* (ed. K.B. Warren), p. 99. Academic Press, New York.
[73] BEHNKE O. (1967) *J. Cell. Biol.,* **34,** 697.
[74] STEPHENS R.E. (1968) *J. Cell. Biol.,* **33,** 517.
[75] STEPHENS R.E. (1969) *Quart. Rev. Biophys.,* **1,** 377.
[76] ALLEN E.R. & PEPE F. (1965) *Amer. J. Anat.,* **116,** 115.
[77] ABRAN D. & KOFFER H. (1964) *J. Mol. Biol.,* **9,** 168.
[78] NEWTON B.A. & KERRIDGE D. (1965) *Symp. Soc. Gen. Microbiol.,* **15,** 225.
[79] LOWY J. & HANSON E.J. (1965) *J. Mol. Biol.,* **11,** 293.
[80] LOWY J. & SPENCER M. (1967) *Symp. Soc. Expl. Biol.,* **22,** 215.
[81] CHAMPNESS J.N. & LOWY J. (1968) In *Symposium on Fibrous Proteins* (Ed. W.G. Crewther), p. 106. Butterworth, London.
[82] CHAMPNESS J.N. (1971) *J. Mol. Biol.,* **56,** 295.
[83] KLUG A. (1967) In *Formation and Fate of Cell Organelles* (Ed. K.B. Warren), p. 1. Academic Press, New York.

[84] ASAKURA S., EGUCHI G. & IINO (1966) *J. Mol. Biol.*, **16**, 302.
[85] FRANKEL-CONRAT H. & WILLIAMS R.C. (1955) *Proc. Nat. Acad. Sci., U.S.*, **41**, 690.
[86] CASPAR D.L.D. (1956) *Nature*, **177**, 475.
[87] CASPAR D.L.D. & KLUG A. (1962) *Cold Spring Harbour Symp. Quant. Biol.*, **27**, 1.
[88] EISERLING F.A. & DICKSON R.C. (1972) *Ann. Rev. Biochem.*, **41**, 467.
[89] CASPAR D.L.D. & HOLMES K.C. (1969) *J. Mol. Biol.*, **46**, 99.
[90] DURHAM A.C.H., FINCH J.T. & KLUG A. (1971) *Nature New Biol.*, **229**, 37.
[91] DURHAM A.C.H. & KLUG A. (1971) *Nature New Biol.*, **229**, 42.
[92] BUTLER P.J.G. & KLUG A. (1971) *Nature New Biol.*, **229**, 47.
[93] KLUG A. (1972) *Fed. Proc.*, **31**, 30.
[94] CASPAR D.L.D. (1963) *Adv. Protein Chem.*, **18**, 37.
[95] PLATO (350 B.C.), *Timaeus*, Penguin Edition, 1965, p. 74 ff. and *Theaetetus*.
[96] KLUG A. (1969) In Ref. [13], p. 425.
[97] KLUG A., FINCH J.T. & FRANKLIN R.E. (1957) *Biochim. Biophys. Acta*, **25**, 242.
[98] HOHN T. (1967) *Eur. J. Biochem.*, **2**, 152.
[99] SUGUJAMA T., HERBERT R.R. & HARTMAN K.A. (1967) *J. Mol. Biol.*, **25**, 455.
[100] BANCROFT J.B. (1970) *Adv. Virus Res.*, **16**, 99.
[101] EDGAR R.S. & EPSTEIN R.H. (1965) *Sci. Amer.* (February).
[102] WOOD W.B. & EDGAR R.S. (1967) *Sci. Amer.* (July).
[103] KELLENBERGER E. (1966) In Ref. [10], p. 192.
[104] KING J. (1971) *J. Mol. Biol.*, **58**, 693.
[105] See e.g. PERUTZ M.F. (1971) *New Scientist*, **50**, 676.
[106] KLOTZ I.M. (1971) In Ref. [14], p. 55.
[107] VAN HOLDE K.E. & VAN BRUGGEN E.F.J. (1971) In Ref. [14], p. 1.
[108] See MATTHEWS B.M. & BERHARD S.A. (1973) *Ann. Rev. Biophys. Bioeng.*, **2**, 257.
[109] GERHART J.C. & SCHACHMAN H.K. (1965) *Biochemistry*, **4**, 1054.
[110] MANGIAROTTI G., APIRION D., SCHLESSINGER D. & SILENGO L. (1968) *Biochemistry*, **7**, 456.
[111] MANGIAROTTI G. & SCHLESSINGER D. (1967) *J. Mol. Biol.*, **29**, 395.
[112] TRAUB P. & NOMURA M. (1968) *Proc. Nat. Acad. Sci., U.S.*, **59**, 777.
[113] NOMURA M. & ERDMANN V.A. (1970) *Nature*, **228**, 744.
[114] MARUTA E., TSUCHIYA T. & MIZUNO D. (1971) *J. Mol. Biol.*, **61**, 123.
[115] MIZUSHIMA S. & NOMURA M. (1970) *Nature*, **226**, 1214.
[116] ENGLEMAN D.M. & MOORE P.B. (1972) *Proc. Nat. Acad. Sci., U.S.*, **69**, 1997.
[117] BLAUROCK A.E. & STOECKENIUS W. (1971) *Nature*, **223**, 152.
[118] GOODENOUGH D.A. & STOECKENIUS W. (1972) *J. Cell Biol.*, **54**, 646.
[119] POLLARD H., MILLER A. & COX G.C. (1973) *J. Supramolecular Structure*, **1**, 295.
[120] FRASER R.D.B., MACRAE T.P., RODGER G.E. & FILSHIE B.K. (1963) *J. Mol. Biol.*, **7**, 90.
[121] CURTIS A.S.G. (1966) *Sci. Progress*, **54**, 61.
[122] CURTIS A.S.G. (1967) *The Cell Surface*. Academic Press, London.
[123] CURTIS A.S.G. (1972) *Sub-cell Biochem.*, **1**, 179.
[124] PARSEGIAN V.A. (1973) *Ann. Rev. Biophys. Bioeng.*, **2**, 221.
[125] JEHLE H. (1970) *Bull. Am. Phys. Soc. Ser. 11*, **15**(11), 1335.
[126] POLLARD E.C. (1973) In *Cell Biology in Medicine* (Ed. E.E. Bittar), p. 357. Wiley, London.

Chapter 5.2
Organelle Development

5.2.1 THE STRUCTURE AND FUNCTION OF CHLOROPLASTS AND MITOCHONDRIA

5.2.1.1 Introduction

There are several reasons why the study of subcellular organelles is important in developmental biology. Although they are comparatively small biological structures, they cannot be formed as a result of self-assembly processes alone (Chapter 5.1) and they, therefore, indicate the limits of self assembly as an explanation of the form of developing systems. They also demonstrate the dependence of cytoplasmic structures on nuclear activity, even though mitochondria and chloroplasts contain their own DNA. This emphasizes the importance of knowledge about the activity of the nucleus during early development (Chapters 1 and 5.3). In addition, chloroplasts and mitochondria provide much of the energy requirements of developing systems and are, therefore, essential components of development.

Photosynthesis and respiration are both complex processes involving the transfer of electrons from a donor to an acceptor along a sequence of intermediate carriers.

In photosynthesis, light energy is used to effect the dissociation of water. This results in the liberation of oxygen and release of electrons, whose subsequent transfer through a series of electron transport components generates ATP and reduced NADP. These products may then be usefully employed in enzyme reactions concerned with the fixation of CO_2.

Respiration is concerned with the efficient utilization of reduced pyrimidine nucleotides and flavoproteins which result from the breakdown of food sources. Electrons removed from the reduced enzyme cofactors are channelled through an electron transport system, finally being handed on to the terminal electron acceptor, which in higher organisms is oxygen. Energy made available as the electrons are transferred down the electrochemical gradient is usefully employed to produce ATP.

Both photosynthesis and respiration occur in prokaryotic as well as eukaryotic organisms. Although some components of these reaction sequences are found in particular regions of the membrane system of bacteria and blue-green algae, they are not compartmentalized into self-contained organelles. This contrasts with the situation in eukaryotic cells, where the processes of photosynthesis and respiration are confined to highly specialized, partly autonomous bodies, within the cell.

These are, of course, the chloroplasts and mitochondria. This chapter is concerned with the way in which these complex structures develop and are maintained. Although functionally quite distinct, a superficial comparison of the structure of mitochondria and chloroplasts shows many parallels. Some of the more obvious similarities should become apparent in the following brief description of their basic morphology.

5.2.1.2 Mitochondria—structure and enzyme localization

Mitochondria consist of two membrane bags, one enclosed within the other. The inner membrane is folded into characteristic invaginations known as cristae. Although the exact number and form of these vary from one tissue to another, they appear to be built on a common plan.

Except where the cristae fold inwards, the contour of the inner membrane closely follows that of the bounding membrane. The two membranes are separated by a gap which extends into the cristae (intercristal space). The area included by both membranes and the invaginations is called the matrix (Fig. 5.2.1). The overall dimensions of mitochondria are also extremely variable depending on tissue type, stage of development and physiological conditions; a 'typical' mitochondrion having dimensions of $1-2\,\mu \times 0.5\,\mu$ which are similar to those of a bacterium, such as *Escherichia coli*.

Most of the evidence concerning the localizations of various enzymes and electron transport components of the mitochondrion comes from experiments in which the organelle is broken down into sub fractions. Three major fractions can be recognized; outer membrane, inner membrane and soluble material. Many workers [6] have suggested that the bulk of readily solubilized tricarboxylic acid cycle enzymes and some of those involved in amino acid metabolism are present in the matrix. This suggestion was based on the results of experiments in which the outer membrane is first stripped off by treatment with detergent, or by osmotic swelling. It is presumed that this releases enzymes present in the intermembrane and intercristal spaces. More vigorous disruption is necessary to damage the inner membrane and release the contents of the matrix.

Needless to say, results achieved with this relatively crude type of approach are conflicting and the belief that many enzymes are of matrix origin is considered by some [63] to be incompatible with the observation that mitochondria in certain

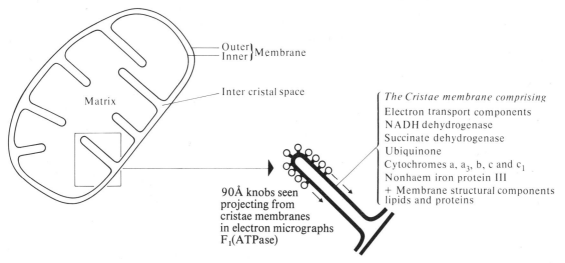

Fig. 5.2.1 Diagram of the basic mitochondrial structure-illustrating the suggested localization of soluble enzymes (in the matrix) and of the electron transport components which are primarily associated with the inner membrane.

Other enzymes, e.g. monoamine oxidase, are found in preparations of the outer membrane (see [6, 107]).

tissues have cristae so densely packed that little or no matrix can be distinguished in electron micrographs. The only obvious alternative location is the intercristal or intermembrane spaces.

The electron transport complexes, ATPase and some of the enzyme complexes of the TCA cycle (e.g. succinate dehydrogenase) are components of the inner membrane and cristae. (See Fig. 5.2.1.)

5.2.1.3 Chloroplasts—Structure and enzyme localization

Chloroplasts and related organelles exhibit a wide range of differentiated forms which reflect a variety of specialized functions, and these include not only photosynthesis but also such activities as the storage of starch (see Section 5.2.2).

Mature chloroplasts, like mitochondria, vary in size and shape [86]. In higher plants they generally appear as elongated discs, the long axis between $3-10 \mu$. They are enclosed by a double membrane structure, called the envelope, which is about $30 \mu m$ in total width. Inside the envelope of mature chloroplasts are to be found further membrane structures which form the basic photosynthetic apparatus—the thylakoids. Each of these forms a flattened sack-like structure with sealed ends. (See Figs 5.2.2a & b.)

Several thylakoids may be stacked on top of each other forming a structure which has been compared with a pile of hollow pennies. However, these stacks of membranous discs, which were originally referred to as grana by light microscopists (as each stack appeared to be a single, solid inclusion) are not quite so regular as this initial description suggests. Many of the

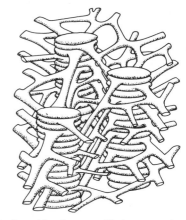

Fig. 5.2.2 (a) Three-dimensional models of grana with interconnecting fretwork system in higher plant chloroplast. By permission [160].

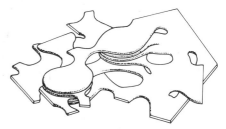

(b) Three-dimensional model of grana with inter-granal connections. By permission [157].

individual membrane discs or thylakoids extend further than their own discrete pile and link up with other thylakoid stacks.

Electron micrographs of sections from different plants have led to several theories about the three-dimensional arrangement of the thylakoid stacks and their interconnections. Two early interpretations were that the grana are connected either by flattened tubular extensions of individual thylakoids [160] or by extensive flattened sheets which extend throughout much of the plastid [157]. (See Figs 5.2.2a & b and Section 5.2.2 for more detailed interpretations.) The remaining volume of the chloroplast not occupied by membranes is known as the stroma which appears in electron micrographs as a granular substance. The membrane elaborations of blue-green algae are very similar to the photosynthetic structures of higher plant chloroplasts (see reference [164]) and in both cases there is some evidence for inter-connections between the thylakoids and the bounding membranes (cytoplasmic membrane in the blue-green algae, or the envelope in chloroplasts). Extreme variations in the exact structure of chloroplasts from this basic pattern have been described. The common features are that mature chloroplasts capable of active photosynthesis consist of interconnected stacks of membranous discs packed in a fluid matrix, the whole structure being surrounded by a double membrane.

As with mitochondria, subfractionation and biochemical investigations have distinguished enzyme activities associated with the membranes and those which are readily solubilized. The light absorbing pigments (chlorophylls and carotenoids) and many of the components associated with electron transfer and concomitant ATP formation are fairly tightly associated with the membranes. It is, therefore, not surprising to find extensive ramification of these structures, which thus provides a large catchment area for light energy.

The 'dark' reactions of photosynthesis, which are concerned with CO_2 fixation, can be performed in essentially membrane and pigment-free supernatants from disrupted chloroplasts. This suggests that many of the enzymes are normally free in the stroma or else, easily liberated on fractionation. In fact it has been suggested on the basis of electron microscopic examination of chloroplast membranes that at least some of the key enzyme involved in CO_2 fixation (ribulose diphospate carboxylase) may be loosely bound to the membrane [5]. However, much of the enzyme activity is found in the water soluble phase after disruption and extraction of chloroplasts. It is now thought that the bulk of the carboxylase enzyme is localized in the stroma.

Thus, in both chloroplasts and mitochondria the basic structure can be divided into the membranous extensions and a ground substance (matrix or stroma). In both, the membranes provide a platform for the organization of electron transport components. Enzymes involved with photosynthesis or energy metabolism, are apparently associated to a varying degree with the membrane fraction. ·Some like ribulose diphosphate carboxylase may be bound *in vivo*, yet liberated on disruption.

Others are thought to exist entirely in the fluid filled spaces of the stroma, matrix or intermembrane spaces.

5.2.1.4 Additional organelle components

The above description has been restricted to the most obvious functions of chloroplasts and mitochondria. However, both organelles contain components which are not directly concerned with photosynthesis and energy metabolism. The enzymes and other machinery necessary to maintain the structure of the organelle are also of considerable importance and these will be dealt with in the next section. Mitochondria are also the central participants in other aspects of cell metabolism. They are, for example, involved in synthesis of haem (δ aminolaevulinic acid synthetase is present in the matrix) and in lipid metabolism (several of the enzymes involved in degradation of fatty acids are associated with both inner and outer membranes).

Nevertheless, even if attention is confined to the components directly associated with the organelles most obvious functions, it is clear that a high degree of structural organization is necessary. How this organization develops, is maintained and what is known about the regulation of the developmental and synthetic processes will be discussed in the following sections.

5.2.1.5 Organelles contain DNA

Both mitochondria and chloroplasts contain DNA, yet convincing biochemical evidence of organelle-specific DNA has only become available during the last decade. So it is only recently that the contribution of this DNA to the regulation of organelle development has become the subject of extensive discussions and investigations.

The overwhelming evidence for the presence of DNA in organelles has accumulated from two major approaches:

(i) visualization by light or electron microscopy;
(ii) characterization of nucleic acids extracted from purified preparations of organelles.

Obviously, one of the main problems in these types of investigation is that of possible confusion of DNA in organelle preparations with nuclear DNA. However, it turns out that organelle DNA can frequently be distinguished from nuclear DNA by base compositional differences as shown by equilibrium centrifugation in CsCl. Furthermore, in many cases, the DNA of chloroplasts and mitochondria has been shown to be circular (see Table 5.2.1).

Another procedure which has proved of great value in distinguishing organelle and nuclear DNA is the measurement of renaturation kinetics. DNA can be denatured (i.e. strands of the double helix separated) by heating, or by exposure to alkali. When returned to more physiological conditions, the rate of reannealing depends on the relative concentration of the com-

Table 5.2.1   The DNA of mitochondria and chloroplasts. (Information from electron microscopy.)

	Length	Form	Approximate molecular weight* (daltons)	Reference
Mitochondria from:				
Most animal cells	5 μ	Circles	10^7	[20, 110, 167]
Tetrahymena	17–18 μ	Linear	$3·5 \times 10^7$	[148]
Yeast	25 μ	Circles	5×10^7	[75]
Plants (pea)	30 μ	Circles	6×10^7	[87]
Chloroplasts from:				
Euglena	~40 μ	Circles	8×10^7	[101]
Spinach	~44 μ	Circles		[100]
Corn	~43 μ	Circles	$8–9 \times 10^7$	[101]

* Estimated from the overall length of the DNA molecule (contour, length for circular DNA). λ phage DNA is often used as a reference. Many organelle genomes were first thought to be linear and were only found to be circular when improved extraction techniques became available. It is probable that those molecules which appear linear on extraction may be particularly fragile.

plementary strands. For a given total amount of DNA in an extract, there will be many more copies of short genomes (i.e. those having coding information for only a few genes) than long, complex genomes, such as those found in nuclei. The speed of strand reassociation can be measured either by changes in UV absorption, or by following the characteristic density shifts of DNA after equilibrium centrifugation. Single-stranded DNA usually bands at positions of higher density than the renatured double stranded helix.

A comparison of the rate of strand reannealing in DNA extracted from organelles and that of DNA from various viruses and bacteria of known genome size, indicates that chloroplasts and mitochondria have DNA of much smaller unique sequence length than that extracted from nuclei. (However, in some cases, confusion may arise as a result of the existence of fast reannealing, repetitious DNA fractions.) It has been possible to make very precise estimates of the amount of coding information available to the organelle using the type of analysis described in Chapter 5.3. Fortunately, the estimates from this type of study agree quite well with calculations based on the assumption that the contour length of circular genomes seen in electron micrographs represents the entire unique sequence length of the organelle genome (see Table 5.2.2).

Given the amount of organelle-specific DNA available, one can get an estimate of how many proteins could be coded for within the organelle using as a rough guide the relationship that 500 base pairs are needed to code for a protein of about mol. wt 20,000. A study of the data presented in Table 5.2.2 shows that informational content of mitochondria and chloroplast DNA is severely restricted. Even these calculations of the number of proteins which could be coded for are maximum estimates. We

shall see in the next section that a considerable fraction of mitochondrial DNA (and to a lesser extent of chloroplast DNA) is necessary to provide a template for ribosomal and transfer RNA.

Table 5.2.2   The genome size of organelle DNA estimated from renaturation kinetics

	Estimated genome size (daltons)*	Reference
Mitochondria from:		
Guinea pig	$1·1 \times 10^7$	[19]
Yeast	$5·0 \times 10^7$	[19]
Pea	$7·4 \times 10^7$	[87]
Chloroplasts from:		
Euglena	$1·8 \times 10^8$	[147]
Lettuce	$1·4 \times 10^8$	[161]
Tobacco	$1·9 \times 10^8$	[152]
Chlamydomonas	$2·0 \times 10^8$	[11]

* A molecular weight of 10^7 daltons corresponds to about 15,000 base pairs.

In several cases the estimates of genome size estimated from renaturation kinetics compare extremely well with those derived from electron microscopic observations, although with chloroplast DNA the figures derived from renaturation data are in general, larger than the estimates from contour length analysis. To obtain a genome size from renaturation kinetics the second order rate constant for the renaturation is determined for the organelle DNA under investigation and a molecular weight for the unique sequence length derived by comparison with the rate constants determined for the DNAs of known molecular weight, e.g. the phages λ or T4. Precise estimates are difficult to obtain because of minor variations probably resulting from differences in base composition (see [19]).

5.2.1.6 Ribosomes and other components required for protein synthesis are also present

As techniques for isolating relatively pure and undamaged organelles became available, it became apparent that chloroplasts and mitochondria could incorporate radioactive amino acids into protein [6, 86, 144, 164]. This suggested that they contained the necessary machinery for protein synthesis (such as ribosomes, rRNA and amino acyl tRNA synthetases).

Most of the initial investigations were directed towards distinguishing protein synthesis in organelles from that of contaminating cytoplasmic protein synthesis. This was made a lot easier by the discovery that mitochondrial and chloroplast ribosomes had several properties which distinguished them from ribosomes found in the rest of the cell.

Ribosomes present in organelles are smaller than the ribosomes of the cytoplasm and in this, as well as in several other properties, show a close parallel to the ribosomes found in bacteria and blue-green algae. The relative sizes of ribosomes from various sources as determined by their rate of sedimentation are presented in Table 5.2.3.

Another difference between the ribosomes or organelles and the cytoplasm which has also proved extremely useful in determining the range of proteins synthesized inside the organelle, as opposed to those which have to be provided from the outside, is that of drug sensitivity. Differences became apparent when known inhibitors of protein synthesis in bacteria, or in higher cell cytoplasm, were tested for effect on the incorporation of amino acids into proteins by isolated organelles. The inhibitors can be divided into three main groups (see Table 5.2.4). It can be seen that some inhibitors have a very wide spectrum of action and presumably affect some stage in protein synthesis common to all ribosome mediated protein synthesis. Puromycin, for example, acts as an amino acyl tRNA analogue.

Other types of inhibitor appear to differentiate between the ribosomes of eukaryote cytoplasm and prokaryotes.

By this criterion of drug sensitivity, the organelle ribosomes seem to resemble those found in bacteria and blue-green algae. Nevertheless, it is important to exercise caution in pressing the analogies between prokaryotic and organelle ribosomes too far. In animal cells the mitochondrial ribosomes (mitoribosomes) have a very much smaller overall size than those of bacteria and represent the smallest class of ribosomes known to function in protein synthesis.

If chloroplasts and mitochondria contain their own genetic information and independent protein synthetic machinery, the question must ultimately arise 'to what extent does this mean that the organelles are independent of the nucleus in controlling their own development?' This rather broad question can be broken down into several aspects. (1) Does the DNA of organelles serve any useful function at all? (2) if the DNA is

Table 5.2.3 The sizes of ribosomes from prokaryotes, organelles and eukaryotic cell cytoplasm

Source		References
Bacteria	70S	[151]
Blue-green algae	70S	[151, 40]
Eukaryotic cytoplasm	80S	[119]
Mitochondria from:		
Yeast	74–80S	[21, 105]
Animal cells e.g. *Hela*	60S	[8]
Chloroplasts	68–70S	[134]

The sedimentation values presented are approximate. Ribosomes from prokaryotes and eukaryotic cytoplasm are conventionally described as belonging to a 70S or 80S class, respectively. In fact, the average value determined by Taylor & Storck [151] for 25 bacteria was 68·4S.

Table 5.2.4   The effect of various inhibitors on protein synthesis in different systems

	Bacteria	Eukaryotic cytoplasm	Yeast mitochondria	Animal mitochondria	Chloroplasts
Antibiotic inhibitor:					
Puromycin	+	+	+	+	+
Cycloheximide	−	+	−	−	−
Emetine	−	+		−	
Chloramphenicol	+	−	+	+	+
Erythromycin	+	−	+	−?	+
Streptomycin	+	−	+	−	+

Symbols

+ Inhibition of protein synthesis.

− No inhibition of protein synthesis at the dose range found to block protein synthesis in other systems. If used at high doses, many of the inhibitors can affect protein synthesis in most systems, often by an indirect mechanism. For additional information on inhibitors which act differently on mitochondrial protein synthesis in yeast and animal cells, see H. Dixon, G.M. Kellerman & A.W. Linnane (1972) *Arch. Biochem. Biophys.*, **152**, 869–75.

functional, does the encoded information differ from that of the nucleus? (3) Where does the organelle protein synthetic machinery originate? If it were made under the direction of the nucleus this would severely limit the expression of organelle independence. In fact, a detailed examination of the last point also goes some way to answering the other two questions.

Protein synthesis inside organelles depends on ribosomes. In bacteria, these usually contain many different proteins [112] and three major classes of RNA (5S, 16S & 23S). The technique of DNA:RNA hybridization makes it possible to discover whether the RNA of organelle ribosomes is coded by nuclear or organelle DNA. In all cases examined, from mitochondria of human cells to tobacco plant chloroplasts, the available evidence indicates that organelles code for their own ribosomal RNA. (Although present in chloroplast ribosomes, 5S RNA does not appear to be a component of mitochondrial ribosomes [see 20, 134].) Hybridization of saturating quantities of ribosomal RNA species with organelle DNA indicates that the organelle genome contains the information for (at least) one 'heavy' and one 'medium' rRNA species [20, 54, 134, 152]. The exact size of the organelle ribosomal RNAs varies somewhat, ribosomal RNA components of animal cell mitochondria being about half the mol. wt. of the equivalent species found in yeast mitochondria and in plant chloroplasts, where they appear to be of similar sizes to those found in bacteria and blue-green algae.

In sharp contrast to the ribosomal RNA, the proteins of organelle ribosomes are made on cytoplasmic ribosomes and most probably are not therefore coded for by the organelle genome. Kuntzel [90] for example, has studied the effect of chloramphenicol, or cycloheximide, on the incorporation of labelled amino acids into mitochondrial ribosomal proteins of *Neurospora* and found the inhibitor of cytoplasmic protein synthesis (i.e. cycloheximide) reduced the labelling by 97%, whereas chloramphenicol had little, or no effect. More recent studies [95] have substantiated these observations and have not provided any evidence for the existence of any mitochondrially made ribosomal protein. Davey & Linnane [43] have shown in yeast that chloramphenicol administration during growth does not interfere with the production of mitochondrial ribosomes as indicated by their activity *in vitro*. Similar types of study have shown that selective inhibitors of chloroplast protein synthesis also have little effect on the accumulation of the organelle ribosomes. Taken together, these results indicate that organelle ribosomal proteins are made in the cytoplasm. However, until more direct experiments support this generalization, the possibility that one or more of the ribosomal proteins are made in, and coded for, by the organelle, cannot be eliminated (see Section 5.2.1.7).

The origin of other components necessary for protein synthesis, e.g. the transfer RNAs, can also be examined directly by the techniques of DNA/RNA hybridization. In all cases examined the 4S (assumed tRNA) found in mitochondria and chloroplasts shows sequence homology with the organelle DNA. In chloroplasts and in the mitochondria of fungi there are enough regions homologous with 4S RNA to provide for at least one copy of each tRNA [20, 134, 154], whereas, in animal mitochondria (e.g. *Xenopus laevis* and human cells in tissue culture, Hela) the information in the genome is limited to about 12 copies of tRNA, clearly not enough to provide one each of the tRNAs needed in 'normal' protein synthesis [3, 7, 44]. It is possible to visualize the actual sites of hybridization in the electron microscope by tagging the 4S RNA with ferritin. Employing this elegant procedure, nine sites have been identified on one strand and three sites on the other strand of Hela mitDNA, the strands first being separated by denaturation and purified by CsCl centrifugation [168]. The, as yet, unanswered question is whether the proteins synthesized in mitochondria of animal cells are unusual in having a restricted number of amino acids or whether additional tRNAs are supplied by the cytoplasm. Some recent experiments [38] seem to suggest that isolated Hela cell mitochondria preferentially incorporate hydrophobic amino acids, but there is no direct evidence of a restricted amino acid composition of mitochondrially made proteins.

Very little information has been accumulated about the other components necessary for efficient protein synthesis such as the factors required in initiation and elongation steps of peptide synthesis. However, it is known that organelles require *N*-formylmethionine tRNA to initiate protein synthesis as do prokaryotic organisms [54, 134, 164].

In yeast it appears that at least some of the additional components for mitochondrial protein synthesis are provided by the cytoplasm. The same holds for the RNA and DNA polymerase activities which have been shown to be present in isolated organelles (see Table 5.2.6).

Table 5.2.5 Some properties of 'petite' and wild type yeast

Wild type (Grande)	Petite
Able to	
Grow on non-fermentable carbon source	Unable to grow
Normal sized colonies formed on fermentable carbon source	Small (petite) colonies
ELECTRON TRANSPORT COMPONENTS	
Cytochromes *a* and *a₃* present	Absent
b present	Absent
c present	Present in elevated quantities
Succinate cytochrome *c* reductase present	Absent
Can develop extensive inner membrane ramifications (cristae formation)	Have lost, almost completely, the ability to form cristae

So, although the original finding of DNA and ribosomes in organelles suggested that their development within the cell might be largely autonomous, perhaps only requiring a switch on, or off, to be triggered by the physiological state of the cell, a more detailed examination of true independence of these components shows that this idea will have to be severely modified.

Different organelles in different organisms show quite large variation in their potential to provide for themselves (in terms of DNA content). Nevertheless, even in the chloroplast which has the most potential and can provide information for all the types of RNA that one might expect in an autonomous body (transfer, ribosomal and message), extensive cooperation with the remainder of the cell is essential even to provide for a complete protein synthesizing apparatus. A picture thus emerges of a complex, enforced cooperation between organelles and nucleus. Whatever the mechanism controlling organelle development, it is clear that only some of the necessary genetic information may be held within the organelle itself (or its developmental precursor). The remainder must be provided by the cell nucleus.

One possible complication, which has been alluded to previously, is that the nucleus itself may be the ultimate source of organelle DNA. The chromosomes might hold a master copy of the organelle genome which could be replicated and transferred to the cytoplasm at some stage during development. This suggestion has little experimental support. Furthermore, the existence of traits which affect organelles and which are transmitted by genetic factors in the cytoplasm strongly argues for the independent nature of the information in organelles (see section 5.2.1.7).

5.2.1.7 Components made on organelle ribosomes

In order to understand the process of development at the subcellular level, it is necessary to know something of the types of component made on organelle ribosomes (presumably under the direction of messages transcribed from organelle DNA) and of those which must be provided by cytoplasmic protein synthesis. Two different approaches have been used in this context, one is typically that of the geneticist and the other that of the biochemist. Organelle biogenesis is one field where both meet on equal terms without, so far, too many conflicting results. The following section contains contributions from both approaches and describes attempts to discover where organelle components other than nucleic acids are synthesized in the cell.

INFORMATION FROM GENETIC STUDIES

Brief mention has already been made of cytoplasmically transmitted traits affecting organelles. A consideration of the nature of male and female gametes and the processes leading to zygote production in higher plants and animals demonstrates the way in which cytoplasmic inheritance can be distinguished from the Mendelian segregation of nuclear genes.

The female gamete usually provides the bulk of the zygote cytoplasm, whereas the contribution from the male gamete is at best minimal and, in many cases, only the nucleus is transmitted.

In animals the male gamete may introduce several mitochondria into the egg. The fate of these mitochondria has been carefully studied in some mammals and it was found that they degenerated during early development [149]. In situations where the maternal and paternal mitDNA can be distinguished by hybridization or by pattern of restriction enzyme cleavage (e.g. in interspecies crosses) only maternal type of mitDNA has been detected in the offspring [43, 78]. The offspring is, therefore, dominated by the complement of organelles present in the female gamete cytoplasm. Genetic determinants which are present in organelles and influencing organelle traits are more or less exclusively transmitted through the female line (see Fig.

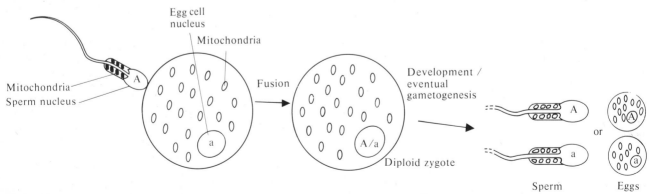

Fig. 5.2.3 Diagrammatic representation of maternal inheritance of mitochondria in animals. Although the mitochondria in the sperm mid-piece may penetrate the egg cell, their descendents have not been detected in the developed offspring (see text). They are possibly degraded [149] and in any case would be vastly outnumbered by the maternally derived organelles.

5.2.3). Any factor transmitted in this way shows 'maternal inheritance'.

The situation in micro-organisms such as yeast and unicellular algae is less straightforward. In the majority of cases, although some traits affecting organelles are transmitted uniparentally in that the offspring resemble only one of the parental type (see Fig. 5.2.4), the contribution to the zygote of cytoplasm from both parental type gametes is equal. One might therefore expect the progeny to show traits resembling both parents. A possible explanation for this enigma has been recently suggested by Sager & Lane for maternally inherited chloroplast traits in *Chlamydomonas* [135]. They propose that the plastid genome of one mating type, although initially present, is broken down soon after fusion of the plastids (which occurs in the zygote), possibly by a mechanism analogous with the modification and restriction of DNA found in bacteria.

In plants, many of the observations can be accounted for by complete elimination of plastids from one or other of the gametes. This has long been known, e.g. for *Spirogyra* [see 86], and is also thought to occur in the male gametes in certain flowering plants. Indeed, no plastids have been found in male

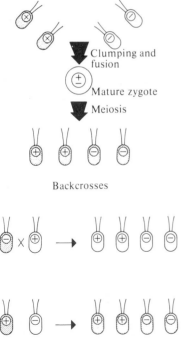

Clumping and fusion

Mature zygote

Meiosis

Backcrosses

Fig. 5.2.4 Uniparental inheritance of streptomycin resistance in the unicellular alga *Chlamydomonas reinhardi*. and reffer to mating type determinants controlled by nuclear genes and which therefore segregate 2:2 in the four meiotic products of the zygote. (Normally only the zygote stage in *Chlamydomonas* is diploid.) Streptomycin resistance (shaded cytoplasm) in the progeny is usually the same as that present in the mating type parent. See text and [135] for possible explanation.

gametes in any of the species so far given a thorough examination [81]. Examples of fern and bryophyte male sperm are known which possess plastids, but which lose them just prior to fertilization [117]. It does not follow, however, that the organelles derived from the male and female gametes will all survive to contribute to the embryo. In certain Gymnosperms a spatial segregation is maintained within the zygote, and it is held that the 'female' plastids degenerate [28, 29, 35].

It is thus possible for one gamete to lose its plastids but even where this occurs, the continuity of plastids is assured by their presence in the cytoplasm of the other. A convincing demonstration of such continuity comes from long term breeding experiments in the genus *Oenothera*, in which one subgenus consists of a number of species formed as a result of combinations of five types of plastid genome and six types of nuclear genome. Since a given type of plastid genome can be recovered unchanged after many successive generations of existence combined with a given type of nucleus, it follows that the plastids carrying that particular genome retained their individuality and were never lost [47, 139].

The most important conclusion to be drawn from the available work on plastid and mitochondrial genetics and the ultrastructure of male and female gametes, is that there is no need to hypothesize a *de novo* origin for organelles. There is no compelling evidence in favour of a *de novo* origin, some genetic evidence against a *de novo* origin, and much evidence in favour of genetic and structural continuity from cell to cell and from generation to generation.

Identification of a cytoplasmically inherited trait affecting organelles does not in itself rigorously establish that the relevant genes are present in organelle-DNA. The bulk of evidence, both circumstantial and direct, however supports the supposition. Detailed investigation at the biochemical level on cytoplasmically inherited alterations to organelles should eventually lead to identification of some organelle DNA-coded products. As we shall see, some limited success has been achieved with this approach.

Most of the cytoplasmically inherited markers, which initially were hopeful candidates for this type of investigation, both in chloroplasts and mitochondria, were mutations conferring resistance to certain antibiotics. Most of the drugs used in these studies were selected because of their inhibitory action on bacterial protein synthesis (see Table 5.2.4), but they also can inhibit growth of eukaryotic cells, presumably by blocking protein synthesis in the organelles. Mutants of yeast and *Chlamydomonas* can be derived spontaneously, or by mutagen treatment, which are resistant to some of these antibiotics. In many cases, genetic analysis demonstrates that the resistance is inherited cytoplasmically. The most detailed studies have been done with mutants resistant to chloramphenicol, erythromycin and streptomycin, all of which can inhibit organelle protein synthesis. This suggests that mutation to resistance is by virtue of an altered organelle ribosome (loss of antibiotic binding site?) or by a change in permeability at the cell or organelle boundary. Therefore, it might be hoped that identification of an altered ribosomal or membrane protein in such mutants would provide clear cut evidence of an organelle DNA-coded component. However, although alterations in the antibiotic binding activity and other physical properties of organelle ribosomes from resistant strains have been reported, only rarely have changes in the ribosomal proteins been implicated, e.g. [104], and there is, as yet, no evidence of an altered primary sequence in a 'mutant' protein [22a]. It may well be that in some mutants the resistance is mediated by the ribosomal RNA, which is known to be coded for by organelle DNA.

There is one class of cytoplasmically inherited trait in yeast, which has received more attention than any other and which has contributed greatly to an appreciation of what the organelle genome can do and what it cannot do, in terms of biogenesis and development. These are the so-called 'petite' mutants; so termed because their colonies are smaller than those of the 'grande' wild type. The mitochondria of 'petites' are severely disrupted both structurally and functionally and for this reason the petite yeast have to be grown on substrates which can provide energy via fermentation, thus avoiding the necessity for normal active respiratory processes.

The term 'petite' embraces a whole variety of lesions but in all cases where the effect is cytoplasmically controlled it can be traced to damaged mitDNA. There are also nuclear inherited traits having similar phenotypic effects but these are excluded from the present discussion.

The extent of alteration to the mitochondrial genome of petite yeast is highly variable. In some cases a few mitochondrial genes capable of expression can be recognized, but, in general, there is apparently extensive deletion and/or rearrangement of the DNA sequence [52, 74]. Extreme cases where all mitDNA is lost have also been described [59]. Therefore, by examining what remains of the mitochondria in extreme petites of this kind, a rough estimate of what is provided by non-mitochondrial systems can be obtained.

A description of the easily observed phenotypic alterations of a petite mutant as compared with the normal wild type yeast is presented in the Table 5.2.5 (see p. 275).

IN VIVO AND *IN VITRO* BIOCHEMICAL
INVESTIGATIONS OF ORGANELLE AUTONOMY

The alternative approach to that of genetics in seeking to establish what the nucleic acids of organelles contribute during development is less rigorous but more widely applicable. It depends on the assumption that the proteins made within the organelle (on organelle ribosomes) are gene products of the chloroplast or mitochondrial DNA. Analysis of the proteins synthesized in organelles can be carried out by following the incorporation of labelled amino acids into proteins of isolated organelles. Alternatively, studies can be done on intact cells by *selectively* blocking cytoplasmic protein synthesis with the drug cycloheximide. Incorporation of amino acids can be followed as with the isolated organelle. Both approaches have severe disadvantages and do not distinguish between proteins made in response to messages of nuclear origin which may be taken up by organelles and those directed by RNA made on organelle genome, unless combined with the use of transcription inhibitors. Nevertheless, both approaches give similar results and (where information is available) are consistent with data from genetic studies. In the Table 5.2.6, information from a wide variety of sources, both genetic and biochemical, have been combined to build up a picture of the components made by the organelle. Information is also presented relevant to those components which are assumed to be particularly important during development, in terms of structure and specialized function, but are made in the cytoplasm. A brief note about evidence on which the table is based is also presented.

Two of the most interesting facts which emerge from the data compiled in Table 5.2.6 are:

(1) Key enzymes which are necessary for the exploitation of the genetic information within the organelle (DNA and RNA polymerases) are coded for by the nucleus.

(2) Even those products apparently coded for by the organelle (e.g. rRNA, tRNA, subunits of cytochrome oxidase, ATPase and ribulose diphosphate carboxylase) are not functional by themselves but have to cooperate with additional components from the cytoplasm (e.g. ribosomal proteins, synthetases, other enzyme subunits) for activity. This suggests a tight interdependence of the organelle and cytoplasmic systems must operate during development. Exceptions to this generalization exist and are discussed in a later section.

5.2.1.8 Mechanism for localization of organelle components made in the cytoplasm during development

One of the puzzling features of organelle biogenesis is that, although some components are manufactured internally, clearly the majority of proteins must come in from the cytoplasm. Therefore, the problem is how these components find their way to their specific subcellular site. An additional complexity is posed by the observation that several mitochondrial enzymes (e.g. malate dehydrogenase and NADP linked isocitrate dehydrogenase) have cytoplasmic equivalents. So two forms of protein exist in the cell, both having similar catalytic properties, yet one of these is found almost exclusively in the mitochondria, the other in the cytoplasm. How is this compartmentalization achieved when both are nuclear gene products* and synthesized on cytoplasmic ribosomes?

---

* In man, genetic studies with somatic cells have indicated that the two malate dehydrogenases are coded for by genes on separate chromosomes [169]. For a general description and discussion of cultured cell genetics and its application to studies on organelle biogenesis, see [39].

Table 5.2.6 Probable origin of major organelle components

Mitochondrial components	Origin	Evidence	Notes
DNA polymerase	Nuclear/cytoplasmic		
RNA polymerase	Nuclear/cytoplasmic		
Ribosomes			
(a) ribosomal RNA	Organelle	H	
(b) ribosomal proteins	Nuclear/cytoplasmic	I	It is difficult to exclude the possibility that one or two of these may be synthesized, or modified, by the mitochondria
Inner membrane proteins			
(a) Some	Organelle	O, I	
(b) Others	Nuclear/cytoplasmic		
T.C.A. Cycle Enzymes	Nuclear/cytoplasmic		
ATPase (yeast)			
(a) 2–3 Subunits	Organelle	I	
(b) Other Subunits	Nuclear/cytoplasmic		
Cytochrome oxidase Some Subunits	Organelle	I	
Other Subunits	Nuclear/cytoplasmic		
Cytochrome b apoprotein	Organelle	I	
Cytochrome c	Nuclear/cytoplasmic		
tRNA (4S RNA)	Organelle	H	Yeast has information for about 20 copies whereas animal cell mitochondria only for about 12
Resistance to Chloramphenicol, Erythromycin, Oligomycin (in yeast)	Organelle	G	Resistance may also be conferred by mutations inherited Mendelianly, i.e. control by nuclear genes
Chloroplast components			
DNA polymerase	Nuclear/cytoplasmic		
RNA polymerase	Nuclear/cytoplasmic		
Ribosomes			
(a) Ribosomal RNA	Organelle	H	
(b) Ribosomal proteins	Nuclear/cytoplasmic	I	Again it is difficult from inhibitor studies to state that all ribosomal proteins are made in the cytoplasm
Cytochrome 559	Organelle		Absent in a mutant of Chlamydomonas lacking organelle ribosomes [see 134]
Cytochrome Q			
tRNA (4S RNA)	Organelle	H	
Ribulose diphosphate carboxylase			
heavy subunit	Organelle	I, G	
light subunit	Nuclear/cytoplasmic		
Resistance to some protein synthesis inhibitors, e.g. Erythromycin	Organelle	G	Resistance can also be conferred by nuclear controlled genes

The evidence for the synthesis of particular components is discussed in the text (see review references). However, the distinction should be borne in mind between those components for which direct evidence implicating the organelle genome in their production (i.e. hybridization or genetic studies) is available and those components for which the evidence only suggests synthesis on organelle ribosomes.

Explanation of symbols used in table

G * Genetic. Alteration to component can be cytoplasmically inherited. (See Bogorad, L. (1975) *Science,* **188,** 891 for recent review.)
H Hybridization. Organelle RNA hybridizes to organelle DNA.
I † Inhibitors. Components not synthesized in presence of chloramphenicol, but not inhibited by cycloheximide.
O Organelle. Component synthesized in the isolated organelle.
   * Although this implicates extranuclear genetic information it does not, in itself, directly identify the genome in question.
   † Indicates the involvement of bacterial like ribosomes (presumably those of the organelle) in the synthesis. Does not directly implicate coding by organelle DNA although it is probable that there is reasonable correlation. A similar argument can be applied to those components manufactured in the isolated organelle.

There are two attractive suggestions that could account for localization. Both have some experimental support. One possibility is that the proteins destined for inclusion within the organelle are synthesized with, or acquire by modification, specific physico-chemical properties. These, or so it is thought, enable recognition and specific uptake at the organelle membrane barrier. The specificity might result from a particular charge configuration, or by addition of a particular carbohydrate group. On this basis, one might, therefore, expect that enzymes found within the organelle would have a similar charge. In fact, some mitochondrial enzymes which have been examined in higher organisms and which are known to possess cytoplasmic equivalents, migrate to similar positions after electrophoresis on starch gel or cellulose acetate [39, 169]. Although this may have no direct relevance to the localization problem, it is of interest to note that similar properties are also found in the analogous enzymes of *Drosophila* and *Paramecium.*

There is some evidence that a localization mechanism similar to the one proposed may hold for lysosomal enzymes. These are normally involved in mobilizing carbohydrate storage products. In a rare human disease several of the lysosomal enzymes are unable to associate properly and integrate with the organelle. Furthermore, they apparently lack specific carbohydrate residues normally associated with the functional enzyme [71]. There are also instances of apparent structural gene mutations resulting in the localization of the altered enzyme at a different subcellular site, e.g. β glucuronidase in mice [114] malate dehydrogenase in *Neurospora* [106].

The second suggestion is that the proteins which eventually find their way into organelles are manufactured on a class of cytoplasmic ribosomes which are actually bound to the organelle surface membranes [138]. It is envisaged that during synthesis the proteins are directed into the organelle, perhaps in an analogous way to the production of proteins made for 'export' on the ribosome studded endoplasmic reticulum. The problem with this idea is that it raises a further question, that of how the different messages differentiate between the ribosomes on the outside of the organelle and any other. Nevertheless, circumstantial evidence for this type of mechanism exists in the electron microscope and biochemical studies which suggest the existence of a group of ribosomes in yeast which appear to associate specifically with the outer membrane [83, 138].

5.2.2 DEVELOPMENT AND BIOGENESIS OF CHLOROPLASTS

5.2.2.1 The variety of forms

The term plastid embraces a variety of differentiated organelle forms (see Figs. 5.2.5–5.2.10). We therefore have to consider not only problems concerned with origin and continuity of plastids but also problems concerned with the development of a wide range of plastid-type organelles, differing amongst themselves in structure and function.

The simplest member of the plastid family is the *proplastid.* Proplastids are small, though usually slightly larger than mitochondria (Fig. 5.2.5). One view which survived well into the electron microscope era was that there is only one class of particle, capable of assuming a variety of disparate forms, including both mitochondria and the more easily recognized forms of plastid [159]. Nowadays, however, it is accepted that the population of small particles in plant cells is heterogeneous and consists of two intermingled populations of genetically discrete entities, mitochondria and proplastids. The proplastids are thought to be transmitted through cell division and reproduction and to be capable of developing into other types of plastid during cell differentiation. They thus act as a plastid stem line.

In general, meristematic cells contain proplastids. As the products of cell division differentiate, so too do their plastids. The types that may appear include *chloroplasts, etioplasts, amyloplasts, chromoplasts* and *leucoplasts.* Chloroplasts are usually green, with photosynthetic functions; etioplasts develop where chloroplast formation is partially blocked by lack of light; amyloplasts store starch in bulk; chromoplasts are pigmented but are non-green and non-photosynthetic; leucoplasts contain little pigment, and this category embraces a fairly heterogeneous assemblage of sub-types, each one within limits typical of the cell in which it is found.

The following sections of the chapter describe aspects of the development of chloroplasts and etioplasts after a brief consideration of the origin and continuity of the lineage.

5.2.2.2 Origin and continuity

It has been established that the plastid genome and the nuclear genome operate in harmony during plastid development. Both are essential, and the current situation can be regarded either as the outcome of a process in which the nucleus has delegated a measure of autonomy to endogenously-generated plastids in the cytoplasm, or as the outcome of gradual genetical assimilation of a one-time independent prokaryotic organism into a fully symbiotic association in which it is now partially controlled by the nucleus of its host cell. For a more detailed account of the various hypotheses concerning the evolutionary origin of chloroplasts and mitochondria, see [14, 30, 91, 128, 133].

In any event, there are very few plant cells that lack plastids. For an exception see p. 277. Meristems in higher plants in general have between 10 and 100 proplastids per cell, and here [4] as in other situations [18] a balance is maintained between the rate of cell division and the rate of plastid multiplication. It is, however, one aspect of the differentiation of certain cell types for this balance to become altered. In cases such as the

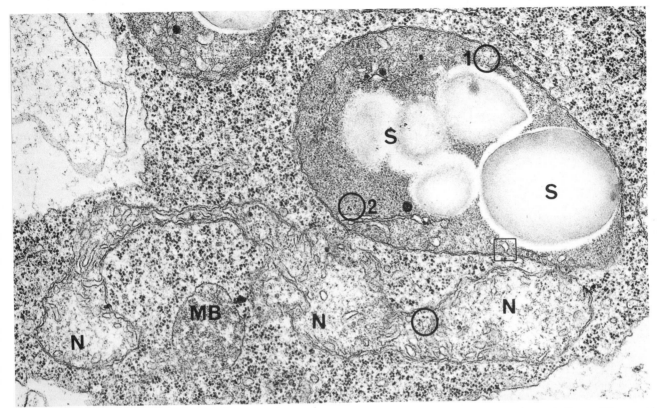

Fig. 5.2.5 Proplastid (upper right) and mitochondrion (lower) in a root tip cell (*Vicia faba*). The proplastid displays several of the general features of plastids: double envelope, nucleoid area, e.g. circle I, ribosomes, e.g. small granules in circle 2, and plastoglobuli (large dense granules). It also contains some starch (S) and a rudimentary internal membrane system, in parts continuous with the inner membrane of the envelope (square). Three nucleoid areas (N) appear in this particular section of the mitochondrion. Its ribosomes, e.g. in circle, like those of the proplastid, can be seen to be smaller than the surrounding cytoplasmic ribosomes. MB-microbody.

formation of photosynthetic leaf mesophyll tissue, plastid division continues long after cell division has ceased, and the population per cell may rise to several hundred [125]. Contrasting with the behaviour of the mesophyll, the plastid population remains at a much lower level in the epidermal cells of the leaf, and usually consists of leucoplasts that are, compared with the mesophyll chloroplasts, poorly developed.

There are pertinent analogies between the behaviour of nuclei and nuclear DNA in dividing and non-dividing cells, and the behaviour of plastids and plastid DNA. In many cases the three processes, nuclear DNA duplication, nuclear division, and cell division, are all closely geared to one another and proceed in step. In others the nuclear DNA can duplicate and reduplicate without concomitant cell division, leading to multinucleate cells or polyploid nuclei depending on whether nuclear division occurs or not. Maintenance of a population of proplastids at a standard level in meristematic cells mimics the most common type of nuclear behaviour. Increasing the population of chloroplasts per cell is like the formation of multinucleate cells. Formation of polyploid nuclei also has its counterpart, in that it is common for chloroplasts to be polyploid, as judged by estimation of the number of nucleoid areas (Fig. 5.2.7) present in them [70], of the extent of incorporation of labelled precursor of DNA [67], and of the amount of DNA present [68]. It seems that the larger the chloroplast the more likely it is to be polyploid. There can be several copies of the plastid DNA per nucleoid, and several nucleoids per plastid [88]. Mitochondria too can have more than one nucleoid area (Fig. 5.2.5) [109], but the assumption of the extensively poly-nucleoid condition in chloroplasts in a leaf cell does not, it seems, necessarily go along with parallel phenomena in the mitochondria of the same cells [88].

The consequences of the polyploidization of plastids, or at any rate chloroplasts, are many. For instance, although there may be very few genes coding for ribosomal RNA in each copy of the plastid genome, by the time the numbers in the population are taken into account, the total number of rRNA genes per cell can rival that in the nucleus [129].

It has been estimated that in *Vicia faba* green leaves, the number of plastid ribosomes approximately equals the number of cytoplasmic ribosomes [49]. Fairly similar figures are available for *Chlamydomonas* chloroplast and cytoplasm [22].

Where the multiplication of plastids does occur as in mesophyll cells, it is clearly a closely controlled process. Little is known about the stimuli that set it in train, or the controls that regulate the rate and duration of population growth. That the nucleus can exert an effect on plastid division, as on so many other aspects of plastid development, has become clear from observing what happens when nuclear DNA duplicates to form a polyploid nucleus [27]. The population of plastids increases by a factor of, on average, 1·7—that is, less than the factor of 2 that would be needed to maintain the level of the population per cell, were the DNA duplication to be followed by nucleokinesis and cytokinesis. It is not necessary that the total nuclear DNA be duplicated in order to bring about an increase in the plastid population per cell: in sugar-beet at least four out of nine different chromosomes carry genes that, if present in three copies instead of the usual two, lead to plastid population growth [27].

Another way of inducing plastid multiplication is convenient for its simplicity and ease of execution as a class exercise. If 'leaves' are pulled off the moss *Funaria* and placed on moist filter paper, it will be found that after a few hours the nuclear DNA will have duplicated, and many of the chloroplasts will have become dumb-bell shaped, preparatory to dividing [56, 57]. Formation of dumb-bells, followed by further constriction and finally fission, seems to be one general mode of plastid division, and another mode that has been claimed to occur involves partitioning of the plastid by an ingrowing septum derived from the chloroplast envelope [41].

As to the factors that stop the population from continuing to grow, current knowledge is extremely hazy. One suggestion is that the nucleus normally inhibits plastid division, and that the plastids can escape from this inhibition when the nucleus divides, or, somewhat less effectively, when the nuclear DNA endoduplicates without nuclear or cell division [27]. Other environmental and hormonal factors may participate in addition [17, 126]. Painstaking measurements of surface areas of mesophyll cells has indicated that multiplication and enlargement of chloroplasts ceases when they cover a certain proportion of the available cell surface, 73% in the case of spinach, and 25% in tobacco [76]. It may therefore be that parameters such as the availability of light, or the ability to transport CO_2 or other molecules across the plasma membrane

and into the cell, can react upon the system that regulates chloroplast development.

Having looked at aspects of plastid multiplication, it remains to consider the reverse phenomenon, namely the reduction in plastid numbers per cell. This is a phenomenon that can occur without profound consequences, as when chloroplast numbers per cell fall in autumn leaves [146], or when temporary fusion occurs in leaf cells [50], but it can also be significant in terms of the inheritance of plastid characters.

In lower plants, a reduction from a many-chloroplast condition to a condition in which there is a single chloroplast per cell, has often been reported for cells that are participating in formation of gametes or even in a vegetative reproductive process [26]. For example, in the moss *Polytrichum*, formation of sporogenous cells involves reduction to one chloroplast per cell, following which mitosis and chloroplast fission keep pace, each of the progeny receiving one nucleus and one chloroplast. Then control of plastid division is altered just prior to meiosis in the spore mother cells. Four are produced instead of the usual two, the four haploid spores that arise following meiosis each receiving one. When the spore germinates, the control system has again altered, and the multi-chloroplast condition is resumed [115]. It has been observed that three processes can contribute to reduction in plastid numbers. One is fusion of plastids, another is cell division without accompanying plastid division [26], and the third is breakdown of entire chloroplasts, including their DNA [99]. It would be of great interest to know whether the reduction in numbers is accompanied by a simplification of the plastid DNA to only a single copy per plastid, for this would mean that the plants arising from the uniplastid spores (or other cells) are genetically uniform with respect to their plastids.

5.2.2.3 Development of the chloroplast's internal membrane system

STRUCTURAL COMPLEXITY

Like other differentiated forms of plastid, chloroplasts arise from proplastids. Plastid multiplication may proceed during the developmental processes; indeed most of the increase in the population during maturation of mesophyll cells seems to occur amongst young chloroplasts, rather than at the earlier, proplastid stage [17, 76, 126].

Chloroplast development includes many syntheses involving both nuclear and plastid genomes. Undoubtedly the change that is most obvious is the formation of the chlorophyll-bearing system of internal membranes. The few fragments of membrane that are present in proplastids (Fig. 5.2.5) first enlarge to form more extensive flattened sacs, or *primary thylakoids,* and then the complex system of *grana* develops upon this foundation. Before examining what happens in more detail it is necessary to enlarge upon the summary of chloroplast ultrastructure presented at the beginning of the chapter.

Some of the earliest representations of chloroplast membranes depicted stacks of flattened, membrane delimited, discs. Each stack was a granum. Some of the discs extended laterally, outwards from the main stack, to contribute to adjacent grana. The membrane system was thus recognized to be in part in the grana, and in part in membranes that traverse the chloroplast stroma. It was later realized that the architecture of the stroma membranes is much more complex. First, they can be highly perforated or fenestrated, so that when imagined in their three dimensions, they resemble ornate carvings, which is why they have been called *frets*. Secondly, the frets interconnect the grana in such a way that no granum discs remain isolated. Where the grana and fretwork system is highly developed, a very high degree of communication is established, with each granum disc possessing *multiple* fret connections, and with frets typically ascending the grana, and sending connections in to each successive disc, in such a way as to generate right handed helices (Fig. 5.2.6) [116]. Every disc loculus is continuous with that of its neighbours within a granum *via* the fret connections, and all the grana are interconnected.

Functionally, the system can be interpreted as a compromise between two paramount requirements—that of optimizing conditions for the light reactions of photosynthesis, and that of maximizing communications between the membranes (the site of the light reactions) and the stroma (the site of the carboxylation and other 'dark' reactions of photosynthesis). Many types of compromise have been thrown up in the course of evolution [55, 86]. The region of the membrane system where adjacent disc membranes lie back to back seems to have special, but as yet incompletely specified, properties in respect of the light reactions. When isolated, these so-called *partitions* are found to be enriched in Photosystem 2 activity, while in the frets, Photosystem 1 predominates [118]. The significance of the partitions may be judged from the fact that they are found throughout all photosynthetic eukaryotes other than the red algae, certain higher plant mutants, and agranal chloroplasts of some C-4 plants which employ additional mechanisms for CO_2 fixation. The techniques of freeze-fracturing, or freeze-etching, shows that a population of large particles exist within the membranes at the partitions. Other, smaller, particles are also present, and unlike the large type, extend away from the partitions into the fret membranes [60, 84, 118].

SELF ASSEMBLY

Not all grana attain the degree of symmetry diagrammed in Fig. 5.2.6. Nevertheless many certainly do [116], and it is extremely difficult to visualize how they could develop upon the primary thylakoids of a young chloroplast. Complex though it is, it

seems that the ability to *self-assemble* is built-in to the system. It has been found that if chloroplast membranes are isolated and suspended in an organic buffer, the partitions very largely fall apart [1, 15, 79, 145]. The intricately interconnected system becomes a series of concentric single thylakoid surfaces, with few or no grana surviving. Although measurements have not

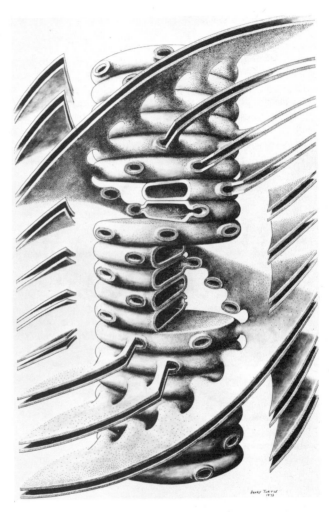

Fig. 5.2.6 An idealized chloroplast granum, with components not accurately to scale, drawn so as to emphasize the symmetry with which fret connections can link the discs and the frets traversing the stroma from granum to granum. Several discs are cut-away to show the disc loculi and the back-to-back membranes constituting the partitions. Most of the frets are cut-away to allow the right-handed helical path of one of them to be shown in full. The multiple fret connections developed by each disc are usually larger than drawn here. Their number per disc can vary, as can the number of discs and the ratio of frets to discs, here shown as one to two. Model based on reconstruction of Paolillo [116], reproduced by permission from [64].

been made, it is likely that no loss of membrane is entailed—merely a change in architecture. The significance from the point of view of development is not so much this experiment, but its sequel: that of adding salts (divalent cations are best). An extraordinary event takes place. Many grana reappear. It is not claimed that the original membrane system is re-created, but the conclusion that chloroplast internal membranes can self-assemble into small grana, given a suitable ionic environment, seems justified. This case can be compared with the other examples of self-assembly in Chapter 5.1.

Further analyses of the self-assembly reactions suggest that ionic and hydrophobic bonds are important in establishing partitions [108], that sulpholipids are required [37] and that special proteins are present in partitions [94]. The latter observation does not mean that the partitions can synthesize proteins on their own, for studies by freeze-etching show that when the original partitions are broken down, their large particles disperse, and when partitions are re-formed, one of the partial processes is the re-aggregation of the large particles [145].

The actual process of granum development in young chloroplasts is less dramatic, but presumably involves elements of the self-assembly reactions discovered in the test-tube experiments described above. Primary thylakoids do not form grana if they are suspended in low salt buffer and then given salt. Evidently they lack materials that are needed in order to form partitions [1]. *In vivo*, such materials soon arise. A primary thylakoid produces an evagination in the form of a small pouch (Fig. 5.2.7). This grows out over the surface of the parent. Its growth ceases when it has attained the shape and diameter of a granum disc—in effect when a granum consisting of two loculi and one intervening partition has been formed. The term *spirocyclic* growth describes subsequent events [156]. Further pouches protrude and grow out over the first granum disc. Each new disc is connected to the underlying disc or to the primary thylakoid at the point where its evagination commenced, and these channels of interconnection, if they are to generate a final form as in Fig. 5.2.6, must lie in a helix. Eventually, a stack of discs, all connected to a helically ascending fret, is produced. The growth process cannot, however, end there. Mature grana have multiple frets, and each disc has multiple fret connections around its periphery [116]. Clearly, in order to establish these, there must be considerable fusions between existing discs and locally growing frets. It may be that establishment of a mature and relatively symmetrical granum is a slow process involving numerous adjustments, many of which can be imagined to occur automatically if it is remembered that in general, symmetry and stability go together. The more order, or 'repeating units' of structure in the system, the less information that the nuclear and plastid genomes need to produce and maintain it. Thus, developmental adjustments which maximize symmetry will be favoured.

Many environmental factors influence the final form of the thylakoid system. Amongst those that affect the size and shapes of grana are: light quality [131], light intensity [10, 98], the photoperiod [23, 73], the CO_2 supply [53], and the supply of mineral nutrients [163], especially forms of nitrogen [92, 127].

MOLECULAR ASPECTS

Turning from the gross conformation of the chloroplast membranes to their development in terms of molecular composition, it seems that the operations involved are similar to those known to occur in the formation of other cell membranes such as the inner mitochondrial membrane and the endoplasmic reticulum [141]. In none of these membranes does the cell manufacture all of the necessary molecular species and then assemble them into an instantly functional membrane. Rather the various types of molecule are added in a multi-step sequence, akin to painting a mosaic one colour at a time, until the final pattern is ready. In order to detect this type of membrane growth it has been necessary to find material in which the thylakoid system of large numbers of cells and chloroplasts is developing synchronously. Mutant algal cells triggered to develop by an abrupt alteration of the nutrient regime have proved valuable, as have populations of etioplasts induced to develop towards the chloroplast conditions by switching on the light. Sequential sampling shows that the concentration ratios of individual pigments, lipids, and proteins change in the course of development, as would be expected from multistep biosynthesis [113, 120].

Electron microscopists have found no means of distinguishing 'young' membrane from 'old'. The fact that the thylakoid system is in places continuous with the inner membrane of the chloroplast envelope has raised suggestions that the latter gives rise to the former [see 131], but until the recent demonstration that the envelope is richly endowed with enzymes that synthesize galactolipids (one important component of thylakoids) [48], the idea lacked supporting evidence. Indeed, by feeding developing chloroplasts in *Chlamydomonas* with radioactive precursors and subsequently

Fig. 5.2.7 'Rough thylakoids' in oat chloroplasts developing from etioplasts. In early stages of greening, the thylakoids dispersing from the prolamellar body are perforated, but are 'smooth', although ribosomes and probably polyribosomes are present in the stroma. (a) Slightly later, the perforations have healed, and ribosome chains lie on the developing thylakoids, e.g. arrows, and free in the stroma. (b) After the lag period of greening has ended, the first granum discs and partitions appear, and polyribosomes are seen free in the stroma as irregular chains, e.g. large arrows, or tight helices (star), as well as on the thylakoids, e.g. small arrows. N-Nucleoid area. (c) (d) By this stage, granum formation is about half complete. The stroma-granum interface, with its associated ribosome chains, e.g. arrows, is seen in profile (d) and face (c) views. (e) Ribosomes remain attached to the young grana even when the chloroplast is broken and the stroma allowed to disperse. Note that in (a), (c) and (d), no ribosomes are seen associated with the chloroplast envelopes. (b) Reproduced by permission from [64].

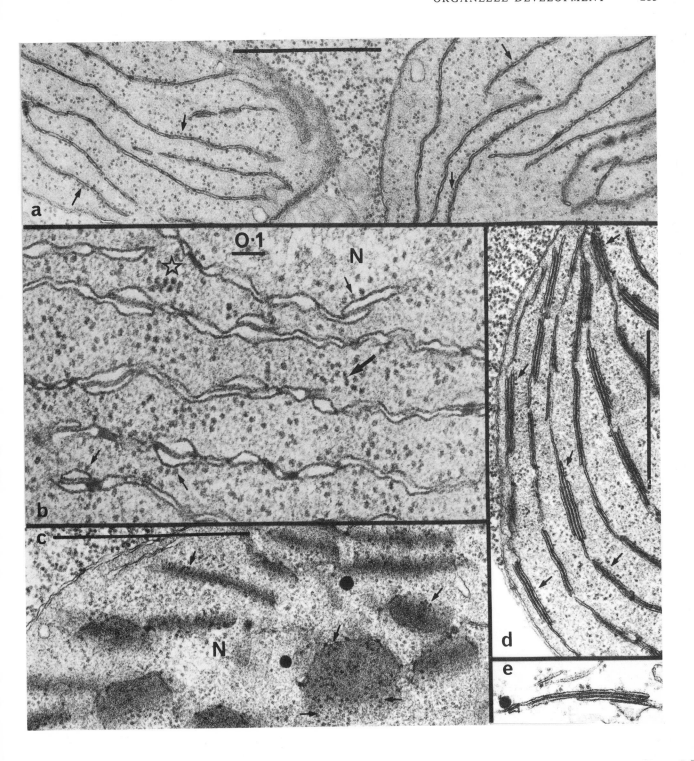

detecting sites of occurrence of labelled material by autoradiography, evidence that *all* parts of the thylakoid surface could grow has been obtained. The key observation is that partitions, frets, and the specialized thylakoids that traverse the pyrenoid, all contain radioactivity, and in different amounts [58]. In view of the known ability of molecules to move in the plane of membranes [142] (one example is the dispersal and re-aggregation of large particles when grana are unstacked and then restacked in the experiments on self-assembly described above), it would be dangerous to assume that the differential labelling implies differential rates of membrane growth.

When thylakoids are being extended rapidly, the microscopist can see (Fig. 5.2.7), and the biochemist can isolate, polyribosomes attached to the membranes [24, 34, 36, 51, 103, 121], i.e. at such times the thylakoids could be described as 'rough', by analogy with the terminology used to describe the endoplasmic reticulum. Treatment with puromycin, which releases nascent proteins from polyribosomes, leaves the proteins that were being assembled at the time of treatment attached to the thylakoids [36, 103]. A process of synthesis and direct incorporation of protein into the growing membrane is inferred. Polyribosomes also occur free in the stroma of chloroplasts (Fig. 5.2.7) [34] which suggests that developing plastids contain different classes of messenger RNA molecules and possess a means of determining whether their translation shall occur free or membrane-bound.

The most reasonable inference from modern work on thylakoid development is that the component molecular species are added independently, and probably not in any strongly localized region of the membrane surface. This is in full accord with much older observations. It has been known since the last century that the familiar ribbon-shaped chloroplast of

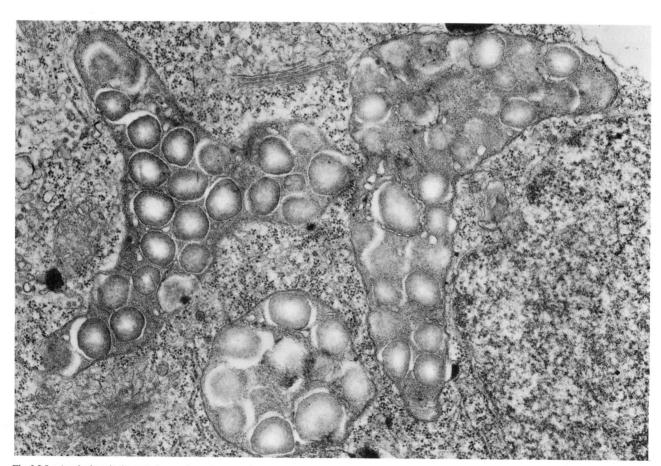

Fig. 5.2.8 Amyloplasts in the central zone of a soybean root cap.

Spirogyra does not have restricted growing points. It extends all along its length, though the tips may grow more than the central region. More recent studies by time-lapse photography (referred to on p. 295) of individual growing chloroplasts of *Nitella* also indicate an even distribution of growth [62]. The conclusion that new material is added to thylakoids all over their surface in units too small to be seen, still holds.

Before closing this section it is worth reiterating that nuclear and plastid genomes collaborate with one another in chloroplast development. Thylakoid assembly is known to involve materials coded for by both systems. The same applies, though at the *intra*-molecular level of protein subunits, rather than the *inter*-molecular level of thylakoids, to the enzyme ribulose diphosphate carboxylase (see Section 5.2.1.7) accumulated in bulk in chloroplasts as they are formed [136, 143]. Very little is known about the intricate signalling systems that must exist to coordinate the activities of the dual transcription and translation systems that work in harmony during chloroplast development. One example will suffice to show the order of complexity that can be anticipated [77]. In the y-1 mutant of *Chlamydomonas reinhardtii*, a major polypeptide of the thylakoid membranes is made on cytoplasmic ribosomes (as judged by the sensitivity of its synthesis to inhibitors specific to this class of ribosome). Despite its extra-chloroplast site of production, its manufacture seems to be geared to the synthesis of chlorophyll *within* the chloroplast. It is thought that a protein that is made in the chloroplast on chloroplast ribosomes regulates the activity of the gene coding for the thylakoid polypeptide. The regulatory protein is in turn influenced by the rate at which chlorophyll is formed, perhaps by interaction with a precursor of the pigment. The harmonous end result is that the membrane protein is made when chlorophyll is being made.

5.2.2.4 Formation of etioplasts

Formation of etioplasts is to some extent an unnatural process, conspicuous only when tissues that would otherwise become green and photosynthetic are forced to develop, or *etiolate*, in darkness. Then, a quasi-crystalline array of membrane in the form of interconnected branched tubules appears within the

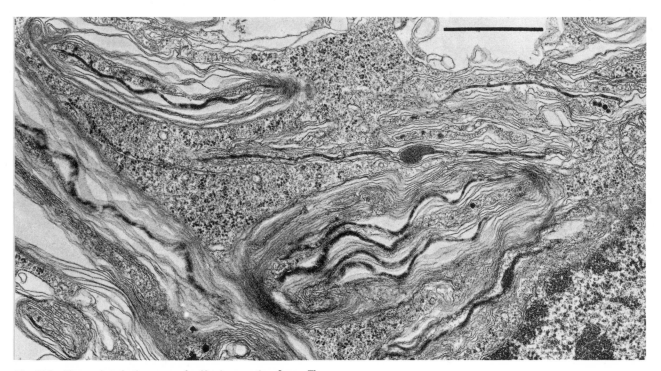

Fig. 5.2.9 Chromoplasts in the corona of a *Narcissus poeticus* flower. The clear areas in them, traversed by undulating thylakoids, represent accumulations of β-carotene.

plastid (Fig. 5.2.10), in place of grana and frets. Because the array of tubules metamorphoses into more conventional thylakoids upon illumination, it is known as the *prolamellar body*. Its production is emphasized during growth in darkness, but darkness is not obligatory; thus small prolamellar bodies appear amongst grana and frets, and connected to them, when leaves develop in low intensity light [66, 131, 158].

Most types of prolamellar body are based upon regularly interconnected tetrahedrally branched tubules [64] (Fig. 5.2.10). Their geometry need not be considered here, but it is worth pointing to one of the major differences between them and grana in chloroplasts. In chloroplasts, a substantial proportion of the total area of membrane is in the partitions, where, because of the back-to-back juxtaposition of membrane, it is not in direct contact with the stroma. In the prolamellar body, on the other hand, *all* of the membrane is bathed (on one surface) by the stroma, which penetrates between the tubules. Although dis-

sected into tubules, the membrane surface is a continuum. It may be argued on rather teleological grounds that if the main *raison d'être* for the prolamellar body is mere storage of membrane, then much more compact forms, e.g. myelin-like arrangements with limited contact with stroma, might be expected. The one hundred percent contact between the prolamellar body membrane—packed with a surface density of up to about 50 μm^2 per μm^3—and the stroma may therefore have as yet obscure functional significance.

Many, but not all, of the proteins and lipids of mature chloroplast thylakoids are present in prolamellar body membranes. There is, however, one major difference. Unless the etioplasts have been exposed at some time to light, the prolamellar body carries not chlorophyll, but its immediate precursor, protochlorophyllide *a*, which undergoes a rapid photochemical reduction to chlorophyllide *a* upon illumination of a dark-grown leaf. Somewhat more slowly, esterification with phytol then gives rise to chlorophyll *a* in a light-independent reaction. These processes are part of the conversion of etioplasts to chloroplasts, to be considered again below. While still in the dark, the protochlorophyllide *a* is associated with a large protein, the protochlorophyll holochrome. Some calculations suggest that the holochrome molecules, each about 10 nm in diameter, that occur in a prolamellar body would cover an area approximately equal to the area of membrane that is present [82]. Whether the protein, with its pigment, does in fact exist in a monolayer in or on the membrane remains to be seen.

Etioplasts contain neither as much pigment nor as much membrane as do chloroplasts. Plastid development is thus not only diverted into the morphologically distinct etioplast condition, but is also held in check. There is considerable evidence that the supply of δ aminolevulinic acid is the limiting factor. This compound is the first biosynthetic intermediate unique to the tetropyrrole pathway that leads to the synthesis of both chlorophylls and haems [12, 13]. If it is supplied to etioplasts in leaves their content of protochlorophyllide is increased, though the addition may not have the same photochemical properties as the original content. In other words, giving an exogenous supply of the precursor overcomes the check that is imposed upon protochlorophyll synthesis. Normally, after a certain amount of development in darkness, during which δ-aminolevulinic acid *is* produced and *is* converted to protochlorophyllide, the pathway is closed down. Illumination will then reopen it. The enzyme or enzymes that produce δ-aminolevulinic acid are relatively unstable, compared with others operating between this compound and protochlorophyllide itself [see 12], and the analysis of a number of mutants of barley plants in which the normal repression system does not function, so that unusually large amounts of protochlorophyllide accumulate, indicate that several nuclear genes operate to provide a control system governing δ-aminolevulinic acid formation [111, 162]. Most probably the controls act at the

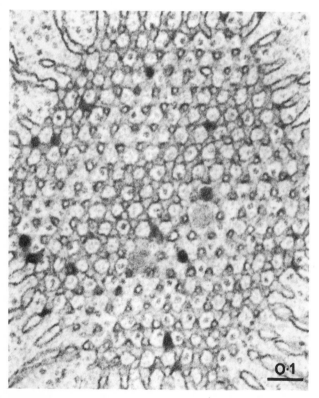

Fig. 5.2.10 A prolamellar body in an oat etioplast. In this particular example, the body was in the form of a hexagonal prism, here sectioned close to the (0001) plane, and made up of tetrahedrally-branched tubules interconnected in the same geometrical relationship as the zinc and sulphur atoms in the mineral wurtzite. The plastid stroma, including many ribosomes, penetrates between the tubules. Reproduced by permission from [40].

level of gene transcription, on the principle that if new enzyme is not continually produced by gene transcription and then translation, the instability of the enzyme will bring the biosynthetic process to a halt.

The first event of the 'greening process' that converts the etioplast to the chloroplast, is the rapid photo-reduction of protochlorophyllide *a* to chlorophyllide *a*. Except in very humid conditions [2] there is then a lag period, lasting a few hours, before rapid synthesis of chlorophylls and thylakoid membrane starts. The initial chlorophyllide *a* becomes phytylated, and the prolamellar body gradually loses its symmetry during the lag period. The tubular form of membrane metamorphoses into a series of relatively flat primary thylakoids which at first are perforated, and devoid of partitions. The perforations disappear (Fig. 5.2.7a), and granum formation, shown in its earliest stages in Fig. 5.2.7b, commences coincident with net synthesis of new chlorophylls, the δ-aminolevulinic acid producing enzymes having been de-repressed. The mechanics of granum formation have already been described for chloroplasts developing from proplastids, and probably are similar during greening.

Some of the early events of the greening process are reversible; thus early primary thylakoids may revert to a prolamellar body configuration if the plant is again darkened [65]. Chlorophyllide does not, however, return to protochlorophyllide. Neither do the grana membranes revert to tubular form. It seems that the stability of the partitions, once formed, preserves them through a period of darkness.

Most of the above account has emphasized developments *from* the proplastid condition, and it is appropriate to close this section of the chapter by noting that plastid differentiation does not seem to result in a modification in the properties of the plastid genome [69, 72]. Thus, when in the formation of reproductive cells, complex forms of plastid become simplified back to a proplastid configuration, the simplification in morphology does not imply any impairment of the developmental potential. Together with the host nuclear DNA, the plastid genome can still provide the new generation with the variety of plastid forms that it will produce in the course of its life.

5.2.3 DEVELOPMENT AND BIOGENESIS OF MITOCHONDRIA

In this section, particular attention will be given to the morphological and biochemical changes which occur during the development of mitochondria from 'precursors' (analogous to proplastids) in yeast cells. The advantage with yeast and some other micro-organisms is that they may be grown under anaerobic conditions when cellular processes can be maintained by the energy produced from fermentation; in these circumstances mitochondria dedifferentiate, as their functions are no longer required. Return to aerobic conditions causes rapid development of active mitochondria and allows the processes of mitochondrial biogenesis to be examined in detail.

A second situation which lends itself to the analysis of mitochondrial biogenesis and development is presented by the rapid changes occurring in these organelles in late stages of insect pupation. Very rapid enlargement and increases in mitochondrial function occur in the flight muscle blocks of insects at times close to emergence from the pupae.

Useful information has also been accumulated from the detailed studies which have been made on the fate of mitochondria during embryogenesis.

5.2.3.1 Development of mitochondria in yeast

Wild type yeast *Saccharomyces* are facultative anaerobes. They are able to grow, in the absence of oxygen, on a carbon source such as glucose, which can yield energy by fermentation pathways. In this situation, the mitochondria are redundant as far as oxygen linked electron transport requirements are concerned. Even in aerobic conditions, as long as a rich supplement of glucose is provided, the cells rely on the fermentative pathways largely to the exclusion of mitochondrial function. High levels of glucose thus leads to a reduction in respiratory activity—the Crabtree effect.

Mitochondria under conditions of glucose repression, or in absence of oxygen, when they are non-functional in terms of electron transport also exhibit considerable structural changes from those normally observed. It was thought at first that mitochondria disappeared altogether under these conditions. More recently it has become recognized that although extremely reduced in structure, mitochondrial equivalents are present; these are referred to as *promitochondria* [137].

There has been considerable controversy surrounding the presence of mitDNA in glucose repressed, or anaerobically grown cells. Original reports suggested that the amount of mitDNA in such cells was considerably less than that found in respiratory active counterparts. However, with the improved procedures developed recently (see Section 5.2.1.5), it is apparent that cells containing inactive and structurally deficient organelles may have the same DNA content (10–20% of total cell DNA) as those cells possessing structurally well developed and active mitochondria. (Note that clear distinction should be borne in mind between yeast which are deficient in mitochondrial profiles because of damage to the organelle DNA (petites) and those resulting from reversible changes in metabolic status).

A dispute also existed for some time concerning the nature of mitochondrial equivalents in anaerobic cells. Some workers were unable to recognize any structures which looked like mitochondria, or their precursors, in electron micrographs, whereas other groups thought they could. It was later discovered that the differences resulted in part from the conditions under which the yeast were grown. Presence of high glucose concentrations and the absence of lipid supplements in the media apparently prevented the accumulation of promitochondria. However, by changing the fixatives used to prepare the sections for electron microscopy, or by use of freeze etching, it was shown that mitochondria-like structures, although poorly developed, are present [42 and see 107]. The mitochondrial DNA found in anaerobically grown cells is associated with these structures.

It would seem then that the structures of promitochondria can vary, depending on the nature of the carbon source and the presence of other metabolic supplements, e.g. lipid, in the medium. Apart from this minor complication an important point has been established, i.e. that some structural precursors are present in all cases. Therefore mitochondrial development on return to aerobic conditions does *not* occur entirely *de novo* nor by transformation of non-mitochondrial precursors.

Promitochondria are much more difficult to isolate from cells than are normal mitochondria. Some success in this direction has been achieved by prefixing the cells with glutaraldehyde before breakage and isolation. The particles obtained by application of this treatment to anaerobic glucose grown cells are smaller than the mitochondria in aerobic cells (being about $0.4\,\mu \times 0.3\,\mu$ c.f. $1.0\,\mu \times 0.75\mu$). Promitochondria obtained by the glutaraldehyde treatment appear to have an extremely dense and granular matrix and the inner membrane is not obviously folded, cristae are absent or very poorly developed [155].

The discovery of these mitochondrial precursors and the availability of techniques for their study, has made it possible to examine in detail the morphological and biochemical changes associated with the development from a non-functional promitochondria into the fully functional organelle.

Although one should obviously expect differences in some detail of the developmental programme, it is generally felt that a similar overall pattern also applies to the development of mitochondria in other organisms. However, it is probable that only in facultative anaerobes do the changes in mitochondrial structure and function fluctuate between such extremes.

ORGANELLE DEVELOPMENT INDUCED BY AERATION

In yeast, the changes which occur on a shift from anaerobic conditions are rapid. Within a few hours the precursors are indistinguishable cytologically from normal mitochondria. Even after one hour of oxygenation some internal membrane organization can be seen in the dense granular matrix of the glutaraldehyde prefixed cells. The proliferation of internal membranes continues and the granular matrix disperses so that by three hours the formation of cristae appears similar to that of aerobically grown cells [155].

The structural changes occur in parallel with the appearance

of a functional organelle protein-synthesizing system and electron transport capability. Some low level of protein synthetic activity may be present in promitochondria particularly if a lipid supplement is provided. In the experiments on the oxygen adaptation of glucose grown, anaerobic yeast, the level of chloramphenicol sensitive, i.e. mitochondrial, protein synthesis increased dramatically after three hours of treatment, reaching one quarter that of aerobically grown cells. The promitochondria retained only vestiges of the functional, coupled-electron transport system, having only cytochrome b_1 and oligomycin sensitive ATPase (see Section 5.2.3.5). Whereas after two hours aeration, the presence of cytochromes c and b could be detected spectroscopically, to be followed shortly afterwards (3 h) by cytochromes a and a_3.

The fatty acid composition of the mitochondria also radically alters during the early stages of adaption, the most notable effects being an increase in long chain unsaturated fatty acids and a concomitant decrease in the percentage of short chain fatty acids. The absolute necessity for unsaturated fatty acids in mitochondrial development has been shown by withholding these from a mutant which required a supplement of unsaturated fatty acids for growth. Respiratory adaption will not take place unless a further addition is made to the medium [93].

Although some disagreement may still exist concerning the extent of dedifferentiation and differentiation on anaerobic-aerobic switching, it is clear that the expression and proliferation of mitochondria is built up on a scaffold of existing precursors. It has been shown by selective radioactive tagging of the promitochondria (^{14}C-leucine + cycloheximide) that it is indeed these precursors which give rise to functional organelles [124].

EFFECT OF PROTEIN SYNTHESIS INHIBITORS
ON MITOCHONDRIAL DEVELOPMENT IN YEAST

Inhibitors preventing either cytoplasmic or mitochondrial protein synthesis restrict the appearance of functional mitochondria as measured by the respiratory capacity of the cells. We have seen though that experiments with selective inhibitors are, at best, only crude attempts to analyse the respective roles of mitochondrial and cytoplasmic protein synthesis in development as the two systems are clearly not independent (for example ribosomal proteins are made on cytoplasmic ribosomes). Nevertheless, fascinating insights into the cooperation between the two systems in establishing mitochondrial activity have recently come to light and are discussed in Section 5.2.3.5.

5.2.3.2 Development of insect sarcosomes

A rather specialized aspect of mitochondrial development can be observed in the changes producing the highly differentiated mitochondria associated with tissues having a high work load. In insect flight muscle, the demands for energy production are met by grossly enlarged mitochondria which have extensive elaboration of the internal membranes. These are called sarcosomes.

In locusts the major development of the flight muscle sarcosomes occurs during a period of about eight days on either side of the imaginal moult, i.e. the final moult which results in the mature adult. In this time period it has been estimated that the mitochondrial membranes increase by a factor of 40-fold

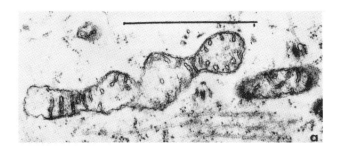

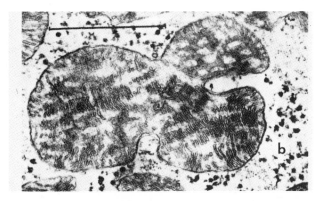

Fig. 5.2.11 Electron micrographs showing the structural development of blowfly sarcosomes. (a) In developing flight muscle three days before emergence. (b) At emergence. (c) Five days post-emergence. (Longitudinal sections; the bar on each print represents 1 μ.) Reproduced from [16] with permission.

[25]. This proliferation of membrane is associated with parallel increases in enzyme activities associated with cristae and matrix. However at all stages, even in the very early muscle precursors (eight days before moulting), the structures exhibit the characteristics of respiratory competent mitochondria both morphologically and in enzyme content. Thus the developmental change in this situation is not from inactive to active, but rather a rapid expansion based on a functionally competent precursor. What the detailed cytological and biochemical investigations have demonstrated in this case is that mitochondria are capable of rapid and extensive growth. From precursors about $1 \cdot 5$ μ long and $0 \cdot 3$ μ wide, mature sarcosomes are produced (6 μ long and 1 μ wide). It is interesting to note that in this case, increasing demand for energy yielding oxidative processes are met by increases in size, rather than in number of organelles.

An analogous situation exists with respect to the developing flight muscle of the blowfly. This has been the subject of intensive investigation by Birt and his colleagues [16]. The flight muscle is built up from an aggregation of large numbers of myoblasts derived from the imaginal discs. The mitochondria of myoblast cells themselves are relatively small (1 μ $\times 0 \cdot 2$ μ) and possess extremely few cristae. It has been estimated that during the development of the muscle blocks and their active involvement in the first days of adult life the volume of the mitochondria increases by about 10-fold and the frequency of the cristae by $2 \cdot 5$-fold (Fig. 5.2.11). The most active period of structural changes occurs from 2—3 days before the pupal hatch (eclosion) to four days after emergence. As in those cases already discussed, the rapid structural changes are also associated with alterations in chemical composition and increasing enzyme activity. Protein/phospholipid ratio increases, as does mitochondrial DNA content. The total organelle protein also builds up dramatically during this period, most of the increase taking place in the 24 h before emergence. Clearly this period represents a time of intense activity, but it has only very recently become recognized that the structural and biochemical changes represent *new* synthesis of material, rather than a redeployment and modification of existing proteins, lipids, etc.

Some original misconceptions, which underemphasized the role of protein synthesis in insect metamorphosis and organelle development were based on: (1) failure to show appreciable incorporation of injected labelled amino acids into mitochondrial proteins; (2) interpretation of electron micrographs which suggested that the sarcosomes were built from structural precursors which pre-existed in the matrix as a disorganized mass.

However, more recent studies [166] on the production of cytochrome *c* and the activity of (mitochondrial) protein synthesis indicate that in the blowfly too, the proliferation of organelle results, at least in part, from *de novo* production of its

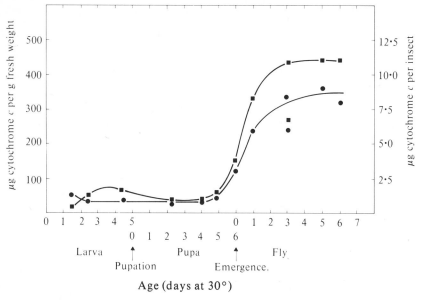

Fig. 5.2.12 Levels of cytochrome *c* in *Lucilia cuprina* (blowfly) during development. Squares = cytochrome *c* per insect; circles = cytochrome *c* per gram fresh weight. Note the rapid increase at about the time of emergence. The increase represents *de novo* synthesis of material (from [166] with permission).

constituents. Fig. 5.2.12 shows the levels of cytochrome *c* during the critical periods of development. All of the observed increase can be accounted for by *de novo* synthesis as indicated by rate of incorporation of labelled amino acid into the functional cytochrome *c*. Cycloheximide but not chloramphenicol block this incorporation.

Williams [165] has also undertaken detailed investigations into the relative contribution of mitochondrial and cytoplasmic protein synthesis during sarcosome development. Incorporation of injected ^{14}-C leucine was analysed at various developmental stages by subsequently isolating the mitochondria from the thorax. These results were compared with the *in vitro* incorporation into aseptically isolated mitochondria. The former gives information concerning the total synthesis of mitochondrial proteins by both mitochondrial and cytoplasmic systems, whereas the latter obviously reflects only the protein synthetic activity of the developing organelle.

The total incorporation of labelled amino acid into the organelle *in vivo* reaches a maximum at emergence. It has been estimated that about 90% of this is accounted for by cytoplasmic protein synthesis and that nearly 40% of the total cell protein synthesis is directed towards making mitochondrial products at this stage in the developing thorax. The isolated mitochondria also show a peak of synthetic activity at the time of emergence (Fig. 5.2.13). After reaching a maximum at

eclosion, the ability to manufacture protein drops rapidly and is virtually undetectable after six days of adult life. Similar results are indicated by the *in vivo* analysis. The pattern appears to be one in which a considerable proportion of the cell's synthetic

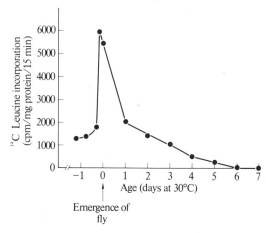

Fig. 5.2.13 Protein synthesis by flight muscle mitochondria isolated from insects at different stages of development. The mitochondria were isolated from thoraces of aseptically reared blowflies and incubated in a complex medium containing ^{14}C leucine (from [165] with permission).

resources both inside and outside the organelle are directed into the rapidly expanding sarcosome at, or just before, emergence from the pupal state. Mitochondrial and cytoplasmic protein synthesis increase and decrease in unison suggesting close cooperation and perhaps interdependency of the two systems. The maximum synthetic activity of the organelle is coincident with the stage of inner membrane proliferation. By the time the insect has reached maturity, protein synthesis directed into organelle material is very much reduced suggesting a slow replacement or turnover of mitochondrial components.

5.2.3.3 Mitochondria and the developing embryo

Studies of mitochondria in developing embryos of several animal and plant species have been carried out. As might be expected, the details of mitochondrial activity and proliferation vary. Some specific examples are discussed below.

THE SOUTH AFRICAN CLAWED TOAD
XENOPUS LAEVIS

The oocytes of *Xenopus laevis* have one hundred thousand-fold more mitochondrial DNA than somatic cells. This massive accumulation of mitochondrial material occurs during oogenesis and the densely packed organelles form a structure below the nucleus which is readily visible under the light microscope [see 9]. The build-up of mitochondrial DNA occurs in the absence of nuclear DNA synthesis indicating that the two processes are independently controlled. This presents the rather unusual situation in that there is more mitochondrial than nuclear DNA present in the oocyte. Studies by Chase & Dawid [33] indicate that the high content of mitDNA and equivalent quantities of mitochondrial protein and RNA, act as a reserve during the early stages of embryogenesis.

Little or no net increase of mitochondrial DNA or protein (e.g. cytochrome oxidase which can be used as a marker for mitochondrial activity) occurs until relatively late stages of development (swimming tadpole stage i.e. 50–65 h). However, incorporation of precursors into mitDNA can be detected at earlier stages. In contrast, nuclear DNA increases from the very earliest stages and keeps apace with cell division [61]. Mitochondrial RNA begins to increase on a per embryo basis at gastrulation and thereafter increases linearly at a slow rate. These basic observations on *Xenopus* suggest that the mitochondrial population does not begin to increase in number (mitDNA), or activity (cytochrome oxidase), until relatively late in development. The original mitochondria are present in sufficient quantity and state of activity to provide the energy requirements for early embryogenesis. The accumulation of mitRNA prior to increases in DNA and protein suggests that assembly of further organelle protein synthetic machinery is a prerequisite for the mitochondrial proliferation which is eventually necessary, when the reserves have been diluted out by cell division.

SEA URCHIN

The eggs of sea urchins also possess large quantities of mitochondrial DNA. However, in contrast to the observations on *Xenopus*, transcription of the DNA resulting in RNA species (identified as rRNA and putative message) begins from the very earliest stages (cleavage) of embryogenesis [32]. These results were obtained by following incorporation of radioactive RNA precursors. Similar labelling patterns were obtained after activation of enucleated eggs. This provides a convincing demonstration that the transcripts are not of nuclear origin and that transcription of mitDNA occurs in sea urchins from the very earliest stages of embryogenesis. Unfortunately it is not yet possible to assign a definite role in embryo development to the mitochondrial RNA produced, other than in providing mitochondrial ribosomes which are presumably necessary for the manufacture of essential proteins required for efficient energy development.

MOUSE EMBRYOS

Experiments with mouse embryos cultured *in vitro* indicate that, as in sea urchins, mitochondrial DNA transcription begins at very early times. There is also considerable cytological evidence that the mitochondria themselves undergo morphological development. Associated with this are concomitant changes in the pattern of energy metabolism and increases in oxygen consumption.

No mitDNA replication has been detected during development up to the blastocyst stage; it is therefore likely that the mitochondria required for early development are synthesized during oogenesis [122].

In two to four cell embryos, the mitochondria are small, vacuolated bodies with a dense matrix and have poorly developed cristae. From this stage they begin to increase in diameter and, at the same time decrease in matrix density. At the 16–32 cell stages more cristae become apparent and by the middle-late blastula stage well developed mitochondria with transverse cristae can be observed, although there is considerable variation in morphology between different mitochondria in the same section [123].

Small electron dense particles (150 Å diam.) which are assumed to be ribosomes become prominent in the mitochondria of expanded blastocysts. It has been estimated that about a fifteen-fold increase in ribosome numbers has occurred by this stage, when compared with the morula.

The contribution of the mitochondrial DNA to development of the organelle and the embryo has been examined by use of ethidium bromide. This selectively inhibits mitochondrial DNA transcription. Mitochondrial protein synthesis can also be selectively blocked with chloramphenicol, as has been discussed. Treatment with such mitochondrial specific transcription and translation inhibitors results in changes in the morphological development of the cristae. They appear much less compact and undergo considerable vesiculation. The increase in ribosome numbers is grossly reduced by ethidium bromide but not by chloramphenicol. This emphasizes the fact that the major contribution of the mitochondrial genome to these structures is the rRNA components. The lack of mitochondrial ribosomes at early stages indicates the relatively small contribution that mitochondrial protein synthesis can make to their own (mitochondrial) development and to that of the early embryo. The application of mitochondrial inhibitors seems to have very little effect on overall developmental processes up to at least the blastocyst stage [123]. It is therefore highly unlikely that the mitochondria export information of components essential to early embryogenesis and the control of cellular processes.

Other workers have also noted the morphological changes and increasing numbers of cristae associated with increases in respiratory activity (oxygen consumption) beginning at 4–8 cell stages of mouse embryos. Biochemical investigations also suggest that at these early stages of development, mitochondria contribute to shifts in metabolic patterns of the developing embryo [89]. At the four-cell stage uptake of tricarboxylic acid cycle intermediates shows a marked increase and may reflect the first stirring of this cycle after a period of relative inactivity.

SUMMARY

Mitochondria serve a vital role during embryogenesis in providing energy, through respiratory activity. The developing embryo meets this by activation and later, proliferation of existing organelles contributed by the oocyte. In some animals very large quantities of mitDNA and protein are present in the zygote and can thus act as a reserve which is diluted out until active mitDNA replication and mitochondrial protein synthesis get underway.

Transcription of mitDNA occurs very early in embryogenesis in some species, one of the probable products being rRNA, as rapid increases in mitoribosomes also occurs. In those species where mitochondrial activity (respiration and RNA synthesis) shows early stimulation, the mitochondria show alterations in morphology, the most characteristic changes being the appearance and proliferation of cristae. Exactly this type of behaviour would be expected to provide the energy to power the rapid cell proliferation and developmental processes. Analogous changes can be seen in other tissues when switching from a relatively inactive state to one dependent on increased respiratory activity.

5.2.3.4 Mitochondrial biogenesis in post-embryonic mammalian tissues

Studies of mitochondrial biogenesis in mammalian systems apart from those during early embryogenesis have been largely confined to the immediate pre- and post-natal stages, or in regenerating tissues (e.g. after partial hepatectomy) [see 54]. As with other systems the requirements for greater energy production, or respiratory metabolism, are met by increased synthesis of mitochondrial components.

Detailed investigations on the synchrony of the appearance of several mitochondrial components (e.g. cytochrome oxidase, succinate dehydrogenase and succinate cytochrome c reductase) in late embryonic and postnatal rat liver cells show that the specific activity of all three increase relative to that of cardiolipin which is a lipid component of the inner membrane exclusively limited to mitochondria. This type of evidence suggests that activation or insertion of the electron transport components into pre-existing membranes can occur [80].

5.2.3.5 The coordination between mitochondrial and cytoplasmic protein synthesis

Before concluding this section on mitochondrial biogenesis there is one aspect which must be discussed because of its relevance to the interaction between organelle and cytoplasmic protein synthesizing systems.

Several mitochondrial enzyme complexes (see p. 271) contain components made by both systems. Cytochrome oxidase and ATPase are made from several subunits, some of which are made on 80S cytoplasmic ribosomes and some on organelle ribosomes. In this type of situation, if anywhere, one might anticipate a close harmony in the two synthetic systems which would ensure a balanced production of subunits. It is not surprising to find that particular attention has been centred on the biogenesis of these complexes.

One of the most informative studies of this type has been done on the ATPase of yeast [154]. The normal ATPase activity of the mitochondria is sensitive to the inhibitor oligomycin (or rutamycin). This property is retained by the complex, even after being solubilized by detergent treatment. Enzyme purified in this way has been shown to consist of 9–10 different polypeptides. On further fractionation, the main ATPase can be separated into three major components. These are:

(1) *ATP hydrolysis activity.* Not sensitive to oligomycin—(5 polypeptides).
(2) *An oligomycin sensitivity conferring protein (OSCP)* which acts as a link between the ATPase activity.
(3) *Proteins of the mitochondrial membrane.* These can be thought of as the part of the membrane to which the ATPase activity is anchored.

The whole complex can be reconstituted from the ATPase, the OSCP and the membrane fraction. It appears that the OSCP itself does not act as the oligomycin target site, but serves to couple the ATPase to the membrane which does contain an oligomycin binding site.

During mitochondrial development in yeast, which has been switched from anaerobic to aerobic conditions, the oligomycin sensitive ATPase activity doubles in 2 h. If chloramphenicol is added, no increase is observed. However, what is of particular relevance is that ATPase activity corresponding to the expected increase is found free in the cell sap. This part of the complex is apparently made there as inhibitors of cytoplasmic protein syntheses (cycloheximide) prevent the accumulation. If now the block on mitochondrial protein synthesis is removed the ATPase activity is then incorporated into the mitochondria. Therefore, in the short term at least there is apparently a disruption of the expected stoichiometric production of ATPase complex components, showing non-obligatory coupling of the two pathways.

More general studies on the increase in respiratory activity of O_2 adapting yeast also provides evidence that the contribution of organelle and cytoplasm to the developing organelle do not have to be exactly coordinated. The exponential rise in respiratory activity normally observed can be completely blocked by cycloheximide (CHI) alone and is severely limited by chloramphenicol (CAP) treatment alone. However, if cells are first exposed to CHI (i.e. allowing only mitochondrial protein synthesis) and then switched from CHI to CAP, a significant respiratory increase can be obtained. Use of the drugs in reverse sequence does not give rise to this increase [132]. The conclusion is, therefore, that some components made by the mitochondria, although not functional on their own, are utilized, or modified by cooperation with the products of cytoplasmic protein synthesis to give rise to functional complexes.

5.2.4 DIVISION OF ORGANELLES

In the early stages of embryogenesis, in the formation of insect flight muscle sarcosomes and in the development of respiratory activity of adapting yeast, a requirement for increased mitochondrial function is met by further elaboration of existing structures. However, in actively dividing cells which already contain mature mitochondria there is an obvious need to increase the number of organelles to keep apace with cell division.

Synthesis of DNA in organelles has been studied in detail and most reports indicate that the replication mechanism is similar to that of bacteria and higher organisms to the extent that it proceeds semiconservatively [see 54, 134].

A fascinating series of electronmicrographs showing the intermediate steps in the replication of DNA in the mitochondria of tissue culture cells have recently been published. The process seems to be partly discontinuous in that the synthesis proceeds from one point on the circular genome, a daughter strand being first formed on one parental strand but not the other. There is no apparent synthesis on the second parental strand until the replicating fork on the first strand has progressed some way round the circle [130]. After initiation the second daughter strand is synthesized in the opposite direction (see Fig. 5.2.14).

It is more difficult to demonstrate division of the whole organelles, even using electron microscopy. The main problem is in the interpretation of what the apparently discrete organelles seen in sections actually represent. Are they really independent of one another and not interconnected in some way?

This is a difficult point to establish without resorting to the tedious process of building up a composite picture from serial sections. When this has been done, the results show that in many cases the organelles (mitochondria in particular) may assume a quite complicated branching structure inside the cell. Nevertheless, fission of the organelle is presumed to provide the mechanism for the approximate portioning of mitochondria and chloroplasts at cell division. Although this is sometimes difficult to demonstrate unequivocally in some higher plants and animals, there is abundant evidence for this type of process in other systems.

Micromonas is a simple flagellate which possesses a single mitochondrion and a single chloroplast each of which divides by fission at cell division [102]. Chloroplast division by fission has also been recorded by time lapse photography in the giant internodal cells of the alga *Nitella* [62]. Nevertheless, it should be borne in mind that the organelles, which appear as discrete bodies in the light microscope, may have interconnections not visible at this level of resolution. There is some evidence which even suggests that mitochondria and chloroplasts of some cells may be interconnected by extensions of their outer membranes. Even so, if one accepts that *de novo* synthesis does not occur, it is obvious that organelle division and separation must take place at some stage in rapidly dividing cells in order to maintain undepleted populations of chloroplasts and mitochondria.

There is strong biochemical evidence that mitochondria found in the bread mould *Neurospora* are descendants of pre-existing structures. *Neurospora* is a convenient organism for genetic studies and many auxotrophic strains have been isolated. One of these requires choline for growth. Choline is incorporated, eventually as lecithin, into the mitochondrial membranes. When *Neurospora* was grown in radioactive choline until half way through the logarithmic phase of growth and then grown for a further three generations on non-labelled supplement, it was found that nearly all of the mitochondria were labelled in radioautograms [97]. This would be expected if

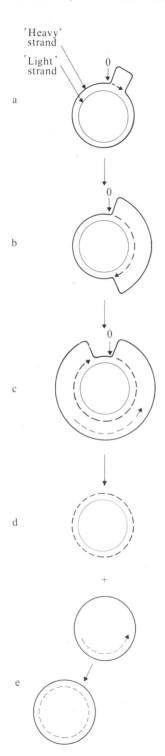

'Heavy' strand

'Light' strand

a

b

c

d

+

e

the new mitochondria were formed by division of pre-existing structures as the radioactive membrane component would be shared out between them. Whereas, if new mitochondria were made more or less from scratch, only those assembled during the early stages would contain the label. The newly synthesized mitochondria would contain only the unlabelled choline.

ORGANELLE TURNOVER

The observation that many cells (e.g. those of the liver) exist for up to 4–5 months raises the question 'what happens to the organelles during this time period?' Once made do they remain functional and stable, or are the component parts turned over? As the vast bulk of cellular protein is apparently 'turned over' in four weeks and the mitochondria represent up to 25% of this total, it is apparent that they are also recycled. Originally it was thought that the mitochondria were broken down and resynthesized as a whole. However, although exact turnover rates are difficult to establish, recent application of a double labelled procedure (by De Bernard, Getz & Rabinowitz [46]) have indicated that the outer membrane is more rapidly recycled than the inner membrane of rat liver. This type of observation has been confirmed by results from independent investigations (see [54, 107]).

Although the studies on organelles made during development have emphasized a coordination in the synthesis of various components, another aspect has become apparent in the studies on the biogenesis of ATPase (and some other respiratory complexes), and from the observations of differential turnover rates for some organelle components. These suggest that some asynchrony exists in the production (and/or integration) of organelle components. This may well indicate a requirement for sequential assembly of some parts of the organelle. The overall process could, therefore, be thought of as an erection and expansion of a basic mitochondrial scaffold, followed by integration of other components from the cytoplasm to provide the structurally organized assembly of respiratory 'enzymes' necessary for efficient electron transport coupled to the production of ATP.

5.2.5 THE PROBLEMS RAISED BY THE PRESENCE OF MULTIPLE COPIES OF ORGANELLE DNA

As we have seen, mitochondria and chloroplasts contain DNA coding for rRNA, tRNAs and probably some proteins

Fig. 5.2.14 Diagram of the model for mitochondrial DNA replication in mouse cells proposed by Robberson *et al.* Note the asynchronous replication of the heavy and light strand. Separation of the parental strands yields one completely replicated molecule and one (comprising the parental heavy strand with an incomplete progeny light strand) which finishes replication later (from [130] with permission).

associated with the organelle membranes. Apart from direct hybridization studies, the most convincing way of showing that a particular function is coded for by the organelle DNA is likely to result from genetic analysis. For this reason, a wide variety of mutagenic treatments have been employed in an attempt to generate identifiable alterations in the organelle whose control can be linked genetically to the organelle genome. Yeast and *Chlamydomonas* have, in particular, been subjected to intensive investigations of this kind and some of the resulting types of organelle mutants have been discussed in the text of this chapter.

Consideration of the unique sequence length of organelle DNA and the total amount present per cell indicate that yeast and animal cells have between 20 and several hundred copies of mitochondrial DNA/cell. Multiplicity also exists in the case of chloroplast DNA. *Chlamydomonas* although having only one structurally discrete chloroplast appears to possess multiple copies of chloroplast DNA. Thus, for any mutation in the organelle DNA to be effective it must either occur in all copies, or be able to exert its effect in the presence of excess copies of the normal wild type. Continued duplication of the mutant DNA copy and its subsequent partitioning among the progeny organelles may generate an organelle containing only mutant DNA copies. This in turn may provide further progeny which, after random segregation and cell division, could give rise to a line of cells entirely populated by mutant organelles. In plants this may generate mutant sectors in the leaves. This and other aspects of the production and occurrence of variegation (many variegations are in fact due to nuclear mutations) are discussed fully elsewhere [85, 86].

There is much yet to be learned about the controls regulating the replication, transcription and translation of organelle DNA and the interaction of components synthesized by the organelle with those provided by the nuclear/cytoplasmic system. This problem of development at the subcellular level constitutes a fascinating topic for future research.

5.2.6 CONCLUSIONS

(1) There are basic similarities in the structural organization of mitochondria and chloroplasts. Both have extensive membrane proliferations folded, or packaged, into characteristic shapes. Many of the complex enzymes and electron transport components concerned with energy production, either by photosynthesis or respiration, are intimately associated with these membranes. Other enzymes and metabolic intermediates are found in the fluid filled spaces (matrix or stroma) and can be liberated by gentle disruption of the organelle. The internal structures of both organelles are separated from the rest of the cell by a bounding membrane.

(2) The high degree of structural organization necessary for efficient energy generation may pose particular problems in developing systems.

(3) The final structure of organelles results from a cooperation between two genetic systems, that of the organelle itself and that of the nucleus.

(4) Mitochondria and chloroplasts contain their own independent protein synthesizing systems, which can be distinguished from those in the cytoplasm of animal or plant cells by their different sensitivity to a range of antibiotic inhibitors. The isolated organelles can also incorporate amino acids into protein.

(5) Biochemical and genetic data have been used independently, or in cooperation, to identify the range of components manufactured in organelles—ribosomal RNA, transfer RNA and a restricted spectrum of membrane associated proteins are thought to be coded for by organelle DNA.

(6) Most of the components identified, cannot operate independently but have to cooperate with cytoplasmically made (nuclear coded) components before becoming fully functional. E.g. organelle ribosomes are compounded from rRNA formed on the organelle DNA template and ribosomal proteins made in the cytoplasm.

(7) One of the most dramatic changes which occurs during chloroplast development from the precursor proplastid is the proliferation of the chlorophyll bearing internal membrane system. Although achieving a rather complex final architecture, there is some evidence that the final structure results in some part from the self-assembly properties of the membrane components.

(8) Other types of plastid may develop from the precursor proplastid. The actual developmental pathway taken depends on the interaction of environmental and nuclear factors. In contrast to the observations on differentiation at the cellular level, most if not all, of the various plastid types are interconvertible.

(9) The development of mitochondria from precursors has been described in two systems. (a) The appearance of active mitochondria in yeast adapting from a fermentation based energy production to one based on respiration. (b) The formation, in insect flight muscle, of elongate sarcosomes which are the equivalents, both structurally and functionally, of giant mitochondria.

(10) Studies on the mitochondria in developing amphibian, sea urchin or mouse embryos indicate that the oocytes contain a stockpile of these organelles which rapidly achieve functional status after fertilization and are subsequently distributed between the rapidly dividing cells of the early embryo. Replication of mitDNA is not observed until the original population is considerably diluted out.

(11) Taken together the information from a wide variety of studies indicates that new organelles are *not* formed *de novo*, but

result from the elaboration of existing precursors, or by division of mature organelles. The synthesis of the various mitochondrial and chloroplast components in development appears to be coordinated under normal conditions, although in some cases, products of cytoplasmic protein synthesis destined for organelle inclusion may accumulate (until sufficient sites for integration are produced).

(12) The genetics of organelles is complicated, not only by the presence of many individual organelles in each cell, but also in many cases by the existence of multiple genome copies in each plastid or mitochondria.

5.2.6 REFERENCES

The references include several review articles and selected research papers and can only represent a small fraction of the vast literature concerning organelle genetics and development.

[1] AKOYUNOGLOU G. & ARGYROUDI-AKOYUNOGLOU J. (1974) Reconstitution of grana thylakoids in spinach chloroplasts. FEBS Letters, 42(2), 135–40.

[2] ALBERTE R.S., THORNBER J.P. & NAYLOR A.W. (1972) Time of appearance of photosystems I and II in chloroplasts of greening jack bean leaves. J. Exp. Bot., 23, 1060–69.

[3] ALONI Y. & ATTARDI G. (1971) Expression of the mitochondrial genome in Hela cells. IV. Titration of mitochondrial genes for 16S, 12S and 4S RNA. J. Molec. Biol., 55, 271–76.

[4] ANTON-LAMPRECHT I. (1968) Anzahl und vermehrung der Zellorganellen im Scheitelmeristem von Epilobium. Ber. dt. Bot. Ges., 80, 747–54.

[5] ARNTZEN C.J., DILLEY R.A. & CRANE F.L. (1969) A comparison of chloroplast membrane surfaces visualized by freeze-etch and negative staining techniques. J. Cell Biol., 43, 16–31.

[6] ASHWELL M. & WORK T.S. (1970) The biogenesis of mitochondria. Ann. Rev. Biochem., 39, 251–90.

[7] ATTARDI G., ALONI Y., ATTARDI B., OJALA D., PICA-MATTOCCIA L., ROBBERSON D.C. & STORIE B. (1970) Transcription of mitochondrial DNA in Hela cells. Cold Spring Harbour Symposium Quant. Biol., 35, 599–619.

[8] ATTARDI G. & OJALA D. (1971) Mitochondrial ribosomes in Hela cells. Nature, New Biology, 229, 133–36.

[9] BALINSKY B.I. (1970) An introduction to embryology. 3rd ed. W.B. Saunders, Philadelphia.

[10] BALLANTINE J.E.M. & FORDE B.J. (1970) The effect of light intensity and temperature on plant growth and chloroplast ultrastructure in soybean. Amer. J. Bot., 57(10), 1150–59.

[11] BASTIA D., CHIANG K.-S., SWIFT H. & SIERSMA P. (1971) Heterogeneity complexity and repetition of the chloroplast DNA of Chlamydomonas reinhardtii. Proc. Natl. Acad. Sci., U.S.A., 68, 1159–61.

[12] BEALE S.I. & CASTELFRANCO P.A. (1974) The biosynthesis of δ-aminolevulinic acid in higher plants. I. Accumulation of δ-aminolevulinic acid in greening plant tissues. Plant Physiol., 53, 291–96.

[13] BEALE S.I. & CASTELFRANCO P.A. (1974) The biosynthesis of δ-aminolevulinic acid in higher plants. II. Formation of ^{14}C-δ-aminolevulinic acid from labelled precursors in greening plant tissues. Plant Physiol., 53, 297–303.

[14] BELL P.R. (1970) Are plastids autonomous? Symp. Soc. Expl. Biol., 24, 109–27.

[15] BERG S., DODGE S., KROGMANN D.W. & DILLEY R.A. (1974) Chloroplast grana membrane carboxyl groups: their involvement in membrane association. Plant Physiol., 53, 619–27.

[16] BIRT L.M. (1971) Structural and enzymic development of blowfly mitochondria and chloroplasts. In Autonomy and biogenesis of mitochondria and chloroplasts (Eds N.K. Boardman, A.W. Linnane and R.M. Smillie), pp. 130–39. North Holland.

[17] BOASSON R., BONNER J.J. & LAETSCH W.M. (1972) Induction and regulation of chloroplast replication in mature tobacco leaf tissue. Pl. Physiol., Lancaster, 49, 97–101.

[18] BOASSON R. & GIBBS S.P. (1973) Chloroplast replication in synchronously dividing Euglena gracilis. Planta (Berl.), 115, 125–34.

[19] BORST P. (1970) Mitochondrial DNA: structure, information content, replication and transcription. Symp. Soc. Exp. Biol., 24, 201–25.

[20] BORST P. (1972) Mitochondrial nucleic acids. Ann. Rev. Biochem., 41, 333–76.

[21] BORST P. & GRIVELL L.A. (1971) Mitochondrial ribosomes. FEBS Letters, 13, 73–88.

[22] BOURQUE D.P., BOYNTON J.E. & GILLHAM N.W. (1971) Studies on the structure and cellular location of various ribosome and ribosomal RNA species in the green alga Chlamydomonas reinhardi. J. Cell Sci., 8, 153–83.

[22a] BOYNTON J.E., BURTON W.G., GILLHAM N.W. & HARRIS E. (1973) Can a non-Mendelian mutation affect both chloroplast and mitochondrial ribosomes? Proc. natl. Acad. Sci., U.S.A., 70, 3463–67.

[23] BRONCHART R., FIRKET H. & SIMAR L. (1964) Ultrastructure du chloroplaste en fonction de la photopériode chez le Perilla nankinesis (Lour.) Decne. (Labiées). C. R. Acad. Sci., Paris, 259, 409.

[24] BROWN F.A.M. & GUNNING B.E.S. (1966) Distribution of ribosome-like particles in Avena plastids. In Biochemistry of chloroplasts. I, pp. 365–73. Academic Press, London & New York.

[25] BUCHER Th. (1965) In Aspects of Insect Biochemistry (Ed. T.W. Goodwin), pp. 15. Academic Press, London.

[26] BURR F.A. (1969) Reduction in chloroplast number during gametophytic regeneration in Megaceros flagellaris. The Bryologist, 72, 200–9.

[27] BUTTERFASS T. (1973) Control of plastid division by means of nuclear DNA amount. Protoplasma, 76, 167–95.

[28] CAMEFORT H. (1968) Cytologie de la fécondation et de la proembryogénèse chez quelques Gymnospermes. Bull. Soc. Bot., Fr., 115, 137–60.

[29] CAMEFORT H. (1969) Fécondation et proembryogénèse chez les Abiétacées (notion de néocytoplasme). Rev. Cytol., et Biol. Vég., 32, 253–71.

[30] CARR N.G. & CRAIG I.W. (1970) The relationship between bacteria blue-green algae and chloroplasts. In Phytochemical phylogeny (Ed. J.B. Harborne), pp. 119–43.

[31] CAVALIER-SMITH T. (1970) Electron microscope evidence for chloroplast fusion in zygotes of Chlamydomonas reinhardi. Nature, London, 228, 333–35.

[32] CHAMBERLAIN J.P. & METZ C.B. (1972) Mitochondrial RNA synthesis in sea-urchin embryos. J. Mol. Biol., 64, 593–607.

[33] CHASE J.W. & DAWID I. (1972) Biogenesis of mitochondria during Xenopus laevis development. Dev. Biol., 27, 504–18.

[34] CHEN J.L. & WILDMAN S.G. (1970) 'Free' and membrane-bound ribosomes, and nature of products formed by isolated tobacco chloroplasts incubated for protein synthesis. Biochem. Biophys. Acta., 209, 207–19.

[35] CHESNOY L. (1969) Sur la participation du gamète mâle à la constitution du cytoplasme de l'embryon chez le Biota orientalis. Endl. Rev. Cytol et Biol. vég., 32, 273–94.

[36] CHUA N.H., BLOBEL G., SIEKEVITZ P. & PALADE G.E. (1973) Attachment of chloroplast polysomes to thylakoid membranes in Chlamydomonas reinhardtii. Proc. Natl. Acad. Sci., U.S.A., 70, 1554–58.

[37] COATS L.W. (1973) Sulfolipid control of grana membrane stacking. J. Cell Biol., 59, 59a.

[38] COSTANTINO P. & ATTARDI G. (1973) A typical pattern of utilization of amino acids for mitochondrial protein synthesis in Hela cells. Proc. Natl. Acad. Sci., U.S.A., 70, 1490–94.

[39] CRAIG I.W. (1975) Cultured Cell Genetics, Proceedings 2nd Int. Symp. on the Genetics of Industrial Micro-organisms. Academic Press.

[40] CRAIG I.W. & CARR N.G. (1968) Ribosomes from the blue-green alga Anabaena variabilis. Arch. Microbiol., 62, 167–77.

[41] CRAN D.V. & POSSINGHAM J.V. (1972) Two forms of division profile in spinach chloroplasts. Nature, New Biol., 235, 142.

[42] CRIDDLE R.S. & SCHATZ G. (1969) Promitochondria of anaerobically grown yeast. I. Isolation and biochemical properties. Biochemistry, 8, 322–34.

[43] DAVEY P.J., YU R. & LINNANE A.W. (1969) The intracellular site of formation of the mitochondrial protein synthetic system. Biochem. Biophys. Res. Commun., 36, 30–4.

[44] DAWID I. (1970) The nature of mitochondrial RNA in oocytes of Xenopus laevis and its relation to mitochondrial DNA. Symp. Soc. Exp. Biol., 24, 227–46.

[45] DAWID I. & BLACKLER A.W. (1972) Maternal and cytoplasmic inheritance of mitochondrial DNA in Xenopus. Dev. Biol., 29, 152–61.

[46] DE BERNARD B., GETZ G.S. & RABINOWITZ M. (1969) The turnover of the protein of the inner and outer mitochondrial membrane of rat liver. Biochem. Biophys. Acta, 193, 58–63.

[47] DIERS L. (1970) Origin of plastids: cytological results and interpretations including some genetical aspects. Symp. Soc. Exp. Biol., 24, 129–45.

[48] DOUCE R. (1974) Site of biosynthesis of galactolipids in spinach chloroplasts. Science, N.Y., 183, 852–53.

[49] DYER T.A. & MILLER R.H. (1971) Leaf nucleic acids. I. Characteristics and role in the differentiation of plastids. J. Exp. Bot., 22, 125–36.

[50] ESAU K. (1972) Apparent temporary chloroplast fusions in leaf cells of Mimosa pudica. Z. Pflanzenphysiol., 67, 244–54.

[51] FALK H. (1969) Rough thylakoids: polysomes attached to chloroplast membranes. J. Cell Biol., 42, 582–87.

[52] FAYE G., FUKUHARA H., GRANDCHAMP C., LAZOWSKA J., MICHEL F., CASEY F., GETZ G.S., LOCKER J., RABINOWITZ M., BOLOTIN-FUKUHARA M., COEN D., DEUTSCH J., DUJON B., NETTER P. & SLONIMSKI P.P. (1973) Mitochondrial nucleic acids in the petite colonie mutants: deletions and repetitions of genes. Biochemie, 55, 779–92.

[53] GERGIS M.S. (1972) Influence of carbon dioxide supply on the chloroplast structure of Chlorella pyrenoidosa. Arch. Mikrobiol., 83, 321–27.

[54] GETZ G.S. (1972) Organelle biogenesis. In Membrane molecular biology (Eds C.F. Fox & A. Keith), pp. 386–434. Sinaver Ass. Inc., Conn.

[55] GIBBS S.P. (1971) The comparative ultrastructure of the algal chloroplast. Ann. N.Y. Acad. Sci., 175, 454–73.

[56] GILES K.L. & TAYLOR A.O. (1971) The control of chloroplast division in Funaria hygrometrica. I. Patterns of nucleic acid, protein and lipid synthesis. Plant and Cell Physiol., 12, 437–45.

[57] GILES K.L. (1971) The control of chloroplast division in Fumaria hygrometrica. II. The effects of kinetin and indoleacetic acid on nucleic acids. Plant and Cell Physiol., 12, 447–50.

[58] GOLDBERG I. & OHAD I. (1970) Biogenesis of chloroplast membranes. V. A radioautographic study of membrane growth in a mutant of Chlamydomonas reinhardi y-1. J. Cell Biol., 44, 572–91.

[59] GOLDRING E.S., GROSSMAN L.I., KRUPNICK D., CRYER D.R. & MARMUR J. (1970) The petite mutation in yeast: loss of mitochondrial DNA during induction of petites with ethidium bromide. J. Mol. Biol., 52, 323.

[60] GOODENOUGH U.W. & STAEHELIN L.A. (1971) Structural differentiation of stacked and unstacked chloroplast membranes. Freeze-etch electron microscopy of wild-type and mutant strains of Chlamydomonas. J. Cell Biol., 48, 594–619.

[61] GRAHAM C.F. (1966) The regulation of DNA synthesis and mitosis in multinucleate frog eggs. J. Cell Sci., 1, 363–74.

[62] GREEN P.B. (1964) Cinematic observations on the growth and division of chloroplasts in Nitella. Am. J. Bot., 51, 334–42.

[63] GREEN D.E., KORMAN E.F., VANDERKOOI G., WAKABAYASHI T. & VALDIVIA E. (1971) Structure and function of the mitochondrial system. In Autonomy and biogenesis of mitochondria and chloroplasts (Eds N.K. Boardman, A.W. Linnane & R.M. Smillie), pp. 1–17. North Holland.

[64] GUNNING B.E.S. & STEER M. (1975) Ultrastructure and the biology of plant cells (Ed. Arnold), pp. 1–300.

[65] HENNINGSEN K.W. & BOYNTON J.E. (1969) Macromolecular physiology of plastids. VII. The effect of a brief illumination on plastids of dark-brown barley leaves. J. Cell Sci., 5, 757–93.

[66] HENNINGSEN K.W. & BOYNTON J.E. (1970) Macromolecular physiology of plastids. VIII. Pigment and membrane formation in plastids of barley greening under low light intensity. J. Cell Biol., 44, 290–304.

[67] HERRMANN R.G. (1970) Multiple amounts of DNA related to the size of chloroplasts. I. An autoradiographic study. Planta, 90, 80–96.

[68] HERRMANN R.G. (1970) Anzahl und Anordnung der genetischen Einheiten (Chloroplastengenome) in Chloroplasten. Ber. Dtsch. Bot. Ges., 83, 359–61.

[69] HERRMANN R.G. (1972) Do chromoplasts contain DNA? II. The isolation and characterization of DNA from chromoplasts, chloroplasts, mitochondria and nuclei of Narcissus. Protoplasma, 74, 7–17.

[70] HERRMANN R.G. & KOWALLIK K.V. (1970) Multiple amounts of DNA related to the size of chloroplasts. II. Comparison of electron-microscope and autoradiographic data. Protoplasma, 69, 365–72.

[71] HICKMAN S. & NEUFELD E.F. (1972) A hypothesis for I-cell disease: defective hydrolases that do not enter lysosomes. Biochem. Biophys. Res. Commun., 49, 992–99.

[72] HINCHMAN R.R. (1971) The DNA of the oat plastid. Argonne Natl. Lab. Ann. Rep., 79–80.

[73] HIRSCHAUER M., REYSS A., SARDA C. & BOURDU R. (1971) Effects of photoperiods on the development of chloroplast lamellae of Lolium multiflorum. 2nd International Congress on Photosynthesis. pp. 2519–23. Stresa.

[74] HOLLENBERG C.P., BORST P., FLAVELL R.A., VAN KREIJL C.F., VAN BRUGGEN E.F.J. & ARNBERG A.C. (1972) Biochem. Biophys. Acta, 277, 44–58.

[75] HOLLENBERG C.P., BORST P. & VAN BRUGGEN E.F.J. (1970) Mitochondrial DNA. V. A 25 μ closed circular duplex DNA molecule in wild-type mitochondria. Structure and genetics complexity. Biochem. et Biophys. Acta, 209, 1–15.

[76] HONDA S.I., HONGLADAROM-HONDA T. & KWANYUEN P. (1971) Interpretations on chloroplast reproduction derived from correlations between cells and chloroplasts. Planta, 97, 1–15.

[77] HOOBER J.K. & STEGEMAN W.J. (1973) Control of the synthesis of a major polypeptide of chloroplast membranes in Chlamydomonas reinhardi. J. Cell Biol., 56, 1–12.

[78] HUTCHINSON C.A., NEWBOLD J.E., POTTER S.S. & EDGELL M.H. (1974) Maternal inheritance of mammalian mitochondrial DNA. Nature, 251, 536–38.

[79] IZAWA S. & GOOD N.E. (1966) Effects of salts and electron transport on the confirmation of isolated chloroplasts. II. Electron Microscopy. Pl. Physiol. Lancaster, 41, 544–52.

[80] JAKOVCIC S., HADDOCK J., GETZ G.S., RABINOWITZ M. & SWIFT H. (1971) Mitochondrial development in liver of foetal and newborn rats. Biochem. J., 121, 341–47.

[81] JENSEN W.A. (1974) Reproduction in flowering plants. In Dynamic aspects of plant ultrastructure (Ed. A.W. Robards), pp. 481–503. McGraw-Hill Book Company (UK) Limited.

[82] KAHN A. (1968) Developmental physiology of bean leaf plastids. II. Negative contrast electron microscopy of tubular membranes in prolamellar bodies. Pl. Physiol. Lancaster, 43, 1769–80.

[83] KELLEMS R.E. & BUTOW R.A. (1972) Cytoplasmic 80S ribosomes associated with yeast mitochondria. J. Biol. Chem., 247, 8043–50.

[84] KIRK J.T.O. (1971) Chloroplast structure and biogenesis. A. Rev. Biochem., 40, 161–96.

[85] KIRK J.T.O. (1972) The genetic control of plastid formation: recent advances and strategies for the future. Sub-cell. Biochem., 1, 333–61.

[86] KIRK J.T.O. & TILNEY-BASSETT R.A.F. (1967) The plastids. W.H. Freeman & Co., London & San Francisco.

[87] KOLODNER R. & TEWARI K.K. (1972) Physicochemical characterization

of mitDNA from pea leaves. *Proc. Natl. Acad. Sci., U.S.A.,* **69,** 1830–34.

[88] KOWALLIK K.V. & HERRMANN R.G. (1972) Variable amounts of DNA related to the size of chloroplasts. IV. Three-dimensional arrangement of DNA in fully differentiated chloroplasts of *Beta vulgaris* L. *J. Cell Sci.,* **11,** 357–377.

[89] KRAMEN M.A. & BIGGERS J.D. (1971) Uptake of tricarboxylic acid intermediates by pre-implantation mouse embryos *in vitro. Proc. Natl. Acad. Sci., U.S.A.,* **68,** 2656–59.

[90] KUNTZEL H. (1969) Proteins of mitochondrial and cytoplasmic ribosomes of *Neurospora crassa. Nature,* **222,** 142–46.

[91] LEDERBERG J. (1952) Cell genetics and hereditary symbiosis. *Physiol. Rev.,* **32,** 403–30.

[92] LEFORT M. & BOURDU R. (1964) Influence de la réanimation nitrique sur l'infrastructure plastidiale de *Bryophyllum Daigremontianum* Berger. *C. R. Acad. Sci., Paris,* **258,** 5031–34.

[93] LEVIN B. (1972) See Getz, G. (1972).

[94] LEVINE R.P., ANDERSON J.M. & DURAM H. (1973) Polypeptides participating in chloroplast membrane stacking. *J. Cell Biol.,* **59,** 192a.

[95] LIZARDI P.M. & LUCK D.J.L. (1972) The intracellular site of synthesis of mitochondrial ribosomal proteins in *Neurospora crassa. J. Cell Biol.,* **54,** 56–74.

[96] LOENING V.E. & INGLE J. (1967) Diversity of RNA components in green plant tissues. *Nature,* **215,** 363–67.

[97] LUCK D.J.L. (1973) Formation of mitochondria in *Neurospora crassa.* A quantitative radioautographic study. *J. Cell Biol.,* **16,** 483–99.

[98] MACHE R. & LOISEAUX S. (1973) Light saturation of growth and photosynthesis of the shade plant *Marchantia* polymorpha. *J. Cell Sci.,* **12,** 391–401.

[99] MANNING J.E. & RICHARDS O.C. (1972) Synthesis and turnover of *Euglena gracilis* nuclear and chloroplast deoxyribonucleic acid. *Biochemistry,* **11,** 2036–43.

[100] MANNING J.E., WOSTENHOLME D.R. & RICHARDS O.C. (1972) Circular DNA molecules associated with chloroplasts of spinach. *Spinacea oleracea. J. Cell Biol.,* **53,** 594–601.

[101] MANNING J.E., WOSTENHOLME D.R., RYAN R.S., HUNTER J.A. & RICHARDS O.C. (1971) Circular chloroplast DNA from *Euglena gracilis. Proc. Natl. Acad. Sci., U.S.A.,* **68,** 1169–72.

[102] MANTON I. (1959) Electron microscopical observations on a very small flagellate: the problem of *Chromulina pusilla* Buchter. *J. Mar. Biol. Ass., U.K.,* **38,** 319–333.

[103] MARGULIES M.M. & MICHAELS A. (1974) Ribosomes bound to chloroplast membranes in *Chlamydomonas reinhardtii. J. Cell Biol.,* **60,** 65–77.

[104] METZ L. & BOGORAD L. (1972) Altered chloroplast ribosomal proteins associated with erythromycin resistant mutants in two genetic systems of *Chlamydomonas reinhardtii. Proc. Natl. Acad. Sci., U.S.A.,* **69,** 3779–83.

[105] MORIMOTO H., SCRAGG A.H., NEKHOROCHEFF J., VILLA V. & HALVORSON H.O. (1971) Comparison of the protein synthesizing systems from mitochondria and cytoplasm of yeast. In *Autonomy and biogenesis of mitochondria and chloroplasts* (Eds N.K. Boardman, A.W. Linnane & R.M. Smillie), pp. 282–92. North Holland.

[106] MUNKRES K.D., BENVISTE K., GORSKI J. & ZUICHES C.A. (1970) Genetically induced subcellular mislocation of *Neurospora crassa* mitochondrial malate dehydrogenase E.C.1.1.1.37. *Proc. Natl. Acad. Sci., U.S.A.,* **67,** 263–70.

[107] MUNN E.A. (1974) *The structure of mitochondria.* Academic Press, London.

[108] MURAKAMI S. & PACKER I. (1971). The role of cations in the organization of chloroplast membranes. *Archs. Biochem. Biophys.,* **146,** 337–47.

[109] NASS M.M.K. (1969) Mitochondrial DNA. I. Intramitochondrial distribution and structural relations of single and double-length circular DNA. *J. Molec. Biol.,* **42,** 521–28.

[110] NASS M.M.K. (1969) Mitochondrial DNA. II. Structure and physicochemical properties of isolated DNA. *J. Mol. Biol.,* **42,** 529–45.

[111] NIELSEN O.F. (1974) Photoconversion and regeneration of active protochlorophyll(ide) in mutants defective in the regulation of chlorophyll synthesis. *Archs. Biochem. Biophys.,* **160,** 430–39.

[112] NOMURA M. (1970) Bacterial ribosome. *Bacterial Rev.,* **34,** 228–77.

[113] OHAD I. (1972) Biogenesis and modulation in chloroplast membranes. In *Role of membranes in secretory processes* (Eds L. Bolis, R.D. Keynes & W. Wilbrandt), pp. 24–51. North Holland, Amsterdam, London.

[114] PAIGEN K. (1971) In *Enzyme synthesis and degradation in mammalian systems,* (Ed. C. Rechcigl), Karger, Basel, pp. 1–46.

[115] PAOLILLO D.J. (1969) The plastids of *Polytrichum.* II. The sporogenous cells. *Cytologia,* **34,** 133–44.

[116] PAOLILLO D.J. (1970) The three-dimensional arrangement of intergranal lamellae in chloroplasts. *J. Cell. Sci.,* **6,** 243–55.

[117] PAOLILLO D.J. (1974) Motile male gametes of plants. In *Dynamic aspects of plant ultrastructure* (Ed. A.W. Robards), pp. 504–31. McGraw-Hill Book Company (U.K.) Limited.

[118] PARK R.B. & SANE P.V. (1971) Distribution of function and structure in chloroplast lamellae. *A. Rev. Pl. Physiol.,* **22,** 395–430.

[119] PETERMANN M.L. (1964) *The physical and chemical properties of ribosomes.* Elsevier Publishing Co., Amsterdam.

[120] DE PETROCELLIS B., SIEKEVITZ P. & PALADE G.E. (1970) Changes in chemical composition of thylakoid membranes during greening of the y-1 mutant of *Chlamydomonas reinhardtii. J. Cell Biol.,* **44,** 618–34.

[121] PHILIPPOVICH I.I., BEZSMERTNAYA I.N. & OPARIN A.I. (1973) On the localization of polyribosomes in the system of chloroplast lamellae. *Expl. Cell Res.,* **79,** 159–68.

[122] PIKO L. (1970) Synthesis of macromolecules in early mouse embryos cultured *in vitrot1: RNA, DNA and a polysaccharide component. Dev. Biol.,* **21,** 257–79.

[123] PIKO L. & CHASE D.G. (1973) Role of the mitochondrial genome during early development in mice. *J. Cell Biol.,* **58,** 357–78.

[124] PLATTNER H., SALPETER M., SALTZGABER J., ROUSLIN W. & SCHATZ G. (1971) Promitochondria of anaerobically grown yeast. In *Autonomy and biogenesis of mitochondria and chloroplasts* (Eds N.K. Boardman, A.W. Linnane & R.M. Smillie), pp. 175–84. North Holland.

[125] POSSINGHAM J.V. & SAURER W. (1969) Changes in chloroplast number per cell during leaf development in spinach. *Planta,* **86,** 186–94.

[126] POSSINGHAM J.V. & SMITH J.W. (1972) Factors affecting chloroplast replication in spinach. *J. Exp. Bot.,* **23,** 1050–59.

[127] PRIOUL J.-L. & BOURDU R. (1968) Utilisation de l'analyse biometrique a l'etude de la dynamique des infrastructures chloroplastiques lors de la levee d'une carence enazote. *J. Microscopie,* **7,** 419–39.

[128] RAFF R.A. & MAHLER H.R. (1972) The non-symbiotic origin of mitochondria. *Science,* **177,** 575–82.

[129] RAWSON J.R.Y. & HASELKORN R. (1973) Chloroplast ribosomal RNA genes in the chloroplast DNA of *Euglena gracilis. J. Molec. Biol.,* **77,** 125–32.

[130] ROBBERSON D.L., KASAMATSU H. & VINOGRAD J. (1972) Replication of mitochondrial DNA. Circular replicative intermediates in mouse L cells. *Proc. Natl. Acad. Sci., U.S.A.,* **69,** 737–41.

[131] ROSINSKI J. & ROSEN W.G. (1972) Chloroplast development: fine structure and chlorophyll synthesis. *Q. Rev. Biol.,* **47,** 160–91.

[132] ROUSLIN W. & SCHATZ G. (1969) Interdependence between promitochondrial and cytoplasmic protein synthesis during respiratory adaption in bakers' yeast. *Biochem. Biophys. Res. Commun.,* **37,** 1002–1007.

[133] SAGAN L. (1967) On the origin of mitasing cells. *J. Theoret. Biol.,* **14,** 225–74.

[134] SAGER R. (1972) *Cytoplasmic genes and organelles.* Academic Press, New York & London.

[135] SAGER R. & LANE D. (1972) Molecular basis for maternal inheritance. *Proc. Natl. Acad. Sci., U.S.A.,* **69,** 2410–13.

[136] SAKANO K., KUNG S.D. & WILDMAN S.G. (1974) Identification of several chloroplast DNA genes which code for the large subunit of *Nicotiana* fraction I proteins. *Molec. Gen. Genet.,* **130,** 91–7.

[137] SCHATZ G. (1965) Subcellular particles carrying mitochondrial enzymes in anaerobically-grown cells of *Saccharomyces cerevisiae. Biochem. Biophys. Acta,* **96,** 342–45.

[138] SCHMITT H. (1969) Characterization of mitochondrial ribosomes from *Saccharomyces cerevisiae*. *FEBS Letters*, **4**, 234–38.

[139] SCHOTZ F. (1970) Effects of the disharmony between genome and plastome on the differentiation of the thylakoid system in *Oenothera*. *Symp. Soc. Exp. Biol.*, **24**, 39–54.

[140] SHANNON J.C. & CREECH R.G. (1973) Genetics of storage polyglucosides in *Zea mays* L. *Ann. N.Y. Acad. Sci.*, **210**, 279–89.

[141] SIEKEVITZ P. (1972) Biological membranes: the dynamics of their organization. *A. Rev. Physiol.*, **34**, 117–39.

[142] SINGER S.J. & NICOLSON G.L. (1972) The fluid mosaic model of the structure of cell membranes. *Science, N.Y.*, **175**, 720–32.

[143] SINGH S. & WILDMAN S.G. (1973) Chloroplast DNA codes for the ribulose diphosphate carboxylase catalytic site on fraction I proteins of *Nicotiana* species. *Molec. Gen. Genet.*, **124**, 187–96.

[144] See 'Biochemical aspects of the biogenesis of mitochondria' (1968) (Eds E.C. Slater, J.M. Tager, S. Papa & E. Quagliariello). Adriatica Editrice, Bari.

[145] STAEHELIN L.A. & MILLER K.R. (1974) Particle movements associated with unstacking and restacking of chloroplast membranes *in vitro*. A freeze-cleave and deep-etch study. 8th International Congress Electron Microscopy, Canberra, II pp. 202–3.

[146] STEARNS M.E. & WAGENAAR E.B. (1971) Ultrastructural changes in chloroplasts of autumn leaves. *Can. J. Genet. Cytol.*, **13**, 550–60.

[147] STUTZ E. (1970) The kinetic complexity of *Euglena gracilis* chloroplast DNA. *FEBS Letters*, **8**, 25–28.

[148] SUYAMA Y. & MIURA K. (1968) Size and structural variations of mitDNA. *Proc. Natl. Acad. Sci., U.S.A.*, **60**, 235–242.

[149] SZOLLOSI D. (1965) The fate of middle-piece mitochondria in the rat egg. *Journal of Experimental Zoology*, **159**, 367–78.

[150] TAYLOR M.M., GLASGOW J.E. & STORCK R. (1967) Sedimentation coefficients of RNA from 70S and 80S ribosomes. *Proc. Natl. Acad. Sci., U.S.A.*, **57**, 164–169.

[151] TAYLOR M.M. & STORCK R. (1964) Uniqueness of bacterial ribosomes. *Proc. Natl. Acad. Sci., U.S.A.*, **52**, 958–65.

[152] TEWARI K.K. & WILDMAN S.G. (1970) Information content in the chloroplast DNA. *Symp. Soc. Exp. Biol.*, **24**, 147–79.

[153] TRIBE M.A. & ASHHURST D.E. (1972) Biochemical and structural variations in the flight muscle mitochondria of ageing blowflies, *Calliphora erythrocephalia*. *J. Cell Sci.*, **10**, 443–69.

[154] TZAGOLOFF A., RUBIN M.S. & SIERRA M.F. (1973) Biosynthesis of mitochondrial enzymes. *Biochem. et Biophys. Acta.*, **301**, 71–104.

[155] WATSON K., HASLAM J.M., VEITCH B. & LINNANE A.W. (1971) Mitochondrial precursors in anaerobically grown yeast. In *Autonomy and biogenesis of mitochondria and chloroplasts* (Eds N.K. Boardman, A.W. Linnane & R.M. Smillie), pp. 162–74. North Holland.

[156] WEHRMEYER W. (1964) Zur Klärung der strukturellen Variabilität der Chloroplastengrana des Spinats in Profil und Aufsicht. *Planta*, **62**, 272–93.

[157] WEHRMEYER W. (1964) Uber membranbildungprozesse im chloroplasten. *Planta*, **63**, 13–30.

[158] WEIER T.E., SJOLAND R.D. & BROWN D.L. (1970) Changes induced by low light intensities on the prolamellar body of eight-day, dark-grown seedlings. *Am. J. Bot.*, **57**, 276–84.

[159] WEIER T.E. & STOCKING C.R. (1952) The chloroplast: structure, inheritance and enzymology. II. *Bot. Rev.*, **18**, 14–75.

[160] WEIER T.E., STOCKING C.R., THOMSON W.W. & DREVER J. (1963) The grana as structural units in chloroplasts of mesophyll of *Nicotiana rustica* and *Phaseolus vulgaris*. *J. Ultrastruct. Res.*, **8**, 122–43.

[161] WELLS R. & BIRNSTIEL M. (1969) Kinetic complexity of chloroplastal deoxyribonucleic acid and mitochondrial deoxyribonucleic acid from higher plants. *Biochem. J.*, **112**, 777–86.

[162] VON WETTSTEIN D., KAHN A., NIELSEN O.F. & GOUGH S. (1974) Genetic regulation of chlorophyll synthesis analysed with mutants in barley. *Science*, **184**, 800–802.

[163] WHATLEY J.M. (1971) Ultrastructural changes in chloroplasts of *Phaseolus vulgaris* during development under conditions of nutrient deficiency. *New Phytol.*, **70**, 725–42.

[164] WHITTON B.A., CARR N.G. & CRAIG I.W. (1971) A comparison of the fine structure and nucleic acid biochemistry of chloroplasts and blue-green algae. *Protoplasma*, **72**, 325–57.

[165] WILLIAMS K.L. (1972) Ph.D. thesis. Australia National University.

[166] WILLIAMS K.L., SMITH E., SHAW D.C. & BIRT L.M. (1972) Studies of the levels and synthesis of cytochrome *c* during adult development of the blowfly *Lucilia cuprina*. *J. Biol. Chem.*, **247**, 6024–30.

[167] WOSTENHOLME D.R. & DAVID I.B. (1968) A size difference between mitDNA molecules of Urodele and Anuran Amphibia. *J. Cell Biol.*, **39**, 222–28.

[168] WU M., DAVIDSON N., ATTARDI G. & ALONI Y. (1972) The relative positions of the 4S RNA genes and of the ribosomal RNA genes in mitochondrial DNA. *J. Mol. Biol.*, **71**, 81–93.

[169] VAN HEYNINGEN V., CRAIG I.W. & BODMER W.F. (1973) Genetic control of mitochondrial enzymes in human-mouse somatic cell hybrids. *Nature*, **242**, 509–12.

Chapter 5.3
Control of Gene Expression during Differentiation and Development

5.3.1 INTRODUCTION

During the last few years, techniques have been developed which allow the identification of RNA molecules soon after transcription. This is an important advance which permits the direct study of the control of gene expression during development and in the future such studies may allow us to understand development in molecular detail.

In Part 1 there is a simple account of the interactions between the nucleus and the cytoplasm during the development of plants and animals. Here we are concerned with the molecules which may be involved in these interactions and it is necessary to discuss the most recent information which is available about the organization of the eukaryotic genome before considering how its expression may be controlled. Ribosomal and transfer RNA molecules are well characterized and so we know more about the cistrons which code for these RNAs than we know about the cistrons which code for messenger RNA. It is also the case that the comparatively easy identification of ribosomal and transfer RNA has permitted more accurate studies of the factors which regulate their synthesis than is possible for messenger RNA. However it seems likely that a study of the control of ribosomal and transfer RNA synthesis may provide a useful model of the control of gene expression in general in eukaryotes, and it is therefore worthwhile describing what is known about them. In addition, the recent characterization of particular messenger RNA molecules makes it possible to decide if other RNA species have their synthesis controlled in a similar manner.

Experiments, which are concerned with the ways in which cells become different from each other, are described in a later section. At present these mechanisms are not understood and it must be remembered that the mechanisms which are eventually discovered must obey the rules of classical embryology (Part 1) and fit with recent studies on pattern formation in the embryo (Chapter 3.6).

ABBREVIATIONS

rRNA	ribosomal ribonucleic acid
mRNA	messenger ribonucleic acid
dRNA	ribonucleic acid with a base composition similar to DNA
SC DNA	single copy deoxyribonucleic acid
IR DNA	intermediate repetitive deoxyribonucleic acid
HR DNA	highly repetitive deoxyribonucleic acid
tRNA	transfer ribonucleic acid.
HnRNA	heterogeneous nuclear ribonucleic acid

5.3.2 THEORIES OF DIFFERENTIATION

Molecular mechanisms governing differentiation, the process whereby cells become recognizably distinct from one another, are of central importance to developmental biology. Three general theories exist which seek to explain differentiation in molecular terms; directed somatic mutation, specific gene amplification and epigenetic theories. Of these only the epigenetic theory survives as a general explanation of differentiation, although both directed somatic mutation and specific gene amplification do occur in special cases.

Directed somatic mutation theory

As its name suggests the directed somatic mutation theory supposes that differentiation is the consequence of irreversible and heritable changes in the DNA of particular cell lines and these changes are responsible for the characteristics by which one cell line is distinct from any other. The phenomena of chromosome elimination and diminution [260] are often cited as examples of such directed changes in the genome. Nuclear transplantation techniques (see Chapter 2.2) provide the means for a direct test of this theory and have failed to produce any evidence in its favour. On the contrary sexually mature frogs have been produced from intestinal cell nuclei [114] of swimming tadpoles and from nuclei of epidermis differentiating in tissue culture [112].

Specific gene amplification theory

Specific gene amplification theory states that differentiation is a consequence of the synthesis of extra copies of genes coding for the proteins necessary for the production of a particular cell type, for instance by this hypothesis differentiating muscle cells would amplify the genes coding for myoglobin, actin, myosin and other muscle specific proteins, while other cells such as brain or kidney cells would not. Support for this theory came from the observation that during oogenesis in many invertebrates and vertebrates the genes for ribosomal RNA (rRNA) are specifically amplified. However recent advances in techniques of nucleic acid hybridization and messenger RNA (mRNA) purification have allowed the theory to be tested directly. It has been shown that the genes for globin in ducks are no more frequent in blood DNA than in liver DNA [14a] and that the genes for silk fibroin in the silk moth are no more frequent in DNA isolated from the silk gland than in DNA isolated from the rest of the animal [239]. Both these observations argue strongly against specific gene amplification as a general mechanism for differentiation.

Epigenetic theory

The epigenetic theory of differentiation states that interaction between cytoplasmic factors and an initially constant genome results in potentially reversible changes of gene activity in particular cell lines such that one cell line becomes recognizably different from another. The main evidence in favour of this hypothesis comes from studies on 'mosaic' egg development, where direct cell lineages are found associated with specific regions of the egg, though there is also considerable evidence from 'regulation' egg development for the presence of cytoplasmic factors which influence specific gene expression. A good example of this type of interaction is seen in germ cell development both in insects and vertebrates where a specific part of the egg cytoplasm, called the germinal plasm, determines that the nuclei entering it become potential germ cells (Chapter 1.3).

Epigenetic control of differentiation may operate at three levels. There may be control of gene expression during transcription, during processing and transport of mRNA to the cytoplasm, and during translation. Elucidation of the relative importance of these three levels of control in determining the differentiated characteristics of particular cell lines is one of the major goals of modern developmental biology.

5.3.3 GENETIC ORGANIZATION IN EUKARYOTES

Before discussing control of genetic activity, it is worth considering what is known of the organization and structure of genes in the eukaryote genome [for recent review see 67]. Genetic fine structure and function analysis in bacteria and bacteriophages provided evidence for the existence of groups of structural genes under common regulatory control and led to the development of the operon model of genetic regulation. To what extent the operon model applies to higher organisms is unknown but several recent theories use it as a basis for models of gene regulation in eukaryotes [67, 96].

Prokaryote chromosomes contain only unique DNA (with

Table 5.3.1   The analytical complexity of the genomes of various prokaryotes and eukaryotes.

Prokaryotes	Genome size (base pairs × 10^{-6})
Phage MS2	0·0005
Phage T4	0·15
Escherichia coli	6·0
Eukaryotes	
Yeast	22·0
Dictyostelium	30·0
Drosophila	120·0
Sea urchin	800·0
Bird	1200·0
Xenopus	2800·0
Triturus	20000·0
Mammal	2800·0

the exception that genes for ribosomal RNAs are several times repeated) which means that no part of the prokaryote genome is represented more than once. By contrast eukaryotes may contain a hundred or a thousand times more DNA per haploid chromosome set than a bacterium such as *E. coli* (Table 5.3.1). Much of this DNA is represented by many hundreds or thousands of copies of one particular family of DNA sequences (Table 5.3.2). This extra DNA may have properties which might

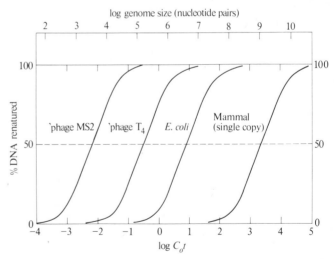

Fig. 5.3.1 Renaturation of DNA from three prokaryotic organisms (bacteriophages MS2 and T4, and bacterium *Escherichia coli*). The DNA has been denatured to form single strands and then under appropriate conditions allowed to renature to form double strands again. The amount of renatured DNA is plotted against the logarithm of Cot (Co is the concentration of single stranded DNA at zero time measured in moles nucleotide per litre; t is the time in seconds from the beginning of the reaction). All three reactions are simple S shaped curves as predicted for ideal second order reactions. The logarithm of the genome size (measured in nucleotide pairs) is plotted along the top of the figure. (Redrawn from data in Britten & Kohne, 1968; ref. [24]).

Table 5.3.2a Examples of structural genes which belong to the intermediate repetitive fraction of DNA

Organism	% of genome (number of copies/haploid genome)				
	rRNA	5S RNA	tRNA	histone	total
E. coli	0·8 (10)	0·002 (10)	0·007 (50)	—	0·81
Yeast	3·1 (200)	0·11 (200)	0·13 (360)	—	3·34
Drosophila	0·38 (200)	0·013 (200)	0·038 (860)	—	0·43
Xenopus	0·12 (450)	0·1 (24 × 10³)	0·025 (9000)	0·02 (500)	0·265
Triturus	0·1 (1·3 × 10⁴)	0·08 (14 × 10⁴)	—	—	0·18
Man	0·05 (280)	0·009 (2000)	0·004 (1300)	0·001 (20)	0·019

Table 5.3.2b Composition of the genome of three eukaryotes in terms of the easily separable kinetic fractions of DNA

Organism	Kinetic fraction				
	Single copy	Intermediate	Highly repetitive	Zero time	Ref.
Drosophila (Insect)					
% DNA	78	15	7	–	265
Reiteration	1	35	2600		
Sequence complexity (base pairs)	9·2 × 10⁷	5·1 × 10⁵	3·1 × 10³		
Strongylocentrotus purpuratus (Sea urchin)					
% DNA	63	27	7	3	102
Reiteration	1	250	6000	–	
Sequence complexity (base pairs)	5·0 × 10⁸	1· × 10⁶	1·3 × 10⁴		
Xenopus (Amphibia)					
% DNA	60	31	6	3	69
Reiteration	1	1600	32000	–	
Sequence complexity (base pairs)	1·7 × 10⁹	6 × 10⁵	6 × 10³		

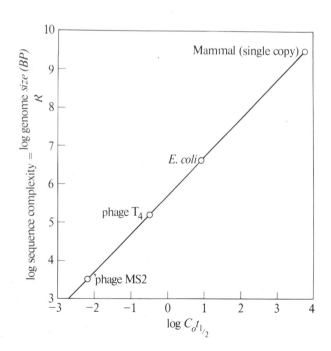

Fig. 5.3.2 Relationship between $Cot_{\frac{1}{2}}$ and the size of the genome for prokaryotic organisms. $Cot_{\frac{1}{2}}$ is the point during a renaturation reaction at which 50% of DNA has become double stranded again (see Fig. 1) and is directly proportional to the size of the genome for prokaryotes which have very little or no sequence repetition in their genomes. When sequence repetition occurs in a kinetically pure fraction of DNA (renaturing with 2nd order kinetics) we can define a parameter, sequence complexity, which is equal to the length of the repeated sequence and may be calculated from the total amount of repeated sequence (analytical complexity measured in nucleotide pairs) divided by the repetition frequency. Thus for any kinetically pure DNA of known analytical complexity it is possible to calculate its reiteration frequency using one of the prokaryote DNAs (say *E. coli*) as a standard:

$$R^u = \frac{AC^u \times Cot_{\frac{1}{2}}^{E.\,coli} \times R^{E.\,coli}}{AC^{E.\,coli} \times Cot_{\frac{1}{2}}^{u}}$$

where R_u is the reiteration frequency of the unknown DNA.

AC_u its analytical complexity and $Cot_{\frac{1}{2}}u$ its $Cot_{\frac{1}{2}}$ for renaturation.

$AC_{E.\,coli}$ is 6×10^6 nucleotide pairs.

$Cot_{\frac{1}{2}E.\,coli}$ is experimentally determined under the same conditions of renaturation as the unknown (usually about 9 mole sec/litre).

$R_{E.\,coli}$ is taken to be one since the *E. coli* genome is not repetitive.

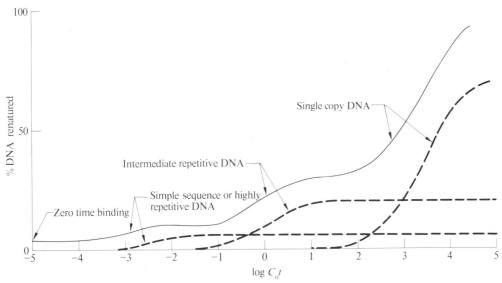

Fig. 5.3.3 Idealized DNA renaturation curve for eukaryote DNA (analytical complexity $2 \cdot 0 \times 10^9$ base pairs) containing four kinetically separable components (4% zero time binding DNA, 6% simple sequence or highly-repetitive DNA, 20% intermediate or middle-repetitive DNA and 70% non-repetitive or unique DNA). The curve is not a second order reaction since it has several components, but it may be analysed in terms of three second order reactions ($---$) and one zero order reaction. Using the relationship derived in Fig. 5.3.2 and using the values given for the *E. coli* genome it is possible to calculate the reiteration frequency and sequence complexity for each of the three kinetically separable components in this figure:

	Observed Cot$\frac{1}{2}$	Fraction of Genome	Corrected Cot$\frac{1}{2}$*	Analytical complexity†	Reiteration frequency‡	Sequence complexity§
1.	3000	0·7	2100	$1 \cdot 4 \times 10^9$	1	$1 \cdot 4 \times 10^9$
2.	1	0·2	0·2	$0 \cdot 4 \times 10^9$	3000	$1 \cdot 3 \times 10^5$
3.	0·003	0·06	0·00018	$0 \cdot 12 \times 10^9$	10^6	120

* Corrected Cot$\frac{1}{2}$ = observed Cot$\frac{1}{2}$ × fraction of the genome, this allows for the dilution of any fraction by the other fractions in the genome.
† Analytical complexity = fraction of the genome × the genome size in base pairs.
‡ Reiteration frequency is calculated according to the formula given in Fig. 5.3.2.
§ Sequence frequency is the analytical complexity divided by the reiteration frequency.

suggest a function for it related to the major increases in organizational complexity (such as chromosome structure, more elaborate processes of development, differentiation, cellular interaction, neurone and brain differentiation) which have occurred during the evolution of eukaryotes.

Britten and Kohne [24] have shown that DNA renaturation should proceed with second order reaction kinetics and provided evidence that this is so or nearly so for DNA of prokaryotes (Fig. 5.3.1). They defined a parameter $C_0 t\frac{1}{2}$ (C_0 is the initial DNA concentration expressed as moles of nucleotide per litre; $t\frac{1}{2}$ is the time in seconds for 50% of the DNA to renature) which is linearly related to genome size (analytical complexity) for prokaryotic organisms (Fig. 5.3.2). However renaturation of DNA from some eukaryotes was more complex but could be explained on the basis of three types of DNA

differing in the degree to which particular DNA sequences are repeated within the genome (Fig. 5.3.3). Slowest renaturing components have the highest $C_0 t\frac{1}{2}$ values consistent with the interpretation that they are mainly sequences present only once per haploid genome (that is so called single copy sequences). The fast and intermediate renaturing components have Cot$\frac{1}{2}$ values consistent with their being composed of highly repeated (greater than 100,000 copies per haploid genome) and moderately repeated (100–100,000 copies per haploid genome) sequences respectively. Representative renaturation curves for various types of eukaryote show that the relative contributions of these three DNA frequency classes to the total genome may vary considerably (Fig. 5.3.4, Table 5.3.2). There is also a fourth component identifiable in renaturation reactions as DNA, which renatures spontaneously at zero time.

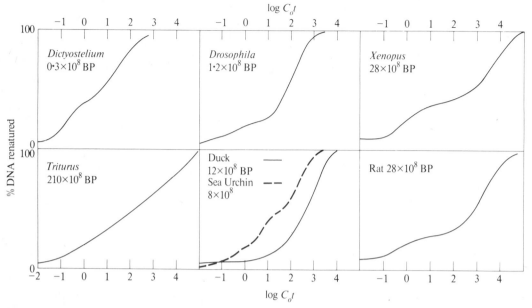

Fig. 5.3.4 Representative renaturation profiles of total DNA from various organsims. Data redrawn from [14a, 68, 84a, 102, 125, 208, 265]. BP = base pairs.

Table 5.3.3a Proteins whose coding sequences occur only once per haploid genome

Protein	Organism	Average reiteration	Reference
Globin	Duck	1–3	14a
	Mouse	1	121
Silk fibroin	Bombyx	1	239
Ovalbumin	Chicken	1	238

Table 5.3.3b Populations of mRNA for many different proteins whose coding sequences occur once per haploid genome

Source of mRNA	Average reiteration (percent RNA)	Reference
Hela cell polysomal poly(A) + RNA	1 (95%)	14
Xenopus and *Triturus* oocyte total poly(A) + RNA	1 (75%)	208
Mouse L cell total polysomal mRNA	1 (80%)	103
Sea urchin gastrula total polysomal mRNA	1 (95%)	88
Dictyostelium total polysomal poly(A) + RNA	1 (95%)	86

Spontaneously renaturing DNA sequences

It has been known for some time that some eukaryotic DNA will renature spontaneously at zero time but the nature of this DNA has received little attention. Explanation of zero time renaturation suggests the existence in eukaryotic DNA of inverted repeated sequences, called palindromes, which would be expected to renature by an intramolecular reaction to form hairpin loops. The rate of hairpin formation would be extremely rapid and proceed at very low DNA concentrations since complementary sequences are not diluted (Fig. 5.3.5). Evidence has

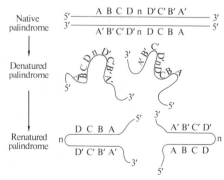

Fig. 5.3.5 An illustration of the way in which an inverted repeated DNA sequence (palindrome) will behave during denaturation and renaturation.

been given which suggests that palindromes are from 300 to 1200 nucleotides pairs in length and that they are located in clusters of two to four, members of a cluster are separated by about 2000 nucleotides though a few may be 25,000 nucleotides apart, and clusters are separated by about 120,000 nucleotides [261]. There is little idea at present about the function of palindromes in eukaryote DNA, but it may be that they provide recognition sites in some nucleic acid-protein interactions since restriction enzymes and the lac repressor, for instance, appear to recognize relatively short palindromes. A further recent suggestion is that palindromes may be important in the replication of the ends of eukaryote chromosomes [41].

Highly repetitive or simple sequence DNAs

Highly repetitive DNA (HR DNA) is composed of tandemly repeated copies of simple short sequences (for reviews see [232, 246, 247]), which do not code for protein and which evolve very rapidly since closely related organisms have quite distinct though related highly repeated DNA components [90, 233]. It has been shown by *in situ* hybridization that highly repeated DNA components, which can be isolated in pure form as satellites on buoyant density gradients, are localized predominantly in centromeric and other regions of constitutive heterochromatin [135, 180]. In organisms where distinct satellites of highly repetitious DNA exist there may be chromosome specific localization [136, 91a]. It may be of course that amounts of satellite DNA sequence below the level of detection by *in situ* techniques exist in other parts of the genome as well.

The function of HR DNA is not well understood but it seems clear that it is not transcribed and that its function is not directly sequence related since the sequence is not rigorously maintained in evolution [90, 232]. Recent theories [60, 182] propose that it may play a structural role in chromosome organization or in the pairing of chromosomes during meiosis [246].

Intermediate or middle repetitive DNA sequences

Some well defined multiple structural genes belong to this kinetic fraction of DNA (Table 5.3.2). They share a basic tandemly repeated structure (Fig. 5.3.6) of transcribed coding and non-coding sequences, and non-transcribed spacer sequences. This may well prove to be a universal model for eukaryote structural genes. Structural studies of ribosomal DNA (rDNA) [34, 70, 255, 256] and 5S rDNA [33] in two closely related species of *Xenopus* (*X. laevis* and *X. mulleri*), have shown that the transcribed sequences are more highly conserved in evolution than the non-transcribed spacer sequences. In the case of rDNA the sequences coding for rRNA are indistinguishable by hybridization or by size, but there is almost no cross-hybridization between their non-transcribed

Structures of repeated genes with known functions

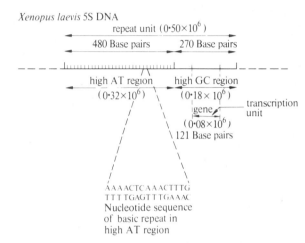

Fig. 5.3.6 An illustration of the structure of two repeated genes, 18 and 28S rRNA and 5S rRNA, in *Xenopus laevis*. Redrawn from data in [36, 255, 256].

spacer sequences which also happen to be approximately the same size. The transcribed spacer sequences are present in the rRNA precursor but absent in the mature 18 and 28S rRNA, and they are also distinguishable by hybridization. 5S rDNA from the two species may be distinguished not only by the absence of cross-hybridization between spacers, but also by spacer size which in *X. mulleri* is twice that of *laevis* [33]. These results suggest that whatever is the function of spacer DNA it is not rigorously related to sequence or size since these two parameters may change quite rapidly in evolution [254]. Recent sequencing studies on the spacer of *X. laevis* 5S rDNA show that it is composed of a basic repeating unit of 15 nucleotides (AAAACTCAAACTTTG) into which mutations have been introduced, a type of sequence organization highly reminiscent of that described for several highly repetitious satellite DNAs [36].

Genes coding for transfer RNA (tRNA) in *Xenopus laevis* are also found in this kinetic fraction of DNA. The average reiteration frequency of each of the 43 basic tRNA sequences distinguishable by nucleic acid hybridization is 200, making a total of about 9000 genes altogether. The numbers of genes for specific aminoacyl-tRNAs varies, for instance four different leucyl-tRNAs distinguishable by hybridization have an average reiteration frequency of 90, while valyl-tRNA with only one species is coded by 240 genes and $tRNA_1^{met}$ (the initiator tRNA) and $tRNA_2^{met}$ are coded by 310 and 170 genes respectively [49]. Relevant to the control of tRNA synthesis is the detailed arrangement of tRNA genes in the genome. Are the genes for different tRNAs intermingled or are they in recognisably separate clusters? Measurement of the buoyant density in neutral CsCl of the sequences homologous to unfractionated tRNA, $tRNA^{val}$, $tRNA_1^{met}$, and $tRNA_2^{met}$ in high molecular weight *X. laevis* DNA, indicate that tRNA genes of the same type are linked to spacer DNA in separate and extensive gene clusters, and that the different clusters contain different spacer DNA sequences [48].

Histone gene coding sequences in Drosophila, echinoderms and amphibians are repeated several hundred-fold per haploid genome [12, 142]. There is evidence that they too are clustered together with spacer DNA, but the detailed arrangement of the genes coding for different histone fractions is not clear at present. Studies of the evolution of histone coding sequences in echinoderms [251] suggest that although the actual amino acid sequences of histones are highly conserved the coding sequence may show considerable non-homology in heterologous hybridization reactions. This implies that use is made of the degeneracy of the genetic code during evolution, perhaps because of the relatively neutral selective forces operating on mutations in the third base (wobble) position of many codons.

Although many well defined types of structural gene belong to intermediate repetitive DNA (IR DNA) the ones so far identified in *Xenopus* account for less than 10% of the total DNA in this fraction (Table 5.3.2). What is the nature and function of the rest of the intermediate repetitive DNA? A suggestion has been made that some intermediate repetitive DNA sequences are derived from highly repetitive sequences by a process of divergence and limited saltatory amplification [247]. Some IR DNA may represent other groups of structural genes, such as those described above, with at present unidentified functions. They may be coding for other proteins which, like histones, might require multiple genes to support their synthesis, for instance ribosomal proteins and immunoglobulins. Certainly a greater proportion of IR DNA is transcribed than can be accounted for by the IR DNAs with known functions and this proportion varies both quantitatively and qualitatively from one cell type to another. However, it is unknown at present whether all IR DNA is transcribed at one time or another during development and differentiation. Except in the well defined cases described above, the function of intermediate repetitive gene transcripts is not known, but there are three possibilities. They may have no function, they may be

messenger RNAs (mRNAs) or they may have a regulatory role in the control of gene expression [23]. Evidence that mRNA other than that for histone is transcribed from intermediate repetitive DNA comes from experiments with rat myoblasts [40], mouse L cells and Hela cells [14, 143a] grown in tissue culture [103], and sea urchin embryos [88], but the nature of the proteins coded for is unknown. The amount of IR DNA involved is quite small and cannot account for more than a few percent of the total. Evidence in favour of a regulatory role of some IR DNA sequences is more circumstantial at present and is derived from experiments on the structure and processing of heterogeneous nuclear RNA (HnRNA), the putative mRNA precursor, and on the intermingling of IR DNA sequences with single copy DNA sequences (see below). The experiments on Hn RNA, to be described in more detail below, show that these large nuclear RNA molecules contain IR DNA transcript covalently linked to single copy DNA transcript. Most HnRNA sequence, perhaps as much as 90%, is degraded within the nucleus but some appears in the cytoplasm as potential mRNA which lacks essentially all IR DNA transcript. It has been suggested that the IR DNA transcript in HnRNA functions during the processing to message by providing specific recognition sites for the processing enzymes. Experiments on intermingling of IR DNA with single copy DNA have been done on a wide variety of organisms and there seems to be general agreement that the majority of single copy DNA sequences are flanked on each side by sequences belonging to the IR DNA kinetic fraction (see next paragraph). Because of this close proximity of IR DNA sequences with single copy DNA sequences, at least some of which have been shown to code for mRNA, it has been suggested that they may play a role in the regulation of protein coding gene activity [23, 67].

Single copy DNA sequences

A fraction of DNA reannealing, with kinetics expected for single copy (SC) sequences, that is sequences which occur only once per haploid genome, has been shown to contain the genes for several different proteins (Table 5.3.3). The percent of total DNA present in this faction is extremely variable (Fig. 5.3.4), indeed in some high *c* value organisms such as the newt *Triturus cristatus carnifex* it is not easily identifiable by optical renaturation methods, but only by the use of radioactive cDNA made by reverse transcriptase on purified mRNA [139]. In other organisms SC DNA may represent as much as 80–90% of the total DNA. It is certain that the sequences present in this fraction are all represented once or a very few times per genome but it is not clear at present what proportion of them code for protein. This is related to the question (see Gene numbers, complementation groups and chromomeres, p. 313) of how many genes are present in eukaryotes. It is clear, however, that if all SC DNA sequences are coding for protein the number of

Table 5.3.4 Extent to which single copy DNA is transcribed in different tissues and at different developmental stages

Organism	Cell or tissue total RNA	% single copy transcribed	Analytical complexity (base pairs)	Ref.
Mouse	Newborn Brain	5·2%	$7·3 \times 10^7$	[34a]
	Adult Brain	8·5%	$12·0 \times 10^7$	
	Newborn Liver	2·5%	$3·5 \times 10^7$	
	2 week old Liver	3·8%	$5·3 \times 10^7$	
	Adult Liver	2·0%	$2·8 \times 10^7$	
Xenopus	Oocytes	0·75%	$1·5 \times 10^7$	[68]
Dictyostelium	Log phase cells	15·1%	$0·9 \times 10^7$	[84a]
	Preaggregation	13·7%	$0·8 \times 10^7$	
	Early slug	17·1%	$1·0 \times 10^7$	
	Culmination	19·3%	$1·1 \times 10^7$	
	All stages combined	28·0%	$1·6 \times 10^7$	

proteins potentially available to an organism is very large indeed (Table 5.3.4).

Estimates of the extent to which SC DNA is transcribed in different tissues are given in Table 5.3.4 together with some data for different developmental stages. It is clear that in no tissue nor at any developmental stage is the whole of the SC DNA transcribed. However it is not possible at present to say what proportion is transcribed during the whole life cycle of an organism. In *Dictyostelium discoideum*, a cellular slime mould the expression of SC DNA has been summed over several developmental stages and gives a value of 56% of which 20% are sequences present at all stages and 36% are stage specific. The sequence complexity of 56% of *Dictyostelium* SC DNA is $16–17 \times 10^6$ nucleotide pairs which is sufficient to code for about 16×10^3 average-sized proteins.

Care must be taken over the interpretation of these results since they define the complexity of SC DNA transcription and not the complexity of SC DNA sequences in actual or potential messenger RNA. Recently it has also been possible to obtain estimates of this parameter (Table 5.3.5) and it is becoming clear that there is considerable discrepancy between the extent of SC DNA sequence transcribed and the amount found in mRNA within one cell type. For instance the sea urchin gastrula appears to transcribe 10–30-fold more SC DNA sequence into nuclear RNA than is present in polysomal RNA [88]. Such an observation clearly implies some qualitative as well as quantitative processing of nuclear RNA.

Intermingling of intermediate repetitive and single copy DNA sequences

Recent experiments from a wide variety of organisms suggests that some fraction of IR DNA is interspersed with a part of the SC DNA [67, 69]. The experiments use short unlabelled DNA at high concentration to 'drive' renaturation reactions with labelled tracer DNA of various defined average lengths from 300 to 10,000 nucleotides. After renaturation to low cot values, such that only repetitive sequences will have renatured, double stranded DNA plus single strand tails is separated from single stranded DNA by binding and elution from hydroxyapatite (HAP) columns (Fig. 5.3.7). Consider a SC DNA sequence covalently linked to an IR DNA sequence. At the end of the low cot renaturation reaction the IR DNA sequence will be double stranded but the SC DNA will remain single stranded. Nevertheless it will bind to HAP because of covalent linkage to a double stranded IR DNA sequence. Such an experiment enables isolation of DNA sequences adjacent to repetitive sequences in quantities sufficient for further analysis. It was found that as the average length of labelled tracer DNA increases so does the amount that will bind to HAP. For instance after renaturation to cot 50 up to 80% labelled *Xenopus* DNA with an average size of 3700 nucleotides will bind to HAP (Fig. 5.3.8).

The curve (Fig. 5.3.8) relating the fraction of tracer DNA bound to HAP after renaturation with excess short (450 nucleotides) DNA to the average size of the tracer DNA shows three things (i) that there is significant binding at very short tracer lengths, (ii) that the binding increases rapidly with length up to a size of 700–1000 nucleotides and (iii) then there is a slower increase up to average tracer lengths of 4000 or greater. Detailed interpretation of this curve requires knowledge of the average size and numbers of repetitive sequences which cause the binding to HAP, proof that they are IR DNA and not HR DNA sequences, and proof that the increasing amounts of

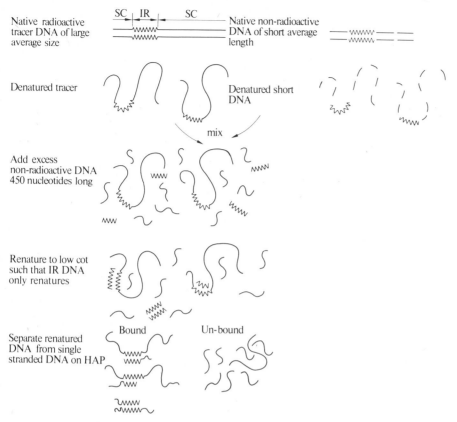

Fig. 5.3.7 An illustration of the way in which excess short (450 nucleotide) non-radioactive DNA will behave when renatured with long radioactive tracer DNA in such a way that intermediate repetitive sequences only become double stranded thus allowing separation on hydroxylapatite columns of any radioactive single copy DNA adjacent to them.

DNA bound is due to longer SC DNA tails. The average size of the IR DNA has been estimated by several techniques (electron microscopy, agarose gel chromatography and hyperchromicity studies) to be approximately 300 nucleotides in both *Xenopus* and *Strongylocentrotus*. For instance after treatment of hybrids with a DNase (S_1 nuclease) specific for single strands, the unrenatured tails can be removed leaving the renatured IR DNA sequence intact. After S_1 nuclease treatment 58% of resistant fragments chromatograph coincident with a 300 nucleotide marker, 25% elute with the void and are greater than 3000 nucleotides in length (presumably composed of clustered repeated DNA sequences), and 17% elutes between the two. Isolated 300 nucleotide fragments from such a column hybridize in DNA excess reactions with the kinetics expected of IR DNA and not HR DNA sequences. Electron microscopy, S_1 nuclease and hyperchromicity studies also indicate that up to tracer fragment lengths of 1000 nucleotides only one repetitive

sequence is present per bound DNA molecule. Proof that the DNA tails are SC DNA is given by isolating HAP bound tracer fragments of 1000 nucleotides average length and shearing them mechanically to a small size (300 nucleotides), which releases the tails intact from the renatured IR DNA sequences. These single stranded tails are now separated from the double stranded IR DNA by HAP chromatography and used as tracer in a cold DNA driven renaturation reaction. It is found that they renature with the kinetics expected of SC DNA.

Qualitative interpretation of the curve (Fig. 5.3.8), which shows how increased amounts of DNA will bind to HAP after low cot renaturation as tracer DNA size increases, can now be given in terms of the interspersion of IR DNA with SC DNA. Consider first the opposite model (Fig. 5.3.9a) in which all IR DNA is in large clusters and not closely interspersed with SC DNA. In such a situation increasing lengths of tracer DNA would not give any more HAP binding after low cot renatura-

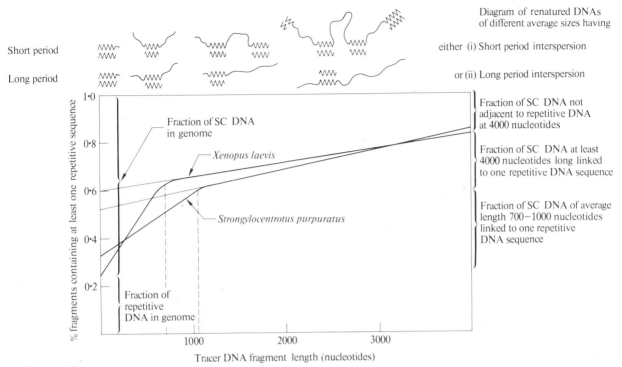

Fig. 5.3.8 An illustration of the relationship between the length of radioactive tracer DNA/and the amount bound to hydroxylapatite by short non-radioactive DNA sequences after renaturation to low Cot values such that only repetitive sequences have renatured for an amphibian *Xenopus laevis* and an echinoderm *Strongylocentrotus purpuratus*. Data redrawn from [69, 102].

tion, since the IR DNAs are not binding a significant proportion of any other type of DNA. The results in Fig. 5.3.8 are clearly not compatible with this type of model. In a second model (Fig. 5.3.9b) a fraction of SC DNA is interspersed with IR DNA in a regular manner such that one IR sequence 300 nucleotides long is found with every 1000 nucleotides of SC DNA. As the fragment size increases so will the amount of SC DNA bound to HAP by renatured IR sequence. At an average tracer DNA length less than the SC DNA length (1000 nucleotides) there will be on average only one IR sequence per molecule and the maximum amount of DNA will bound only when the average size is greater than the size of the SC DNA sequence. At these greater lengths there will be two or more IR sequences per molecule but the amount of DNA bound will not increase since all interspersed SC DNA will be bound. The curve will plateau at the length of the SC DNA sequence and the plateau value will be a measure, after subtraction of the IR DNA amount (i.e. binding at zero fragment length), of the proportion of the SC DNA in this interspersed arrangement. Finally in a third model (Fig. 5.3.9c) IR sequences are distributed partly as in model 6 and partly in a longer period interspersion (10,000 nucleotides) with SC DNA. Part of the SC DNA has no

detectable IR DNA associated with it. It is clear that as the tracer DNA size increases beyond the length of the SC DNA in short period interspersion, the amount of DNA binding to HAP will still increase up to a fragment length approaching 10,000 nucleotides as a result of the binding of increasing amounts of the long period interspersed SC DNA.

Quantitative interpretation of the curves in Fig. 5.3.8 are based on the type of model (model 3) which gives the best qualitative fit. It shows that about 25% of the *Xenopus* genome is in repetitive sequences and 75% is single copy or low frequency. At least 60% of the repetitive elements are sequences 300 nucleotides long, which by hybridization kinetics have been identified as IR DNA sequences. These IR sequences are interspersed with about 75–80% of the SC DNA at either short (700–1000 nucleotides) or long (greater than 4000 nucleotides) distances. Up to 20% of SC DNA may be organized into quite large blocks without repetitive sequences, and up to 40% of repetitive sequences (both HR and IR) may be organized into tandemly repeating structures without SC DNA over large distances. A similar model with slight quantitative differences has been derived from *Strongylocentrotus* [67, 102].

Although these experiments clearly show that the major part

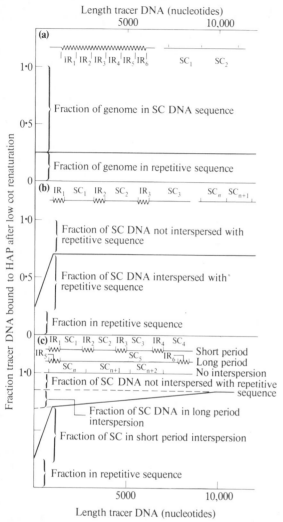

Fig. 5.3.9 Three models for the distribution of repetitive and single copy sequences in the genome. (a) No interspersion, (b) short period interspersion and (c) short period and long period interspersion. The graphs show the relationship between length of tracer DNA and the amount bound to hydroxylapatite after renaturation with excess cold driver DNA which is predicted for each model.

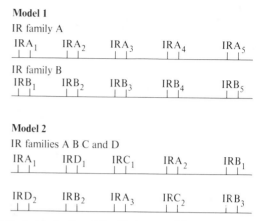

Fig. 5.3.10 Two models for the relationship between different families of interspersed intermediate repetitive DNA sequences. (a) Members of one family are more closely linked to each other than to members of another family. (b) Members of one family are more closely linked to members of another family than to each other.

end of the genes for globin in ducks there is a repetitive DNA sequence which is not poly(dA . dT) [13].

A further question relates to the distribution of IR DNA sequences which belong to specific families. For instance in *Xenopus laevis* there are about 1750 different IR DNA families with an average reiteration frequency of 1600 sequences 300 nucleotides long per family. Are the members of one family of IR sequences more closely linked within the genome than they are to members of another family? Two extreme models may be envisaged (Fig. 5.3.10) however at present there is no evidence which might help to discriminate between them.

The master-slave hypothesis and DNA circle formation

In order to account for the large differences in total DNA (*c* value) between closely related organisms and also many genetic and cytological observations on the structure of eukaryote chromosomes, Callan [39a] proposed a specific model of gene organization called the master-slave hypothesis. In essence this states that all eukaryote genes are composed of identical tandemly repeating DNA coding sequences. The obvious genetic problems posed by such a model, such as how do multiple genes mutate, crossover and evolve as single units, were accommodated by a special mechanism in which one of the sequences (the master) in each multiple gene unit serves as a template for the correction of any mutations which have occurred in the other (slave) sequences since the last correction event. Mutations occurring in the master sequence are propagated to the slaves at the same time, and crossing over is allowed only between master sequences. Such a mechanisms would in theory allow a gene composed of many identical sequences to behave genetically as a single unit. The frequency

of SC DNA is intermingled with IR DNA they do not show that any of the SC DNA in this class is protein coding. The coding potential of the SC DNA which is not intimately mixed with repetitive DNA is very considerable for instance in *Xenopus* 20% of single copy DNA will code for 300,000 different average sized proteins. This is a very large number and is at least an order of magnitude greater than the coding potential of mRNA so far isolated from any single cell type (Table 5.3.4). Experimental evidence has recently been provided that at the 3′

of the correction event would be every generation (or it may possibly occur less often) at some point during meiosis.

In principle the master-slave hypothesis can be partially verified by showing that all structural genes in higher organisms are composed of identical tandemly repeated DNA sequences. The demonstration that much of the DNA of higher organisms is single copy and that at least some coding sequences for well defined proteins occur once or a few times per genome argues against this as a general phenomenon (see Single copy DNA sequences, p. 308). Nevertheless a significant proportion of the structural genes may be organized in this way.

Several years ago it was shown that Salmon sperm DNA could be made to form circular molecules (slipped circles) by denaturation and renaturation [242a]. These circles must be derived from tandem repeated DNA sequences (see Fig. 5.3.11). More recently it has been possible to prepare circles from native DNA [153a, 194a] using enzymes (exonucleases) which digest one strand of a DNA double helix from either 3'OH or 5'PO$_4$ ends depending on their specificity. Use of these enzymes will generate DNA molecules with single strand tails (Fig. 5.3.11). If these tails are base complementary, as they would be for

tandem repeated sequences, they will form duplex during renaturation reactions causing circular molecules to form. The proportion of rings to linear forms after correction for efficiency of ring formation, is a measure of the fraction of DNA molecules with some repeating structure in them. The figures estimated show considerable variation from organism to organism but is usually in the region of 20–40% [153a, 194a].

There has been considerable debate over the exact nature of the DNA which forms these rings but there seems to be general agreement now that they are derived almost exclusively from the highly repetitious simple-sequence DNAs (p. 307). A small fraction (less than 10%) of the intermediate repetitive DNA, which may be exclusively genes such as those for rRNA, 5S RNA, tRNA and histones, will also form rings and are known from other studies to be tandemly repeated. The evidence that some but not all DNA is arranged in large blocks of tandemly repeating sequences is now very good but their relationship to chromosome structure and function is not understood.

Existence of tandemly repeating sequences does not prove that the special mechanisms proposed by the Master-Slave hypothesis also exist. Clear evidence that they do is not available for any particular set of tandem repeating sequences. On the contrary the recent demonstration of sequence and length heterogeneity within such well defined repetitive genes as the rRNA and 5S RNA genes of *Xenopus* [36, 254] suggest that correction mechanisms of the Master-Slave type are not operating very frequently within these gene blocks.

Gene numbers, complementation groups and chromomeres

All eukaryote genomes have an enormous protein coding potential (see Single copy DNA sequences, p. 308), and it is of considerable importance to understand whether this potential is likely to be realized since the complexity of genetic control systems must be reflected in the number of genes to be controlled.

Early attempts (for refs to early work, see [138]) to define the number of genes in *Drosophila* were based on comparison of natural or X-ray induced X linked lethal mutation rates per locus compared with the lethal mutation rate for the X chromosome as a whole. They gave figures in the region of 800–1000 loci for this chromosome, which, since it represents one-fifth to one-sixth of the total genome, would mean a total of about 5–6000 lethal producing loci in *Drosophila*, a figure equivalent to the number of bands (chromomeres) seen in salivary gland chromosomes [8]. Recent extensive analysis by deletion mapping of lethal and semi-lethal mutations in the *white-zeste* region of the *Drosophila* X chromosome has shown that the number of complementation groups is equivalent to the number of cytologically visible bands (or interbands, since the mapping techniques had not sufficient resolution to discriminate between them) in this region [137, 138]. Furthermore the

(a) Formation of slipped rings

Native repeated DNA molecule $A_1A_2A_3A_4A_5A_6A_7A_8A_9A_{10}$

Denatured

Renatured with slipping to form ring structure

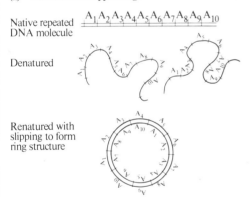

(b) Formation of DNA rings after exonuclease treatment

Native repeated DNA molecule $A_1A_2A_3A_4A_5A_6A_7A_8A_9A_{10}$

Resect with exonuclease $A_1A_2A_3A_4A_5A_6A_7A_8$ A_9A_{10}

Renature to form circles

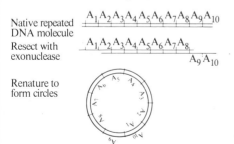

Fig. 5.3.11 An illustration of the formation of DNA rings (a) by denaturation and renaturation, and (b) after exonuclease treatment.

mutants of any one complementation group are equivalent in respect of the developmental stage at which they die, suggesting they all affect one particular function [220]. Since the analytical complexity of the *Drosophila* genome is approximately $1\cdot5 \times 10^8$ base pairs and there are about 5000 visible bands in polytene chromosomes the average amount of DNA per band plus interband is sufficient to code for 30 proteins with average molecular weight (40,000 daltons). An interband contains much less DNA than a band perhaps sufficient to code for one average-sized protein.

Genes which are able to give rise to lethal mutations may not be representative of all genes, so that estimates of gene numbers based on numbers of lethally mutating genes may be underestimates. Evidence that lethal mutations may be in a special class of genes comes from studies which show that they usually involve deletions of whole loci as defined by complementation analysis [46]. Also there are several structural genes for defined protein products in *Drosophila* whose expression is suppressed by non-lethal mutations (null mutations) [178]. This suggests that there are two classes of gene, one which codes for dispensable functions and another which codes for indispensable functions. It may be that lethal mutation identifies only genes for indispensable functions, such as controlling elements and essential proteins, but that these genes may be

intimately associated with genes for dispensable functions, such as code for non-essential proteins.

Recently [164] studies of *Drosophila* DNA sequence organization have shown that it contains approximately 3000 IR DNA sequences of about 6000 nucleotides average length split between 40 different families with an average repetition frequency of 70. The total number of IR DNA sequences is so closely similar to the number of chromomeres that it is tempting to suppose that there is one per chromomere. Since the average length of DNA per chromomere in *Drosophila* is in the region of 40,000 nucleotides, this would imply a model for the interspersion of IR DNA with SC DNA at distances of about 34,000 nucleotides. There is evidence [164] that some IR DNA sequences are interspersed with SC DNA at distances greater than 13,000 nucleotides, although a shorter period interspersion for other IR sequences is not excluded [265]. This model stands in contrast to that for *Xenopus* and *Strongylocentrotus* [67, 69, 102] which show predominantly short period interspersion. If we put the genetic and biochemical information together we can generate a model for the chromomere in *Drosophila* (Fig. 5.3.12) in which there is one IR sequence and sufficient SC sequence to code for about 30 average size proteins some of which may be dispensable and some indispensable. Such a structure might be transcribed in two ways, either (i) to give a

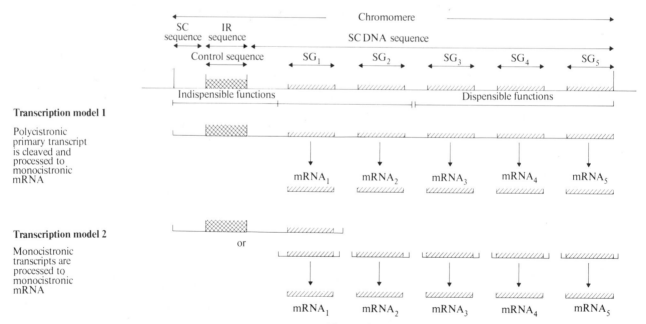

Fig. 5.3.12 An illustration of a possible model for chromomere structure in *Drosophila* together with two models for its transcription. See text above for further explanation.

Table 5.3.5 Extent of single copy DNA transcript present in polysome bound mRNA and total nuclear RNA from Sea Urchin Gastrula (ref. [88]).

Source of RNA	% single copy transcribed	Analytical complexity (base pairs)
total polysomal	2·7%	$1·7 \times 10^7$
total nuclear	20–30%	$12–18 \times 10^7$

Table 5.3.6 DNA dependent RNA polymerase activities in eukaryotes.

Polymerase	Intracellular location	Alpha-amantinin	Product
I	Nucleolus	Resistant	rRNA precursor
II	Nucleoplasm	Sensitive	HnRNA
III	Nucleoplasm	Sensitive	4 and 5S RNA

polycistronic primary transcript, for which there is no evidence in eukaryotes, and which would have to be processed to give monocistronic messages, or (ii) to give monocistronic precursors which may or may not require processing to functional mRNA. There does not seem to be any evidence at the moment which will distinguish these possibilities.

Two further points relevant to this discussion have been brought into focus recently by estimates of mRNA and total transcriptional sequence complexity (Tables 5.3.5 & 5.3.6). These measurements clearly show that within particular cell types the number of average size proteins coded by mRNA is of the order of 5–15,000 depending on organism and cell type. They do not yet tell us about the total mRNA sequence complexity required for the whole life cycle of an organism. Estimates of *total* RNA sequence complexity suggest that at least a ten-fold greater amount of DNA sequence is transcribed than is required in cytoplasmic mRNA and this clearly indicates some form of processing and selection of RNA sequences after transcription (see Single copy DNA sequences, p. 308).

5.3.4 TRANSCRIPTION

Transcription of DNA is mediated by enzymes known as DNA dependant RNA polymerases. Since they are involved in the primary events of gene expression it is important to test the hypothesis that control of gene activity involves the availability of polymerase molecules with specificity for particular regions of the genome. In bacterial and bacteriophage systems it now seems clear that such positive control is exerted by polymerase core enzyme which derives its specificity for initiation of RNA synthesis by interaction with different modifying sigma factors and for termination by interaction with modifying rho factors.

The DNA dependent RNA polymerases

Work on eukaryotic RNA polymerase has been hampered by technical difficulties such as its lack of stability, relatively low concentration in cells, and tight binding to chromatin. These problems now seem to have been overcome and the existence of three chromatographically distinct DNA dependent RNA polymerase activities has been described (Table 5.3.6) [17, 204]. Polymerase I is located in the nucleolus [205], is probably responsible for synthesis of 18S and 28S rRNA, and is resistant to high concentrations of α-amanitin, a cyclic polypeptide which is a major active agent in poisoning by the mushroom *Amanita phalloides*. Polymerase II is sensitive to α-amanitin at low (0·01 µg/ml) concentrations, is located in the nucleoplasm [205] and is possibly responsible for synthesis of HnRNA the putative mRNA precursor (see p. 318). RNA polymerase III is also located in the nucleoplasm [205], is resistant to α-amanitin at low concentrations, but sensitive at concentrations above 1 µg/ml, and is involved in synthesis of 4S and 5S RNA [253]. More detailed chromatographic and structural studies have revealed further heterogeneity of polymerase activity within these fractions [44]. For instance two polymerase activities are separable from polymerase III fraction of a mouse myeloma [165]. These observations suggest that regulation of genetic activity in eukaryotes is based at least partly on RNA polymerases with different template specificities.

Functional studies of these enzymes has been made difficult by the fact that most purified eukaryote DNAs contain a substantial number of single-strand nicks which promote non-specific initiation of transcription *in vitro*. Also assays for the fidelity of *in vitro* transcription by comparison with *in vivo* synthesized RNA in nucleic acid hybridization competition experiments has been complicated by the high sequence complexity of a eukaryote DNA. The genes for rRNA, which are easily isolated, present a favourable system in which to approach the question of whether the specificity of *in vivo* initiation can be simulated *in vitro* [203]. Contrary to expectation, purified *Xenopus* rDNA was not preferentially transcribed in comparison with total DNA by polymerase I, as predicted from its nucleolar location *in vivo*, and polymerase II was equally good at transcribing rDNA sequences. Moreover neither polymerase showed any greater preference than *E. coli* RNA polymerase to initiate RNA synthesis specifically in any one of the well defined regions (Fig. 5.3.6) of rDNA, nor to transcribe it with any greater fidelity from one strand [196]. Greater strand specificity has been reported for *Micrococcus luteus* polymerase, when transcribing *Xenopus* rDNA with very high single strand molecular weight. In this study there was an indication of preferential initiation in the 18S rRNA region. It would seem that the undoubted specificity of transcription observed *in vivo* is not a unique property of either the purified polymerase nor of the DNA sequences but presumably resides in the interaction of these two components with other factors, such as chromosomal proteins or sigma-like factors, an idea which has received some support recently in studies of transcription using reconstituted chromatin (see Biochemical effects of cytoplasmic localization, p. 338).

Lack of correlation between availability of polymerase I and rate of rRNA synthesis in early development

During early development in most organisms there is a progressive increase in the rate of RNA synthesis per embryo (see Section 5.3.6) and also a progressive qualitative change in the type of RNA synthesized. One reasonable prediction would be that the type and amount of each different polymerase would change during development according to the amount and type of RNA synthesis. Early cleavage stages of *Xenopus* development, with few cells and little if any RNA synthesis, contain only 4—5-fold less total RNA polymerase activity than do swimming tadpoles with thousands more cells and high rates of RNA synthesis [203]. This observation shows that there is no absolute correlation between total amount of polymerase and the rate of RNA synthesis. It may be that assays of total polymerase activity mask more subtle changes in the relative amounts of each type of polymerase. For instance at the onset of rRNA synthesis during late blastula we might expect to see a change in the amount of polymerase I activity, however this is not the case [203]. Moreover the anucleolate mutant of *Xenopus laevis*, which has no rRNA genes and synthesizes no detectable rRNA, has normal levels of polymerase I activity. Thus there is no correlation between total polymerase I activity and rRNA synthesis in either the presence or absence of rRNA genes. Sea urchin embryos similarly show the presence of polymerase I in nuclei which do not synthesize rRNA and in which nucleoli are absent [17, 206]. Furthermore, there is no change in polymerase activity, either quantitative or qualitative, during cell differentiation in the amoebo-flagellate *Naegleria gruberi* even though this requires new RNA synthesis [230]. Such observations lead to the conclusion that regulation of the level of polymerase I availability is unlikely to be a major factor in the control of rRNA synthesis during early development, and reinforce the idea that control of transcription is mediated via the interaction of specific factors with the polymerases and/or the genes themselves.

Correlation of polymerase activity with RNA synthesis

In certain situations the level of specific polymerases may be correlated with changes in type and amount of RNA synthesis. For instance in sea urchin development, cleavage stage embryos as well as their isolated nuclei synthesize largely HnRNA, and over 90% of RNA synthesis at this stage is α-amanitin sensitive. The relative amount of HnRNA synthesis declines several fold during development and is paralleled by a decrease in the level of polymerase II [17, 206]. Similarly in situations such as partial hepatectomy and the action of oestrogen on the uterus, where the amount and type of RNA synthesis is altered experimentally, changes in the levels of both polymerases I and II can be detected [17]. These observations suggest that in some situations polymerase availability may become rate limiting for RNA synthesis.

Is initiation of transcription rate limiting for RNA synthesis?

If initiation of transcription is rate limiting for RNA synthesis polymerase will be sparsely packed on the DNA coding sequence, while if it is not limiting polymerase will be maximally packed. Thus determination of the degree to which polymerase is packed on to specific DNA sequences should give some indication of the extent to which initiation limits transcription. Electron microscopy of amphibian oocyte nucleoli and of nuclei from Hela cells using special techniques for the preparation and spreading of samples led to the visualization of 'active' genes for rRNA [170]. These genes have a well-defined 'fir tree' structure of a single central DNA duplex with regularly spaced RNP (ribonucleoprotein) fibrils at right angles to the main axis. At the base of each fibril, at its point of insertion on to the axis, is a globular protein molecule with the dimensions expected for an enzyme the size of RNA polymerase. These pictures allowed certain conclusions about the structure of ribosomal genes, about the direction of transcription in adjacent genes, and an estimate of the packing of polymerase. Confirmation was provided that rRNA genes are tandem repeats of a well-defined length sufficient to account for transcription of a molecule the size of the rRNA precursor. These repeats were spaced by apparently untranscribed regions of sufficient length to account for the untranscribed spacer DNA. Transcription, judged by the polarity given to the 'fir tree' by RNP fibrils of different lengths, is in the same direction in adjacent transcription units. In both *Triturus* and *Xenopus* the number of such fibrils is 80—100 per repeat unit, which means that for a polymerase protein of diameter $12 \cdot 5$ μm, approximately half of the total transcription unit is occupied by polymerase. In any event this must represent greater than 50% of maximum packing and may well represent maximum possible packing of polymerase. In Hela cells the presumptive rRNA genes have 100—150 polymerases per repeat unit which is 5—7 times greater than the number predicted from biochemical parameters but represents less dense packing than the amphibian genes since the mammalian transcription unit is about twice the size. The conclusions from these observations seem to be that rRNA genes are maximally packed with polymerase, or nearly so, and that if inactive rRNA genes occur they must be inactivated in blocks which would not be distinguishable by these techniques.

5.3.5 THE PRODUCTS OF TRANSCRIPTION

There are six major classes of RNA in cells of eukaryotes, 18 and 28S ribosomal RNA, 5S ribosomal RNA, transfer (4S) RNA, messenger RNA, heterogeneous nuclear RNA and low molecular weight nuclear RNA, and this list may not be complete. It now seems likely that none of these classes is synthesized in its final functional form, but rather as a larger precursor molecule which is processed (shortened) and chemically modified in various ways. These processing and modification mechanisms are not well understood at present, but clearly they may be important candidates for regulating the rate of appearance of mature gene products. In addition gene expression may be controlled by selective degradation, and by transport to the cytoplasm of some molecules and not others (for recent reviews, see [59, 64, 159]).

The structure and processing of RNA

THE STRUCTURE AND PROCESSING OF rRNA

The main steps in the processing and modification of ribosomal RNA are well known (Fig. 5.3.13) and seem, with minor modifications, to be a general feature of all eukaryotes (for recent review, see [59]). However the detailed mechanisms and enzymes involved are not well characterized. The initial transcript, known as ribosomal precursor RNA, varies in size between different taxonomic groups ($45S, 4 \cdot 1 \times 10^6$ daltons in mammals, $40S, 2 \cdot 7 \times 10^6$ daltons in amphibians and plants). Recent secondary structure mapping of the various components of the rRNA processing series has confirmed and extended the original work based on the results of polyacrylamide gel electrophoresis [252, 255]. Secondary structure maps clearly

show that the 28S RNA is located at the 5′ end of the precursor, that the 18S RNA is towards the 3′ end but separated from both the 28S RNA and the 3′ end by transcribed spacer RNA which is lost during processing.

Chemical modification (methylation and pseudo uridine formation) of rRNA occurs twice; soon after synthesis of the precursor RNA and late in maturation on the 18S RNA. Methylation of ribose occurs rapidly on the precursor at about 105 sequence specific sites, and a further 5 or 6 methylations of bases are also observed [160]. These methylated sequences are all conserved during processing and appear in the mature 18 and 28S RNA [160]. Later during maturation six base methylations occur on the 18S molecule [210].

This late methylated sequence occurs in the smaller rRNA of mammals, yeast and E. coli and this suggests that it may have some general role in ribosomal maturation or function. Methylation of the sequence is not however essential, since there is a viable mutant of E. coli in which the sequence is not methylated. Experiments on methionine (the donor of methyl groups in RNA methylation) starvation have shown that methylation is essential for processing and maturation of rRNA [161], and that under methylated precursor can be rescued by restoration of methionine.

Ribosome assembly begins in the nucleolus (for reviews and references, see [59, 159]) with the association of the rRNA precursor and a group of specific proteins, some of which are found in mature cytoplasmic ribosomes and some of which are recycled within the nucleolus. It is not surprising therefore that there is a close coupling of rRNA and protein synthesis. However synthesis of ribosomal proteins will continue for some time in the absence of rRNA synthesis, but the ribosomal proteins made under these conditions are not utilized in ribosome formation when rRNA synthesis is resumed. There is

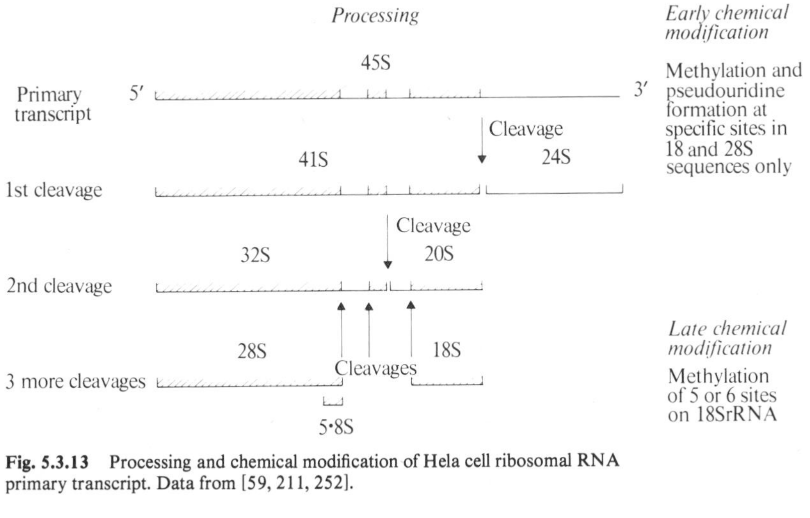

Fig. 5.3.13 Processing and chemical modification of Hela cell ribosomal RNA primary transcript. Data from [59, 211, 252].

a strict coupling of rRNA synthesis to protein synthesis but not *vice versa*, and recent experiments suggest that the processing and transport to the cytoplasm of large, but not small, ribosome subunits is dependant on the continued transcription of rDNA [146].

5S RNA (42,000 daltons MW) is a small RNA which is found mainly in association with the large ribosome subunit or its precursors. The complete nucleotide sequence is known from many organisms (bacteria, animals and plants), and it contains no chemically modified bases or ribose moieties. There is an indication for bacteria [82] and *Xenopus* [71] that 5S RNA is synthesized in precursor form with a few extra nucleotides at the 5' and 3' end respectively, which are removed before or during the incorporation of 5S RNA into the 50S ribosome precursor. The 5' terminal residue in eukaryotes may be recovered in part as the tetraphosphate pppNp suggesting that the initial transcript is intact at this end.

It has recently been shown that as in bacteria there is heterogeneity of 5S RNA sequence in *Xenopus* and *Triturus* [84, 249]. There is tissue specific control of the expression of these sequences since somatic cells, tissue culture cells, and oocytes show very different proportions of the various types of 5S RNA sequence. It is not known at present whether these types of 5S RNA are functionally different or whether the base changes observed result from neutral mutations in the multiple genes coding for 5S RNA.

Transfer RNA molecules (about 25,000 daltons MW, 75–90 nucleotides) play a very well defined role in protein synthesis. Specific tRNAs accept specific amino acids and act as adaptor molecules in locating their amino acid at the correct codon during translation of mRNA. The specificity of location on the correct codon is determined by the base sequence of part of the tRNA molecule called the anticodon. The specificity for acceptance of a particular amino acid is determined by the interaction of tRNA with its specific aminoacyl tRNA synthetase enzyme. For each amino acid there is at least one tRNA and usually more than one. Different tRNAs accepting the same amino acid are known as isoacceptors. The nucleotide sequences of about 60 different tRNA molecules are known and all of them can be formed by virtue of base pairing between different regions of the molecule) into the well known two dimensional structure called the cloverleaf [7]. Recently X-ray crystallography has produced three-dimensional models for tRNA structure [200]. Nucleotide sequencing work has made it quite clear that tRNAs contain a large number (greater than 30) of rare nucleotides which are produced by secondary chemical

modification after transcription [7]. There is a large literature on the post-transcriptional modifications of tRNA and the variability of tRNA populations in various physiological and developmental conditions [95, 116]. It is clear that the degeneracy of the genetic code and the occurrence of isoacceptors which may vary in concentration might allow some fine control of the rates of synthesis of specific proteins. It is also of importance to know how different cells manage to control the levels of isoacceptors and tRNAs in such a precise manner. Part of the answer to this sort of question may be provided by studies on the tRNA genes themselves (see Intermediate or middle repetitive DNA sequences, p. 307).

It is clear that the initial tRNA transcript is larger than the final mature tRNA in both pro- and eukaryotes [9, 37]. An elegant series of experiments [3, 4] has led to the isolation of a specific tRNA precursor for tRNAsuIII in *E. coli*. This molecule has the complete tRNAsuIII sequence including the −CCA 3' terminus together with an extra 41 nucleotides at the 5' end and an extra three nucleotides at the 3' end. A purified ribosome bound nuclease (RNase P) removes the extra nucleotides from the 5' end as a fragment 41 nucleotides long exposing the 5' end of the mature tRNA but leaves the 3' end of the precursor intact. Although such a detailed study of the processing of eukaryote tRNA precursor is not available, it is clear that enzymes are present in eukaryote cells which will process tRNA precursors in a similar way [37].

It is convenient to consider the structure of HnRNA and mRNA together since there is considerable circumstantial evidence, though no direct proof, that at least some fraction of HnRNA is a precursor to mRNA (for recent reviews, see [22, 65, 87, 250]). Rapidly labelled nuclear RNA with large (greater than 40S) heterogeneous sedimentation coefficients has been known for many years; while mRNA from cytoplasmic polyribosomes is less rapidly labelled and has a much smaller (less than 40S) heterogeneous sedimentation coefficient. Early kinetic experiments suggested that most (>80%) rapidly labelled nuclear RNA is degraded within the nucleus and never reaches the cytoplasm [117]. Recent experiments have shown, in both mammalian (mouse L cells) tissue culture cells and sea urchin embryos, that HnRNA decays with an apparent half life of 23 and 7 min respectively and represents about 7% of total cellular RNA [20, 21, 190]. Messenger RNA represents about 5% of total cell RNA and in mouse L cells has an apparent half life of 10 h [20]. The proportion of HnRNA which enters the cytoplasm is about 2% in mouse L cells and 10% in sea urchin embryos though it is not clear at present whether this difference is of phylogenetic or developmental significance [20, 21, 190].

The hypothesis that HnRNA is precursor to mRNA was based on several observations at the time of its discovery [124]. Its base composition resembles that of mRNA and DNA, it is apparently much larger than mRNA, and although mRNA must be synthesized in the nucleus no mRNA sized precursor has been found there. Recent observations have added strength to this idea [65]. For instance it has been found that a large fraction of HnRNA is transcribed from single copy DNA and contains a post-transcriptionally added poly(A) sequence at the 3' end like mRNA [87, 133, 144, 174]. Also some viral and specific protein mRNA sequences have been identified in HnRNA, although there is considerable debate as to whether this result is due to the fortuitous aggregation of mRNA sequences with HnRNA or to the mRNA sequences being in a covalently intact high molecular weight molecule [131, 158, 169, 237].

The occurrence of a poly(A) fragment at the 3' end of some HnRNA and mRNA molecules has made possible the isolation of these molecules by affinity chromatography on poly(U) sepharose or oligo (dT) cellulose columns (Fig. 5.3.14). It also allows the isolation of both the 5' and 3' regions of these molecules, since if purified HnRNA or mRNA is broken in half by limited treatment with alkali and re-run through an affinity column 5' ends will not be bound but 3' ends will (Fig. 5.3.14). Poly(A) allows quantitation of mRNA by hybridization with radioactive poly(U) [15, 104] and the synthesis of a DNA copy of at least a part of the mRNA molecule by an RNA dependant

DNA polymerase (reverse transcriptase) which requires an oligo (rAdT) primer on which to initiate [139]. The DNA copy (called cDNA) may be made radioactive and is extremely useful for nucleic acid hybridization reactions involving both DNA and RNA [244].

As well as the similarities between HnRNA and mRNA there are many chemical differences apart from the apparent size difference. An oligo (U) fragment, 35–40 nucleotides long, containing about 80% uridylic acid, has been isolated from HnRNA [175, 176]. It seems to occur on average once per molecule and is located towards the 5' end of the molecule and is therefore probably coded for by the genome. A similar sequence does not occur in mRNA. HnRNA in mammals and sea urchins is transcribed partly from single copy DNA and partly from intermediate repetitive DNA, the proportions of each are about 3:1 and may vary from system to system. It has been shown that the bulk of IR DNA transcript is intermingled with SC DNA transcript [102, 125, 228] in a manner reminiscent of the intermingling of IR and SC DNA sequences in the genome (see Intermingling of intermediate repetitive and single copy DNA sequences, p. 309). Although in some cells small (5–10%) fraction of mRNA may be transcribed from IR DNA [14, 140] it seems that these molecules are not covalently linked to SC DNA sequence transcripts and may represent mRNA for proteins coded by multiple genes like those for histone. Although the IR DNA transcript in HnRNA is intermingled with SC DNA transcript, it is located pre-

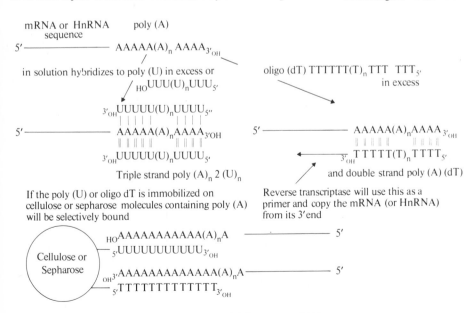

Fig. 5.3.14 An illustration of the way in which poly(A) fragments on RNA molecules enables their quantitation [15], purification and reverse transcription to give c(copy) DNA.

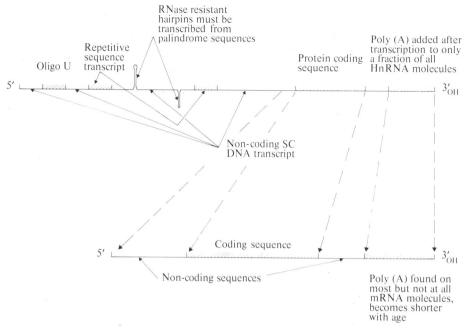

Fig. 5.3.15 Illustration of a possible general model for heterogeneous nuclear RNA and its relationship to a general model for messenger RNA.

dominantly towards the 5′ end of the molecule [134, 175]. Apart from the poly(A) tails, HnRNA also contains a significant amount of double stranded RNA which may be isolated by its RNase resistance. The RNase resistant fragments appear to be hairpin loops less than 500 nucleotides in length and located towards the 5′ end of the HnRNA molecule and their hybridization properties suggest they are transcribed from IR DNA [134]. It is possible to make a general model for HnRNA structure based on these observations ([175], Fig. 5.3.15) but it is clear that the details may vary from one phylogenetic group to another [87].

Messenger RNA structure is now also much better known mainly due to work on purified mRNAs coding for known proteins (Fig. 5.3.15). Most eukaryote mRNAs have been shown to contain at their 3′ end a poly(A) sequence of a few hundred residues which is added post-transcriptionally [15, 87] and is therefore not coded by the genome [87]. Histone mRNAs, and there may be others [169a, 177a], are an exception to this since they do not have 3′ poly(A) sequences [20]. In fact since cytoplasmic poly(A) becomes shorter with age it may be that there is a subpopulation of most messages which do not contain poly(A) [221]. Messenger RNA contains a protein coding sequence of a length defined by the number of amino acids in the protein. All of the purified mRNAs for known proteins also contain a variable number of extra non-

coding nucleotides partly at the 5′ and partly at the 3′ end of the the coding sequence. For instance globin mRNA contains about 150 nucleotides of non-poly(A) non-coding sequences; of these about 40 are located between the 3′ end of the coding sequence and the poly(A) fragment, while the other 110 are at the 5′ end [173, 194]. The function of these non-coding sequences in mRNA is unknown at present [126, 194] but the availability in the near future of non-coding sequences in other purified mRNAs may give some indication of their importance [173, 194]. Recent work with mRNA isolated from *Xenopus* embryos and insect cells has suggested that some of them contain at their 5′ ends sequences transcribed from the IR DNA fraction [74, 130]. It is not exactly clear from this work what proportion of mRNA molecules contain such a sequence nor what proportion of the IR DNA fraction is represented. Results with newly synthesized sea urchin gastrula polysomal mRNA suggest on the contrary that less than 3% of mRNA molecules contain recognizable IR DNA transcripts [20a]. Chemically modified nucleotides also occur in mRNA but at a very much lower level than in either rRNA or tRNA [189].

LOW MOLECULAR WEIGHT NUCLEAR RNA

The existence in nuclei of metabolically stable low molecular weight RNA has been known for some time [for references, see

250]. High resolution polyacrylamide gel electrophoresis has demonstrated the presence of as many as eleven discrete types of small RNA molecule in nuclei [177]. Two of these RNAs are clearly nucleolar in origin, but most of them are found in the extranucleolar fraction of nuclei. These RNA molecules are also extensively modified by methylation, and nucleotide sequencing studies have shown that they are unique species of RNA. They occur very widely and although there are phylogenetic differences in terms of the molecular weights and sequences of individual components, the proportion of each molecular species seems to be quite similar in the cells examined so far. Some of these RNAs are coded for in by multiple genes. Their detailed functions are unknown but it would seem likely that the nucleolar associated molecules play some role in ribosome formation and some of the others may be involved in the control of DNA synthesis [50].

5.3.6 THE CONTROL OF RNA SYNTHESIS IN DEVELOPMENT

RNA synthesis may vary both quantitatively and qualitatively. One extreme model for the control of transcription in development and differentiation supposes that all genes are transcribed in all cells but that the relative amounts of specific direct gene product is controlled by varying the rate of transcription. The other extreme model supposes that some gene transcripts are required by all or most cells but that genes specifying differentiated cell functions are active only in those cells having these functions. It is likely that neither model is entirely sufficient to account for the control of gene activity in all systems and that both may have some validity. We will now examine the evidence for both quantitative and qualitative control of transcription.

HnRNA synthesis in development and differentiation

HnRNA SYNTHESIS DURING OOGENESIS

During the diplotene phase of oogenesis, the chromosomes have a characteristic lampbrush like structure [39]. They have been seen in a wide variety of organisms but have been best studied in animals with high c values (DNA content) such as the Urodele amphibians [67]. Lampbrushes are pairs of homologous chromosomes held together at points by chiasmata. Each chromosome is formed of a pair of chromatids along which the DNA is condensed at intervals into chromomeres separated by uncondensed DNA. Extending laterally from each chromomere is a single stranded DNA loop. These lateral loops have been shown to be sites of RNA synthesis [91] and to contain attached ribonucleoprotein (RNP) particles [163, 170]. They have recently been visualized in electron micrographs in a manner similar to that used for observation of nucleolar rRNA

genes [163, 170]. The RNP fibrils, like the nascent ribosome precursors, become longer from one end of the lampbrush loop to the other, but grow to much greater lengths, and probably to the full length of the loop itself [170]. The average size of lampbrush loops in *Triturus viridescens* is 50 μm and some are 200 μm long. Thus the nascent RNA of lampbrush chromosomes in *Triturus* may be on average 150,000 nucleotides (5×10^7 daltons) in length and fall outside the size range of HnRNA molecules seen in mammals and birds ($1-2 \times 10^7$ daltons). It has been estimated that 5% of the total genome is in loop DNA at any one time, the bulk of the rest being condensed into chromomeres. This means that a minimum expression of 5% of the genome is possible at any one time during oogenesis, and it may be more since loop DNA may move from one side of a chromomere to the other [229]. We will come back to this point below when considering the proportion of the genome expressed (sequence complexity) in oogenesis.

Isolation of these lampbrush loop particles from *Triturus cristatus* oocytes has been achieved by differential and sucrose gradient centrifugation [231]. They are found to have about 3% RNA and 97% protein. The RNA sediments faster than 40S, has a base composition approaching that of *Triturus* DNA, is labelled with RNA precursors during the lampbrush phase of oogenesis and is present but not synthesized in mature oocytes. The proteins are of several different kinds [218, 231a] unlike the relatively few proteins extracted from nuclear non-ribosomal RNP particles in other cells [185]. One of the proteins has been shown to be localized specifically on one particular lampbrush loop [218] while others are more generally distributed. Further details concerning the structure [163], stability, function and sequence complexity of the RNA components make it likely that they represent particles containing HnRNA in oocytes.

HnRNA IN DEVELOPMENT

Autoradiographic and biochemical experiments with labelled RNA precursors established that most RNA synthesized in the nucleus during early development of several organisms never reaches the cytoplasm [66]. Recently biochemical studies involving the measurement of absolute rates of RNA synthesis in sea urchin embryos have shown that very high rates of HnRNA synthesis occur during cleavage but that the rate on a per nucleus basis declines during subsequent development [5, 21]. Newly labelled RNA appears to fall into two classes on the basis of half life measurements [20, 21]. One class has an apparent half life of 5–10 min and the other one of 60–90 min, and these half lives do not alter significantly between blastula and pluteus stage [21].

Isolation of the newly synthesized nuclear RNA from sea urchin embryos, followed by fractionation on sucrose gradients or in denaturing formamide gels, has shown that it is of high molecular weight (up to 3×10^6 daltons) [148]. The presence of a small (3%) proportion of HnRNA molecules containing poly(A) has been shown in the hatching blastula of sea urchins [148]. This is in contrast to the situation in *Dictyostelium* where the majority of pulse labelled nuclear non-ribosomal RNA molecules contain post transcriptionally added poly(A) [87]. It is also different to the situation in Hela cells in which about 30% of the largest and 15% of the smallest HnRNA molecules have poly(A) sequences [65]. It may be that these differences are of developmental significance suggesting a lower level of processing of HnRNA to mRNA in sea urchin embryos compared with other cells. Electrophoresis of poly(A) containing HnRNA from these embryos reveals a few major discrete size components with molecular weights between $1-2\cdot5\times10^6$ daltons and a background of molecules ranging in size from $0\cdot5-5\cdot0\times10^6$ daltons, quite unlike the very broad heterogeneous size distribution of total sea urchin embryo HnRNA and suggests that poly(A) is added selectively to HnRNA molecules and not to a random sample of the whole population [148]. RNA with molecular weights of the same order as HnRNA has been found in cytoplasmic RNA preparations from early sea urchin embryos. It has been argued that this is not due to leakage from nuclei during cell fractionation but to the presence of a large proportion of cells in mitosis [99]. During mitosis HnRNA is released from nuclei, but whether it remains there during the subsequent interphase or returns to the nucleus is not known.

Sea urchin gastrula HnRNA has been shown to contain 8% sequences transcribed from IR DNA and the rest (92%) from SC DNA [102] similar to HnRNA from other cell types [103]. Measurement of the proportion of total SC DNA sequence represented in sea urchin gastrula HnRNA gives figures of between 10 and 30% [88]. Messenger RNA present in sea urchin gastrula polyribosomes is transcribed from about $2\cdot8\%$ of the total SC DNA coding for about 14,000 average sized proteins [88]. Clearly this represents a developmental situation in which the number of diverse SC DNA sequence transcripts present in cytoplasmic polyribosomes is at least four-fold and probably ten-fold restricted compared with the complexity of nuclear RNA [88]. There is no evidence at present which suggests that any of the nuclear transcripts of SC DNA in sea urchin embryos are related to any of the cytoplasmic SC DNA transcripts as many, if not all, of the cytoplasmic mRNAs may have been maternally inherited (see mRNA in development, p. 324). However, since gastrulation is the time when new gene expression becomes essential for further development [66], it would seem likely that at least some nuclear and cytoplasmic SC DNA transcripts are identical. These observations may be interpreted in two ways depending on whether all SC DNA transcripts can be regarded as potential protein coding sequences or not. If it is true that much of the SC DNA is not protein coding then these results may be interpreted to indicate transcription of 4 to 10 times more non-protein coding than coding SC DNA reflecting two functionally distinct classes of SC DNA. However, the possibility still remains that some selective processing of potential mRNA sequences does occur. Although there is no such detailed study of HnRNA in early development of any other organism it is likely that HnRNA synthesis is a general feature of early development.

HnRNA SYNTHESIS IN DIFFERENTIATION

Particularly favourable material for the study of gene activity in differentiation is presented by the polytene chromosomes found in many tissues of Dipteran (Insect) larvae. These chromosomes have a characteristic banding pattern, each band representing the lateral alignment of about one thousand identical chromomeres formed during lateral amplification of chromosomes after pairing of homologues. The majority of bands are apparently inactive in RNA synthesis since they incorporate little if any RNA precursors as judged autoradiographically [186]. A few bands, become enlarged (puffed) and show intense incorporation of RNA precursors [6, 8]. Puffing patterns are tissue and developmental stage specific and may be altered by a variety of hormone and physiological treatments [6, 8]. Since there is a strong indication that genes, identified by complementation analysis, are equivalent to chromomeres (bands) (see p. 313) it is very likely that puffing represents the transcriptional activity of a particular complementation group. It is clear that the big differences in transcription rates between puffs and non-puffs represent at least a quantitative control of gene activity, and if one accepts that some non-puffs are entirely inactive in RNA synthesis represents a qualitative control as well [6, 8].

Recent studies of a particularly large puff (Balbiani ring 2 = BR2) in *Chironomus* salivary glands have shown that the primary transcript is of high molecular weight ($15-35\times10^6$ daltons) and may represent total transcription of Balbiani ring DNA. If BR2 DNA is of average size for *Chironomous* it would have a double stranded molecular weight of 60×10^6 daltons [62]. Nucleic acid hybridization studies suggest that BR2 RNA hybridizes with the kinetics expected for RNA transcribed from sequences repeated 200-fold, but it hybridizes to the BR2 band only [152]. Clearly the BR2 RNA must be internally repetitive since it is unlikely that 200 sequences of the length of the BR2 RNA could be accommodated within the DNA of the BR2 region with which it hybridizes. Thus, BR2 RNA may not be typical of primary gene transcripts. RNA transcribed from two other Balbiani rings (BR1 and 3) show that they have different sequence specificity, though BR1 RNA also seems to be internally repetitive. If these RNAs code for the secretory proteins produced by the salivary gland [108], they are probably transcribed from multiple genes [152] and transcription from them may not be representative of transcription from single copy coding sequences.

Messenger RNA synthesis in development and differentiation

MESSENGER RNA SYNTHESIS IN OOGENESIS

There is much convincing evidence that protein synthesis during early development in animals and germination in plants is dependent to a large extent on mRNA synthesized and stored during oogenesis [29, 107, 195]. The observation that maximum genome activity occurs during early diplotene of oogenesis in Urodeles when lampbrush chromosomes are fully developed suggests that most of this mRNA may be synthesized at this time [66, 91], although incorporation of RNA precursors into lampbrush loops persists for some time after the maximum phase of activity [66, 91]. This conclusion was also supported by early work using various techniques, such as template activity in cell free protein synthesis and nucleic acid hybridization to SC DNA, to assay for the appearance of informational RNA during oogenesis in *Xenopus* (for review of early literature, see [66]). These studies also suggested that there was no accumulation of either template active or new SC DNA sequence transcript after the lampbrush stage [66].

There is evidence for sea urchin oogenesis [106], which suggests that heterogeneous non-ribosomal RNA is synthesized at least three months prior to ovulation and that synthesis continues at least until the last few days. There is no reason to suppose that this RNA is all mRNA but at least there is a good possibility that some of it is. If this is the case it is likely that there are differences between phylogenetically distinct groups in the timing of information RNA synthesis in oogenesis.

The fact that most mRNA molecules are known to contain a 3' poly(A) sequence which may be used both for quantitation and purification of mRNA has allowed a study of the appearance of poly(A) containing RNA in oogenesis [207]. It was shown that accumulation of poly(A) containing RNA during oogenesis in *Xenopus* occurs during pre- and early vitellogenesis, coinciding with the period of maximum lampbrush chromosome activity. As predicted from earlier work no accumulation was detected during the later vitellogenic stages. Of course estimation of total poly(A) containing RNA does not exclude the possibility that substantial turnover of these molecules occurs and that a steady state in which synthesis and degradation are equalized is established during later stages of oogenesis.

Poly(A) containing RNA from *Xenopus* oocytes has a mean sedimentation value of 18S (*ca.* 2000 nucleotides) but is very heterogeneous with some molecules sedimenting at greater than 40S. The poly(A) fragment is about 60–70 nucleotides long which is approximately half the size of poly(A) from *Xenopus* tissue culture cells. It is not known whether this size difference has any developmental importance but a similar effect is observed in sea urchin eggs [226]. The poly(A) containing RNA is template active in cell free protein synthesizing systems [134a] and gives rise to a large number of different polypeptides. This result agrees with the result that cDNA (produced by reverse transcription of purified oocyte poly(A) containing RNA) hybridizes almost' exclusively to SC DNA (similar results were also obtained for *Triturus* oocytes) and that total unique sequence transcript in mature *Xenopus* oocytes hybridizes to at least 0·9% of the *Xenopus* genome [68]. On the other hand poly(A) containing RNA, which most likely represents a subset of informational RNA [169a, 177a], is transcribed from only 0·3 − 0·6% of the *Xenopus* genome [208]. It is not clear whether the approximately two-fold difference in these figures, which were derived by completely different methods, is a real difference suggesting that some unique sequence transcripts are not present in poly(A) containing RNA, or whether it is fortuitous. The *Xenopus* genome contains $2·25 \times 10^9$ base pairs of SC DNA thus total unique sequence transcript in mature oocytes is sufficient to code for 10^4 different proteins with mRNA of 2000 nucleotides, while the poly(A) containing RNA contains about half this number. Although the *Triturus* genome is seven times the size of *Xenopus* it expresses only 0·1% as poly(A) containing RNA during oogenesis which will code for approximately the same number of 18S sized mRNA sequences as found in *Xenopus*. The conclusion from this is that the extra DNA found in *Triturus* has no function in coding for poly(A) containing RNA during oogenesis.

Complementary DNA (cDNA) may be hybridized back on to the template RNA from which it was synthesized. When this is done using an excess of template RNA to drive the reaction and a trace of radioactive cDNA to follow it, the reaction will proceed at a rate determined by the sequence complexity of the template and by the relative abundance of different sequences [14]. For instance if a population of mRNA molecules of roughly the same size is made up 50% of one species and 1% each of fifty species, reaction of cDNA made using this population (assuming all species have been copied in their correct proportions by the reverse transcriptase) will proceed 50 times faster with the most abundant than with the least abundant. The course of the reaction (Fig. 5.3.16) will follow a two-step pattern, the first step (transition) will represent the reaction of the most abundant species and the second transition the reaction of the least abundant mRNA species. A standard reaction system between a defined mRNA species of known sequence complexity (e.g. Hb α chain mRNA) and its cDNA enables the calculation of the numbers of different sequences reacting in the two transitions of the unknown reaction [14]. Assuming that all mRNAs are copied by reverse transcriptase with equal efficiency and that mRNA molecules are not internally redundant, then the plateau level of each transition gives the proportion of mRNA molecules in the population which form hybrids during the course of that transition. From measurements of this sort it is possible to calculate the numbers of different mRNA sequences and their frequency of occurrence in any mRNA population.

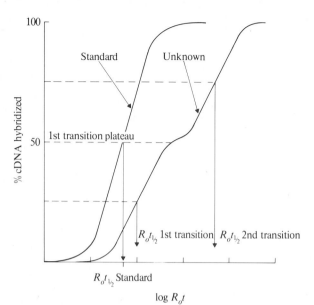

Fig. 5.3.16 Hybridization of copy DNA to its own mRNA template for a single purified standard mRNA and for a population of mRNAs, composed of 50% of one mRNA species and 50% of 50 species. See text, p. 323, for further explanation.

The reaction of *Xenopus* and *Triturus* poly(A) containing RNA with its homologous cDNA [208] suggests the existence of two classes of mRNA, an abundant class representing about 25% of mRNA molecules in 150–300 different kinds and a complex class representing 75% of mRNA molecules in 5–10,000 different kinds. Knowing that poly(A) containing RNA is about 0·5%–1% of total *Xenopus* oocyte RNA (i.e. 0·02–0·04 µg poly(A) containing RNA per oocyte) it is calculated that individual sequences of the abundant class are present in 2×10^7 copies per oocyte while the different sequences of the complex class are present in about 2×10^6 copies per oocyte. Assuming that all the mRNAs are transcribed from SC DNA (as indicated by the reaction of cDNA with DNA in excess [208]) and that the maximum rate of transcription (based on a polymerase packing of 30 per 2000 nucleotides and transcription time of 1 min per 2000 nucleotides) is 30 mRNA molecules per minute per gene then it will take 100 days to synthesize the abundant mRNA species and 10 days to synthesize the complex mRNA species on the four copies of each gene present in a meiotic cell. These figures fall well within the time scale of oogenesis in *Xenopus laevis* (about 1 year).

MESSENGER RNA IN DEVELOPMENT

Studies of mRNA synthesis during early stages of development are complicated by the presence of a large store of maternally inherited mRNA molecules [107]. Sequence complexity studies indicate that there may be of the order of 10,000 different mRNA sequences in *Xenopus* and *Triturus* eggs [208] and 14,000 in sea urchin gastrula [88]. Within this population, only two mRNAs of defined function have so far been identified, those for histone proteins [107, 224] and for microtubule protein [195]. These two mRNAs differ in that there is also new transcription of histone mRNA from very early cleavage stages [142a, b], but no new transcription of microtubule mRNA has been found at this time [195a]. Transcription of histone genes in early cleavage is the only example at present of the synthesis of mRNA with a defined function in early development. It must however be emphasized that the control of histone mRNA synthesis may be unusual in that it appears to have no nuclear precursor, has a very rapid exit time from the nucleus, and no 3^1OH poly(A) sequence [2], and histone synthesis is normally closely correlated with DNA synthesis in the cell cycle [38]. Other studies [128] in sea urchins using labelling of heterogeneous RNA from cytoplasmic polyribosomes, in the virtual absence of new rRNA synthesis, leads to the general conclusion that new mRNA is synthesized very early in development. How far this is generally true is not obvious at present except that there is evidence in mammals [213, 215] that new poly(A) containing RNA is synthesized by the 16-cell stage. Since it is well known from enucleation and RNA synthesis inhibition experiments that development in many organisms will proceed through cleavage to late blastula, the significance of new mRNA synthesis at this time is hard to judge. It would appear likely that some new mRNAs, like those for histones, are dispensable because they already exist in the maternal mRNA store in sufficient quantities for the needs of the embryo until blastula. Other new mRNAs may code for proteins which are dispensable because there is a sufficient store of the protein itself, and still others may be synthesized but not used until blastula or later stages.

In a situation where no rRNA synthesis takes place, it is possible to easily study mRNA synthesis. A study of the synthesis of heterogeneous RNA with DNA-like base composition during early development of both normal and anucleolate *Xenopus* embryos provides such a situation and has shown several interesting things [28, 32]. The anucleolate mutants synthesize no detectable rRNA and present a unique opportunity to study the synthesis of other high molecular weight RNAs in development. During ovulation it has been shown that at least 1 ng of heterogeneous high molecular weight RNA with a DNA-like base composition (dRNA) is synthesized [29]. The total amount of dRNA increases during cleavage relative to DNA content and then remains constant, that is it doubles with every doubling in the amount of DNA between gastrula and early swimming stages [32]. It was not clear from this study what proportion of dRNA is purely nuclear and how much is cytoplasmic. In normal embryos the measurements of new dRNA were obscured after late blastula stage by increasing amounts of new rRNA. A study of dRNA synthesis in anucleolate mutants overcame this difficulty and confirmed the earlier findings with normal embryos on accumulation of dRNA in *Xenopus* development; however, attempts to define its cellular location were unsuccessful. A further point did emerge from the study on anucleolate embryos suggesting that the larger dRNA molecules synthesized during gastrulation are fairly stable until tailbud at which point there is a loss of large dRNA (greater than 20S) and the appearance of small dRNA (10–20S). By making use of microinjected ^{3}H-guanosine, which has a relatively small intracellular pool in *Xenopus* embryos, effective pulse-chase conditions were obtained allowing the study of the fate of newly synthesized dRNA. Calculations indicated that if the large dRNA gives rise to the 10–20S component not more than 10% can do so and it takes many hours for completion. The small dRNA was found to be as stable as the new 4S RNA synthesized at the same time. These results are consistent with the interpretation that the large dRNA represents HnRNA and the small dRNA, cytoplasmic mRNA. If this is the correct interpretation it provides evidence for the synthesis and stabilization of mRNA precursor with processing to active mRNA at a much later time. It also points to a difference between HnRNA in sea urchin and amphibian embryos in terms of their half lives, sea urchin HnRNA turning over much more rapidly [21, 79].

Several early studies of the nucleic acid hybridization-competition type were done in an attempt to show that informational RNA synthesis changes in development (see, for review [66]). However these early measurements probably failed to discriminate the SC DNA transcripts and are likely to refer only to IR DNA transcripts. They did, however, succeed in showing that stage specific transcription of IR DNA occurs. It is not possible to say from these experiments how they relate if at all to the synthesis of HnRNA or mRNA.

MESSENGER RNA SYNTHESIS IN TERMINAL CELL DIFFERENTIATION

A considerable number (Table 5.3.7) of mRNA molecules coding for specific proteins have been isolated from terminally differentiated cells, that is cells which have become committed to the synthesis of a large amount of one or a few polypeptides. They may, as in the case of mammalian reticulocytes, lack a nucleus and be incapable of further cell division, and are generally thought to to be incapable of further differentiation to another cell type.

There are five general points about studies on the synthesis of these mRNAs (for review, see [210a]). The first is that they are found only in the cytoplasm of one cell type. It may be that this is not absolutely valid and that more sophisticated nucleic acid hybridization and cell fractionation techniques will be able to set an upper limit on say the number of globin mRNAs found in

Table 5.3.7   Specific messenger RNA molecules from terminally differentiated cells.

Protein	Cell type	Reference
Globin	Reticulocyte	[14a], [121], [194]
Silk fibroin	Silk gland	[239]
Ovalbumin	Oviduct	[238]
Crystallin	Lens	[9a]
Myosin Actin	Muscle	[181]
Immunoglobulin light and heavy chains	Myeloma	[173], [194], [237]

brain or kidney cell cytoplasm. Secondly one would like to be able to conclude, but cannot at present through lack of evidence, that they also do not occur in nuclear RNA of cells other than their own end cell. It is still an open question whether for instance the globin gene sequences are transcribed in brain or kidney cell nuclei and are subsequently eliminated by processing before entering the cytoplasm. This is a hard question to answer, but it may be possible to put an upper limit on the number of globin sequences present in the nuclear RNA of other cells by hybridization of globin cDNA to nuclear RNA. Thirdly it may be generally true that these end cell specific mRNAs are synthesized initially during a period of cell differentiation and are then stored and are not immediately translated [210a]. For instance during reticulocyte maturation [184, 263], pancreas differentiation [257], lens cell differentiation [179], and enzyme formation in early development (see p. 336 [258]), such storage may occur, and synthesis of these proteins becomes resistant to RNA synthesis inhibitors, such as Actinomycin D, some time before the proteins are actually produced. Great care must be taken in the interpretation of such experiments since the Actinomycin D may not eliminate the synthesis of mRNA, and it has recently been shown that although the bulk of RNA synthesis in slime moulds is inhibited by this drug nevertheless some mRNA synthesis continues [85].

Fourthly mRNA specific to terminal cell differentiation appears to be unusually stable [142]. For instance the half life of the message for cocoonase appears to be 100 h compared to 2–3 h for other cell mRNAs [140]. Finally there is now evidence which excludes the possibility that the control of their synthesis involves a process of specific gene amplification [14, 238, 239]. Indeed it has been calculated [141], given their stability and making reasonable assumptions of transcription time and polymerase loading, that one gene copy per a genome is sufficient to allow for the synthesis of the amounts of mRNA necessary to support the rates of protein synthesis observed in these terminally differentiated cells and therefore the requirement for specific gene amplification is eliminated.

Ribosomal RNA synthesis in development and differentiation

RIBOSOMAL RNA SYNTHESIS IN OOGENESIS

Cytochemical and autoradiographic techniques have shown that oocytes of a wide variety of plants and animals accumulate large quantities of RNA [66]. The RNA is usually the product of oocyte genome activity though in some orders of insects a number of nurse cells is responsible for synthesis of the major portion of oocyte RNA which enters the oocyte via intercellular cytoplasmic bridges. Although oocytes of all organisms are also associated with somatic cells (follicle cells) which are active in RNA synthesis there is no evidence to suggest that they make any contribution to the complement of oocyte RNA. In fact there is functional evidence in *Xenopus* against the idea that there is any effect of the maternal or paternal gonadal environment on the developmental or differentiation potential of germ cells [16].

The major component of oocyte RNA is ribosomal RNA [30, 100]. The amount of rRNA present in the mature egg is highly variable from one phylogenetic group to another, but is roughly correlated with the size of the egg and with the length of development before the embryo is able to obtain nutrition from its environment. For instance mammals have small eggs with little rRNA and obtain nutrients from the mother from the one cell stage onwards, whereas amphibians with large amounts of rRNA have large eggs and develop for several days before feeding. Clearly the rRNA synthetic capacity required by the oocytes of different phylogenetic groups must vary. In many organisms in which the demand for ribosomes during oogenesis is great, it has been shown that the genes for rRNA are selectively amplified in the oocyte ([26] Differential synthesis of ribosomal RNA genes, p. 328). It has often been argued that ribosomal gene amplification alleviates a potential rate limiting step in rRNA synthesis by providing more gene copies. Clearly it is important to know how and when these extra genes become active in rRNA synthesis. One approach to this problem is to measure the absolute rate of rRNA synthesis at different stages of oogenesis after rDNA amplification. Such experiments, which involve the use of radioactive RNA precursors and the estimation of specific activities of intracellular precursor pools, have recently been done for two stages of *Xenopus* oocyte development [151]. The results suggest that although the total incorporation of ^{3}H guanosine into RNA is much greater in stage 4 (vitellogenic) oocytes than in stage 6 (full grown) oocytes, yet when allowance is made for differences in specific activity of precursor pools, it turns out that the stage 6 oocytes make more RNA (1·1–1·6 ng/oocyte/h) than the stage 4 oocytes (0·73–1·39 ng/oocyte/h). These are rates of synthesis which would allow accumulation of the complete stage 6 RNA content (4·5 μg) in 3–5 months assuming complete stability of the rRNA. This result is unexpected since it was generally thought that the full grown *Xenopus* oocyte is much less active

in RNA synthesis than the stage 4 oocyte [66, 212], and emphasizes the caution required when interpreting incorporation data in the absence of knowledge about the specific activity of precursor pools. Estimates of total RNA in full grown oocytes [29, 30, 82a, 212] have shown that there is no accumulation of RNA above about 4·5 μg RNA per oocyte. Since RNA synthesis is continuing, one is forced to conclude that either degradation of RNA is occurring at the same rate so that there is no net synthesis, or that the techniques of *in vitro* culture used for the experiment have caused a reactivation of RNA synthesis normally quiescent *in vivo*, or that the specific activities of the extractable precursor pools are not good estimates of the specific activities of the pools used for RNA synthesis in these oocytes.

A different approach to the same problem has been to follow the accumulation of RNA during the growth of oocytes after partial ovariectomy and removal of all the vitellogenic oocytes [212]. This study shows that 4 μg of RNA may be accumulated within a period of 40 days, which is a rate (4·1 ng/oocyte/h) of accumulation 3–5 times that calculated from the incorporation data [151]. Further evidence is also available which supports the idea of a change in the rate of RNA accumulation during oogenesis in *Xenopus* at about the onset of vitellogenesis [82a]. During the first 10–15 days after the operation, RNA is accumulated at a rate of 0·7 ng/oocyte/h, while over the next 25 days RNA accumulation is at 6·7 ng/oocyte/h representing a 10-fold increase in rate. The conclusion from this is that full use of the amplified rRNA genes is not made immediately after amplification but only later and suggests a gradual recruitment of genes for rRNA synthesis during the transition from pre-vitellogenic to vitellogenic stages.

Cytological and autoradiographic observations show that nucleoli which incorporate labelled RNA precursors are present both prior to and after amplification. However the densely staining rDNA cap characteristic of late pachytene nuclei only gradually disperses and the full complement of about 1500 nucleoli is only reached at the beginning of vitellogenesis [157a], consistent with the view that the amplified genes are not all immediately transcribed.

Amplification, though necessary to support the high rate of rRNA synthesis during vitellogenesis, is clearly not sufficient to cause its onset. Other factors must be involved, such as the dispersion of the condensed pachytene rDNA cap and the assembly of all the proteins and enzyme systems necessary for nucleolar function.

THE SYNTHESIS OF rRNA DURING EARLY DEVELOPMENT

Post-fertilization development in most organisms is characterized by a period of intense DNA synthesis and rapid cell division. During this time many quantitative and qualitative changes in genetic activity have been recorded. One of the first descriptions of such a change concerned the regulation of rRNA synthesis [27, 43]. Onset of detectable rRNA synthesis occurs at the late blastula stage in amphibians [29, 150], echinoderms [79, 217] and several other chordate and invertebrate phyla [66, 101, 216]. Mammals appear exceptional in this respect as new rRNA synthesis begins at the 2–4 cell stage [144], which may be related to the small size of the egg, its low ribosome content, and the early demand for protein synthesis. New ribosome synthesis is correlated with the appearance of nucleoli, for example nucleoli appear at 2–4 cell stage in mammals and late blastula in amphibians and echinoderms. Appearance of nucleoli may be used as a guide to the likely developmental stage of rRNA gene activation in organisms which have not been analysed biochemically. For example, many molluscs show nucleoli during cleavage which may be correlated, as in mammals, with small eggs and rapid development, compared with other molluscs with larger eggs and slower development in which nucleoli do not appear until blastula [66]. Plants also show a lag after the onset of germination before new ribosome synthesis is observed [43]. Although the exact point in development at which new ribosome synthesis starts is variable from one organism to another it does seem to be generally true that at least a few cell divisions occur without detectable rRNA synthesis.

A careful recent study has argued that the onset of rRNA synthesis at blastula stage in sea urchins is more apparent than real [79]. Measurements of absolute amounts of rRNA and dRNA synthesized during early sea urchin development, suggests that the amount of rRNA synthesized per nucleus does not change from the time it is first detected (blastula in this study). However there is a great excess (10–100-fold) in the amount of DNA-like RNA over rRNA synthesized per nucleus during cleavage and blastula stages. By the pluteus stage the rates of dRNA and rRNA synthesis are approximately equal. It was suggested that the small amount of rRNA which might be synthesized during cleavage will be rendered undetectable by the 100-fold excess of dRNA also synthesized at this time. However an increase of rRNA synthesis per nucleus has been seen in another species (*Paracentrotus*) of sea urchin [217] and in *Xenopus* [150] after gastrulation. There is thus still considerable uncertainty about whether the control of rRNA synthesis in early development is an example of a qualitative mechanism (i.e. some stages without any rRNA transcription and some other stages with rRNA transcription) or whether it is a quantitative control, with early stages making less rRNA per nucleus than later stages. If it is the case that all stages make rRNA, then clearly rRNA synthesis is uncoupled from nucleolus formation. Rapid cell division (every hour in sea urchins, and every 20–30 min in *Xenopus*) may place a severe restriction on nucleolus formation without affecting transcription of ribosomal genes. A further point is that if the synthesis of rRNA is

regulated during the cell cycle then restrictions may also be placed on rRNA synthesis in early development. Evidence suggests that only 4S and 5S RNA are made during mitosis in tissue culture cells [266], and if the same is true for embryonic cells and since mitosis occupies at least 50–60% of the total cycle time in these cells, then at least half the time between fertilization and blastula stage is unavailable for rRNA synthesis. Therefore, it is expected that the rate of rRNA synthesis per nucleus during cleavage would be reduced two-fold on this basis alone. Similar consideration can be given to the processing of rRNA, which takes about 20 min for 18S and 50 min for 28S RNA in Hela cells, if processing is interrupted during mitosis. These considerations may provide an explanation for the observed changes in rates of rRNA synthesis per nucleus in early development without providing any grounds for believing rRNA synthesis is controlled in any special way during this period.

EFFECT OF GENE DOSAGE ON THE RATE OF rRNA SYNTHESIS IN EARLY DEVELOPMENT

Wild type populations of the two amphibians (*Bufo marinus* [9] and *Xenopus laevis* [171, 172]) are known to be composed of individuals with differing numbers of rRNA genes. Wild type (+/+nu) *Xenopus* have 900 to 1200 rRNA genes per diploid nucleus, and two large nucleoli; the anucleolate mutant (o/o nu) which is a recessive lethal has less than 5% of this number and no true nucleoli [27]. A second nucleolar mutant of *Xenopus* has only a partial nucleolus [172], contains only 25% the wild type number of rRNA genes and synthesizes rRNA at 25%–40% of the normal rate. Such a mutant of genotype p^1/o nu is made by crossing a heterozygote for the partial nucleolar mutation (p^1/+ nu) with an animal having only one nucleolar organizer (+/o nu). The four types of offspring all distinguishable by their nucleolar cytology are +/+, +/o, +/p^1 and p^1/o nu. The first three genotypes all synthesize rRNA at the same rate and develop normally, while at all stages of development the p^1/o nu embryos show reduced rRNA synthesis, which is not due to an error in processing of precursor rRNA [145]. There are similarities between these partial deletions of rDNA in *Xenopus* and the bobbed mutants of *Drosophila melanogaster*, which are known to contain less than the normal amounts of rRNA genes [198] and in which it has also been proposed that the reduction in rRNA synthesis is responsible for lower growth rates and lethality. It is instructive to examine the average rate of rRNA synthesis per a gene in these various *Xenopus* nucleolar genotypes (Table 15.3.8). There is a two-fold variation in rate which clearly means that either the rate of transcription per gene may change or that the number of active genes is altered in each genotype. Only when the number of genes is less than half the normal wild type number is rRNA synthesis reduced to an extent incompatible with development beyond the swimming tadpole stage.

Table 5.3.8 Average rate of rRNA synthesis per gene in various nucleolar genotypes of *Xenopus laevis*. [Ref. 145]

	Nucleolar Genotype				
	+/+	+/p^1	+/o	p^1/o	o/o
Relative amount of rDNA	100	60	45	23	6
Relative rate of rRNA synthesis	100	100	100	25–40	<2
Relative rate of rRNA synthesis per gene	100	166	222	112–175	<33

DIFFERENTIAL SYNTHESIS OF RIBOSOMAL RNA GENES

It is evident that over or under replication of limited parts of the genome might provide a general mechanism for controlling differentiation. The rRNA genes provide a particularly good system for studying this problem because they are relatively easily assayed.

Over-replication

One situation in which rRNA genes would be rate limiting for rRNA synthesis if special mechanisms had not been evolved to specifically amplify them is in oogenesis. In *Xenopus* for instance it can be calculated that it would take about 100 years to make the required number (10^{12}) of ribosomes on the 2000 genes which would be present in the meiotic nucleus if amplification did not occur, and similar calculations may be applied to a wide variety of other organisms. Although the phenomenon of gene amplification was first described and is best understood in *Xenopus laevis* [89] it is widespread occurring in some insects, echiuroid worms, molluscs, crustaceans and echinoderms, though probably not in mammals [26]. Amplification is often associated (amphibians, echinoderms) with the production of supernumerary extra-chromosomal nucleoli, but this is not obligatory since both *Urechis* and *Spisula* oocytes have only single very large nucleolus [26]. It is not clear whether this reflects a fundamental difference in the nature of the amplified genes related to whether or not they behave as episomes on the one hand or as an integral part of the chromosome on the other, or if it has a trivial explanation based on the fusion of nucleoli.

Under-replication

Independent control of rRNA gene number also occurs in polytene chromosomes of *Drosophila melanogaster* [234]. During polytenization of salivary gland chromosomes, homologous chromosomes pair and the euchromatin undergoes nine rounds of replication resulting in the formation of 1024 fibres (chromatids). Heterochromatin does not replicate during this time but remains condensed in the chromocentre formed by the association of heterochromatin from all chromosomes [91a]. In *Drosophila* ribosomal RNA genes are localized in the heterochromatin of X and Y chromosomes, but are nevertheless active in transcription unlike the rest of the heterochromatin. Analysis of the amount of DNA complementary to rRNA in both diploid tissue (brain and imaginal discs) and polytene tissue (salivary gland) shows that relative to total DNA the rRNA genes are six-fold under-replicated in polytene cells, but are still represented by 150–200 times the normal diploid content [234]. This means that during polytenization while the euchromatin replicates through nine doublings, the rDNA replicates through only six or seven, and the heterochromatin containing predominantly simple sequence highly reiterated DNA is not replicated at all [91a]. Such results strongly suggest that the replication of these three types of DNA sequence is under independent control.

rRNA gene dosage compensation

In a situation where an organism has only one of a pair of nucleolar organizers, it is possible to ask whether rDNA is amplified to compensate for the absence of one organizer. There are mutants of both *Xenopus* and *Drosophila* which allow this to be tested. In *Xenopus* (see Effect of gene dosage on the rate of rRNA synthesis in early development, p. 327) when two animals of genotype +/o nu are mated the progeny are of three kinds +/+, +/o and o/o nu distinguishable by having two, one and no nucleoli. During amplification of rDNA the +/o nu embryos produce as much rDNA as the +/+ nu animals [187]. In *Drosophila* a similar situation can be engineered by producing animals with either one or two X chromosomes. Diploid cells of XX individuals contain twice the amount of rDNA as diploid cells of XO individuals. However during polytenization the amount of rDNA in both genotypes becomes the same [234]. These results suggest that only one nucleolar organizer is involved in amplification, or amplification is completely independent of chromosomal rDNA (for instance, an episome) or that the amplification mechanism is sensitive to the final absolute amount of rDNA and thus independent of the amount of rDNA present prior to amplification.

The mechanism of rDNA amplification in Xenopus laevis

Although as we have seen amplification of rDNA is phylogenetically widespread [26], studies of the molecular mechanisms involved are largely confined to *Xenopus*. Detailed knowledge of such mechanisms are of general interest to developmental biologists since they provide the means for specifically replicating a limited portion of the genome. It has often been suggested that such mechanisms with other specificities might play an important role in the differentiation of other cell types.

In *Xenopus* rDNA amplification occurs at or about metamorphosis, during the pachytene stage of oocyte meiosis. Limited amplification also occurs at the same stage in spermatogenesis [10]. Amplification takes about three weeks to complete [10] and during that time 75% of the oocyte's nuclear DNA becomes rDNA, representing a 4000-fold increase over the haploid amount of rDNA. The amplified rDNA is extra chromosomal and during diplotene disperses into thousands of nucleoli which actively engage in rRNA synthesis. We can define three major problems concerning the mechanism of rDNA amplification:

(1) Is amplification dependent on the inheritance of non-chromosomal rRNA genes (that is of a rDNA episome) [25]?
(2) If amplification is dependent on chromosomal genes (that is on a chromosome copy mechanism) what is the mechanism which produces the first extrachromosomal copy?
(3) What is the mechanism involved in replicating extrachromosomal rDNA?

rDNA amplification is by a chromosome copy mechanism. The chromosome copy mechanism predicts chromosomal inheritance for rDNA amplification, but the episome hypothesis predicts maternal inheritance since it is specifically postulated that the amplified rDNA of the egg is sequestered by the primordial germ cells in early development and provides the template for rDNA amplification in the next generation [25]. In a situation where it is possible to distinguish maternal and paternal rDNA contributions it will be clear whether maternal inheritance is occurring. Such a situation is presented by interspecies hybrids of *Xenopus laevis* and *X. mulleri* [25]. A study using nucleic acid hybridization and CsCl buoyant density techniques to distinguish *laevis* and *mulleri* rDNA amplification in the ovaries of hybrid females has shown that only the *laevis* rDNA is amplified regardless of whether *laevis* was the maternal or paternal parent and in spite of the fact that

both *laevis* and *mulleri* rDNA is present in somatic cells. Such a result is clearly incompatible with the episome model as stated above, however it does not distinguish another form of episome model which states that all cells contain a small rDNA episome for use in rDNA amplification during oogenesis and does not require maternal inheritance.

Amplification of *mulleri* rDNA is clearly suppressed in the presence of *laevis* DNA although it is not obvious at present whether this is due specifically to the presence of *laevis* rDNA or to some other component of the *laevis* genome. In principle this can be distinguished in hybrids between *mulleri* (m^+/m^+) and *laevis* ($1^+/1^-$) with one nucleolus since half the progeny will have only one *mulleri* nucleolus and be lacking any *laevis* rDNA ($m^+/1^-$). If amplification of *mulleri* rDNA proceeds in females of this genotype a direct effect of *laevis* rDNA in suppression of *mulleri* rDNA amplification is suggested.

Origin of the first non-chromosomal rDNA copy. As we have seen it is likely that amplification proceeds by a chromosome copy mechanism, and we are faced with the problem of the origin of the first extra-chromosomal rDNA copy. There are three solutions to this problem:

(1) Crossing over within the rRNA gene cluster would release a circular DNA molecule containing from one to most of the rRNA genes (Fig. 5.3.17a). One consequence of this mechanism is that the chromosome from which the circle is derived is now deficient in rRNA genes and these would presumably have to be replaced by crossing over between an amplified circle and the chromosome at a later time [36a].
(2) Semiconservative DNA replication of a part or the whole of a rDNA locus followed by excision at specific points and circularization by crossing over or ligation (Fig. 5.3.17b).
(3) An RNA transcript of the whole rRNA gene including the normally non-transcribed spacer DNA sequence is copied by an RNA dependent DNA polymerase (reverse transcriptase) into DNA which may then circularize by crossing over or ligation (Fig. 5.3.17c).

The last two models involve mechanisms which do not destroy the chromosomal rDNA locus like the first and so do not require reintroduction of rDNA later in oogenesis.

Discovery of RNA dependent DNA polymerases in certain animal RNA viruses and in RNA virus infected cells [240] laid the foundation for the third hypothesis given above. A firm prediction of this hypothesis is the synthesis of a complete RNA transcript of rDNA containing both the normal transcription unit (40S, $2 \cdot 5 \times 10^6$ daltons MW) and the normally non-transcribed spacer, such an RNA molecule would have a molecular weight of $4 \cdot 3 \times 10^6$ daltons (47S). This prediction follows from the knowledge that the amplified rDNA is indistinguishable from chromosomal rDNA on the basis of the length of its repeating unit [61]. Evidence has been provided in

Model 1

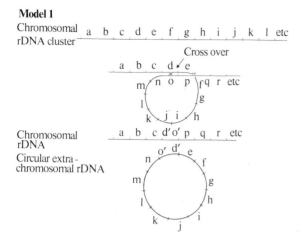

Model 2

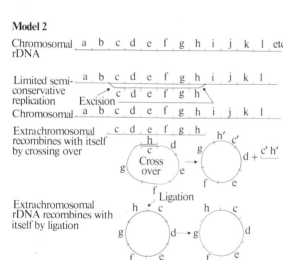

Model 3

Chromosomal rDNA gives full repeat size RNA transcript, which gives full repeat size DNA by reverse transcription, which circularizes by ligation.

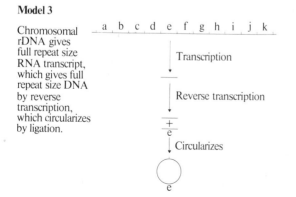

Model 4

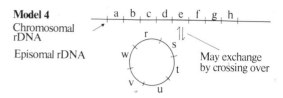

Fig. 5.3.17 Four possible models for the origin and replication of extra-chromosomal ribosomal DNA for amplification during oogenesis. (a) Crossover within rDNA produces extra-chromosomal circular DNA which replicates by a rolling circle mechanism. (b) Limited semi-conservative replication followed by excision and circularization by crossing over or ligation. (c) Reverse transcription of a full repeat size RNA transcript. (d) Episomal inheritance of ribosomal DNA with rolling circle replication during oogenesis.

amplifying ovaries for the existence of a 47S RNA with sequences complementary to the normally non-transcribed spacer [61]. Further predictions of the model such as the presence of an RNA/DNA complex as intermediate in the synthesis of rDNA, the specific inhibition of rDNA synthesis but not chromosomal DNA synthesis by a rifampicin derivative known to inhibit viral reverse transcriptase, and the isolation of a rifampicin sensitive RNA dependent DNA polymerase

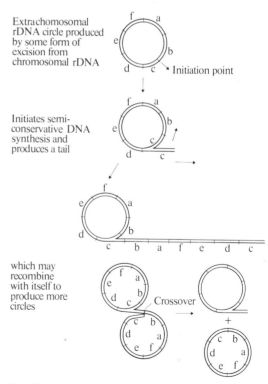

Fig. 5.3.18 The rolling circle mechanism for ribosomal DNA amplification.

activity from amplifying ovaries were all met [35]. Such an overwhelming agreement between the predictions of the model and experimental observation might have led to its acceptance were it not for the fact that the same experiments repeated using techniques of greater sophistication and refinement failed to confirm several of them [11]. It is difficult to reconcile the two sets of experiments but the differences may be due to analysis of slightly different stages in the amplification process occurring at different times in oogenesis. If this is so then it may be that production of a few rDNA molecules by reverse transcription of a complete RNA copy are the first events in amplification. This stage is followed by a period of DNA dependent DNA synthesis templated by the free rDNA formed by reverse transcription.

Such a suggestion is also compatible with recent evidence that amplification proceeds by way of a rolling circle intermediate (Fig. 5.3.18) [126, 201]. Electron microscopy of rDNA isolated from amplifying ovaries shows that although most molecules are linear 2–5% are circles or rolling circles (i.e. circles with tails). Partial denaturation mapping proved that both circles and rolling circles contained a repeating unit indistinguishable from rDNA [126]. By using radioactive thymidine and deoxycytidine to label replicating DNA prior to isolation it was shown that rolling circles do indeed contain newly replicated DNA [201]. It seems that rDNA appears exponentially during amplification [157a] suggesting that newly synthesized rDNA will act as a template for further synthesis.

rRNA SYNTHESIS IN DIFFERENTIATION

The control of rRNA synthesis in various fully differentiated or terminally differentiating tissues has been studied but it is not well understood (for reviews, see [59, 159]). For instance in normally non-dividing fully differentiated tissues such as liver or kidney the rate of ribosome precursor synthesis appears to be several times in excess of that required by the known rate of ribosome degradation if a steady state is to be maintained [45]. In mammalian liver the half life of ribosomes is about 100 h [45, 195]. Turnover of ribosomes appears to be related to the protein synthetic capacity of the particular cell type under investigation [159]. In liver with a high rate of protein synthesis per cell the ribosome turnover is quite high but turnover is lower in the kidney which also has a lower rate of protein synthesis [159]. In non-dividing cells, such as unstimulated lymphocytes [56], or in cells moving from a rapidly dividing phase into a non-dividing phase such as lens epithelial cells [179] giving rise to stationary phase fibre cells, or proliferating myoblasts fusing to give non-dividing myotubes [53], there appears to be an over-production of rRNA precursor and consequent wastage of rRNA during processing. In lymphocytes stimulated by phytohaemaglutinin [56] this wastage is abolished and establishes a higher rate of ribosome formation. Similarly in regenerating liver [45], or in a

kidney compensating by growth for the experimental removal of the other [1], there is a dramatic increase in the rate of rRNA synthesis which probably involves increasing the number of ribosomes being made at one time rather than decreasing the time of synthesis.

The synthesis of 5S RNA in development and differentiation

Since 5S RNA is found in equal molar amount with 18S and 28S rRNA in ribosomes, it is reasonable to suppose that these molecules will be synthesized in equimolar amounts. It is clear that 18S and 28S rRNA are usually synthesized in equimolar amounts because they share a common precursor; in situations in which they do not appear to be synthesized in this way (see previous page), it is possible to account for the discrepancy by the different stability of the two molecules. Not only does 5S RNA not share a common precursor with 18S and 28S rRNA but it is synthesized on separate genes which are usually not closely linked to the main ribosomal RNA genes. Early experiments suggested that 5S RNA is synthesized coordinately with 18S and 28S rRNA during oogenesis, development and differentiation [31]. This coupling of synthesis could not be based on equal numbers of 5S and 18S and 28S rRNA genes since in every eukaryote which has been studied so far the 5S genes are in excess of the other rRNA genes (Table 5.3.2).

It is now established that separate RNA polymerase molecules are responsible for 5S and 18S and 28S RNA synthesis, and the mechanism of coordination is not based on common transcriptional controls but must presumably involve some more subtle and as yet unknown relationship between 5S and 18S and 28S RNA synthesis [253].

In tissue culture cells it is possible to uncouple 5S RNA from 18 and 28S RNA synthesis by treatment with low doses of Actinomycin D, which inhibits the synthesis of the larger rRNAs but allows synthesis of 5S to continue [188]. A similar situation occurs naturally during mitosis when 5S synthesis continues at a reduced rate but 18 and 28S synthesis ceases [266]. Another natural situation in which 5S synthesis is uncoupled from synthesis of the larger rRNAs is in oogenesis of amphibia [83, 162]. In Xenopus for example 5S RNA accumulates during previtellogenesis, a period of oogenesis after rDNA amplification but before vitellogenesis, when maximal rRNA synthesis starts. These three instances of uncoupling clearly show that continued 5S synthesis is not dependent on continued 18 and 28S synthesis and is taken to indicate that the coupling normally seen is not related to an effect of rRNA synthesis on 5S synthesis. It is clearly of interest to try to find a situation in which synthesis of rRNA continues in the absence of 5S RNA synthesis but so far none is known.

The synthesis of 5S RNA during oogenesis in Xenopus presents several other interesting features. The amount of 5S

RNA accumulated by early vitellogenesis is approximately half the requirement of the fully grown oocyte, and represents about 45% of the total RNA of previtellogenic oocytes. The 5S genes are not amplified like the rRNA genes but there are about 80,000 of them (4 × haploid value of 20,000) [26] and since after amplification the oocyte has some 2,000,000 rRNA genes, this represents a 25-fold excess over 5S gene, the reverse of the situation in somatic cells where there is a 40-fold excess of 5S genes. RNA sequencing studies on the 5S RNA accumulated by oocytes shows that the genes are heterogeneous with respect to sequence [84, 249].

Comparison with the sequence of 5S RNA accumulated by somatic cells suggests that most of the 5S genes expressed in oocytes are inactive in somatic cells [84]. The mechanism of this differential expression of similar but not identical genes is clearly of importance in understanding control of gene activity but so far very little is known about it.

Recent experiments have shown that in *Xenopus* 5S RNA synthesis is first detected in early development at the blastula stage; this is at a time before new 18S and 28S rRNA synthesis is detectable and suggests that these molecules are not synthesized coordinately at this time [170a]. 5S RNA synthesis also occurs during the development of the anucleolate mutant of *Xenopus* and this provides a further example of the uncoupling of the synthesis of these two types of RNA [170a].

The synthesis of tRNA in development and differentiation

THE SYNTHESIS OF tRNA IN OOGENESIS

Synthesis of tRNA is usually related to a cell's capacity for protein synthesis and there are normally 15–20 molecules of tRNA per ribosome. Oocytes and early embryos of a wide variety of organisms (except perhaps mammals) show low levels of protein synthesis relative to their ribosome content and have of the order of 1–2 tRNA molecules per ribosome [31]. An examination of the accumulation of tRNA during oogenesis in *Xenopus* has shown that it parallels 5S rRNA synthesis, and it is tempting to suppose that this may be due to the fact that although they are synthesized by distinct polymerases [165] they are both synthesized by polymerases of type III [253]. About half of the tRNA required by the full grown *Xenopus* oocyte is synthesized during the previtellogenic phase and at this time represents about 45% of the total oocyte RNA. The tRNAs synthesized at this time may be distinguished from those found in a *Xenopus* tissue culture cell line, but whether these differences are due to secondary modifications or to different primary sequences has yet to be determined [83a]. In pre-vitellogenic oocytes transfer RNA accumulates in the cytoplasm in association with 5S RNA and specific proteins forming a particle which sediments at 42S [83], less than 5% of

tRNA is in the soluble fraction at this stage. The molar ratio of 5S to tRNA in this particle is 1:3 and it contains about 75% protein and 25% RNA. The particle does not persist in oogenesis and in the fully grown oocyte all the tRNA is found in the soluble fraction. Although the details of tRNA synthesis during oogenesis may vary from organism to organism it is clear that in *Xenopus* it is not related to the cell's demands for protein synthesis and is closely correlated with 5S RNA synthesis.

SYNTHESIS OF tRNA IN DEVELOPMENT

Synthesis of tRNA in development stands in contrast to the synthesis of rRNAs and comparison of the two provides one of the clearest examples of the differential control of the rate of gene transcription. The tRNA content of most cells is constant with respect to DNA content and in actively growing cells there are about 20 tRNA molecules per ribosome. We have seen that in oogenesis in *Xenopus*, tRNA accumulates during the pre-vitellogenic phase in the absence of DNA or rRNA synthesis to give tRNA to ribosome ratios of 100–200 [83]. By the end of oogenesis the ratio has fallen to about 1 or 2 to 1 as the direct result of increased rRNA synthesis during vitellogenesis. Thus the embryo inherits a low tRNA to ribosome ratio, but a very high content relative to DNA. One might predict that if tRNA to ribosome ratio was important for controlling tRNA synthesis that it would start immediately at fertilization to try to redress the balance. If on the other hand tRNA to DNA ratio was important tRNA synthesis would not start until later in development when tRNA to DNA ratio has fallen as a direct consequence of rapid DNA synthesis during cleavage. In fact the latter case is seen in *Xenopus* development. Transfer RNA synthesis is first detected in late cleavage and increases in rate relative to DNA synthesis until late neurula after which it declines to reach a constant level by hatching. In contrast rRNA synthesis is first detected in late blastula, increases only gradually relative to DNA synthesis during gastrulation and plateaus at a constant high level after the embryo hatches. The net result of these changing patterns of gene activity is that the amount of tRNA relative to DNA decreases rapidly during cleavage until gastrulation and from then on the embryo doubles its tRNA content for every doubling in the amount of DNA. By the hatching stage the number of tRNA molecules per ribosome has increased to 15 since the total content of ribosomes increases only slightly during the same period.

Such a detailed description of tRNA synthesis does not appear to exist for any other organism, but it is probably generally true. Mammals may be an exception since tRNA precursor is first synthesized at the two cell stage but may not be processed until the four cell stage when the first new 4S synthesis is detected [144]. Since rRNA synthesis also starts at the two cell stage in mammals there is no temporal separation of tRNA and rRNA gene activation.

Interest in the synthesis of tRNA during differentiation stems from two general points (for review, see [155]). The first is that both the rate and type of protein synthesis on a particular population of mRNAs may be profoundly altered by changes in the levels of different tRNAs or their isoacceptors, and secondly it has been shown that many tRNAs contain substituted purines (called cytokinins) which are known to be plant hormones which have considerable influence on basic developmental events such as cell division and differentiation [116]. The existing literature is very large [for review and references, see [155]) and the general points will be illustrated by only a few of the numerous examples.

Terminally differentiating cells committed to the synthesis of large amounts of one or a few proteins, which may have unusual amino acid compositions, in general adapt their tRNA populations to the amino acid composition of these proteins. For instance the silk glands of the moth *Bombyx mori* produces the silk protein, containing large amounts of glycine and alanine, and shows elevated levels of glycyl and alanyl tRNAs. During embryogenesis and differentiation in several plant and animal systems there are descriptions of changes in the patterns of isoaccepting tRNAs. Hormones will alter the pattern of tRNAs seen in their target organs. Malignant cells and cells infected with virus also show differences in tRNA patterns relative to normal or uninfected cells. Although the evidence correlating changing patterns of tRNAs with particular differentiating systems is overwhelming it is not entirely clear in any system whether this is a cause or an effect of the differentiated property.

5.3.7 TRANSLATION

Translation of mRNA is the second main process in the expression of genetic information and it has three major steps: (i) initiation, (ii) elongation, and (iii) termination. The general features of these steps are quite well understood (for recent review, see [122]) but important details are still being worked out [63, 132]. For the present discussion it is worth distinguishing between a rate limiting step during initiation and a rate limiting step during either elongation or termination since it is likely, by analogy with other control systems, that control of translation will operate at the first step in a sequence of events. If elongation or termination are the rate limiting steps in protein synthesis mRNA will become maximally packed with ribosomes, since the instant an initiation site becomes vacant another initiation event will occur. If on the other hand initiation

is rate limiting the packing of ribosomes on mRNA will be inversely related to the rate of elongation and termination. High rates of elongation will lead to small numbers of ribosomes per unit length of message and vice versa.

Evidence for translational control of gene expression

We have seen that gene expression may be most efficiently controlled by regulation of transcription, however since in eukaryotes translation takes place at a site remote from the site of transcription it may quite often play a modulating role in regulating the amount of particular proteins finally produced. There may be quantitative controls, that is the number of protein molecules made per message by two different mRNAs coexisting within one cell or by the same message in two different cells or in the same cell at two different stages of its development and differentiation might vary; and there may be qualitative controls, that is some cells might contain specific messages which are not translated at one time but are at another. We will now consider some of the evidence for these two kinds of translational control.

Favourable material for studying the quantitative control of translation is provided by the mammalian reticulocyte synthesizing α and β globin chains. There is considerable evidence which suggests that α and β globin synthesis may be regulated independently at the genetic level. For instance, α chain synthesis starts before β chain synthesis in development, the mammalian foetus having a haemoglobin $(\alpha_2 \delta_2)$ in which β globin is replaced by a δ chain. Beta globin synthesis starts at around birth just as δ chain synthesis declines. The switch is based on the production of a new population of cells synthesizing α and β rather than α and δ chains and is not a change from δ to β synthesis within one population of cells. α and β globin synthesis may also be regulated at the translational level. The relative amounts of α and β chain made by cell-free protein synthesizing systems are clearly dependent on the ionic composition of the reaction mixture. Since the ionic conditions prevailing in cells vary from type to type and also probably during the differentiation of one cell they may have an important though relatively unspecific effect on protein synthesis. Also although reticulocytes accumulate equal quantities of α and β chains they have about 1·4 times more ribosome associated α globin than β globin mRNA [129a]. Initiation is less efficient (65%) on α mRNA than it is on β [157] and given equal rates of elongation and termination this result accounts both for the equality of chains synthesized and for the fact that α globin is made on smaller polyribosomes (3–4 mers) than β globin (4–5 mers) [129] since the size of the coding sequence is the same for both messages. Here we have a good case for believing that different rates of initiation on two distinct mRNAs within one cell result in different numbers of protein molecules being produced per message molecule. It is not clear from these experiments which component of the initiation reaction causes the variation in rates, for instance, it may be that mRNA secondary structure is important [156], or there may be α and β specific initiation factors.

Purified globin mRNAs have also been translated after microinjection into living *Xenopus* oocytes (for review, see [110, 149]). In this situation it was found that five molecules of α globin are synthesized for every molecule of β globin [110]. The ratio could be restored to equality by injection of haemin [97], a factor known to act during initiation events of translation in a reticulocyte cell-free protein synthesizing system [154]. Although it is not known at what point haemin is acting on translation in oocytes it clearly shows some message specificity. Further evidence for the existence of message specific initiation factors is suggested by work comparing myosin with globin synthesis [123], and globin with viral [259] specific protein synthesis in cell-free systems to which specific proteins isolated from ribosome salt washes may be added. These factors may be important in the quantitative regulation of protein synthesis but are probably not relevant to qualitative control since no absolute requirement for them has ever been observed.

EVIDENCE FOR UNTRANSLATED mRNA IN CELLS

There is much evidence (for recent reviews, see [105, 227]) which suggests that protein synthesis following fertilization in most animals (possibly except mammals) and following germination in plants [75, 214, 248] and fungi [153] is templated by mRNA molecules synthesized previously during oogenesis, plant embryo or fungal spore formation and stored in an inactive form until activated by the events of fertilization or germination [51, 153, 214]. The types of experiment which have suggested this conclusion involve suppression of nuclear and mitochondrial RNA synthesis either by drug treatment [219] or mechanical enucleation experiments [72, 73]. Experiments using drugs suffer two general difficulties one is that although most RNA synthesis is inhibited synthesis of specific RNAs may continue to be made at rates sufficient to account for the observed effects, this problem has been seen in studies with Actinomycin D on Slime mould differentiation [85]. Secondly it may be that the drugs used have unknown side effects which might be important in producing the specific effects observed. Enucleation experiments provide the strongest evidence for lack of direct genomic control over protein synthesis in early sea urchin and amphibian development.

In sea urchins following fertilization there is a period during which the absolute rate of protein synthesis increases by as much as fifteen times [80]. This increase is correlated with numerous other changes in for instance the properties of ribosome salt wash factors [92, 167], activation and availability of tRNA [42], the appearance of polyribosomes [197] and the adenylation of pre-existing RNA by a cytoplasmic enzyme system [225, 262]. A careful study [127] has emphasized that fertilization in sea urchins results in increased utilization of mRNA by recruitment of pre-existing untranslated molecules rather than by increased rates of initiation or elongation on mRNA already committed to protein synthesis. An increased rate of elongation will reduce the transit time of a ribosome on a message, but no change in this parameter is observed at fertilization. Since there is no change in the average size of proteins produced after fertilization the rate of elongation is unlikely to have altered. Similarly if fertilization increases the rate of initiation it would lead to greater loading of ribosomes on to messages, which should be observed as a shift from small to large polysomes, but no such change in polysome size is observed and hence the rate of initiation does not alter detectably. If neither initiation nor elongation rates change at fertilization then protein synthesis can only increase by recruitment of previously untranslated messages.

Using hybridization of poly(U) to detect poly(A) containing RNA (presumably largely mRNA by analogy with other systems) it has been possible to demonstrate the presence of a significant proportion of total egg mRNA sedimenting in sucrose gradients as ribonucleoprotein (RNP) particles between 40 and 100S in both sea urchins and amphibians [207, 225]. More specifically both histone and microtubule protein mRNAs have been identified by cell-free translation and nucleic acid hybridization in the subribosomal region of gradients in which RNP particles from sea urchin eggs have been separated [107, 195]. In amphibians these particles are not ribosome bound since they sediment in the same way under conditions in which ribosomes and polysomes are disrupted [207]. In sea urchins after fertilization poly(A) containing RNA shifts into a ribosome associated compartment presumably active in protein synthesis [225]. The detailed analysis of poly(A) containing RNA at fertilization will help us understand more clearly the mechanisms controlling the increase of protein synthesis at this time and also may help in clarifying the role of poly(A) in translation. Recent experiments suggest that there is no absolute dependence of post-fertilization increases in translation rates on continued polyadenylation [166a]; this is perhaps not too surprising in the light of other experiments which show that mRNAs which have been stripped of their poly(A) tails are translated both in cell-free systems and after injection into oocytes [126, 223], but that their prolonged activity as messages is severely limited.

Fertilization and germination provide good examples for the existence of quantitative controls on translation but the detailed mechanism involved in the mobilization of stored mRNA for protein synthesis is still not clear. It seems likely from experiments in which mRNAs coding for known proteins are injected into *Xenopus* eggs and oocytes that both these cell types have an excess translational capacity [110]. Since they also have a store of potential mRNA which is untranslated it must be inactivated either by (i) sequestration within a specific cellular compartment, or (ii) by packaging with specific proteins, or (iii) by the lack of a specific component necessary to allow the mRNA to function. Activation after fertilization would then be due to either (i) breakdown of the cell compartment, or (ii) the removal of packaging proteins, or (iii) the provision of the specific component necessary for mRNA function.

EVIDENCE FOR DELAYED TRANSLATION

Maternally inherited mRNA discussed above is a special case of a more general phenomenon in which specific mRNAs are synthesized in advance of the requirement for their translation products (see p. 325 and p. 336). Phenomena of this kind have been observed in several developing and differentiating systems. The evidence that it occurs, usually involves enucleation or treatment with drugs to suppress RNA synthesis in order to define a period during which continued RNA synthesis is required for subsequent new protein synthesis. Such a period of sensitivity often precedes a period in which new transcription is not required for new enzyme or protein synthesis to continue. Some examples of such systems are given on p. 325, (messenger RNA synthesis in terminal cell differentiation). In none of these systems is a complete analysis available (this would involve estimation of rates of specific mRNA synthesis, of nuclear mRNA precursor pools, of the rate of processing of mRNA precursor, of pool sizes of actual and potential mRNA, and of rates of specific protein synthesis and degradation). However, there is a strong indication that terminally differentiating cells and some developmental systems may synthesize specific mRNAs in advance of a requirement for their translation, and such cases possibly present examples of translational control processes in these cells.

LACK OF EVIDENCE FOR QUALITATIVE CONTROL OF TRANSLATION

A demonstration that qualitative control of translation is occurring requires proof that specific mRNAs for proteins not being made at that time coexist within one cell alongside other mRNAs which are being translated, and that at some later time these inactive mRNAs become translated. At present there is no example of such a system though it may turn out that there are specific mRNAs stored during oogenesis or seed embryo formation which code for proteins not made at the time but only later after fertilization or germination. Two maternally inherited mRNAs for well defined proteins, those for histones and microtubule protein in sea urchin embryos [107, 195], are

known but both histones and microtubule protein are also made during oogenesis [1a], and so these only qualify as examples of quantitative control of translation.

5.3.8 CYTOPLASMIC FACTORS IMPORTANT IN CONTROLLING GENE ACTIVITY IN DEVELOPMENT

Localized factors and fate maps

Classical embryology provides many examples in both vertebrates and invertebrates of cytoplasmic factors, often localized in specific regions of the egg, which are spatially correlated with the appearance of particular differentiated regions of the embryo (for reviews, see [66, 260]). Localized factors become segregated during embryonic cleavage into particular cell lines and it has often been suggested, though the detailed biochemical mechanisms are not clear, that they control the expression of the genes relevant to the development and differentiation of these cell lines. Evidence showing that cytoplasmic components are at least in part responsible for the developmental potential of particular embryonic cell lines has come from three kinds of experiments: (i) alteration of the spatial arrangement of cytoplasmic components by centrifugation, (ii) removal of particular regions of the cytoplasm, and (iii) rescue experiments which restore cytoplasmic components either to regions of the embryo which do not normally have them or to embryos which have had components removed or destroyed (see Chapters 1 and 3.6).

Biochemical effects of cytoplasmic localization

POLAR LOBE OF ILYANASSA

During the first few cleavages of *Ilyanassa* (Mollusca) development a particular form of cytoplasmic sequestration occurs. An anucleate polar lobe forms at the first cleavage, giving the embryo the superficial appearance of having three cells (called the trefoil stage), and remains associated with one (the CD cell) of the two cells formed. The polar lobe forms at the second cleavage and its contents enter only one (the D cell) of the four cells produced. It appears at several subsequent cleavages but always remains associated with the D cell lineage which gives rise to mesodermal cells of the adult. Removal of the polar lobe at the first or subsequent divisions results in embryos deficient in mesodermal cells and since no nuclear material is removed at the same time presents a case of cytoplasmically localized developmental information.

Studies with lobeless embryos show that they incorporate significantly (though not dramatically) less amino acid into protein [54] and uridine into RNA [54] but accumulation of rRNA is unaffected [55]. Moreover isolated lobes incorporate amino acid into protein though possibly at a lower rate per unit protein than lobeless embryos [52]; of course, since the lobe contains no nucleus it synthesizes no nuclear RNA. These results suggest that the lobe cytoplasm may play a role in determining the rates of protein and RNA synthesis, but whether its effect is also qualitative has yet to be determined.

TISSUE SPECIFIC ENZYMES IN THE DEVELOPMENT OF *CIONA INTESTINALIS*

Early observations on the development of the Tunicate *Styela* showed that different regions of the embryo were easily recognizable by natural differences in colour of some cytoplasmic components, and although these components are not superficially apparent in the unfertilized egg they do appear early in development [57]. A recent study of the appearance of two tissue specific enzymes in the development of another Tunicate, *Ciona,* has given a clearer insight into the importance of regional differences in eggs in the control of specific enzyme synthesis [258]. One of these enzymes, acetylcholine esterase, which catalyses the hydrolysis of acetylcholine, is first detected histochemically in presumptive muscle cells of the early neurula at eight hours after fertilization. The other enzyme, dopa oxidase, which catalyses a reaction important in melanin (pigment) synthesis, appears slightly later at nine hours and is first detected in presumptive pigment cells. Three kinds of experiments were done in order to determine whether (i) new RNA synthesis, (ii) new protein synthesis and (iii) a pre-determined number of cell divisions has to take place before either enzyme will appear. New RNA synthesis was blocked to greater than 70% with Actinomycin D and it was found that prior to the appearance of both enzymes there is a sensitive period (Fig. 5.3.19) during which new RNA synthesis is required. After this period enzyme production is relatively resistant to inhibition of RNA synthesis. Inhibition of protein synthesis, using puromycin, at the time when enzyme activity is normally appearing completely prevents it but has no short term effect on enzyme activity after this has been fully expressed suggesting that puromycin is not inhibiting the enzyme directly, but is likely to be preventing its synthesis or the synthesis of a cofactor necessary for its activity. These results suggests that both enzymes require new RNA and new protein synthesis for their appearance, and the simplest explanation, though there is no proof is that mRNA specific for each enzyme is made during the actinomycin D sensitive period prior to the synthesis of new enzyme protein. If this is the correct explanation then the cytoplasmic factor controlling the cell specific production of these enzymes cannot be either mRNA or inactive enzyme.

Cleavage arrested embryos, produced by cytochalasin B or colcemid treatment, show both new enzyme activities at the normal time after fertilization, though of course the embryos will have fewer than normal cells. By arresting cleavage at

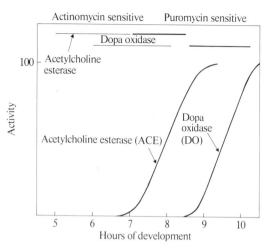

Fig. 5.3.19 Diagram of the relationship between Actinomycin D and puromycin sensitive periods and the appearance of acetylcholine esterase and dopa oxidase activity during development of *Ciona intestinalis* [258].

different times after fertilization and growing the embryos until the time of enzyme appearance it was possible to trace the cells containing the information specific for enzyme production right back to the fertilized egg (single cell) stage. It was found that segregation of specific enzyme forming capacity occurs very early in development and follows the cell lineage specific for either muscle or pigment cells. Normal development will occur in centrifugally stratified eggs and in eggs from which as much as 50% of the cytoplasm has been removed [57]. For the faithful distribution of cytoplasmic developmental information to occur in such situations it must be unaffected by the experimental treatment. In view of this it seems that the cell membrane or the stiff cortical layer immediately beneath it are good repositories for such information, since they are not easily altered by centrifugation or by withdrawal of cytoplasm, and it is relatively easy to see how particular regions of membrane or cortex are apportioned to particular cells during cleavage. In conclusion these experiments indicate regional localization of enzyme forming capacity which is probably not based on storage of enzyme specific mRNA nor inactive enzyme, but on some other component which activates new gene expression at a particular stage of development. Timing of new enzyme synthesis is not related to the number of cell divisions and must be due to a developmental clock which counts neither cell divisions nor cells.

CYTOPLASMIC CONTROL OF RNA AND DNA SYNTHESIS

In a situation where it is possible to introduce the nucleus of one cell type into the cytoplasm of another it is possible to ask questions about the effect of this experimental manoeuvre on the synthesis of RNA or DNA within that nucleus. Situations of this kind have been produced by two general procedures (i) by microinjection [110] and (ii) by cell fusion [81]. In general the nuclear activities of the recipient cell are imposed on the donor nucleus. For instance when a chicken erythrocyte nucleus, normally inactive in RNA or DNA synthesis is introduced into the cytoplasm of a Hela cell it will eventually resume both RNA and DNA synthesis [18, 118]. This reactivation is accompanied by an increase in volume and dry mass of the nucleus which precedes DNA synthesis but is roughly proportional to, but not dependent on, new RNA synthesis [18]. Reactivation of specific chicken genes also occurs since new cell surface antigens and a specific enzyme hypoxanthine: guanine phosphoribosyl transferase also appear after reactivation. However, these genes had previously been active in erythrocyte development and since new chicken globin synthesis is not detectable after fusion not all previously active erythrocyte genes are reactivated [119, 120].

The effect of a particular type of cytoplasm on the activity of a recipient nucleus has been studied by microinjection of adult frog brain nuclei which synthesize a little RNA but DNA only occasionally into either (i) ovarian oocytes; (ii) ovulated oocytes undergoing maturation or (iii) unfertilized eggs artificially activated by the injection procedure [109]. These three types of cell differ in respect of DNA and RNA synthesis. The full grown ovarian oocyte which synthesizes RNA but no DNA induces brain nuclei to enlarge and synthesize RNA but not DNA. Unfertilized eggs which synthesize nuclear DNA after activation induce DNA but not RNA synthesis, while oocytes undergoing maturation and in which chromosomes are condensed on spindles and do not synthesize RNA or DNA induce brain nuclei to condense chromosomes which appear on cytoplasmic division spindles. Each type of recipient cell imposes its own synthetic regime on the donor nucleus [109, 113], presumably by diffusible factors which the donor nucleus accumulates from the recipient, rather than by selective loss of factors by the donor nucleus. There is evidence for the entry of newly synthesized protein into injected nuclei during maturation and early development of both *Xenopus, Rana* and sea urchins (see below), though there is evidence against loss of protein from donor nuclei [227].

CYTOPLASMIC CONTROL OF RNA SYNTHESIS IN DEVELOPMENT

There is a well-defined pattern of specific synthesis in early development of many animals and plants (see 5.3.6). In *Xenopus* this pattern is controlled partly by cytoplasmic factors since nuclei transplanted from neurula cells, in which they were synthesizing rRNA, 4S RNA and HnRNA, into enucleate eggs will give transplant embryos in which the unfolding pattern of RNA synthesis is indistinguishable from normal embryos [115].

These cytoplasmic factors are species specific since *Xenopus* neurula nuclei injected into enucleate *Discoglossus* (a related genus of frogs) eggs will show activation of only some types of RNA synthesis [115, 264].

Reversible repression of rRNA synthesis in early development has been suggested to be due to the presence of an inhibitory substance up to blastula stage and its dilution, removal or destruction during gastrulation [111, 222, 222a]. Extraction of a factor from *Xenopus* blastulae which reduces rRNA synthesis in *Xenopus* neurula cells, but has little effect on 4S RNA, HnRNA or DNA synthesis has been reported [222a]. There is an inverse correlation between the stages and regions of embryos from which this factor may be isolated and the rate of rRNA synthesis which occurs in them [245]. For instance, in gastrulating embryos, rRNA synthesis occurs only in animal halves of embryos and the inhibitor is only extracted from vegetal halves where there is little rRNA synthesis [245]. These results have been repeated recently using different extraction procedures and show that an acid soluble component of *Xenopus* blastula but not neurula cells will reduce rRNA synthesis relative to 4S RNA by an order of magnitude [152a]. Such experiments represent a start in the search for specific regulatory substances in development but at the moment there is no proof that they are normally active repressors of rRNA synthesis in development.

PASSAGE OF NEWLY SYNTHESIZED PROTEINS FROM CYTOPLASM TO NUCLEUS IN DEVELOPMENT

There is ample evidence that some proteins synthesized during oocyte maturation or during early embryonic development will accumulate within nuclei [76, 143, 166, 227]. Perhaps this is not too surprising since it is well known that nuclei do contain specific proteins which are synthesized in the cytoplasm [143] the histones are a good example. A recent careful study [1a] of histone synthesis in oocytes, eggs and embryos of *Xenopus* has shown that they all synthesize F2A1, F2A2, F2B and F3 types of histone but only at late blastula is F1 histone synthesis detected. A stage specific shift in type of histone synthesis has also been shown in sea urchin embryos [209, 210] and is probably due to a change in the type of histone gene transcription rather than to the recruitment of new previously masked maternally inherited histone mRNA [209]. In *Xenopus* oocytes and unfertilized eggs no nuclear DNA synthesis occurs and so, unlike most cells growing in tissue culture which show coordinated DNA and histone synthesis [38] there is an uncoupling of histone and DNA synthesis. It is argued that in *Xenopus* as in sea urchins the function of the high rate of histone synthesis in early development is to provide for the extremely rapid rates of DNA synthesis which occur late in cleavage [1a]. Since labelled histone injected into oocytes accumulates in the nucleus to many times its natural concentration [110] it is likely that the histones normally synthesized by oocytes and early embryos also accumulate in nuclei.

The conclusion from this section is that cytoplasmic proteins may enter nuclei and it may be that some of these proteins act as specific regulators of gene activity in development. So far there is no experimental evidence to suggest that a specific protein entering a nucleus has altered its activity in respect of either RNA or DNA synthesis in a specific way.

5.3.9 NUCLEAR PROTEINS

There is a general argument which leads us to believe that gene activity is controlled at least in part by proteins which associate with DNA in the nucleus. We have seen in the previous section how experimentally induced changes in nuclear activity are correlated with nuclear uptake of protein from the cytoplasm. It is therefore of importance to learn as much as possible about nuclear proteins and their properties in the hope of also gaining knowledge of the role of these proteins in regulating gene activity. There are two main classes of nuclear protein: (i) the histones, and (ii) the nuclear acidic proteins (for recent review, see [78]).

Histones

Histones represent 70% of nuclear protein and are normally in equal quantities by weight with DNA. They have been implicated in gene regulation for some time [235] but are generally thought, by virtue of their limited heterogeneity, not to have sufficient specificity for fine control of gene transcription. Recently models have proposed that they may be important in the gross control of differentiation [23, 60, 67, 96, 182], or that they may gain greater specificity by association with specific RNA molecules [87a] or by acting in groups to allow greater flexibility [243]. There is some evidence to suggest that different histone fractions show preferential binding to regions of DNA with either high or low GC content [47]. Opinion now seems to favour some general structural role for histones which may be important in overall regulation of gene activity [60, 67, 182]. In spite of their limited heterogeneity histones do show tissue and developmental stage specificity [1a], which cannot be wholly ascribed to the vagaries of post-translational modification such as phosphorylation, methylation and acetylation [78]. Changes in the levels of these post-translational modifications have been correlated with particular developmental or differentiation processes [78]. So far, however, there is no real indication as to whether they are casual factors or merely part of the expression of the differentiated state intself.

Acidic nuclear proteins

Non-histone or acidic nuclear proteins represent about 30% of nuclear protein and are extremely heterogeneous [94, 168].

They have received attention because they are reported to be required in chromatin reconstitution experiments for the retention of transcriptional specificity [98, 183, 242]. They may undergo post-translational modification by phosphorylation [242] but show only limited tissue specificity, which may reflect inadequate sensitivity of the techniques used to detect differences among regulatory molecules which may be present down to one or a few copies per nucleus. There are both qualitative and quantitative changes in acidic nuclear protein synthesis during stimulation of rat uterus by oestrogen [241], in phytohaemagglutinin stimulated lymphocytes [154a], during polytene chromosome formation in *Drosophila* [77] and during the phases of the cell cycle but it is not clear to what extent these are related to changes in gene activity which occur at the same time [185]. Although this evidence may be suggestive of a role for nuclear acidic proteins in gene regulation there is no example in eukaryotes of a defined purified regulator molecule such as the lac repressor or CAP protein in *E. coli* [19]. It is likely that the techniques used so far to fractionate non-histone nuclear proteins are insufficiently sensitive to discriminate molecules which might be present in only a few copies per nucleus.

5.3.10 CONCLUSIONS

In the absence of adequate genetic techniques which have proven so useful in studies of gene control mechanisms in prokaryotic systems, we have tried to show how many diverse types of experiment are beginning to throw light on the fundamental question of control of gene expression in development and differentiation of eukaryotes. In particular the sequence organization of eukaryote DNA suggests a distribution of some classes of repetitive sequence amongst the single copy DNA sequences some of which are known to code for proteins. Transcripts of repetitive sequences appear in nuclear RNA intermingled on the same molecules with unique sequence transcripts, but are not found in cytoplasmic messenger RNA populations. Nuclear RNA has considerably greater coding potential (sequence complexity) than cytoplasmic RNA suggesting the preferential elimination of some single copy sequences within the nucleus along with the repetitive sequence transcript. Although differential gene transcription is seen as the predominant determining factor in differentiation and development the possible modulating effects of differential processing of primary gene transcripts or differential translation of messenger RNAs are also considered. The processes which control differential gene activity in development and differentiation involve nuclear interactions with molecules which may be located initially in specific regions of the egg cytoplasm. While mechanisms which maintain the differentiated state probably involve interactions between specific nuclear proteins and DNA and these interactions may be a consequence of the process of differentiation itself.

5.3.11 REFERENCES

[1] AB G. & MALT R.A. (1970) Metabolism of ribosomal precursor ribonucleic acid in kidney. *J. Cell Biol.*, **46**, 362–69.

[1a] ADAMSON E.D. & WOODLAND H.R. (1974) Histone synthesis in early amphibian development: histone and DNA syntheses are not coordinated. *J. molec. Biol.*, **86**, 263–85.

[2] ADESNIK M. & DARNELL J.E. (1972) Biogeneis and characterization of histone mRNA in Hela cells. *J. molec. Biol.*, **67**, 397–406.

[3] ALTMAN S. (1971) Isolation of tyrosine tRNA precursor molecules. *Nature, New Biol.*, **229**, 19–21.

[4] ALTMAN S. & SMITH J.O. (1971) Tyrosine tRNA formed *in vitro* by specific nucleolytic cleavage of a precursor molecule. *Nature, New Biol.*, **233**, 35–39.

[5] ARONSON A.I. & WILT F.H. (1969) Properties of nuclear RNA in sea urchin embryos. *Proc. nat. Acad. Sci. (Wash.)*, **62**, 186–93.

[6] ASHBURNER M., CHIHARA C., MELTZER P. & RICHARDS G. (1973) Temporal control of puffing activity in polytene chromosomes. *Cold Spr. Harb. Symp. quant. Biol.*, **38**, 655–62.

[7] BARRELL B.G. & CLARK B.F.C. (1974) *Handbook of nucleic acid sequences*. Joynson-Bruvvers.

[8] BEERMANN W. (1972) Chromomeres and genes. In *Developmental studies on giant chromosomes: results and problems in cell differentiation*, Vol. 4 (Ed. W. Beermann). Springer-Verlag, Berlin.

[9] BERNHARDT D. & DARNELL J.E. (1969) tRNA synthesis in Hela cells. A precursor to tRNA and the effects of methionine starvation on tRNA synthesis. *J. molec. Biol.*, **42**, 43–56.

[9a] BERNS A.J., KRAAIKAMP M., BLOEMENDAL H. & LANE C.D. (1972) Calf crystallin synthesis in frog cells: the translation of lens-cell 14S RNA in oocytes. *Proc. nat. Acad. Sci. (Wash.)*, **69**, 1606–9.

[10] BIRD A.P. & BIRNSTIEL M.L. (1971) A timing study of DNA amplification in *Xenopus laevis* oocytes. *Chromosoma, Berl.*, **35**, 300–9.

[11] BIRD A., ROGERS E. & BIRNSTIEL M.L. (1973) Is gene amplification RNA directed? *Nature, New Biol.*, **242**, 226–30.

[12] BIRNSTIEL M.L., TELFORD J., WEINBERG E. & STAFFORD D. (1974) Isolation and some properties of the genes coding for histone proteins. *Proc. nat. Acad. Sci. (Wash.)*, **71**, 2900–4.

[13] BISHOP J.O. & FREEMAN K.B. (1973) DNA sequences neighbouring the duck haemoglobin genes. *Cold Spr. Harb. Symp.*, **38**, 707–16.

[14] BISHOP J.O., MORTON J.G., ROSBASH M. & RICHARDSON M. (1974) Three abundance classes in Hela cell mRNA. *Nature (Lond.)*, **250**, 199–204.

[14a] BISHOP J.O. & ROSBASH M. (1973) Reiteration frequency of duck haemoglobin genes. *Nature, New Biol.*, **241**, 204–7.

[15] BISHOP J.O., ROSBASH M. & EVANS D. (1974) Polynucleotide sequences in eukaryotic DNA and RNA that form ribonuclease-resistant complexes with polyuridylic acid. *J. molec. Biol.*, **85**, 75–86.

[16] BLACKLER A.W. (1972) Functional interactions between the amphibian oocyte and general ovarian cells. In *Oogenesis* (Eds J.D. Biggers & A.W. Schuetz), pp. 321–37. Butterworths, University Park Press.

[17] BLATTI S.P., INGLES C.J., LINDELL T.J., MORRIS P.W., WEAVER R.F., WEINBERG F. & RUTTER W.J. (1970) Structure and regulatory properties of eukaryotic RNA polymerase. *Cold Spr. Harb. Symp. Quant. Biol.*, **35**, 649–57.

[18] BOLUND L., RINGERTZ N.R. & HARRIS H. (1969) Changes in the cytochemical properties of erythrocyte nuclei reactivated by cell fusion. *J. Cell Sci.*, **4**, 71–87.

[19] BOURGEOIS S. (1972) Methods for studying protein nucleic acid interaction. *5th Karolinska symposium on research methods in reproductive endocrinology*, pp. 178–202.

[20] BRANDHORST B.P. & McCONKEY E.H. (1974) Stability of nuclear RNA in mammalian cells. *J. molec. Biol.*, **85**, 451–63.

[21] BRANDHORST B.P. & HUMPHRIES T. (1971) Synthesis and decay rates of major classes of deoxyribonucleic acid like ribonucleic acid in sea urchin embryos. *Biochemistry*, **10**, 877–81.

[22] BRAWERMAN G. (1974) Eukaryotic messenger RNA. *Ann. Rev. Biochemistry*, **43**, 621–42.

[23] BRITTEN R.J. & DAVIDSON E.H. (1969) Gene regulation for higher cells: a theory. *Science*, **165**, 349–57.

[24] BRITTEN R.J. & KOHNE D.E. (1968) Repeated sequences in DNA. *Science*, **161**, 529–40.

[25] BROWN D.D. & BLACKLER A.W. (1972) Gene amplification proceeds by a chromosome copy mechanism. *J. molec. Biol.*, **63**, 75–83.

[26] BROWN D.D. & DAWID I. (1968) Specific gene amplification in oocytes. *Science*, **160**, 272–80.

[27] BROWN D.D. & GURDON J.B. (1964) Absence of ribosomal RNA synthesis in the anucleolate mutant of *Xenopus laevis*. *Proc. Nat. Acad. Sci. (Wash.)*, **51**, 139–46.

[28] BROWN D.D. & GURDON J.B. (1966) Size distribution and stability of DNA like RNA synthesized during development of anucleolate embryos of *Xenopus laevis*. *J. molec. Biol.*, **19**, 399–422.

[29] BROWN D.D. & LITTNA E. (1964) RNA synthesis during the development of *Xenopus laevis*, the South African Clawed Toad. *J. molec. Biol.*, **8**, 669–87.

[30] BROWN D.D. & LITTNA E. (1964) Variations in the synthesis of stable RNAs during oogenesis and development of *Xenopus laevis*. *J. molec. Biol.*, **8**, 688–95.

[31] BROWN D.D. & LITTNA E. (1966) The synthesis and accumulation of low molecular weight RNA during embryogenesis of *Xenopus laevis*. *J. molec. Biol.*, **20**, 95–112.

[32] BROWN D.D. & LITTNA E. (1966) The synthesis and accumulation of DNA-like RNA during embryogenesis of *Xenopus laevis*. *J. molec. Biol.*, **20**, 81–94.

[33] BROWN D.D. & SUGIMOTO K. (1973) The 5S DNAs of *Xenopus laevis* and *Xenopus mülleri*: the evolution of a gene family. *J. molec. Biol.*, **78**, 397–415.

[34] BROWN D.D., WENSINK P.C. & JORDAN E. (1972) A comparison of the ribosomal DNAs of *Xenopus laevis* and *Xenopus mülleri*: the evolution of tandem genes. *J. molec. Biol.*, **63**, 57–73.

[34a] BROWN I.R. & CHURCH R.B. (1972) Transcription of non-repeated DNA during mouse and rabbit development. *Devl. Biol.*, **29**, 73–84.

[35] BROWN R.D., MATTOCCIA E. & TOCCHINI-VALENTINI G.P. (1972) On the role of RNA in gene amplification. *5th Karolinska Symposium on Research Methods in Reproductive Endocrinology*, pp. 307–16.

[36] BROWNLEE G.G., CARTWRIGHT E.M. & BROAN D.D. (1974) Sequence studies of the 5S DNA of *Xenopus laevis*. *J. molec. Biol.*, **89**, 703–48.

[36a] BUONGIORNO-NARDELLI M., AMALDI F. & LAVA-SANCHEZ P.A. (1972) Amplification as a rectification mechanism for the redundant rRNA genes. *Nature (Lond.)*, **238**, 134–37.

[37] BURDON R.H. (1971) Ribonucleic acid maturation in animal cells. *Prog. Nuc. Acid Res. Mol. Biol.*, **11**, 33–79.

[38] BUTLER W.B. & MUELLER G.C. (1973) Control of histone synthesis in Hela cells. *Biochim. biophys. Acta (Amst.)*, **294**, 481–96.

[39] CALLAN H.G. (1963) The nature of lampbrush chromosomes. *Int. Rev. Cytol.*, **15**, 1–34.

[39a] CALLAN H.G. (1967) The organization of genetic units in chromosomes. *J. Cell Sci.*, **2**, 1–8.

[40] CAMPO M.S. & BISHOP J.O. (1974) Two classes of messenger RNA in cultured rat cells: repetitive sequence transcripts and unique sequence transcripts. *J. molec. Biol.*, **90**, 649–64.

[41] CAVALIER-SMITH T. (1974) Palindromic base sequences and replication of eukaryotic chromosome ends. *Nature (Lond.)*, **250**, 467–70.

[42] CECCARINI C., MAGGIO R. & BARBATA G. (1967) Aminoacyl sRNA synthetases as possible regulators of protein synthesis in the embryo of the sea urchin *Paracentrotus lividus*. *Proc. nat. Acad. Sci. (Wash.)*, **58**, 2235–39.

[43] CHEN D., SCHULTZ G. & KATCHALSKI E. (1971) Early ribosomal RNA transcription and appearance of cytoplasmic ribosomes during germination of the wheat embryo. *Nature, New Biol.*, **231**, 69–72.

[44] CHAMBON P., GISSINGER F., KEDINGER C., MANDEL J.L., MEILHAC M. & NURET P. (1972) Structural and functional properties of three mammalian nuclear DNA dependent RNA polymerases. *5th Karolinska Symp. on Research Methods in Reproductive Endocrinology*, pp. 222–46.

[45] CHANDHURI S. & LIEBERMAN I. (1968) Time required by the normal and

regenerating rat liver cell to make a ribosome. *J. molec. Biol.*, **33**, 323–26.

[46] CHOVNICK A., FINNERTY V., SCHALET A. & DUCK P. (1969) Studies on genetic organization in higher organisms. I. Analysis of a complex gene in *Drosophila melanogaster*. *Genetics*, **62**, 145–60.

[47] CLARK R.J. & FELSENFELD G. (1972) Association of arginine-rich histones with GC-rich regrons of ENAin chromatin. *Neture, New Biol.*, **240**, 22y–29.

[48] CLARKSON S.G., BIRNSTIEL M.L. & PURDOM I.F. (1973) Clustering of transfer RNA genes of *Xenopus laevis*. *J. molec. Biol.*, **79**, 411–29.

[49] CLARKSON S.G., BIRNSTIEL M.L. & SERA V. (1973) Reiterated transfer RNA genes of *Xenopus laevis*. *J. molec. Biol.*, **79**, 391–410.

[50] CLASON A.E. & BURDON R.H. (1969) Synthesis of small nuclear RNAs of mammalian cells in relation to the cell cycle. *Nature (Lond.)*, **223**, 1063–64.

[51] CLEGG J.S. & GOLUB A.L. (1969) Protein synthesis in *Artemia salina* embryos. II. Resumption of RNA and protein synthesis upon cessation of dormancy in the encysted gastrula. *Devl. Biol.*, **19**, 178–200.

[52] CLEMENT A.C. & TYLER A. (1967) Protein synthesizing activity of anucleate polar lobe of the mud snail *Ilyanassa obsoleta*. *Science*, **158**, 1457–58.

[53] CLISSOLD P. & COLE R.J. (1973) Regulation of ribosomal RNA synthesis during mammalian myogenesis in culture. *Expl. Cell Res.*, **80**, 159–69.

[54] COLLIER J.R. (1961) Nucleic acid and protein metabolism of the *Ilyanassa* embryo. *Expl. Cell Res.*, **24**, 320–26.

[55] COLLIER J.R. (1965) RNAs of the *Ilyanassa* embryo. *Science*, **147**, 150–51.

[56] COOPER H. (1969) Ribosomal ribonucleic acid production and growth regulation in human lymphocytes. *J. biol. Chem.*, **244**, 1946–52.

[57] CONKLIN E.G. (1931) The development of centrifuged eggs of Ascidians. *J. expl. Zool.*, **60**, 1–120.

[58] CRAIG N.C. (1971) On the regulation of the synthesis of ribosomal proteins in L cells. *J. molec. Biol.*, **55**, 129–34.

[58a] CRAIG N.C. & PERRY R.P. (1971) Persistent cytoplasmic synthesis of ribosomal proteins during selective inhibition of ribosomal RNA synthesis. *Nature, New Biol.*, **229**, 75–80.

[59] CRAIG N.C. (1974) Ribosomal RNA synthesis in eukaryotes and its regulation. In *MTP International Review of Science, Biochemistry of Nucleic Acids*, Series one (Eds. K. Burton). Butterworths, University Park Press. Vol. 6, Chapter 9.

[60] CRICK F.H.C. (1971) General model for chromosomes of higher organisms. *Nature (Lond.)*, **234**, 25–27.

[61] CRIPPA M. & TOCCHINI-VALENTINI G.P. (1971) Synthesis of amplified DNA that codes for ribosomal RNA. *Proc. Nat. Acad. Sci. (Wash.)*, **11**, 2769–72.

[62] DANEHOLT B. & HOSICK H. (1973) The transcription unit in Balbiani ring 2 of *Chironomus tentans*. *Cold Spr. Harb. Symp. Quant. Biol.*, **38**, 629–35.

[63] DARNBROUGH C.H., LEGON S., HUNT T. & JACKSON R.J. (1973) Initiation of protein synthesis. Evidence for messenger RNA independant binding of methionyl-transfer RNA to the 40S ribosomal subunit. *J. molec. Biol.*, **76**, 379–403.

[64] DARNELL J.E. (1968) Ribonucleic acids from animal cells. *Bact. Rev.*, **32**, 262–90.

[65] DARNELL J.E., JELINEK W.R. & MOLLOY G.R. (1973) Biogenesis of mRNA: genetic regulation in mammalian cells. *Science*, **181**, 1215–21.

[66] DAVIDSON E.H. (1968) *Gene activity in development*. Academic Press, New York and London.

[67] DAVIDSON E.H. & BRITTEN R.J. (1973) Organization, transcription and regulation in the animal genome. *Quart. Rev. Biol.*, **48**, 565–613.

[68] DAVIDSON E.H. & HOUGH B.R. (1971) Genetic information in oocyte RNA. *J. molec. Biol.*, **56**, 491–506.

[69] DAVIDSON E.H., HOUGH B.R., AMENSON C.S. & BRITTEN R.J. (1973) General interspersion of repetitive and non-repetitive sequence elements in DNA of *Xenopus*. *J. molec. Biol.*, **77**, 1–23.

[70] DAWID I.B., BROWN D.D. & REEDER R.H. (1970) Composition and structure of chromosomal and amplified ribosomal DNAs of *Xenopus laevis*. *J. molec. Biol.*, **51**, 341–60.

[71] DELANGE R.J. & SMITH E.L. (1971) Histones: structure and function. *A. Rev. Biochem.*, **40**, 279–314.

[71a] DENIS H. & WEGNEZ M. (1973) Recherches biochimiques sur l'oogenese 7. Synthese et maturation du RNA 5S dans les petits oocytes de *Xenopus laevis*. *Biochimie*, **55**, 1137–51.

[72] DENNIS-SMITH L. & ECKER R.E. (1969) Role of the oocyte nucleus in physiological maturation in *Rana pipiers*. *Devl. Biol.*, **19**, 281–309.

[73] DENNY P.C. & TYLER A. (1964) Activation of protein biosynthesis in non-nucleate fragments of sea urchin eggs. *Biochem. biophys. Res. Commun.*, **14**, 245–49.

[74] DINA D., CRIPPA M. & BECCARI E. (1973) Hybridization properties and sequence arrangement in a population of mRNAs. *Nature, New Biol.*, **242**, 101–5.

[75] DURE L. & WATERS L. (1965) Long-lived messenger RNA: evidence from cotton seed germination. *Science*, **147**, 410–12.

[76] ECKER R.E. & SMITH L.O. (1971) The nature and fate of *Rana pipiens* proteins synthesized during maturation and early cleavage. *Devl. Biol.*, **24**, 559–76.

[77] ELGIN S.C.R. & HOOD L.E. (1973) Chromosomal proteins of *Drosophila* embryos. *Biochemistry*, **12**, 4984–85.

[78] ELGIN S.C.R., FROEHNER S.C., SMART J.E. & BONNER J. (1971) The biology and chemistry of chromosomal proteins. *Adv. Cell Mol. Biol.*, **1**, 1–57.

[79] EMERSON C.P. & HUMPHREYS T. (1970) Regulation of DNA-like RNA and the apparent activation of ribosomal RNA synthesis in sea urchin embryos: quantitative measurements of newly synthesized RNA. *Devl. Biol.*, **23**, 86–112.

[80] EPEL D. (1967) Protein synthesis in sea urchin eggs: a 'late' response to fertilization. *Proc. nat. Acad. Sci. (Wash.)*, **57**, 899–906.

[81] EPHRUSSI B. (1972) *Hybridization of somatic cells*. Oxford University Press.

[82] FEUNTEUN J., JORDAN B.R. & MONIER R. (1972) Study of the maturation of 5S RNA precursors in *Escherichia coli*. *J. molec. Biol.*, **70**, 465–74.

[82a] FORD P.J. (1972) Ribonucleic acid synthesis during oogenesis in *Xenopus laevis*. In *Oogenesis* (Eds J.D. Biggers & A.W. Schuetz), pp. 167–92. Butterworths.

[83] FORD P.J. (1971) Non-coordinated accumulation and synthesis of 5S ribonucleic acid by ovaries of *Xenopus laevis*. *Nature (Lond.)*, **233**, 561–64.

[83a] FORD P.J. & CLARKSON S.G. (in preparation).

[84] FORD P.J. & SOUTHERN E.M. (1973) Different sequences for 5S RNA in kidney cells and ovaries of *Xenopus laevis*. *Nature, New Biol.*, **241**, 7–12.

[84a] FIRTEL R.A. (1972) Changes in the expression of single-copy DNA during development of the cellular slime mould *Dictyostelium discoideum*. *J. molec. Biol.*, **66**, 363–77.

[85] FIRTEL R.A., BAXTER L. & LODISH H.F. (1973) Actinomycin D and the regulation of enzyme biosynthesis during development of *Dictyostelium discoideum*. *J. molec. Biol.*, **79**, 315–27.

[86] FIRTEL R.A., JACOBSON A. & LODISH H.F. (1972) Isolation and hybridization kinetics of messenger RNA from *Dictyostelium discoideum*. *Nature, New Biol.*, **239**, 225–28.

[87] FIRTEL R.A. & LODISH H.F. (1973) A small nuclear precursor of messenger RNA in the cellular slime mould *Dictyostelium discoideum*. *J. molec. Biol.*, **79**, 295–314.

[87a] FRENSTER J.H. (1965) A model of specific de-repression within interphase chromatin. *Nature (Lond.)*, **206**, 1269–70.

[88] GALU G.A., BRITTEN R.J. & DAVIDSON E.H. (1974) A measurement of the sequence complexity of polysomal messenger RNA in sea urchin embryos. *Cell*, **2**, 41–52.

[89] GALL J.G. (1969) The genes for ribosomal RNA during oogenesis. *Genetics*, **61**, supplement 1, 121–32.

[90] GALL J.G. & ATHERTON D.D. (1974) Satellite DNA sequences in *Drosophila virilis*. *J. molec. Biol.*, **85**, 633–64.

[91] GALL J.G. & CALLAN H.G. (1962) H³ Uridine incorporation in lambrush chromosomes. *Proc. Nat. Acad. Sci. (Wash.),* **48**, 562–70.

[91a] GALL J.G., COHEN E.H. & POLAN M.L. (1971) Repetitive DNA sequences in *Drosophila. Chromosoma,* **33**, 319–44.

[92] GAMBINO R., METAFORA S., FELICETTI L. & RAISMAN J. (1973) Properties of the ribosomal salt wash from unfertilized and fertilized sea urchin eggs and its effect on natural mRNA translation. *Biochim. biophys. Acta,* **312**, 377–91.

[93] GAREL J.P., MANDEL P., CHAVANCY G. & DAILLIE J. (1971) Functional adaptation of tRNAs to protein biosynthesis in a highly differentiated cell system. III. Induction of isoacceptor tRNAs during secretion of fibroin in the silk gland of *Bombyx mori. FEBS Letters,* **12**, 249–52.

[94] GARRARD W.T. & BONNER J. (1974) Changes in chromatin proteins during liver regeneration. *J. biol. Chem.,* **249**, 5570–79.

[95] GEFTER M.L. & RUSSELL R.L. (1969) Role of modifications in tyrosine RNA: a modified base affecting ribosome binding. *J. molec. Biol.,* **39**, 145–57.

[96] GEORGIEV G.P. (1969) On the structural organization of operon and the regulation of RNA synthesis in animal cells. *J. theor. Biol.,* **25**, 473–90.

[96a] GEORGIEV G.P., RYSKOV A.P., COUTELLE C., MANTIEVA V.L. & AVAKYAN E.R. (1972) On the structure of transcriptional unit in mammalian cells. *Biochim. biophys. Acta,* **259**, 259–82.

[97] GIGLIONI B., GIANNI A.M., COMI P., OTTOLENGHI S. & RUNGGER D. (1973) Translational control of globin synthesis by haemin in *Xenopus* oocytes. *Nature, New Biol.,* **246**, 99–102.

[98] GILMOUR R.S. & PAUL J. (1969) RNA transcribed from reconstituted nucleoprotein is similar to natural RNA. *J. molec. Biol.,* **40**, 137–40.

[99] GIUDICE G., SCONZO G., RAMIREZ F. & ALBONESE I. (1972) Giant RNA is also found in the cytoplasm in sea urchin embryos. *Biochim. biophys. Acta,* **262**, 401–03.

[100] GOULD M.C. (1969) RNA and protein synthesis in unfertilized eggs of *Urechis caupo. Devl. Biol.,* **19**, 460–81.

[101] GOULD M.C. (1969) A comparison of RNA and protein synthesis in fertilized and unfertilized eggs of *Urechis caupo. Devl. Biol.,* **19**, 482–97.

[102] GRAHAM D.E., NEUFELD B.R., DAVIDSON E.H. & BRITTEN R. (1974) Interspersion of repetitive and non-repetitive DNA sequences in the sea urchin genome. *Cell,* **1**, 127–37.

[103] GREENBERG J.R. & PERRY R.P. (1971) Hybridization properties of DNA sequences directing the synthesis of mRNA and heterogeneous nuclear RNA. *J. Cell Biol.,* **50**, 774–81.

[104] GREENBERG J.R. & PERRY R.P. (1972) Relative occurrence of poly adenylic acid sequences in messenger and heterogeneous nuclear RNA of L cells as determined by polyU-hydroxylapatite chromatography. *J. molec. Biol.,* **72**, 91–8.

[105] GROSS P.R., GROSS K.W., SKOULTCHI A.I. & RUDERMAN J.V. (1973) Maternal mRNA and protein synthesis in the embryo. *6th Karolinska Symp. on Research Methods in Reproductive Endocrinology,* pp. 244–58.

[106] GROSS P.R., MALKIN L.I. & HUBBARD M. (1965) Synthesis of RNA during oogenesis in the sea urchin. *J. molec. Biol.,* **13**, 463–81.

[107] GROSS K., RUDERMAN J., JACOBS-LORENZA M., BAGLIONI C. & GROSS P.R. (1973) Cell-free synthesis of histones directed by messenger RNA from sea urchin embryos. *Nature, New Biol.,* **241**, 272–74.

[108] GROSSBACH U. (1973) Chromosome puffs and gene expression in polytene cells. *Cold Spr. Harb. Symp. quant. Biol.,* **38**, 619–27.

[109] GURDON J.B. (1968) Changes in somatic cell nuclei inserted into growing and maturing amphibian oocytes. *J. Embryol. exp. Morph.,* **20**, 401–14.

[110] GURDON J.B. (1974) *The control of gene expression in animal development.* Oxford University Press.

[111] GURDON J.B. & BROWN D.D. (1965) Cytoplasmic regulation of RNA synthesis and nucleolus formation in developing embryos of *Xenopus laevis. J. molec. Biol.,* **12**, 27–35.

[112] GURDON J.B. & LASKEY R.A. (1970) The transplantation of nuclei from single cultured cells into enucleate frogs' eggs. *J. Embryol. exp. Morph.,* **24**, 227–48.

[113] GURDON J.B. & SPEIGHT V.A. (1969) The appearance of cytoplasmic DNA polymerase activity during the maturation of amphibian oocytes into eggs. *Expl. Cell Res.,* **55**, 253–56.

[114] GURDON J.B. & UEHLINGER V. (1966) 'Fertile' intestine nuclei. *Nature (Lond.),* **210**, 1240–41.

[115] GURDON J.B. & WOODLAND H.R. (1969) The influence of the cytoplasm on the nucleus during cell differentiation with special reference to RNA synthesis during amphibian cleavage. *Proc. R. Soc. B,* **173**, 99–111.

[116] HALL R.H., DYSON W.H., CHHEDA G.B., DUTTA S.P. & HONG C.I. (1972) Modified components of tRNA: their possible role in the process of differentiation. *FEBS Symposium,* **23**, 131–45.

[117] HARRIS H. (1963) Rapidly labelled ribonucleic acid in the cell nucleus. *Nature (Lond.),* **198**, 184–85.

[118] HARRIS H. (1967) The reactivation of the red cell nucleus. *J. Cell Sci.,* **2**, 23–32.

[119] HARRIS H. & COOK P.R. (1969) Synthesis of an enzyme determined by an erythrocyte nucleus in a hybrid cell. *J. Cell Sci.,* **5**, 121–33.

[120] HARRIS H., SIDEBOTTOM E., GRACE D.M. & BRAMWELL M.E. (1969) The expression of genetic information: a study with hybrid animal cells. *J. Cell Sci.,* **4**, 499–526.

[121] HARRISON P.R., BIRNIE G.D., HELL A., HUMPHRIES B., YOUNG B.D. & PAUL J. (1974) Kinetic studies of gene frequency. I. Use of a DNA copy of reticulocyte RNA to estimate globin gene dosage in mouse tissues. *J. molec. Biol.,* **84**, 000–000.

[122] HASELKORN R. & ROTHMAN-DENES L.B. (1973) Protein synthesis. *A. Rev. Biochem.,* **42**, 397–438.

[122a] HECHT R.M. & BIRNSTIEL M.L. (1972) Integrity of the DNA template, a prerequisite for the faithful transcription of *Xenopus* rDNA *in vitro. Eur. J. Biochem.,* **29**, 489–99.

[123] HEYWOOD S.M. (1970) Formation of the initiation complex using muscle messenger RNAs. *Nature (Lond.),* **225**, 696–98.

[124] HOLMES D.S. & BONNER J. (1973) Preparation, molecular weight, base composition and secondary structure of giant nuclear ribonucleic acid. *Biochemistry,* **12**, 2330–38.

[125] HOLMES D.S. & BONNER J. (1974) Interspersion of repetitive and single-copy sequences in nuclear ribonucleic acid of high molecular weight. *Proc. nat. Acad. Sci. (Wash.),* **71**, 1108–12.

[126] HOURCADE D., DRESSLER D. & WOLFSON J. (1973) The amplification of ribosomal RNA genes involves a rolling circle intermediate. *Proc. nat. Acad. Sci. (Wash.),* **70**, 2926–30.

[126a] HUEZ G., MARBAIX G., HUBERT E., LECLERQ M., NUDEL U., SOREQ H., SALOMON R., LEBLEU B., REVEL M. & LITTAUER U.Z. (1974) Role of the polyadenylate segment in the translation of globin messenger RNA in *Xenopus* oocytes. *Proc. nat. Acad. Sci. (Wash.),* **71**, 3143–46.

[127] HUMPHREYS T. (1969) Efficiency of translation of messenger RNA before and after fertilization in sea urchins. *Devl. Biol.,* **20**, 435–58.

[128] HUMPHREYS T. (1971) Measurements of messenger RNA entering polysomes upon fertilization of sea urchin eggs. *Devl. Biol.,* **26**, 201–08.

[129] HUNT T., HUNTER T. & MUNRO A. (1968) Control of haemoglobin synthesis: distribution of ribosomes on the messenger RNA for α and β chains. *J. molec. Biol.,* **36**, 31–45.

[129a] HUNT T., HUNTER T. & MUNRO A. (1969) Control of haemoglobin synthesis: rate of translation of mRNA for the α and β chains. *J. molec. Biol.,* **43**, 123–33.

[130] ILAN J. & ILAN J. (1973) Sequence homology at the 5' termini of insect messenger RNAs. *Proc. nat. Acad. Sci. (Wash.),* **70**, 1355–58.

[131] IMAIZUMI T., DIGGELMANN H. & SCHERRER K. (1973) Demonstration of globin messenger sequences in giant nuclear precursors of messenger RNA a avian erythroblasts. *Proc. nat. Acad. Sci. (Wash.),* **70**, 1122–26.

[132] JAY G. & KAEMPFER R. (1974) Sequence of events in initiation of translation: a role for initiator tRNA in the recognition of messenger RNA. *Proc. nat. Acad. Sci. (Wash.),* **71**, 3199–203.

[133] JELINEK W., ADESNIK M., SALDITT M., SHEINESS D., WALL R., MOLLOY G., PHILIPSON L. & DARNELL J.E. (1973) Further evidence on the

nuclear origin and transfer to the cytoplasm of polyadenylic acid sequences in mammalian cell RNA. *J. molec. Biol.*, **75**, 515–32.

[134] JELINEK W., MOLLOY G., FERNANDEZ-MUNOZ R., SALDITT M. & DARNELL J.E. (1974) Secondary structure in heterogeneous nuclear RNA: involvement of regions from repeated DNA sites. *J. molec. Biol.*, **82**, 361–70.

[134a] JENKINS N., TAYLOR M.W. & RAFF R.A. (1973) *In vitro* translation of oogenetic messenger RNA of sea urchin eggs and picornavirus RNA with a cell-free system from sarcoma 180. *Proc. nat. Acad. Sci. (Wash.)*, **70**, 3287–91.

[135] JONES K.W. (1970) Chromosomal and nuclear location of mouse satellite DNA in individual cells. *Nature (Lond.)*, **225**, 912–15.

[136] JONES K.W. & CORNEO G. (1971) Location of satellite and homogeneous DNA sequences on human chromosomes. *Nature, New Biol.*, **233**, 268–71.

[137] JUDD B.H., SHEN M.W. & KAUFMAN T.C. (1972) The anatomy and function of a segment of the X chromosome of *Drosophila melanogaster*. *Genetics*, **71**, 139–56.

[138] JUDD B.H. & YOUNG M.W. (1973) An examination of the one cistron: one chromomere concept. *Cold Spr. Harb. Symp. quant. Biol.*, **38**, 573–79.

[139] KACIAN D.L., SPIEGELMAN S., BANK A., TERADA M., METAFORA S., DOW L. & MARKS H.A. (1972) *In vitro* synthesis of DNA components of human genes for globins. *Nature, New Biol.*, **235**, 167–69.

[140] KAFATOS F.C. (1972) The Cocoonase zymogen cells of silk moths: a model of terminal cell differentiation for specific protein synthesis. In *Current topics in developmental biology* (Eds A.A. Moscona & A. Monroy), Vol. 7, pp. 125–91.

[141] KAFATOS F.C. (1972) mRNA stability and cellular differentiation. *Karolinska Symposium on Research Methods in Reproductive Endocrinology*, **5**, 319–41.

[142] KEDES L.A. & BIRNSTIEL M.L. (1971) Reiteration and clustering of DNA sequences complementary to histone mRNA. *Nature, New Biol.*, **230**, 165–69.

[142a] KEDES L.H. & GROSS P.R. (1969) Identification in cleaving embryos of three RNA species serving as templates for the synthesis of nuclear proteins.. *Nature (Lond.)*, **223**, 1335–39.

[142b] KEDES L.H. & GROSS P.R. (1969) Synthesis and function of messenger RNA during early embryonic cleavage. *J. molec. Biol.*, **42**, 559–75.

[143] KEDES L.H., GROSS P.R., GOGNETTI G. & HUNTER A.L. (1969) Synthesis of nuclear and chromosomal proteins on light polyribosomes during cleavage in the sea urchin embryo. *J. molec. Biol.*, **45**, 337–51.

[143a] KLEIN W.H., MURPHY W., ATTARDI G., BRITTEN R.J. & DAVIDSON E.H. (1974) Distribution of repetitive and non-repetitive sequence transcripts in Hela mRNA. *Proc. nat. Acad. Sci (Wash.)*, **71**, 1785–89.

[144] KNOWLAND J.S. & GRAHAM C.F. (1972) RNA synthesis at the two cell stage of mouse development. *J. Embryol. exp. Morph.*, **27**, 167–76.

[145] KNOWLAND J. & MILLER L. (1970) Reduction of ribosomal RNA synthesis and ribosomal RNA genes in a mutant of *Xenopus laevis* which organizes only a partial nucleolus. I. Ribosomal RNA synthesis in embryos of different nucleolar types. *J. molec. Biol.*, **53**, 321–28.

[146] KUMAR A. & WU R.S. (1973) Role of ribosomal RNA transcription in ribosome processing in Hela cells. *J. molec. Biol.*, **80**, 265–76.

[147] KUMAR A. & WARNER J.R. (1972) Characterization of ribosomal precursor particles from Hela cell nucleoli. *J. molec. Biol.*, **63**, 233–46.

[148] KUNG C.S. (1974) On the size relationship between nuclear and cytoplasmic RNA in sea urchin embryos. *Devl. Biol.*, **36**, 343–56.

[149] LANE C.D., MARBAIX G. & GURDON J.B. (1971) Rabbit haemoglobin synthesis in frog cells: the translation of reticulocyte 9S RNA in frog oocytes. *J. molec. Biol.*, **61**, 73–91.

[150] LANDESMAN R. (1972) Ribosomal RNA synthesis in pre- and post-gastrula embryos of *Xenopus laevis*. *Cell Differentiation*, **1**, 209–13.

[151] LAMARCA M.J., SMITH L.D. & STROBEL M.C. (1973) Quantitative and qualitative analysis of RNA synthesis in stage 6 and stage 4 oocytes of *Xenopus laevis*. *Devl. Biol.*, **34**, 106–18.

[152] LAMBERT B. (1973) Repeated nucleotide sequences in a single puff of *Chironomus tentans* polytene chromosomes. *Cold Spr. Harb. Symp. quant. Biol.*, **38**, 637–44.

[152a] LASKEY R.A., GERHART J. & KNOWLAND J.S. (1973) Inhibition of ribosomal RNA synthesis in neurula cells by extracts from blastulae of *Xenopus laevis*. *Devl. Biol.*, **33**, 241–47.

[153] LEAVER C.J. & LOVETT J.S. (1974) An analysis of protein and RNA synthesis during encystment and outgrowth (germination) of *Blastocladiella zoospores*. *Cell Differentiation*, **3**, 165–92.

[153a] LEE C.S. & THOMAS C.A., Jr. (1973) Formation of rings from *Drosophila* DNA fragments. *J. Molec. Biol.*, **77**, 25–55.

[154] LEGON S., JACKSON R.J. & HUNT T. (1973) Control of protein synthesis in reticulocyte lysates by haemin. *Nature, New Biol.*, **241**, 150–52.

[154a] LEVY R., LEVY S., ROSENBERG S.A. & SIMPSON R.T. (1973) Selective stimulation of non-histone chromatin protein synthesis in lymphoid cells by phytohemagglutinin. *Biochemistry*, **12**, 224–28.

[155] LITTAUER U.Z. & INOUYE H. (1973) Regulation of tRNA. *A. Rev. Biochem.*, **42**, 439–70.

[156] LODISH H.F. (1970) Secondary structure of b?cteriophage f2 ribonucleic acid and the initiation of *in vitro* protein synthesis. *J. molec. Biol.*, **50**, 689–702.

[157] LODISH H.F. (1971) Alpha and beta globin messenger ribonucleic acid: different amounts and rates of initiation of translation. *J. biol. Chem.*, **246**, 7131–38.

[157a] MACGREGOR H.C. (1972) The nucleolus and its genes in amphibian oogenesis. *Biol. Rev.*, **47**, 177–210.

[158] MACNAUGHTON M., FREEMAN K.B. & BISHOP J.O. (1974) A precursor to haemoglobin mRNA in nuclei of immature duck red blood cells. *Cell*, **1**, 117–25.

[159] MADEN B.E.H. (1971) The structure and formation of ribosomes in animal cells. *Prog. in biophys. and molecular biology*, **22**, 127–77.

[160] MADEN B.E.H., SALIM M. & SUMMERS D.F. (1972) Maturation pathway for RNA in the Hela cell nucleus. *Nature, New Biol.*, **237**, 5–9.

[161] MADEN B.E.H. & VAUGHAN M.H. (1968) Synthesis of ribosomal proteins in the absence of ribosome maturation in methionine deficient Hela cells. *J. molec. Biol.*, **38**, 431–35.

[162] MAIRY M. & DENIS H. (1971) Recherches biochimiques sur l'oogenèse. Synthèse et accumulation du RNA pendant l'oogènese du crapaud sud-africain *Xenopus laevis*. *Devl. Biol.*, **24**, 143–65.

[163] MALCOLM D.B. & SOMMERVILLE J. (1974) The structure of chromosome-derived ribonucleoprotein in oocytes of *Triturus cristatus carnifex* (Laurenti). *Chromosoma*, **48**, 137–58.

[164] MANNING J.E., SCHMID C.W. & DAVIDSON N. (1975) Interspersion of repetitive and non-repetitive DNA sequences in the *Drosophila melanogaster* genome. *Cell*, **4**, 141–55.

[165] MARZLUFF W.F., MURPHY E.C. & HUANG R.C.C. (1974) Transcription of genes for 5S ribosomal RNA and transfer RNA in isolated mouse myeloma cell nuclei. *Biochemistry*, **13**, 3689–96.

[166] MERRIAM R.W. (1969) Movement of cytoplasmic proteins into nuclei induced to enlarge and initiate DNA or RNA synthesis. *J. Cell Sci.*, **5**, 333–49.

[166a] MESCHER A. & HUMPHREYS T. (1974) Activation of maternal mRNA in the absence of poly(A) formation in fertilized sea urchin eggs. *Nature (Lond.)*, **249**, 138–39.

[167] METAFORA S., FELICETTI L. & GAMBINO R. (1971) The mechanism of protein synthesis activation after fertilization of sea urchin eggs. *Proc. nat. Acad. Sci. (Wash.)*, **68**, 600–4.

[168] MACGILLIVRAY A.J. & RICKWOOD D. (1974) The heterogeneity of mouse chromatin non-histone proteins as evidenced by two dimensional polyacrylamide gel electrophoresis and ion exchange chromatography. *Eur. J. Biochem.*, **41**, 181–90.

[169] MCKNIGHT G.S. & SCHIMKE R.T. (1974) Ovalbumin messenger RNA: evidence that the initial product of transcription is the same size as polysomal ovalbumin messenger. *Proc. nat. Acad. Sci. (Wash.)*, **71**, 4327–31.

[169a] MILCAREK C., PRICE R. & PENMAN S. (1974) The metabolism of a poly(A) minus mRNA fraction in Hela cells. *Cell*, **3**, 1–10.

[170] MILLER O.L. & BAKKEN A.H. (1972) Morphological studies of transcription. *5th Karolinska Symposium on Research Methods in Reproductive Endocrinology*, pp. 155–73.

[170a] MILLER L. (1974) Metabolism of 5S RNA in the absence of ribosome production. *Cell*, **3**, 275–81.

[171] MILLER L. & BROWN D.D. (1969) Variation in the activity of nucleolar organisers and their ribosomal gene content. *Chromosoma*, **28**, 430–44.

[172] MILLER L. & KNOWLAND J.S. (1970) Reduction of ribosomal RNA synthesis and ribosomal RNA genes in a mutant of *Xenopus laevis* which organises only a partial nucleolus. I. Ribosomal RNA synthesis in embryos of different nucleolar types. *J. molec. Biol.*, **53**, 321–28.

[173] MILSTEIN C., BROWNLEE G.G., CARTWRIGHT E.M., JARVIS J.M. & PROUDFOOT N.J. (1974) Sequence analysis of immunoglobulin light chain messenger RNA. *Nature (Lond.)*, **252**, 354–59.

[174] MOLLOY G.R. & DARNELL J.E. (1973) Characterization of the poly adenylic acid regions and the adjacent nucleotides in heterogeneous nuclear RNA and messenger RNA from Hela cells. *Biochemistry*, **12**, 2324–30.

[175] MOLLOY G., JELINEK W., SALDITT M. & DARNELL J.E. (1974) Arrangement of specific oligonucleotides within poly(A) terminated HnRNA molecules. *Cell*, **1**, 43–53.

[176] MOLLOY G.R., THOMAS W.L. & DARNELL J.E. (1972) Occurrence of uridylate rich oligonucleotide regions in heterogeneous nuclear RNA of Hela cells. *Proc. nat. Acad. Sci. (Wash.)*, **69**, 3684–88.

[177] MORIYAMA Y., HODNETT J.L., PRESTAYKO A.W. & BUSCH H. (1969) Studies on the nuclear 4–7S RNA of the Novikoff Hepatoma. *J. molec. Biol.*, **39**, 335–49.

[177a] NEMER M., GRAHAM M. & DUBROFF L.M. (1974) Co-existence of non-histone messenger RNA species lacking and containing polyadenylic acid in sea urchin embryos. *J. molec. Biol.*, **89**, 435–54.

[178] O'BRIEN S.J. (1973) On estimating functional gene number in eukaryotes. *Nature, New Biol.*, **242**, 52.

[179] PAPACONSTANTINOU J. & JULKA E. (1969) The regulation of ribosomal RNA synthesis and ribosomal assembly in vertebrate lens. *J. Cell Physiol.*, **72**, supplement 1, 161–78.

[180] PARDUE M.L. & GALL J.G. (1970) Chromosomal localization of mouse satellite DNA. *Science*, **168**, 1356–58.

[181] PATERSON B.M., ROBERTS B.E. & YAFFE D. (1974) Determination of actin messenger RNA in cultures of differentiating embryonic chick skeletal muscle. *Proc. nat. Acad. Sci. (Wash.)*, **71**, 4467–71.

[182] PAUL J. (1972) General theory of chromosome structure and gene activation in eukaryotes. *Nature (Lond.)*, **238**, 444–46.

[183] PAUL J., CARROLL D., GILMOUR R.S., MORE J.A.R., THRELFALL G., WILKIE M. & WILSON S. (1972) Functional studies on chromatin. *5th Karolinska Symposium on Research Methods in Reproductive Endocrinology*, pp. 277–95.

[184] PAUL J. & HUNTER J.A. (1969) Synthesis of macromolecules during induction of haemoglobin synthesis by erythroprotein. *J. molec. Biol.*, **42**, 31–41.

[185] PEDERSON T. (1974) Gene activation in eukaryotes: are nuclear acidic proteins the cause or the effect? *Proc. nat. Acad. Sci. (Wash.)*, **71**, 617–21.

[186] PELLING C. (1959) Chromosomal synthesis of ribonucleic acid as shown by incorporation of uridine labelled with tritium. *Nature (Lond.)*, **184**, 655–56.

[187] PERKOWSKA E., McGREGOR H.C. & BIRNSTIEL M.L. (1968) Gene amplification in the oocyte nucleus of mutant and wild type *Xenopus laevis. Nature (Lond.)*, **217**, 649–50.

[188] PERRY R.P. & KELLEY D.E. (1968) Persistent synthesis of 5S RNA when production of 28S and 18S ribosomal RNA is inhibited by low doses of Actinomycin D. *J. Cell Physiol.*, **72**, 235–46.

[189] PERRY R.P. & KELLEY D.E. (1974) Existence of methylated messenger in mouse L cells. *Cell*, **1**, 37–42.

[190] PERRY R.P., KELLEY D.E. & LA TORRE J. (1974) Synthesis and turnover of nuclear and cytoplasmic polyadenylic acid in mouse L cells. *J. molec. Biol.*, **82**, 315–31.

[191] PRESTAYKO A.W. & BUSCH H. (1968) Low molecular weight RNA of the chromatin fraction from the novikoff hepatoma and rat liver nuclei. *Biochem. biophys. Acta*, **169**, 327–37.

[192] PRESTYAKO A.W., TONATO M. & BUSCH H. (1970) Low molecular weight RNA associated with 28S nucleolar RNA. *J. molec. Biol.*, **47**, 505–15.

[193] PRESTYAKO A.W., TONATO M., LEWIS B.C. & BUSCH H. (1971) Heterogeneity of nucleolar U_3 ribonucleic acid of the Novikoff Hepatoma. *J. biol. Chem.*, **246**, 182–87.

[194] PROUDFOOT N.J. & BROWNLEE G.G. (1974) Sequence at the 3' end of globin mRNA shows homology with immunoglobulin light chain mRNA. *Nature (Lond.)*, **252**, 259–62.

[194a] PYERITZ R.E. & THOMAS C.A., Jr. (1973) Regional organization of eukaryotic DNA sequences as studied by the formation of folded rings. *J. molec. Biol.*, **77**, 57–73.

[195] RAFF R.A., COLOT H.V., SELVIG S.E. & GROSS P.R. (1972) Oogenetic origin of messenger RNA for embryonic synthesis of microtubule proteins. *Nature (Lond.)*, **235**, 211–14.

[195a] RAFF R.A., GREENHOUSE G., GROSS K.W. & GROSS P.R. (1971) Synthesis and storage of microtubule proteins by sea urchin embryos. *J. Cell Biol.*, **50**, 516–27.

[196] REEDER R.H. & BROWN D.D. (1970) Transcription of the ribosomal RNA genes of an amphibian by the RNA polymerase of a bacterium. *J. molec. Biol.*, **51**, 361–77.

[197] RINALDI A.M. & MONROY A. (1969) Polyribosomal formation and RNA synthesis in early post fertilization stages of the sea urchin egg. *Devl. Biol.*, **19**, 73–86.

[198] RITOSSA F.M., ATWOOD K.C. & SPIEGELMAN S. (1966) A molecular explanation for the bobbed mutants of *Drosophila* as partial deficiencies of 'ribosomal' DNA. *Genetics*, **54**, 819–34.

[199] ROBERTSON H.D., ALTMAN S. & SMITH J.D. (1972) Purification and properties of a specific *Escherichia coli* ribonuclease which cleaves a tyrosine transfer ribonucleic acid precursor. *J. biol. Chem.*, **247**, 5243–51.

[200] ROBERTUS J.D., LADNER J.E., FINCH J.T., RHODES D., BROWN R.S., CLARK B.F.C. & KLUG A. (1974) Structure of yeast phenylalanine tRNA at 3 Å resolution. *Nature (Lond.)*, **250**, 546–51.

[201] ROCHAIX J.D., BIRD A. & BAKKEN A. (1974) Ribosomal RNA gene amplification by rolling circles. *J. molec. Biol.*, **87**, 473–87.

[202] RO CHOI T.S., MORIYAMA Y., CHOI Y.C. & BUSCH H. (1970) Isolation and purification of a nuclear 4–5S ribonucleic acid of the Novikoff Hepatoma. *J. biol. Chem.*, **245**, 1970–77.

[203] ROEDER R.G., REEDER R.H. & BROWN D.D. (1970) Multiple forms of RNA polmerase in *Xenopus laevis*: their relationship to RNA synthesis *in vivo* and their fidelity of transcription *in vitro*. *Cold Spr. Harb. Symp. quant. Biol.*, **35**, 727–35.

[204] ROEDER R.G. & RUTTER W.J. (1969) Multiple forms of DNA-dependent RNA polymerase in eukaryotic organisms. *Nature (Lond.)*, **224**, 234–37.

[205] ROEDER R.G. & RUTTER W. (1970) Specific nucleolar and nucleoplasmic RNA polymerases. *Proc. nat. Acad. Sci. (Wash.)*, **65**, 675–82.

[206] ROEDER R.G. & RUTTER W.J. (1970) Multiple RNA polymerases and RNA synthesis during sea urchin development. *Biochemistry*, **9**, 2543–53.

[207] ROSBASH M.R. & FORD P.J. (1974) Polyadenylic-acid containing RNA in *Xenopus laevis* oocytes. *J. molec. Biol.*, **85**, 87–101.

[208] ROSBASH M.R., FORD P.J. & BISHOP J.O. (1974) Analysis of the *c* value paradox by molecular hybridization. *Proc. nat. Acad. Sci. (Wash.)*, **71**, 3746–50.

[209] RUDERMAN J.V., BAGLIONI C. & GROSS P.R. (1974) Histone mRNA and histone synthesis during embryogenesis. *Nature (Lond.)*, **247**, 36–8.

[210] RUDERMAN J.V. & GROSS P.R. (1974) Histones and histone synthesis in sea urchin development. *Devl. Biol.*, **36**, 286–98.

[210a] RUTTER W.J., PICTET R.L. & MORRIS P.W. (1973) Toward molecular

mechanisms of developmental processes. *A. Rev. Biochem.*, **42**, 601–46.

[211] SALIM M. & MADEN B.E.H. (1973) Early and late methylations in Hela cell ribosome maturation. *Nature (Lond.)*, **244**, 334–36.

[212] SCHEER U. (1973) Nuclear pore flow rate of ribosomal RNA and chain growth rate of its precursor during oogenesis of *Xenopus laevis*. *Devl. Biol.*, **30**, 13–28.

[213] SCHULTZ G.A. (1973) Characterization of polyribosomes containing newly synthesized messenger RNA in preimplantation rabbit embryo. *Expl. Cell Res.*, **82**, 168–74.

[214] SCHULTZ G.A., CHEN D. & KATCHALSKI E. (1972) Localization of a messenger RNA in a ribosomal fraction from ungerminated wheat embryos. *J. molec. Biol.*, **66**, 379–90.

[215] SCHULTZ G., MANES C. & HAHN W.E. (1973) Synthesis of RNA containing polyadenylic acid sequences in preimplantation rabbit embryos. *Devl. Biol.*, **30**, 418–26.

[216] SCHWARZ M.C. (1970) Nucleic acid metabolism in oocytes and embryos of *urechis caupo*. *Devl. Biol.*, **23**, 241–60.

[217] SCONZO G. & GIUDICE G. (1971) Synthesis of ribosomal RNA in sea urchin embryos. V. Further evidence for an activation following hatching blastula stage. *Biochim. biophys. Acta*, **254**, 447–51.

[218] SCOTT S.E.M. & SOMMERVILLE J. (1974) Location of nuclear proteins on the chromosomes of newt oocytes. *Nature (Lond.)*, **250**, 680–82.

[219] SELVIG S.E., GROSS P.R. & HUNTER A.L. (1970) Cytoplasmic synthesis of RNA in the sea urchin embryo. *Devl. Biol.*, **22**, 343–65.

[220] SHANNON M.P., KAUFMAN T.C., SHEN M.W. & JUDD B.H. (1972) Lethality patterns and morphology of selected lethal and semi-lethal mutations in the zeste-white region of *Drosophila melanogaster*. *Genetics*, **72**, 615–38.

[221] SHEINESS D. & DARNELL J.E. (1973) Polyadenylic acid segment in mRNA becomes shorter with age. *Nature, New Biol.*, **241**, 265–68.

[222] SHIOKAWA K. & YAMANA K. (1967) Pattern of RNA synthesis in isolated cells of *Xenopus laevis* embryos. *Devl. Biol.*, **16**, 368–88.

[222a] SHIOKAWA K. & YAMANA K. (1967) Inhibitor of ribosomal RNA synthesis in *Xenopus laevis* embryos. *Devl. Biol.*, **16**, 389–406.

[223] SIPPEL A.E., STAVRIANOPOULOS J.G., SCHUTZ G. & FEIGELSON P. (1974) Translational properties of rabbit globin mRNA after specific removal of poly(A) with ribonuclease H. *Proc. nat. Acad. Sci. (Wash.)*, **71**, 4635–39.

[224] SKOULTCHI A. & GROSS P.R. (1973) Maternal histone messenger RNA: detection by molecular hybridization. *Proc. nat. Acad. Sci. (Wash.)*, **70**, 2840–44.

[225] SLATER I., GILLESPIE D. & SLATER D.W. (1973) Cytoplasmic adenylation and processing of maternal RNA. *Proc. nat. Acad. Sci. (Wash.)*, **70**, 406–11.

[226] SLATER D.W., SLATER I. & GILLESPIE D. (1972) Post fertilization synthesis of polyadenylic acid in sea urchin embryos. *Nature (Lond.)*, **240**, 333–37.

[227] SMITH L.D. & ECKER R.E. (1970) Regulatory processes in the maturation and early cleavage of amphibian eggs. *Current Top. in Devel. Biol.*, **5**, 1–38.

[228] SMITH M.J., HOUGH B.R., CHAMBERLIN M. & DAVIDSON E.H. (1974) Repetitive and non-repetitive sequence in sea urchin heterogeneous nuclear RNA. *J. molec. Biol.*, **85**, 103–26.

[229] SNOW M.H.L. & CALLAN H.G. (1969) Evidence for a polarized movement of the lateral loops of newt lampbrush chromosomes during oogenesis. *J. Cell Sci.*, **5**, 1–25.

[230] SOLL D.R. & FULTON C. (1974) The constancy of RNA-polymerase activities during a transcriptionally dependent cell differentiation in *Naegleria*. *Devl. Biol.*, **36**, 236–44.

[231] SOMMERVILLE J. (1973) Ribonucleoprotein particles derived from lampbrush chromosomes of newt oocytes. *J. molec. Biol.*, **78**, 487–503.

[231a] SOMMERVILLE J. & HILL R.J. (1973) Proteins associated with heterogeneous nuclear RNA of newt oocytes. *Nature, New Biol.*, **245**, 104–06.

[232] SOUTHERN E. (1974) Eukaryotic DNA. In *MTP international review of science biochemistry of nucleic acids, series one* (Ed. K. Burton), Volume 6, Chapter 4.

[233] SOUTHERN E.M. (1970) Base sequence and evolution of guinea pig α satellite DNA. *Nature (Lond.)*, **227**, 794–98.

[234] SPEAR B.B. & GALL J.G. (1973) Independent control of ribosomal gene replication in polytene chromosomes of *Drosophila melanogaster*. *Proc. nat. Acad. Sci. (Wash.)*, **70**, 1359–63.

[235] STEDMAN E. & STEDMAN E. (1947) The chemical nature and functions of the components of cell nuclei. *Cold Spr. Harb. Symp. quant. Biol.*, **12**, 224–36.

[236] STEIN G., CHANDHURI S. & BASERGA R. (1972) Gene activation in WI-38 fibroblasts stimulated to proliferate: role of non-histone chromosomal proteins. *J. biol. Chem.*, **247**, 3918–22.

[237] STEVENS R.H. & WILLIAMSON A.R. (1973) Isolation of nuclear pre-mRNA which codes for immunoglobulin heavy chain. *Nature, New Biol.*, **245**, 101–04.

[238] SULLIVAN D., PALACIOS R., STAVNEZER J., TAYLOR J.M., FARAS A.J., KIELY M.L., SUMMERS N.M., BISHOP J.M. & SCHIMKE R.T. (1973) Synthesis of a deoxyribonucleic acid sequence complementary to ovalbumin messenger ribonucleic acid and quantification of ovalbumin genes. *J. biol. Chem.*, **248**, 7530–39.

[239] SUZUKI Y., GAGE L.P. & BROWN D.D. (1972) The genes for silk fibroin in *Bombyx mori*. *J. molec. Biol.*, **70**, 637–49.

[240] TEMIN H.M. & MIZUTANI S. (1970) RNA-dependent DNA polymerase in virus of *Rous sarcoma* virus. *Nature (Lond.)*, **226**, 1211–13.

[241] TENG C.S. & HAMILTON T.H. (1969) Role of chromatin in estrogen action in the uterus II hormone-induced synthesis of non-histone acidic proteins which restore histone-inhibited DNA dependent RNA synthesis. *Proc. nat. Acad. Sci. (Wash.)*, **63**, 465–72.

[242] TENG C.S., TENG C.T. & ALLFREY V.G. (1971) Studies of nuclear acidic proteins. Evidence for their phosphorylation, tissue specificity, selective binding to DNA and stimulatory effects on transcription. *J. biol. Chem.*, **246**, 3597–609.

[242a] THOMAS C.A., Jr., HAMKALO B.A., MISRA D.N. & LEE C.S. (1970) Cyclization of eukaryotic DNA fragments. *J. molec. Biol.*, **51**, 621–31.

[243] TSANEV R. & SENDOV B.I. (1971) Possible molecular mechanism for cell differentiation in multicellular organisms. *J. theor. Biol.*, **30**, 337–93.

[244] VERMA I.M., TEMPLE G.F., FAN H. & BALTIMORE D. (1972) *In vitro* synthesis of DNA complementary to rabbit reticulocyte 10S RNA. *Nature, New Biol.*, **235**, 163–66.

[245] WADA K., SHIOKAWA K. & YAMANA K. (1968) Inhibitor of ribosomal RNA synthesis in *Xenopus laevis* embryos. I. Changes in activity of the inhibitor during development and its distribution in early gastrulae. *Expl. Cell Res.*, **52**, 252–60.

[246] WALKER P.M.B. (1971) Origin of satellite DNA. *Nature (Lond.)*, **229**, 306–308.

[247] WALKER P.M.B. (1971) Repetitive DNA in higher organisms. *Prog. Biophys. and Molecular Biol.*, **23**, 147–90.

[248] WEEKS D.P. & MARCUS A. (1971) Preformed messenger RNA of quiescent wheat embryos. *BBA*, **232**, 671–84.

[249] WEGNEZ M., MONIER R. & DENIS H. (1972) Sequence heterogeneity of 5S RNA in *Xenopus laevis*. *FEBS Letters*, **25**, 13–20.

[250] WEINBERG R.A. (1973) Nuclear RNA metabolism. *A. Rev. Biochem.*, **42**, 329–54.

[251] WEINBERG E.S., BIRNSTIEL M.L., PURDOM I.F. & WILLIAMSON R. (1972) Genes coding for polysomal 9S RNA of sea urchins: conservation and divergence. *Nature (Lond.)*, **240**, 225–28.

[252] WEINBERG R.A. & PENMAN S. (1970) Processing of 45S nucleolar RNA. *J. molec. Biol.*, **47**, 169–78.

[253] WEINMANN R. & ROEDER R.G. (1974) Role of DNA-dependent RNA polymerase III in the transcription of the tRNA and 5S RNA genes. *Proc. nat. Acad. Sci. (Wash.)*, **71**, 1790–94.

[254] WELLAUER P.K., REEDER R.H., CARROLL D., BROWN D.D., DEUTCH A., HIGASHINAKAGAWA T. & DAWID I.B. (1974) Amplified ribosomal DNA from *Xenopus laevis* has heterogeneous spacer lengths. *Proc. nat. Acad. Sci. (Wash.)*, **71**, 2823–27.

[255] WELLAUER P.K. & DAWID I.B. (1974) Secondary structure maps of

ribosomal RNA and DNA. I. Processing of *Xenopus laevis* ribosomal RNA and structure of single stranded ribosomal DNA. *J. molec. Biol.*, **89**, 379–95.

[256] WENSINK P.C. & BROWN D.D. (1971) Denaturation map of the ribosomal DNA of *Xenopus laevis*. *J. molec. Biol.*, **60**, 235–48.

[257] WESSELLS N.K. & WILT F.H. (1965) Action of Actinomycin D on exocrine pancreas cell differentiation. *J. molec. Biol.*, **13**, 767–79.

[258] WHITTAKER J.R. (1973) Segregation during ascidian embryogenesis of egg cytoplasmic information for tissue specific enzyme development. *Proc. nat. Acad. Sci. (Wash.)*, **70**, 2096–2100.

[259] WIGLE D.T. & SMITH A.E. (1973) Specificity in initiation of protein synthesis in a fractionated mammalian cell free system. *Nature, New Biol.*, **242**, 136–40.

[260] WILSON E.B. (1925) *The cell in development and heredity.* Macmillan.

[261] WILSON D.A. & THOMAS C.A., Jr. (1974) Palindromes in chromosomes. *J. molec. Biol.*, **83**, 115–45.

[262] WILT F.H. (1973) Polyadenylation of maternal RNA of sea urchin eggs after fertilization. *Proc. nat. Acad. Sci. (Wash.)*, **70**, 2345–49.

[263] WILT F.H. (1965) Regulation of initiation of chick embryo haemoglobin synthesis. *J. molec. Biol.*, **12**, 331–41.

[264] WOODLAND H.R. & GURDON J.B. (1969) RNA synthesis in an Amphibian nuclear transplant hybrid. *Devl. Biol.*, **20**, 89–104.

[265] WU J.R., HURN J. & BONNER J. (1972) Size and distribution of the repetitive segments of the *Drosophila* genome. *J. molec. Biol.*, **64**, 211–19.

[266] ZYLBER .E.A. & PENMAN S. (1971) Synthesis of 5S and 4S RNA in metaphase arrested Hela cells. *Science*, **172**, 947–49.

Conclusions to Part 5

Self-assembly

1. BIOLOGICAL STRUCTURES OFTEN CONSIST OF AGGREGATES OF LIKE MOLECULES

Collagen, myosin, tubullin, and flagellin aggregate to form long fibres and phospholipids aggregate to form membranes. The advantage of building biological structures from like molecules is that only a few genes are required, and it is probably for this advantage that many viruses only consist of aggregates of two types of molecule (ribonucleic acid and many identical proteins).

2. THE FORM OF SUCH STRUCTURES USUALLY DEPENDS ON THE PROPERTIES OF THE MOLECULES

In some cases the form of a molecule is likely to be determined by its amino acid sequence, e.g. ribonuclease, tropcollagen, tropomyosin. Molecules are also capable of self-assembly into larger biological structures under particular ionic conditions; this means that a cell could provide for the formation of a microtubule or a tendon by synthesizing either tubullin or collagen in a controlled environment.

3. BACTERIOPHAGES AND RIBOSOMES ASSEMBLE IN A SEQUENCE

The dissociated components of a ribosome can self-assemble into an active ribosome. The assembly requires warming and occurs far more quickly if the components (16S ribonucleic acids and 21 different proteins for the 30S subunit) are added in a particular order. Similarly most parts of a bacteriophage can be assembled from the components of the mature organism presented in a particular sequence; however in this case a catalytic gene product not present in the mature organism is required for one of the assembly steps. It is therefore likely that at this level of biological complexity, a cell must not only synthesize the structural proteins, but must also specify the sequence of their inclusion into the structure and provide catalysts. In life, biological structures are formed in pre-existing cells and their organization may provide templates and constraints which do not immediately depend on gene activity.

Cell organelles

4. MITOCHONDRIA AND CHLOROPLASTS ARE COMPLEX AND CANNOT SELF-ASSEMBLE

Mitochondria and chloroplasts are more complex (contain more different molecules) than ribosomes and bacteriophages, and they always develop from pre-existing structures despite the self-assembly properties of some chloroplast components. They contain DNA and are maternally inherited in animals and inherited through either or both germ lines in plants.

5. THE ORGANELLES DEPEND ON NUCLEAR GENES FOR REPLICATION AND ACTIVITY

The nucleus codes for the enzymes required for organelle replication (DNA polymerase) and for transcription of organelle genes (RNA polymerase). The organelle genome indeed codes for a variety of products (ribonucleic acids, subunits of cytochrome oxidase, ATPase and ribolase diphospate carboxylase) but these are not functional by themselves and require additional components from the cell's cytoplasm (e.g. ribosomal proteins, synthetases, other enzyme subunits).

Control of Gene Expression

6. ANIMAL GENOMES CONTAIN MORE DNA AND ARE MORE COMPLEX THAN THOSE OF CELL ORGANELLES AND BACTERIA

Insect, amphibian, and mammalian genomes contain 20 to 3,000 times more DNA than a bacterium. Much of this DNA is in repeating sequences: highly repetitive sequences (about 7% of total) are not transcribed; intermediate repetitive sequences (between 15 and 31% of total) include genes which are transcribed (e.g. r-RNA, t-RNA, 5S RNA, and histones) and non-transcribed regions. Single copy sequences, which include the structural genes which code for proteins, make up to 80% of the rest of the genome. The single copy DNA (mammalian) has a sequence complexity 1000 times greater than that of bacteria and its expression must be controlled as cells develop and synthesize different proteins.

7. DIFFERENCES BETWEEN DIFFERENTIATED CELL TYPES

A comparison of the variety of proteins synthesized by different cell types indicates the precision of protein synthesis control which is required for development. Most cell types have in common the many proteins needed for cell maintenance and respiration and only qualitatively differ in the synthesis of a few specialized products (e.g. globin in reticulocytes, silk fibroin in the silk gland). Cells which do specialize in the production of a few proteins, contain large quantities of stable messenger RNA and t-RNA populations suitable for the translation of these messages (silk gland). The problem is then to discover how these

particular RNA populations arise and are maintained in different cell types.

8. CONTROL OF TRANSCRIPTION

The amount of RNA transcribed by different genes varies widely; this is shown by autoradiographic studies on insect polytene chromosomes and by studying the amounts of different classes of RNA synthesized during oögenesis and development. In the latter case it can be shown that the cytoplasm controls r-RNA synthesis; there is no direct correlation between the rate of r-RNA synthesis and either the number of gene copies or the amount of the special r-RNA polymerase and so this control is exercised through an unknown factor.

9. CONTROL OF RNA PROCESSING

In the sea urchin gastrula, transcripts of 20 to 30% of the single copy DNA are found in the nucleus while only one-tenth of these are found in polysomes. The implication is that only a selection of the messenger RNA molecules produced in the nucleus may reach the protein synthesizing machinery.

The most certain information about RNA processing comes from studies on r-RNA. The genes for 18 and 28S r-RNA are grouped so that one pair of 18 and 28S genes are separated from another pair by nontranscribed spaces. The initial transcript contains both the 18 and 28S sequence; these are cleaved from each other with the loss of some of the initial transcript and some of their bases are modified by methylation.

Single copy DNA transcripts (presumably including messenger RNA) are probably processed similarly. Single copy sequences in the genome are interspersed with intermediate repetitive sequences and many of their transcripts in the nucleus are attached to transcripts of intermediate repetitive regions; the latter appear to be reduced as the molecules move to the cytoplasm. Poly-A is added to some of the nuclear transcripts and most messenger RNA molecules in the cytoplasm contain poly-A. It is therefore possible that control of processing could determine which messenger RNA molecules are available for translation in the cytoplasm; certainly much of the RNA synthesized in the nucleus during early development never reaches the cytoplasm.

10. CONTROL OF TRANSLATION

Messenger RNA molecules, such as those for histones and microtubular protein, are stored in organisms in preparation for rapid growth; long lived messenger RNA is therefore found in the unfertilized eggs of animals and the ungerminated seeds and spores of plants and fungi. Following fertilization or germination these messenger RNAs are translated but the mechanism of this activation, which is probably not messenger RNA specific, is not known.

The frequency with which different messages are translated seems to depend on the efficiency of chain initiation which may be influenced by initiation factors or co-factors (e.g. haemin in globin synthesis). However these factors are unlikely to provide a basis for the control of specific protein synthesis by particular cell types. It is much more likely that this is achieved by control of transcription and processing.

11. DEVELOPMENTAL CONTROL OF PROTEIN SYNTHESIS

Several experiments demonstrate that the cytoplasm and membranes control gene expression in development (e.g. nuclear transplant in amphibia and studies on localization in the eggs). The molecular basis of this control is not understood in any eukaryotic system, but the changing pattern of gene expression which occurs in development provides suitable situations for solving this problem. The strategy of control must conform to the embryological rules described in other parts of the book.

Part 6
Environmental Control of Development

Chapter 6.1
Environmental Control
of Development

6.1.1 INTRODUCTION

So far we have paid little attention to the influences of environmental factors in development, yet in many organisms, particularly plants and insects, environmental stimuli play a vital role in controlling key steps in development. For several reasons, growth and development in plants appear to be more profoundly affected by environmental influences than in animals. Firstly, the 'open ended' pattern of development of plants, resulting from their mode of growth by apical meristems, renders the plant body much more susceptible to modification by such external factors as light, temperature, water supply and mineral nutrition. Secondly, being sedentary organisms, land plants can take little action to avoid environmental stresses and hence they have tended to become adapted to *tolerate* adverse conditions, whereas many animals are able to *avoid* seasonal stresses by migration, hibernation and so on.

It is useful to distinguish between (1) *direct,* non-adaptive effects of environmental factors, such as freezing temperatures or drought, on the organism and (2) *indirect,* adaptive responses of the organism to an environmental factor, such as unilateral illumination or gravity, in which the plant may show an active, 'programmed' response, e.g. a growth curvature. In such cases the environmental factor acts as a stimulus to which the organism responds in an adaptive manner. Among the most interesting examples of environmental control of development are adaptive responses to seasonal climatic changes, and we shall confine our attention to such cases in this introductory section.

The seasonal alternation between warm summers and cold winters in temperate regions has had a profound effect upon the life-cycles of many plants and insects, and many species show an alternation of active and dormant phases which are synchronized with the seasonal climatic changes. To achieve this coordination such organisms need to be able to 'tell the seasonal time', and many do so by responding to seasonal changes in daylength, which vary extremely regularly and predictably. It is well known, of course, that a variety of developmental responses in a wide range of organisms, including not only plants but also insects, birds and even mammals, are controlled by daylength, a phenomenon known as *photoperiodism.*

6.1.2 PHOTOPERIODIC CONTROL OF FLOWERING

One of the best studied processes controlled by daylength is flowering in plants. (For an introduction to this subject see [10], and for more detailed treatment see [9].) In some plant species flowering is promoted by short days ('short-day plants'), in others by long days ('long-day plants'), while many other species are relatively insensitive to variations in daylength and are said to be 'day neutral'. In general, short-day plants are either autumn flowering species of temperate regions (e.g. the garden chrysanthemum) or plants of tropical regions (e.g. sugar cane, maize, rice), while long-day plants are summer-flowering plants of temperate regions.

The transition from the vegetative to the reproductive phase involves a profound change in the structure of the shoot apical meristems. The structure of the vegetative meristem of flowering plants has been described earlier (p. 35). During the change to a flowering apex, the tunica, corpus and subsidiary zones become obliterated and cell division becomes limited to the outer layers, giving rise to a 'meristematic mantle' overlying a central pith of vacuolated cells (Fig. 6.1.1). As a result of this radical change in organization, the apex loses its capacity for indefinite growth and gives rise to the flower which is a structure of 'determinate' growth. Thus, it is clear that the transition of a shoot apex from the vegetative to the reproductive state involves a radical switch in developmental pathways. Evidently, in plants in which flowering is sensitive to daylength this developmental switch is environmentally controlled. We have already seen (p. 48) that in many other plant species flower initiation is induced by winter chilling, thus providing another example of environmental control of development in plants.

Although flower initiation involves changes at the shoot apex, it is controlled by the daylength conditions to which the *leaves* are exposed. Thus, flower initiation must involve the transmission of some stimulus from the leaves to the shoot apex, and there is much circumstantial evidence that this stimulus must be of a hormonal nature. However, attempts to isolate the hypothetical 'flower hormone' have so far proved abortive. There is, indeed, considerable difference of opinion as to whether there is a specific flowering stimulus, or whether flowering is controlled by variations in the levels of non-specific

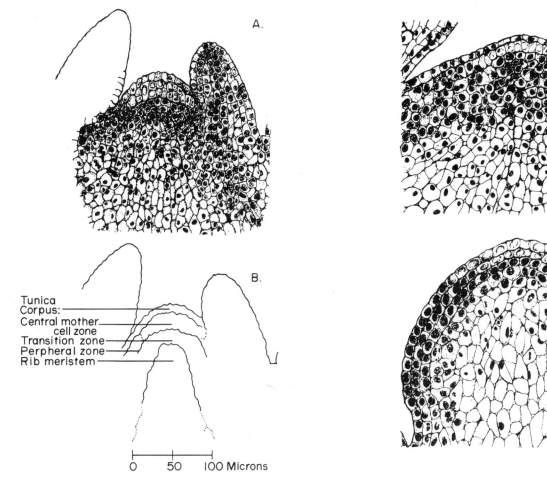

Tunica
Corpus:
Central mother
        cell zone
Transition zone
Perpheral zone
Rib meristem

0   50   100 Microns

Fig. 6.1.1 Histological changes during transition from the vegetative to the reproductive condition in *Xanthium strumarium*. A. Median section through vegetative apex. B. Diagram showing zonation of A. C. Beginning of the transition to the floral bud. D. Floral stage, showing the meristematic mantle overlying the enlarged rib meristem. (From F.B. Salisbury *The Flowering Process,* Pergamon Press, Oxford, 1963.)

hormones, such as gibberellins or cytokinins. It is of interest that external application of gibberellic acid will induce flowering of some long-day plants under non-inductive daylength conditions and there is evidence that the levels of endogenous gibberellins and cytokinins are affected by daylength. Moreover, gibberellic acid also stimulates flowering in some species which normally require winter chilling. However, there is other evidence that the flowering stimulus is identical in both short-day and long-day plants, and yet gibberellins are generally not effective in stimulating flowering in short-day plants. Hence, it would seem that the flowering stimulus is not a gibberellin.

Whatever the nature of the flowering stimulus, it appears that in short-day plants flower-promoting processes occur during the dark period. Thus, flowering in short-day plants will only occur when the daily dark period exceeds a certain length ('critical dark period') and the effect of the dark period is completely nullified (i.e. flowering is inhibited) if it is interrupted by as little as 1 min of light. The most effective part of the spectrum in this 'night break' effect is the red region (660 nm), and the effect of a short interruption by red light can be reversed by far-red irradiation (in the region of 730 nm), thus indicating that the photo-receptor phytochrome (p. 353) is involved. As will be shown in Chapter 6.2 phytochrome can exist in two alternative states and on exposure to red light is converted from a red-absorbing form (Pr) into a far-red absorbing form (Pfr), with the reverse process occurring on exposure to far-red. That is:

$$Pr \underset{\text{Far Red}}{\overset{\text{Red}}{\rightleftharpoons}} Pfr$$

Since red light given during the dark period inhibits flowering in short-day plants, it would appear that Pfr is inhibitory to the flower-promoting processes occurring during the dark period and that it is necessary for the levels of Pfr to decay to a certain level during the early part of the dark period. It is postulated that the length of the critical dark period is determined by some time-measuring process ('internal clock'), and that flower hormone synthesis proceeds when this critical dark period is exceeded.

These ideas are summarized in the following diagram:

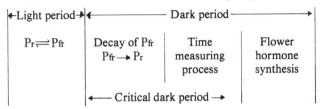

Whether or not this scheme is valid, it seems clear that certain steps in the processes involved in flowering are phytochrome-controlled. Moreover, it may be of significance that short periods of red light effect rapid increases in the levels of endogenous gibberellins and cytokinins in plant tissues (p. 367).

6.1.3 PHOTOPERIODISM AND REPRODUCTION IN ANIMALS

The successful reproduction of animals depends on abundant food supplies; food is required for females which either carry nutritive eggs (e.g. birds and insects) or which carry embryos in the uterus (e.g. mammals), and food is also required to sustain the rapid growth of the young. For this reason, the period of reproduction is usually related to food abundance; rapidly developing animals such as insects, may start to reproduce at the time that plant food becomes abundant while slowly developing animals, such as mammals, must anticipate food abundance by mating at a time that allows their young to benefit from a readily available food supply. Plants and animals therefore tend to reproduce and develop in seasons related to the total energy flow into the ecosystem (mainly light energy) and it is not surprising that the photoperiod is an important proximate controller of reproduction in animals as it is in plants.

It is not certain which features of the photoperiod are important for controlling breeding and development. In some cases the absolute length of the photoperiod is of prime importance (reviewed [8]), while in other situations the rate of change of the photoperiod on successive days seems to be the important factor (reviewed [2]). It may also happen that in animals the organism is measuring the time between first light and last light in a 24 h period (e.g. [6]). It is certainly difficult to think of any single biochemical mechanism which could account for this variety of responses to different features of the photoperiod.

There is no doubt that the photoperiod can control both the onset of breeding and the products of reproduction in many animals. The most impressive effect of the photoperiod is on the breeding of aphids. These insects alternate cycles of breeding parthenogenetically with cycles of breeding sexually. Usually sexual reproduction is initiated by the shortening day-length of autumn and eggs are produced for overwintering at this time. If the insect is maintained on long day-lengths artificially, then it continues to reproduce parthenogenetically and one aphid has been taken through 248 parthenogenetic generations in this way (reviewed [5]). The young which reproduce parthenogenetically may be morphologically quite different from the young which reproduce sexually and the effect of the photoperiod appears to be mediated through the mother which bears the eggs (reviewed [7]). This is then a case where the photoperiod controls not only the timing of breeding but also the path of development.

Many mammals and birds can be induced to breed outside their normal season by artificial changes in the photoperiod; thus animals which breed in nature in the spring (e.g. many birds, rodent and carnivore mammals) may be taken into premature breeding by a lengthening photoperiod while animals which breed in the autumn (e.g. sheep and deer) will breed in response to artificially shortened day-length.

All these animals must be able to detect features of the photoperiod. It appears that the eyes are not required for detection because blinded ducks, rats and aphids will undergo reproductive changes in response to light directed at the brain (reviewed [3, 4]). It has been supposed that porphyrins inside the brain may be involved in the response and in some cases the pineal gland of vertebrates may also be implicated. There is little doubt that the effect of light on the brain is rapidly translated into changes in hormone levels in the body via either the corpora allata in insects or the hypothalamic-pituitary system in mammals.

Here we have stressed the importance of the photoperiod in animals so that these studies can be related to those in plants. One should also be aware that numerous social and nutritive factors which can also control the onset and consequences of breeding in animals.

6.1.4 ENVIRONMENTAL CONTROL OF DORMANCY AND DIAPAUSE

Reproduction is not the only phase in the life-cycle which may be environmentally controlled. Thus many organisms of temperate regions enter a phase of dormancy during the winter.

Plants may survive the winter as seeds or form resting organs such as the winter resting buds of trees, underground rhizomes, bulbs, tubers etc. Likewise many insects enter into the state of diapause, which is analogous to dormancy in plants, in which development is arrested at the embryo, larval or pupal stage. They survive the winter in this state and further development to the adult stage proceeds in the following spring or summer. In the state of dormancy in plants or of diapause in insects the tissues of the organism undergo certain biochemical changes which render them more resistant to freezing and other adverse conditions. The onset of dormancy in woody plants and the formation of resting organs such as tubers of potato are promoted by short days in the autumn. Similarly diapause in insects is, in many species, induced by short days. Thus, dormancy and diapause provide further examples of the environmental control of development. As in reproduction, there is a great deal of evidence that the environmental control of dormancy is mediated through the hormone-system of the organism, in both plants and animals. Dormancy in seeds and other resting organs of plants appears to be regulated by growth promoting hormones, such as gibberellins and cytokinins, and growth inhibitors, such as abscisic acid. The reader is referred to other works for detailed treatments of the hormonal control of dormancy and diapause [1, 11, 12].

The following chapter of this section deals with the role of light in plant development, a topic which has been chosen because rapid advances have been made in this field in recent years and it provides a good example of environmental control, but it is well to be aware of the profound influences of other environmental factors in the development of a wide variety of organisms and of the intermediary role of hormones in many such responses.

6.1.5 REFERENCES

[1] CHAPMAN R.F. (1969) *The Insects. Structure and Function*. English Universities Press, London.

[2] CLARKE J.R. (1972) Seasonal breeding in female mammals. *Mammal Review*, 1, 217–29.

[3] DE WILDE J. & DE LOOF A. (1973) Reproduction—Endocrine control. In *The physiology of insecta*, Vol. 1, 97–157 (Ed. M. Rockstein). Academic Press.

[4] DONOVAN B.T. (1970) *Mammalian neuroendocrinology*. McGraw-Hill, London.

[5] ENGELMANN F. (1970) *The physiology of insect reproduction*. Pergamon Press, Oxford.

[6] FOLLETT B.K. & SHARP P.J. (1969) Circadian rhymicity in photoperiodically induced gonadotrophin release and gonadal growth in the quail. *Nature, Lond.*, 223, 968–71.

[7] LEES A.D. (1966) The control of polymorphism in insects. In *Advances in insect physiology*, Vol. 3, 207–77 (Eds. J.N.L. Beament, J.E. Treherne & V.B. Wigglesworth). Academic Press.

[8] SADLIER R.M.F.S. (1972) Environmental effects. In *Reproduction in mammals. 4 Reproductive patterns*, pp. 69–93 (Eds. C.R. Austin and R.V. Short). Cambridge University Press.

[9] VINCE-PRUE D. (1975) *Photoperidosm in plants*. McGraw-Hill, London.

[10] WAREING P.F. & PHILLIPS I.D.J. (1970). *The Control of growth and differentiation in plants*. Pergamon Press, Oxford.

[11] WAREING P.F. & SAUNDERS P.F. (1971) Hormones and Dormancy. *Ann. Rev. Plant Physiol.*, 22, 261–88.

[12] WIGGLESWORTH V.B. (1970) *Insect Hormones*. Oliver & Boyd, Edinburgh.

Chapter 6.2
External Factors
Controlling Development:
The Photocontrol of
Plant Development

6.2.1 THE PLASTICITY OF PLANT DEVELOPMENT

The form and function of all organisms is adapted to the environment in which they live; these genotypically determined characteristics are the result of the selection pressures exerted upon the species during the course of its evolution, and are invariant features of each normal individual of the species. All organisms also possess, to varying degrees, the ability to redirect their developmental processes in response to environmental pressures. This capacity, although obviously genetically determined in common with all other functions of the organism, is often termed phenotypic plasticity, since it allows the organism developmental flexibility such that each individual, rather than the species, may become well-suited to the particular habitat it happens to occupy. Thus, organisms possess the ability to detect specific environmental parameters and to direct their cellular development (or metabolism) accordingly.

As we have seen, there are basically two strategies by which an organism may cope with adverse environmental conditions; it may escape, or it may become adapted to tolerate the conditions. During evolution, the alternative strategy of physiological adaptation has been refined to an extremely high level at which the higher plant, in particular, is able to detect and respond to almost every changeable aspect of the environment. The various phases of development in higher plants may be sensitive to light, temperature, gravity, wind, humidity, magnetic and electrostatic fields and various subtle combinations or sequences of these environmental stimuli. It is the reactions of the plant to such environmental stimuli that ensure its survival to the reproductive stage.

Thus the appearance and behaviour of any plant is determined by both genetic and physiological mechanisms. The characteristic anatomy and morphology of the species is arrived at by a genetically determined ground plan or programme of development. Environmental conditions can modulate, over a wide range, the rate at which the basic programme is followed, and even, in special cases, cause the selection or preselection of specific sub-programmes of development. In this way the basic developmental processes of cell division, cell enlargement and cell differentiation can be both quantitatively and qualitatively altered in response to changes in the environment. It is with the mechanisms which sense the environment and transduce

environmental stimuli into developmental processes, and which confer on higher plants their uniquely high degree of developmental plasticity, that this chapter is concerned. It is clearly not possible in a single chapter to cover adequately all of the ways in which plants respond to the range of environmental stimuli mentioned above. However, since the general principles of the environmental control of development are shared by all of the different responses, it is quite logical to restrict coverage to only one stimulus, and the most intensively investigated is light.

In all cases where an aspect of the environment controls development, the overall process can be divided into three fundamental steps:

(a) perception of the environmental stimulus,
(b) transduction of the environmental stimulus into a biological stimulus, and
(c) control of differentiation and development in response to the biological stimulus.

Clearly in order fully to understand the whole process it is necessary to integrate information on the physical nature of the environmental stimulus, the properties of the sensing mechanism, and the mechanisms responsible for the ultimate changes in growth and development. In the special case of the photocontrol of development, these considerations mean that real progress can only be made by the cooperation of physicists, biochemists and plant physiologists. Fortunately, this multi-disciplinary attack has been successfully mounted and we now know a great deal about the mechanisms whereby plants respond to changes in the light environment. As will be seen, the area of greatest current ignorance lies within the central question of the internal control mechanisms of development.

6.2.2 THE PHENOMENOLOGY OF THE PHOTOCONTROL OF PLANT DEVELOPMENT

Light is certainly the most important environmental factor as far as the regulation of plant development is concerned. Solar radiation provides the plant with the energy it requires to manufacture and maintain cellular materials, and it is not surprising that plants have evolved mechanisms to sense the light environment and regulate development in such a way that optimum utilization of solar energy can be achieved.

Photocontrol of development has been observed in every known class of plants, including fungi, and in higher plants, which have been the most intensively investigated, virtually every phase of development may, under certain circumstances, be controlled by light. An encyclopaedic coverage of such a wide-ranging topic is not possible, but Table 6.2.1 should give some idea of the comprehensive nature of the photocontrol of plant development.

It is usual to define the various types of photocontrol in terms of the nature of the stimulus required to elicit a response. In this way we can define three major categories of developmental photocontrol and these are known as phototropism, photoperiodism and photomorphogenesis.

Phototropism

Phototropic responses are elicited only by a *directional* light stimulus, and they result in growth curvature of the stimulated organ in a direction determined by the direction of the incident irradiation. Phototropism was first intensively studied by Charles Darwin and his son Francis [22] and led to their realization that transmissible influences are present in plants which can control development from a distance—the first indication of hormone action in plants. Phototropism is of great importance in the natural environment in orienting the growth of the plant so that it is able to optimize its interception of the available solar energy for photosynthesis.

The orientation of leaves to achieve this optimum arrangement for light interception is, in many species controlled by growth curvatures of the stem and petioles. However, there are other diurnal movements in the orientation of leaves which, although they are affected by light and dark, are responses to a non-directional light stimulus and are said to be *photonastic*. Such 'sleep movements' may be brought about by growth curvatures but in some species, such as clover, they are due to changes in turgor of the cells on opposite sides of the petiole.

Photoperiodism

As was explained in the preceding introductory chapter, photoperiodic phenomena are developmental responses to natural light regimes in which there is a daily alternation of light and darkness, with a total cycle length of 24 hours. Thus, in photoperiodism the response is to a *periodic* stimulus. The responses are controlled both by the length of the light period and, even more markedly, by the length of the dark period. Moreover, it is the *absolute* lengths of the light and dark periods which are critical, not their relative lengths. In a number of plant species there appears to be an endogenous rhythm in photoperiodic sensitivity, with a daily alternation of 'photophile' and 'skotophile' phases in which the responses of the plant are promoted by light and by darkness, respectively. Thus

photoperiodism provides the plant with a time-measuring system and enables it to 'tell the seasonal time', and thereby to adapt to seasonal climatic changes. The time-measuring system is subject to genetical variation which allows for the existence of photoperiodic ecotypes of individual species which have become specifically adapted to the latitudes in which they live.

Photomorphogenesis

Photomorphogenic phenomena are those developmental processes which are initiated by light stimuli that are not necessarily either directional or periodic. Photomorphogenic responses occur throughout the whole life of the plant and may be responsible for controlling seed germination, stem elongation, leaf expansion, the formation of stomata, the formation of leaf hairs, the synthesis of chlorophyll, the development of chloroplasts and the synthesis of a wide range of secondary products including carotenoids, flavonoids and alkaloids (see Table 6.2.1). As will be detailed later, these photomorphogenic responses can all be shown to be quantitative rather than qualitative changes. This point is of importance to a consideration of the mechanisms of the photocontrol of development.

The most profound photomorphogenic changes occur when plants are sown and grown in darkness for some days, then

Table 6.2.1   Processes that are known to be under photomorphogenic control (after Smith [105]).

NON-VASCULAR LOWER PLANTS	ANGIOSPERMAE (cont.)
Spore germination	Coleoptile elongation
Cell number per protonema	Hypocotyl hook opening
Thallus growth rate	Leaflet movement (photonasty)
Chloroplast size	Geotropic reactivity
Chloroplast division	Phototropic reactivity
Chloroplast movement	Root initiation
	Leaf primordia initiation
PTERIDOPHYTA	Leaf dry weight
Spore germination	Seedling protein content
Rhizoid growth	Cytoplasmic viscosity
Filament growth	Sucrose translocation
Filament differentiation	Starch degradation
	Anthocyanin synthesis
	Flavonoid synthesis
GYMNOSPERMAE	Chlorophyll synthesis
Seed germination	Ascorbic acid synthesis
Hypocotyl hook formation	Various enzyme activities
Stem extension growth	Gibberellin formation/release
	Cytokinin formation/release
	Auxin metabolism
	Polysome formation
ANGIOSPERMAE	Membrane permeability to water
Seed germination	Membrane permeability to ions
Plant height (growth rate)	ATP/ADP conversions
Leaf number increase	Electric potential establishment
Leaf area increase	

exposed to light, where upon very large changes in the development of the different organs occurs, as is shown. in Fig. 6.2.1. Growth in darkness produces etiolated seedlings in which stem elongation is extreme and leaf expansion restricted. The exposure to light reverses these abnormal developmental extremes. Another process often subject to photomorphogenic control is seed germination, some seeds being stimulated to germinate and others being inhibited by exposure to white light. The study of these phenomena have probably provided most of the really valuable evidence on the molecular mechanisms whereby plants perceive and respond to a light stimulus.

Thus, there are three principal ways in which light controls plant development:*

(a) a directional control over the orientation of plant parts (phototropism);
(b) a long-term, often qualitative, control over the whole pattern of development (photoperiodism); and
(c) a shorter term, quantitative control over the rates of developmental processes (photomorphogenesis).

Since photomorphogenesis is probably better understood at the present time than either phototropism or photoperiodism, the detailed analysis that follows will emphasize photomorphogenesis. Furthermore, there is good reason to believe that the mechanisms responsible for photomorphogenesis may also be involved in the other two phenomena, and where appropriate, these cross-connections will be brought out.

6.2.3 THE PHOTOMORPHOGENIC RESPONSE SYSTEMS

The perception of a light stimulus can only be achieved by the absorption of light by a specific substance which is thereby 'photoactivated', in which state it can initiate the biochemical reactions that lead to the observed developmental changes. An absorbing molecule absorbs light by the whole energy of a photon being captured by a single electron. The results of this capture depend on the energy of the absorbed photon. In the wavelengths below c. 350 nm, each individual photon contains so much energy that the absorbing electron is, in effect, ejected from the molecule, which is left in an ionized and thereby highly reactive state. Ionized molecules have major deleterious effects

* It is important to realize that although plants utilize light as their sole source of energy which is captured through the process of photosynthesis, nevertheless the photosynthetic machinery is not normally involved in *developmental* responses to light. Clearly, photosynthesis is important in growth and development in providing a supply of chemical energy for the necessary biosynthetic processes involved, but absorption and transformation of light energy by the photosynthetic machinery is not responsible for any *regulation* of development.

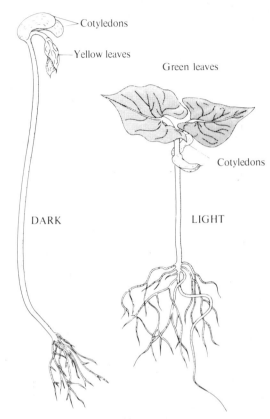

Fig. 6.2.1 Dark grown and light grown bean plants showing the developmental effects of light; (left) bean seedling which has been sown in darkness and grown in darkness for six days; (right) a genetically almost identical bean seedling which has been grown for six days under high intensity white fluorescent light.

on living processes and are largely responsible for the radiation damage which occurs upon exposure to such short-wave radiation. In the infra-red and longer wavelengths, the energy of each photon is very low and is only sufficient to increase the vibrational and rotational energy of the electron. This is exhibited as an increase in kinetic energy, but not as a photo-activated state of the molecule.

In the visible region, however, the energy of each photon is sufficient to raise individual electrons within a molecule to an orbital of higher energy. As the chemical properties of a substance depend upon its electronic distribution, then raising an electron to a higher orbital by light absorption results in the formation of a different chemical species. Thus the molecule achieves a so-called excited state in which it can enter into reactions impossible for the same molecule in its ground state. The absorption of a photon can only occur if the energy of the photon is exactly the correct amount to raise the electron to one of its precisely quantised energy levels. This means, in its turn, that a photoreceptive substance will only absorb light of specific wavelengths, thus providing a characteristic absorption spectrum for the molecule. In practice, the absorbing electrons in complex organic molecules have a rather wide range of possible energy levels associated with each excited state thus giving relatively broad absorption spectra.

From the foregoing, it should be clear that every photo-biological process must be mediated by a specific photo-receptor, one of whose properties would be a characteristic absorption spectrum. One of the major aims of investigation is to identify, isolate and characterize the responsible photo-receptors for the photomorphogenic responses. The fact that each photoreceptor must have a characteristic absorption spectrum allows a direct approach to the question of its identity. This is to construct an action spectrum (a plot of relative biological activity against wavelength) which theoretically should be similar to the absorption spectrum. This approach has been used with great effect in the study of photomorphogenesis but, as will be seen, although much information has been generated by the construction of action spectra, it has in no case led to the direct identification of the photoreceptor. Other information is also necessary, particularly if the photoreceptor is present only in very small amounts. Useful evidence comes from other characteristics of the photoresponse, in particular energy and timing requirements.

Over the past 40 years or so intensive study of the wavelength dependence of photomorphogenic responses has led to the conclusion that at least three distinguishable types of response occur, mediated by probably two different photo-receptors. The responses may be described as:

(a) the low energy, red/far-red photoreversible response
(b) the high energy, far-red response
(c) the high energy, blue response.

The low energy red/far-red photoreversible response

The first meaningful attempt to analyse the wavelength dependence of a photomorphogenic response was by Flint and McAllister in 1935 and 1937. Working with a variety of lettuce seed that required light for maximum germination but which gave about 30% germination in darkness, they showed that red light (i.e. 600–700 nm) was the active wavelength region for the stimulation of germination [28]. They also found that near infra-red (i.e. *c.* 700–800 nm) actually inhibited the germination which normally occurred in dark imbibed seeds [27]. (This region of the spectrum [*c.* 700–800 nm] is now known as *far-red* by plant physiologists.) Thus red light stimulated germination whilst far-red inhibited it. In this response only a short period of light (*c.* 5 min) is necessary; germination then

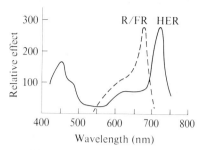

Fig. 6.2.2 The action spectra of low energy red/far-red photoreversible responses (R/FR) and the high energy reaction (HER) (after Mohr[76]).

takes place in darkness over a period of hours. This shows that only low energies of light are required.

In the early 1940s a group of research workers at the United States Department of Agriculture Research Station at Beltsville, Maryland, began a painstakingly thorough investigation of a very wide range of photomorphogenic responses. This group, led by Dr Harry A. Borthwick and Dr Sterling B. Hendricks, during the subsequent 20 years or so, produced an enormous amount of information upon which the current theories of photomorphogenesis are firmly based. They first produced action spectra of several aspects of de-etiolation (e.g. stem growth in etiolated barley seedlings [9], stem and leaf growth in etiolated pea seedlings [82], of the germination of lettuce seeds [11], and of the effect of a short break of light in the middle of a long dark period on the photoperiodic induction of flowering in cocklebur plants [81]. The action spectra for these diverse responses were all remarkably similar, each exhibiting maximum action at or near 660 nm (Fig. 6.2.2). It could thus be postulated that the responsible photoreceptor would have an absorption spectrum with a maximum at 660 nm. This information, on its own, would probably not have been sufficient to allow the identification of the photoreceptor as there are several substances present in plants, including proto-chlorophyll and chlorophyll which have absorption bands near 660 nm.

In 1952 a simple but critically important discovery was made without which further progress would have been most difficult. The Beltsville team resurrected Flint and McAllister's observation that lettuce seed germination, although promoted by red light, was inhibited by far-red light, and they proceeded to test the effects of sequences of red and far-red light on germination. Their results are shown in Table 6.2.2. Clearly, red and far-red light have antagonistic effects on development and, after a sequence of treatments with the two wavelengths, the ultimate developmental response is determined by the nature of the last irradiation treatment. On the basis of these findings it was proposed that the photoreceptor (P) existed in two photo-

Table 6.2.2 Control of lettuce seed germination by red and far-red light (from Borthwick *et al.* [10]).

Irradiation	Percentage germination
Red	70
Red/Far-red	6
Red/Far-red/Red	74
Red/Far-red/Red/Far-red	6
Red/Far-red/Red/Far-red/Red	76
Red/Far-red/Red/Far-red/Red/Far-red	7
Red/Far-red/Red/Far-red/Red/Far-red/Red	81
Red/Far-red/Red/Far-red/Red/Far-red/Red/Far-red	7

convertible forms, one which absorbed maximally in the red regions of the spectrum (Pr) and one which absorbed maximally in the far-red (Pfr) in accordance with the following scheme:

$$Pr \underset{far\text{-}red}{\overset{red}{\rightleftarrows}} Pfr$$

Subsequently, it was found that all of the phenomena which could be shown to be initiated by brief red light treatment, could also be reversed by immediate subsequent irradiation with brief far-red light. Action spectra for the reversal of a red light effect all showed maximum activity between 710–750 nm (Fig. 6.2.2). Since developmental changes could not be initiated by far-red light alone, and far-red only had its effect in plants which had already been irradiated with red light, it was further proposed that Pfr was the biologically active form of the hypothetical photoreceptor. The photoresponses could thus be described by the following simple model:

$$Pr \underset{far\text{-}red}{\overset{red}{\rightleftarrows}} Pfr \rightarrow Biological\ action$$

On the basis of purely physiological data, it could therefore be predicted that plants should contain a photoreceptive substance, existing in two forms, one absorbing maximally at *c.* 660 nm, and the other at 710–750 nm each being mutually interconvertible upon absorbing radiant energy. It was argued further that the spectral changes which such a photoconvertible substance would exhibit upon sequential irradiation with red and far-red light should be detectable in intact plants with a sensitive spectrophotometer. Thus in 1959, Butler *et al.* published *in vivo* spectrophotometric data with etiolated maize seedlings demonstrating the predicted spectral changes [19]. Red irradiated seedlings absorbed rather less light at 660 nm

and rather more at 730 nm than did dark-grown or far-red irradiated seedlings. The physical detection of the photoreceptor in tissue was supported by the finding that partially purified protein extracts from etiolated tissues also showed similar spectral changes upon irradiation with red and far-red light. These findings elegantly validated the predictions made from the physiological observations. At about this time (*ca.* 1959) the photoreceptor was half-jokingly (according to Borthwick [8]), given the name *phytochrome* by W.L. Butler.

Thus, during the 1950s the concept of a low-energy, red/far-red photoreversible photomorphogenic response became accepted. This response system has been shown to have a controlling influence on a wide range of developmental and metabolic processes (see Table 6.2.1) and has been studied in great detail. It should be borne in mind, however, that the responses as studied in the laboratory bear no directly obvious relationship to the control of development and metabolism in plants growing in the natural environment; treatments with short periods of red or far-red light are necessarily involved in experimental work and such conditions do not occur in nature. On the other hand, a reductionist approach to photomorphogenesis is necessary to obtain sufficient understanding of the partial processes for a useful synthesis eventually to take place.

The low-energy, red/far-red photoreversible response manifests itself as a syndrome of specific characteristics. The responses typically require only short periods of irradiation, are red/far-red photoreversible, exhibit an action maximum for induction at 660 nm and for reversion at or near 730 nm, are saturated at relatively low total energies, and display a logarithmic relationship between energy and response. In practice, demonstration of red/far-red photoreversibility is normally accepted as an adequate and sufficient criterion for the involvement of phytochrome.

ENERGY REQUIREMENTS

The energy requirements vary between different plants and responses, and reported saturation energies range from 0·1 Joules cm^{-2} for the red-light mediated opening of the epicotyl hook in *Phaseolus* [62] to about 2000 Joules cm^{-2} for the red-light inhibition of flowering by light break in the short-day plant *Pharbitis nil* [109]. Below the saturation energy the relationship between energy and response is normally log-linear as shown by the red-light mediated increase in barley leaf width in Fig. 6.2.3.

PHOTOREVERSIBILITY

Dark-grown (i.e. etiolated) plants usually exhibit no developmental responses to a brief irradiation with far-red light, but far-red given immediately after red light negates the inductive action of the red light. If a period of time in darkness is interposed between the red and far-red irradiation treatments a

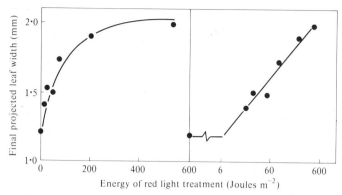

Fig. 6.2.3 Dose-response for the control of barley leaf unrolling by red light; (left) linear plot; (right) semi-log plot.

reduction in the degree to which the far-red will reverse the red effect is observed. This can be considered as an escape from far-red reversibility, probably indicating that Pfr interacts with some other component in the biological system, and once it has done so, removal of Pfr by far-red irradiation can no longer negate the response. On this basis it ought to be possible to measure the time taken for Pfr to carry out its primary action by estimating the time course of escape from far-red reversibility. Unfortunately, there is enormous variation in the escape times that have been reported, e.g. the $t_{\frac{1}{2}}$ (time for 50% loss of photoreversibility) for the light break effect on the photoperiodic induction of flowering in *Pharbitis* is 1·5 min [30], the $t_{\frac{1}{2}}$ for flavonoid synthesis in peas is 30 min [49], and the $t_{\frac{1}{2}}$ for rice coleoptile growth is 8 h [86]. This extreme variability probably means that the processes with which Pfr interacts are themselves limited by factors not related to the photoreceptive mechanism. It is, however, interesting that loss of photoreversibility usually occurs well before any developmental effect of the red-light treatment can be detected. As mentioned below,

some phytochrome mediated responses occur very rapidly indeed, and in these cases far-red reversibility is difficult to demonstrate for purely practical reasons. With the normal long term developmental effects, however, it seems quite clear that Pfr does not itself direct the actual processes of development; it must set in train a sequence of steps which ultimately leads to the modification of development rates, the later steps not requiring the presence of Pfr.

TIME-COURSE OF PHYTOCHROME RESPONSES

The early experiments upon which the concept of a low energy phytochrome response was based were concerned with developmental effects that were not manifested until several hours or even days after the light stimulus. In recent years much emphasis has been laid on determining just how rapidly phytochrome can act, and several non-developmental responses have been found to be controlled by light through phytochrome with very short lags indeed. Most of these rapid phenomena, as is discussed in detail below (page 366) appear to be associated with changes in membrane properties. A selected list of observed lag periods (i.e. the time between the onset of red light treatment and the first detection of a response) is given in Table 6.2.3. Thus phytochrome can act to alter bioelectric potentials within 15 s, although the lag periods for the developmental responses lie between 1 and 4 h. One interpretation of these observations is that the primary biological action of phytochrome is completed very quickly, but the ensuing processes linking the primary action with the developmental changes are relatively slow.

The high-energy reactions

Exposure of etiolated seedlings to brief red light as described above initiates developmental processes which tend to reverse the symptoms of etiolation. The extent of the red light initiated

Table 6.2.3 Lag periods of some selected low-energy phytochrome responses (after Smith [105]).

Response	Lag-period	Plant materials
Change in electric potentials	15 s	Oat coleoptiles [80]
Adhesion of roots to glass	30 s	Mung bean [109b]
ATP formation	1 min	Oat mesocotyls [97]
Gibberellin increases	10 min	Wheat leaf [68a]
Geotropic sensitivity increase	30 min	Oat coleoptiles [117]
Ascorbic acid synthesis	60 min	Mustard hypocotyls [100]
Phenylalanine ammonia-lyase increase	90 min	Pea buds [3]
Inhibition of stem growth	90 min	Peas [68]
Stimulation of leaf expansion	4 h	Peas [33]

responses, however, is relatively small compared with the changes wrought by exposing etiolated plants to long periods of continuous high irradiance white light. In other words, the low energy phytochrome response, even when light saturated, is not capable of completely 'de-etiolating' etiolated seedlings. Moreover, the greater magnitude of the continuous white light effects indicate that one or more additional photoresponsive mechanisms are operative in the etiolated seedling. The first demonstration of a distinct photoresponse was reported in 1957 from observations on hypocotyl expansion and anthocyanin synthesis in *Sinapis alba* (white mustard). In this work, Mohr [74] showed that cotyledon expansion was stimulated by red light in a manner consistent with the operation of the low energy phytochrome response. The red light effect was reversed by brief exposure to low-energy far-red light, but if the exposure to far-red was continued for a period of 2–3 h or longer, no reversal was seen and in fact a stimulation of expansion was observed. The response to long periods of far-red was greater with higher irradiances and thus Mohr coined the term 'High Energy Reaction'. The high energy reaction has since then been implicated in a very wide range of developmental processes including seed germination, stem and leaf expansion, anthocyanin synthesis, nucleic acid and protein synthesis, and changes in the levels of specific enzymes.

Many attempts have been made to construct accurate action spectra of high energy reactions, a task which is clearly much more difficult for a response requiring long periods of irradiation than for one elicited by brief light treatments. The most detailed action spectrum yet published is that constructed by Hartmann [42] for the photo-inhibition of hypocotyl extension in etiolated lettuce seedlings (Fig. 6.2.4). There are obviously two spectral regions which inhibit hypocotyl extension, the blue and the far-red, and this duality of action maxima is common throughout most of the high energy responses that have been investigated.

Some authorities (e.g. Hartmann [41]) believe that the responses to blue and far-red light may be mediated by the same photoreceptor whilst others (e.g. Vince [115]) are of the opinion

that two wholly separate photoreceptors are necessary. There is considerable physiological evidence to suggest that blue and far-red light operate *via* different mechanisms. For example, blue light and far-red light both inhibit hypocotyl extension in gherkin seedlings, but blue light causes growth inhibition almost immediately, whereas the inhibition caused by far-red light has a lag phase of about 30 min [72]. Similarly, both blue and far-red light cause increased anthocyanin synthesis in etiolated turnip seedlings. However, if a red light treatment is given prior to irradiation with far-red light, a marked reduction in anthocyanin synthesis is observed; if a red light treatment is given before the blue light, no such reduction is seen [38]. Although such complex physiological experiments are difficult to interpret, it seems inescapable that blue and far-red light must act through different mechanisms. The simplest, but not the only, way in which this might occur is for there to be two separate and distinct photoreceptors. A major question therefore is the identity of the photoreceptors responsible for the two action maxima of the High Energy Reaction.

The high-energy far-red response

The action maximum for the photocontrol of lettuce hypocotyl extension by continuous irradiation is at precisely 716 nm. Since this is within the spectral regions absorbed by phytochrome, it is obviously possible that phytochrome is the responsible photoreceptor. The first suggestion that this might be so came from Hendricks and Borthwick [46] in 1959 when they proposed that high energy reactions with action maxima in the red and far-red spectral regions could be mediated by the simultaneous absorption of light by Pr and Pfr and their consequent simultaneous excitation. When it is realized that Pr and Pfr have broad overlapping absorption spectra (page 364) it can be seen that irradiation with continuous light within the spectral regions absorbed by both Pr and Pfr must lead to the establishment of a dynamic equilibrium between the two forms. Thus at any wavelength absorbed by the two forms, a so-called 'photostationary state' will be established in which the rate of formation of Pfr is exactly balanced by the rate of formation of Pr, and in which the steady-state proportions of Pr and Pfr will be determined by the wavelength of the actinic radiation. It is, therefore, reasonable to propose that high energy reactions may depend on maintaining a certain ratio of Pfr:Pr for a sufficiently long period of time, and that this can be achieved by irradiation with monochromatic light at some wavelength between the absorption maxima of Pr and Pfr.

This hypothesis was most elegantly tested by Hartmann [41] in 1965 by the use of two separate beams of red and far-red light. By simultaneously irradiating lettuce seedlings with monochromatic light at various red and far-red wavelengths, he was able to show that maximum inhibition occurred when the irradiation conditions set up a photostationary state in which

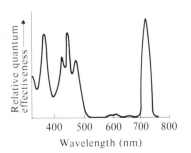

Fig. 6.2.4 A detailed action spectrum for the high energy inhibition of hypocotyl extension in lettuce seedlings (after Hartmann [42]).

3% of the total phytochrome was present as Pfr. This value was consistent with the action spectrum since 716 nm light (the action maximum) also established 3% of total phytochrome as Pfr. The dual wavelength approach has since been applied to several high energy reactions, and in each case, irrespective of the wavelength of the actinic sources, maximum action occurred only at a certain critical Pfr/Ptotal ratio. The critical Pfr/Ptotal value is not necessarily the same for different photoresponses, but in any one photoresponse, the establishment of Pfr/Ptotal values either lower, or higher, than the critical value results in a reduced developmental effect.

This evidence implies very strongly that phytochrome is the photoreceptor for the far-red action maximum of the high energy reaction. If this is so, the manifestations of phytochrome action in the high energy mode are very different from those resulting from phytochrome action in the low energy mode. High energy reactions are typically irradiance dependent over a wide range, with a direct linear relationship between irradiance and response. It is extremely difficult to account for these relationships. If we take the scheme on page 357 as the basis for phytochrome action in the low energy mode, then the biological action is dependent on the availability of Pfr. With continuous irradiation, however, a photostationary state is achieved and maintained in which the proportions of Pr and Pfr remain constant with respect to each other. Increasing the irradiance of the actinic radiation will have no effect on the proportions of Pr and Pfr; the only result will be an increased rate of cycling between Pr and Pfr. Thus the irradiance dependence of the high energy reaction must be related in some way to the rate of cycling between Pr and Pfr, and not to the steady state concentration of Pfr *per se*.

Another thorny question is how the developmental effects of continuous irradiation are mediated by necessarily low proportions of Pfr:Ptotal, whereas brief red light effects are maximal with high proportions of Pfr:Ptotal. Several possible explanations have been advanced to account for this anomaly, all of which are still totally speculative.

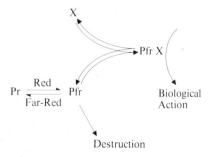

Fig. 6.2.5 Hartmann's scheme for the operation of phytochrome in the high energy reaction (after Hartmann [41]).

Hartmann [41] suggested that Pfr, in order to initiate developmental changes, must combine with an unknown reactant, X, and thus it is PfrX, and not Pfr *per se* that is responsible for the changes. If Pfr is unstable, then a competition would exist between the formation of PfrX and the degradation of Pfr. If Pfr was present in high concentrations (e.g. with irradiation with red light) whilst X was only present in low concentrations, then Pfr degradation would be favoured, and over a period of time, large amounts of phytochrome would be lost, thus reducing the effectiveness of red light. On this basis, far-red light is effective because it maintains Pfr at an adequate level to combine with X, but not at a sufficient level for significant Pfr degradation to occur. This model is shown diagrammatically in Fig. 6.2.5. It is known from *in vivo* spectrophotometry that Pfr is very much more unstable than Pr.

Schopfer and Mohr [101] have extended Hartmann's hypothesis by suggesting that, during the photoconversion of Pr to Pfr, an excited state of Pfr is produced (denoted by Pfr*). Pfr* is thought to be the effective molecule in the high energy reaction and it is proposed that the steady-state concentration of Pfr* is irradiance dependent. If Pfr* is an intermediate between Pr and Pfr it would be expected that its concentration would increase with increased irradiance as cycling between Pr and Pfr increased. As a corollary to this proposal, it was also suggested that $Pfr_{ground\ state}$, produced from Pfr* and relatively long-lived compared to Pfr*, has reduced effectiveness in biological terms compared to Pfr*. Thus $Pfr_{ground\ state}$ (formally identical to Pfr in the scheme on page 353) is thought to be responsible for the responses to brief irradiation with red light, whilst Pfr* brings about the more extensive responses to continuous high energy irradiation. Although this idea is attractive, it is wholly speculative and, moreover, difficult to test.

A further possibility, put forward by Smith [103], is that phytochrome is a membrane bound transport factor with anisotropic binding sites for a critical metabolite X. Thus Pr would bind X on one side of a membrane and Pfr on the other. Conversion of Pr to Pfr or *vice versa* would cause the transport of X across the membrane in one direction or the other. It can be shown on thermodynamic grounds (see [105]) that under cycling conditions, net transport of X would be from Pr to Pfr if the photostationary state was low in Pfr (see Fig. 6.2.6). If cycling is increased by increased irradiance, then the rate of transport of X across the membrane would be increased. Thus, this model can also account for both the dependence on low Pfr/Ptotal values, and the irradiance dependence of the high energy reaction. This hypothesis is also purely speculative, but it implies that (a) phytochrome should be a membrane component, and (b) it should have binding sites for the critical metabolite—whatever that might be!

Irrespective of the mechanism, however, it seems reasonably certain that phytochrome is the photoreceptor for the far-red

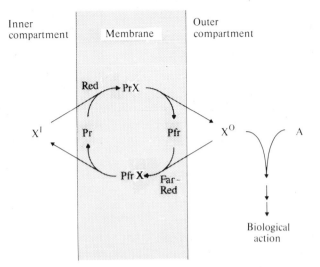

Fig. 6.2.6 Smith's model for the mechanism of phytochrome action in both the low-energy and high-energy modes (after Smith [103]).

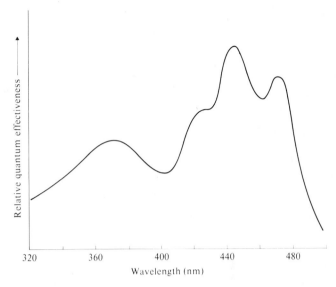

Fig. 6.2.7 An action spectrum for the phototropic curvature of oat coleoptiles.

action maximum of the high energy reaction although there has been some dissent on this matter.

THE HIGH ENERGY BLUE RESPONSE

Examination of the action spectrum in Fig. 6.2.4 will show that the response to blue light is characterized by multiple action maxima. It is instructive to compare this action spectrum with that for phototropism in *Avena* coleoptiles (Fig. 6.2.7). It is clear that a great deal of similarity exists, suggesting that the same, or at least a related, substance is the photoreceptor in each case. Much research has been expended on attempts to identify the phototropism photoreceptor. Currently, there appear to be two major contenders, a carotenoid, or a flavin. Both of these substances have absorption spectra which are similar to, but not identical with, the action spectra for phototropism and the high energy reaction. It has not been possible, however, to make a categorical decision as to which is the responsible substance. At present, several research groups are trying to unravel this difficult question, and the best guess is that the photoreceptor will turn out to be a flavoprotein.

Conclusions

There appear to be three distinct photoreactive response systems responsible for the direct effects of light on plant development. These are:

(a) the low energy phytochrome system;
(b) the high energy phytochrome system;
(c) the high energy blue system.

The simplest conclusion is that only two photoreceptors are necessary to explain the observed data, namely phytochrome, and the as yet unidentified blue-absorbing photoreceptor. It should be stressed, however, that this is only the simplest conclusion and it is only too likely that other photoreactive response systems will be defined in the future. It is certainly true, for example, that fungi exhibit photomorphogenic responses that do not fit into any of the above categories. As we know very little about the blue-absorbing photoreceptor, the rest of this chapter will concentrate on the properties and mode of action of phytochrome.

6.2.4 STRUCTURE AND PROPERTIES OF PHYTOCHROME

Phytochrome is now known to be a protein with a covalently attached chromophoric prosthetic group which gives the molecule its specific absorption properties. As will be seen, it is a most unusual, if not unique, molecule, of great intrinsic interest to physical and biological chemists. As often happens with scientific discovery, however, the realization that phytochrome is a protein, was based almost wholly on an intelligent, but probably incorrect, guess—that phytochrome is an enzyme.

Detection and measurement of phytochrome

The isolation and purification of any substance is dependent on the existence of a sensitive and specific assay. In the case of phytochrome, the only known method of detection utilizes the photoreversible changes in absorption properties. This assay is wholly specific to phytochrome, but in comparison with, for example, an enzyme assay, is not particularly sensitive. The principle of the assay is to measure the absorption of the tissue or extract at 660 nm and 730 nm (the absorption maxima of Pr and Pfr respectively) and to assess the changes in absorption at these two wavelengths upon irradiation with actinic light in the red and far-red regions. Normally, a dual wavelength spectrophotometer is used in which rapidly alternating beams of 660 nm and 730 nm light are passed through the sample to a photomultiplier where the difference in absorption at the two wavelengths is computed electronically. The sample is then irradiated with high energy red light to photoconvert Pr to Pfr, after which the difference in absorption at 660 nm and 730 nm is again measured. Subsequently the sample is photoconverted from Pfr to Pr with high energy far-red light and the absorption difference measured. The overall difference in absorption between maximum Pfr and maximum Pr is a relative measure of total phytochrome content. By various other sequences the proportions of Pr and Pfr in the sample can also be determined. The principle of the method can be seen from Fig. 6.2.8, which shows actual absorption spectra of etiolated tissue with phytochrome present either totally in the Pr form or mainly in the Pfr form.

Isolation and purification of phytochrome

Phytochrome may be isolated by the normal methods of protein extraction and purification. It is, however, a difficult protein to purify—it is present in very low concentrations, it is unstable to conditions which oxidize sulphydryl groups, the Pfr form is very unstable especially in the presence of metal ions, and at least in some plants there appears to be a specific protease which attacks phytochrome, is co-purified with it, and is activated by sulphydryl reagents! Several groups, however, have successfully purified phytochrome from a range of plants, but the greatest success has been achieved with rye and oats. The latest extraction procedure, worked up by Briggs and coworkers at Harvard [35] is shown in flow sheet form in Table 6.2.4. This procedure takes two to three days to complete, but yields a product which has been shown to be a single homogeneous protein by acrylamide gel electrophoresis. Full discussions of extraction methods will be found in reference [94].

The protein moeity

The molecular weight of purified phytochrome varies with the source of the preparation, the way it was purified and the manner in which the molecular weight is determined. Most of the earlier work gave values of around 60,000 daltons, although

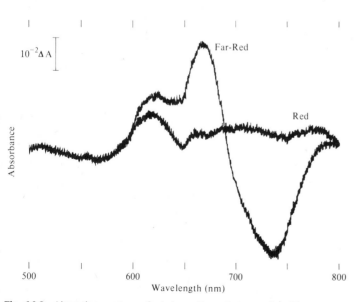

Fig. 6.2.8 Absorption spectrum of whole seedlings of *Amaranthus* either treated with saturating red light (Red), or with saturating far-red light (Far-Red). (Unpublished data of R.E. Kendrick.)

Table 6.2.4 The Briggs procedure for isolating phytochrome (compiled from Gardner *et al.* [35]).

---

1 kg oats, chilled and harvested
Grind in 1 l buffer: 50 mM tris + 0·7% 2-mercaptoethanol
Extract in Waring blender
↓
Centrifuge at *ca.* 3000 rpm, 15 min
↓ discard pellet
Make supernatant 10 mM in CaCl$_2$
↓
Centrifuge at *ca.* 3000 rpm, 15 min
↓ discard pellet

Add sample to Brushite column (1 l brushite per kg fresh tissue) equilibrated with two column volumes 10 mM KPB (KPB = potassium phosphate buffer), pH 7·0, followed by two volumes 10 mM KPB + 0·7% 2-mercaptoethanol, pH 7·0

After sample addition wash column with one volume 10 mM KPB + 0·7% 2-mercaptoethanol, pH 7·0, followed by one volume 15 mM KPB, pH 7·0
↓
CRUDE PHYTOCHROME precipitated with 33% (NH$_4$)$_2$ SO$_4$
↓ centrifuge
Dissolve precipitate in buffer and dialyse 1 h against same buffer
↓
Apply to DEAE-cellulose column equilibrated with 10 mM KPB, pH 7·4
↓
Elute with a convex gradient of 500 ml 0.3 M KCl in 10 mM KPB, pH 7·4 into 250 ml 10 mM KPB, pH 7·4
↓
Active fractions concentrated and precipitated with 40% (NH$_4$)$_2$SO$_4$ pH 7·8
↓ discard supernatant
Redissolve precipitate and dialyse against 10 mM KPB, pH 7·5
↓
PHYTOCHROME
↓
Further purification by chromatography on Hydroxylapatite, Agarose and Sephadex G-200

---

there were reports ranging from 12,000 to 180,000. Briggs and Rice in a scholarly review of phytochrome [16] have suggested that the differences are due to varying degrees of proteolysis and to the methods of determining molecular weights. In rye and oat seedlings, a specific protease exists which appears to cleave native phytochrome into units of approximately 60,000 daltons [84]. If the protease is inhibited by PMSF (phenylmethyl-sulphonyl fluoride) a much larger molecular weight form can be obtained. When the molecular weight of the large form was measured by shape-independent methods (e.g. equilibrium density gradient centrifugation in caesium chloride, or electrophoresis in acrylamide gels containing the detergent sodium dodecyl sulphate, which denatures the protein and allows it to migrate as a linear polypeptide chain) the value was either 120,000 or 240,000 daltons. With other measures of molecular weight which are dependent on the shape of the molecule, quite unrelated values were obtained.

On the basis of this elegant work Briggs and Rice [16] proposed that phytochrome is not a simple globular protein. They suggest that the native molecule probably has a molecular weight of about 120,000 daltons, and is cleaved by the specific protease into two 60,000 M.W. portions. The 240,000 M.W. units observed are probably aggregation states.

The chromophore

The visible absorbance of phytochrome and its photoreversible changes are properties of a specific chromophore attached to the protein and much effort has been devoted to elucidating the chemical identity of the chromophore. In 1950, nine years before the first isolation of phytochrome, Borthwick, Parker & Hendricks [12] suggested that the pigment responsible for the photomorphogenic changes might be a linear tetrapyrrole, similar to the chromophore of the algal pigment phycocyanin. Later it was shown that the absorption spectrum of phytochrome bears very close similarities to those of C-phycocyanin and allophycocyanin [102] and this served to direct further work along the lines of pyrrole chemistry.

One of the major difficulties has been the preparation of sufficient amounts of the chromophore for a detailed chemical analysis. The chromophore is extremely tightly bound to the apoprotein and even after denaturation in trichloroacetic acid and refluxing in methanol, only 5–10% of the chromophore was released [102]. An intriguing calculation by Kroes [64] shows that from 25 kilogrammes of oat coleoptile material, one might expect to produce only 67 microgrammes of chromophore! Nevertheless Siegelman *et al.* [102] were able to carry out a series of chromatographic tests on purified oat chromophore which proved that the chromophore was a linear tetrapyrrole.

The most conclusive evidence on the structure, however, comes from a different approach in which the chromophore is oxidized by chromic acid *in situ* on the protein. This is a classical method of investigating the structure of bile pigments (which are also proteins with linear tetrapyrrole prosthetic groups) and yields a range of identifiable products which can be traced to the various pyrrole rings of the native molecule. Using this method, Rüdiger [95] has been able to postulate almost a complete structure for both Pr and Pfr (Fig. 6.2.9). It is still not certain whether the covalent attachment to the protein is via Ring III (as shown here) or Ring II, as both would yield the same products on oxidation. Furthermore, there is some question as to whether or not Ring I is also covalently attached to the protein in addition to probably having an electrostatic binding with it in the Pfr form.

Spectral properties

Purified phytochrome exhibits the expected absorption bands (λ_{max} 660 nm for Pr; λ_{max} *ca.* 730 nm for Pfr). However, the

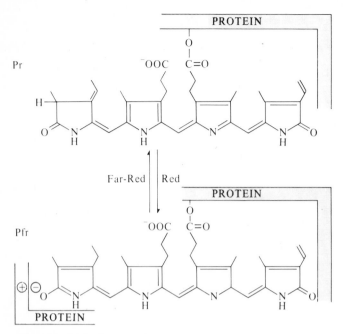

Fig. 6.2.9 Rüdiger's scheme for the chromophore structure of Pr and Pfr (after Rüdiger [95]).

two absorption spectra overlap at all wavelengths below *ca.* 720 nm and upon irradiation with light below 720 nm, a photoequilibrium is established in which the conversion of Pr to Pfr is balanced by the conversion of Pfr to Pr. At any wavelength absorbed by both forms, the established photoequilibrium is characterized by the proportions of Pr and Pfr; thus at 660 nm, 81% of Ptotal is Pfr and 19% is Pr [18]. Thus it is not possible to prepare phytochrome totally in the Pfr form and the absorption spectrum of Pfr can only be deduced by calculations from the spectra of photoequilibria. Representative spectra are shown in Fig. 6.2.10.

The establishment of photoequilibria with continuous irradiation is very important for the interpretation of the high energy reaction and for an understanding of the role of phytochrome in the natural environment. The relationship between photoequilibria of phytochrome and wavelength is shown in Fig. 6.2.11.

The nature of the photoconversions

INTERMEDIATES BETWEEN PR AND PFR

The process of photoconversion of Pr to Pfr and *vice versa* has been studied in great detail and it is known that there are several short-lived molecular intermediates in both directions. The earlier observations were made with the use of flash

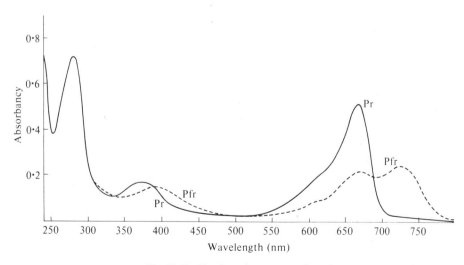

Fig. 6.2.10 The absorption spectrum of pure large rye phytochrome in the Pr and Pfr form. Note that Pfr in this case actually means an equilibrium mixture of 80% Pfr and 20% Pr, thus the absorption spectrum has a large shoulder in the red region (after Rice & Briggs [91a]).

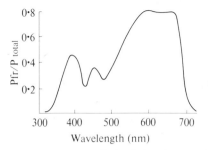

Fig. 6.2.11 Photoequilibria of phytochrome under continuous irradiation at various wavelengths (after Hartmann [41]).

spectroscopy [66] and suggested that different intermediates were involved in the photoconversions in the two directions. Furthermore, it was also suggested that several alternative and parallel pathways existed between Pr and Pfr and *vice versa*. More recent work using spectrophotometric methods, has shown that relatively long-lived intermediates accumulate under cycling conditions, i.e. when both Pr and Pfr are continuously absorbing quanta [14, 15]. Indeed, Kendrick & Spruit [59] have shown that under continuous irradiation with high energy white light, a significant proportion of the Ptotal (up to 30% at 0°C) is present in the form of one particular intermediate whose rate of conversion to Pfr is relatively slow. From this type of work a picture has been built up of a series of thermochemical reactions initiated by a single photochemical reaction as shown in diagrammatic form in Fig. 6.2.12. As can be seen, the various intermediates are recognized by their specific absorption characteristics.

It has been suggested by Kroes [64] that the initial photochemical reaction causes an isomerization of the Pr chromophore to yield an unstable transition state. The extra energy of this intermediate induces a local conformational change in the protein to produce, *via* a series of partially changed intermediates, a new stable complex between protein and chromophore, i.e. Pfr. The reverse transformation is energetically easier since the Pfr protein-chromophore complex is thought to be less stable.

PROTEIN CONFORMATIONAL CHANGES

There is considerable evidence that a change in protein conformation does occur during the transformation, although the extent of the change appears to be rather small. It is known, for example, that Pfr is more susceptible to denaturation by urea than is Pr [20]. Since denaturation involves the unfolding of the polypeptide, this difference can best be explained by a difference in the tertiary structure of the two forms. Similarly, Pr was more sensitive to glutaraldehyde fixation than Pfr and, out of a total of 27 lysine residues in highly-purified 60,000 M.W. phytochrome, 13 reacted with glutaraldehyde when Pr was fixed, but only 11 when Pfr was fixed [92]. This suggests that the availability of two lysines to glutaraldehyde is different in Pr and Pfr.

All these data are consistent with a small protein conformational change accompanying the photoconversions. If the protein moiety of phytochrome has some form of catalytic or transport function (see later), clearly the conformational change could be associated with an activation mechanism.

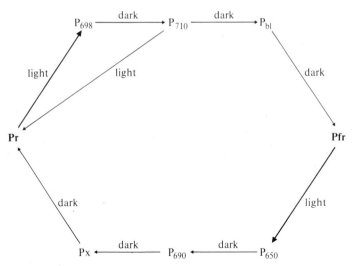

Fig. 6.2.12 Scheme for the photoconversions of Pr and Pfr showing the known intermediates (after Kendrick & Spruit [60]).

THE CHEMICAL NATURE OF THE PHOTOCONVERSIONS

Currently it is not possible to state exactly what chemical changes occur in the chromophore and its local protein environment during the photoconversions. The best available evidence, however, supports the scheme of Rüdiger [95] in which Ring I is involved in a proton exchange with the protein (Fig. 6.2.9). Thus a proton acceptor group must exist on the protein in the vicinity of Ring I and this must stabilize Pfr. Furthermore, the photochemically induced change in electron distribution in Ring I should only enable proton transfer when the acceptor group is in its vicinity, i.e. the local protein conformation must change with the photoconversions. This elegant scheme therefore fits the observed facts concerning the properties of Pr and Pfr, but it does not attempt to take into account the intermediates.

Conclusions

The last decade has seen an enormous increase in our knowledge and understanding of the chemistry of phytochrome. One of the most striking aspects to emerge is the great similarity between phytochrome, the major photoreceptor for photomorphogenic responses in plants, and rhodopsin, one of the photoreceptors for vision in animals. Both are chromoproteins, both can exist in relatively stable, photoconvertible mesomeric forms, both undergo similar types of chromophore isomerization during photoconversion (even though the chromophores are quite different), and both undergo protein conformational changes. It will be seen that this remarkable parallelism of evolution appears to extend to the molecular mechanisms through which the two photoreceptors act.

6.2.5 THE PRIMARY ACTION OF PHYTOCHROME

The establishment of a specific photoequilibrium between Pr and Pfr that is characteristic of the spectral energy distribution of the actinic light can be said to represent the first of the three fundamental processes of photomorphogenesis—namely, the perception of the environmental stimulus. The second fundamental process is the transduction of the environmental stimulus into a biological stimulus. By this, it is meant that the specific Pr:Pfr ratio must elicit a primary internal change, presumably biochemical in nature, which ultimately causes the observed developmental effects. Before the nature of this primary action of phytochrome can be considered in detail, several important aspects of the properties of phytochrome within the plant need to be raised.

Intracellular location of phytochrome

Phytochrome has been detected by *in vivo* dual wavelength photometry in all the representative tissues of etiolated seedlings. It is, however, generally more concentrated in regions of high growth rates, e.g. the meristematic growing tips of roots and shoots and the cambial zones of stems. This correlation with growth activity also extends to temporal relationships, since it has been observed that the early periods of rapid growth in etiolated barley and maize seedlings coincide with an increase in total phytochrome content per seedling.

The localization of phytochrome within cells, on the other hand, is rather more difficult to determine. Several different techniques, however, do suggest that phytochrome is preferentially located in or close to various cellular membranes.

CELL FRACTIONATION STUDIES

When plant tissues are homogenized and fractionated by centrifugation, the distribution of phytochrome depends on the pH of the extraction buffer. If the pH is below about $6·5-7·0$, phytochrome appears to be non-specifically precipitated on to all the particulate cell constituents. Above pH $8·0$, on the other hand, phytochrome is highly soluble, and appears only in the final high-speed supernatant. It is thus very difficult to determine whether or not phytochrome is a component of membranes which is released during high pH extraction, or alternatively, is present in soluble form within the cell, and precipitated on to membranes during low pH extraction. Nevertheless, there are some indications that at least a fraction of the cellular phytochrome is either a component of membranes, or binds to membranes *in vivo*.

Rubinstein *et al.* [93] found in 1969 that a small fraction (*ca.* 5%) of oat phytochrome could be recovered in a fraction sedimenting between 1500 and 40,000 × g, even when extraction was carried out at pH $7·4$ and the pellet washed at pH $8·0$. The phytochrome could be dissociated from the pellet by treatment with the detergent Triton X-100. Pelletable phytochrome exhibited different properties from those of the 95% of the total phytochrome which was soluble. In particular, soluble phytochrome was quite stable over a 24 h period at 0°C whereas pelletable phytochrome was lost during that period.

More recently Quail *et al.* [88] have confirmed these observations by showing that between pH $6·8$ and $8·0$, from 5% to 25% of the total phytochrome of maize or pumpkin tissues could be pelleted at $20,000 × g$. These results are particularly striking, since the proportion of phytochrome that can be pelleted is affected by whether it was in the Pr or Pfr state in the tissue prior to extraction.

In an extension of this work, Briggs and colleagues [70, 71] have been able to find conditions under which a very high proportion (*ca.* 40%) of the total phytochrome may be pelleted from a homogenate of zucchini squash hypocotyls. Very little phytochrome was pelletable (*ca.* 5%) if dark-grown tissues were used, and maximum pelletability occurred with red-irradiated hypocotyls. The pelletability was very dependent on the Mg^{++} concentration, a maximum being achieved at 10 mM. The pelleted structures appear to be membranous, as they contain phospholipids and other membrane constituents. Under certain conditions (i.e. varying Mg^{++} concentration) the pelleted phytochrome is reversibly released *in vitro*. The authors conclude that the membrane possesses binding sites for phytochrome and imply that *in vivo* Pfr is bound to the membrane and Pr is soluble.

A slightly different approach using cell fractionation methods is to attempt to demonstrate not phytochrome itself, but phytochrome-mediated processes in isolated cell fractions. This type of investigation is only just beginning but already evidence for the presence of phytochrome in etioplasts, and possibly in mitochondria has been reported. Welburn & Welburn [116] were able to isolate etioplasts (i.e. plastids from etiolated seedlings) in such a state that they were capable of a certain degree of ultrastructural development *in vitro*. By a rather complex scoring procedure, they were able to quantify the developmental changes they observed, and showed that red and far-red light had the expected effect on etioplast development *in vitro*. This suggests either that phytochrome is present within the etioplasts or is attached to the outer membrane. In work described in more detail below, Manabe & Furuya [69] have recently shown that specific enzyme activities of cell particles sedimenting between 1000 and 7000 g are controllable *in vitro* by red and far-red light. Although they did not come to a decision regarding the nature of the particulate fraction, mitochondria may have been present even though the sedimentation speed was rather low. In this case the authors also demonstrated the presence of phytochrome by the classical photoreversible spectral changes.

IMMUNOCYTOCHEMICAL LOCALIZATION

An ingenious approach to the localization question is the use of antibodies to phytochrome that can be visualized histochemically. Pratt & Coleman [87] raised rabbit antibodies to purified phytochrome and challenged paraffin embedded thin sections of oat shoot tissues with the antiserum. The section was then sequentially challenged with a sheep antiserum to rabbit immunoglobulin, and a rabbit antiperoxidase-peroxidase complex. The outcome of this complicated procedure is the binding of each phytochrome molecule to a large number of peroxidase molecules, which can be readily visualized by histochemical techniques.

So far as intracellular location is concerned, the results were not very clear cut, although they did indicate that phytochrome was present on most of the membranous components of the cell,

in addition to being present within the cytoplasm. It is intriguing, however, that certain cells appeared to contain no phytochrome whatsoever; in the lower regions of the oat coleoptile, for example, only the epidermal cells contained phytochrome, whereas in the mesocotyl, it was apparently restricted to the procambial cells. These findings might well be due to unpredictable differences in the penetration of the various antisera into the different cells.

EVIDENCE FROM POLARIZED LIGHT EXPERIMENTS

Indirect physiological experiments carried out on a small number of single-celled plant materials have led to the conclusion that phytochrome is located at or near the plasmalemma, and that the pigment molecules are oriented with respect to each other. The prime example of this type of approach concerns the phytochrome-controlled rotation of the single chloroplast in the cells of the filamentous green alga *Mougeotia*. Over the past decade, Haupt and colleagues have shown by the use of microbeams of red and far-red light that the molecules of phytochrome which control this response are located not in the chloroplast itself, but in the outer regions of the cytoplasm near the plasmalemma. Furthermore, the use of polarized light has demonstrated that phytochrome molecules in the Pr form are oriented with their absorbing axes (i.e. the chromophores) parallel with the surface of the cell, whereas Pfr molecules are oriented with their chromophores perpendicular to the cell surface. Thus, upon photoconversion, the photoreceptor molecules must flip over. The *Mougeotia* experiments have been comprehensively reviewed by Haupt & Schönbaum [43, 44, 45].

Similar dichroic orientation of phytochrome molecules at or near the plasmalemma has been observed in the filamentous germ tubes of the fern *Dryopteris filix-mas* [25]. Such orientation of photoreceptor molecules implies that they are connected to each other either directly, by being bound together, or indirectly, by being associated with some other structure which has either planar or linear characteristics. The obvious and perhaps most likely explanation is that the phytochrome, or at least a fraction of the phytochrome, is either bound to, or is a component of, the plasmalemma. Other possibilities exist however. It is known, for example, that plant cells have microtubules running around the cell just within the plasmalemma. Microtubules, although not planar structures, have linear axes and are usually oriented with respect to each other. In the case of *Mougeotia*, Schönbohm [99] has shown that the actual movement of the chloroplast is mediated by contractile microfibrils attaching the chloroplast edge to the parietal cytoplasm. It is not known whether these fibrillar elements are oriented with respect to each other. If so, either the microfibrils or the microtubules could be sites of phytochrome localization.

From all these data, the most prudent general conclusion is that at least some of the cell phytochrome is associated with either membranous or particulate components of the cell. As will be seen, several other lines of evidence suggest that the primary action of phytochrome involves a change in the properties of certain cell membranes.

In vivo phytochrome interconversions

As described previously, the two forms of phytochrome are interconvertible upon exposure to actinic irradiation. *In vivo,* however, a number of non-photochemical conversions are known to occur. The two most important are (a) phytochrome *destruction* (formerly known as decay) in which the spectral photo-reversibility is lost completely; (b) *reversion,* in which Pfr is converted to Pr in darkness. In addition, the overall levels of phytochrome appear to be regulated by a balance between synthesis and degradation.

PHYTOCHROME DESTRUCTION

Etiolated tissues contain relatively large amounts of spectrally assayable phytochrome, but if such tissues are irradiated with red light, and then left in darkness, the total phytochrome decreases quite rapidly. The loss of total phytochrome in darkness following red light can be halted at any point by a brief irradiation with far-red. Thus, in etiolated tissues, phytochrome appears to be quite stable as Pr and unstable as Pfr. This property has its analogy in the greater reactivity and sensitivity to denaturation of Pfr *in vitro*. The process of loss of total phytochrome was originally known as 'decay', but the alternative term 'destruction' is now preferred. Very little is known of the mechanism of destruction and it should be stressed that it only represents the loss of photoreversible spectral changes, which could be caused by a wide range of chemical alterations, ranging from complete degradation of the whole molecule, to a small but irreversible change in the electronic distribution of the protein-chromophore complex. Destruction appears to be a normal metabolic process as it is virtually prevented at 0°C and is severely inhibited by anaerobiosis and poisons such as azide and cyanide.

In most tissues examined, destruction is a first-order reaction in which the rate constant is proportional to the Pfr:Pr ratio. Thus although destruction is not light-dependent, it occurs most rapidly ($t_{\frac{1}{2}} \sim 30$ to 60 min at 25°C) in tissues continuously irradiated with red light which maintains the maximum Pfr:Pr ratio. Because of this, it is normally impossible to detect by *in vivo* spectrophotometry, the presence of phytochrome in plants which have become de-etiolated. Furthermore, such tissues normally have relatively large amounts of chlorophyll which effectively masks the weak absorption due to the residual phytochrome present. Indeed Grill [39] has suggested that screening by newly-formed chlorophyll might cause the loss of apparent photoreversibility in continually illuminated tissues. If this is so, then destruction would appear to be an artefact of the method of measurement. It is difficult, however, to see how chlorophyll screening could account for phytochrome destruction in tissues which have only been briefly irradiated with red light and returned to darkness. In such tissues, significant amounts of chlorophyll are not formed, and thus the loss of photoreversibility cannot be explained by chlorophyll screening.

The destruction process is currently somewhat of a mystery, both in terms of its mechanism, and its function. It is perhaps possible that destruction merely represents normal turnover which only takes place with Pfr, and thus is prevented in dark-grown tissues which only contain Pr. (See [29] for review of destruction.)

DARK REVERSION

As a result of purely physiological experiments, the slow dark reversion of Pfr to Pr was first postulated before phytochrome was discovered. In spite of the physiological evidence, however, it is now known that reversion of Pfr to Pr only occurs in a restricted range of plants. It has not been detected spectrophotometrically in any monocotyledonous species, and is also absent from certain families of the Dicotyledonae (e.g. the Centrospermae [58]). However, dark reversion of Pfr to Pr definitely does occur in several dicotyledonous plants, and in some tissues dark reversion occurs in the complete absence of destruction (e.g. cauliflower head tissue). (See [29] for review).

SYNTHESIS OF PHYTOCHROME

When the level of total phytochrome is reduced by destruction in seedlings which have received a single red light irradiation, the level of total P eventually begins to increase again. This process of 'apparent resynthesis' begins when about 20% of the original phytochrome is left, and leads to a significant increase in the amounts of assayable phytochrome [21]. As only the spectral changes can be measured, it is prudent to use the conditional term 'apparent synthesis' when describing the increases.

Recently, however, direct demonstration of the synthesis of phytochrome in etiolated, and light-treated seedlings has been reported [89]. In these experiments seedlings were incubated in heavy water (D_2O) so that any newly-synthesized proteins would incorporate deuterium instead of hydrogen. Any protein incorporating deuterium would have a higher buoyant density, and could thus be differentiated from the native protein by centrifugation to equilibrium in a density gradient of caesium chloride solution. This technique of density labelling has been used with great effect to demonstrate the *de novo* synthesis of a range of enzymes in plant tissues [26].

Quail *et al.* [89] were able to density label phytochrome in the hypocotyls and cotyledons of pumpkin seedlings both in darkness and after a brief irradiation with red light. They were also able to show that turnover of the total phytochrome in the tissues occurred. Phytochrome was synthesized in the Pr form, and both Pr and Pfr appeared to turnover, although the rate constant of degradation of Pfr could be up to 100 times greater than that of Pr.

CONCLUSIONS

Thus the levels of Pr and Pfr in a cell appear to be regulated by two photochemical reactions, two non-photochemical inter-conversions and by synthesis and turnover. The simple scheme of phytochrome reactions used previously therefore needs considerable modification as follows:

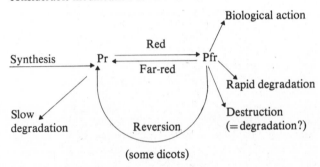

The primary action of phytochrome

There are two main lines of thought concerning the mechanism whereby the environmental stimulus is transduced to a biological stimulus. In 1966 Mohr [75] proposed that phytochrome acts by a direct effect of Pfr on the expression of certain genes. In 1967 Hendricks & Borthwick [47] proposed that phytochrome acts by altering the permeability of cellular membranes. These two hypotheses are not necessarily mutually exclusive, and there is no reason to suppose that they cannot operate together, or at least sequentially.

THE GENE-EXPRESSION HYPOTHESIS

In its original formulation, Mohr's gene expression hypothesis proposed that Pfr interacted directly with the mechanisms controlling the synthesis of messenger-RNA on the genome. Evidence for this view came principally from rather negative experiments in which certain photomorphogenic responses could be shown to be prevented, or partially inhibited by applications of actinomycin D, an antibiotic which, under ideal

conditions, specifically inhibits RNA synthesis [65,78]. The concept was that the genome of a particular cell could be divided into four categories: (a) genes which are active and not subject to Pfr control; (b) genes which are inactive and not subject to Pfr control; (c) genes which are inactive but capable of being switched on by Pfr; and (d) genes which are active, but capable of being switched off by Pfr. In this form, the hypothesis represents an attempt to account for the *primary* action of Pfr. In later publications [76,77] Mohr has implied that Pfr need not *itself* interact directly with the genome, and that some product or products of Pfr primary action might be responsible for controlling gene expression. In this form, the hypothesis does not seek to account for the primary action of Pfr.

As pointed out in the next section, it is now almost a truism to say that developmental control implies the regulation of gene expression, and it does seem likely, although it is as yet unproven, that the ultimate effects of Pfr action will be found to involve the induction and repression of enzyme synthesis. However, there is overwhelming evidence (as Mohr has himself pointed out [77]) against the original concept that Pfr interacts directly with the genome.

Foremost amongst this evidence is the knowledge that phytochrome can act very quickly, both in non-developmental

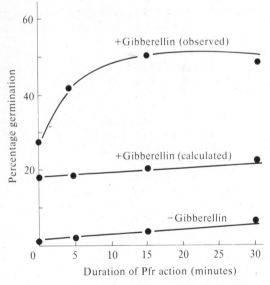

Fig. 6.2.13 Rapid action of Pfr in the germination of lettuce seeds. Imbibed seeds were given red light followed by far-red such that Pfr was able to act for the stated period. In the absence of added gibberellin little or no stimulation of germination occurred (lower curve). With gibberellin, interaction between Pfr and the hormone was apparent (upper curve). The middle curve shows the expected germination if Pfr and gibberellin were to act additively (after Bewley, Black & Negbi [6]).

and developmental processes. Table 6.2.3 shows that the time lag of several phytochrome displays is less than 5 min, and that a response can be detected within 15 s in the case of the alteration of electric potentials in *Avena* coleoptiles [80]. Even in developmental responses, Pfr can often be shown to operate very quickly, although the actual display may take several hours. In the case of lettuce seed germination, for example, it has been shown that Pfr can act within a few minutes of being formed [6]. If seeds are given red light (setting up 80% Pfr) in the presence of very low concentrations of the plant hormone gibberellic acid (GA$_3$), and then given far-red (which removes the Pfr) only 5 min later, a significant increase in eventual germination is obtained when compared with dark controls (Fig. 6.2.13). Thus Pfr and GA$_3$ interact in stimulating germination, and action of Pfr present for only a few minutes may be detected as long as low levels of GA$_3$ are available. It seems extremely unlikely that a mechanism involving the synthesis of new messenger-RNA, new enzymes and the action of these enzymes, could account for such rapid effects of Pfr. It is known, for example, that the substrate induction of nitrate reductase in plants, which apparently involves all these processes, has a lag of several hours before increases in enzyme activity can be detected [107].

THE MEMBRANE-PERMEABILITY HYPOTHESIS

In view of the known, very rapid effects of Pfr, Hendricks & Borthwick [47] proposed a mechanism involving the control of membrane permeability as the primary function of phytochrome. Their selection of membrane permeability stemmed from the realization that many of the rapid effects of red light appeared to involve the displacement of ions and the consequent establishment of electric charges. This is particularly evident for the changes in electric potential of the *Avena* coleoptile [80] and the mung bean root tip [51]. Other responses similarly obviously associated with membrane properties include the photoreversible electrostatic adhesion of barley and mung bean root tips to a negatively charged glass surface [109a, 109b] and the control of the photonastic movements of *Mimosa* and *Albizzia* leaflets [34]. In all these cases, it has subsequently been shown that the immediate effect of phytochrome is to cause the redistribution of cations, probably K$^+$, within the tissues, thus affecting either the electrical properties or the osmotic potentials of the cells (see [104a] for review). Hendricks & Borthwick [47] however, did not attempt to specify the mechanism whereby phytochrome might regulate membrane permeability, and Mohr [77] has rightly pointed out that such a very general hypothesis has little heuristic value.

At least one attempt, however, has been made to render the membrane permeability hypothesis more specific and amenable to investigation. Jaffe, working with mung bean root tips, observed that red light treatment caused a rapid (maximum 4 min) increase in the concentration of acetylcholine within the tips [52]. Acetylcholine, a mammalian neurohumour whose pharmacological action is to activate ion transport across nerve cell membranes, had only rarely been observed in plants and no suggestions had previously been made as to its function. Jaffé also observed that acetylcholine could mimic red light in causing the root tips to adhere electrostatically to a negatively charged glass surface and thus he proposed that the primary action of Pfr was to cause the release of acetylcholine from a storage compartment, or alternatively to stimulate its synthesis. Although the action of acetylcholine on root tips appears to be due to its being a monovalent cation [110], and although exogenous acetylcholine appears to have no effect in other photomorphogenic phenomena [56,98], nevertheless the increased internal levels after red light are most interesting and need further examining. Jaffe has produced a speculative model [53] in which acetylcholine is considered as a second messenger, released by Pfr, and which moves throughout the cell to initiate a wide range of metabolic events including changed membrane permeability. It is only fair to say that this proposal has met considerable scepticism, largely due to the reluctance of plant physiologists to consider that plants might have even the most rudimentary elements of a nervous system; however, as pointed out by another reviewer [16] the onus is on the sceptics to obtain further information on the possible role of acetylcholine in plants.

The membrane permeability hypothesis, either in its general form or as postulated by Jaffe, does not necessarily imply that phytochrome is a component of cell membranes. A great deal of evidence, however, points to the concept that at least part of the phytochrome of the etiolated plant is membrane associated. It was this concept that led to the formulation by Smith in 1970 of a variant of the membrane permeability hypothesis in which phytochrome is considered to be a transport factor within the membrane, catalysing the transport of a critical metabolite across the membrane [103]. This model, already shown in Fig. 6.2.5 implies that Pfr *per se* is not the active form of phytochrome, but that the action is associated with photoconversions of Pr to Pfr. This model unfortunately suffers from the problem of being difficult to test, although it does imply that phytochrome will have affinity for a critical metabolite.

Finally, an allosteric model has recently been proposed for the effect of phytochrome on membrane properties [71]. This is based on the observed photoreversible binding of phytochrome to membrane fractions both *in vivo* and *in vitro*, and suggests that phytochrome acts as an effector to modify the conformation of the membrane to which it is bound. Thus, several molecules of phytochrome might act cooperatively to bring about relatively enormous changes in membrane conformation, possibly causing changed permeability and transport properties.

CONCLUSIONS

It is impossible to state precisely at the present moment how the photoconversion of phytochrome is transduced into a biochemical stimulus within the cell. Certainly the most acceptable idea is that the initial event involves membrane properties in some way, either by relatively unspecific effects on the permeability of membranes to ions such as K^+, or more specifically by controlling the transport or release of a second messenger. It would seem extremely important to clear up the question of whether a fraction of phytochrome is a component of cell membranes, or whether it is merely found there as a result of an artifact of preparation. Further investigation of acetylcholine as a potential second messenger would also seem most desirable. So far as gene expression is concerned, most would agree that it is futile to search for the primary action of phytochrome at this level. Nevertheless, as is detailed below, the ultimate developmental events, perhaps mediated by the postulated second messenger, seem certainly to involve the regulation of gene expression, perhaps via several distinct mechanisms.

Possible role of hormones as second messengers for phytochrome action

If the hypothesis that phytochrome acts via a second messenger is accepted the question of whether or not any of the known plant hormones may function as the second messenger ought to be considered. There is, in fact, evidence linking most of the known hormones to phytochrome although much of it is extremely difficult to reconcile with a simple scheme.

AUXINS AND ETHYLENE

Early work has shown that irradiation of etiolated seedlings with red light leads to a fall in the content of free auxin in the stem tissues [7, 48, 83, 113]. So far as is known, however, these responses have time lags of the order of several hours and thus are unlikely to be related to the primary action of phytochrome. In addition, although the depression of free auxin levels is associated with a reduced growth rate of the tissues, in no case could the growth inhibition be reversed by application of exogenous auxin. However, bean hypocotyl hook opening is stimulated by red light and inhibited by auxin, which correlates well with the observed drop in auxin content [63].

The production of ethylene, a gaseous plant hormone whose action appears to be tied up with auxin in some as yet unexplained manner, is also known to be controlled by phytochrome. Goeschl et al. in 1957 showed that red light treatment of pea seedlings caused a marked drop in the rate of ethylene evolution which exactly paralleled the decreased growth rate of the stems [37]. More recently Kang & Ray [55] have shown that red light treatment of etiolated bean seedlings depressed ethylene production and stimulated CO_2 evolution. Applications of exogenous CO_2 caused opening of the hypocotylar hook in a manner similar to the red light response; ethylene, on the other hand, antagonized the red light induction of hook opening. Thus, Kang & Ray concluded that ethylene acted in vivo as a regulator of hook opening, and that phytochrome acts through the control of ethylene formation. In addition they proposed an ingenious biochemical hypothesis for the control of ethylene production via a phytochrome-mediated increase of polyphenol synthesis. In common with the auxin observations, however, the effects on ethylene levels appear to be separated from the primary action of phytochrome by at least a few hours, and thus it seems unlikely that ethylene could be a general second messenger for phytochrome action.

GIBBERELLINS

Dwarf varieties exist of several species of crop plants (e.g. peas and maize) and in many cases the dwarfism is only apparent in seedlings which have received light. Thus the growth rates of etiolated seedlings of tall and dwarf varieties are similar, but the dwarf varieties experience a dramatic and permanent inhibition of stem extension rate upon exposure to a brief period of red light. The growth of tall varieties is also inhibited by such treatment but the effect is much smaller and is only transient. In 1965, Lockhart [67] showed that gibberellic acid (GA_3) could prevent the red light inhibition of dwarf pea stem extension. This led him to postulate [68] that phytochrome acts by depressing the rate of gibberellin synthesis. Although this hypothesis is supported by other cases in which exogenous gibberellin overcomes the red light effects (e.g. the stimulatory effects of red light on hook opening in beans [61], and on flavonoid synthesis in peas [96], are prevented by GA_3) nevertheless, there are several instances in which phytochrome action is insensitive to gibberellin application. Thus, the hypothesis cannot be considered as being generally true.

Attempts to measure changes in gibberellin levels in light-treated seedlings of dwarf and tall peas have to date been unsuccessful. Kende & Lang [57] for example, showed that two fractions with gibberellin activity (probably GA_1 and GA_5) occur in pea seedlings, neither of which is affected by light treatment. However, the sensitivity of the growth of pea stems to exogenous GA_5 is decreased by prior red light treatment. Later it was shown that light-treated stem sections were able to bind much less radioactive-GA_5 than etiolated tissues, suggesting that the availability of specific binding sites for GA_5 was reduced in the light-treated tissues [79]. Very recent work has shown that plant tissues contain highly specific gibberellin-binding proteins which may be involved in the growth responses to gibberellin [108]. The phytochrome-mediated changes in the ability of the pea stem tissues to bind radioactive GA_5 had a lag phase of about three hours, so here again we are not dealing with primary actions of phytochrome.

In the case of cereal leaves, however, red light has been shown to cause very rapid changes in the levels of endogenous gibberellins. As mentioned earlier, the expansion rate of many leaves is reduced in darkness and increased significantly by red light treatment. This is most readily observed in dicotyledonous plants, but an analogous phenomenon occurs in cereals in which the leaves, which are tightly rolled around a longitudinal axis in darkness, open out after light treatment by expansion of the upper epidermal cells. This process is a highly sensitive phytochrome-controlled phenomenon which can be studied in leaf sections floating on water in a petri dish.

In 1968, Reid, Clements & Carr [90] showed that red light caused a very rapid, transient surge in gibberellin levels in sections of barley leaves. Later, Wareing and colleagues [5, 68a] demonstrated a similar increase in wheat leaf sections. Maximum gibberellin levels occurred at about 15 minutes after the red light treatment. Subsequently, it was observed that an exactly analogous response occurred in homogenates of barley leaves [91]. A typical time course of the increases in free and bound (acid-hydrolysable) gibberellin in a red-light treated homogenate is shown in Fig. 6.2.14. There appears to be a rapid release of free gibberellin which peaks at about five minutes after the onset of the light period, followed by a slower, but quantitatively much greater rise in bound gibberellin. There has been some slight controversy over whether or not the initial increase in gibberellin levels is due to synthesis, or to release from a bound form, or from a subcellular compartment. The rapidity of the response suggests that release is more likely than synthesis, although the later effects on bound gibberellin may involve synthesis.

These results are consistent with a gibberellin acting as a second messenger for phytochrome action, in cereal leaves at least. In addition, application of exogenous gibberellin to leaf sections in darkness causes opening of the leaves in an analogous manner to that caused by red light [5]. However, gibberellins are not the only plant hormones which can do this. Cytokinins will also cause wheat leaf unrolling, although the effect is quantitatively not as great as with the gibberellins [5]. One cannot be certain therefore that the phytochrome-mediated surges in endogenous gibberellins are necessarily causally involved in the subsequent growth responses. A major problem is that the red-light mediated unrolling of cereal leaves is still subject to significant far-red reversibility for up to two hours after the light treatment. Thus far-red can still reverse the growth effects some time after the transient changes in free gibberellin levels have been completed.

CYTOKININS

Although cytokinins also mimic red light in the cereal leaf opening response, no light-mediated changes in endogenous cytokinin levels could be detected in barley leaves [73]. However, Hewett & Wareing [48] have recently shown that brief red light treatment of mature poplar leaves (taken from the

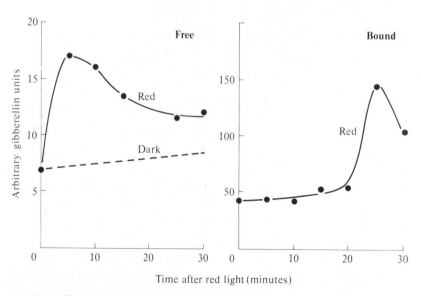

Fig. 6.2.14 Time course of the changes in free and bound (acid-hydrolysable) gibberellin after red light treatment of homogenates of dark-grown barley leaves (unpublished data of Miss Audrey Evans).

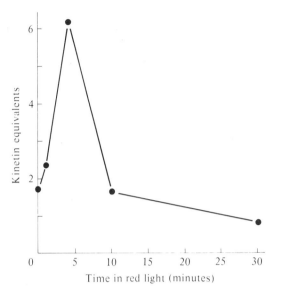

Fig. 6.2.15 Time course of the changes in cytokinin levels of poplar leaves during red light treatment (replotted from Hewett & Wareing [48]).

field in the middle of the night) caused large and rapid increases in cytokinin levels (Fig. 6.2.15). Unfortunately, far-red reversibility tests were not reported.

In a related investigation van Staden & Wareing [114] have shown that brief red light treatment of dark-imbibed *Rumex obtusifolius* seeds caused a very large increase in extractable cytokinins within about 10 minutes. This effect was reversible by subsequent far-red light. The authors concluded that the massive increase in endogenous cytokinins may mediate the control of germination by phytochrome, although they could not replace red light by exogenous application of either synthetic or natural cytokinins.

CONCLUSIONS

There seems to be undeniable evidence that the major plant hormones are implicated in some phytochrome controlled growth responses. It is not possible, however, to suggest a simple hypothesis in which one of the hormones mediates the action of phytochrome in all its multivalent displays. Conclusions of this nature are unfortunately only too common when the role and functions of plant hormones are being considered. Unlike most of the animal hormones, plant hormones do not have specific functions in differentiation. The molecular mechanisms of the individual hormones are incompletely understood, although it is known that the auxins, the gibberellins and ethylene control cell enlargement under specific conditions, whilst the cytokinins are necessary for cell

division. In a multicellular environment, however, it appears to be the balance between these, and possible other, hormones which determines the fate of any particular cell in differentiation. Thus, it would be naive to expect to be able to understand the mechanism of phytochrome action in terms of a single plant hormone. We are forced to the conclusion that the observed changes in hormone levels, even the rapid ones, are themselves secondary events of phytochrome action which may or may not be causally related to the ultimate developmental events. If a single second messenger for phytochrome action exists, then it is most unlikely to be one of the known plant hormones, although one of its actions would presumably be to mediate the observed changes in hormone levels. A corollary of this conclusion is that a second messenger need not be present, or active, for more than a very short period of time; in barley leaves for example it would only need to be available for 10–15 minutes in order to bring about the increase in free gibberellin. If all the subsequent steps were set in train by the free gibberellin, then the continued presence of the second messenger would be irrelevant. If this is so, it is going to mean that the search for the postulated second messenger is likely to be long and tedious.

Rapid metabolic effects of phytochrome action

One approach towards the elucidation of phytochrome action is to monitor metabolic changes during the first few minutes after the establishment of Pfr in tissues. With hindsight, it seems quite remarkable that so little is, in fact, known of the immediate biochemical consequences of the photoactivation of phytochrome. In common with other aspects of phytochrome studies, there has been little direction of effort, and most investigations have employed different species, different tissues and even different irradiation conditions. However, progress is beginning to be made and glimpses of some tantalizing possibilities can already be discerned.

As stated above, the earliest manifestation of phytochrome action is the establishment of an electric potential along the *Avena* coleoptile [80], a response which has a lag of about 15 s. This is clearly a membrane-associated response. In the same species, however, an almost equally rapid, metabolic event has been observed which is not obviously connected with membranes. Sandmeier & Ivart [97] showed that irradiating *Avena* mesocotyl tissue with red light caused an almost stoichiometric conversion of ADP to ATP. The peak of ATP concentration occurred at about one minute after the onset of irradiation, and considerable action had occurred within 30 seconds.

Indeed, there seems to be growing evidence that phytochrome has a modulating effect on the pyridine nucleotide complement of plant cells, and perhaps through these changes, on the energy metabolism and biosynthetic activities of the cells. Tezuka & Yamamoto in 1969 showed that irradiation of *Pharbitis nil* cotyledons with red light rapidly increased the levels of NADP; the increase could be partially reversed with far-red light [111]. Subsequently, they showed that partially purified pea phytochrome had NAD kinase activity, and that red light treatment given *in vitro* increased the activity of the NAD kinase. The NAD kinase and the phytochrome were not, however, part of the same protein, as they could be separated by calcium phosphate gel chromatography [112]. Although the NAD-kinase findings have failed to be independently confirmed to date [50], other workers have shown increased amounts of NADPH in oat tissues after red-light treatment [32]. Tezuka & Yamamoto [112] were able to demonstrate a range of metabolic alterations as a result of applications of NADP, and they proposed that phytochrome acts by modulating the cellular concentration of NADP.

Recently, a further *in vitro* phytochrome effect on pyridine nucleotide levels has been reported [69]. In this work a subcellular particulate fraction sedimenting between 1000 and 7000 g was obtained containing both phytochrome and a NADP-linked glucose-6-phosphate dehydrogenase. If the particulate fraction was given a red light treatment before the addition of $NADP^+$ and glucose-6-phosphate, the rate of formation of NADPH was between 1·5 and 2 times that of a sample given either far-red light or darkness. The authors conclude that phytochrome photoconversion induces either a conformational change in the glucose-6-phosphate dehydrogenase, or an increased rate of transport of $NADP^+$ into the particles comprising the active fraction.

Since NADPH is involved in a large range of mainly biosynthetic metabolic reactions, it is quite clear that a hypothesis could be built up around $NADP^+$ or NADPH as an intermediary in the action of phytochrome. On purely *a priori* grounds, however, it is difficult to see $NADP^+$/NADPH having the necessary specificity to act as a second messenger in the sense used previously. Clearly, much remains to be learnt about the photocontrol of pyridine nucleotide levels, and indeed, about any other metabolic processes that can be shown to occur very soon after the photoconversion of Pr to Pfr.

6.2.6 MOLECULAR MECHANISMS OF DEVELOPMENTAL CONTROL

One of the current dogmas of developmental biology is that the regulation of development necessarily involves selective control of the expression of the genetic information. The total acceptance of this view occurred when the alternative of the selective loss of 'unnecessary' genetic information throughout development was shown to be wholly untenable. It is now considered a truism that when a cell changes its function dramatically, it must be expressing parts of its information store

which had not hitherto been expressed. In spite of all this, it is necessary to examine all specific developmental processes very closely before precipitously concluding that control of gene expression is involved. Developmental processes are extremely complex, and it is quite possible that apparently dramatic changes may not be due to the selection of new activities, but to the integrated acceleration and deceleration of existing activities. The distinction here is between *induction* and *modulation*, and as will be seen is crucially important to a study of photomorphogenesis.

Induction—or, more colloquially, 'switching-on'—of a new activity obviously must involve the selection of new parts of the genetic information for expression. Modulation, on the other hand, does not imply such selection, as it is concerned merely with the rates of on-going activities. Thus, it is important to distinguish between qualitative (inductive) events and quantitative (modulatory) events. As briefly mentioned previously, all known strictly photomorphogenic events can be shown to be quantitative in nature, even though extremely profound overall changes are involved. It is not possible in a single chapter to catalogue the evidence which leads to this

Table 6.2.5 Enzymes whose extractable activities are regulated by phytochrome (see Smith [104a] for references).

Enzyme	Plant material
Intermediary metabolism	
NAD kinase (*in vitro*)	Pea
Lipoxygenase	Mustard
Amylase	Mustard
Ascorbic acid oxidase	Mustard
Galactosyl transferase	Mustard
NAD^+-glyceraldehyde-3-phosphate dehydrogenase	Bean
Nitrate reductase	Pea
Nucleic acid and protein metabolism	
RNA polymerase (nuclear)	Pea
Ribonuclease	Lupin
Amino acid activating enzymes	Pea
Photosynthesis and Chlorophyll Synthesis	
Ribulose-1,5-diphosphate carboxylase	Bean
Transketolase	Rye
$NADP^+$-glyceraldehyde-3-phosphate dehydrogenase	Bean
Alkaline fructose-1,6-diphosphatase	Pea
Inorganic pyrophosphatase	Maize
Adenylate kinase	Maize
Succinyl CoA synthetase	Bean
Peroxisome and Glyoxisome Enzymes	
Peroxidase	Mustard
Glycollate oxidase	Mustard
Glyoxylate reductase	Mustard
Isocitrate lyase	Mustard
Catalase	Mustard
Secondary Product Synthesis	
Phenylalanine ammonia lyase	Peas (and many others)
Cinnamate hydroxylase	Peas

conclusion, but it will be seen that the quantitative nature of phytochrome action is quite consistent with what little is known of the regulatory mechanisms involved.

Developmental processes, in common with other activities, result from the concerted and integrated functioning of a large number of biochemical reactions, and since enzymes are responsible for the catalysis and control of these reactions, it is logical to direct attention at particular enzymes and the way in which they respond to a light stimulus. A large number of enzyme activities have been shown to be affected by light acting via phytochrome (see Table 6.2.5), but unfortunately only a very small number have been investigated in sufficient detail to allow the derivation of meaningful conclusions. The best examples are phenylalanine ammonia-lyase and ascorbic acid oxidase, both of which have been studied in mustard seedlings.

Phenylalanine ammonia-lyase

Phenylalanine ammonia-lyase (PAL) catalyzes the removal of ammonia from the amino acid phenylalanine to yield cinnamic acid which is subsequently used to form part of the structure of the red/blue flower pigments known as anthocyanins. In a large number of seedlings, PAL activity increases in response to light treatment. In mustard seedlings, continuous far-red light gives maximum response indicating that phytochrome is operating in the high energy mode [23] (Fig. 6.2.16). The important question is whether the increase in activity is due to enzyme synthesis or activation.

Various experiments with protein synthesis inhibitors (e.g. cycloheximide) suggest that enzyme synthesis is involved, but such experiments are always rather dubious. Recently, a direct attempt to demonstrate *de novo* synthesis has been made using the method of density labelling briefly described above. Attridge *et al.* [2] grew mustard seedlings in H_2O, in darkness, transferred them to D_2O and gave some far-red light whilst leaving others in darkness as a control. In this technique, enzyme synthesis is indicated if the extracted enzyme can be shown to have a higher buoyant density than the native H-enzyme. This can only happen if deuterium is incorporated into the protein during synthesis. Table 6.2.6 gives the results from a

Table 6.2.6 Buoyant densities of phenylalanine ammonia-lyase (PAL) and ascorbic acid oxidase (AO) in the dark and after saturating far-red light (data from Attridge [1, 2])

Enzyme	Treatment	Buoyant densities
PAL	Dark H_2O	1·295
	Far-Red H_2O	1·294
	Dark D_2O	1·321
	Far-Red D_2O	1·314
AO	Dark H_2O	1·303
	Far-Red H_2O	1·302
	Dark D_2O	1·306
	Far-Red D_2O	1·315

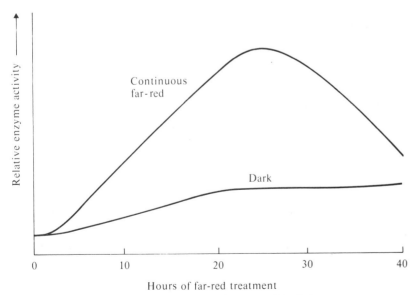

Fig. 6.2.16 Time course of phenylalanine ammonia lyase activity during continuous irradiation of mustard seedlings with far-red light (after Schopfer & Mohr [101]).

typical experiment showing that the enzyme from dark grown D_2O-treated plants has a much higher buoyant density than that from far-red D_2O, treated plants, even though actual enzyme activity was several fold higher in the latter case.

These results are somewhat difficult to interpret as the final buoyant density in situations involving a transfer from H_2O to D_2O depends on the balance between synthesis and turnover. Thus the decreased buoyant density in the far-red treated plants could be due either to a massive reduction in PAL degradation, or to the activation of pre-existing inactive enzyme molecules.

In related experiments carried out with gherkin seedlings, evidence has been put forward for a large pool of inactive PAL existing in dark grown seedlings [4]. Furthermore, a macromolecular inactivator of PAL has been isolated from gherkin seedlings [31].

Whichever way the density labelling is interpreted, there is no evidence for an actual increase in the rate of PAL synthesis under the influence of far-red light. Indeed, it could be concluded that the rate of PAL synthesis is considerably greater in the absence of light. The quantitative nature of this response is quite evident and it would be very difficult to account for the increase in enzyme activity by invoking control at the transcription level.

Ascorbic acid oxidase

Ascorbic acid oxidase is also increased in activity in mustard seedlings upon exposure to far-red light. The time course is rather similar to that for PAL, but in this case the density labelling evidence is quite different. Attridge [1] has carried out analogous experiments to those described for PAL and typical results are shown in Table 6.2.6. In this case the enzyme from the far-red treated seedlings has a much higher buoyant density than that from the dark controls. These results are consistent with an increased rate of ascorbic acid oxidase synthesis under far-red light, but the complications of possible alterations in turnover rate should be borne in mind.

Thus a very precise technique used by the same workers, in the same tissue, on different enzymes, produced in one case evidence for an effect of phytochrome on enzyme activation, and in the other, evidence for an effect on synthesis. At the very least, this suggests that the ultimate mechanisms through which phytochrome controls enzyme levels can be different even within the same tissues. In addition, it shows that the phytochrome control of development, as exemplified by enzyme activities, does not always operate through the regulation of gene expression at the level of transcription or translation.

Nucleic acid synthesis

Several investigators have looked into the possibility of increased rates of nucleic acid and protein synthesis. Most of

these attempts have merely shown that any increases that are detectable, occur rather slowly and thus are not likely to be causal steps in the overall developmental changes. Unfortunately, techniques for isolating and measuring the rates of synthesis of individual messenger-RNA molecules are not yet available, and thus a study of the most interesting fractions of RNA cannot be undertaken. Nevertheless, there are some indications that messenger-RNA availability may increase after red light treatment.

These come from studies of phytochrome controlled changes in the levels of polyribosomes in etiolated bean leaves. Pine & Klein [85] in 1972 first showed that red light treatment led to a significant increase in polyribosome proportions in bean leaves, after a lag of 2 to 3 hours. Unfortunately, the methods available at that time did not prevent ribonuclease action and accurate measurements could not be made. Subsequently, better methods were published enabling Pine & Klein's results to be extended by Smith [106]. With ribonuclease completely inhibited, it was possible to show that polyribosome proportions in etiolated bean leaves increased very significantly well within 1 h of irradiation, after which they stayed at the new high level for several hours (Fig. 6.2.17). Increased formation of polyribosomes implies an increase in the availability of messenger-RNA. In this case it is not yet known whether the increased availability of messenger-RNA is due to its increased synthesis, or to its sudden release from some form of storage. This area of phytochrome study has been surprisingly neglected and should form a valuable field of study during the next few years.

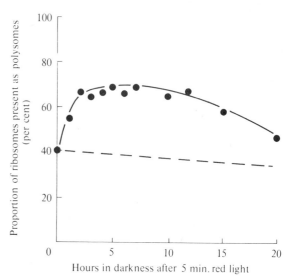

Fig. 6.2.17 Red light mediated increases in the proportions of polyribosomes in etiolated bean leaves (unpublished data).

Conclusions

The very limited number of examples quoted in this section have been been selected to show that evidence can be adduced in favour of phytochrome controlling development at several different levels. Polyribosome studies suggest that messenger-RNA availability may be controlled, the ascorbic acid oxidase evidence suggests that the rate of enzyme synthesis may be controlled, whilst the PAL evidence suggests that enzyme turnover or activation may be controlled. None of this evidence can be considered conclusive at this point. Nevertheless, it is impossible to propose a neat and simple hypothesis to cover all of the available information. Perhaps the best that can be done at the present time, and it is a very weak best, is to extend the second messenger concept referred to earlier along the lines of the multiple roles of the mammalian second messenger cyclic-AMP, which is known to be capable of controlling several diverse and quite separate cellular functions [54]. On this basis, perhaps we can imagine that phytochrome, operating at the cell membrane, causes the release of a second messenger, X, which proceeds to modulate the rates of a wide range of cellular activities, perhaps not all in the same cell. This concept is shown in Fig. 6.2.18 in which the lengths of the arrows are meant to indicate roughly the time lags of the various responses. If this scheme has any virtue other than to draw a set of divergent observations together, it can only be in concentrating ideas on the nature of the postulated second messenger. It could perhaps

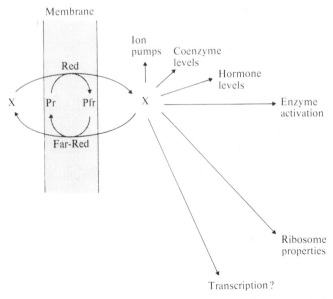

Fig. 6.2.18 A hypothetical second messenger scheme for the control of development by phytochrome.

be acetylcholine or NADPH. It could even be cyclic-AMP, which has recently been positively identified in plant tissues [17] although there is no acceptable evidence that its concentration responds to light treatment. There is, however, no compelling reason to believe that a plant second messenger need necessarily be chemically related to an animal second messenger, although it would be comforting to think so. Thus, it must be concluded that the area of greatest ignorance concerning phytochrome lies in the central question of the molecular mechanisms controlling developmental processes.

6.2.7 THE ROLE OF PHYTOCHROME IN THE NATURAL ENVIRONMENT

The highly successful reductionist analysis of phytochrome and its properties has been accompanied by an unaccountably ruthless exclusion of the main question: what is phytochrome there for? Hardly any major review has considered this question in the 15 years or so since phytochrome was first isolated. But when some thought is given to it, it quickly begins to appear as if phytochrome is rather too complicated for its apparent role. Plants in their natural environment are not exposed to brief periods of red or far-red light, nor are they adapted to continuous irradiation with monochromatic light. It is obvious that the detection of light is of great importance to the germinating seed or the developing seedling, so that the activities of the organism may be appropriate for growth below or above the soil surface. But what advantage does the red/far-red photoreversibility of phytochrome confer on the organism so far as the mere detection of light is concerned? Plants contain other photoreceptive substances, for example the various forms of chlorophyll and the as yet unidentified blue-absorbing photoreceptor, which have quantitatively greater effects on the metabolism and growth than phytochrome. In order to approach an understanding of the special role of phytochrome in plants growing in nature, it is necessary to consider the important parameters of the natural radiation environment and to see whether or not they are correlatable with the known properties of phytochrome.

Plants in the natural environment are exposed to long periods of irradiation from a broad spectrum source, the sun. Phytochrome in such plants will be continually cycling between Pr and Pfr, with specific proportions of the two forms present as characterized by the amounts of red and far-red incident on the plants. The spectral distribution of daylight (on a sunny day in mid-England!) is given in Fig. 6.2.19a and it can be seen that roughly equal amounts of red and far-red light are present on a quantum basis. When this mid-day sunlight is filtered through a vegetation canopy, however, an enormous change in spectral energy distribution takes place (Fig. 6.2.19b). The chlorophyll in the leaves causes the absorption of almost all the red and blue

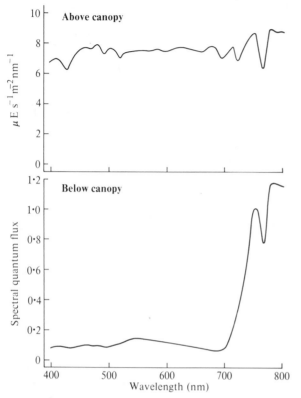

Fig. 6.2.19 Spectral energy distribution of daylight (direct sun, mid-summer, mid-England) and of daylight filtered through the canopy of a wheat crop (unpublished data of A. McCartney & M.G. Holmes).

Table 6.2.7 Phytochrome photoequilibria under natural conditions. The data represent the range of Pfr/Ptotal proportions measured in etiolated *Avena* coleoptiles exposed either to midday sunlight, or to sunlight filtered through sugar beet leaves (after Smith [104]).

Light source	Pfr/Ptotal (per cent)
Midday sunlight	53·0–60·0
Sunlight through one leaf	17·0–26·0
Sunlight through two leaves	5·0–14·0

light, but the leaves are relatively transparent to wavelengths above about 700 nm. Thus the amount of far-red filtering through a vegetation canopy is very much greater than the red light. Consequently, although phytochrome will still be cycling, the steady-state proportions of Pr and Pfr will be very different indeed. Table 6.2.7 gives some actual data of the Pfr/Ptotal values observed in *Avena* coleoptiles irradiated either with summer sunlight, or with the same source filtered through one or two sugar beet leaves.

Thus, phytochrome may be responsible for the plant's ability to detect shading from neighbouring plants and to direct its development accordingly. Stephen Hales in 1727 was probably the first to draw attention to the fact that shaded plants grow much taller than non-shaded plants [40]. This is partly an as yet unexplained irradiance effect, as it occurs to some extent with neutral density shading in which the spectral energy distribution of the light is not altered. However, very profound effects of added far-red light on development have been shown to occur. Downs *et al.* [24] in 1957 showed that stem elongation in a number of species could be substantially increased by brief supplemental treatment with far-red light given immediately after the end of the daily light period.

Table 6.2.8 shows the striking effects of differing proportions of red and far-red light in otherwise white light on the growth of

Table 6.2.8 Developmental changes in *T. maritimum* and *C. album* over 15 days at $25 \pm 1°C$ under fluorescent (low far-red) and incandescent (high far-red) light sources.

	T. maritimum		*C. album*	
	Fluorescent	Incandescent	Fluorescent	Incandescent
Height (cm)	29·7 ± 1·2	59·4 ± 1·4	15·0 ± 1·	28·4 ± 1·5
Internode length (cm)	0·84 ± 0·09	3·53 ± 0·18	—	—
Leaf dry weight (g)	0·483 ± 0·029	0·464 ± 0·024	0·338 ± 0·016	0·310 ± 0·016
Stem dry weight (g)	0·329 ± 0·017	0·583 ± 0·032	0·104 ± 0·006	0·197 ± 0·009
Leaf: stem dry weight ratio	1·47	0·80	3·25	1·57
Leaf area (cm²)	—	—	107·1 ± 1·2	78·5 ± 0·9
Chlorophyll $a + b$ (mg per g fresh weight)	75·7 ± 1·9	67·0 ± 1·7	112·3 ± 1·8	93·8 ± 4·5

The light sources were adjusted to produce equal quantum fluxes in the 400–700 nm waveband. *T. maritimum* seedlings were grown in a glasshouse until 9 weeks old, then transferred to fluorescent lighting in a growth chamber for 1 week before starting the treatments which were carried out in 8-h days. *C. album* were collected from an open situation in the field, potted and grown under fluorescent lighting for 10 days. The plants were then divided equally between the fluorescent and incandescent sources; 16 h photoperiods were used.

two common weeds of cereals in England, *Tripleurospermum maritimum*, (mayweed) and *Chenopodium album* (fat-hen). Seedlings were grown in either fluorescent or incandescent light adjusted to give equal amounts of photosynthetically active light. The only difference between the two sources is therefore in spectral energy distribution; the fluorescent source cuts off at about 700 nm with virtually no far-red, whereas the incandescent source has relatively large amounts of far-red. The two sources set up 70% and 54% Pfr/Ptotal respectively. The very marked effects of the differing Pfr/Ptotal proportions are clearly evident from the results. In nature, the weeds are capable of growing very tall and of competing with the cereal crop when inside the canopy; on open ground they rarely grow to more than 20–30 cm high.

It now seems likely that a large number of metabolic and developmental changes occur in plants subject to shading from others. The most important function of phytochrome, therefore, may not be to control de-etiolation and seed germination, but to enable the plant growing in its natural environment to adapt to the proximity of other plants.

ACKNOWLEDGEMENTS

The author gratefully acknowledges the permission of Dr R.E. Kendrick of the University of Newcastle-upon-Tyne, Dr T.H. Attridge, Miss Audrey Evans, Mr A. McCartney, Mr C.B. Johnson and Mr M.G. Holmes of the University of Nottingham, to include their unpublished results.

6.2.8 REFERENCES

[1] ATTRIDGE T.H. (1974) Phytochrome mediated synthesis of ascorbic acid oxidase in *Sinapis alba*. *Biochim. Biophys. Acta*, **362**, 258–265.

[2] ATTRIDGE T.H., JOHNSON C.B. & SMITH H. (1974) Density labelling evidence for the phytochrome-mediated activation of phenylalanine ammonia-lyase in mustard cotyledons. *Biochim. Biophys. Acta*. (In press.)

[3] ATTRIDGE T.H. & SMITH H. (1967) A phytochrome-mediated increase in the level of phenylalanine ammonia-lyase activity in the terminal buds of *Pisum sativum. Biochim. Biophys. Acta*, **148**, 805–7.

[4] ATTRIDGE T.H. & SMITH H. (1973) Evidence for a pool of inactive phenylalanine ammonia-lyase in *Cucumis sativus* seedlings. *Phytochemistry*, **12**, 1569–74.

[5] BEEVERS L., LOVEYS B., PEARSON J.A. & WAREING P.F. (1970) Phytochrome and hormonal control of expansion and greening of etiolated wheat leaves. *Planta*, **90**, 286–94.

[6] BEWLEY J.D., BLACK M. & NEGBI M. (1967) Immediate action of phytochrome in light stimulated lettuce seeds. *Nature*, **215**, 648–49.

[7] BLAAUW-JANSEN G. (1958) The influence of red and far-red light on the growth and phototropism of the *Avena* seedling. *Acta Botanica Neerl.*, **8**, 1–39.

[8] BORTHWICK H.A. (1972) History of phytochrome. In *Phytochrome* (Eds K. Mitrakos & W. Shropshire, Jr.), pp. 3–22. Academic Press, London.

[9] BORTHWICK H.A., HENDRICKS S.B. & PARKER M.W. (1951) Action spectrum for the inhibition of stem growth in dark-grown seedlings of albino and non-albino barley (*Hordeum vulgare*). *Botan. Gaz.*, **113**, 95–105.

[10] BORTHWICK H.A., HENDRICKS S.B., PARKER M.W., TOOLE E.H. & TOOLE V.K. (1952) A reversible photoreaction controlling seed germination. *Proc. Nat. Acad. Sci.*, **38**, 662–66.

[11] BORTHWICK H.A., HENDRICKS S.B., TOOLE E.H. & TOOLE V.K. (1954) Action of light on lettuce seed germination. *Bot. Gaz.*, **115**, 202–25.

[12] BORTHWICK H.A., PARKER M.W. & HENDRICKS S.B. (1950) Recent developments in the control of flowering by photoperiod. *Am. Soc. Natur.*, **84**, 117–34.

[13] BRIGGS W.R. (1963) Red light, auxin relationships, and the phototropic responses of corn and oat coleoptiles. *Amer. J. Bot.*, **50**, 196–207.

[14] BRIGGS W.R. & FORK D.C. (1969) Long-lived intermediates in phytochrome transformation. I. *In vitro* studies. *Plant Physiol.*, **44**, 1081–8.

[15] BRIGGS W.R. & FORK D.C. (1969) Long-lived intermediates in phytochrome transformation. II. *In vitro* and *in vivo* studies. *Plant Physiol.*, **44**, 1089–94.

[16] BRIGGS W.R. & RICE H.V. (1972) Phytochrome: chemical and physical properties and mechanism of action. *Ann. Rev. Plant Physiol.*, **23**, 293–334.

[17] BROWN E.G. & NEWTON R.P. (1973) Occurrence of adenosine 3′:5′-cyclic monophosphate in plant tissues. *Phytochemistry*, **12**, 2683–5.

[18] BUTLER W.L., HENDRICKS S.B. & SIEGELMAN H.W. (1964) Action spectra of phytochrome *in vitro*. *Photochem. Photobiol.*, **3**, 521–28.

[19] BUTLER W.L., NORRIS K.H., SIEGELMAN H.W. & HENDRICKS S.B. (1959) Detection, assay, and preliminary purification of the pigment controlling photoresponsive development of plants. *Proc. Nat. Acad. Sci.*, **45**, 1703–8.

[20] BUTLER W.L., SIEGELMAN H.W. & MILLER C.O. (1964) Denaturation of phytochrome. *Biochemistry*, *t33*, 851–57.

[21] CLARKSON D.T. & HILLMAN W. (1967) Stability of phytochrome concentration in dicotyledonous tissues under continuous far-red light. *Planta*, **75**, 286–90.

[22] DARWIN C. (1880) *The power of movement in plants* (assisted by F. Darwin). John Murray, London.

[23] DITTES L., RISSLAND I. & MOHR H. (1971) On the regulation of enzyme levels (phenylalanine ammonia-lyase) in different organs of a plant (*Sinapis alba* L.). *Z. Naturforsch.*, **26b**, 1175–80.

[24] DOWNS R.J., HENDRICKS S.B. & BORTHWICK H.A. (1957) Photoreversible control of elongation of pinto beans and other plants under normal conditions of growth. *Bot. Gaz.*, **118**, 119–208.

[25] ETZOLD H. (1965) Der Polarotropismus und Phototropismus der Chloronemen von *Dryopteris filix mas* (L.) Schoff. *Planta*, **64**, 254–80.

[26] FILNER P. & VARNER J.E. (1967) A test for *de novo* synthesis of enzymes; density labelling with H_2O^{18} of barley α-amylase induced by gibberellic acid. *Proc. Nat. Acad. Sci.*, **58**, 1520–26.

[27] FLINT L.H. & MCALISTER E.D. (1935) Wavelengths of radiation in the visible spectrum inhibiting the germination of light-sensitive lettuce seed. *Smithsonian Misc. Collection*, **94**(5), 1–11.

[28] FLINT L.H. & MCALISTER E.D. (1937) Wavelengths of radiation in the visible spectrum promoting the germination of light-sensitive lettuce seed. *Smithsonian Misc. Collection*, **96**(2), 1–9.

[29] FRANKLAND B. (1972) Biosynthesis and dark transformations of phytochrome. In *Phytochrome* (Eds K. Mitrakos & W. Shropshire, Jr.), pp. 195–225. Academic Press, London.

[30] FREDERICQ H. (1964) Conditions determining effects of far-red and red irradiations on flowering response of *Pharbitis nil. Plant Physiol.*, **39**, 812–6.

[31] FRENCH C.J. & SMITH H. (1975) An inactivator of phenylalanine ammonia-lyase from gherkin hypocotyls. *Phytochemistry*, **14**, 963–966.

[32] FUJII T. & KOUDO N. (1969) Changes in the levels of nicotinamide nucleotides and in activities of NADP-dependent dehydrogenases after a brief illumination with red light. *Develop. Growth Differ.*, **11**, 40–45.

[33] FURUYA M. & THOMAS R.G. (1964) Flavonoid complexes in *Pisum sativum*. II. Effects of red and far-red light on biosynthesis of kaempferol complexes and on growth of etiolated plumules. *Plant Physiol.*, **39**, 634–42.

[34] GALSTON A.W. & SATTER R.L. (1972) A study of the mechanism of phytochrome action. In *Recent advances in phytochemistry* (Eds V.C. Runeckles & T.C. Tso), **5**, 51–80. Academic Press, New York.

[35] GARDNER G., PIKE C.S., rice H.V. & BRIGGS W.R. (1971) 'Disaggregation' of phytochrome *in vitro*–a consequence of proteolysis. *Plant Physiol.*, **48**, 686–93.

[36] GARNER W.W. & ALLARD H.A. (1920) Effect of the relative length of day and night and other factors of the environment on growth and reproduction in plants. *J. Agric. Res.*, **18**, 553–606.

[37] GOESCHL J.D., PRATT H.K. & BONNER B.A. (1967) An effect of light on the production of ethylene and the growth of the plumular portion of etiolated pea seedlings. *Plant Physiol.*, **42**, 1077–80.

[38] GRILL R. (1969) Photo-control of anthocyanin formation in turnip seedlings. *Planta*, **85**, 42–56.

[39] GRILL R. (1972) The influence of chlorophyll on *in vivo* difference spectra of phytochrome. *Planta*, **108**, 185–202.

[40] HALES S. (1727) *Statical Essays*, **1**, 334.

[41] HARTMANN K.M. (1966) A general hypothesis to interpret 'high energy phenomena' of photomorphogenesis on the basis of phytochrome. *Photochem. Photobiol.*, **5**, 349–66.

[42] HARTMANN K.M. (1967) Ein Wirkungsspektrum der Photomorphogenese unter Hochenergie Bedingungen und seine Interpretation auf der Basis des Phytochroms. (Hypokotylwachstumshemmung bei *Lactuca sativa* L.) *Z. Naturforschg.*, **22b**, 1172–75.

[43] HAUPT W. (1972) Short-term phenomena controlled by phytochrome. In *Phytochrome* (Eds K. Mitrakos & W. Shropshire, Jr.), pp. 349–68. Academic Press, London.

[44] HAUPT W. (1973) Localization of phytochrome within the cell. In *Phytochrome* (Eds K. Mitrakos & W. Shropshire, Jr.), pp. 553–69. Academic Press, London.

[45] HAUPT W. & SCHONBOHM E. (1970) Light-oriented chloroplast movements. In *Photobiology of microorganisms* (Ed. P. Holdal), pp. 283–307. Wiley, Chichester.

[46] HENDRICKS S.B. & BORTHWICK H.A. (1959) Photocontrol of plant development by simultaneous exitation of two interconvertible pigments. *Proc. Nat. Acad. Sci., USA*, **45**, 344–49.

[47] HENDRICKS S.B. & BORTHWICK H.A. (1967) The function of phytochrome in regulation of plant growth. *Proc. Nat. Acad. Sci.*, **58**, 2125–30.

[48] HEWETT E.W. & WAREING P.F. (1973) Cytokinins in *Populus × robusta*: changes during chilling and bud burst. *Physiologia Plantarum*, **28**, 393–400.

[49] HILLMAN W.S. & GALSTON A. W. (1957) Inductive control of indoleacetic acid oxidase activity by red and near infrared light. *Plant Physiol.*, **32**, 129–35.

[50] HOPKINS D.W. & BRIGGS W.R. (1973) *Plant Physiol.* (suppl.)

[51] JAFFE M.J. (1968) Phytochrome-mediated bioelectric potentials in mung bean seedlings. *Science*, **162**, 1016–17.

[52] JAFFE M.J. (1970) Evidence for the regulation of phytochrome-mediated processes in bean roots by the neurohumor, acetylcholine. *Plant Physiol.*, **46**, 768–77.

[53] JAFFE M.J. (1972) Acetylcholine as a native metabolic regulator of phytochrome-mediated processes in bean roots. In *Recent advances in phytochemistry* (Eds V.C. Runneckles & T.C. Tso), Vol. 5, pp. 80–104. Academic Press, New York.

[54] JOST J.P. & RICKENBERG H.V. (1971) Cyclic AMP. *Ann. Rev. Biochem.*, **40**, 741–74.

[55] KANG B.G. & RAY P.M. (1969) Role of growth regulators in the bean hypocotyl hook opening response. *Planta*, **87**, 193–205.

[56] KASEMIR H. & MOHR H. (1972) Involvement of acetylcholine in phytochrome mediated processes. *Plant Physiol.*, **49**, 453–54.

[57] KENDE H. & LANG A. (1964) Gibberellins and light inhibition of stem growth in peas. *Plant Physiol.*, **39**, 435–40.

[58] KENDRICK R.E. & HILLMAN W.S. (1971) Absence of phytochrome dark reversion in seedlings of the Centrospermae. *Am. J. Bot.*, **58**, 424–28.

[59] KENDRICK R.E. & SPRUIT C.J.P. (1972) Light maintains high levels of phytochrome intermediates. *Nature, New Biol.*, **237**, 281–82.

[60] KENDRICK R.E. & SPRUIT C.J.P. (1973) Phytochrome properties and the molecular environment. *Plant Physiol.*, **52**, 327–31.

[61] KLEIN W.H. (1959) Interaction of growth factors with photoprocess in seedling growth. In *Photoperiodism* (Ed. R.B. Withrow), pp. 207–15. Washington American Ass Adv. Science.

[62] KLEIN W.H., WITHROW R.B. & ELSTAD V.B. (1956) Response of the hypocotyl hook of bean seedlings to radiant energy and other factors. *Plant Physiol.*, **31**, 289–94.

[63] KLEIN W.H., WITHROW R.B., ELSTAD V.B. & PRICE L. (1957) Photocontrol of growth and pigment synthesis in the bean seedling as related to irradiance and wavelength. *Am. J. Bot.*, **44**, 15–19.

[64] KROES H.H. (1970) A study of phytochrome, its isolation, structure and photochemical transformations. *Meded. Landbouwhoogeschool, Wageningen*, **70**, 18, 1–112.

[65] LANGE H. & MOHR H. (1965) Die Hemmung der Phytochrominduzierten Anthocyansynthese durch Actinomycin D und Puromycin. *Planta*, **67**, 107–21.

[66] LINSCHITZ H., KASCHE V., BUTLER W.L. & SIEGELMAN H.W. (1966) The kinetics of phytochrome conversion. *J. Biol. Chem.*, **241**, 3395–403.

[67] LOCKHART J.A. (1956) Reversal of light inhibition of pea stem growth by gibberellins. *Proc. Nat. Acad. Sci., USA*, **42**, 841–48.

[68] LOCKHART J.A. (1959) Studies on the mechanisms of stem growth inhibition by visible radiation. *Plant Physiol.*, **34**, 457–60.

[68a] LOVEYS B.R. & WAREING P.F. (1971) The red light controlled production of gibberellin in etiolated wheat leaves. *Planta*, **98**, 109–16.

[69] MANABE K. & FURUYA M. (1973) A rapid phytochrome-dependent reduction of NADP in particle fraction from etiolated bean hypocotyl. *Plant Physiol.*, **51**, 982–83.

[70] MARMÉ D., BOISARD J. & BRIGGS W.R. (1973) Binding properties *in vitro* of phytochrome to a membrane fraction. *Proc. Nat. Acad. Sci., USA*, **70**, 3861–65.

[71] MARMÉ D., MACKENZIE Jr, J.M., BOISARD J. & BRIGGS W.R. (1974) The isolation and partial characterization of membrane vesicles containing phytochrome. *Plant Physiol.*, **54**, 263–271.

[72] MEIJER G. (1968) Rapid growth inhibition of gherkin hypocotyls in blue light. *Acta Bot. Neerl.*, **17**(1), 9–14.

[73] MENHENETT R. & CARR D.J. (1973) Cytokinins in etiolated barley leaves. *Aust. J. Biol. Sci.*, **26**, 1073–80.

[74] MOHR H. (1957) Der Einfluss monochromatischer Strahlung auf das Langenwachstum des Hypocotyls und auf die Anthocyanbildung bei Keimlingen von *Sinapis alba* L. *Planta*, **49**, 389–405.

[75] MOHR H. (1966) Differential gene activation as a mode of action of phytochrome 730. *Photochem. Photobiol.*, **5**, 469–83.

[76] MOHR H. (1969) Photomorphogenesis. In *Physiology of plant growth and development* (Ed. M.B. Wilkins), pp. 509–56. McGraw-Hill, London.

[77] MOHR H. (1972) *Lectures in Photomorphogenesis*. Springer-Verlag, Berlin.

[78] MOHR H. & BIENGER I. (1967) Experimente zur Wirkung von Actinomycin D auf die durch Phytochrom bewirkte Anthocyansynthese. *Planta*, **75**, 180–94.

[79] MUSGRAVE A., KAYS S.E. & KENDE H. (1959) *In vivo* binding of radioactive gibberellins in dwarf pea shoots. *Planta*, **89**, 165–77.

[80] NEWMAN I.A. & BRIGGS W.R. (1972) Phytochrome-mediated electric potential changes in oat seedlings. *Plant Physiol.*, **50**, 687–93.

[81] PARKER M.W., HENDRICKS S.B., BORTHWICK H.A. & SCULLY N.J. (1946) Action spectra for photoperiodic control of floral initiation in short-day plants. *Bot. Gaz.*, **108**, 1–26.

[82] PARKER M.W., HENDRICKS S.B., BORTHWICK H.A. & WENT F.W. (1949) Spectral sensitivities for leaf and stem growth of etiolated pea seedlings and their similarity to action spectra for photoperiodism. *Am. J. Bot.*, **36**, 194–204.

[83] PHILLIPS I.D.J., VLITOS J.A.J. & CUTLER H. (1959) The influence of gibberellic acid upon the endogenous growth substances of the Alaska pea. *Contribs Boyce Thompson Inst.*, **20**, 111–20.

[84] PIKE C.S. & BRIGGS W.R. (1972) Partial purification and characterization of a phytochrome-degrading neutral protease from etiolated oat shoots. *Plant Physiol.*, **49**, 521–30.

[85] PINE K. & KLEIN A.O. (1972) Regulation of polysome formation in etiolated bean leaves by light. *Dev. Biol.*, **28**, 280–189.

[86] PJON C.J. & FURUYA M. (1967) Phytochrome action in *Oryza sativa* L. I. Growth responses of etiolated coleoptiles to red, far-red and blue light. *Plant and Cell Physiol.*, **8**, 709–18.

[87] PRATT L.H. & COLEMAN R.A. (1971) Immunocytochemical localization of phytochrome. *Proc. Nat. Acad. Sci.*, **68**, 2431–35.

[88] QUAIL P., MARMÉ D. & SCHÄFER E. (1973) Particle-bound phytochrome from maize and pumpkin. *Nature, New Biol.*, **245**, 189–90.

[89] QUAIL P.H., SCHÄFER E. & MARMÉ D. (1973) *De novo* synthesis of phytochrome in pumpkin hooks. *Plant Physiol.*, **52**, 124–27.

[90] REID D.M., CLEMENTS J.B. & CARR D.J. (1968) Red light induction of gibberellin synthesis in leaves. *Nature*, **217**, 580–82.

[91] REID D.M., TUING M.S., DURLEY R.C. & RAILTON I.D. (1972) Red light enhanced conversion of tritiated gibberellin A₉ into other gibberellin-like substances in homogenates of etiolated barley leaves. *Planta*, **108**, 67–75.

[91a] RICE H.V. & BRIGGS W.R. (1973) Partial characterization of oat and rye phytochrome. *Plant Physiol.*, **51**, 927–38.

[92] ROUX S.J. & HILLMAN W.S. (1969) The effect of glutaraldehyde and two monoaldehydes on phytochrome. *Arch. Biochem. and Biophys.*, **131**, 423–29.

[93] RUBINSTEIN B., DRURY K.S. & PARK R.B. (1969) Evidence for bound phytochrome in oat seedlings. *Plant Physiol.*, **44**, 105–9.

[94] RUDIGER W. (1972) Isolation and purification of phytochrome. In *Phytochrome* (Eds K. Mitrakos & W. Shropshire, Jr.), pp. 107–28. Academic Press, New York.

[95] RUDIGER W. (1972) Chemistry of phytochrome chromophore. In *Phytochrome* (Eds K. Mitrakos & W. Shropshire, Jr.), pp. 129–41. Academic Press, London.

[96] RUSSELL D.W. & GALSTON A.W. (1969) Blockage by gibberellic acid of phytochrome effects on growth, auxin responses, and flavonoid synthesis in etiolated pea internodes. *Plant Physiol.*, **44**, 1211–16.

[97] SANDMEIER M. & IVART J. (1972) Modification du taux des nucleotides adenyliques (ATP, ADP et AMP), par un eclairment de lumiere rouge-clair (660 nm). *Photochem. Photobiol.*, **16**, 51–59.

[98] SATTER R.L., APPLEWHITE P.B. & GALSTON A.W. (1972) Phytochrome-controlled nyctinasty in *Albizzia julibrissin*. V. Evidence against acetylcholine participation. *Plant Physiol.*, **50**, 5231–25.

[99] SCHÖNBOHM E. (1973) Kontraktile Fibrillen als aktive Elemente bei der Mechanik der Chloroplastenverlagerung. *Ber. Deutsch Bot. Ges.*, **86**, 407–22.

[100] SCHOPFER P. (1966) Der Einfluss von Phytochrom auf die statilnain Konzentrationen von Ascorbinsaure und Dehydroascorbinsaure bein Senfkeimlingen. *Planta*, **69**, 158–77.

[101] SCHOPFER P. & MOHR H. (1972) Phytochrome-mediated induction of phenylalanine ammonia-lyase in mustard seedlings. A contribution to eliminate some misconceptions. *Plant Physiol.*, **49**, 8–10.

[102] SIEGELMAN H.W., TURNER B.C. & HENDRICKS S.B. (1966) The chromophore of phytochrome. *Plant Physiol.*, **41**, 1289–92.

[103] SMITH H. (1970) Phytochrome and photomorphogenesis in plants. *Nature*, **227**, 665–68.

[104] SMITH H. (1973) Light quality and germination: ecological implications. In *Seed ecology* (Ed. W. Heydecker), pp. 219–31. Butterworths, London.

[104a] SMITH H. (1975) The biochemistry of photomorphogenesis. In *Plant biochemistry, International review of biochemistry*, Vol. 13 (Ed. D.H. Northcote), pp. 159–97. (MTP: London.)

[105] SMITH H. (1974) *Phytochrome and photomorphogenesis*. McGraw-Hill, London.

[106] SMITH H. (1975) Phytochrome mediated control of polysome levels in bean leaves. *Eur. J. Biochem.* (In press.)

[107] STEWART G.R. (1968) The effect of cycloheximide on the induction of nitrate and nitrite reductase in *Lemna minor* L. *Phytochemistry*, **7**, 1139–42.

[108] STODDART J.L., BREIDENBACH W., NEDEAU R. & RAPPAPORT L. (1974) Selective binding of ³H-gibberellin A₁ by protein fractions from pea epicotyls. *Proc. Nat. Acad. Sci., USA.* **71**, 3255–59.

[109] TAKIMOTO A. & HAMNER K.C. (1965) Studies on red light interruption in relation to timing mechanisms involved in the photoperiodic response of *Pharbitis nil. Plant Physiol.*, **40**, 852–54.

[109a] TANADA T. (1968) Substances essential for a red, far-red light reversible attachment of mung bean root tips to glass. *Plant Physiol.*, **43**, 2070–71.

[109b] TANADA T. (1968) A rapid photoreversible response of barley root tips in the presence of 3-indoleacetic acid. *Proc. Nat. Acad. Sci., USA*, **59**, 376–80.

[110] TANADA T. (1972) On the involvement of acetylcholine in phytochrome action. *Plant Physiol.*, **49**, 860–61.

[111] TEZUKA T. & YAMAMOTO Y. (1969) NAD kinase and phytochrome. *Botan. Mag., Tokyo*, **82**, 130–33.

[112] TEZUKA T. & YAMAMOTO Y. (1972) Photoregulation of nicotinamide adenine dinucleotide kinase activity in cell-free extracts. *Plant Physiol.*, **50**, 458–62.

[113] VAN OVERBEEK J. (1936) Growth hormone and mesocotyl growth. *Rec. Trav. bot., Neerl.*, **33**, 333–40.

[114] VAN STADEN J. & WAREING P.F. (1972) The effect of light on endogenous cytokinin levels in seeds of *Rumex obtusifolius. Planta*, **104**, 126–33.

[115] VINCE D. (1964) Photomorphogenesis in plant stems. *Biol. Rev.*, **39**, 506–36.

[116] WELLBURN F.A.M. & WELLBURN A.R. (1973) Response of etioplasts *in situ* and in isolated suspensions to pre-illumination with various combinations of red, far-red and blue light. *New Phytol.*, **72**, 551–60.

[117] WILKINS M.B. & GOLDSMITH H.M. (1964) The effects of red, far-red, and blue light on the geotropic response of coleoptiles of *Zea mays. J. Exp. Bot.*, **15**, 600–15.

[118] WITHROW R.B., KLEIN W.H. & ELSTAD V. (1957) Action spectra of photomorphogenic induction and photoinactivation. *Plant Physiol.*, **32**, 453–62.

Conclusions to Part 6

Photomorphogenesis in plants was chosen as the topic to illustrate environmental control of development because it provides one of the most intensively studied examples of this phenomenon. All organisms need to be able to sense a range of environmental signals and, as we have seen earlier, many plants and animals need to relate their cycle of development to both periodic and aperiodic changes in the environment. Clearly, in plants, phytochrome plays a central role as a universal sensing device for information on the light environment, and the variety of developmental responses it controls and initiates is remarkable (Table 6.2.1). Moreover, phytochrome can operate either in the 'low energy' or the 'high energy' mode and can apparently interact with other photosystems, such as one involving the blue region of the spectrum.

When we come, however, to consider the mode of action of phytochrome in bringing about such a wide variety of responses, we are faced with the same type of problem as was presented by the multiplicity of effects of a single hormone, such as gibberellic acid in plants or thyroxine in amphibia. Indeed, it is not clear whether there is a single primary mode of action of phytochrome or possibly several. We have seen that several pieces of experimental evidence, such as the very rapid effects on bioelectric potentials, seem to indicate that phytochrome is located in or near the cell membranes, such as the plasmalemma, and that it acts by changing the properties of the membranes. On the other hand, phytochrome conversion resulting from exposure to red light leads to increased activity of several types of hormone and some of the diverse developmental responses are probably mediated by hormones acting as 'second messengers'. If this is so, then a major conceptual synthesis, involving both phytochrome and the hormone synthesis, will have been achieved.

Phytochrome conversion also leads to increased activity of a range of enzymes (Table 6.2.5). It is still uncertain as to whether this increased activity is due to activation of pre-existing enzymes or to the synthesis of new enzyme proteins. However, the varied developmental responses controlled by phytochrome would seem to involve selective gene expression, and the question at issue is one which has been a recurring problem throughout this work—namely how is selective gene expression controlled during development? Is it controlled directly in the nucleus through selective transcription of specific parts of the genome, or in the cytoplasm at the translation of messenger-RNA in protein synthesis? Or is it possible that development does not involve activation of new parts of the genome, but changes in the ratio of existing activities ('modulation'), as has been suggested in the preceding chapter? These remain some of the major problems of development. Information on the structure of the genome in eucaryotes is increasing steadily and it may not be too long before we understand how gene activity and gene expression are regulated in higher organisms. We will then be in a much better position to tackle some of the many unanswered questions raised in this book.

And so we have come full circle, because the problems raised by the mode of action of phytochrome are essentially the same as those with which we began our discussion of development in Part I.

Index

Italics indicate a figure